Advances in Aerospace Guidance, Navigation and Control

Joël Bordeneuve-Guibé · Antoine Drouin
Clément Roos
Editors

Advances in Aerospace Guidance, Navigation and Control

Selected Papers of the Third CEAS Specialist Conference on Guidance, Navigation and Control Held in Toulouse, France, in April 2015

Springer

Editors
Joël Bordeneuve-Guibé
ISAE
University of Toulouse
Toulouse
France

Clément Roos
ONERA The French Aerospace Lab
Toulouse
France

Antoine Drouin
ENAC
University of Toulouse
Toulouse
France

ISBN 978-3-319-17517-1 ISBN 978-3-319-17518-8 (eBook)
DOI 10.1007/978-3-319-17518-8

Library of Congress Control Number: 2015937022

Springer Cham Heidelberg New York Dordrecht London

Printed on acid-free paper

Springer International Publishing AG Switzerland is part of Springer Science+Business Media (www.springer.com)

About the Book

For the 3rd CEAS Specialist Conference on Guidance, Navigation and Control, the International Program Committee conducted a formal review process. Each paper was reviewed in compliance with good journal practice by at least two independent and anonymous reviewers. The papers published in this book were selected based on the results and recommendations from the reviewers.

The active members of the International Program Committee are:

Sébastien Aubry	ONERA, France
José Raul Azinheira	IST, Portugal
Samir Bennani	ESA, The Netherlands
Paolo Castaldi	Bologna University, Italy
Daniel Choukroun	Ben Gurion University, Israel
Qiping Chu	TU Delft, The Netherlands
John Crassidis	University of Buffalo, USA
François Defaÿ	ISAE, France
Daniel Delahaye	ENAC, France
Fabrice Demourant	ONERA, France
Boguslaw Dolega	University of Rzeszów, Poland
Chris Edwards	University of Leicester, UK
Aleksander Efremov	Moscow Aviation Institute, Russia
Patrick Fabiani	ISAE, France
Pierre Fabre	Airbus, France
Nicolas Fezans	DLR, Germany
Walter Fichter	University of Stuttgart, Germany
Christopher Fielding	RAeS, United Kingdom
Benoit Frapard	Astrium, France
Hagström Martin	Defense Research Agency, Sweden
Florian Holzapfel	TU Munchen, Germany
Eric Johnson	Georgia Institute of Technology, USA
Frank Jouhaud	ONERA, France

Erik-Jan van Kampen	TU Delft, The Netherlands
Karl-Heinz Kienitz	Instituto Tecnológico de Aeronáutica, Brazil
Youdan Kim	Seoul National University, South Korea
Aymeric Kron	Bombardier Aerospace, Canada
Christophe Louembet	LAAS/CNRS, France
Marco Lovera	Politecnico di Milano, Italy
Mark Lowenberg	University of Bristol, UK
Robert Luckner	Berlin Technical University, Germany
Bob Mulder	TU Delft, The Netherlands
Janusz Narkiewicz	University of Varsow, Poland
Alexander Nebylov	IIAAT, Russia
Radhakant Padhi	Indian Institute of Science, India
Christelle Pittet	CNES, France
Han Ping	Civil Aviation University of China, China
Lorenzo Pollini	University of Pisa, Italy
Fausto de Oliveira Ramos	IAE, Brazil
Alexandre Carlos Brandão Ramos	UNIFEI, Brazil
Arthur Richards	University of Bristol, UK
David Alexandre Saussié	EPM, Canada
Stephan Theil	DLR, Germany
Frank Thielecke	TU Hamburg Harburg, Germany
Andrzej Tomczyk	University of Rzeszów, Poland
Coen de Visser	TU Delft, The Netherlands
James Whidborne	Cranfield University, UK
Rama Yedavalli	Ohio State University, USA
Ali Zolgadri	IMS, France

Preface

The two first CEAS (Council of European Aerospace Societies) Specialist Conferences on Guidance, Navigation and Control (CEAS EuroGNC) were held in Munich, Germany in 2011 and in Delft, The Netherlands in 2013. ONERA The French Aerospace Lab, ISAE (Institut Supérieur de l'Aéronautique et de l'Espace) and ENAC (Ecole Nationale de l'Aviation Civile) accepted the challenge of jointly organizing the 3rd edition. The conference, chaired by Daniel Alazard and Felix Mora-Camino, took place on April 13–15, 2015, at ISAE-SUPAERO, Toulouse, France, one of the leading aerospace engineering schools in Europe. The Organizing Committee composed of Christelle Cumer and Nadine Barriety, and the International Program Committee composed of about 50 eminent scientists and engineers, strongly contributed to the success of this event. About a hundred papers were selected for presentation at the conference and this book contains the forty best contributions. The topics addressed here represent the most actively researched areas in guidance, navigation and control.

It is well known that the challenges are often more demanding in aerospace than in many other fields. The control of aerospace vehicles remains a difficult task because of ever larger flight domains, more complex and coupled dynamics, and wider variety of flying vehicles. Among the most promising control techniques, adaptive control has gained significant interest due to recent developments ensuring fast adaptation to environmental changes while preserving robust stability. A renewed interest in robust control is also observed. Recent advances in non-smooth optimization and developments of efficient softwares have contributed to bridge the gap between theory and practice, allowing these techniques to be used in many industrial applications. It is now possible for example to design very simple controllers such as PIDs using H_∞ based techniques.

Visual servoing, also known as vision-based control, has emerged more recently with the development of small, accurate and affordable cameras. This technique uses feedback information extracted from vision sensors to control the motion of a vehicle or a robot. The ever-growing computer power makes it now possible to process the rich information provided by these sensors, which is an essential step towards the control and the guidance of vehicles with fast dynamics. Many theoretical and practical results have already been presented, but solid mathematical analyses and proofs, real-time issues and efficient hardware implementations of image processing algorithms still deserve to be further investigated.

Then before flight testing, each aerospace vehicle has to go through a rigorous certification and qualification process to prove to the authorities that the flight control system is safe and reliable. Currently significant time and money is spent by the aeronautical industry on this task. Monte-Carlo simulations are used in most cases, but it is often difficult to isolate worst case scenarios or to confidently assert that no such scenario exists. Fortunately, many stability, performance, loads and comfort criteria can be reformulated as robustness analysis problems. Promising techniques such as multi-objective optimization under uncertainty using for example evolutionary algorithms can thus be applied to determine parameters/inputs/flight conditions for which the criteria are violated or poorly satisfied. A considerable effort is currently underway to enhance these techniques, motivated by the increase in computer power and the advent of multi-core processors, which allow to perform parallel computing at a reasonable cost.

These topics are all the more important that they are significant to both traditional aerospace vehicles as well as to emerging ones such as small Unmanned Aerial Vehicles (UAVs). Originating in the 60's as a military tool, UAVs have evolved from expensive and complicated military tools into inexpensive, relatively easy to use machines that are accessible to most people. Revenues generated by the activity have seen a tremendous growth as underlined by the initial public offering (IPO) of the DJI company.

This class of vehicles introduces many new challenges in term of control and navigation. Compared to classical flying vehicles, they perform a wider variety of missions, many of which including tightly space-constrained evolutions requiring high dynamics trajectories. Another particularity of those vehicles is that they are often operated closer to one another as well as to obstacles, requiring accurate and reactive navigation. As many of the applications intended for UAVs are motivated by their affordability, the costly validation and certification techniques traditionally used in aerospace can not be directly transposed. Their small mass and velocity imply limited consequences in case of crash and could justify relaxed regulations.

An additional advantage of this relaxed certification is the possibility of using the latest generation of microprocessors. The vast processing power available allows to experiment with a new class of algorithms that were previously inapplicable on other vehicles. The cost constraint limits the quality of sensors used on those vehicles and motivates new challenges for navigation algorithms, as does the use of innovative sensors like vision. Last but not least about unmanned systems, integration with Air Traffic Management has become a key issue that needs to be urgently tackled.

More recently, the same trend can be seen with CubeSats. Those small low cost satellites promise to offer a whole new range of applications once the technical and regulatory issues differentiating them from their full sized counterparts are be solved.

Obviously the papers presented at the conference and selected in this book do not suffice to fully cover all these challenging fields. However they represent an excellent source of information for those seeking the latest theoretical and practical developments in guidance, navigation and control of aerospace vehicles.

The organization of the CEAS EuroGNC 2015 would have been impossible without the strong support of many people and communities. On behalf of the Local Organization Committee of CEAS EuroGNC 2015, we would like to thank all contributors to the conference. These contributors are: Council of European Aerospace Societies (CEAS), ONERA The French Aerospace Lab, ISAE, ENAC, American Institute of Aeronautics and Astronautics (AIAA), Fondation ISAE-SUPAERO, all members of the CEAS EuroGNC 2015 International Program Committee and all reviewers of technical papers.

The book is divided into four chapters : Guidance and Control, Navigation and Estimation, Atmospheric Applications and Space Applications. We hope you will enjoy reading this overview as much we enjoyed assembling it.

Toulouse, France
April, 2015

Joël Bordeneuve-Guibé
Antoine Drouin
Clément Roos

Editorial Committee

Joël Bordeneuve-Guibé
Université de Toulouse
ISAE
10 avenue Edouard Belin
31055 Toulouse
France

Antoine Drouin
Université de Toulouse
ENAC
7 avenue Edouard Belin
31055 Toulouse
France

Clément Roos
ONERA
The French Aerospace Lab
2 avenue Edouard Belin
BP 74025
31055 Toulouse
France

Biographies

Dr. Clément Roos graduated from ISAE-Supaéro (The French Superior Institute of Aeronautics and Space) in 2004. He holds a PhD in Automatic Control from ISAE-Supaero, for which he received two awards. He joined ONERA (The French Aerospace Lab) as a research scientist in 2007. He often takes part in industrial projects with Airbus and Dassault, and was notably involved in the European projects GARTEUR-AG17 and COFCLUO. His research interests focus on aircraft modeling, robustness analysis and control laws validation, as well as nonlinear design based on robustified dynamic inversion schemes and anti-windup synthesis. He is the author or co-author of several papers, book chapters, teaching documents and Matlab toolboxes. He also gives courses on automatic control, signal processing and flight mechanics at ISAE-Supaéro.

Dr. Antoine Drouin obtained his MSc degree in 1997 from ENAC (French Civil Aviation University) in Aerospace Engineering. He received his PhD in 2012 from the university Toulouse Paul Sabatier. The topic of his PhD research was the development of adaptive guidance laws for micro UAVs. For 6 years, Dr. Drouin worked for the French air navigation service provider as a software engineer specializing in high availability software. He worked specifically on the redundancy system for radar and flight plan applications. From 2003 to 2006 he worked for the French air traffic control research center on the development of new generation graphical user interface and simulation code. Since 2007, Dr. Drouin has been a member of the faculty at ENAC, in the automatic control laboratory as assistant professor. Since 2003, he has been instrumental in the development of Paparazzi, an open source UAV system. He is involved in collaborations with foreign universities including the Georgia Institute of Technology and companies such as Joby Aviation. In 2000, Dr. Drouin began teaching at ENAC, initially in the field of operating systems and computer programming, and later in dynamical systems in addition to numerical programming. His research interests include navigation and nonlinear control theory.

Dr. Joël Bordeneuve-Guibé received the PhD degree in Automatic Control in 1990 from the Paul Sabatier University of Toulouse. The topic of the PhD was the multivariable predictive and adaptive control of industrial thermal processes. From 1991 to 1992, Dr. Bordeneuve-Guibé worked as a Research Assistant at the Department of Automatic Control of the Polytechnic University of Valencia, Spain (UPV). Since September 1992, Dr. Bordeneuve-Guibé is a research fellow at the Department of Mathematics, Computer Science and Control Design, Institut Supérieur de l'Aéronautique et de l'Espace (ISAE-Supaéro), Toulouse. His research interests are in adaptive and robust control, predictive control, spacecraft control and guidance. Dr. Bordeneuve-Guibé is currently associate professor of Automatic Control at ISAE-Supaéro.

Contents

Part I: Guidance and Control

Robust Lateral Control of Future Small Aircraft 3
Thaddäus Baier, Matthias Heller

Flexible Launch Vehicle Control Using Robust Observer-Based Controller Obtained through Structured H_∞ Synthesis 23
Emmanuel Chambon, Pierre Apkarian, Laurent Burlion

Lyapunov-Based Three-Dimensional Terminal Angle Constrained Guidance Laws 39
Mingu Kim, Yongwoo Lee, Seokwon Lee, Youdan Kim

Application of Optimization-Based Worst Case Analysis to Control Law Assessment in Aerospace 53
Hans-Dieter Joos

Robust Output Tracking of a *3DOF* Helicopter via High-Order Sliding Mode Observers 67
Alejandra Ferreira de Loza, Jérome Cieslak, David Henry, Ali Zolghadri, Leonid Fridman

Nonlinear Quadrotor Control with Online Model Identification 81
Peng Lu, Erik-Jan van Kampen, Qiping P. Chu

Comparison of L1 Adaptive Augmentation Strategies for a Differential PI Baseline Controller on a Longitudinal F16 Aircraft Model 99
Fabian Hellmundt, Andreas Wildschek, Rudolf Maier, Robert Osterhuber, Florian Holzapfel

Flight Path Management System of EOLE UAV 119
Frank Jouhaud

Fault Tolerant $\mathcal{L}_1$ Adaptive Control Based on Degraded Models 135
Toufik Souanef, Walter Fichter

$\mathcal{L}_1$ Adaptive Control of Systems with Disturbances of Unknown Bounds 151
Toufik Souanef, Ahsene Boubakir, Walter Fichter

Differential Games Based Autonomous Rendezvous for Aerial Refueling 167
Ezra Tal, Tal Shima

Nonlinear and Fault-Tolerant Flight Control Using Multivariate Splines 187
H.J. Tol, C.C. de Visser, E. van Kampen, Qiping P. Chu

Rotor State Feedback in the Design of Rotorcraft Attitude Control Laws 205
Simone Panza, Marco Lovera

Cooperative 2-On-1 Bounded-Control Linear Differential Games 227
Shmuel Y. Hayoun, Tal Shima

A New Impact Angle Control Guidance Law to Reduce Sensitivity on Initial Errors 247
Hyo-Sang Shin, Jin-Ik Lee, Antonios Tsourdos

Part II: Estimation and Navigation

On-Line Safe Flight Envelope Determination for Impaired Aircraft 263
Thomas Lombaerts, Stefan Schuet, Diana Acosta, John Kaneshige

A Sigma-Point Kalman Filter for Remote Sensing of Updrafts in Autonomous Soaring 283
Martin Stolle, Yoko Watanabe, Carsten Döll

Multiple-Model Adaptive Estimation of Time-Varying Residual Magnetic Moment for Small Satellites 303
Halil Ersin Soken, Shin-ichiro Sakai

Sliding Mode Observers for Fault Estimation in Multisensor Avionics Systems 323
Jérome Cieslak, Alejandra Ferreira de Loza, David Henry, J. Dávila, Ali Zolghadri

CubeSat Attitude Estimation via AUKF Using Magnetometer Measurements and MRPs 343
Francesco Sanfedino, Marco Scardino, Jérémie Chaix, Stéphanie Lizy-Destrez

Sensor Fault Detection and Estimation for Quadrotors Using Kinematic Equations 363
Peng Lu, Laurens Van Eykeren, Erik-Jan van Kampen, Qiping P. Chu

An Image Processing Algorithm for Ground Navigation of Aircraft 381
Kevin Theuma, David Zammit Mangion

A New Observer for Range Identification in Perspective Vision Systems 401
Victo Gibert, Laurent Burlion, Abdelhamid Chriette, Josep Boada-Bauxell, Franck Plestan

Part III: Atmospheric Applications

Flocking Algorithm for Fixed-Wing Unmanned Aerial Vehicles 415
Cezary Kownacki, Daniel Ołdziej

Airborne Doppler LiDAR Sensor Parameter Analysis for Wake Vortex Impact Alleviation Purposes 433
Jana Ehlers, Nicolas Fezans

Traffic Management along Air Streams through Space Metering 455
Mastura Ab Wahid, Hakim Bouadi, Antoine Drouin, Benjamas Panomruttanarug, Felix Mora-Camino

Integrated Design and Control of a Flying Wing Using Nonsmooth Optimization Techniques 475
Yann Denieul, Joël Bordeneuve, Daniel Alazard, Clément Toussaint, Gilles Taquin

Structured Control Law Design and Robustness Assessment for the Automatic Launch of Small UAVs 491
Jan Bolting, Jean-Marc Biannic, François Defaÿ

Behavior Trees with Stateful Tasks 509
Andreas Klöckner

Functional Interior Point Programming Applied to the Aircraft Path Planning Problem 521
Stephane Puechmorel, Daniel Delahaye

Nonlinear Visual Servoing Control for VTOL UAVs with Field of View Constraint 531
Henry de Plinval, Laurent Burlion

Automatic Landing of a High-Aspect-Ratio Aircraft without Using the Thrust 549
Maxim Lamp, Robert Luckner

Part IV: Space Applications

Bearings-Only Rendezvous with Enhanced Performance 571
Jonathan Grzymisch, Walter Fichter, Damiana Losa, Massimo Casasco

Structured Accelero-Stellar Estimator for Microscope Drag-Free Mission 591
Christelle Pittet, Pascal Prieur

Lidar-Aided Camera Feature Tracking and Visual SLAM for Spacecraft Low-Orbit Navigation and Planetary Landing 605
Franz Andert, Nikolaus Ammann, Bolko Maass

Camera-Based Tracking for Rendezvous and Proximity Operation of a Satellite 625
Nassir W. Oumer, Giorgio Panin

Linear Dynamic Modeling of Spacecraft with Open-Chain Assembly of Flexible Bodies for ACS/Structure Co-design. 639
Jose Alvaro Perez, Daniel Alazard, Thomas Loquen, Christelle Cumer, Christelle Pittet

Mechanical-Attitude Controller Co-design of Large Flexible Space Structures 659
Hari Hara Sudhan Murali, Daniel Alazard, Luca Massotti, Finn Ankersen, Chiara Toglia

Probabilistic Collision Avoidance for Long-term Space Encounters via Risk Selection 679
Romain Serra, Denis Arzelier, Mioara Joldes, Aude Rondepierre

Motion Planning and Control of a Space Robot to Capture a Tumbling Debris 699
Vincent Dubanchet, David Saussié, Daniel Alazard, Caroline Bérard, Catherine Le Peuvédic

Cross-Entropy Based Probabilistic Analysis of VEGA Launcher Performance 719
Anusha Mujumdar, Prathyush Purushothama Menon, Christophe Roux, Samir Bennani

Author Index 739

Part I
Guidance and Control

Robust Lateral Control of Future Small Aircraft

Thaddäus Baier and Matthias Heller

Abstract. It is a well-known fact that the General Aviation (GA) sector exhibits a significant higher accident rate compared to common transport aircraft (airliners). This is caused by two major reasons: First, pilots of General Aviation Aircrafts commonly show a relatively low training level and a small number of flight hours compared to airliner (ATPL) pilots. Thus, their flight experience and hazard awareness is in general strictly limited. Second, General Aviation Aircraft usually are not equipped with various well-proven safety enhancing assistance systems like an active Fly-by-Wire Flight Control System (FbW FCS), as it is state of the art in current transport aircraft. The use of active FbW FCS supports the pilot by providing excellent Flying and Handling Qualities and thus, reducing pilot's workload at the same time by directly generating appropriate control deflections, dependent on the pilot's commands and (measured) flight condition. Unfortunately, this valuable safety increasing systems did not find their way into the General Aviation sector, due to the tremendous costs of typical Fly-by-Wire control technology. The continuation of the project "Future Small Aircraft (FSA)" at the Institute of Flight System Dynamics of the Technische Universität München comprises the development of active FbW FCS with the primary objective to provide excellent Flying and Handling Qualities to yield best possible pilot's assistance for General Aviation Aircraft. In this paper, the development process of robust lateral flight control algorithms and the proof of robustness, both perfectly tailored to the specific

---

Thaddäus Baier
Institute of Flight System Dynamics (FSD), Technische Universität München (TUM), Boltzmannstraße 15, D-85748 Garching, Germany
e-mail: th.baier@tum.de

Matthias Heller
Rudolf Diesel Industry Fellow of TUM-IAS,
TUM, Institute for Advanced Study (IAS) and
Institute of Flight System Dynamics (FSD), Technische Universität München (TUM), Boltzmannstraße 15, D-85748 Garching, Germany
e-mail: matthias.heller@tum.de

J. Bordeneuve-Guibé et al. (eds.), *Advances in Aerospace Guidance, Navigation and Control*, DOI: 10.1007/978-3-319-17518-8_1

needs of manufacturers of small and medium-sized planes, are presented. The robustness proof is mandatory for GA aircraft due to the present considerable model uncertainty owing to the lower affordable modeling efforts.

1 Introduction

By comparison of the accident statistics of General Aviation (GA) versus common transport aircraft (airliners) it becomes immediately obvious that the General Aviation sector exhibits a significant higher rate of accidents (Ref. [1]). This fact is not new and even though the sources are manifold, two main reasons behind may be identified. On the one hand, General Aviation Pilots mostly hold a Private Pilot License (PPL) only and hence, their number of flight hours and thus their experience is commonly considerably limited in contrast to airliner pilots possessing an Airline Transport Pilot License (ATPL). On the other hand, current modern transport aircraft feature a noteworthy higher technical standard providing various beneficial pilot assistant systems in order to support the pilot to fly the aircraft safely and to reduce pilot's workload extensively.

The most important and effective assistant systems, which are well-known as active Fly-by-Wire Flight Control Systems (FbW FCS) intervene directly and actively into the aircraft's control and besides improving and homogenizing flying and handling qualities considerably, they offer a wide range of functionalities including pilot input monitoring, provision of warnings plus limitations and advanced protections. Hence, they increase the passenger, crew and aircraft comfort and safety competently towards the overall ultimate objective "carefree handling".

This valuable safety increase, which has become standard in current transport planes and modern business jets, unfortunately did not find its way into the general aviation sector due to the tremendous cost of typical Fly-by-Wire control technology always requiring complex redundancy and reversionary systems in order to fulfill the strict certification requirements and specifications. However, to achieve a corresponding safety enhancement for GA aircraft and thus, to depress their high accident rates and to protect human lives, the well-proven beneficial functionalities of active Flight Control Systems unquestionably have to be made available and especially affordable for this small airplane class. This applies particularly in the context of the expected significant rise in the number of GA planes and movements (Ref. [2]), which otherwise would inevitably come along with a further increase in accidents and victims, and this has to be prevented certainly. Consequently, specifically tailored active FbW FCS technology suitable and in particular affordable for GA airplanes has to be matured or developed, respectively, and made available immediately. For instance, recent technological achievements and trends in the field of (miniaturized, efficient and reliable) actuators, sensors and flight control computers (FCC) offer the potential to design more cost-effective active assistance systems to be utilized within future GA planes.

A substantial contribution to this subject is one major objective of the ambitious Technology Research Program “Future Small Aircraft (FSA)” of the Austrian aircraft manufacturer Diamond Aircraft Industries in cooperation with the Institute of Flight System Dynamics of the Technische Universität München.

Within this joint multinational Research program concerning upcoming Future Small Aircraft, first a hybrid control concept for longitudinal dynamics has been proposed (Ref. [3]), which currently is extended by the development of an appropriate lateral flight control system (refer to Ref. [4]). Although the control law layout is aimed for the provision of homogeneous excellent handling & flying qualities and thus, highly effective pilot's assistance, another main focus is set on the elaboration of special processes, tools and hardware requirements/solutions supporting the design of control algorithms which are perfectly adapted to the specific needs of manufacturers of small and medium-sized planes.

Summarized, the development process applied comprises the following main topics:

- Elaboration of a generalized controller structure for lateral dynamics suitable for all classical airplane configurations (fixed-wing aircraft with empennage) irrespective of whether a full Fly-by-Wire or "Hybrid Flight Control System" will be implemented, see Ref. [4].
- Provision of a corresponding appropriate gain layout methodology, denoted "Model Reference Direct Eigenstructure Assignment" (*MR DEA*), which is directly adapted to the generalized controller structure previously introduced in order to determine the gain sets/tables along the entire envelope (Ref. [4]).
- Development of a controller analysis tool for robustness assessment based on a detailed multivariable model uncertainty description and evaluation by means of the *Advanced Structured Singular Value* $\mu_{\Delta\Gamma}$ in order to enable a (affordable) strict certification process (main focus of this paper).

Verification, validation and test of the whole development chain will be accomplished by implementation of the lateral controller designed into an state-of-the-art general aviation aircraft simulator (D-SIM-42) and finally, by actually flying the active FbW FCS proposed on the dedicated flying testbed "*Fliegender Erprobungsträger*", a research aircraft based on a DA-42 NG MPP airframe which was projected, developed and modified at the Institute of Flight System Dynamics of Technische Universität München.

2 Development, Integration and Test Environment (along) with Flight Dynamics Modeling

A small aircraft featuring an active Fly-by-Wire Flight Control System (FbW FCS) represents a new class of general aviation airplanes. By utilizing the benefits of such a system, some basic aircraft design objectives (e.g. inherent static stability and naturel damping characteristics) may be shifted to other focuses and thus, the aircraft concept may differ from today's conventional shape. Nevertheless, the novel active FbW FCS must be implemented, demonstrated and tested out within a suitable flying testbed before designing an entire new aircraft configuration relying on such a novel active Flight Control System technology.

For this purpose, the Institute of Flight System Dynamics owns a fully-fledged Flight Control System Development and Integration Environment:

- A DA-42 Flight Training Device (D-SIM-42 FTD simulator) with extensive capabilities to simulate malfunctions of multiple aircraft systems for design and validation of control laws and pilot in the loop verification.
- A DA-42 Airframe and Control System Iron Bird for component tests and verification, integration tests for research aircraft and hardware in the loop simulation in connection with the Flight Training Device
- And in particular a research aircraft Diamond DA-42 MPP NG (Multi Purpose Platform New Generation), particularly dedicated as in-flight simulator with an Experimental Fly-by-Wire (EFbW) control system, see Fig. 1.

Fig. 1 Research Flying Testbed (DA-42MPP NG) featuring EFbW FCS

With this continuous "end-to-end" development and integration chain at hand, it suggests itself to utilize the Diamond DA-42 aircraft as reference configuration for the control design methodology application and demonstration. For this type of aircraft the feasibility, advantages and reliability of the active FbW FCS for general aviation aircraft will be analyzed, verified and finally (in-flight) proven.

Regarding the functional layout and development of the control system, an appropriate high-fidelity flight simulator for functional testing is of vital importance. The available D-SIM-42 Flight Training Device is equipped with an original glass cockpit, *Garmin G1000 Avionics package* with primary flight display (*PFD*) as well as a multi-function display in order to achieve the most realistic cockpit environment. The DA-42 simulator thus represents a perfect tool for functional controller pilot-in-the-loop testing under "real world" conditions. Unfortunately, the D-SIM-42 flight dynamics model as well as all internal signal processing are completely capsuled "black boxes" and entirely isolated from external inputs. Both is adverse for the use of the simulator as design and implementation platform regarding the active FCS to be developed.

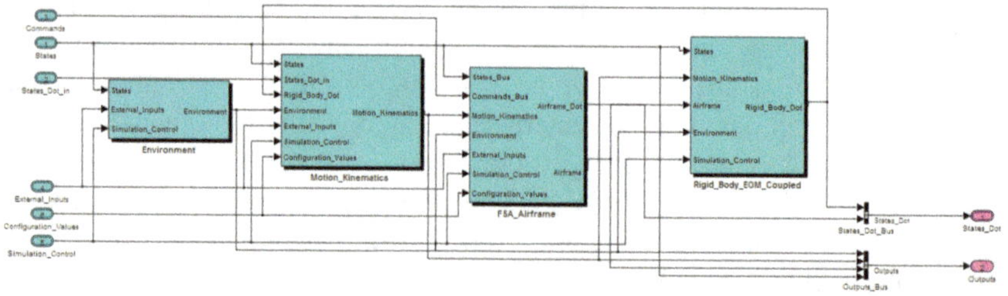

Fig. 2 FSD DA-42/FSA Nonlinear 6 DOF Flight Dynamics & Simulation Model

For this reason, an independent full-fledged nonlinear six 6 DoF model called "FSD DA-42/FSA 6 DOF Flight Dynamics & Simulation Model" has been implemented using *MATLAB / Simulink* and the corresponding top-level block structure is depicted in Fig. 2. Particular emphasis was placed on a high-fidelity physical modeling and on a exact reproduction of the dynamics of the D-SIM-42 FTD simulator.

The validation of the FSD DA-42/FSA 6 DOF flight dynamics model was accomplished by systematic simulator flight tests conducted in the D-SIM-42 device (as detailed within Ref. [3]) assuring a excellent matching versus the D-SIM-42 FTD.

3 Lateral Flight Control Design Methodology

3.1 Objectives, Requirements and Derived "Design Philosophy"

The primary objective of the FSA lateral controller design is to provide excellent homogenous flying/handling qualities along the entire flight envelope in order to reduce pilot's workload significantly when flying the airplane manually. Related intensions are to increase the safety and comfort of passengers and crew together with the efficiency of the aircraft (fuel consumption).

To achieve these goals, it is required to improve and homogenize the stability and command characteristics to shape them intuitive and predictable over the operational envelope and to accomplish an effective gust load rejection. Additionally, the effects of aircraft configuration changes (e.g. one engine inoperative) should automatically be compensated and a further reduction of the pilot's workload should be achieved by partial automation of secondary controls, e.g. spoilers, flaps or thrust setting.

Consequently, the following primary design objectives may be deduced and appointed in terms of our "design philosophy":

- Modification of the aircraft's stability characteristics:
  The basic stability characteristics and thus the flying qualities shall be modified by assignment of "optimal" damping and natural frequency or time constant, respectively, each to the different eigenmodi of the lateral motion. Furthermore, an effective decoupling of roll and yaw axes shall be applied.
- Augmentation of the command behavior:
  - Rate Command/Attitude Hold (*RC/AH*) characteristics w.r.t. roll axis versus Angle of Sideslip (*AOS*) Command/Zero Lateral Load behavior (yaw axis):
    The provision of "velocity vector roll" and angle of sideslip command has shown to be very intuitive and predictable for the pilot in combination with attitude hold and zero lateral load factor (n_y=0) characteristics, respectively, when the inceptor is released.

  - Feed Forward Path Augmentation:
    By implementation of a "direct link" feed forward branch each for yaw/roll axis, the aircraft's control sensitivity will be improved by means of "integrator feedback pole" cancellation out of the corresponding command transfer functions in order to obtain a "crisp" common control response.
- Decoupling of command inputs (introduction of decoupled "auxiliary control effectors"):
  To manually initiate a velocity vector roll, the pilot would have to apply coordinated aileron and rudder command and vice versa a pure buildup of sideslip would require adding adequate aileron deflections to the pedal input. This maneuver coordination concerning the decoupling of experimental roll and yaw axes shall be accomplished automatically by means of so-called "auxiliary control effectors". A respective control axes decoupling is not only convenient for the pilot, it also makes the subsequent controller layout much more intuitive and straightforward.

3.2 Controller Design

As described within the previous section, the controller to be developed has to comply with several different objectives. The fulfillment of the specific design goals is thereby typically associated with the different branches of the controller structure. The overall controller structure is depicted in Fig. 3, a detailed description can be found in Ref. [4].

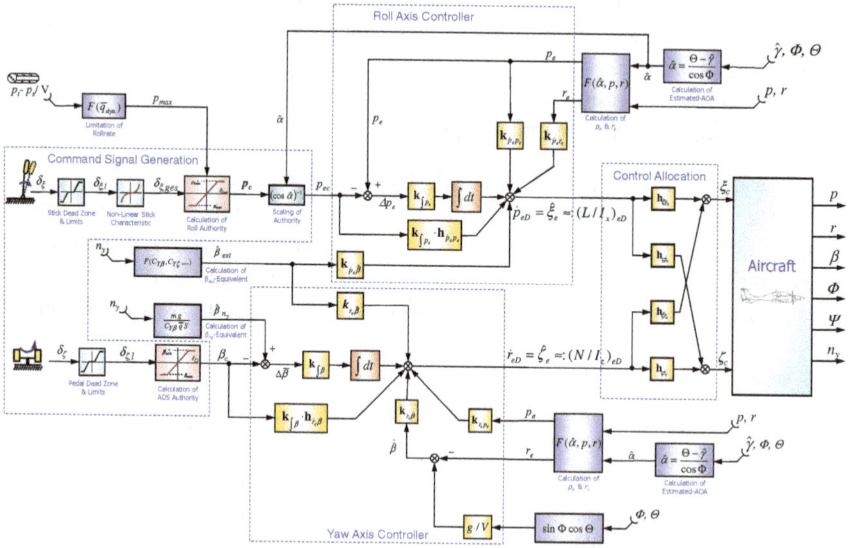

Fig. 3 Controller Structure

Generally, the controller consists of the *Command Signal Generation* providing roll rate and angle of sideslip command signals, the decoupling *Yaw* and *Roll Axis Feedback Controller* relying on the calculated commands plus sensor feedbacks generating generalized input variables in terms of demanded roll and yaw control acceleration (normalized control moment w.r.t. the experimental axes) as well as the *Control Allocation* converting the generalized input variables to equivalent allocated rudder/aileron control surface deflection commands.

The *Command Signal Generation* path scales, limits and processes the pilot's stick and pedal inputs to yield/provide intuitive stick and pedal characteristics. The *Yaw* and *Roll Axis Controller* each basically represents a classical *PI*-scheme augmented by a direct link path. Accordingly, the relative and absolute damping as well as the natural frequency of the dutch roll, the roll and spiral time constants are adjusted to meet so-called "optimal Level 1*" flying qualities derived from Ref. [5] and a decoupling of the roll and yaw motion (with regard to the experimental frame, i.e. velocity vector roll / pure sideslip buildup) is realized. The direct link paths are used to ensure a "crisp" roll and angle of sideslip response. The *Control Allocation* provides a coordinated deflection of rudder and aileron according to the *Yaw* and *Roll Axis Controller* outputs $\dot{p}_{eD}$ and $\dot{r}_{eD}$ to ensure a decoupled command behavior relating to the experimental frame.

To ensure proper controller function, high and low frequency measurements have to be available appropriately, as listed in the following table (whereas the "hat-variables" represent not directly measured but estimated signals).

Table 1 Measurement Signals

High Frequency Measurements	*Low Frequency Measurements*
n_y, p, r	Φ, Θ, $(\hat{\gamma})$

The layout of the specific controller gains (feedback, feedforward plus cross-feeds) for a representative set of flight conditions along the envelope is performed by applying *Model Reference Direct Eigenstructure Assignment* (*MR DEA*). A detailed description of *MR DEA* can be found in Ref. [4].

Appropriate controller gain tables holding for the entire flight range were established as follows: For a dense mesh of discrete trim points along the envelope corresponding state space models are generated for each of which a gain set has been computed by the *MR DEA* method. Based on these gain matrices, a gain-scheduled controller covering the whole admissible flight envelope is designed. Furthermore, in case of an air data (dynamic pressure) failure, a "reversionary mode" fixed- gain back-up controller is developed.

4 Robustness Considerations and Proof

To ensure safe and reliable operation of the controlled aircraft along the entire admissible envelope, it is mandatory to proof/verify/demonstrate that the minimum required closed-loop stability characteristics are guaranteed for all deviations (to be expected/specified) between the real (perturbed) system and the nominal plant plus controller development model. The differences (deviations/errors) between a real system dynamics and the nominal models used for control system design are typically denoted by the term *model uncertainty*. A controlled system is called robust, if sufficient stability margins for all perturbed plants up to the worst-case model uncertainty are guaranteed.

In the past until today, robustness verification for flight control systems within the governmental permission and certification process is adduced via single-input-single-output (*SISO*) phase and gain margins (Nyquist criterion or Nichols "disk margins", respectively), compare Ref. [6]. However, this is also done for multiple-input-multiple-output (*MIMO*) systems, e.g. the lateral aircraft dynamics, by applying so-called "bottleneck" single loop cuts (with all other loops closed). Nevertheless, *Doyle* showed by counterexample (Ref. [7]), that a single-cut (*SISO*) robustness proof of a *MIMO* system is basically not sufficient. In contrast, sufficient and non-conservative evidence of robustness for a *MIMO* system can be provided by means of the "Structured Singular Value (*SSV*, μ-Analysis) see Ref. [8]. Hence, an innovative advancement of the μ-Analysis in order to prove "*robust compliance*" w.r.t. user-defined specific "stability characteristics borders", will be introduced.

4.1 Introduction of Robust Compliance and $\mu_{\Delta\Gamma}$-Analysis

The μ-Analysis provides a sufficient evidence of *robust stability* for a controlled system (Ref. [8]). That is, via the structured singular value μ it can be assured that for a specified set of model uncertainties no pole/eigenvalue of a nominal stable system reaches or crosses the imaginary axis of the complex plane and thus, the system will remain stable.

From a flight dynamics point of view, the imaginary axis is commonly not decisive for sufficient flying qualities and hence, not relevant for adequate stability and robustness characteristics. For example, the spiral mode of an aircraft with a control and stability augmentation system (CSAS) should preferably be neutral stable for excellent flying qualities ("MIL-Level 1*). However, sufficient for MIL-Level 1 flying qualities, also a instability featuring a long time-to-double of the spiral mode is permitted (Ref. [5]). A verification of robustness with respect to a neutral or bounded unstable spiral mode using classical μ-Analysis is not possible due to the nominal stability precondition and the imaginary axis reference. Therefore, in order to demonstrate adequate robustness properties w.r.t. sufficient flying and handling qualities, e.g. allowing an unstable spiral pole meeting the time-to-double limit, the *robust compliance* and the $\mu_{\Delta\Gamma}$-Analysis is proposed.

The $\mu_{\Delta\Gamma}$-Analysis does not refer to the imaginary axis as stability border, but evaluates the stability characteristics in relation to a user-defined border Γ within the complex plane. In direct analogy to the imaginary axis as stability border for *robust stability*, we define (*robust compliance*): If all poles/eigenvalues of a linear time-invariant system are to the left of the user specified border Γ for all perturbed plants, the system fulfills the required robust stability characteristics and is called *robust compliant*. If at least one pole/eigenvalue lies on (or to the right of) the specified robustness bound, the system is *not robust compliant, i.e. non-compliant*. In Fig. 4, the concept of *compliance* and *non-compliance* in analogy to *stable* an *unstable* are illustrated.

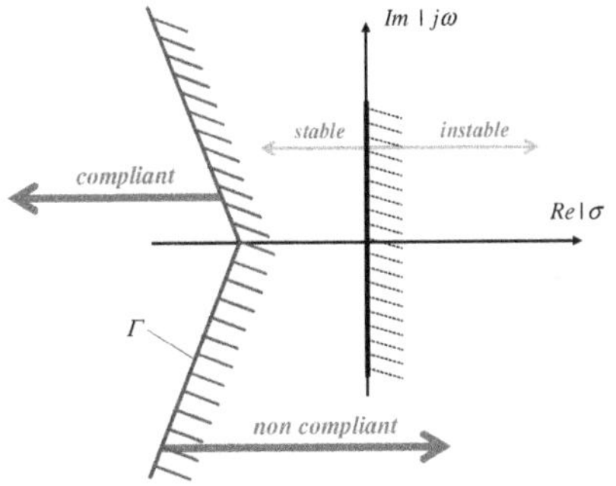

Fig. 4 Compliant and non-compliant

The execution of a robustness analysis for a *MIMO* linear time-invariant dynamic system via the $\mu_{\Delta\Gamma}$-Analysis is based on a standard structure of the robust control theory, the $\boldsymbol{M\Delta}$-Structure (Fig. 5), compare Ref. [8], [9] et al. The $\boldsymbol{M\Delta}$-Structure yields the partition of known system parts (plant and controller) $\boldsymbol{M}(\boldsymbol{G},\boldsymbol{K})$ and uncertainties $\boldsymbol{\Delta}$.

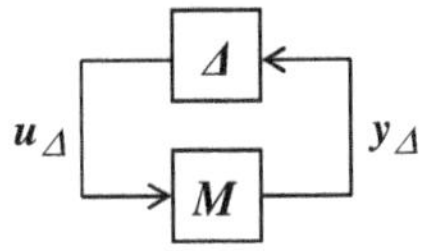

Fig. 5 **MΔ**-Structure

In correspondence to the definition of the structured singular value μ (Ref. [8]), $\mu_{\Delta\Gamma}$ for the analysis of *robust compliance* (*RC*) is defined as follows:

Definition $\mu_{\Delta\Gamma}$*: If* Γ *is the user-defined bijective compliance border and if for the* $\boldsymbol{M\Delta}$*-Structure is true:*

- $\boldsymbol{M}_{(s_\Gamma)}$ *and* $\boldsymbol{\Delta}_{(s_\Gamma)}$ *are complex matrixes*
  for every point of the compliance border Γ *(Index* s_Γ*)*
- $\boldsymbol{\Delta}_{(s_\Gamma)} \in \underline{\boldsymbol{\Delta}}$ *with*
  $\underline{\boldsymbol{\Delta}} = \{diag(\delta_1 \boldsymbol{I}_1, \dots, \delta_k \boldsymbol{I}_k, \Delta_1, \dots, \Delta_k) \mid \delta_i \in \mathbb{C},\ \Delta_j \in \mathbb{C}^{m_j x m_j}\}$,

the structured singular value $\mu_{\Delta\Gamma}$ *of the matrix* $\boldsymbol{M}$ *is for every point of the compliance border* $s_\Gamma = \sigma_\Gamma(\omega) + j\omega$ *with* $\omega \in (-\infty; +\infty)$ *(for the analysis of robust compliance) defined as follows:*

$$\mu_{\Delta\Gamma}(\boldsymbol{M}_{(s_\Gamma)}) := \begin{cases} \left(\min\{\bar{\sigma}(\boldsymbol{\Delta}_{(s_\Gamma)}) \mid \det(\boldsymbol{I} - \boldsymbol{M}_{(s_\Gamma)}\boldsymbol{\Delta}_{(s_\Gamma)}) = 0\}\right)^{-1} \\ 0 \mid \det(\boldsymbol{I} - \boldsymbol{M}_{(s_\Gamma)}\boldsymbol{\Delta}_{(s_\Gamma)}) \neq 0,\ \forall\, \boldsymbol{\Delta} \in \underline{\boldsymbol{\Delta}} \end{cases} \quad (1)$$

The proof of *robust compliance* via the structured singular value $\mu_{\Delta\Gamma}$ of a nominal compliant linear time-invariant dynamic system is as follows:

Robust Compliance via $\mu_{\Delta\Gamma}$*: If* Γ *is the user-defined unique compliance border and if for the* $\boldsymbol{M\Delta}$*-Structure is true:*

- $\boldsymbol{M}(s)$ *and* $\boldsymbol{\Delta}(s)$ *are complex transfer matrixes, which are both compliant, i.e. all poles/eigenvalues of* $\boldsymbol{M}(s)$ *and* $\boldsymbol{\Delta}(s)$ *lie to the left of the compliance border and*
- $\boldsymbol{\Delta}(s)$ *is a structure block diagonal uncertainty matrix conform with* $\bar{\sigma}(\boldsymbol{\Delta}(s)) \leq 1,\ \forall\, s = \sigma + j\omega \ \ mit \ \ \sigma = \sigma_\Gamma(\omega)$,

the closed $\boldsymbol{M\Delta}$*-Structure is robust compliant (fulfills robust compliance) for all permitted uncertainties, if*

$$\mu_{\Delta\Gamma}\big(\boldsymbol{M}_{(s_\Gamma)}\big) < 1 \ \ \forall\omega. \tag{2}$$

According to the SSV μ the direct calculation of $\mu_{\Delta\Gamma}$ is not possible, but an upper and lower bound of $\mu_{\Delta\Gamma}$ can be determined based on the properties of $\mu_{\Delta\Gamma}$. This is analog to the μ-Analysis (Refs. [8], [9]).

A simplified descriptive interpretation of $\mu_{\Delta\Gamma}$ and *robust compliance* is as follows: The roots of $\det\big(\boldsymbol{I} - \boldsymbol{M}_{(s_\Gamma)}\,\boldsymbol{\Delta}_{(s_\Gamma)}\big)$ are identical with the poles/eigenvalues of the closed $\boldsymbol{M\Delta}$-Structure. If for every point of the compliance border, the equation $\det\big(\boldsymbol{I} - \boldsymbol{M}_{(s_\Gamma)}\,\boldsymbol{\Delta}_{(s_\Gamma)}\big)$ is solved, it can be identified if a pole/eigenvalue lies on the borderline. Regarding a nominal *compliant* system (i.e. all poles/eigenvalues lie to the left of the compliance border) and considering a successively growing uncertainty $\boldsymbol{\Delta}$, the smallest $\boldsymbol{\Delta}$ for which $\det\big(\boldsymbol{I} - \boldsymbol{M}_{(s_\Gamma)}\,\boldsymbol{\Delta}_{(s_\Gamma)}\big) = 0$ holds, can be identified. Thereby $\boldsymbol{\Delta}$ is measured by means of the maximum singular value $\bar{\sigma}\big(\boldsymbol{\Delta}_{(s_\Gamma)}\big)$. That means that at least one pole/eigenvalue has reached the compliance border. As a result, the structured singular value $\mu_{\Delta\Gamma}\big(\boldsymbol{M}_{(s_\Gamma)}\big)$ is a measure for the crossing of the compliance border by a pole/eigenvalue in relation to the uncertainty $\boldsymbol{\Delta}$. A large value of $\mu_{\Delta\Gamma}\big(\boldsymbol{M}_{(s_\Gamma)}\big)$ means "worse robustness characteristics"; a small value of $\mu_{\Delta\Gamma}\big(\boldsymbol{M}_{(s_\Gamma)}\big)$ means "good robustness properties". According to the small gain theorem (Ref. [8, 9]) and under the constraint of allowed uncertainties $\bar{\sigma}\big(\boldsymbol{\Delta}_{(s_\Gamma)}\big) \leq 1$, the requirement for *robust compliance* by $\mu_{\Delta\Gamma}\big(\boldsymbol{M}_{(s_\Gamma)}\big) < 1$ follows.

A detailed and comprehensive derivation and proof of the *robust compliance* as well as the structured singular value $\mu_{\Delta\Gamma}$ including its properties is done by the authors, but is outside the scope of this paper. The above introduction has no claims of completeness and is only intended to provide a short introduction for the better understanding of the following.

4.2 Uncertainty Modeling

The difference between the real system dynamics and the model used for system analysis and controller design (nominal system) is captured in the form of uncertainties, against which *robust compliance* has to be shown.

For a sufficient proof of *robust compliance,* it is necessary to verifiy robustness not only for a dense grid of discrete trim points over the entire envelope, but also for a continuous close description of the system dynamics over the admissible envelope. For this purpose, the trim point uncertainty δ_{Env} is established, which represents the dynamic pressure. In dependence of the dynamic pressure, the primary change of the dynamic behavior of the aerodynamic forces and as a result of the flight dynamics and aircraft characteristics can be described over the complete envelope. For that reason, all essential changes over the flying range of the matrix elements $\boldsymbol{A}$ and $\boldsymbol{B}$ of the system-describing state space model and also the gain scheduling of the controller are represented as a function of δ_{Env}. Thus it is possible to describe the essential behavior modification of the controlled aircraft over the entire flight envelope as a continuum by one state space model. A detailed explanation of such a proceeding can be found in Ref. [10]. In the case that it is not possible to find an adequate description of the system dynamics changes over the envelope by one parameter (here the dynamic pressure and δ_{Env}), more envelope and aircraft parameters, for example flying speed, flight altitude and aircraft mass, and consequently additional trim point uncertainty δ_i, should be used.

The difference between the real system dynamics and the nominal system, for example caused by wrongly identified aerodynamic parameters like the derivatives C_{Lp} or $C_{n\beta}$, is measured by parametric uncertainties. These parametric uncertainties are considered for every matrix derivative (e.g. N_p or L_ξ) of the state space model.

For a better understanding of the uncertainty modeling of the aircraft dynamics, the derivative L_p is exemplarily considered, as shown in equation (3). The first part, summarized as $L_{p_{Env\,nom}}$, represents the nominal value, which changes over the flight envelope, dependent on the dynamic pressure. The change of the nominal $L_{p_{Env\,nom}}$ along the envelope is captured with good approximation as a second degree polynomial in dependence of the envelope uncertainty δ_{Env}. The second part Δ_{L_p} represents the parametric uncertainty of the particular flight condition, which describes the difference between the real system dynamics and the nominal model (nominal system).

$$L_p = \underbrace{\left(\nu_{1/22}\,\delta_{Env}{}^2 + \nu_{2/22}\,\delta_{Env} + \nu_{3/22}\right)}_{L_{p_{Env\,nom}}} \cdot \Delta_{L_p} \tag{3}$$

For the proof of *robust compliance*, a dynamic uncertainty of the actuators is also taken into account, consequently the uncertain transfer function of the lateral control units are shown in the following equation.

$$G_{i\,uc} = G_{i\,nom}\,(1 + \Delta_{Gi} W_{Gi}) \qquad i = \xi, \zeta \qquad \text{mit} \quad \|\Delta_{Gi}\|_{\infty\Gamma} \le 1 \tag{4}$$

In equation (4), Δ_{Gi} is the norm-bounded uncertainty and W_{Gi} the dynamic uncertainty weight, which allows considering uncertainties in dependence of phase and amplitude.

4.3 *Robustness Analysis and Assessment*

Before presenting the results of *robust compliance analysis*, the specified compliance border of the executed $\mu_{\Delta\Gamma}$-Analysis will be introduced.

Extracted from Ref. [5], there are three demands from which the compliance boundary of the minimum stability characteristics can be defined to ensure adequate controllability and stability properties of the aircraft, taking into account the most adverse combination of the assumed uncertainties. First, there is the request for a maximum permissible instability of the spiral mode. For the compliance border, the constraints of the spiral mode that ensure so called MIL-Level 1 flying qualities ($\lambda_{SPmax} = 0{,}0346\ [rad/s]$) are taken into account, to fulfill the requirements of the allowed but bounded "in"-stability of the spiral mode. Second, there is the request for a minimum absolute damping of the dutch roll mode. Since this is a "stable" demand, the requirements for the MIL-Level 2 flying qualities are taken into account ($\sigma_{DRmin} = -0{,}05\ [rad/s]$). Third, there is the demand for a minimum relative damping of the dutch roll mode, which is also based on the MIL-Level 2 request ($\zeta_{DRmin} = 0{,}02\ [-]$). The roll mode is not decisive for the compliance border, because of its position in the left half-plane.

Based on the selected minimal requirements, that have to be fulfilled in consideration of the assumed uncertainties, a compliance border can be established. To merge the requirements for the spiral and the dutch roll mode, a tapered region along the real axis is inserted. The top of tapered area lies on the real axis and is congruent with the requirements for the spiral mode. The real axis values of the endpoints of the v-notch correspond to the required minimal absolute damping of the dutch roll. For the specification of the imaginary axis values, a theoretically possible constellation of a conjugate complex pair of poles/eigenvalues with a minimal time of oscillation of about $62\ [s]$ is considered acceptable, because of its slow dynamic motion. The remaining part of the compliance border fulfills the relative and absolute damping of the dutch roll. In Fig. 6, the finale compliance border that is used for the robustness proof is pictured.

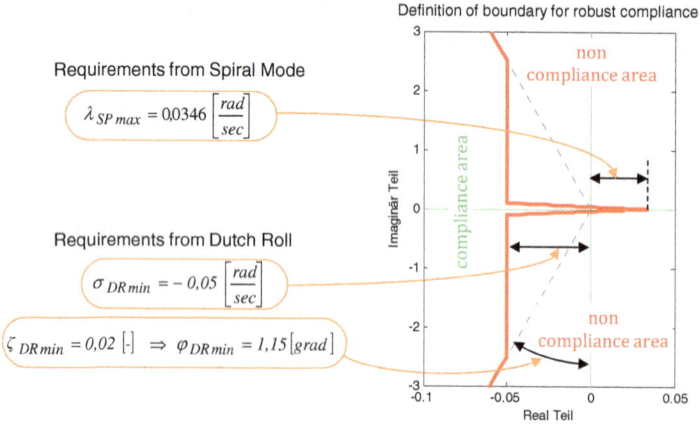

Fig. 6 Compliance border

The result of the $\mu_{\Delta r}$-Analysis for the defined compliance border considering the below specified uncertainty conditions is pictured in Fig. 7. The aircraft-describing state space model is equipped with a ± 6 percent parametric uncertainty of every derivative of the $\boldsymbol{A}$ and $\boldsymbol{B}$ matrix and is taking the trim point uncertainty δ_{Env} for a continuous robustness analysis over the admissible envelope into account. The actuators of aileron and ruder, modeled as PT_2-transferfunctions, are performed with dynamic uncertainties, see Eq. (4). The scheduling of the controller over the flight envelope is specified in dependence of the trim point uncertainty δ_{Env}.

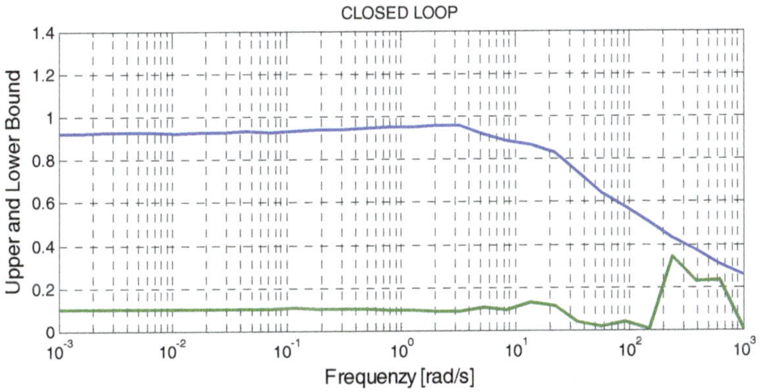

Fig. 7 $\mu_{\Delta r}$-Analysis of the complete System (gain-scheduled controller, trim point uncertainty)

The progression of the upper boundary in Fig. 7 of the structured singular value $\mu_{\Delta r}$ does not reach the value 1, which means that no pole/eigenvalue of the nominal stable system hits the compliance border, considering the specified uncertainty model. The system fulfills *robust compliance* for the admissible flight envelope.

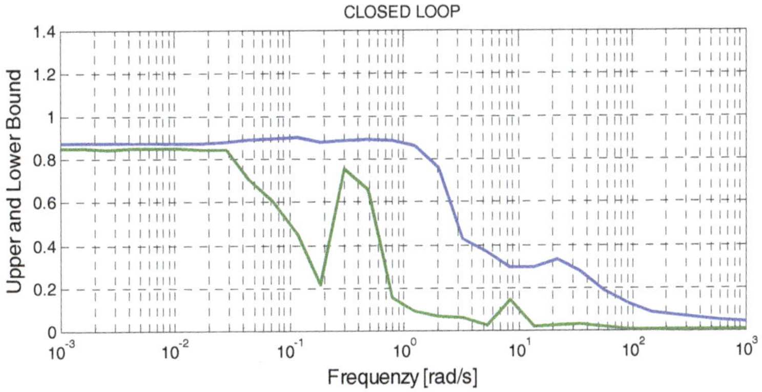

Fig. 8 $\mu_{\Delta r}$-Analysis of the complete System (fixed gain controller, trim point uncertainty)

In addition, a fixed gain back-up controller, which can be used in the case of breakdown of the dynamic pressure signal, is elaborated. The flying qualities reached with the fixed gain controller are less adequate than the gain scheduled controller, but significantly better than those of the uncontrolled aircraft. In Fig. 8, the results of the $\mu_{\Delta\Gamma}$-Analysis for the *robust compliance* of the total system consisting of the aircraft, the actuator system and the fixed gain controller, are shown. The uncertainty model is the same as in the previous case, but without the trim point uncertainty for the controller. As a result, the total system with the fixed gain controller also fulfills *robust compliance* for the complete flight envelope.

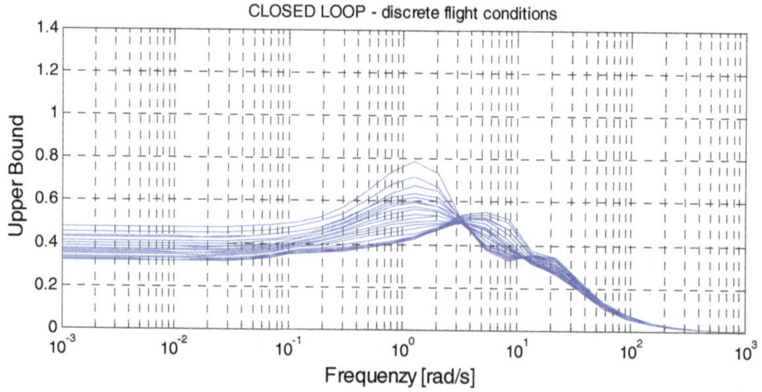

Fig. 9 $\mu_{\Delta\Gamma}$-Analyses for discrete trim points (gain-scheduled controller)

Supplementary to the robustness proof in form of a continuum over the admissible flight envelope, a $\mu_{\Delta\Gamma}$-Analysis of discrete trim points, is worked out according to the today commonly accepted SISO robustness analysis by Nyquist Stability Criterion for discrete flight conditions.

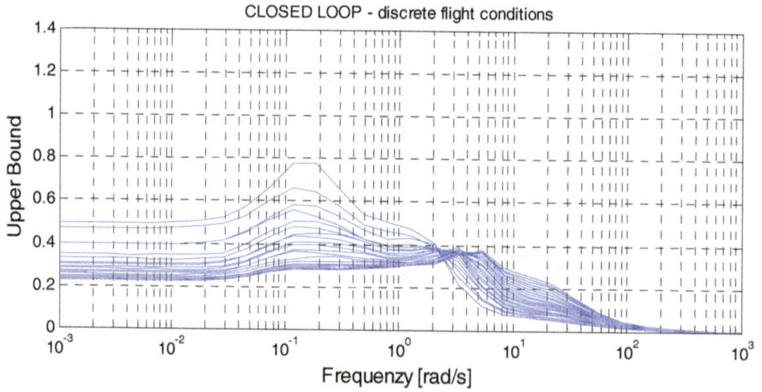

Fig. 10 $\mu_{\Delta\Gamma}$-Analyses for discrete trim points (fixed gain controller)

For a dense mesh of discrete trim points consisting of state space models, $\mu_{\Delta\Gamma}$-Analyses are done. The uncertainty model of the aircraft contains the same dynamic uncertainty for the actuators as for the continuous analysis and parametric uncertainties for every derivative of the $\boldsymbol{A}$ and $\boldsymbol{B}$ matrix of $\pm 30\%$ with the gain scheduled controller and $\pm 20\%$ with the fixed gain controller. Every considered trim point fulfills *robust compliance* for the above defined compliance border and the assumed uncertainty model. In Fig. 9, the upper boundaries of the structured singular value $\mu_{\Delta\Gamma}$ of all analyzed trim points with the gain-scheduled controller are shown and no curve meets the value 1. In addition, Fig. 10 shows the same for the fixed gain controller.

The presented $\mu_{\Delta\Gamma}$-Analyses show sufficient *robust compliance* of the analyzed system, considering the defined compliance border. Therefore, the controlled aircraft fulfills the specified requirements and exhibits proofed robust stability characteristics (*robust compliance*).

The reason, that the system, analyzed as a continuum, allows a smaller deviation ($\pm 6\%$) compared to the $\mu_{\Delta\Gamma}$-Analysis of the discrete trim points with a parametric uncertainty of ($\pm 30\%$), can be motivated by the conservative effects of the curve fitting capturing the change of the ***A*** and ***B*** matrix elements over the flight range, compare eq. (3).

5 Linear and Nonlinear Analysis

Finally, a brief linear analysis of the closed loop system is presented to verify the dynamic characteristics of the controlled aircraft and to show the improvement compared to the uncontrolled airplane. The final evaluation of the complete

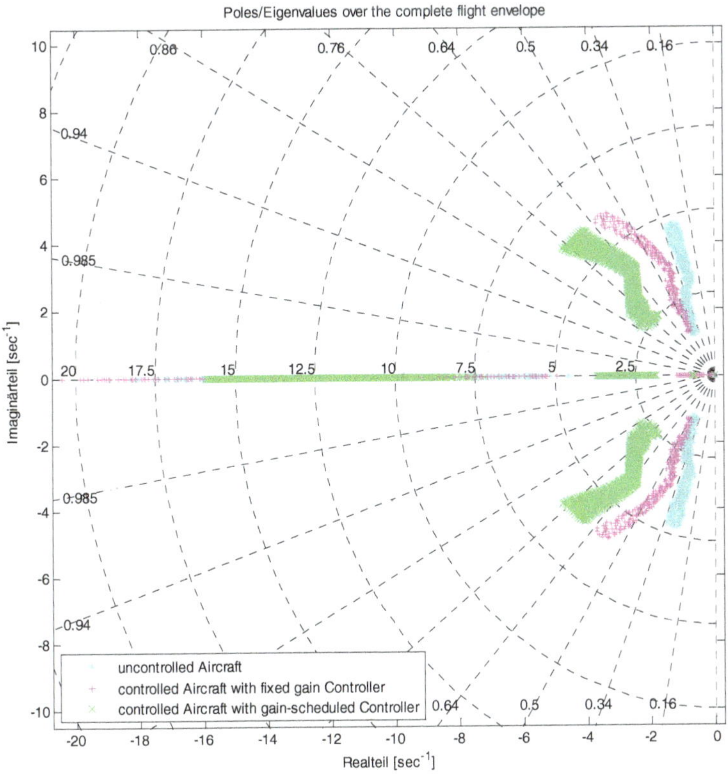

Fig. 11 Poles / eigenvalues in comparison

system including actuator dynamics and dead time for digital processing is done with a nonlinear numerical simulation, which will also be introduced.

In Fig. 11, the poles/eigenvalues of the controlled aircraft with the gain-scheduled controller, of the controlled aircraft with the fixed gain controller and in addition of the uncontrolled aircraft, are illustrated.

The variation of the roll mode over the envelope is reduced to an acceptable range that provides intuitive flying properties and in particular the high absolute damping, which causes too large control sensitivity (Ref. [11]), is reduced measurably. The position of the dutch roll poles/eigenvalues is changed by the gain scheduled controller to receive an optimal absolute and relative damping and also a higher natural frequency, so that the dutch roll fulfills MIL-Level 1* properties. In addition, the fixed gain controller improves the dutch roll to get high damping and therefore better flying properties compared to the uncontrolled aircraft. The spiral mode of the uncontrolled aircraft that has stable characteristics is modified by the controller in order to be neutral stable as is required for excellent flying and handling qualities of an aircraft equipped with a control and stability augmentation system (CSAS).

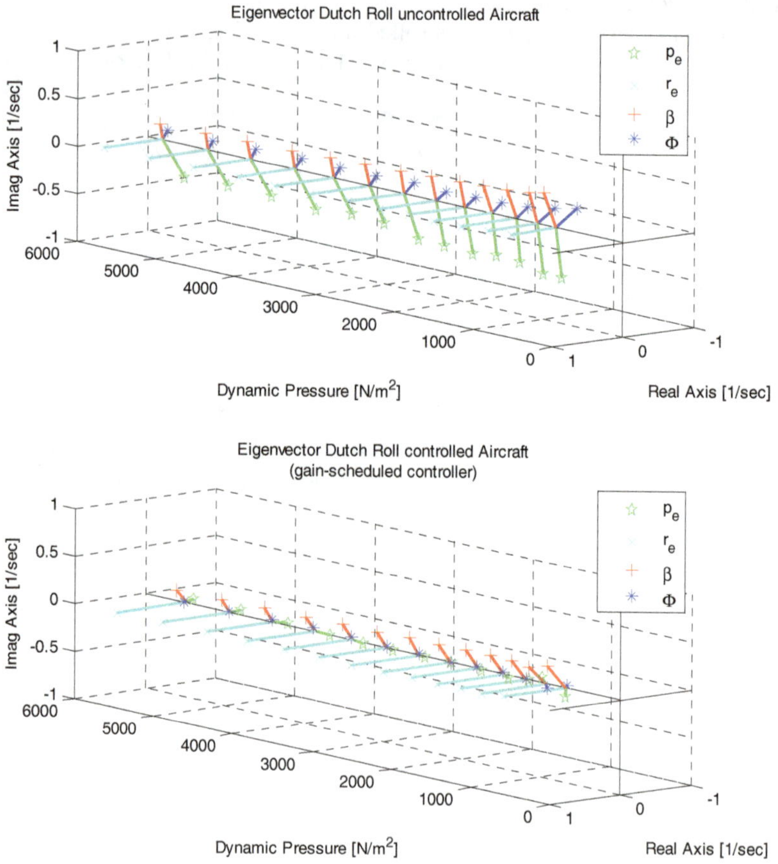

Fig. 12 Eigenvector dutch roll

The requested decoupling of the roll and yaw dynamic, which is an important requirement to generate optimal flying characteristics of aircrafts with a CSAS, is also realized by the gain-scheduled controller. To demonstrate this fact, the dutch roll eigenvector will be analyzed: In Fig. 12, the dutch roll eigenvector of the uncontrolled aircraft and of the aircraft with the gain-scheduled controller is pictured in dependence on the dynamic pressure. The figure clearly shows that the part of the roll rate (p_e) and the bank angle (ϕ) of the controlled aircraft is reduced to a minimum and thus the yaw dynamics are decoupled from the roll axis states, as required.

It has been proofed that, in analogy to the dutch roll, the part of the angle of sideslip both of the roll and the spiral mode is also reduced to a minimum. As a result, the roll axis dynamic is decoupled from the angle of sideslip.

The analysis of eigenvalues and eigenvectors demonstrates clearly that the requirements concerning excellent flying properties, particularly the decoupling of the lateral motion, is fulfilled adequately in consideration of the linear model.

Finally, in the following, the fulfillment of these requirements will be confirmed in a nonlinear analysis:

The nonlinear numerical simulation of the controlled aircraft using the gain-scheduled controller is performed with the above explained *FSD DA-42/FSA 6 DOF* simulation model. The aircraft characteristics will be analyzed first for a command in the roll axis (p_{ec}), second for an atmospheric disturbance from the side. The analysis of the controlled aircraft is always executed in comparison to the uncontrolled one.

The dynamic behavior of the controlled aircraft with a disconnected square-wave signal of p_{ec} is shown in the left diagram line of Fig. 13. In comparison to the uncontrolled aircraft, the controlled one performs a coordinated motion in building up and reducing the bank angle as well as in the stationary turn flight. As a result of the automatic coordinated flight, there are almost no effects to the angle of sideslip as required. This is caused by the harmonized controller-demanded rudder and aileron deflection. As is evident, the uncontrolled aircraft (right diagram line of Fig. 13) does not execute a coordinated flight motion for a lateral stick input, thus the pilot workload to generate suchlike behavior is very high.

After the analysis for a p_{ec}-command has been shown, the atmospheric disturbance will be examined:

The shape of the side wind gust has a "$1 - cos$" form, see Ref. [5]. The dynamic performance of the aircraft with the gain-scheduled controller is pictured in the left diagram line of Fig. 14. The uncontrolled aircraft behavior caused by the gust is presented in the right diagram line of Fig. 14. In addition, in the lowest diagrams of Fig. 14 the crosswind gust velocity v_{Wind} is shown.

The controller performs coordinated deflection of the ruder and the aileron, which improves the dynamic characteristics of the yaw axis states r_e and β significantly, because of the better transient response. Furthermore, the automatic surface deflection of the controlled aircraft reduces considerably the effects to the roll axis. The roll rate p_e is reduced to a minimum and thus the bank angle is almost zero in contrast to the bank angle of the uncontrolled airplane.

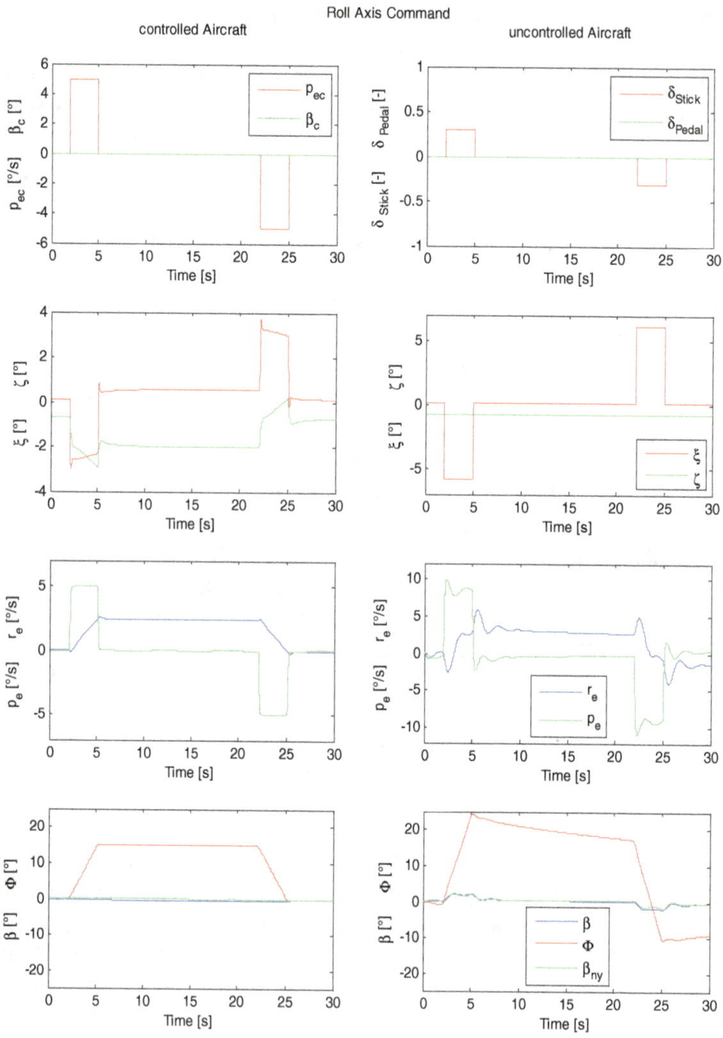

Fig. 13 Comparison of controlled (gain-scheduled controller) and uncontrolled aircraft for roll command

The analysis of the disturbance behavior shows that the stability characteristics of the controlled aircraft are superior compared to the characteristics of the uncontrolled one. Because of the improved disturbance response the pilot workload of the controlled aircraft is reduced immensely.

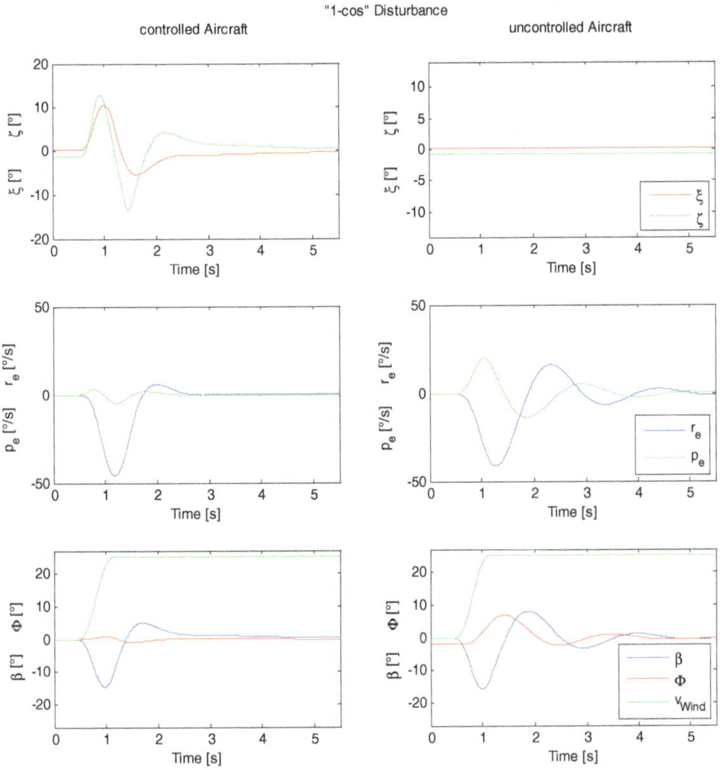

Fig. 14 Comparison of controlled (gain-scheduled controller) and uncontrolled aircraft for a "1-cos" side wind gust

6 Summary / Conclusion

The development of a novel Fly-by-Wire Control System designed to provide excellent flying qualities and pilot`s assistance for the lateral dynamics of Future Small (GA) Aircraft has been presented focusing on a sufficient proof of robust compliance.

After introducing the high safety risks in the general aviation sector and the need for improvement by establishing active Flight Control Systems, a controller structure, specifically tailored for excellent flying and handling qualities, as well as a corresponding design methodology has been proposed. Subsequently, a robust compliance verification of the presented control system, which is the main focus of this paper, is presented.

The conventional robustness analysis based on Nichols Charts (single loop cuts) does not guarantee sufficient stability characteristics. For that reason, the $\mu_{\Delta\Gamma}$-Analysis, which is an enhancement of the μ-Analysis allowing for a comprehensive robustness proof w.r.t. a user-defined boundary of compliant stability characteristics, is proposed. By means of the $\mu_{\Delta\Gamma}$-Analysis, a robustness assessment of the controlled aircraft considering an extensive uncertainty model executed and *robust compliance* has been verified along the entire envelope.

A linear and nonlinear analysis (simulation) of the controlled aircraft in comparison to the non-augmented aircraft dynamics reveals the potential of the concept presented. The analyses presented highlight the significant improvements concerning the flight and command behavior/characteristics of the aircraft with the active FbW Flight Control System.

Acknowledgments. This paper was prepared with the support of the Technische Universität München - Institute for Advanced Study, funded by the German Excellence Initiative.

References

1. Emma, K.: Australia probes high fatal accident rate in GA sector. In: Flight International, July 19-25, p. 21. Reed Business Information Ltd. (2011)
2. U.S. Department of Transportation - Federal Aviation Administration - Aviation Policy and Plans, FAA Aerospace Forecast Fiscal Years 2010-2030 (January 22, 2013), `https://www.faa.gov/data_research/aviation/aerospace_forecasts/2010-2030/`
3. Heller, M., Schuck, F., Peter, L., Holzapfel, F.: Hybrid Control System for a Future Small Aircraft. In: AIAA Guidance, Navigation, and Control Conference, Portland, Oregon, AIAA-2011-6635 (August 2011)
4. Heller, M., Baier, T., Schuck, F.: Lateral Fly by Wire Control System Dedicated to Future Small Aircraft. In: Advances in Aerospace Guidance, Navigation and Control, pp. 353–372. Springer, Heidelberg (2013)
5. Department of Defense, MIL-F-8785C, Military Specification, Flying Qualities of Piloted Airplanes, Washington (1980)
6. Department of Defense, Flying Qualities of Piloted Airplanes. Handbook, MIL HDBK 1797A, U.S. Department of Defense, Washington (1997)
7. Doyle, J., Redondo Beach, C.: lecture notes, Internal Report, Caltech, Pasadena, CA (1986)
8. Skogestad, S., Postlethwaite, I.: Multivariable Feedback Control, Analysis and Design, 2nd edn. John Wiley & Sons Ltd., West Sussex (2007)
9. Zhou, K., Dolye, J.C.: Robust and optimal control. Prentice Hall, Upper Saddle River (1996)
10. Herrenberger, M., Heller, M., Leitner, R., Sachs, G.: Fortschrittliche Unsicherheitsmodellierung und Robustheitsanalyse für den Basisregler eines modernen Trainers. DGLR Jahrestagung, Braunschweig (2006)
11. Heller, M.: Untersuchung zur Steuerung und Robusten Regelung der Seitenbewegung von Hyperschall-Flugzeugen; Dissertation, München: Herbert Utz Verlag GmbH (1999)

Flexible Launch Vehicle Control Using Robust Observer-Based Controller Obtained through Structured H_∞ Synthesis

Emmanuel Chambon, Pierre Apkarian, and Laurent Burlion

Abstract. Control of a flexible launch vehicle in the atmospheric ascent phase is highly challenging as it involves multiple concurrent design requirements. This ranges from reduction of the angle of attack in face of wind, minimum gain-phase and parametric margins as well as flexible modes attenuation. In this work, we discuss recently available non-smooth optimization techniques as a central tool to solve this problem. We consider designing an observer-based controller based on a Kalman filter suitably augmented with Dryden wind dynamics. We suggest a non-conservative approach to handle model uncertainties based upon multiple models of the launcher. This preliminary work aims at testing the potential of non-smooth controller tuning on a generic launcher model. It will serve as a stepping stone for a more in-depth study of the benchmark developed by M. Ganet at *Airbus Defence and Space*.

Keywords: flexible launch vehicle, robust control, observer-based controller, structured H_∞ synthesis, multiple models, unknown input.

Nomenclature

0_m	Null vector of length m
$0_{m\times n}$	$m \times n$ null matrix
α	Angle of Attack
β	Thruster angle relatively to launch vehicle body
G_l	Dryden wind gusts forming filter
G_m	Gain margin
$\mathscr{F}_u(M,K)$	Upper *LFT* of M with respect to K

Emmanuel Chambon · Pierre Apkarian · Laurent Burlion
Onera – The French Aerospace Lab, F-31055 Toulouse, France
e-mail: {Emmanuel.Chambon, Pierre.Apkarian, Laurent.Burlion}@onera.fr

J. Bordeneuve-Guibé et al. (eds.), *Advances in Aerospace Guidance, Navigation and Control*, DOI: 10.1007/978-3-319-17518-8_2

$\mathscr{F}_l(M,K)$	Lower *LFT* of M with respect to K
LFT	Linear Fractional Transformation
LPV	Linear Parametric Varying
ψ	Attitude (Euler angle)
PBH	Popov-Belevitch-Hautus (Lemma)
ϕ_m	Phase margin
Φ_l	Dryden wind gusts spectral density
PSD	Power Spectral Density
$\mathbb{R}$	Set of real numbers
$T_{w \Rightarrow z}$	Transfer from input w to output z
V	Launcher relative velocity
v_z	Lateral velocity
w	Wind disturbance input
$\|\| \ \|\|$	H_∞ or H_2 norm possibly restricted to frequency intervals of interest

1 Introduction

As far as launch vehicle atmospheric flight is concerned, wind speed variation is critical for dimensioning control laws. A key variable considered for control design to minimize aerodynamic load is the angle of attack (AoA), that is the angle between the launcher main axis and the relative speed of wind vector:

$$\alpha = \psi + \frac{v_z - w}{V} \tag{1}$$

where ψ is the attitude Euler angle, v_z the lateral deviation, w the unknown wind speed and V is the launcher relative speed to wind. These notations are illustrated in Fig. 1. To ensure minimal aerodynamic load, $|\alpha(t)|$ must remain below a given positive value α_{max} at every instant of the atmospheric flight. Due to the presence of flexible modes at specific frequencies which interfere with thruster control β, this problem is a difficult one.

There is a rich literature on launch vehicle control design. As far as research studies are concerned the following articles can be mentioned, [17] uses structured H_∞ synthesis in parallel with tuned elliptic filters to filter flexible modes; [9] introduces the use of *guardian maps* to analyse closed-loop stability; [1] exploits the Normalized Coprime Factorization approach combined with LQG/LTR additions for gain scheduling problem; [12] investigates the design of *LPV* control law with industrial application to flexible launcher. Note that in solutions making use of H_2 or H_∞ synthesis, deterministic information on the wind model is brought by weighting filters in the synthesis step. On an industrial application viewpoint, the following can be cited as an example [15] which deals with the Ares I control system design and the chosen implementation.

In this study, we propose to enhance wind disturbance rejection by means of an unknown input estimator as part of the controller. For a comprehensive discussion on unknown disturbance observers, we refer the reader to [18]. Such techniques try

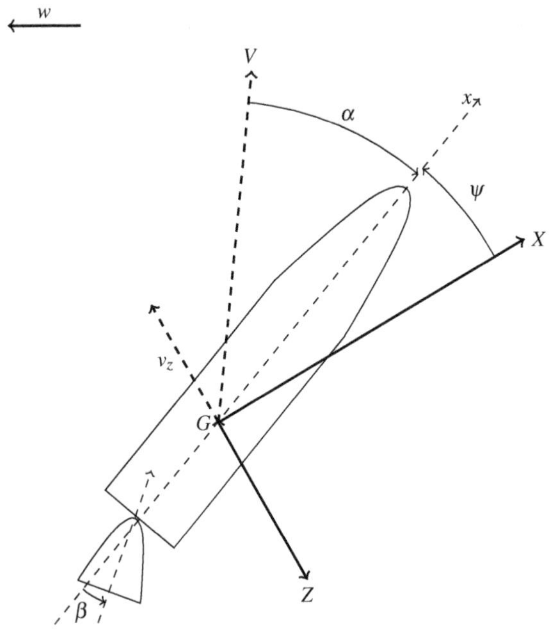

Fig. 1 Launch vehicle frame (longitudinal motion) and notations

to estimate both disturbances but also unmodelled plant dynamics. Unknown Input Observer (UIO) theory [13] makes assumptions on the unknown input model to estimate the state associated with disturbance. At first, our approach is similar in the way we augment the existing plant representation by a disturbance model before using Kalman formalism. Uncertainties are handled using multi-model synthesis. The purpose of using Kalman formalism are the possibilities

- to model wind in homogeneous meteorological conditions by a stochastic process where the forming filter can be used to augment base model, and
- to externalize estimation gain and express Kalman variance requirements in terms of H_2 norm which can be used in H_∞ synthesis methods.

Wind modelling is difficult and mainly relies on confrontation between physical theories and on-site measurements such as Jimspheres and smoke-trail. Obviously, choices need to be made to capture necessary deterministic information [22]. Two main approaches are used depending on the considered type of wind speed variation anomaly, namely turbulence in clear sky or wind shears.

The former are modelled using von Kármán [23] or Dryden [8, 19] stochastic models which can be expressed as filters driven by unbiased Gaussian white noises whose statistical characteristics are tabulated with respect to altitude and general meteorological conditions (from light to severe). The latter are discrete anomalies and as such are modelled using a completely different approach based on synthetic winds first described in [21]. The discrete approach is highly dependent on wind profile measurements. Dimensioning parameters are thus tabulated for a given measurements campaign at a given launching site [14]. Of course, stochastic models are not designed for capturing wind shears. Assumptions need to be made depending on the type of anomaly we wish to capture.

In this paper, we use a Kalman filter to estimate the unknown wind input and to give an observer-based structure to the controller. State augmentation is obtained using deterministic information brought by making an assumption on the wind speed variation mathematical nature. Note that the plant model is uncertain and multiple critical models are thus considered in the synthesis. The main contribution of this paper is to propose a method to synthesize a robust observer-based controller over multiple critical models using non-smooth structured H_∞ synthesis as presented in [5, 4].

The paper is structured as follows. Sect. 2 presents the formalism of unknown input estimation using Kalman filtering and describes the obtained observer-based controller. Assumptions on the unknown input are made in this section. Based on this prerequisite, Sect. 3 develops the main result of the simultaneous tuning of observer-based controller gains, augmented by Youla parameter. Sect. 4 presents an application to a generic flexible launch vehicle model with five flexible modes, one unknown input and uncertainties. Concluding remarks are then given.

2 Observer-Based Controller

An observer-based controller is a dynamic controller structured using mixed observer in tandem with state feedback loops. Such observer-based structures are also present in the so-called Youla parametrization of stabilizing controllers. These structures are used within controllers to enable a physical interpretation of the controller variables which are directly linked to the plant states through an observer stage. It also reduces the number of tuning parameters while benefiting from a dynamic structure. Moreover it is expected to improve gain-scheduling results by smoothing transition between controllers as experienced in [3]. This however is far from the spectrum of this article.

In this section we describe this structure and consider an augmentation relative to unknown input modelling. The observer stage is designed as a stationary Kalman filter. A structure useful for simultaneous tuning of all parameters is finally obtained.

2.1 *Unknown Input Modelling*

Without loss of generality, let consider our plant is subject to a unique unknown input w. Kalman formalism expects centered Gaussian white noises as inputs [2]. Since we also want to account for biased unknown disturbance, a little reformulation is required. Let $m_w = \mathbb{E}[w(t)]$ be the mean disturbance and b a unit variance white noise. Signal b is used to feed a forming filter $G_l(s)$ so that we obtain a coloured noise with Power Spectral Density (PSD) $\Phi_{ww}(s) = G_l(-s)G_l^\top(s)$. Such a decomposition leads to a Markov representation of w to be determined with state variable x_l. Illustration is provided in Fig. 2 and resulting state space representation is:

$$(G_w)\begin{cases} \dot{x}_l = A_l x_l + B_l b \\ w = C_l x_l + m_w \end{cases} \tag{2}$$

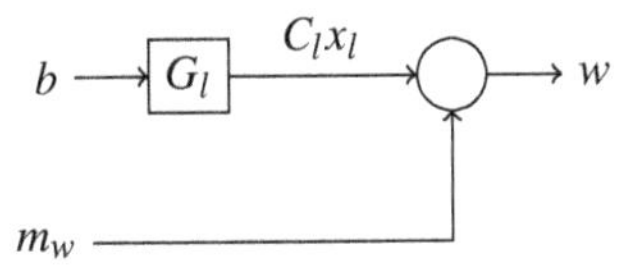

Fig. 2 Unknown input model G_w

In case mean disturbance $m_w(t)$ is not known, as is the case in our application, we proceed to state augmentation to estimate m_w. The bias is modelled by a new state $\dot{x}_b = A_b x_b$ where A_b value is discussed in Sect. 2.2.2 and $\widehat{m}_w = x_b$:

$$\begin{pmatrix} \dot{x}_l \\ \dot{x}_b \end{pmatrix} = \begin{bmatrix} A_l & 0 \\ 0 & A_b \end{bmatrix} \begin{pmatrix} x_l \\ x_b \end{pmatrix} + \begin{bmatrix} B_l \\ 0 \end{bmatrix} b \tag{3}$$

$$w = \begin{bmatrix} C_l & 1 \end{bmatrix} \begin{pmatrix} x_l \\ x_b \end{pmatrix} \tag{4}$$

In case of multiple disturbances, each unknown input requires dynamic augmentation to model coloured white noises and biases.

2.2 *Estimation Stage*

We use Kalman formalism as presented in [2] for the observer structure of our controller. A state-space representation of the more conservative stationary Kalman filter is finally derived.

2.2.1 Plant Model

A multi-model approach means considering multiple plant models. We need to provide the Kalman filter with one of them. In our case, the choice is guided by physical considerations as discussed in 4.2. Let this model be a linear model with state vector $x \in \mathbb{R}^{n_x}$ and measurements $y \in \mathbb{R}^{n_y}$. Inputs are divided into known deterministic inputs $u \in \mathbb{R}^{n_u}$ and an unknown disturbance $w \in \mathbb{R}$ standing for wind input. The state space representation is given as:

$$\begin{pmatrix} \dot{x} \\ y \end{pmatrix} = \begin{bmatrix} A & B_w & B_u \\ C_y & D_{yw} & D_{yu} \end{bmatrix} \begin{pmatrix} x \\ w \\ u \end{pmatrix} \tag{5}$$

where all matrices are of appropriate dimensions.

2.2.2 Kalman Model

In Sect. 2.1, Eq. (4) has been derived from all deterministic information contained in our modelling of input w. Suppose our measurements are also perturbed by a Gaussian white noise $v \in \mathbb{R}^{n_y}$. The nominal perturbed plant can be described by a state space representation called the Kalman model [2]:

$$\begin{cases} \dot{x}_a &= A_a x_a + B_a u + Mb, \text{ state equation} \\ y_m &= C_a x_a + D_a u + v, \quad \text{measurement equation} \end{cases} \tag{6}$$

where $x_a = (x, x_l, x_b)^\top \in \mathbb{R}^{n_x+2}$. Let Cov_b and Cov_v be the state and measurement noise covariance matrices. We have:

$$A_a = \begin{bmatrix} A & B_w C_l & B_w \\ 0_{1\times n_x} & A_l & 0 \\ 0_{1\times n_x} & 0 & A_b \end{bmatrix}, B_a = \begin{bmatrix} B_u \\ 0_{2\times n_u} \end{bmatrix}, M = \begin{bmatrix} 0_{n_x} \\ B_l \\ 0 \end{bmatrix},$$
$$C_a = \begin{bmatrix} C_y \; D_{yw} C_l \; D_{yw} \end{bmatrix}, D_a = D_{yu} \tag{7}$$

One can see that A_b is inevitably a model eigenvalue. An important property of the Kalman model to be valid is that the pair (A, C) be detectable [2]. We use the Popov-Belevitch-Hautus (PBH) Lemma [20] to easily discard unacceptable A_b values. We then use the Kalman model to obtain an expression of the Kalman filter.

2.2.3 Stationary Kalman Filter

The Kalman filter is fed with the perturbed measurements y_m and the deterministic input u. It outputs the plant augmented state estimate $\widehat{x}_a$. We have the following state space representation [2]:

$$\begin{cases} \dot{\widehat{x}}_a &= A_a \widehat{x}_a + B_a u + K_f (y_m - C_a \widehat{x}_a) \\ y_{kf} &= \widehat{x}_a \end{cases} \tag{8}$$

where $K_f \in \mathbb{R}^{(n_x+2)\times n_y}$ is the stationary Kalman filter gain. Its automated tuning along with the command feedback and Youla parameter using observer-based structure in Sect. 2.4 is the core of this article.

2.3 *State-Feedback and Youla Parameter*

Stationary Kalman filter outputs estimated augmented state $\widehat{x}_a$. A static state-feedback $K_v \in \mathbb{R}^{1\times(n_x+2)}$ is then added to the estimation stage to obtain an observer-based structure (see Fig. 3).

Given the complexity of the problem, the available degrees of freedom (static K_f and K_v) may not be enough to satisfy specified requirements. The order of the controller can be augmented using Youla Q-parametrization theory. Let consider a plant P and a stabilizing observer-based controller J_0:

$$J_0 = \left[\begin{array}{c|c} A_a - B_a K_v - K_f C_a & K_f \\ \hline -K_v & 0_{1\times 2} \end{array}\right] \tag{9}$$

Let Q a stable transfer function of order n_Q. Then $C_v = \mathscr{F}_l\,(J^\star, Q)$ is a stabilizing control law for P where:

$$J^\star = \left[\begin{array}{c|cc} A_a - B_a K_v - K_f C_a & K_f & B_a \\ \hline -K_v & 0_{1\times 2} & 1 \\ -C_a & I_2 & 0_{2\times 1} \end{array}\right] \tag{10}$$

Hence the controller is artificially augmented of n_Q states, thus providing more degrees of freedom.

2.4 Observer-Based Structure

The complete observer-based structure obtained from Kalman filter in Sect. 2.2.3 and state-feedback considerations in Sect. 2.3 is shown in Fig. 3. The controller C_v is obtained through $C_v = \mathscr{F}_l\,(J^\star, Q)$ where $J^\star$ is defined as in Eq. (10). In this observer-based structure, the parameters to tune are K_v, K_f and Q.

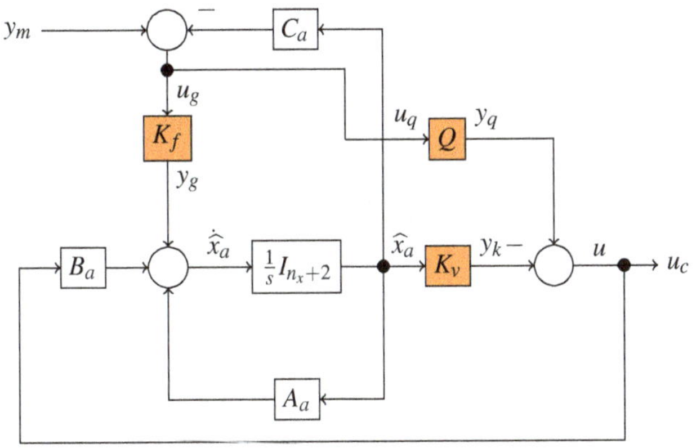

Fig. 3 Observer-based controller C_v

3 Simultaneous Tuning

Having obtained a detailed expression of our controller, H_∞ synthesis can be used to tune the components K_f, K_v and Q. A synthesis model is presented where these components are externalized. Multiple models synthesis is considered to account for requirements to be enforced even when uncertainties are considered, as is the case with our application.

3.1 H_∞ Synthesis

Nominal H_∞ synthesis solutions have been proposed using algebraic Riccati equations [7] or semidefinite programming (SDP) [10]. When structural constraints are expressed on the controller, the H_∞ synthesis problem is no longer convex and solutions and algorithms were proposed in [5, 6]. In this work, we use the non-smooth approach developed in [5, 4, 11] for tuning the structured controller. The considered problem is to find a controller $C(s,p)$ – where $p \in \mathbb{R}^n$ are the tunable parameters – answering to problem 1:

Problem 1

$$\min_{p} \max_{i=1,\ldots,n_{soft}} \{\| T_{w_i \to z_i}(C(s,p)) \|\} \tag{11}$$

subject to $\| T_{w_j \to z_j}(C(s,p)) \| \leq 1,\ j = 1,\ldots,n_{hard}$.

3.2 Synthesis Model

Simultaneous synthesis consists in tuning the Kalman gain K_f, feedback K_v and Youla parameter Q using a unique synthesis model shown in Fig. 4. The transfers highlighted by the synthesis model G_{syn} are thus used to express requirements both on the estimation stage and on the feedback. An inner view of this model is provided in Fig. 5. In this figure, G_{plant} is an augmented model of the plant where we also output the augmented state x_a so as to obtain the estimation error ε on which variance requirements will be expressed to obtain a minimal error variance estimator. F is the LFT-form of observer-based corrector C_v with additional output $\widehat{x}_a$. As mentioned in Sect. 2.2.2, b and v are Gaussian white noises with respective covariance matrices Cov_b and Cov_v. They indirectly define our confidence in the wind model and in the measurements.

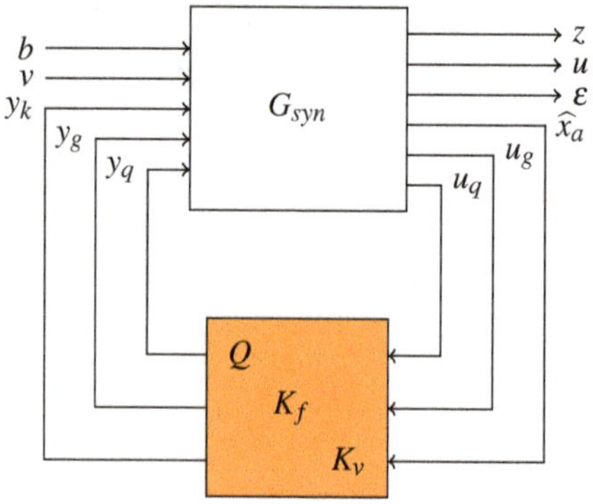

Fig. 4 Simultaneous synthesis model I/Os; see Fig. 5 for inner G_{syn}

In case the controller should stabilize multiple models a multi-model approach is used. Synthesis model then consists of a collection of models where G_{plant} is chosen so as to enforce specific plant properties. H_∞ synthesis requirements can thus be expressed on all or specific models. In Fig. 5, this is symbolized by y_Δ and u_Δ I/Os where uncertainties can be injected using upper *LFT* of G_{plant} with respect to a matrix Δ: $\mathscr{F}_u\left(G_{plant}, \Delta\right)$.

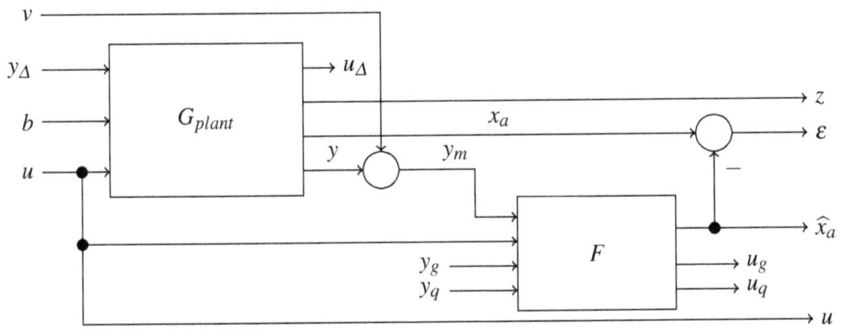

Fig. 5 Synthesis model G_{syn} inner view

4 Application

In this section, we use a generic launcher linear model to test the potential of the proposed structured controller. We use multiple uncertain models to account for the approximate knowledge of flexible modes characteristics, especially frequency and damping.

4.1 Wind Model

Using a Dryden approach, the wind is approximated by a slow time-varying signal m_w perturbed by random gusts. The longitudinal wind gusts are modelled by a random noise of spectral density [16]:

$$\Phi_l(s) = \frac{2L_l\sigma_l^2}{V\pi} \frac{1}{1-\left(\frac{L_l s}{V}\right)^2} \tag{12}$$

where L_l and σ_l are tabulated with respect to altitude and are thus fixed for a given working point. We neglect the other wind directions as well as angular rates since the wind influence is already projected in our longitudinal model. As shown in Sect. 2.1, this spectral density can be recovered using a transfer function $G_l(s)$ with input a white noise of unit variance and output a random signal of spectral density $\Phi_l(s) = G_l(-s)G_l^\top(s)$. In our case, we have:

$$G_l(s) = \frac{\sqrt{2k}}{s+a} \tag{13}$$

where $a = \frac{V}{L_l}$ and $k = \frac{a\sigma_l^2}{\pi}$.

The mean wind is non-zero so we use the Markov representation given in Eq. (4). Using the PBH Lemma, we notice that the Kalman model Eq. (6) pair (A,C) is not detectable if $A_b = 0$. To avoid similar pathological cases and to account for mean wind slow time variation, we made the choice of a stabilizing $A_b = -0.01$. Fig. 6 shows a stochastic (severe) wind profile generated using a Dryden model.

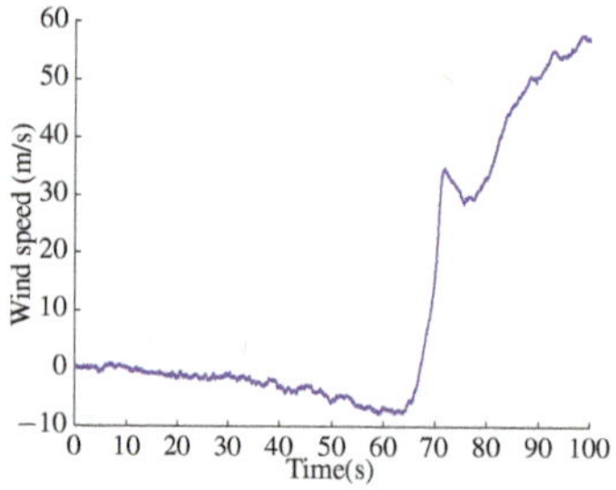

Fig. 6 Dryden wind gust stochastic model around a mean wind profile

These filters are used to augment the plant Kalman model. This enables us to use the input b as a design input during synthesis with information on its white noise nature. A Gaussian white noise is applied to this input during simulations.

4.2 *Launcher Model*

In this application, we only consider longitudinal motion as illustrated in Fig. 1. We use the generic launcher model from Eq. (14) which is a linearised model around a given working point. In this model, $r \in \mathbb{N}$ uncertainties are injected through y_Δ by upper-*LFT* $\mathscr{F}_u(G_{plant}, \Delta)$ with $\Delta = \mathrm{Diag}(\{\Delta_i\}_{1 \le i \le r})$ where $\forall i,\ \Delta_i \in \{-1, 0, 1\}$.

$$(G_{plant}) \left\{ \begin{pmatrix} \dot{x} \\ u_\Delta \\ y \end{pmatrix} = \left[\begin{array}{c|c|cc} A & B_\Delta & B_w & B_u \\ \hline C_\Delta & D_{\Delta\Delta} & D_{\Delta w} & D_{\Delta u} \\ \hline C_y & D_{y\Delta} & D_{yw} & D_{yu} \end{array}\right] \begin{pmatrix} x \\ y_\Delta \\ w \\ \beta \end{pmatrix} \right. \tag{14}$$

In the presence of 5 flexible modes and considering second-order sensor and actuator dynamics, A is of dimension 17×17 and $r = 27$ uncertainties are considered:

- 2 uncertainties on the rigid dynamics,
- 5 uncertainties for each flexible mode especially uncertainties on damping and frequency of the associated second-order dynamics. Other uncertainties pertain to coefficients in measurements y which depend on flexible states.

We have $y = (\alpha, v_z, \psi, \dot{\psi})^\top$ where only the attitude and its rate $(\psi, \dot{\psi})$ are measured. We command the system through thruster angle $u = \beta$. The angle of attack α is a critical dimensioning variable for our control law as shown in Sect. 4.3. In our application, we consider 5 critical models where diagonal blocks $\{\Delta_i\}_{1\leq i\leq 27}$ are varied to consider worst cases. Bode plots of transfers from system command β to measurement ψ are shown in Fig. 7 in the 5 considered critical cases. The presence of peaks in the magnitude due to the dependency of measurements on flexible modes can lead to instability in the case of an ill-dimensioned control law hence the need of specifications and the use of robust control synthesis techniques.

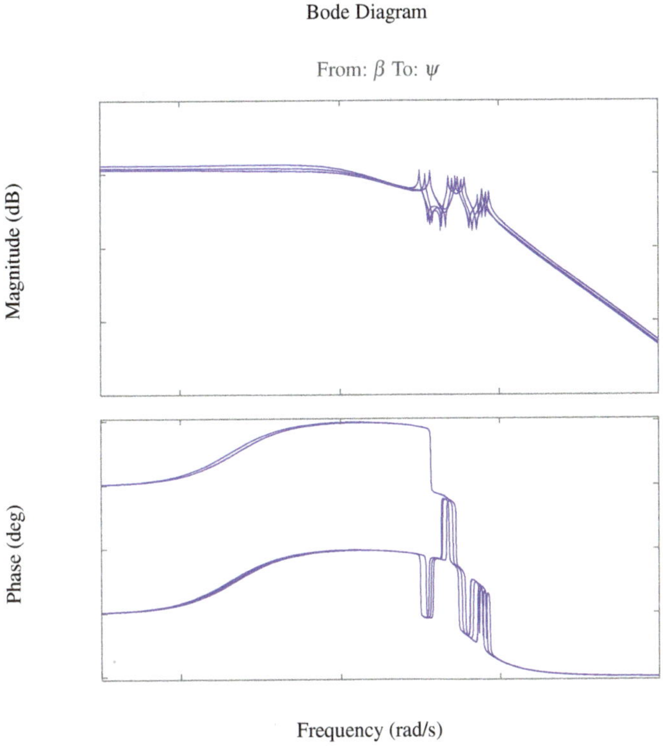

Fig. 7 Magnitude and phase of transfer from command β to measurement ψ (open loop) in the 5 considered critical cases

4.3 Specifications

To minimize aerodynamic load and ensure flexible modes do not disturb control, different specifications stated in the frequency- or time-domain are issued. During synthesis, one must follow these objectives:

- ensure stable closed-loop,
- ensure guaranteed values for gain G_m and phase ϕ_m margins,
- phase-control low frequency flexible modes,
- gain-control high frequency flexible modes through roll-off,

- ensure minimal angle of attack ($|\alpha|$ kept below α_{max}) – in other words: minimize wind gusts effects on AoA,
- minimize thruster actuation consumption.

These specifications are captured through the definition of appropriate weighting functions used to emphasize certain frequency ranges.

4.4 Synthesis

As mentioned above, we use five models in our synthesis, each differing by the model used in G_{plant} (see Fig. 5). These models are obtained by upper *LFT* $\mathscr{F}_u(G_{syn}, \Delta)$ as shown in Fig. 8. Requirements are expressed on transfers between: the stochastic b (state noise) and v (measurements noise) inputs, and: regulated output z (angle of attack) and estimation error ε (see Fig. 8). The objective is to reduce the impact of input noises on estimation stage and performance. Margins and roll-off requirements are expressed at the plant input. Gain and phase margins results are presented in Table 1.

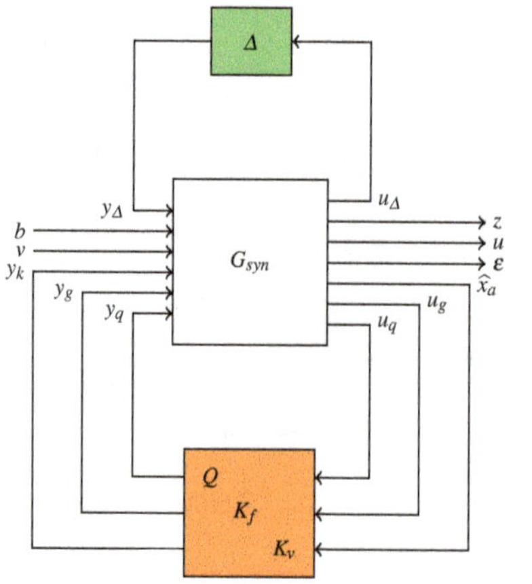

Fig. 8 Synthesis model considering uncertainties in *M*-Δ *LFT* form

Table 1 Closed-loop phase and gain margins over critical models

Model	G_m (dB)	ϕ_m (°)
Nominal	3.85	27.39
Critical 1	2.27	15.22
Critical 2	2.27	24.87
Critical 3	1.57	16.84
Critical 4	1.57	17.77

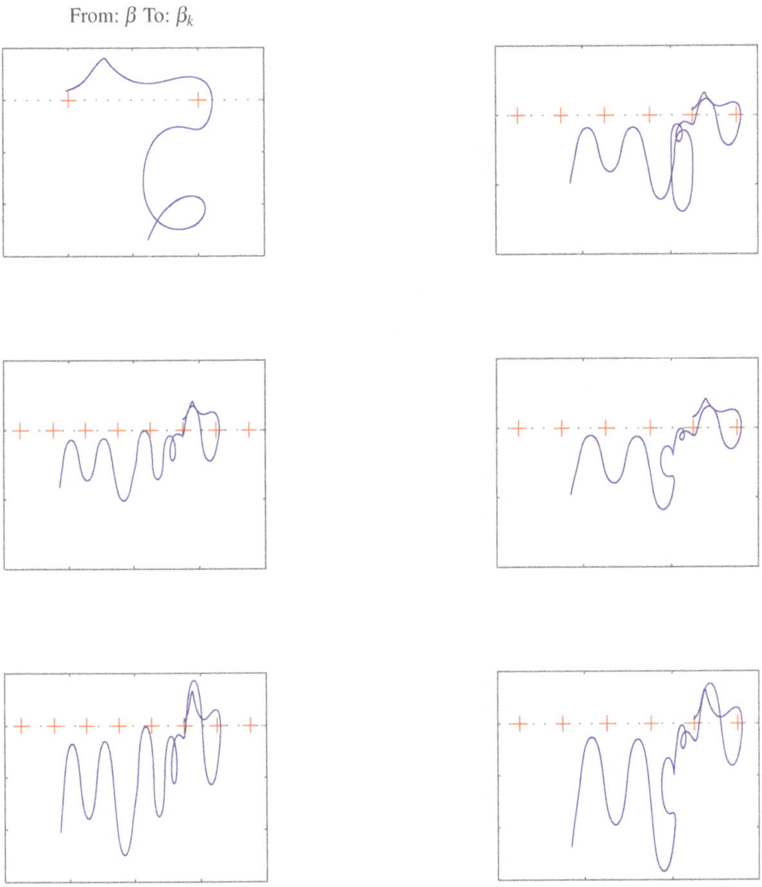

Fig. 9 Nichols plots with G_{plant} chosen respectively equal to the rigid model (no flexible modes) and to the 5 critical models (left to right and top to bottom), loop opening is at controller output $u_c = \beta_k$ (see Fig. 3)

4.5 Simulations and Validations

Simulations are performed using all critical models in closed-loop with synthesized controller. We essentially consider the angle of attack α in these simulations. It is to be kept minimal so as to minimize dynamic load during atmospheric flight.

Table 2 Simulation of maximal AoA over critical models

Model	Max. AoA (°)
Nominal	3.30
Critical 1	3.36
Critical 2	3.36
Critical 3	3.24
Critical 4	3.25

[a] where T is final simulation time.

Simulation results over all critical models are shown in Table 2. In all cases, maximal AoA $\overline{|\alpha|}$ satisfies $\overline{|\alpha|} < \alpha_{max}$ as specified. The angle of attack evolution over time is shown in Fig. 10. The corresponding thruster actuation is shown on Fig. 11. Note that some critical models require more thruster angular corrections to satisfy requirements on AoA.

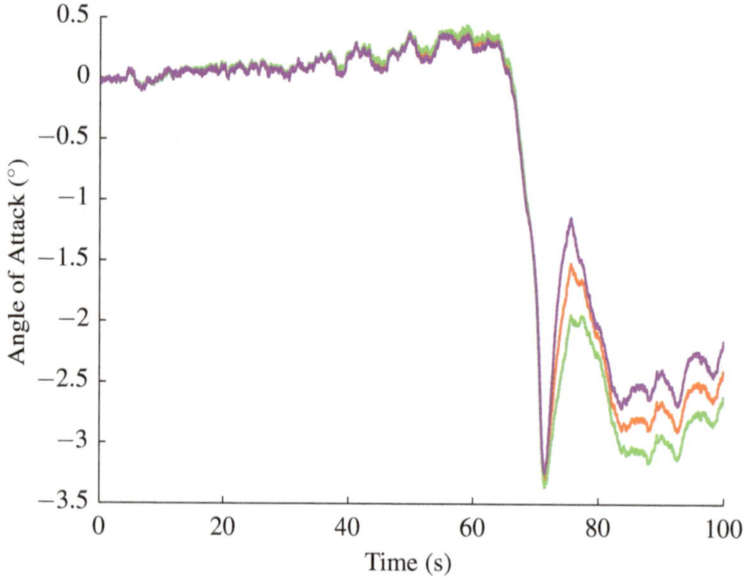

Fig. 10 Angle of Attack α response after simulation over the five critical models, maximal value is $\alpha_{max} = 3.4°$

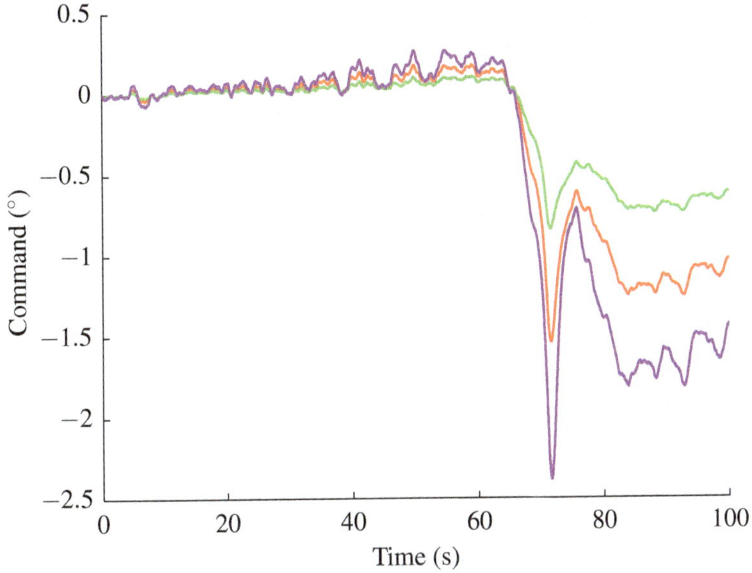

Fig. 11 Command β required to satisfy specified requirements

5 Conclusions

In this article, we have explored the design of an observer-based controller under multiple frequency- and time- domain specifications using multi-model non-smooth H_∞ synthesis techniques. This approach has been applied to a generic uncertain flexible launcher model. In this application, we have considered augmentation of the observer by a stochastic Dryden wind gusts model to account for unknown wind disturbance.

Despite good results – similar to those obtain with different approaches [22] – fine tuning of parameters especially of the wind model remains critical which may hamper finding a suitable solution.

This preliminary study indicates that non-smooth optimization approaches hold promises to design an observer-based controller while accounting for parametric variations in the plant model. Further research will be dedicated to reduce sensitivity to the modelling of the unknown wind input.

References

1. Abbas-Turki, M., Duc, G., Clément, B., Theodoulis, S.: Robust gain scheduled control of a space launcher by introducing LQG/LTR ideas in the NCF robust stabilisation problem. In: 46th IEEE Conference on Decision and Control, New Orleans, LA, USA, pp. 2393–2398 (December 2007)
2. Alazard, D.: Introduction to Kalman filtering. ISAE – Institut Supérieur de l'Aéronautique et de l'Espace (December 2011)

3. Alazard, D., Imbert, N., Clément, B., Apkarian, P.: Launcher attitude control: some additional design and optimization tools. In: CNES/EADS Conference on Launcher Technology, Madrid (November 2003)
4. Apkarian, P.: Tuning controllers against multiple design requirements. In: American Control Conference Proceedings, Washington, USA, pp. 3888–3893 (June 2013)
5. Apkarian, P., Noll, D.: Nonsmooth H_∞ synthesis. IEEE Transactions on Automatic Control 51(1), 71–86 (2006)
6. Burke, J.V., Henrion, D., Lewis, A.S., Overton, M.L.: HIFOO - a MATLAB package for fixed-order controller design and H_∞ optimization. In: 5th IFAC Symposium on Robust Control Design, Toulouse, France (August 2006)
7. Doyle, J., Glover, K., Khargonekar, P., Francis, B.A.: State space solutions to standard h2 and H_∞ control problems. IEEE Transactions on Automatic Control 34, 831–847 (1989)
8. Dryden, H.L., Kuethe, A.M.: Turbulence in wind tunnel measurements. Technical Report 342, National Advisory Commitee for Aeronautics, Washington (August 1929)
9. Dubanchet, V., Saussié, D., Bérard, C., Saydy, L., Gourdeau, R.: Robust control of a launch vehicle in atmospheric ascent based on guardian maps. In: American Control Conference, ACC 2012, Fairmont Queen Elizabeth, Montréal, Canada, pp. 938–943 (June 2012)
10. Gahinet, P., Apkarian, P.: A linear matrix inequality approach to H_∞ control. International Journal of Robust and Nonlinear Control 4, 421–448 (1994)
11. Gahinet, P., Apkarian, P.: Structured H_∞ synthesis in MATLAB. In: Proceedings of the 18th IFAC World Congress, Milan, pp. 1435–1440 (2011)
12. Ganet-Schoeller, M., Ducamp, M.: LPV control for flexible launcher. In: AIAA Guidance, Navigation, and Control Conference, Toronto, Ontario Canada (July 2010)
13. Hostetter, G.H., Meditch, J.S.: Brief paper on the generalization of observers to systems with unmeasurable unknown inputs. Automatica 9, 721–724 (1973)
14. Jacquemin, B., Voiron, T.: La détermination du profil du vent pour les activités spatiales. La Météorologie (33), 45–55 (May 2001)
15. Jang, J.-W., Alaniz, A., Hall, R., Bedrossian, N., Hall, C., Ryan, S., Jackson, M.: Ares I flight control system design. In: Proceedings of AIAA Guidance, Navigation, and Control Conference (2010)
16. Keegan, W.B.: Terrestrial environment (climatic) criteria handbook for use in aerospace vehicle development. Technical report (August 2000)
17. Knoblauch, M., Saussié, D., Bérard, C.: Structured H_∞ control for a launch vehicle. In: American Control Conference, ACC 2012, Fairmont Queen Elizabeth, Montréal, Canada, pp. 967–972 (June 2012)
18. Radke, A., Gao, Z.: A survey of state and disturbance observers for practitioners. In: American Control Conference, Minnesota, USA (June 2006)
19. Taylor, G.I.: Statistical theory of turbulence. Proceedings of the Royal Society of London. Series A, Mathematical and Physical Sciences 151(873), 444–454 (1935)
20. Terrell, W.J.: Stability and Stabilization: An Introduction. Cloth (2009)
21. Vaughan, W.W., Scoggins, J.R., Smith, O.E.: Role of applied meteorology in the development of large space vehicles. Bulletin of American Meteorological Society 44, 157–174 (1963)
22. Voinot, O., Alazard, D., Piquereau, A.: A robust multi-objective synthesis applied to launcher attitude control. In: 15th IFAC Symposium on Automatic Control in Aerospace (September 2001)
23. von Kármán, T., Howarth, L.: On the statistical theory of turbulence. Proceedings of the Royal Society of London. Series A, Mathematical and Physical Sciences 164(917), 192–215 (1938)

Lyapunov-Based Three-Dimensional Terminal Angle Constrained Guidance Laws

Mingu Kim, Yongwoo Lee, Seokwon Lee, and Youdan Kim

Abstract. Three-dimensional nonlinear guidance laws are proposed considering terminal angle constraints. Unlike conventional two-dimensional guidance laws, the three-dimensional geometry is considered without the assumption that the yaw channel and the pitch channel are decoupled. It is shown that the states converge to the desired values by using Lyapunov stability theory and LaSalle's invariance theorem. Numerical simulation results are presented to demonstrate the performance of the proposed guidance laws.

1 Introduction

Lots of studies have been performed to improve the performance or survivability of vehicles including missiles or unmanned aerial vehicle (UAV) systems. Impact angle constraint is especially important to improve the performance of missiles when striking a vulnerable area of a target. Also, in the UAV system, the concept of an impact angle can be utilized to consider an arrival angle, an incident angle, or an approach angle for precise landing or waypoint approaching. For this reason, the research on impact angle control problem has been widely performed.

Kim and Grider studied an impact angle constrained problem for terminal guidance schemes of reentry vehicles [1]. Linear quadratic optimization technique was used to design the terminal guidance law. Song et al. developed an impact angle control guidance (IACG) law in terminal phase for planar engagement by solving a minimum energy problem and combined with a suboptimal filter for estimating target acceleration [2]. Ryoo et al. proposed an optimal IACG law by solving a minimum energy problem and considered the first order autopilot [3]. Lee et al. generalized the formulation of optimal IACG laws [4]. Cho et al. developed an optimal IACG law against maneuvering targets using zero-effort collision triangle [5].

Mingu Kim · Yongwoo Lee · Seokwon Lee · Youdan Kim
Department of Mechanical and Aerospace Engineering, Seoul National University,
1 Gwanak-ro, Gwanak-gu, Seoul, 151-744, Korea
e-mail: {mingukim, leeyw17, blueswl, ydkim}@snu.ac.kr

J. Bordeneuve-Guibé et al. (eds.), *Advances in Aerospace Guidance, Navigation and Control*, DOI: 10.1007/978-3-319-17518-8_3

Proportional Navigation Guidance (PNG) was also used to consider an impact angle constrained problem against a stationary target by controlling the navigation constant [6]. It was extended to nonstationary nonmaneuvering target cases [7]. State-Dependent Riccati Equation(SDRE) was utilized to solve the IACG problem against a stationary target [8]. The backstepping control method was used to design an IACG law [9]. Simple Lyapunov candidate functions were introduced to design a pursuit-like IACG law [10].

Most of the conventional guidance laws have been designed in two-dimensional space under the assumption that the pitch channel and the yaw channel are decoupled. In reality, however, the two channels are highly coupled, and therefore, the consideration of the coupling effect should be considered to improve the performance of missiles or UAVs. The performance and the capturability of three-dimensional pure PNG law were analyzed using Lyapunov-like approach [11, 12].

Recently, several studies on impact angle control problem in the three-dimensional environment have been performed. Lyapunov stability theory was used to design a three-dimensional IACG law [13]. The concept of a reference circle on a moving coordinate frame fixed to a target was used to solve a three-dimensional impact angle control problem [14]. Model predictive static programming technique was applied to design a three-dimensional guidance law with impact angle constraints for air-to-ground missiles [15]. The two-dimensional near-optimal guidance law and the concept of a maneuvering plane were used for midcourse guidance to a predicted interception point with a three-dimensional impact angle constraint [16]. A partial integrated guidance and control scheme was utilized to solve an impact angle control problem in the three-dimensional environment [17]. A sliding mode control scheme was used to design three-dimensional IACG laws [18, 19].

In this study, new three-dimensional nonlinear guidance laws considering terminal angle constraints are proposed. Two Lyapunov candidate functions are introduced to design nonlinear guidance laws with terminal angle constraints. The longitudinal acceleration command and the lateral acceleration command are derived considering the three-dimensional geometry without the assumption that the pitch channel dynamics and the yaw channel dynamics are decoupled. The convergence to the desired states is shown using Lyapunov stability theory and LaSalle's invariance theorem.

This paper is organized as follows: In Section 2, a three-dimensional geometry is presented to formulate the problem. In Section 3, two three-dimensional nonlinear guidance laws are proposed considering terminal angle constraints based on Lyapunov stability theory. The stability of the proposed guidance laws is also analyzed. Section 4 provides the results of numerical simulations to demonstrate the performance of the proposed three-dimensional terminal angle constrained guidance laws. Finally, conclusions are given in Section 5.

2 Problem Formulation

Let us consider a three-dimensional geometry of a point mass vehicle model as shown in Fig. 1. In Fig. 1, (x_I, y_I, z_I), (x_L, y_L, z_L), and (x_M, y_M, z_M) are reference coordinate system, Line-of-Sight(LOS) coordinate system, and body coordinate system, respectively, θ_L and ψ_L are elevation and azimuth angles of LOS coordinate system with respect to the reference coordinate system, respectively, θ_M and ψ_M are Euler angles from the LOS coordinate system to the body coordinate system, respectively, and λ_y and λ_z are y_L and z_L components of the LOS vector, respectively. Note that the direction of the LOS vector is the x-axis of the LOS coordinate system.

To formulate the three-dimensional kinematic equations, the following assumptions are required. First, the vehicle is considered as a point mass. Second, the autopilot and seeker/sensor dynamics are faster than the vehicle dynamics. Third, the speed of the vehicle is constant. Fourth, the angle-of-attack of the vehicle is very small so that it can be neglected. Fifth, the target is stationary. Using the above assumption, the kinematic equations can be obtained as [11].

$$\dot{R} = -\mathrm{V}_M \cos\theta_M \cos\psi_M \tag{1}$$

$$\dot{\lambda}_y = \frac{\mathrm{V}_M}{R} \sin\theta_M \tag{2}$$

$$\dot{\lambda}_z = -\frac{\mathrm{V}_M}{R} \cos\theta_M \sin\psi_M \tag{3}$$

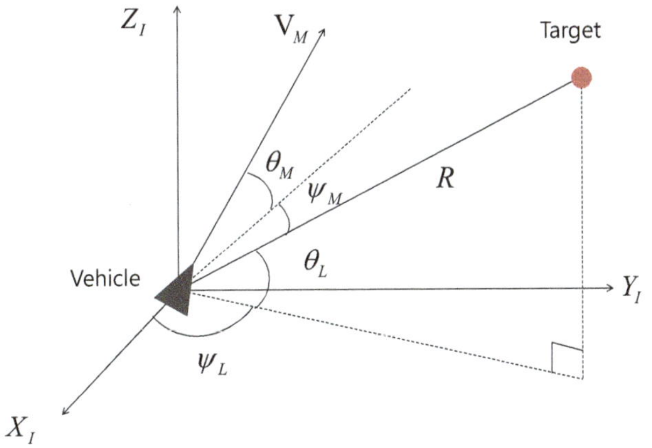

Fig. 1 Three-dimensional geometry of a point mass vehicle model

$$\dot{\theta}_M = \frac{a_{zM}}{V_M} + \frac{1}{R} V_M \sin^2 \psi_M \tan \theta_L \cos \theta_M + \frac{1}{R} V_M \cos \psi_M \sin \theta_M \tag{4}$$

$$\begin{aligned} \dot{\psi}_M = & \frac{a_{yM}}{V_M \cos \theta_M} - \frac{1}{R} V_M \sin \theta_M \sin \psi_M \cos \psi_M \tan \theta_L \\ & + \frac{1}{R \cos \theta_M} V_M \sin^2 \theta_M \sin \psi_M + \frac{1}{R} V_M \cos \theta_M \sin \psi_M \end{aligned} \tag{5}$$

where R is the distance between the vehicle and the target, V_M is the speed of the vehicle, and a_{y_M} and a_{z_M} are yaw and pitch accelerations of the vehicle, respectively.

To achieve the terminal angle constraints, let us define variables θ_{L_d} and ψ_{L_d} [18].

$$\theta_{L_d} = \theta_{L_0} - \lambda_{y_d} \tag{6}$$

$$\psi_{L_d} = \psi_{L_0} - \lambda_{z_d} \tag{7}$$

where θ_{L_0} and ψ_{L_0} are the initial elevation and azimuth angles, respectively, and λ_{y_d} and λ_{z_d} are the desired LOS rotation angles, respectively.

To accomplish an interception mission or a waypoint passing mission, the velocity vector of a vehicle should be coincided with the LOS vector. Therefore, the final conditions can be constructed as follow.

$$\dot{\lambda}_y = 0, \dot{\lambda}_z = 0 \tag{8}$$

$$\theta_M = 0, \psi_M = 0 \tag{9}$$

Note that the terminal angle errors can be substituted as the total LOS rotation error as follows.

$$e_{\lambda_y} = \lambda_y - \lambda_{y_d} \tag{10}$$

$$e_{\lambda_z} = \lambda_z - \lambda_{z_d} \tag{11}$$

3 Guidance Laws

In this section, two three-dimensional terminal angle constrained guidance laws are proposed under the assumption that $\theta_L \neq \pm\pi/2$, $|\theta_M| < \pm\pi/2$, and $|\psi_M| < \pm\pi/2$. In the physical point of view, $\theta_L = \pm\pi/2$ means the vertical flight, such as a high diving maneuver, and $|\theta_M| < \pi/2$, and $|\psi_M| < \pi/2$ means the flight toward the target. In other words, the high diving maneuver is not considered in this study. Now, let us introduce two Lyapunov candidate functions to design the guidance laws.

3.1 Proportional Navigation(PN)-Like Guidance Law

Using the final conditions, (8), and the terminal angle errors, (10) and (11), let us consider a following Lyapunov candidate function to obtain the guidance law considering terminal angle constraints in the three-dimensional space.

$$f_L = \frac{1}{2}\left(R\dot{\lambda}_y\right)^2 + \frac{1}{2}\left(R\dot{\lambda}_z\right)^2 + \frac{1}{2}c_1 V_M^2 e_{\lambda_y}^2 + \frac{1}{2}c_2 V_M^2 e_{\lambda_z}^2 \tag{12}$$

where c_1 and c_2 are positive constants. The time derivative of (12) can be expressed as

$$\begin{aligned}\dot{f}_L =& V_M^2 \sin\theta_M \cos\theta_M \cos^2\psi_M \dot{\theta}_M + V_M^2 \cos^2\theta_M \sin\psi_M \cos\psi_M \dot{\psi}_M \\ &+ c_1 V_M^2 e_{\lambda_y} \dot{\lambda}_y + c_2 V_M^2 e_{\lambda_z} \dot{\lambda}_z\end{aligned} \tag{13}$$

Substituting (2)–(5) into (13) yields

$$\begin{aligned}\dot{f}_L =& \frac{1}{2} V_M^2 \sin 2\theta_M \cos^2\psi_M \left(\frac{a_{zM}}{V_M}\right) + \frac{1}{2} V_M^2 \cos^2\theta_M \sin 2\psi_M \left(\frac{a_{yM}}{V_M \cos\theta_M}\right) \\ &+ \frac{1}{2} V_M^2 \sin 2\theta_M \cos^2\psi_M \left(\frac{V_M}{R}\cos\theta_M \tan\theta_L \sin^2\psi_M + \frac{V_M}{R}\cos\psi_M \sin\theta_M\right) \\ &- \frac{1}{2} V_M^2 \cos^2\theta_M \sin 2\psi_M \left(\frac{V_M}{2R}\sin\theta_M \tan\theta_L \sin 2\psi_M\right) \\ &+ \frac{1}{2} V_M^2 \cos^2\theta_M \sin 2\psi_M \left(\frac{V_M}{R\cos\theta_M}\sin^2\theta_M \sin\psi_M + \frac{V_M}{R}\cos\theta_M \sin\psi_M\right) \\ &+ c_1 e_{\lambda_y} \frac{V_M}{R}\sin\theta_M - c_2 e_{\lambda_z} \frac{V_M}{R}\cos\theta_M \sin\psi_M\end{aligned} \tag{14}$$

Now, let us propose following acceleration commands.

$$a_{yM} = -\frac{V_M^2}{R}\sin\psi_M + c_2 \frac{V_M^2}{R}\frac{e_{\lambda_z}}{\cos\psi_M} - c_4 \frac{V_M^2}{R}\frac{\sin\psi_M}{\cos\theta_M \cos\psi_M} \tag{15}$$

$$a_{zM} = -\frac{V_M^2}{R}\sin\theta_M \cos\psi_M - c_1 \frac{V_M^2}{R}\frac{e_{\lambda_y}}{\cos\theta_M \cos^2\psi_M} - c_3 \frac{V_M^2}{R}\frac{\sin\theta_M}{\cos\theta_M \cos^2\psi_M} \tag{16}$$

where c_3 and c_4 are positive constant guidance gains. Substituting the guidance commands (15) and (16) into (14) yields

$$\begin{aligned}\dot{f}_L &= -c_3 \frac{V_M^2}{R}\sin^2\theta_M - c_4 \frac{V_M^2}{R}\sin^2\psi_M \\ &\le 0\end{aligned} \tag{17}$$

The Lyapunov candidate function (12) is a continuously differentiable function, such that for some $r > 0$, $\Omega_r = \{x_L \in \Re^n | f_L(x_L(t)) < r\}$ is bounded, and its first time derivative is negative semi-definite, where $x_L = \left\{\theta_M, \psi_M, e_{\lambda_y}, e_{\lambda_z}\right\}$. Let D be the set of all points satisfying $\dot{f}_L(t) = 0$ within Ω_r. And let us define M as the largest invariant set in D.

$$D = \{x_L \in \Re^n | \theta_M(t) = 0, \psi_M(t) = 0\} \tag{18}$$

$$M = \left\{x_L \in \Re^n | \theta_M(t) = 0, \psi_M(t) = 0, e_{\lambda_y}(t) = 0, e_{\lambda_z}(t) = 0\right\} \tag{19}$$

Assuming that M contains a point with $e_{\lambda_y}(t) \neq 0$ and $e_{\lambda_z}(t) \neq 0$, following equations can be obtained.

$$\begin{aligned}
\lim_{\theta_M,\psi_M \to 0} \dot{\theta}_M(t) &= \lim_{\theta_M,\psi_M \to 0} \left(\frac{a_{zM}}{V_M} + \frac{V_M}{R}\sin^2\psi_M \tan\theta_L \cos\theta_M + \frac{V_M}{R}\cos\psi_M \sin\theta_M\right) \\
&= \lim_{\theta_M,\psi_M \to 0} \left(-\frac{V_M}{R}\sin\psi_M + c_2\frac{V_M}{R}\frac{e_{\lambda_z}}{\cos\psi_M} - c_4\frac{V_M}{R}\frac{\sin\psi_M}{\cos\theta_M\cos\psi_M}\right) \\
&+ \lim_{\theta_M,\psi_M \to 0} \left(\frac{V_M}{R}\sin^2\psi_M \tan\theta_L \cos\theta_M + \frac{V_M}{R}\cos\psi_M \sin\theta_M\right) \\
&\neq 0
\end{aligned} \tag{20}$$

$$\begin{aligned}
\lim_{\theta_M,\psi_M \to 0} \dot{\psi}_M(t) &= \lim_{\theta_M,\psi_M \to 0} \left(\frac{a_{yM}}{V_M\cos\theta_M} - \frac{V_M}{2R}\sin\theta_M \tan\theta_L \sin 2\psi_M\right) \\
&+ \lim_{\theta_M,\psi_M \to 0} \left(\frac{V_M}{R\cos\theta_M}\sin^2\theta_M \sin\psi_M + \frac{V_M}{R}\cos\theta_M \sin\psi_M\right) \\
&= \lim_{\theta_M,\psi_M \to 0} \left(-\frac{V_M}{R}\tan\theta_M \cos\psi_M + c_1\frac{V_M}{R}\frac{e_{\lambda_y}}{\cos^2\theta_M\cos^2\psi_M}\right) \\
&+ \lim_{\theta_M,\psi_M \to 0} \left(c_3\frac{V_M}{R}\frac{\sin\theta_M}{\cos^2\theta_M\cos^2\psi_M} - \frac{V_M}{R}\sin\theta_M \tan\theta_L \sin\psi_M \cos\psi_M\right) \\
&+ \lim_{\theta_M,\psi_M \to 0} \left(\frac{V_M}{R\cos\theta_M}\sin^2\theta_M \sin\psi_M + \frac{V_M}{R}\cos\theta_M \sin\psi_M\right) \\
&\neq 0
\end{aligned} \tag{21}$$

Therefore, if the trajectory moves out of D, then it will also move out of M. This is the contradiction to the definition of the invariant set M.

$$\dot{f}_L = 0 \Rightarrow \theta_M = 0, \psi_M = 0 \Rightarrow \dot{\theta}_M = 0, \dot{\psi}_M = 0 \Rightarrow e_{\lambda_y} = 0, e_{\lambda_z} = 0 \Rightarrow f_L = 0 \tag{22}$$

In other words, according to LaSalle's invariance theorem, the proposed acceleration commands make the system be asymptotically stable at the equilibrium point, i.e., $\theta_M(t) = 0$, $\psi_M(t) = 0$, $e_{\lambda_y}(t) = 0$, and $e_{\lambda_z}(t) = 0$.

3.2 *Pursuit-Like Guidance Law*

Using the final conditions, (9), and the terminal angle errors, (10) and (11), let us consider another Lyapunov candidate function to obtain the three-dimensional guidance law considering terminal angle constraints.

$$f_L = 2\sin^2\frac{\theta_M}{4} + 2\sin^2\frac{\psi_M}{4} + \frac{1}{2}k_1 e_{\lambda_y}^2 + \frac{1}{2}k_2 e_{\lambda_z}^2 \tag{23}$$

where k_1 and k_2 are positive constant. The time derivative of (23) can be expressed as

$$\dot{f}_L = \sin\frac{\theta_M}{4}\cos\frac{\theta_M}{4}\dot{\theta}_M + \sin\frac{\psi_M}{4}\cos\frac{\psi_M}{4}\dot{\psi}_M + k_1 e_{\lambda_y}\dot{\lambda}_y + k_2 e_{\lambda_z}\dot{\lambda}_z \tag{24}$$

Substituting (2)–(5) into (24) yields

$$\begin{aligned}\dot{f}_L =& \frac{1}{2}\sin\frac{\theta_M}{2}\left(\frac{a_{zM}}{V_M} + \frac{V_M}{R}\cos\theta_M\tan\theta_L\sin^2\psi_M + \frac{V_M}{R}\cos\psi_M\sin\theta_M\right)\\ &+ \frac{1}{2}\sin\frac{\psi_M}{2}\left(\frac{a_{yM}}{V_M\cos\theta_M} - \frac{V_M}{2R}\sin\theta_M\tan\theta_L\sin 2\psi_M\right)\\ &+ \frac{1}{2}\sin\frac{\psi_M}{2}\left(\frac{V_M}{R\cos\theta_M}\sin^2\theta_M\sin\psi_M + \frac{V_M}{R}\cos\theta_M\sin\psi_M\right)\\ &+ k_1 e_{\lambda_y}\frac{V_M}{R}\sin\theta_M - k_2 e_{\lambda_z}\frac{V_M}{R}\cos\theta_M\sin\psi_M\end{aligned} \tag{25}$$

Let us propose following acceleration commands.

$$\begin{aligned}a_{yM} =& -\frac{V_M^2}{R}\sin\psi_M + \frac{V_M^2}{4R}\sin 2\theta_M\tan\theta_L\sin 2\psi_M\\ &- k_4\frac{V_M^2}{2R}\cos\theta_M\sin\frac{\psi_M}{2} + 4k_2 e_{\lambda_z}\frac{V_M^2}{R}\cos\theta_M\cos\frac{\psi_M}{2}\end{aligned} \tag{26}$$

$$\begin{aligned}a_{zM} =& -\frac{V_M^2}{R}\sin\theta_M\cos\psi_M - \frac{V_M^2}{R}\cos\theta_M\tan\theta_L\sin^2\psi_M\\ &- k_3\frac{V_M^2}{2R}\sin\frac{\theta_M}{2} - 4k_1 e_{\lambda_y}\frac{V_M^2}{R}\cos\frac{\theta_M}{2}\end{aligned} \tag{27}$$

where k_3 and k_4 are positive constant guidance gains. Substituting the guidance commands (26) and (27) into (25) yields

$$\begin{aligned}\dot{f}_L &= -k_3\frac{V_M}{4R}\sin^2\frac{\theta_M}{2} - k_4\frac{V_M}{4R}\sin^2\frac{\psi_M}{2}\\ &\le 0\end{aligned} \tag{28}$$

Similarly, the Lyapunov candidate function (23) is a continuously differentiable function, such that for some $r > 0$, $\Omega_r = \{x_L \in \Re^n | f_L(x_L(t)) < r\}$ is bounded, and its first time derivative is negative semi-definite, where $x_L = \left\{\theta_M, \psi_M, e_{\lambda_y}, e_{\lambda_z}\right\}$. Let D be the set of all points satisfying $\dot{f}_L(t) = 0$ within Ω_r. And let us define M as the largest invariant set in D.

$$D = \{x_L \in \Re^n | \theta_M(t) = 0, \psi_M(t) = 0\} \tag{29}$$

$$M = \left\{x_L \in \Re^n | \theta_M(t) = 0, \psi_M(t) = 0, e_{\lambda_y}(t) = 0, e_{\lambda_z}(t) = 0\right\} \tag{30}$$

Assuming that M contains a point with $e_{\lambda_y}(t) \neq 0$ and $e_{\lambda_z}(t) \neq 0$, the following equations can be obtained.

$$\begin{aligned}\lim_{\theta_M,\psi_M \to 0} \dot{\theta}_M(t) &= \lim_{\theta_M,\psi_M \to 0} \left(\frac{a_{zM}}{V_M} + \frac{V_M}{R}\sin^2\psi_M \tan\theta_L \cos\theta_M + \frac{V_M}{R}\cos\psi_M \sin\theta_M\right) \\ &= \lim_{\theta_M,\psi_M \to 0} -\left(k_3 \frac{V_M}{2R}\sin\frac{\theta_M}{2} + 4k_1 e_{\lambda_y}\frac{V_M}{R}\cos\frac{\theta_M}{2}\right) \\ &\neq 0\end{aligned} \tag{31}$$

$$\begin{aligned}\lim_{\theta_M,\psi_M \to 0} \dot{\psi}_M(t) &= \lim_{\theta_M,\psi_M \to 0} \left(\frac{a_{yM}}{V_M \cos\theta_M} - \frac{V_M}{R}\sin\theta_M \tan\theta_L \sin\psi_M \cos\psi_M\right) \\ &\quad + \lim_{\theta_M,\psi_M \to 0} \left(\frac{V_M}{R\cos\theta_M}\sin^2\theta_M \sin\psi_M + \frac{V_M}{R}\cos\theta_M \sin\psi_M\right) \\ &= \lim_{\theta_M,\psi_M \to 0} \left(-k_4 \frac{V_M}{2R}\sin\frac{\psi_M}{2} + 4k_2 e_{\lambda_z}\frac{V_M}{R}\cos\frac{\psi_M}{2}\right) \\ &\neq 0\end{aligned} \tag{32}$$

Hence, if the trajectory moves out of D, then it will also move out of M. This is the contradiction to the definition of the invariant set M.

$$\dot{f}_L = 0 \Rightarrow \theta_M = 0, \psi_M = 0 \Rightarrow \dot{\theta}_M = 0, \dot{\psi}_M = 0 \Rightarrow e_{\lambda_y} = 0, e_{\lambda_z} = 0 \Rightarrow f_L = 0 \tag{33}$$

Therefore, according to LaSalle's invariance theorem, the proposed acceleration commands make the system be asymptotically stable at the equilibrium point, i.e., $\theta_M(t) = 0$, $\psi_M(t) = 0$, $e_{\lambda_y}(t) = 0$, and $e_{\lambda_z}(t) = 0$.

4 Numerical Simulations

Numerical simulations are performed to demonstrate the performance of the proposed three-dimensional terminal angle constrained guidance laws. In the simulations, an interception scenario against a stationary target and a carrier landing scenario are considered. The performance of the proposed guidance laws are compared with that of the three-dimensional sliding mode control(SMC)-based terminal angle constrained guidance law [19].

4.1 Interception Scenario

In this simulation, three missiles fired at same locations are considered. However, the guidance laws used for each missile are different. The speed of missiles is

$250m/s$, and the initial position of missiles is $(0,\ 0,\ 0)\,m$. The position of the target is $(8{,}000,\ 6{,}000,\ 0)\,m$, and θ_{M0} and ψ_{M0} of the missiles are $10deg$ and $40deg$, respectively. The limit of the acceleration command is chosen as $10g$, where g is a gravitational acceleration. The desired impact angles, θ_{L_d} and ψ_{L_d}, are -30deg and 30deg, respectively.

Figure 2 shows the three-dimensional flight trajectories of the missiles. In Fig. 2, the solid line is the trajectory of the missile using the three-dimensional pursuit-like guidance law with terminal angle constraints, the dashed line is the trajectory of the missile using the three-dimensional PN-like guidance law with terminal angle constraints, and the dashed-dot line is the trajectory of the missile using the three-dimensional SMC terminal angle constrained guidance law. All missiles intercepted the target successfully while satisfying the given terminal angle constraints. The absolute values of θ_M and ψ_M of each missile do not exceed $\pm\pi/2$ as shown in Fig. 3.

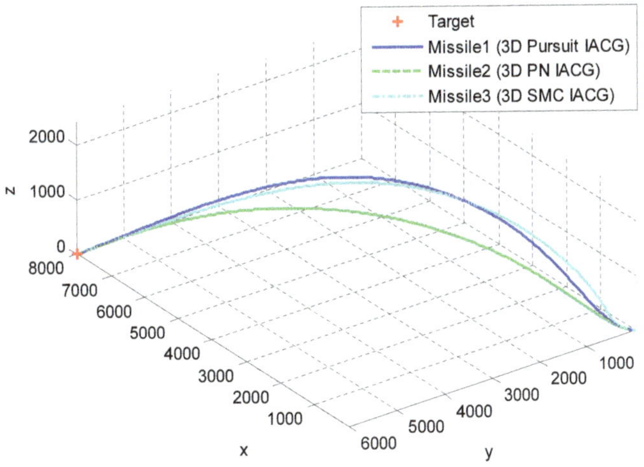

Fig. 2 Three-dimensional flight trajectory: an interception mission

To compare the performance of three guidance laws, the following performance indexes, the final impact angle errors $\Delta\theta_L$ and $\Delta\psi_L$, the fuel consumption J_F, and the energy consumption J_E, are calculated.

$$\Delta\theta_L = \left|\theta_{L_f} - \theta_{L_d}\right| \tag{34}$$

$$\Delta\psi_L = \left|\psi_{L_f} - \psi_{L_d}\right| \tag{35}$$

$$J_F = \int_{t_0}^{t_f} \left(\left|a_{y_M}\right| + \left|a_{z_M}\right|\right) dt \tag{36}$$

$$J_E = \int_{t_0}^{t_f} \left({a_{y_M}}^2 + {a_{z_M}}^2\right) dt \tag{37}$$

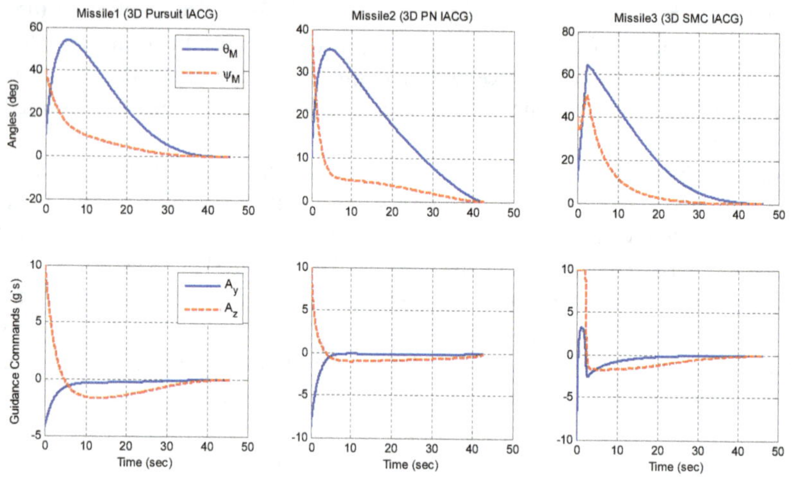

Fig. 3 Time history of θ_M and ψ_M, and acceleration commands (Interception mission)

The final impact angle errors and the final miss distance of each missile are summarized in Table 1. The final impact angle errors and the final miss distance of each missile are smaller than 1 *deg* and 0.5*m*, respectively. The fuel consumption and the energy consumption of each missile are summarized in Table 2. In the view of the fuel consumption and the energy consumption, the three-dimensional PN-like terminal angle constrained guidance law shows the best performance as shown in Table 2.

Table 1 The final impact angle error and the final miss distance: an interception mission

Missile	The final impact angle error (*deg*) ($\Delta\theta_L$ and $\Delta\psi_L$)	The final miss distance(*m*)
Missile 1 (3D pursuit-like IACG)	0.15, 0.09	0.23
Missile 2 (3D PN-like IACG)	0.16, 0.06	0.11
Missile 3 (3D SMC IACG)	0.67, 0.09	0.21

Table 2 The fuel consumption and the energy consumption: an interception mission

Missile	The fuel consumption (J_F)	The energy consumption (J_E)
Missile 1 (3D pursuit-like IACG)	693.59	1.7082×10^4
Missile 2 (3D PN-like IACG)	562.00	1.3532×10^4
Missile 3 (3D SMC IACG)	863.58	3.2609×10^4

4.2 Carrier Landing Scenario

In this simulation, three unmanned aerial vehicles(UAVs) trying to land on a nonmaneuvering, i.e., moving with a constant velocity, carrier are considered. Again, each UAV uses different a guidance law. The speed of UAVs is $50m/s$, and the initial position of UAVs is $(0,\ 0,\ 500)m$. The initial position, the speed, and the moving direction of the carrier are $(4{,}000,\ 2{,}000,\ 0)m$, $10m/s$, and $45deg$, respectively. Therefore, the desired approach angles θ_{L_d} and ψ_{L_d} are $0deg$ and $45deg$ in this scenario. The limit of the acceleration command is chosen as $2g$. Other simulation conditions are same as the previous scenario.

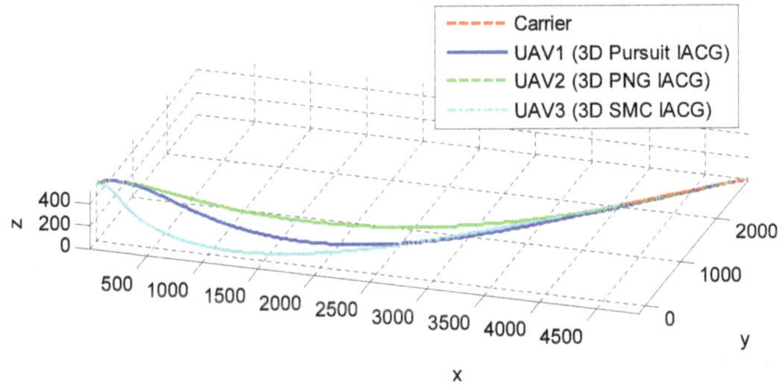

Fig. 4 Three-dimensional flight trajectory: an carrier landing mission

Figure 4 shows the three-dimensional flight trajectories of UAVs. In Fig. 4, the solid line is the trajectory of the UAV using the three-dimensional pursuit-like guidance law with terminal angle constraints, the dashed line is the trajectory of the UAV using the three-dimensional PN-like guidance law with terminal angle constraints, the dashed-dot line is the trajectory of the UAV using the three-dimensional SMC

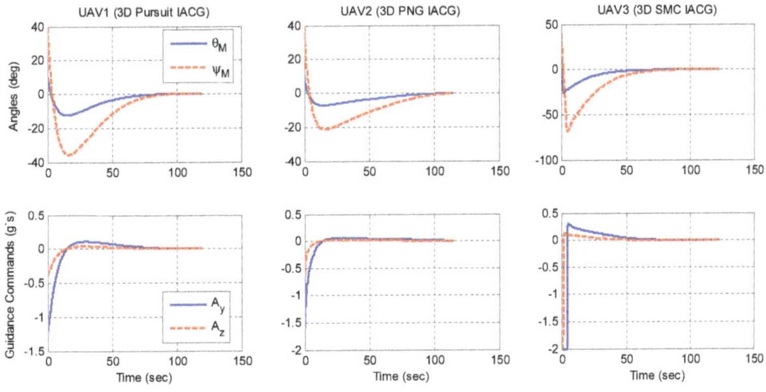

Fig. 5 Time history of θ_M, ψ_M, and acceleration commands: an carrier landing mission

terminal angle constrained guidance law, and the red dashed line is the trajectory of the carrier. All UAVs landed on the carrier successfully while satisfying the given terminal angle constraints. The absolute values of θ_M and ψ_M of each UAV do not exceed $\pm\pi/2$ as shown in Fig. 5.

To compare the performance of three guidance laws considering terminal angle constraints, the same performance indexes considered in the previous simulation are considered. The final approach angle errors and the final miss distance of each missile are summarized in Table 3. All guidance laws have the good performance of the final approach angle errors and the final miss distance as shown in Table 3. The fuel consumption and the energy consumption of each UAV are summarized in Table 4. In the view of the fuel consumption and the energy consumption, the three-dimensional PN-like terminal angle constrained guidance law has the best performance as shown in Table 4.

Table 3 The final approach angle error and the final miss distance: an carrier landing mission

UAV	The final approach angle error (deg) ($\Delta\theta_L$ and $\Delta\psi_L$)	The final miss distance(m)
UAV 1 (3D pursuit-like IACG)	0.04, 0.12	0.03
UAV 2 (3D PN-like IACG)	0.08, 0.24	0.03
UAV 3 (3D SMC IACG)	0.001, 0.004	0.01

Table 4 The fuel consumption and the energy consumption: an carrier landing mission

UAV & The fuel consumption (J_F) & The energy consumption (J_E)		
UAV 1 (3D pursuit-like IACG)	141.62	532.32
UAV 2 (3D PN-like IACG)	111.51	434.12
UAV 3 (3D SMC IACG)	217.64	2,405.2

5 Conclusions

In this paper, three-dimensional terminal angle constrained guidance laws were proposed. The three-dimensional geometry was considered without the assumption that the pitch channel and the yaw channel are decoupled in this study. Two Lyapunov candidate functions were introduced to develop the guidance laws considering terminal angle constraints. Lyapunov stability theory and LaSalle's invariance theorem were used to analyze the stability of the proposed guidance laws. Numerical simulations were performed to demonstrate the performance of the proposed guidance laws and to compare the performance of the three-dimensional terminal angle constrained guidance laws. The future work will include the singularity avoidance of three-dimensional terminal angle constrained guidance laws.

Acknowledgements. This study was supported by Guidance/Control Study for Take-off and Landing on a ship program through the Agency for Defense Development (ADD) of KOREA (UD1130053JD).

References

1. Kim, M., Grider, K.V.: Terminal Guidance for Imapct Attitude Angle Constrained Flight Trajectories. IEEE Transactions on Aerospace Electron Systems 9(6), 852–859 (1973)
2. Song, T., Shin, S., Cho, H.: Impact Angle Control for Planar Engegements. IEEE Transactions on Aerospace Electron Systems 35(4), 1439–1444 (1999)
3. Ryoo, C., Cho, H., Tahk, M.: Optimal Guidance Law with Terminal Impact Angle Constraint. Journal of Guidance, Control, and Dynamics 28(4), 724–732 (2005)
4. Lee, C., Tahk, M., Lee, J.: Generalized Formulation of Weighted Optimal Guidance Laws with Impact Angle Control. IEEE Transactions on Aerospace Electron Systems 49(2), 1317–1322 (2013)
5. Cho, H., Ryoo, C., Tsourdos, A., White, B.: Optimal Impact Angle Control Guidance Law Based on Linearization About Collision Triangle. Journal of Guidance, Control, and Dynamics 37(3), 958–964 (2014)
6. Ratnoo, A., Ghose, D.: Impact Angle Constrained Interception of Stationary Targets. Journal of Guidance, Control, and Dynamics 31(6), 1816–1821 (2008)
7. Ratnoo, A., Ghose, D.: Impact Angle Constrained Guidance Against Nonstationary Nonmaneuvering Targets. Journal of Guidance, Control, and Dynamics 33(1), 269–275 (2010)
8. Ratnoo, A., Ghose, D.: State-Dependent Riccati-Equation Based Guidance Law for Impact-Angle-Constrained Trajectories. Journal of Guidance, Control, and Dynamics 32(1), 320–325 (2009)
9. Kim, K., Jung, B., Kim, Y.: Practical Guidance Law Controlling Impact Angle. Journal of Aerospace Engineering, Proceedings on the Institution of Mechanical Engineers Part G 221(1), 755–774 (2007)
10. Kim, M., Kim, Y.: Lyapunov-Based Pursuit Guidance Law with Impact Angle Constraint. In: Proc. of the 19th IFAC World Congress, Cape Town, South Africa, pp. 2509–2514 (2014)
11. Song, S., Ha, I.: A Lyapunov-Like Approach to Performance Analysis of 3-Dimensional Pure PNG Laws. IEEE Transactions on Aerospace Electron Systems 30(1), 238–248 (1994)
12. Oh, J., Ha, I.: Capturability of the 3-Dimensional Pure PNG Law. IEEE Transactions on Aerospace Electron Systems 35(2), 491–503 (1999)
13. Ma, P., Zhang, Y., Ji, J., Zhang, X.: Three-Dimensional Guidance Law with Terminal Impact Angle Constraint. In: Proc. of the IEEE International Conference on Mechatronics and Automation, Changchun, China, pp. 4162–4166 (2009)
14. Yoon, M.: Relative Circular Navigation Guidance for Three-Dimensional Impact Angle Control Problem. Journal of Aerospace Engineering 23(4), 300–308 (2010)
15. Oza, H., Padhi, R.: Impact-Angle-Constrained Suboptimal Model Predictive Static Programming Guidance of Air-to-Ground Missiles. Journal of Guidance, Control, and Dynamics 35(1), 153–164 (2012)

16. Indig, N., Ben-Asher, J., Farber, N.: Near-Optimal Spatial Midcourse Guidance Law with an Angular Constraint. Journal of Guidance, Control, and Dynamics 37(1), 214–223 (2014)
17. Wang, X., Wang, J.: Partial Integrated Guidance and Control for Missiles with Three-Dimensional Impact Angle Constraints. Journal of Guidance, Control, and Dynamics 37(2), 644–657 (2014)
18. Lee, Y., Kim, Y.: Three-Dimensional Impact Angle Control Guidance Law for Missiles Using Dual Sliding Surfaces. In: Proc. of the 19th IFAC Symposium on Automatic Control in Aerospace, Wurzburg, Germany, pp. 137–142 (2013)
19. Kumar, S.R., Ghose, D.: Three-Dimensional Impact Angle Constrained Guidance Law using Sliding Mode Control. In: Proc. of the 2014 American Control Conference, Portland, OR, pp. 2474–2479 (2014)

Application of Optimization-Based Worst Case Analysis to Control Law Assessment in Aerospace

Hans-Dieter Joos

Abstract. The flight control law design assessment problem can be formulated as a robustness analysis problem, where a set of suitably defined assessment criteria must be checked to lie within certain limits for all admissible variations of vehicle parameters, external inputs and all flight conditions. Optimization based worst case analysis can be used to find those parameters/inputs/flight conditions for which the criteria are violated or poorly satisfied. The potential of this approach is its general applicability to any kind of simulation models and scenarios including complex non-linearity in control laws. But in order to confidently assert that no violation of assessment criteria exists, a global optimization problem has to be solved. Especially in case of many assessment criteria, global worst case search can lead to a huge computational effort. However, solving worst case problems as a multi-objective problem can help to reduce the number of computations since all or some of the assessment criteria can be considered simultaneously. Optimization-based approaches can also be used to detect parameter sensitivities on the assessment criteria and can help to find safe parameter regions.

1 Introduction

When applying traditional approaches like Monte-Carlo simulations or parameter studies for flight control system (FCS) assessment it is problematic to find confidently weak or even worst cases or to confidently assert that those cases do not exist [14]. Along with the normally huge computational effort necessary for these

Hans-Dieter Joos
German Aerospace Center (DLR), Institute of Robotics and Mechatronics,
82234 Wessling, Germany
e-mail: dieter.joos@dlr.de

J. Bordeneuve-Guibé et al. (eds.), *Advances in Aerospace Guidance, Navigation and Control*, DOI: 10.1007/978-3-319-17518-8_4

approaches, this motivated research in [13] to explore the benefits of several new analysis techniques for the assessment analysis of flight control laws, like stability analysis of linear parameter varying systems or optimization based worst case search.

In the latter approach, the assessment problem is formulated as a robustness analysis problem, where a set of suitably defined assessment criteria must be checked to lie above (below) certain bounds for all admissible variations of aircraft parameters, external inputs and all flight conditions. Optimization tools can then be used to find the minimum (maximum) solution indicating whether these bounds are violated or not and the global extremum found is called the worst case. Such an approach was already formulated in [2].

Hence the difficulty in using the optimization based approach does not lie in finding a parameter combination such that a criterion is not satisfied, i.e. to demonstrate that the FCS is not validated, but to confidently assert that a criterion is verified in all cases. That means one has to solve a global optimization problem and the global solution should be found with some reasonable confidence level. A strategy for approaching this is proposed in [12].

There exist several investigations on optimization based assessment regarding different applications and usage of different single objective optimization algorithms, see e.g. [4, 7, 9, 10, 11, 14]. The results clearly demonstrate the high potential of the optimization-based approach in reliably solving assessment problems with many simultaneous uncertain parameters and criteria. The reasons for that are in particular the generality and reliability of the approach without adding conservatism. These advantages also hold for the multi-criteria approach. Since the assessment process normally requires demonstrating system performance for a whole set of criteria one can expect that the multi-objective approach will reduce computational effort substantially since only one optimization run is necessary to get the required results for all criteria under consideration.

Not only worst-case problems can be solved by means of optimization but also the effect of individual parameter variations on assessment criteria violation can be examined. For this a norm optimization problem has to be formulated with the assessment conditions as constraints.

Moreover using the niching concepts introduced for genetic algorithms [5], regions in the parameter space can be detected where assessment criteria are violated or the system is of weak performance. These parameter regions may be spread over the whole admissible parameter range not only concentrated on the region of convergence to the worst case parameter combination.

In this paper the formulation of assessment and sensitivity problems as optimization problems will be given in sect. 2. The control law assessment process based on this optimization tasks is demonstrated by two aerospace applications. In sect. 3, multi-objective optimization is used to validate eight control law performance criteria of a nonlinear six degrees of freedom time domain simulation model of the VEGA launcher [8]. For this application also the effects of the more than eighty parameters are investigated applying the proposed optimization approaches. The second application is to find regions of reduced performance in the flight envelope of a research FCS for a large civil transport aircraft regarding load limits exceedance (Sect. 4). The development of a fault tolerant FCS including protections is one of the benchmark problems formulated in the FP7 Project RECONFIGURE (Reconfiguration of Control in Flight for Integral Global Upset Recovery) [15].

2 Formulation of Assessment Problems as Optimization Tasks

2.1 *Worst Case Search*

Following the approaches in [2, 4, 7, 9, 10, 11, 13] for utilising optimisation methods the assessment problem has to be expressed as a scalar objective function $c(p,d)$ with optimisation parameters p that are uncertain or varying during operation (e.g. aerodynamic coefficients, aircraft mass, inertia, speed, height or wind parameters, etc.) and discrete conditions d (e.g. aircraft configuration, landing gear settings, etc.). The components of p are assumed to be bounded and continuously varying over known intervals, defining a hyper-box P. The assessment problem can now be formulated as a minimization problem. Let $c_0(d)$ a lower acceptable value of c then

$$c_{min}(d) = \min_{p \in P}(c(p,d)) \tag{1}$$

is a measure for the assessment performance. The assessment requirement is fulfilled for condition d if $c_{\min}(d) \geq c_0(d)$, otherwise the criterion is not cleared.

This formulation can immediately be extended to several criteria using the theory of Pareto optimality [16, 17]. The following optimization problem has now to be solved

$$\text{minimize } [c_i(p)] \; for\; i = 1, \dots, m, s.t.\; p_j^L \leq p_j \leq p_j^U \,,\; j = 1, \dots n. \tag{2}$$

If $[c_i^*(p^*)]$, $p^* \in P_{Pareto}$ is a solution point of the Pareto-optimal set, the worst case of criterion c_i^{wc} is given by the minimum over all p^*:

$$c_i^{wc} = \min_{p^* \in P_{Pareto}}\left(c_i^*(p^*)\right). \tag{3}$$

Generally the two most common approaches to solve multiple objective problems are: (1) combine them into a single objective function and obtain a single solution such as in the cases of the weighted sum method or maximum utility functions, or (2) obtain a set of non-dominated Pareto-optimal solutions directly. The purpose of this study is to demonstrate the usefulness of approach (2) for worst case search problems, where only the extreme values of the Pareto-optimal solution set are of primary interest.

A sophisticated algorithm called NSGA-II based on non-dominated sorting was published by Deb [1]. The objective of the NSGA algorithm is to improve the fit of candidate solutions to a Pareto front with respect to a set of objective functions based on the concept of Pareto-ranking [3] where the fitness of an individual is measured by the number of individuals which dominate it. An individual (set of parameters) *A* is dominated by an individual *B* if the all criteria values corresponding to *B* are less than the criteria values corresponding to *A*. The Pareto-rank of an individual is then defined as the number of dominating individuals plus one.

The algorithm itself is an instance of an evolutionary algorithm using evolutionary operators including selection, genetic crossover, and genetic mutation. The population is sorted into a hierarchy of sub-populations based on Pareto ranking. The diversity of the members along a Pareto-front is improved by fitness sharing concepts.

The computations of the assessment-study of this paper have been performed by means of DLR's optimization environment MOPS [6]. The algorithm for multi-objective optimization implemented there is a combination of a global exploration mechanism based on non-dominated sorting with a local search algorithm based on gradient-free pattern search. In such a hybrid approach global multi-objective search alternates iteratively with short local single-objective optimization. The following advantages can be expected:

- A deterministic search for each single criterion can provide rapidly good solutions marking the minimum solutions of each criterion.
- Improved minima of the individual objectives widen the range of the Pareto-front. This normally speeds up the search of elements of the Pareto-front between these points by means of genetic algorithms.
- The minimum value for each criterion is found more precise by means of local search algorithms. This is especially useful for worst case search.

Experiences showed that a multi-objective WCS costs about the same amount on criteria evaluations as a WCS with a single objective only especially when using the hybrid optimization technique. Hence in case of usually several assessment criteria multi-objective WCS can help to reduce computational effort since all or at least some of the assessment criteria can be considered simultaneously in one run.

2.2 Regions of Compliance

The problem of finding regions of compliance is of great interest in aerospace in order to get information where a flight envelope should be restricted. In practice, one has to find ranges in the parameter space where system performance is confidently satisfied is of great interest in aerospace in order to get information where a flight envelope should be restricted. A general formulation of this problem at least for simply connected parameter ranges as an optimization problem is:

$$\max\left[p_i^U - p_i^L\right] , \ s.t. \ c_j(p) \geq c_j^0, \ for\ all\ p \epsilon \left[p^L, \ p^U\right] \tag{4}$$

A multi-objective optimization problem has to be solved with the length of parameter intervals as optimization criteria subject to the constraint that the assessment criteria are satisfied. In general this global optimization problem cannot be solved with reasonable effort at least for many parameters.

Without loss of generality we can assume that the optimization parameters are normalized to $[-1, 1]$. Restricting now the range of compliance problem to search for centered equal bounded parameter intervals the following practical formulation of the optimization problem can be made assuming that for $\bar{p} = 0$ all assessment criteria are satisfied:

$$\max(\|\bar{p}\|_\infty), \ s.t. \ c_j \geq c_j^0 , \ \bar{p} \ni [-1, 1] \tag{5}$$

Now a single objective optimization has to be solved under the constraint that the assessment criteria are satisfied.

Evolutionary computation offers another concept to find regions of non-compliance, i.e. regions where at least one assessment criterion might be not satisfied. These regions may be disconnectedly spread over the entire parameter range. With classical optimization techniques, which normally converge in some way to the optimal solution, disconnected parameter regions with weak performance cannot be detected safely. The niching (or crowding distance) concepts developed for genetic algorithms allow (i) to keep multiple, highly fit, but significantly different solutions in a generation, and (ii) help to avoid premature convergence to local minima [5].

When niching is applied, the fitness of an individual is calculated as a properly scaled ratio of the objective function value and minimal distance to other individuals in the parameter space. In our optimization environment niching is applied only for single objective problems.

2.3 *Parameter Effects on Criteria*

The knowledge of the parameter effectivity on the assessment criteria is of great interest for the design of robust control laws. The importance of individual parameters for cases where assessment criteria are violated can be estimated by means of the following optimization problem:

$$\min(\|\bar{p}\|_1), \ s.t. \ c_j < c_j^0 , \ \bar{p} \ni [-1, 1] \tag{6}$$

Again it is assumed that for zero parameter values all assessment criteria are satisfied. Since all components of parameter vector $\bar{p}$ contribute to the 1-norm, a minimum can only be achieved when all values of ineffective or less effective parameters tend to zero. Due to the constraints, the more effective parameters achieve nonzero values. The minimum must not be unique. At the global minimum all ineffective parameters are expected to be zero.

3 Application Control Law Assessment of a Launcher Vehicle

The assessment analysis performed in this study is based on a standard industrial nonlinear, six degree of freedom simulation model of a launch vehicle which was also the benchmark model in the ESA project "Robust Flight Control System Design Verification and Validation Framework" (ESA AO/1-6322/09/NL/JK). The model describes the VEGA launcher, a new European small launch vehicle developed under the responsibility of ESA, including non-linear actuator dynamics, flight mechanics, aerodynamics, wind input, inertia system and sensors. It is equipped with a full guidance navigation and control system for thrust vector control (TVC) and roll and attitude control (RACS). For assessment studies the model is provided as a black-box simulation model allowing no access to structure and states.

However, the model allows external access to a large number of uncertain parameters regarding aerodynamics, wind, inertial reference system, thrust scattering, mass, center of gravity, inertia, thrust offset and misalignment, atmosphere and separation. In this study the first flight phase with height between 35m and 60km is considered, utilizing 84 uncertainty parameters for assessment. All parameters are scaled to [-1, 1] for external access.

The assessment criteria considered here represent the traditional requirements for TVC and RACS during first flight phase. Eight optimization criteria have been extracted from the compliance matrix regarding lateral control performance and load requirements during flight as well as separation conditions:

- Load requirements are validated via the product of dynamic pressure and angle of attack, $Q\alpha$, depending on Mach (criterion C1)
- Drift of position and speed in y- and z-direction are measured within the entire altitude range of 35m to 60km (criteria C2, C3, C4, and C5).
- Limits on transversal angular acceleration have to be kept during the whole flight phase (criterion C6)
- Transverse angular velocity at separation must be within some limits (criterion C7)
- Total angle of attack at separation (criterion C8)

All criteria are normalized to their limits or envelop bounds such that a violation of the requirement occurs when the criteria values are less than minus one. For a given set of uncertain parameters the first phase of the flight is simulated until separation. The criteria are then calculated from the various outputs provided by the high-fidelity simulation model.

The applicability of multi-objective optimization to such a multi-criteria assessment problem has already been demonstrated in [12]. As a consequence of these results the control law has been retuned in order to get better performance especially for loads criteria C1 making the worst case search much more ambitious even though hybrid optimization techniques have been used now.

Fig. 1 shows the criteria results of one optimization run for the parameter combinations with Pareto rank 1, i.e. those points which are candidate for the Pareto front. It shows that all criteria have improved performance compared to the control law assessed in [12]. However, there are still violations of the assessment bound of -1 for criteria 1 and 7. Moreover it can be seen that there exist parameter combinations where both criteria are violated at the same time.

Following the assessment strategy outlined in [12] to assert that the global worst cases are reached reliably the WCS has been repeated 10 times with different initial and statistical conditions. Applying 10 runs and assuming a success rate of 0.7 for sufficiently approaching the (global) worst cases by one optimization run a confidence level of 99% is achieved.

In all runs no WCS could violate criteria C2 to C6 which therefore can be considered as verified. However, criteria C1, C7 and C8 are violating the assessment bounds at least slightly (C8) in all runs. For these criteria the assessment requirements are therefore still not fulfilled.

The hybrid optimization was performed applying a population size of 200 over 100 generations, yielding 20000 simulations and criteria evaluations for global search only. The iteratively activated local searches of the hybrid algorithm require on average almost the same amount on evaluations. Such an optimization run with approximately 40000 evaluations takes about 4 hours of computation time on a desktop PC equipped with two Intel® Xeon® Processors X5550 (2.66GHz) when the parallelization features of MOPS were used utilizing 4 workers.

For the overall assessment problem with 8 criteria and 84 uncertain parameters a region of non-compliance was looked for by means of the norm optimization problem formulated by (5). The max-norm optimization results in the maximum parameter interval around zero, normally the nominal case, inside which the assessment criteria are not violated for all parameter combinations, provided that the global minimum was found. Fig. 2 shows the result of an exhaustive optimization run with about 57.000 evaluations, using the genetic optimization method implemented in MOPS. The blue bars indicate that parameters are allowed to deviate up to about 75% from nominal before a violation of any of the assessment criteria occurs.

Parameter sensitivity was investigated by a follow on optimization run now minimizing the 1-norm of the uncertainties where the parameter range is restricted to the maximum magnitude of 0.75 as found before. It is expected that all parameters which have no or only less effect on the criteria will tend to zero. The result is shown by the green bars. Several parameters represented by a large blue bar are not accompanied by green ones; see for example parameters numbered as 19 and 21. This indicates that these parameters have no or only small effect on criteria violations. However, since the parameter region is restricted, a deviation of 75% from nominal is still necessary for more than 30 parameters in order to violate at least one criterion.

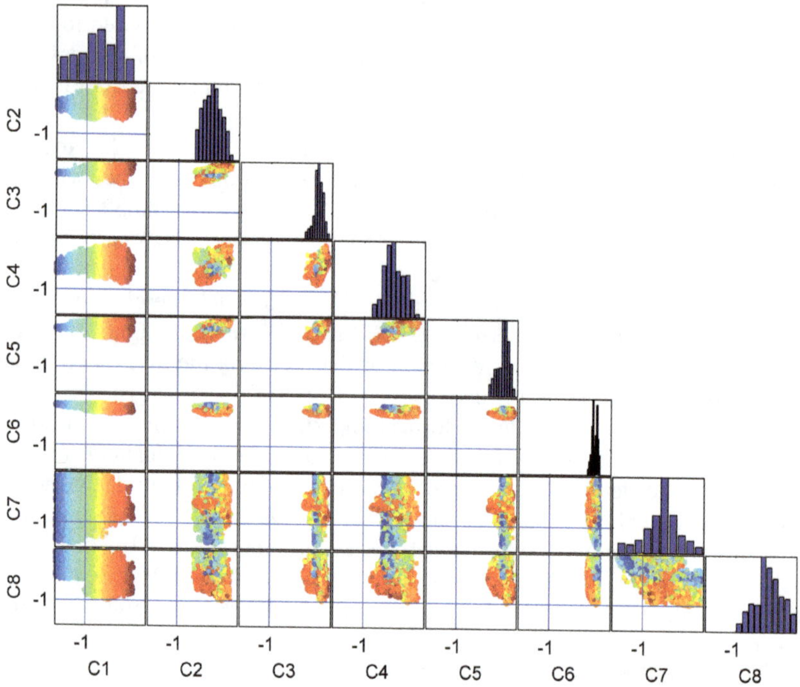

Fig. 1 Two-dimensional section planes of the 8-dimensional Pareto-front for all criteria pairs. The diagrams in the diagonal show the histogram of a single criterion.

When deviations of the whole parameter range are allowed, the optimization problem (6) gives an answer to the question which parameters are the most sensitive ones. An example result of such an optimization is depicted by the brown bars in Fig. 2. Only 10 parameters show a significant deviation from nominal. Hence parameter deviations of those 10 most sensitive parameters will result in violations of at least one assessment criterion. However, the deviation must be up to 100% for most of the parameters.

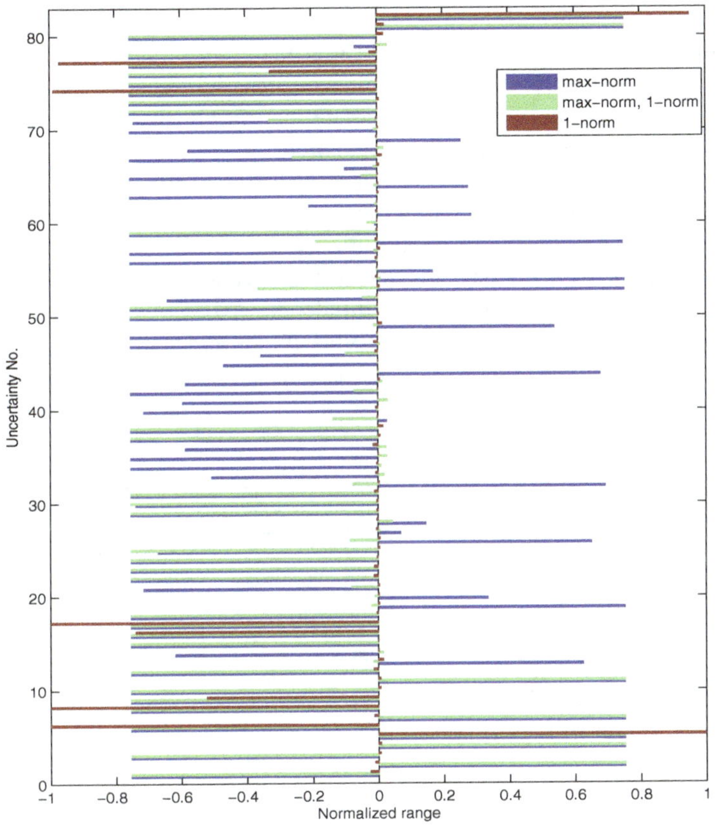

Fig. 2 Range of non-compliance found by max-norm optimization and parameter sensitivity found by 1-norm optimization for the launcher vehicle.

4 Application of the Niching Concepts to Detect Regions of Non-compliance

For a large commercial aircraft one of the assessment requirements is that normal load should not exceed 2.5g. This property is validated using a high-fidelity industrial aircraft simulation model which is augmented by an experimental FCS including non-linear protection laws to prevent load factor exceedance and to comply with angle of attack, attitude angle or speed limits. WCS is able to keep with such a highly non-linear assessment problem without approximations or simplifications.

The diagrams in Fig. 3 and Fig. 4 show the results of two different WC searches. Both diagrams depict the points visited during optimization in the normalized flight domain. Points that violate load factor assessment bounds are marked with a red cross. About 200 evaluations have been performed for each optimization run. Fig.3 shows the search result without using niching concepts. The optimizer converges rapidly to a minimum indicating bad performance in a single region around the normalized altitude of 0.6 and high speed. In Fig.4 where the results with niching are depicted a second region of weak performance can be recognized (normalized altitude about 0). However, the minimum found by the run applying niching concepts is not as small as the one found without niching.

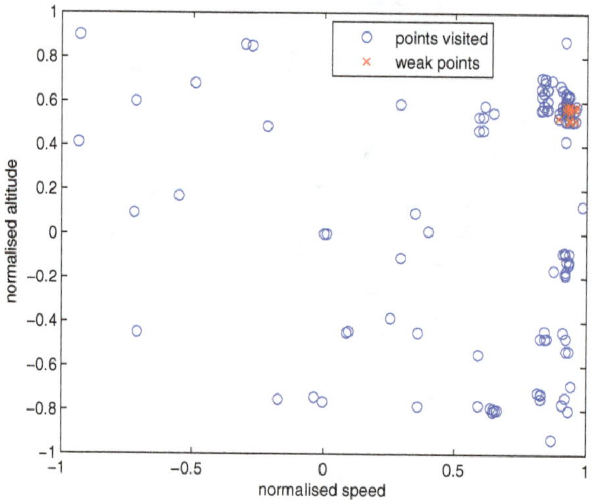

Fig. 3 WCS results indicating points in the normalized flight envelope with weak performance (x) regarding load factor limits of an experimental FCS applying an implementation of a genetic algorithm without niching concepts.

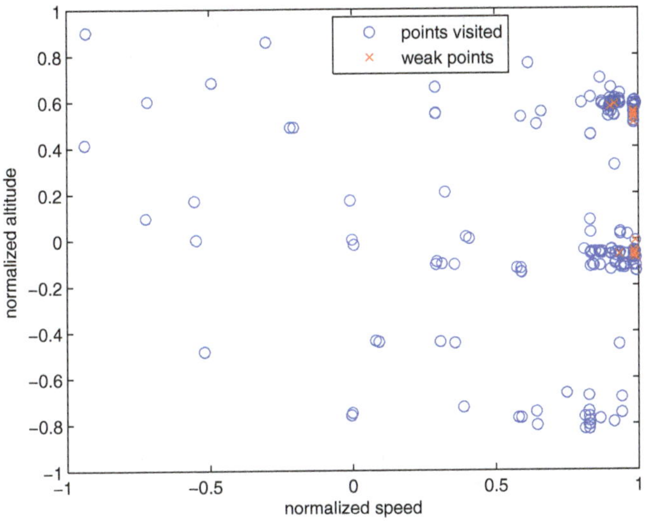

Fig. 4 WCS results indicating points in the normalized flight envelope with weak performance (x) regarding load factor limits of an experimental FCS; application of a genetic algorithm including niching concepts. Two disconnected regions of noncompliance can be detected.

5 Conclusions

From these observations and results obtained for the benchmark problems, one can conclude that multi-objective optimization-based approaches can be a valuable part of assessment analysis procedures. While several criteria can be examined in one step, the computational effort does not increase compared to single objective assessment. Furthermore, the detection whether several assessment criteria can be violated simultaneously can only be made by multi-objective considerations.

The concepts of norm minimization with assessment constraints and the niching concept can be effectively used to find regions of noncompliance or to provide information about regions of safe operation. Moreover, the knowledge of parameter effects on assessment criteria can give useful advice for designing a FCS.

The direct formulation of assessment criteria from simulation results can lead to noisy or even discontinuous objective functions with multiple minima which are difficult to detect. However, genetic search techniques are insensitive to such kind of disturbed criteria and hence are able to solve the worst case search effectively and reliably.

Optimization based worst case search seems to be very 'aggressive' in the sense that it can take advantage out of any possibly incomplete or incorrect modelling of the validation criterion. Hence, unsatisfactory criterion values are not necessarily due to control law weaknesses only. Before identifying a criterion not to be cleared the preconditions have to be verified carefully such as: validity of parameter ranges; completeness and correctness of the models involved (aircraft, controller, criterion); reasonable and realistic maneuvers and according implementations as simulation tasks. Since worst case search often operates at extreme flight conditions in the flight envelope, the validity of the aerodynamic models in those regions is of great importance. Therefore worst case search can not only be applied for assessment of flight control laws but also for verification of design and simulation models.

References

1. Dep, K., Pratrap, A., Agarwal, S., Meyarivan, T.: A Fast and Elitist Multiobjective Genetic Algorithm: NSGA-II. IEEE Transactions on Evolutionary Computation 6(2) (2002)
2. Fielding, C., Varga, A., Benani, S., Sellier, M. (eds.): Advanced Techniques for Clearance of Flight Control Laws. LNCIS, vol. 283. Springer, Heidelberg (2002)
3. Fonseca, C.M., Fleming, P.J.: An overview of evolutionary algorithms in multiobjective optimization. Evolutionary Computing 3(1), 1–16 (1995)
4. Forssell, L.S.: Flight clearance analysis using global nonlinear optimization-based search algorithms. In: AIAA Guidance, Navigation, and Control Conference and Exhibit, Austin, Texas, pp. 1023–1030 (2003)
5. David, E.: Genetic Algorithms with sharing for multimodal function optimization. In: Grefenstette, J.J. (ed.) Proceedings of the Second International Conference on Genetic Algorithms, Lawrence Erlbaum Associates, Hillsdale (1987)
6. Joos, H.-D., Bals, J., Looye, G., Schnepper, K., Varga, A.: A multi-objective optimization based software environment for control systems design. In: Proceedings of the 2002 IEEE International Conference on Control Applications and International Symposium on Computer Aided Control Systems Design, CCA/CACSD, Glasgow, Scotland, U.K (2002)
7. Joos, H.-D.: Worst-case parameter search based clearance using parallel nonlinear programming methods. In: Varga, A., Hansson, A., Puyou, G. (eds.) Optimization Based Clearance of Flight Control Laws. LNCIS, vol. 416, pp. 149–159. Springer, Heidelberg (2012)
8. Andres, M., Christophe, R., Max, R., Hans-Dieter, J., Samir, B., Felipe, P.L., Augusto, C.: The V&V problematic for launchers: current practise and potential advantages on the application of modern analysis techniques. In: 8th International ESA Conference on Guidance, Navigation & Control Systems Proceedings of ESA GNC, Czech Republic, Karlovy Vary (2011)
9. Menon, P.P., Kim, J., Bates, D.G., Postlethwaite, I.: Clearance of nonlinear flight control laws using hybrid evolutionary optimization. IEEE Transactions on Evolutionary Computation 10(6), 689–699 (2006)
10. Menon, P.P., Bates, D., Postlethwaite, I.: Optimization-based flight control law clearance. In: Bates, D., Hagström, D.M. (eds.) Nonlinear Analysis and Synthesis Techniques for Aircraft Control. LNCIS, vol. 365, pp. 259–300. Springer, Heidelberg (2007)

11. Menon, P.P., Bates, D., Postlethwaite, I.: Nonlinear Robustness Analysis of Flight Control Laws for Highly Augmented Aircraft. Journal of Control Engineering Practice 15(6), 655–662 (2007)
12. Pfiffer, H., Joos, H.-D.: Robust Flight Control System Design Verification and Validation by Multiobjective Worst-case Search. In: AIAA GNC/AFM/MST/ASC, Minneapolis, USA (2012)
13. Varga, A.: Optimization-based Clearance. In: Fielding, C., Varga, A., Benani, S., Sellier, M. (eds.) Advanced Techniques for Clearance of Flight Control Laws. LNCIS, vol. 283, pp. 107–117. Springer, Heidelberg (2002)
14. Varga, A., Hansson, A., Puyou, G. (eds.): Optimization Based Clearance of Flight Control Laws. LNCIS, vol. 416. Springer, Berlin (2011)
15. Varga, A., Ossmann, D., Joos, H.-D.: A fault diagnosis based reconfigurable longitudinal control system for managing loss of air data sensors for a civil aircraft. In: IFAC, World Congress, Cape Town, South Africa (2014)
16. Sawaragi, Y., Nakayama, H., Tanino, T.: Theory of Multiobjective Optimization. Mathematics in Science and Engineering, vol. 176. Academic Press Inc., Orlando Florida (1985)
17. Zitzler, E.: Evolutionary Algorithms fo Multiobjective Optimization: Methods and Applications, PhD thesis, Swiss Federal Institute of Technology (ETH), Zurich, Switzerland (November 1999)

Robust Output Tracking of a *3DOF* Helicopter via High-Order Sliding Mode Observers

Alejandra Ferreira de Loza, Jérome Cieslak, David Henry,
Ali Zolghadri, and Leonid Fridman

Abstract. This paper tackles the output tracking problem of a MIMO system subjected to actuator faults and unmatched perturbations. The proposed methodology is based on high order sliding mode observation and identification techniques. A dynamic sliding surface is proposed using a nested-backward design strategy in order to counteract the effects of the unmatched perturbations. A super-twisting control is used to steer the state to the sliding surface. The identified value of the fault is injected to alleviate the control gain while accomplishing fault accommodation. As a consequence, the chattering is attenuated. A simulation example for a $3-DOF$ helicopter highlights the efficiency of the present method.

1 Introduction

Fault Tolerant Control (FTC) is used to avoid a potentially hazardous, out-of-tolerance or dangerous behaviour of the controlled system. In the open literature, one can find a great number of publications on various strategies for FTC, see for example [1, 2, 4, 20] and many references therein. In this context, sliding mode

Alejandra Ferreira de Loza
Instituto Politecnico Nacional, Avenida del Parque 1310, Mesa de Otay, Tijuana, Mexico
e-mail: da_ferreira@yahoo.com

Jérome Cieslak · David Henry · Ali Zolghadri
Université de Bordeaux, CNRS, IMS-lab, 351 cours de la Libération,
33405 Talence cedex, France
e-mail: {jerome.cieslak,david.henry,
ali.zolghadri}@ims-bordeaux.fr

Leonid Fridman
Universidad Nacional Autonoma de Mexico (UNAM), Department of Control,
Engineering Faculty. C.P. 04510. Mexico D.F.
e-mail: lfridman@servidor.unam.mx

J. Bordeneuve-Guibé et al. (eds.), *Advances in Aerospace Guidance, Navigation and Control*, DOI: 10.1007/978-3-319-17518-8_5

control (SMC) has interesting characteristics thanks to its insensibility to matched perturbations, i.e. those which appear implicitly at the input channels [7]. This property has been taken into account to withstand with actuator faults in a wide class of applications, [3, 8, 9, 10]. In [8], fault detection and isolation is achieved by means of a residual generation scheme and thresholds definition. The fault accommodation takes place using the healthy redundant actuators. The robustness of the method may be spoiled if a fault occurrence is disregarded due to an inappropriate threshold selection. This is not the case of [9, 10], where the fault information appears as a given weighting matrix which contains the effectiveness level of the actuators. Thus a control allocation is proposed to accommodate the control among the actuators. Such strategies achieve the proposed goal assuming full state information, whereas in [10] the results are extended considering only output information an using integral sliding mode control scheme. Still, the aforementioned approaches are tarnished by the so-called chattering effect. To overcome this issue, High Order Sliding Mode (HOSM) techniques have been considered. Strategies based on fault reconstruction and compensation using HOSM schemes are developed in [11, 12, 13]. In [11] the actuator fault is reconstructed using a HOSM differentiator and then the identified signal is injected in the control law in order to compensate its effects. Nevertheless, it cannot stand for disturbances affecting the sub-actuated dynamics.

Strategies explicitly designed to cope with unmatched perturbations, i.e. those which are not implicit in the input channels, have been proposed in [12, 13, 14, 15]. In [15] the unmatched perturbations and faults are estimated by means of an adaptive mechanism whereas a backstepping technique is applied to compensate its effects. The performance of the closed-loop system is affected by the convergence rate of the adaptation mechanism. HOSM identification based strategies circumvent this issue owing to its finite time convergence capabilities. In this direction, in [12] a smooth backstepping control based on exact identification is proposed. This technique consider full state measurements and exploits a HOSM differentiator to identify the perturbations at each coordinate. Moreover, additional HOSM differentiators are involved to facilitate the control law computation. In [14, 13] output regulation strategies are introduced. While a priory knownlege of the perturbation dynamics is assumed in [14], in [13] perturbations are tackled as unknown inputs. Thus, a dynamic sliding surface is designed to cope with the unmatched perturbations. While a discontinuous control counteracts the matched perturbations. As a consequence chattering may occurs.

The goal of the paper is to design an observed-based control for a $3DOF$ helicopter system affected by unmatched perturbations and subjected to a class of actuator faults. The effect of the perturbations and actuator faults are thus handle as additive time-varying unknown inputs that can be identified by a HOSM observer [16, 17]. To achieve the control design, first a finite time convergent HOSM observer is used to estimate the state and identify the actuator faults as well as the unmatched perturbations and its successive derivatives based on the available output measurements [18]. Then, a dynamic sliding surface is designed to accomplish the desired control goal (tracking) despite the unmatched perturbations using a backstepping approach. Finally, a super-twisting based control law steers the state trajectories to the sliding surface guaranteeing that there remains despite the perturbations, the identified value of the fault is injected to alleviate the controller gain accomplishing fault accommodation.

In [19] the regulation problem of a $3-DOF$ helicopter subjected to matched uncertainties was carried HOSM observation and compensation. However the case of unmatched uncertainties was not considered.

Contribution. In this work the output tracking problem of a $3DOF$ helicopter system affected with actuator faults and unmatched perturbations is achieved. The proposed solution possesses fault accommodation capability against additive time-varying actuator faults. Moreover, the effects of the unmatched perturbations are counteracted.

The paper is organized as follows. In Section II, some preliminaries are given; the state estimation and perturbation identification by a HOSM observer are summarized. In Section III the problem formulation and control challenge are presented. Section IV presents the control strategy. The performance of the proposed technique is illustrated through numerical simulations for a 3-DOF Helicopter in Section 5. Finally some concluding remarks are given in Section 6.

2 Preliminaries

The following notation is used. For a matrix $X \in \mathbb{R}^{n\times m}$ with $rank(X) = r$, the matrix $X^{\perp} \in \mathbb{R}^{n-r\times n}$ with $rank(X^{\perp}) = n-r$ is defined such that $X^{\perp}X = 0$. If $X \in \mathbb{R}^{n\times m}$ with $rank(X) = m$, the matrix $X^{+} = (X^TX)^{-1}X^T$ is defined as the pseudo-inverse of X. If $X \in \mathbb{R}^{n\times m}$ with $rank(X) = m$. For any $x \in \mathbb{R}$ the symbol $|x|$ denotes its absolute value. If $x \in \mathbb{R}^n$ then $|x|$ states for the Euclidean norm. If $X \in \mathbb{R}^{n\times m}$ is a matrix, the symbol $|X|$ denotes the corresponding induced norm.

Consider the particular class of faulty systems that can be modeled as a linear time invariant system affected by unknown inputs

$$\dot{x} = Ax + B(u+f) + Gv \tag{1}$$

$$y = Cx \tag{2}$$

where $x \in \mathbb{R}^n$, $y \in \mathbb{R}^p$ $(1 \le p < n)$ represent the system's state and measured outputs respectively. The vector $u \in \mathbb{R}^m$ represents the input and $v \in \mathbb{R}^q$ $(q \le n-m)$ are the perturbations. The matrices A, B, G and C are of the corresponding dimensions. $f \in \mathbb{R}^m$ f refers to the (actuator) faults and behaves to the particular class of additive faults, see [20] for more details on modeling faults. The following is assumed henceforth: $A1)$ the system is strongly observable, or equivalently the triplet $(A, C, [B\ G])$ has no zeros; $A2)$ $span(G) \subset span(B^{\perp})$; $A3)$ v and its derivatives up to order r are bounded, i.e. $|v| \le \check{v}_0$ as well as $|v^{(i)}| \le \check{v}_i$ for $i = \overline{1,r}$, for all $t \ge 0$; $A4)$ f and its derivatives up to order $r-2$ are bounded, i.e. $|f| \le \check{f}_0$ as well as $|f^{(i)}| \le \check{f}_i$ for $i = \overline{1,r-2}$, for all $t \ge 0$.

Notice that assumption $A2$ distinguishes between faults and perturbations which belong to different subspaces. In this way, the smoothness characteristics of faults signals are relaxed $A3$ w.r.t. perturbations signals $A4$.

Following [18], a HOSM observer is introduced to reconstruct the state vector and identify the actuator faults, the perturbations as well as the perturbations successive derivatives.

2.1 High Order Sliding Mode Observer

First, a dynamic auxiliary system is proposed to bound the observation error, i.e. $\dot{\tilde{x}} = A\tilde{x} + Bu + L(y - C\tilde{x})$, $\tilde{x} \in \mathbb{R}^n$, the gain L is designed such that $\tilde{A} := A - LC$ is Hurwitz. Let $e = x - \tilde{x}$ whose dynamics follows

$$\dot{e} = \tilde{A}e + Dw \tag{3}$$

where $D = [\,B\;G\,]$ and $w = [\,f^T\;v^T\,]^T$ and $y_e = Ce$. Thus, under the assumptions $A2$ and $A3$, the unknown inputs vector w, and its successive derivatives, are also bounded. Thus, it is well known that e will have a bounded norm, i.e. $|e| < \check{e}$ for all $t > t_e$.

The error vector will be represented as an algebraic expression of the output and its derivatives. To this aim, a decoupling algorithm is involved in order to get rid of the effects of the unknown input vector w.

Starting with $M_1 := C$ and $J_1 := (M_1 D)^{\perp}$, let M_κ be defined in a recursive way in the following form,

$$M_\kappa = \begin{bmatrix} (M_{\kappa-1}D)^{\perp} M_{\kappa-1}\tilde{A} \\ M_1 \end{bmatrix} \quad J_{\kappa-1} = (M_{\kappa-1}D)^{\perp} \begin{bmatrix} J_{\kappa-2} & 0 \\ 0 & I_p \end{bmatrix}$$

Due to $A1$, there exists a unique positive integer $\kappa \le n$ such that the matrix M_κ generated recursively satisfies the condition $rank(M_\kappa) = n$ (see [18]). Therefore, the following algebraic expression can be constructed

$$M_\kappa e = \frac{d^{\kappa-1}}{dt^{\kappa-1}} \begin{bmatrix} J_{\kappa-1} & 0 \\ 0 & I_p \end{bmatrix} \begin{bmatrix} y_e \\ \vdots \\ y_e^{[\kappa-1]} \end{bmatrix} = \frac{d^{\kappa-1}}{dt^{\kappa-1}} Y \tag{4}$$

where $y_e^{[i]}$ represents the $i-th$ anti-differentiator of y_e, that is, $y_e^{[i]} = \int_0^t \cdots \int_0^{\tau_i} y_e d\tau_i \ldots dt$. From the above equation a solution for e exists, i.e. $e = M_\kappa^{+} \frac{d^{\kappa-1}}{dt^{\kappa-1}} Y$.

Thus, a real time HOSM differentiator is used to provide exact and finite time differentiation of Y [21]. It is given by

$$\begin{aligned}
\dot{z}_{i,0} &= \lambda_0 \Upsilon^{\frac{1}{\ell}} |z_{i,0} - Y_i|^{\frac{\ell}{\ell+1}} sign(z_{i,0} - Y_i) + z_{i,1} \\
\dot{z}_{i,j} &= \lambda_j \Upsilon^{\frac{1}{\ell-j}} |z_{i,j+1} - \dot{z}_{i,j-1}|^{\frac{\ell-j}{\ell}} sign(z_{i,j} - \dot{z}_{i,j-1}) + z_{i,j+1} \\
\dot{z}_{i,\ell} &= \lambda_\ell \Upsilon |z_{i,\ell} - \dot{z}_{i,\ell-1}|^{\frac{1}{2}} sign(z_{i,\ell} - \dot{z}_{i,\ell-1})
\end{aligned} \tag{5}$$

with $j = \overline{1, \ell - 1}$, ℓ is the differentiator order. The differentiator input Y_i for $i = \overline{1, n}$ represents the components of Y. A positive sequence of λ_j can be selected as in [21]. The gain Υ is a Lipschitz constant of $Y^{(\ell)}$. Due to assumption $A3$, the higher differentiator order may be $\ell = \kappa + r - 3$, i.e. $|e^{(r-1)}| < \Upsilon$. Therefore, $\Upsilon \geq |\tilde{A}|^{r-1}\breve{e} + \Sigma_{j=1}^{r-2} |\tilde{A}|^j (|B|\breve{f}_j + |G|\breve{v}_j)$.

In [21], it was shown that with the proper choice of the constants λ_j, and Υ for all $j = \overline{0, \ell}$ there is a finite time $t \geq t_o$ such that $z_j = \frac{d^j}{dt^j} Y$ is fulfilled for all $j = \overline{0, \ell}$, $z_j = [z_{1,j} \; \dots \; z_{n,j}]^T$. Hence, the vector e in (4) is recovered from the $(\kappa - 1) - th$ sliding dynamics, i.e. $e = M_\kappa^+ z_{\kappa-1}$ holds for $t \geq t_o$. Consequently the next expression holds

$$\hat{x} := M_\kappa^+ z_{\kappa-1} + \tilde{x} \tag{6}$$

where $\hat{x} \in \mathbb{R}^n$ is the estimated value of x for all $t \geq t_x$.

2.2 Faults and Perturbations Identification

The perturbations and actuator faults vectors v and f can be recovered from (3), where $\dot{e}$ can be obtained from the HOSM differentiator (5), i.e. the equality $z_\kappa = \dot{e}$ is accomplished for all $t > t_x$. Hence, working out (3) it yields to

$$\hat{f} = (G^\perp B)^+ G^\perp (z_\kappa - \tilde{A} z_{\kappa-1}) \tag{7}$$

where $\hat{f} \in \mathbb{R}^m$ represents the identified value of f. Whereas for the perturbations and its successive derivatives it yields to

$$\hat{v}^{(j)} = (B^\perp G)^+ B^\perp (z_{\kappa+j} - \tilde{A} z_{\kappa-1+j}) \tag{8}$$

for $j = \overline{0, r-3}$, where $z_{\kappa+j}$ comes from (5). Therefore, $\hat{v}^{(j)} \in \mathbb{R}^q$ represents the estimate of $v^{(j)}$ for all $t > t_x$. The next derivative $\hat{v}^{(r-2)}$ may be achieved from $\hat{v}^{(r-2)} = (B^\perp G)^+ B^\perp (e^{(r-1)} - \tilde{A} z_{\kappa+r-2})$ where $B^\perp e^{(r-1)}$ can be recovered by means of $B^\perp e^{(r-1)} = B^\perp \frac{d}{dt} z_{\kappa+r-2}$. In order to do that, the Lipschitz constant may be calculated from $|B^\perp e^{(r)}| < \Upsilon_2$, i.e. $\Upsilon_2 \geq |B^\perp \Gamma| + |B^\perp G| \breve{v}_{r-1}$. Thus, with the proper choice of λ_i it follows from [21] $\hat{v}^{(r-2)} = v^{(r-2)}$ for all $t > t_v > t_x$.

3 Problem Statement

Consider a strict feedback form representation of system (1), see [22, 23]. Due to the pair (A, B) is controllable, system (1) can be always reduced to the block-controllable form, it is composed into r connected sub-systems, i.e

$$\dot{x}_1 = A_1 x_1 + B_1(x_2 + \Gamma_1 \nu) \tag{9}$$
$$\dot{x}_i = A_i \bar{x}_i + B_i(x_{i+1} + \Gamma_i \nu) \tag{10}$$
$$\dot{x}_r = A_r \bar{x}_r + B_r(u + f) \tag{11}$$

for $i = \overline{2, r-1}$, where $\bar{x}_i = [x_1^T \ \dots \ x_i^T]^T$, $x_i \in \mathbb{R}^{n_i}$, $n_i = rank(B_i)$, $\sum_{i=1}^{r} n_i = n$. The state vector $x \in \mathbb{R}^n$ is given by $\bar{x} = [x_1^T \ x_i^T \ \dots \ x_r^T]^T$. The measured output signal is given by (2) while the controlled output is given by $x_1 \in \mathbb{R}^{n_1}$. The sub-system (9)-(10) for $i = \overline{2, r-1}$ represents the sub-actuated dynamics and (11) corresponds to the actuated dynamics thus, $x_r \in \mathbb{R}^m$. The matrices A_i, Γ_i are of the corresponding dimensions. Further details can be found in [24].

Control Aim. The goal is to design an output feedback continuous sliding mode controller u which allows the coordinate x_1 to track a smooth signal x_d in spite of system perturbations ν and the occurrence of the faults f. ■

The sliding control design relies on the availability, in finite time, of the exact estimation of the state and the identification of the faults as well as perturbations and its successive derivatives until $r-2$-th order. Thus, from (6), after $t > t_o$, the identities $\hat{x}_1 = x_1, \dots, \hat{x}_r = x_r$ are certainly obtained.

4 Control Design

First, the design of the sliding surface is achieved in $r-1$ steps, it is carried such that the sub-actuated dynamics (9)-(10) accomplishes the desired behavior in spite of the perturbations. Then, at the $r-th$ step, the control law is conceived such that the state trajectories reach the sliding surface and there remain. It is assumed that the successive derivatives of the desired trajectory x_d are available.

Step 1. Starting with sub-system (9), the coordinate x_2 can be exploited as a fictitious control input $x_2 := \phi_1$,

$$\phi_1 = -\Gamma_1 \hat{\nu} - B_1^{\dagger}(A_1 x_1 - \hat{A}_1(x_1 - x_d) - \dot{x}_d) \tag{12}$$

where $\hat{A}_1 \in \mathbb{R}^{n_1}$ is a Hurwitz matrix containing the desired convergence performance of x_1 towards x_d. Notice that since $rank(Bi) = n_i$ then $B_i B_i^T$ is invertible, thus $B_i^{\dagger} = B_i^T (B_i B_i^T)^{-1}$ is the right inverse matrix of B_i.

Step i. The coordinate x_{i+1} is a fictitious control input for (10), it is $x_{i+1} := \phi_i$

$$\begin{aligned} \phi_i = & -\Gamma_i \hat{\nu} - B_i^{\dagger}(A_i \bar{x}_i - \hat{A}_i(x_i - \phi_{i-1}) \\ & + X_{i-1}(x_{i-1} - \phi_{i-2}) - \dot{\phi}_{i-1}) \end{aligned} \tag{13}$$

for all $i = \overline{2, r-1}$, $\hat{A}_i \in \mathbb{R}^{n_i}$ is a Hurwitz matrix and $X_{i-1} = P_i^{-1} B_i^T P_{i-1}$, with P_i a positive definite matrix satisfying $P_i \hat{A}_i + \hat{A}_i^T P_i = -I$.

Step r. Finally, the sliding surface s is designed like

$$s = x_r - \phi_{r-1} \tag{14}$$

with $s \in \mathbb{R}^m$, $s = [\,s_1 \,\ldots\, s_m\,]^T$ and the control law becomes

$$u = -B_r^{\dagger}\left(A_r\bar{x}_r - \dot{\tilde{\phi}}_r - B_{r-1}^T(x_{r-1} - \phi_{r-2}) + \upsilon\right) - \hat{f} \tag{15}$$

where $\dot{\tilde{\phi}}_r$ is a linear combination of the state x and $\hat{v},\ldots,\hat{v}^{(r-2)}$. The control $\upsilon \in \mathbb{R}^m$ is a super-twisting cóntrol (see [25]) given by

$$\upsilon = K_1\Psi_1(s) + K_2\int_{t_o}^{t}\Psi_2(s) \tag{16}$$

with the matrices $K_j = diag\left(k_{j1},\ldots,k_{jm}\right)$, for $j = \overline{1,2}, \Psi_1 = diag(\psi_{11},\ldots,\psi_{1m})$ and $\Psi_2 = diag(\psi_{21},\ldots,\psi_{2m})$, where $\psi_{1j} = |s_j|^{\frac{1}{2}} sign(s_j)$ and $\psi_{2j} = sign(s_j)$ for $j = \overline{1,m}$.

Following [26], the components of the matrices gains K_1 and K_2 are given as

$$k_{1j} > 0 \qquad k_{2j} > 3\breve{\omega} + 2\left(\frac{\breve{\omega}}{k_{j1}}\right)^2 \tag{17}$$

for $j = \overline{1,m}$, the constant $|B_{r-1}^{\dagger}\ldots B_2^{\dagger}\Gamma_1 v^{(r)}| \leq \breve{\omega}$ ■

Now, the stability of the closed loop system is achieved in two stages. First, the convergence of the trajectories to the sliding surface $s = 0$ is shown. Then, a composite Lyapunov function is designed to demonstrate the convergence of the sub-actuated dynamics.

For the sake of simplicity, the convergence of the individual components of the sliding surface is studied. Differentiating (14) and applying the control law (15) it yields to $\dot{s}_j = -k_{j1}\psi_{1j} - k_{j2}\int_{t_o}^{t}\psi_{2j} - \omega_j$ where $j = \overline{1,m}$, ω_j, s_j, represent the components of vectors s, ω. The vector $\omega = -B_{r-1}^{\dagger}\ldots B_2^{\dagger}\Gamma_1 v^{(r-1)}$ is an uncertainty due to the effects of the unmatched perturbation in the actuated dynamics. Defining $\mu_j = -k_{j2}\int_{t_o}^{t}\psi_{2j} + \omega_j$, yields to

$$\dot{s}_j = -k_{j1}\psi_{1j} + \mu_j \qquad \dot{\mu}_j = -k_{j2}\psi_{2j} + \dot{\omega}_j \tag{18}$$

with $\dot{\omega} = B_{r-1}^{\dagger}\ldots B_2^{\dagger}\Gamma_1 v^{(r)}$ which under $A3$ is bounded, i.e. there exist a constant $\breve{\omega}$ such that $|\dot{\omega}| \leq |\breve{\omega}|$. In [26], it was stated that the auxiliary system (18) is stabilized and achievied $s = 0$ by selecting the controller gains as in (17). From (14), it follows that $x_r \equiv \phi_{r-1}$.

Now, the sub-actuated dynamics is examined for the second part of the convergence exposition. Notice that $x_{i+1} = \phi_i$ is exploited as a fictitious control for the coordinate x_i. First, the error variables $\sigma_1 = x_1 - x_d$, $\sigma_i = x_i - \phi_{i-1}$ for $i = \overline{2, r-1}$ are defined. Second, consider x_2 as a pseudo-control to achieve $x_1 \to x_d$. Then, a Lyapunov function candidate given by $V_1 = \sigma_1^T P_1\sigma_1$ is proposed. Differentiating it along the time yields to $\dot{V}_1 = 2\sigma_1^T(A_1x_1 + B_1(x_2 + \Gamma_1 v) - \dot{x}_d)$. Since $x_2 = \phi_1 + \sigma_1$ and in virtue of (12), it yields to $\dot{V}_1 \leq -|\sigma_1|^2 + 2\sigma_1 P_1 B_1\sigma_2$. In the next step, the second term of the Lyapunov function derivative will be taked into account. To this aim, equations (9)-(10) are rewritten as:

$$\begin{aligned}\dot{\sigma}_1 &= \hat{A}_1\sigma_1 + B_1\sigma_2\\ \dot{\sigma}_2 &= A_2\bar{x}_2 + B_2(x_3 + \Gamma_2 v) - \dot{\phi}_1\end{aligned}$$

where x_3 can be seen as a pseudo-control to stabilize σ_1, σ_2 at zero. Consider the Lyapunov candidate $V_2 = V_1 + \sigma_2^T P_2 \sigma_2$ whose derivative with respect to time is given by $\dot{V}_2 = \dot{V}_1 + \sigma_2^T P_2 \left(A_2\bar{x}_2 + B_2(x_3 + \Gamma_2 v) - \dot{\phi}_1\right)$. Given that $x_3 = \phi_2 + \sigma_2$ and in virtue of (14), it yields to $\dot{V}_2 \leq -|\sigma_1|^2 - |\sigma_2|^2 + 2\sigma_2 P_2 B_2 \sigma_3$. A similar procedure is achieved for each coordinate. In the step $r-1$, the equations (9)-(10) are rewritten as:

$$\begin{aligned}\dot{\sigma}_1 &= \hat{A}_1\sigma_1 + B_1\sigma_2\\ \dot{\sigma}_i &= \hat{A}_i\bar{x}_i + B_i\sigma_{i+1} - X_{i-1}\sigma_{i-1}\\ \dot{\sigma}_{r-1} &= A_{r-1}\bar{x}_{r-1} + B_{r-1}(x_r + \Gamma_{r-1}v) - \dot{\phi}_{r-2}\end{aligned}$$

As a result, the composite Lyapunov function candidate for the entire sub-actuated error dynamics is given by $V_\sigma = \sum_{i=1}^{r-1}\sigma_i^T P_i \sigma_i$. Differentiating it produces $\dot{V}_\sigma = -\sum_{i=1}^{r-2}|\sigma_i|^2 + 2\sigma_{r-1}^T B_{r-2} P_{r-2} \sigma_{r-2} + 2\sigma_{r-1}^T (A_{r-1}\bar{x}_{r-1} + B_{r-1}(x_r + \Gamma_{r-1}v) - \dot{\phi}_{r-2})$. Now, as the state trajectories are already on the sliding surface, i.e., $s \equiv 0$ thus $x_r \equiv \phi_{r-1}$ and the composite Lyapunov function yields to $\dot{V}_\sigma = -\sum_{i=1}^{r-2}|\sigma_i|^2$. As a result, σ_i for $i = \overline{r-1, 1}$ converges asymptotically to zero. Consequently, $x_i \to \phi_{i-1}$ for $i = \overline{r-1, 2}$. Therefore, $x_1 \to x_d$.

5 Example

Consider a 3-DOF helicopter, see [29]. The system is composed by two arms: a small arm with a propeller at each end connected with an arm swinging in a fixed base (see Fig. 1). The system can rotate freely around three axes. A linearized model around $\varepsilon_2^* = 0$ is given by

$$\begin{aligned}\ddot{\varepsilon}_1 &= 0.45\left((u_1 + f_1) + (u_2 + f_2)\right)\\ \ddot{\varepsilon}_2 &= 3.05\left((u_1 + f_1) - (u_2 + f_2)\right) \quad \ddot{\varepsilon}_3 = -0.49\varepsilon_2 + v\end{aligned} \tag{19}$$

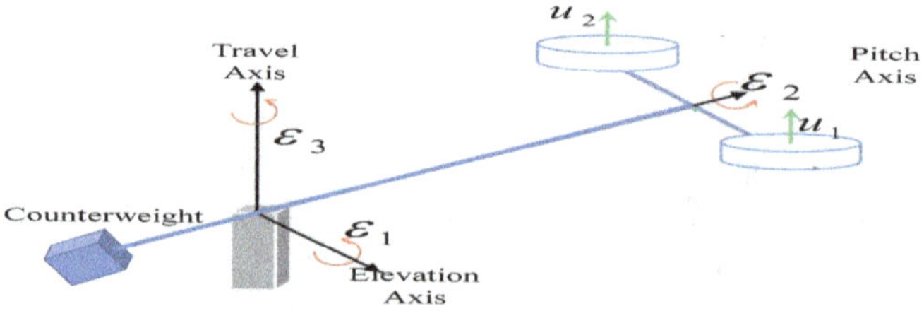

Fig. 1 Schematic diagram of a 3-DOF helicopter.

where $\varepsilon_1, \varepsilon_2, \varepsilon_3 \in R$ represent respectively the elevation, pitch and travel angles and $u \in R^2$. The system is subjected to actuator faults f_1, f_2 while a disturbance signal v is acting on the travel dynamics. The measured outputs correspond to the angular positions, i.e. $y = [\,\varepsilon_1\ \varepsilon_2\ \varepsilon_3\,]^T$.

The goal is that travel position ε_3 tracks a desired trajectory $x_d = 5\sin(1.1t)$ in spite of faults and disturbances. The simulation trial consists in four stages: (a) for $0 \le t < 60$ the system is working free of faults and disturbances (i.e. $f = [0\ 0]$, $v = 0$); (b) during $60 \le t$ a disturbance appears $v = 2\sin 3t + 1.5$; (c) in $100 \le t$ a type of liquid oscillatory fault (oscillatory fault is added to the nominal signal [27]) occurs in the first actuator $f_1 = 5\cos(1.5t) + 5$; (d) for $140 \le t$, a drifting fault (i.e. $f_2 = \beta t : \beta > 0$) appears in the second actuator. Notice that both faults represent realistic faults that can affect actuator servo-loops in flight systems, see [20]. The simulation sampling time is $100[\mu s]$.

Observer design. Firstly, the state vector is selected as $x = [\,\varepsilon_1\ \varepsilon_2\ \varepsilon_3\ \dot\varepsilon_1\ \dot\varepsilon_2\ \dot\varepsilon_3\,]^T$. Then, it can be easily shown that for the obtained state space representation, the triplet (A, C, D) is strongly observable. Thus, following Section 2.1 an auxiliary system matrix $\tilde{A}$ with eigenvalues $-\{1.6, 1.5, 1.8, 1.9, 1.7, 1.1\}$ is considered. Subsequently, for $\kappa = 2$ iterations a full column rank matrix $M_2 = [-21\ 0\ 0\ 1\ 0\ 0;\ 0\ -13\ 0\ 0\ 1\ 0;\ C]$ is found.

The differentiator in (5) has an order $\ell = 3$, such that $\hat{x}$, $\hat{f}$, $\hat{v}$ and $\hat{v}^{(1)}$ can be estimated. The bounds are $\breve{v} = 3.5$, $\breve{v}_1 = 6$, $\breve{v}_2 = 18$, $\breve{f}_1 = 10$. $\breve{f}_2 = 7.5$. Then, $\Upsilon = 150$, $\lambda_i = \{1.1, 1.5, 3, 5\}$.

Control design. The goal is $\varepsilon_3 \to x_d$. Moreover, the elevation angle is kept at $\varepsilon_o = 15[deg]$ so, the variables $\tilde{\varepsilon}_1 := \varepsilon_1 - \varepsilon_o$ and $\varepsilon_4 = \int \tilde{\varepsilon}_1$ are introduced. As a consequence, an extended state vector is considered $x = [\,\tilde{\varepsilon}_1\ \varepsilon_2\ \varepsilon_3\ \dot\varepsilon_1\ \dot\varepsilon_2\ \dot\varepsilon_3\ \varepsilon_4\,]^T$. This extended system can be transformed into (9)-(11). Following the coordinates transformation given in [24] with $x_1 = \varepsilon_3$, $x_2 = [\,\dot\varepsilon_3\ \varepsilon_4\,]^T$, $x_3 = [\,0.98\ 0.14;\ -0.14\ -0.98\,][\,\tilde{\varepsilon}_1\ \varepsilon_2\,]^T$, $x_4 = [\,\dot\varepsilon_1\ -\dot\varepsilon_2\,]^T$ it yields to

$$\dot{x}_1 = B_1 x_2 \qquad \dot{x}_2 = B_2(x_3 + \Gamma_2 v) \tag{20}$$

$$\dot{x}_3 = B_3 x_4 \qquad \dot{x}_4 = B_4(u + f) \tag{21}$$

where $B_1 = [\,1\ 0\,]$, $B_2 = [\,0.49\ 0.07;\ -0.14\ 0.98\,]$, $B_3 = [\,0.14\ 0.98;\ -0.98\ 0.14\,]$, $\Gamma_2 = [\,1.99\ -0.29\,]^T$, $B_4 = [\,-0.45\ -0.45; -3.05\ 3.05\,]$.

The fictitious controls ϕ_i for $i = \overline{1,3}$ are computed following (12)-(14). The sliding surface is thus designed as $s = x_4 - \phi_3$. The control signal (15) has $\hat{A}_1 = -1.8$, $\hat{A}_2 = diag(-2, -3)$, $\hat{A}_3 = diag(-4, -5)$, $\breve{\omega} = 160$. The super-twisting gains (17) are $k_{1i} = 10$, $k_{2i} = 600$.

For comparison purposes, a LQR control law $u = -K_n x$ that involves full state measurements is considered. The LQR controller is designed for a nominal system without faults and disturbances, see [29]. The LQR control keeps the elevation at $15[deg]$ while the travel coordinate tracks the desired trajectory x_d. Augmenting the state with $\tilde{\varepsilon}_3 = \varepsilon_3 - x_d$ and $\varepsilon_5 := \int \tilde{\varepsilon}_3$, i.e. $x = [\,\tilde{\varepsilon}_1\ \varepsilon_2\ \tilde{\varepsilon}_3\ \dot\varepsilon_1\ \dot\varepsilon_2\ \dot{\tilde{\varepsilon}}_3\ \varepsilon_4,\ \varepsilon_5\,]^T$. The controller gain yields to $K_n = [\,15.6\ 9\ -7.8\ 9.7\ 3.6\ -20.5\ 2.2\ -1;\ 15.6\ -9\ 7.8\ -9.7\ -3.6\ 20.5\ 2.2\ 1\,]$. Fig. 2 shows the tracking performance. Both controllers achieve the desired travel

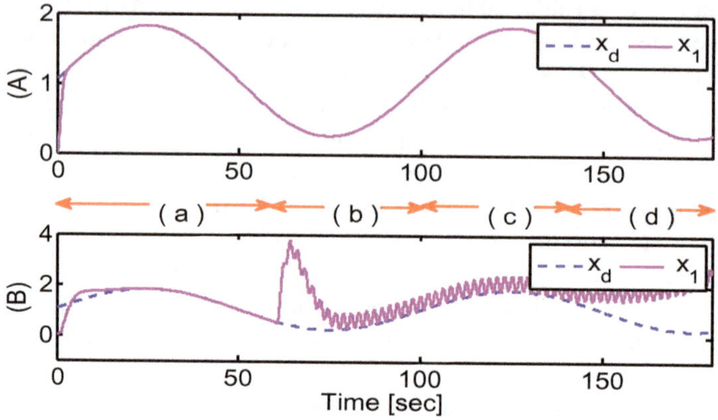

Fig. 2 Travel tracking performance: (A) proposed control and (B) LQR control.

goal when neither faults nor disturbances are present, see stage (a). However, when disturbances and/or faults are present, see stages (b)-(d), only the proposed controller in Fig. 2.A is capable to attain the prescribed goal. Fig.3 shows the controllers' capability to maintain the prescribed elevation level (i.e. $15[deg]$). Fig. 4.B shows the continuous control signals. Fig. 4.A. depicts the sliding surfaces. Fig. 5 exhibits the disturbance (A) and the faults (B) acting on the system (solid line). The identified values $\hat{v}, \hat{f}_1, \hat{f}_2$ are also depicted (dotted line).

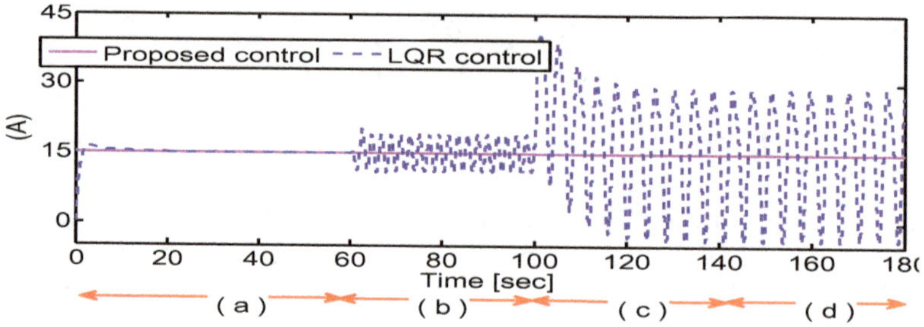

Fig. 3 Elevation level.

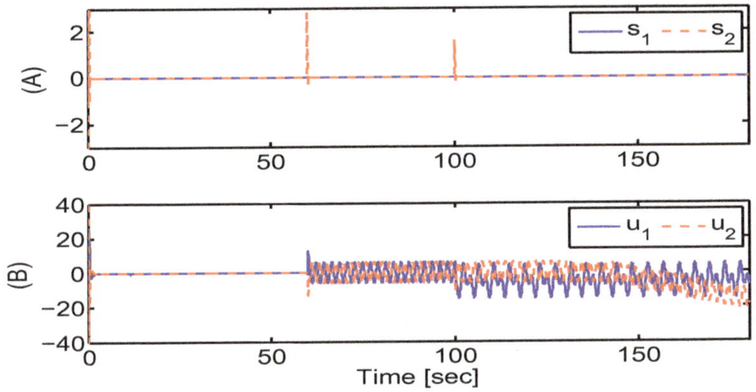

Fig. 4 (A) Sliding surfaces; (B) control signals.

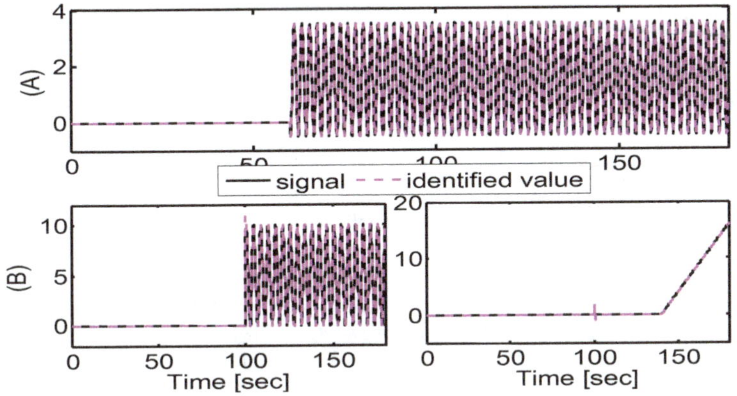

Fig. 5 Disturbance (A) and faults (B) signals acting on the system (solid line) and their identified values (dotted line).

6 Conclusions

An output tracking scheme for a $3-DOF$ helicopter subjected to unmatched perturbations and actuator faults was proposed based on HOMS observation and identification techniques. The unmatched perturbations compensation was carried through the sliding surface. To this aim, a dynamic sliding surface was designed. A super-twisting control drives the state trajectories to the sliding surface in spite of the perturbations effects. Moreover, fault accommodation is achieved by means of the fault identification alleviating this way the controller gain. Simulations are considered to validate the method.

References

1. Edwards, C., Lombaerts, T., Smaili, H.: Fault Tolerant Flight Control: A Benchmark Challenge. Springer (2010)
2. Berdjag, D., Cieslak, J., Zolghadri, A.: Fault diagnosis and monitoring of oscillatory failure case in aircraft inertial system. Control Engineering Practice 20, 1410–1425 (2012)
3. Ferreira de Loza, A., Cieslak, J., Henry, D., Dvila, J., Zolghadri, A.: Sensor fault diagnosis using a non-homogeneous high-order sliding mode observer with application to a transport aircraft. IET Control Theory and Applications (2015), doi:10.1049/iet-cta.2014.0226
4. Cieslak, J., Efimov, D., Henry, D.: Transient management of a supervisory fault-tolerant control scheme based on dwell-time conditions. Int. J. Adapt. Control Signal Process (2014), doi:10.1002/acs.2465
5. Efimov, D., Cieslak, J., Henry, D.: Supervisory faulttolerant control with mutual performance optimization. Int. J. Adapt. Control Signal Process. 27, 251–279 (2013)
6. Cieslak, J., Henry, D., Zolghadri, A.: Fault tolerant flight control: from theory to piloted flight simulator experiments. IET Control Theory and Applications 4, 1451–1464 (2010)
7. Utkin, V.I.: Sliding modes in control and optimization. Springer, Berlin (1992)
8. Corradini, M.L., Monteriu, A., Orlando, G.: An actuator failure tolerant control scheme for an underwater remotely operated vehicle. IEEE Trans. on Control Systems Technology 19, 1036–1046 (2011)
9. Alwi, H., Edwards, C.: Fault Detection and Fault-Tolerant Control Using Sliding Modes. AIC. Springer, New York (2011)
10. Hamayun, M.T., Edwards, C., Alwi, H.: A fault tolerant control allocation scheme with output integral sliding modes. Automatica 49, 1830–1837 (2013)
11. Efimov, D., Zolghadri, A., Rassi, T.: Actuator fault detection and compensation under feedback control. Automatica 47, 1699–1705 (2011)
12. Davila, J.: Exact tracking using backstepping control design and high-order sliding modes. IEEE Trans. on Automat. Control 58, 2077–2081 (2013)
13. Ferreira, A., Fridman, L., Punta, E., Bartolini, G.: Output nested backward compensation of unmatched effects of unknown inputs. In: Proc. of the IEEE Conference on Decision and Control, pp. 1448–1453 (2010)
14. Utkin, V.I.: Sliding modes in control and optimization. Springer, Berlin (1992)
15. Wang, W., Wen, C.: Adaptive compensation for infinite number of actuator failures or faults. Automatica 47, 2197–2210 (2011)
16. Basin, M.V., Loukianov, A.G., Hernandez-Gonzalez, M.: Joint state and parameter estimation for uncertain stochastic nonlinear polynomial systems. Int. J. Systems Science 44, 1200–1208 (2013)
17. Bejarano, F.J., Figueroa, M., Pacheco, J., De Jesus Rubio, J.: Robust fault diagnosis of disturbed linear systems via a sliding mode high order differentiator. Int. J. of Control 85, 648–659 (2012)
18. Fridman, L., Davila, J., Levant, A.: High-order sliding-mode observation for linear systems with unknown inputs. Nonlinear Analysis: Hybrid Systems 5, 337–347 (2011)
19. Ferreira de Loza, A., Rios, H., Rosales, A.: Robust regulation for a 3DOF helicopter via sliding-mode observation and identification. Journal of the Franklin Institute 349, 700–718 (2012)

20. Zolghadri, A., Henry, D., Cieslak, J., Efimov, D., Goupil, P.: Fault Diagnosis and Fault-Tolerant Control and Guidance for Aerospace Vehicles: From theory to application. Advances in Industrial Control. Springer (2014) ISBN 978-1-4471-5312-2
21. Levant, A.: Higher-order sliding modes, differentiation and output feedback control. Int. J. of Control 76, 924–941 (2003)
22. Loukyanov, A., Utkin, V.: Methods of reducing equations for dynamic systems to a regular form. Automatic and Remote Control 42, 413–420 (1993)
23. Drakunov, S.V., Izosimov, D.B., Lukyanov, A.G., Utkin, V.I.: Block control principle I. Automation and Remote Control 51, 601–609 (1990)
24. Polyakov, A.: Nonlinear feedback design for fixed-time stabilization of linear control systems. IEEE Transactions on Automatic Control 57, 2106–2110 (2012)
25. Levant, A.: Sliding order and sliding accuracy in sliding mode control. Int. J. of Control 58, 1247–1263 (1993)
26. Moreno, J.A., Osorio, M.: Strict Lyapunov functions for the supertwisting algorithm. IEEE Trans. on Automat. Control 57, 1035–1040 (2012)
27. Cieslak, J., Efimov, D., Zolghadri, A., Henry, D., Goupil, P.: Design of a non-homogeneous differentiator for actuator oscillatory failure case reconstruction in noisy environment. Proc. IMechE Part I: J Systems and Control Engineering (2014), doi:10.1177/0959651814561091
28. Krstic, M., Kanellakopoulos, I., Kokotovic, P.V.: Nonlinear and adaptive control design. Wiley, New York (1995)
29. Quanser Inc. 3-DOF Helicopter Reference Manual, Document Number 644, Revision 2.1

Nonlinear Quadrotor Control with Online Model Identification

Peng Lu, Erik-Jan van Kampen, and Qiping P. Chu

Abstract. This paper proposes a new Fault Tolerant Control (FTC) system for quadrotors which are subjected to actuator loss of effectiveness faults. The proposed FTC system is composed of three subsystems: the state estimation, the loss of effectiveness estimation and the Backstepping (BS) controller. A new method to estimate the loss of effectiveness online is proposed, which can provide fault information for the controller to achieve fault tolerant control. The performance of the FTC system is validated using two different simulations: position control of the quadrotor in the presence and absence of actuator faults. The simulation results show that the proposed system can enable the quadrotor to maintain the flight even all the four rotors fail consecutively, which demonstrate its satisfactory performance.

1 Introduction

Recently, autonomous Unmanned Aerial Vehicles (UAV)s have attracted considerable attention due to their strong autonomy and ability to fulfill complex tasks without the aid of the human. Quadrotor is one type of UAV which is an easy to build and fly system. It is able to take off and land vertically, and hover at a fixed point. The purpose of these quadrotors are various, ranging from scientific exploration and data collection, to provision of commercial services, military reconnaissance and intelligence gathering [6]. Since these vehicles operates in an environment subjected to a high degree of uncertainties and disturbances, the problem of precise and accurate control and estimation of these vehicles is difficult and requires advanced control and estimation techniques [6]. Furthermore, due to the increasing requirement for control systems to be more secure and reliable, Fault Detection and Diagnosis (FDD) and FTC are becoming more and more critical and significant [6].

Peng Lu · Erik-Jan van Kampen · Qiping P. Chu
Delft University of Technology, Delft, 2600 GB, The Netherlands
e-mail: {P.Lu-1,E.vanKampen,q.p.chu}@tudelft.nl

J. Bordeneuve-Guibé et al. (eds.), *Advances in Aerospace Guidance, Navigation and Control*, DOI: 10.1007/978-3-319-17518-8_6

A number of control approaches have been applied to the quadrotors in the absence of failures [11, 13, 14, 3]. However, only few papers consider the actuator failures of the quadrotor [1, 17, 16]. When there are actuator faults, the effectiveness of the actuator changes, which could lead to a Loss Of Effectiveness (LOE) fault. In this situation, the controller of the quadrotor should have tolerance against the actuator faults to prevent the quadrotor from crashing.

There are some papers which deal with the actuator fault detection [6]. However, there are few papers which estimate the actuator faults [1, 4]. Most of these papers use estimation techniques to estimate the faults. Cen et. al[4] uses a adaptive Thau observer to estimate the actuator faults. Amoozgar et. al [1] applies the Kalman filter for the detection and diagnosis of the actuator faults.

The contribution of this paper consists of proposing a novel FTC strategy based on a novel online model identification strategy. The FTC system consists of three subsystems: a state estimation subsystem, a LOE estimation subsystem and a nonlinear control system. The state estimation subsystem is responsible for correcting the biases and drifts of sensors and estimating the state of the quadrotor. The LOE estimation subsystem is based on a novel LOE estimation strategy which makes use of the information provided by the state estimation subsystem to estimate the actuator LOE factors. The LOE estimation is used by the nonlinear control system to achieve FTC. The control approach used in this paper is a BS [7] controller.

The performance of the proposed FTC system is validated using two different simulations: nonlinear quadrotor control with and without actuator faults. Both simulation results show that the proposed FTC system is able to control the quadrotor even in the presence of actuator faults. This demonstrates the effectiveness of the FTC system.

The structure of the paper is as follows: Sect. 2 gives a detailed description of the quadrotor, its actuator and the measurement. The modeling of actuator LOE faults is also modeled. Sect. 3 presents the new method which estimates the LOE factors online. The whole FTC system structure is given in Sect. 4. The performance of the proposed FTC is validated in two different situations, whose result is given in Sect. 5. Sect. 6 concludes the paper.

2 Modeling of the Quadrotor, The Actuator Faults and the Measurement

This section first introduces the model of the quadrotor, which includes the translational and rotational dynamics. The assumptions which are required are also presented. Then, the modeling of actuator and actuator faults are introduced. Finally, the measurement model is given.

2.1 Dynamics of the Quadrotor

Define an earth frame $\{\Sigma_e\}(O_e, x_e, y_e, z_e)$ and $\{\Sigma_b\}(O_b, x_b, y_b, z_b)$ in which O_b is fixed to the quadrotor (see Fig. 1). The earth frame is the North East Down (NED)

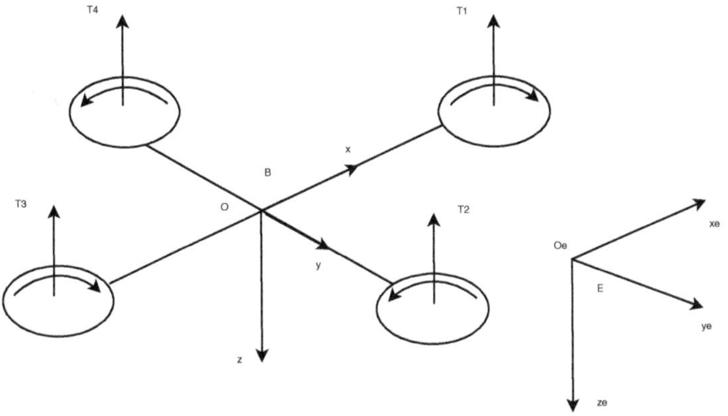

Fig. 1 Block diagram for the MMAE scheme

frame in which the z_e axis points down. The z_b axis of the body frame also points down. The rotation of the body frame with respect to the earth frame is denoted by the following rotation matrix **R**:

$$\mathbf{R} = \begin{bmatrix} \cos\theta\cos\psi & \sin\phi\sin\theta\cos\psi - \cos\phi\sin\psi & \cos\phi\sin\theta\cos\psi + \sin\phi\sin\psi \\ \cos\theta\sin\psi & \sin\phi\sin\theta\sin\psi + \cos\phi\cos\psi & \cos\phi\sin\theta\sin\psi - \sin\phi\cos\psi \\ -\sin\theta & \sin\phi\cos\theta & \cos\phi\cos\theta \end{bmatrix} \tag{1}$$

where ϕ, θ and ψ will be defined later.

Before introducing the equations of motion of the quadrotor, the following assumptions have to be made [2]:

1. The structure is supposed rigid.
2. The structure is supposed symmetrical.
3. The center of Gravity and the origin of the body fixed frame are assumed to coincide.
4. The propellers are supposed rigid.
5. The thrust and the drug are proportional to the square of the propeller's rotational speed.

Let η denote the Euler angles and ω denote the angular rates of the quadrotor:

$$\eta = [\phi, \theta, \psi]^T \tag{2}$$

$$\omega = [p, q, r]^T \tag{3}$$

The Euler angles ϕ and θ are assumed to satisfy the following conditions in this paper:

$$\phi \in [-\pi/2, \pi/2] \tag{4}$$
$$\theta \in (-\pi/2, \pi/2) \tag{5}$$

The position of the quadrotor is denoted as $d = (x, y, z)$ while the velocity of the quadrotor expressed in the inertial frame is denoted as $V = (V_x, V_y, V_z)$. The total thrust generated by the four rotors are denoted as T. Therefore, the translational dynamics of the quadrotor can be described by the following equations:

$$\dot{x} = V_x \tag{6}$$
$$\dot{y} = V_y \tag{7}$$
$$\dot{z} = V_z \tag{8}$$
$$\dot{V}_x = -(\cos\phi \sin\theta \cos\psi + \sin\phi \sin\psi)T \tag{9}$$
$$\dot{V}_y = -(\cos\phi \sin\theta \sin\psi - \sin\phi \cos\psi)T \tag{10}$$
$$\dot{V}_z = mg - (\cos\phi \cos\theta)T \tag{11}$$

where $g = 9.81$ m/s^{-2} is the gravity constant, m is the total mass of the quadrotor.

The relationship between the derivative of the Euler angles and the angular rates can be described as follows:

$$\dot{\phi} = p + q\sin\phi \tan\theta + r\cos\phi \tan\theta \tag{12}$$
$$\dot{\theta} = q\cos\phi - r\sin\phi \tag{13}$$
$$\dot{\psi} = q\frac{\sin\phi}{\cos\theta} + r\frac{\cos\phi}{\cos\theta} \tag{14}$$

Since the vehicle is assumed to be rigid and symmetrical, the inertia matrix of the quadrotor can be denotes as

$$I = \begin{bmatrix} I_x & 0 & 0 \\ 0 & I_y & 0 \\ 0 & 0 & I_z \end{bmatrix} \tag{15}$$

Let ω_i, $i = 1,2,3,4$ denote the rotational speed of the four rotors. Using the well-known rigid-body dynamics, the rotational dynamics of the quadrotor including the propeller gyro effect can be described as follows:

$$\dot{p} = \frac{1}{I_x}[(I_y - I_z)qr + \tau_x + I_r q\omega_r] \tag{16}$$
$$\dot{q} = \frac{1}{I_y}[(I_z - I_x)rp + \tau_y - I_r p\omega_r] \tag{17}$$
$$\dot{r} = \frac{1}{I_z}[(I_x - I_y)pq + \tau_z + I_r\dot{\omega}_r] \tag{18}$$

where I_r is the inertia of the propeller and ω_r is defined as

$$\omega_r = -\omega_1 + \omega_2 - \omega_3 + \omega_4 \tag{19}$$

The thrust generated by the rotor is given by the following equation [5]:

$$T = C_T \rho A (\omega R)^2 \tag{20}$$

$$\frac{C_T}{\sigma a} = (\frac{1}{6} + \frac{\mu^2}{4}\theta_0) - (1+\mu^2)\frac{\theta_{tw}}{8} - \frac{\lambda}{4} \tag{21}$$

where A, R are the parameters of the rotor. λ and μ are the inflow ratio and the advance ratio of the rotor respectively. For simplicity, it is assumed that the relationship between the thrust and torque is linear such that the following equation holds:

$$\begin{bmatrix} T \\ \tau_x \\ \tau_y \\ \tau_z \end{bmatrix} = \begin{bmatrix} b_r & b_r & b_r & b_r \\ 0 & -Lb_r & 0 & Lb_r \\ Lb_r & 0 & -Lb_r & 0 \\ -d_r & d_r & -d_r & d_r \end{bmatrix} \begin{bmatrix} \omega_1^2 \\ \omega_2^2 \\ \omega_3^2 \\ \omega_4^2 \end{bmatrix} \tag{22}$$

where L is the distance between the center of the mass and the rotor. b_r and d_r are the thrust and torque coefficient which varies with the flight condition. For controller design without model identification, the thrust and torque coefficient are assumed to be constant such that the following equation is used for controller design:

$$\begin{bmatrix} T \\ \tau_x \\ \tau_y \\ \tau_z \end{bmatrix} = \begin{bmatrix} b & b & b & b \\ 0 & -Lb & 0 & Lb \\ Lb & 0 & -Lb & 0 \\ -d & d & -d & d \end{bmatrix} \begin{bmatrix} \omega_1^2 \\ \omega_2^2 \\ \omega_3^2 \\ \omega_4^2 \end{bmatrix} \tag{23}$$

where b and d are the thrust coefficient and torque coefficient during the hovering of the quadrotor, respectively.

2.2 Modeling of the Actuator and Actuator Faults

The actuator dynamics are modeled by a simple first-order filter. It is assumed that there are no sensors to measure the rotational speed of the actuators. The rotational speed of the rotors and the commanded rotational speed are described as:

$$\omega_i = \frac{K}{\tau s + 1}\omega_{ci}, \quad i = 1,2,3,4 \tag{24}$$

where ω_{ci} are the commanded rotational speed to the actuator. τ is the time constant of the actuator and K is the gain of the actuator.

In case of actuator faults, the actuator effectiveness will decrease. Denote l_i, $i = 1,2,3,4$ as the LOE factor of the four actuators. Then Eq. (23) should be adapted to

$$\begin{bmatrix} T \\ \tau_x \\ \tau_y \\ \tau_z \end{bmatrix} = \begin{bmatrix} b & b & b & b \\ 0 & Lb & 0 & -Lb \\ Lb & 0 & -Lb & 0 \\ -d & d & -d & d \end{bmatrix} \begin{bmatrix} \omega_1^2 & 0 & 0 & 0 \\ 0 & \omega_2^2 & 0 & 0 \\ 0 & 0 & \omega_3^2 & 0 \\ 0 & 0 & 0 & \omega_4^2 \end{bmatrix} \begin{bmatrix} (1-l_1) \\ (1-l_2) \\ (1-l_3) \\ (1-l_4) \end{bmatrix} \tag{25}$$

It can be seen that $l_i = 0,\ i = 1,2,3,4$ when the quadrotor is in hover. When the quadrotor is not in hover or there is an actuator fault in the ith rotor, b_r and d_r are not constant. In these conditions, $l_i \neq 0$.

2.3 Measurement Model

The position and velocity of the quadrotor is assumed to be measured by a optic track system which is composed of cameras. It is assumed that there are no biases in the measurement of the position and velocity. The Euler angles are measured by the attitude reference system and the angular rates are measured by the Inertial Measurement Unit (IMU). The biases in the IMU are considered. Therefore, the complete measurement model is as follows:

$$x_m = x + \nu_x \tag{26}$$
$$y_m = y + \nu_y \tag{27}$$
$$z_m = z + \nu_z \tag{28}$$
$$v_{xm} = v_x + \nu_{vx} \tag{29}$$
$$v_{ym} = v_y + \nu_{vy} \tag{30}$$
$$v_{zm} = v_z + \nu_{vz} \tag{31}$$
$$\phi_m = \phi + \nu_\phi \tag{32}$$
$$\theta_m = \theta + \nu_\theta \tag{33}$$
$$\psi_m = \psi + \nu_\psi \tag{34}$$
$$p_m = p + \nu_p + \lambda_p \tag{35}$$
$$q_m = q + \nu_q + \lambda_q \tag{36}$$
$$r_m = r + \nu_r + \lambda_r \tag{37}$$

where λ_p, λ_q and λ_r are the biases in the IMU measurement. These biases maybe time-varying and will be removed using the state estimation.

3 Online Model Identification of the Quadrotor

This section presents the novel model identification for the quadrotor. It should be noted that there are some papers which estimate the actuator fault using filters or observers. This section proposes a new method to estimate the LOE factors. This method consists of two steps. The first step is state estimation, whose objective is to correct the biases of the IMU sensors and estimate the states of the system. The second step is to estimate the LOE factors using the estimated states. One advantage of using this method is that there is no need to choose the process noise covariance matrix which is required by the filter to estimate the LOE factors. Another advantage is that it can both cope with sensor biases and actuator fault estimation.

3.1 State Estimation

Since there are biases in the IMU measurement, they will also be estimated in order to get a more accurate state estimation. The idea is to use the IMU measurements as the input to the system and use other measurements as the output.

The state vector of the estimation is

$$\mathbf{x} = [x, y, z, v_x, v_y, v_z, \phi, \theta, \psi] \tag{38}$$

The input vector of the system is

$$\mathbf{u} = [A_{xm}, A_{ym}, A_{zm}, p_m, q_m, r_m] \tag{39}$$

The measurement vector of the estimation is

$$\mathbf{y} = [x_m, y_m, z_m, v_{xm}, v_{ym}, v_{zm}, \phi_m, \theta_m, \psi_m] \tag{40}$$

The equations used for the state estimation are Eqs. (6) - (11) and Eqs. (12) - (14).

It should be noted that a number of people have worked with this problem. Mulder et al. [12] and Lombaerts [9] used Extended Kalman Filter (EKF) and Iterated Extended Kalman Filter (IEKF) to cope with the problem. Van Eykeren et al. [15] dealt with the problem using Augmented Extended Kalman Filter (AEKF) while Lu et al. [10] used Augmented Unscented Kalman Filter (AUKF). For details, the reader is referred to these references.

3.2 LOE Factor Estimation

In this section, the state estimated by the first step is applied to estimate the LOE factors. Combining Eqs. (11), (16), (17) and (18), it follows

$$\hat{T} = m(g - \dot{V}_z)/(\cos\phi_E \cos\theta_E) \tag{41}$$
$$\hat{\tau}_x = [I_x\dot{p} - (I_y - I_z)q_E r_E] \tag{42}$$
$$\hat{\tau}_y = [I_y\dot{q} - (I_z - I_x)r_E p_E] \tag{43}$$
$$\hat{\tau}_z = [I_z\dot{r} - (I_x - I_y)p_E q_E] \tag{44}$$

where subscript "E" denotes the estimated states. There are a number of approaches which are suitable to compute the derivatives of the angular rates such as Sliding Mode Differentiator (SMD) [8]. The details are omitted here.

Using Eq. (25), it is possible to compute the LOE factors. The estimated LOE factors are

$$\begin{bmatrix} \hat{l}_1 \\ \hat{l}_2 \\ \hat{l}_3 \\ \hat{l}_4 \end{bmatrix} = \begin{bmatrix} 1 \\ 1 \\ 1 \\ 1 \end{bmatrix} - \begin{bmatrix} \omega_1^2 & 0 & 0 & 0 \\ 0 & \omega_2^2 & 0 & 0 \\ 0 & 0 & \omega_3^2 & 0 \\ 0 & 0 & 0 & \omega_4^2 \end{bmatrix}^{-1} \begin{bmatrix} b & b & b & b \\ 0 & Lb & 0 & -Lb \\ Lb & 0 & -Lb & 0 \\ -d & d & -d & d \end{bmatrix}^{-1} \begin{bmatrix} \hat{T} \\ \hat{\tau}_x \\ \hat{\tau}_y \\ \hat{\tau}_z \end{bmatrix} \tag{45}$$

The estimated LOE factors will be used by the controller to achieve adaptive control in the presence of actuator faults.

4 Nonlinear Adaptive Controller Structure

In this section, the complete controller structure will be presented. The controller is implemented using the BS approach since it can provide a guaranteed stable control law for nonlinear systems.

The BS controller is a multi-loop controller which control the position and altitude of the quadrotor. The control input of the BS controller is the thrust and torques generated by the rotors. The BS controller design details are omitted here and the readers are referred to [7].

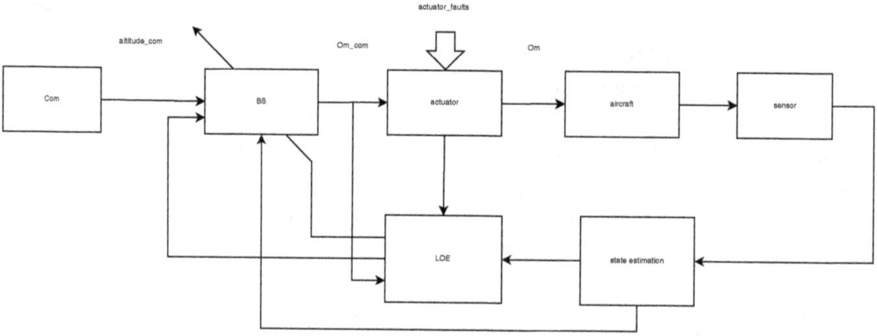

Fig. 2 Block diagram for the FTC scheme

As can be seen from Fig. 2, the complete FTC system is composed of three subsystems: the state estimation, the LOE estimation and the BS controller. The state estimation system deals with the biases of the IMU sensors and estimate the states of the quadrotor. The estimated information is used to estimate the LOE factors. Both the information of the state estimation and the LOE estimation is used by the BS controller to achieve actuator FTC. The performance of this adaptive control system will be demonstrated in the following section.

5 Simulation Results

In this section, the performance of the proposed adaptive controller is shown. The different simulation situations are shown as follows:

1. Without actuator faults
2. With actuator faults

The performance will be compared to the normal BS controller without online model identification. The objective is to control the position and altitude of the quadrotor with and without actuator faults.

5.1 Simulation without Actuator Faults

This subsection will compare the performance of the adaptive controller with the BS controller without adaptation, in the absence of actuator faults. The result using the normal BS is shown in Fig. 3. As can be seen from the figure, the control performance is good despite of little delay which is caused by the actuator dynamics. The position and altitude of the quadrotor can track the reference well. A small steady-state error is observed in the response.

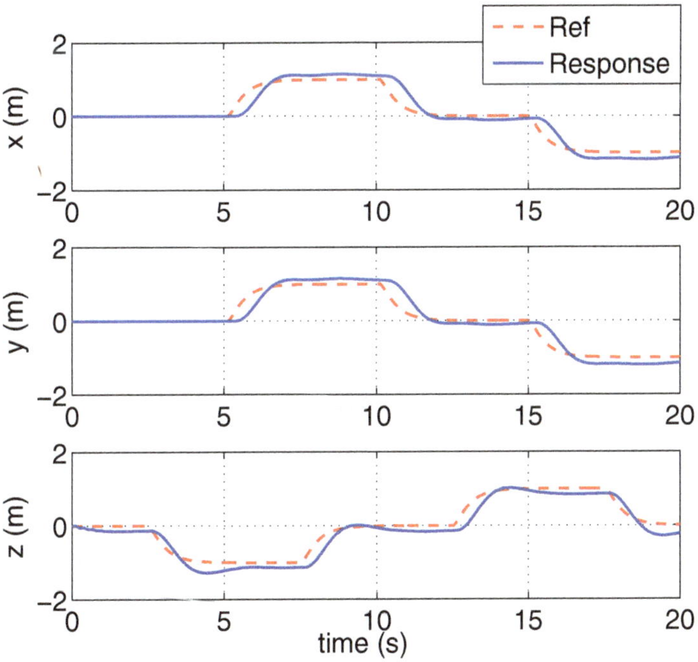

Fig. 3 Position response using normal BS without online model identification, in the absence of actuator faults

The result using the proposed FTC system is shown in Figs.4 - 8. The position performance is shown in Fig. 4. It can be seen that the performance is good. There is also some delay in the response. However, a slightly bigger overshoot is observed from the response. The estimation of the LOE factors introduces additional slight delay which leads to a bigger overshot. However, when the quadrotor reaches hovering condition, the steady-state error is eliminated. This is better than the normal BS controller without the online model identification system. The reason is that during the process, the control effectiveness of the actuator does not remain constant, which introduces uncertainties to the BS controller. Therefore, the normal BS controller without the model identification will suffer from performance degradation while the FTC system does not since the control effectiveness is estimated online.

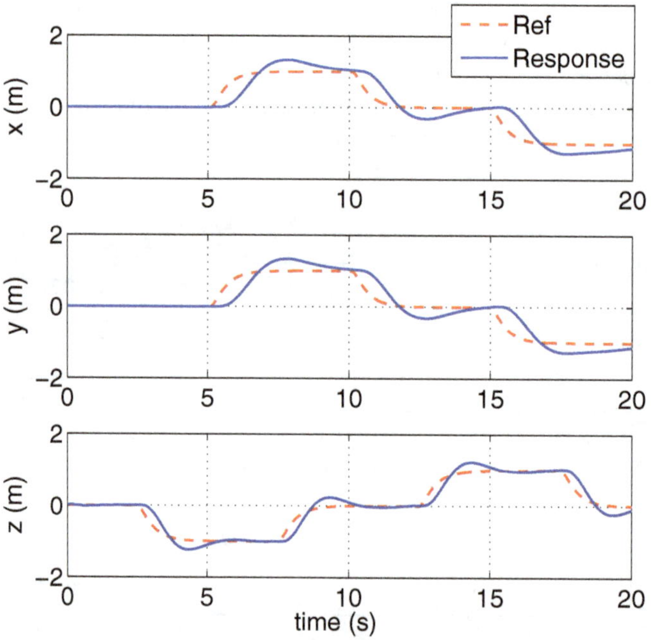

Fig. 4 Position response using the FTC system in the absence of actuator faults

The velocity response and its reference of the quadrotor using the FTC system is shown in Fig. 5. As can be seen, the velocity can track the reference well. The Euler angles of the quadrotor is shown in Fig. 6. The roll angle and pitch angle can track the reference well while the yaw angle of the quadrotor cannot fully track the reference. However, the discrepancy between the response and the reference is small (the maximum deviation is 0.01 rad). This is caused by the maneuvers of the quadrotor. The angular rates of the quadrotor can also track the reference except there is a discrepancy between the yaw rate response and its reference, as shown in Fig. 7.

The LOE factor estimation result using the FTC system is presented in Fig. 8. The red dotted lines denote the real LOE factors caused by the faults while the blue

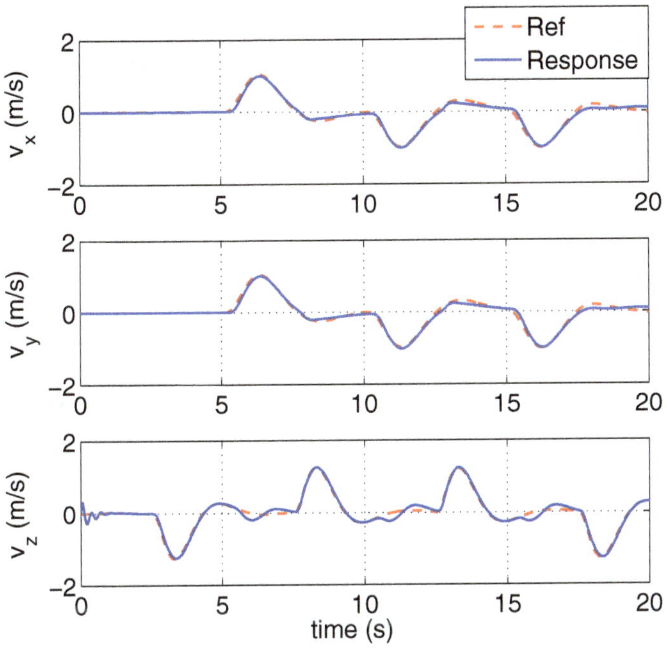

Fig. 5 Velocity response using the FTC system in the absence of actuator faults

solid lines represent the LOE caused by both the faults and the maneuver of the quadrotor. It is obvious that the LOE factors are not zero even when there are no actuator faults. They only remain constant during $1\ \text{s} < t < 2\ \text{s}$ when the quadrotor is in hover. As long as the quadrotor is not in hover, the LOE factors varies with the rotational speed of the rotors. This is caused by the changing inflow velocity and advancing ratio of the quadrotor. Since these speed-varying LOE factors are estimated online, the control performance will be better as long as the estimation converges. This is the reason why the steady-state error using the FTC system is better than that using the normal BS without online model identification.

5.2 *Simulation with Actuator Faults*

This section compares the performance of the FTC system with the normal BS without online model identification. The fault type is LOE fault which means that the quadrotor losses some effectiveness probably caused by the loss of part of its propellers. It is assumed that all of the four rotors experience LOE faults consecutively. The fault scenario is given in Table 1.

The result of using the normal BS without the online model identification is shown in Fig. 9. It can be seen that immediately after the first fault, the controller shows no ability to recover the control and the quadrotor begins to loss control.

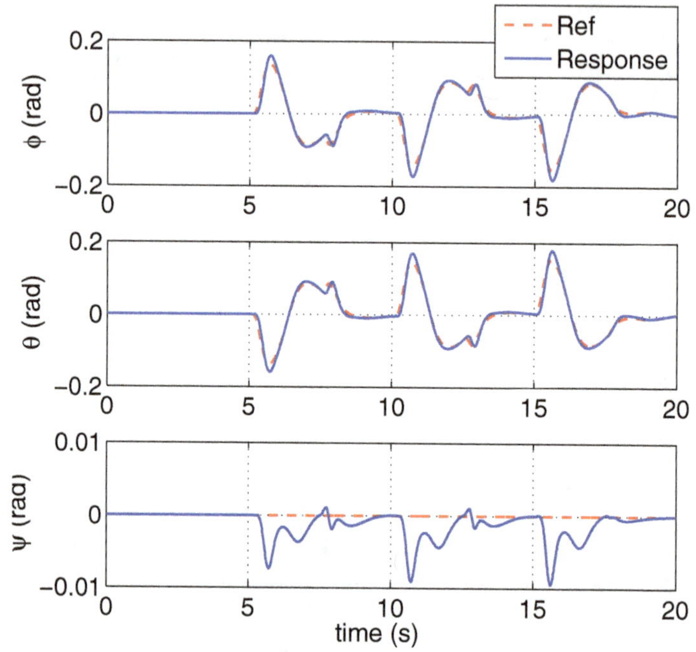

Fig. 6 Euler angle response using the FTC system in the absence of actuator faults

Table 1 Actuator LOE fault scenario

Rotor	Failure time	Fault Type	Fault Magnitude
1	$t = 5$ s	LOE	0.2
2	$t = 10$ s	LOE	0.2
3	$t = 15$ s	LOE	0.2
4	$t = 20$ s	LOE	0.2

The result using the FTC system is demonstrated in Fig. 10. The FTC system successfully recovers the control in the presence of four consecutive actuator LOE faults. The position response of the quadrotor degrades which is caused by the actuator faults. However, the controller can maintain the flight after it is recovered. The altitude response is good despite of the four consecutive faults. The reason is that the altitude is directly controlled by the thrust which is less affected by the actuator faults than the torques generated by the four rotors.

The velocity response is shown in Fig. 11. There is also discrepancy between the response and its references. The vertical speed can track the reference though. The Euler angle response is shown in Fig. 12. It is noticed that the yaw angle error is larger than the situation where there are no actuator faults. One of the reasons is that

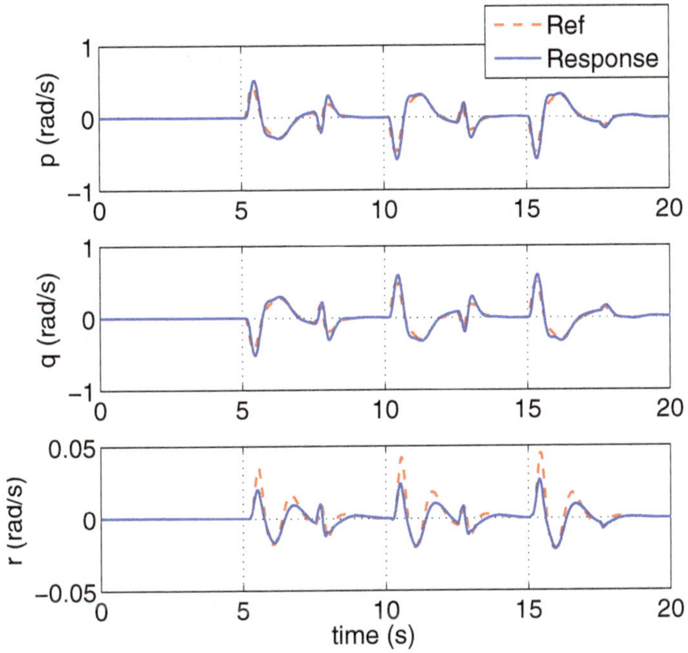

Fig. 7 Rate response using the FTC system in the absence of actuator faults

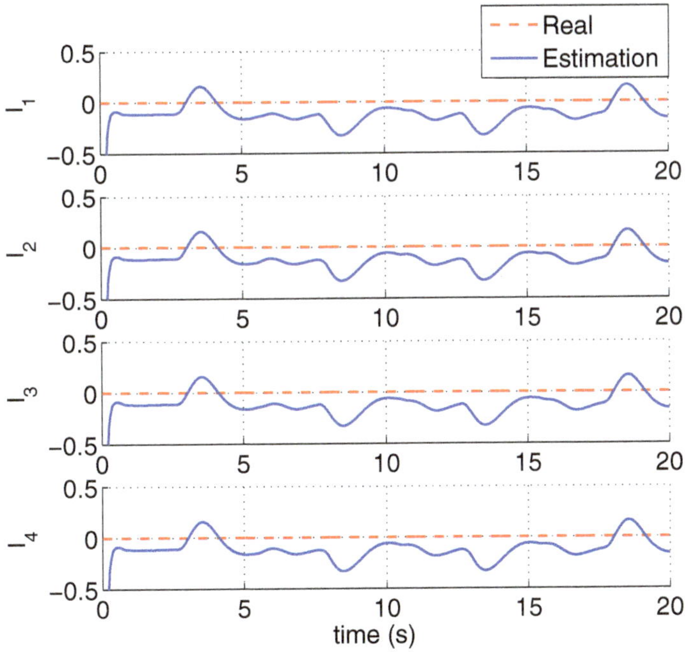

Fig. 8 LOE estimation using the FTC system in the absence of actuator faults

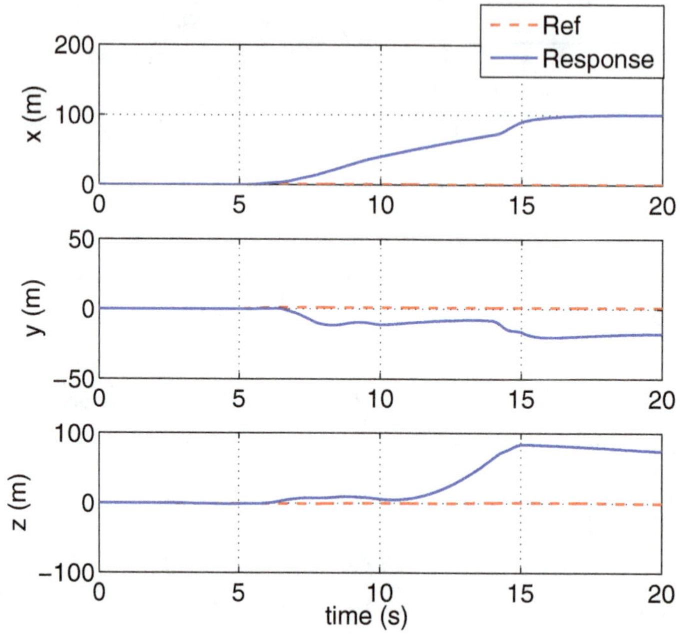

Fig. 9 Position response using the normal BS without online model identification, in the presence of actuator faults

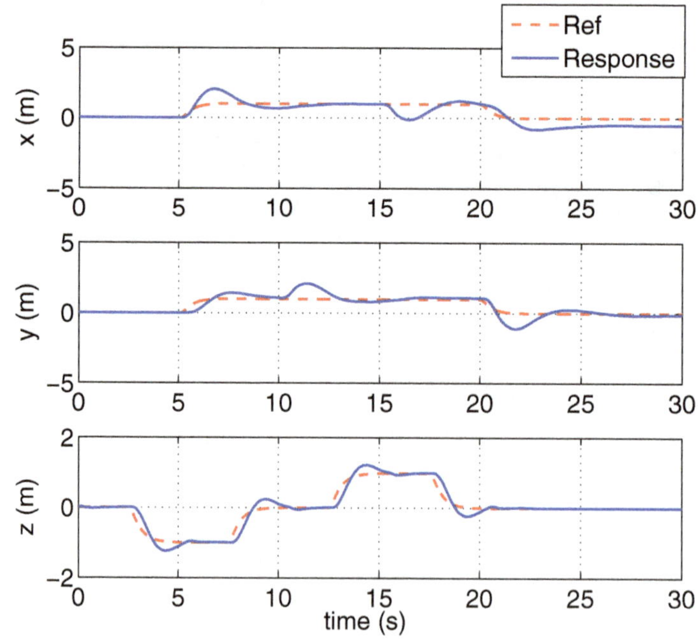

Fig. 10 Position response using the FTC system in the presence of actuator faults

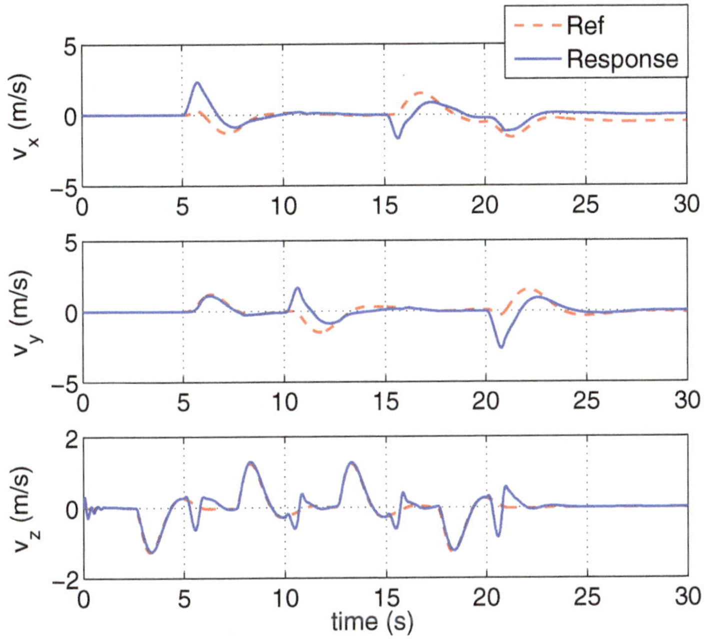

Fig. 11 Velocity response using the FTC system in the presence of actuator faults

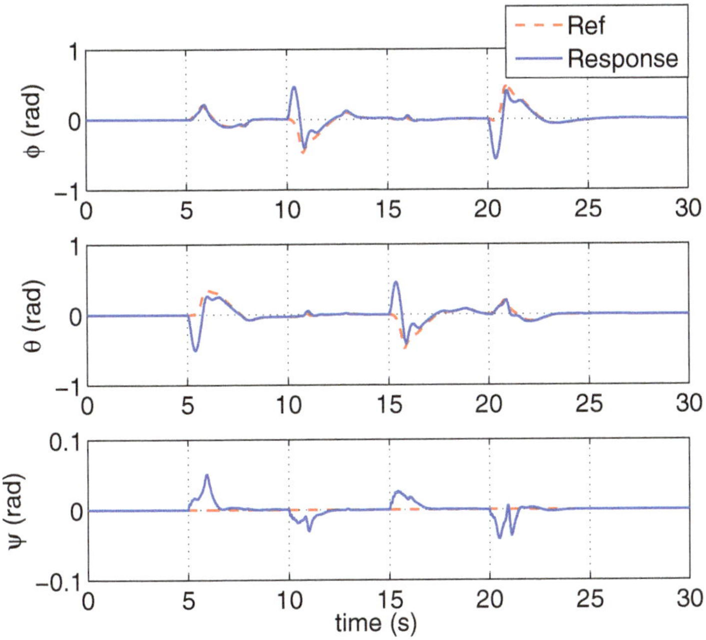

Fig. 12 Euler angle response using the FTC system in the presence of actuator faults

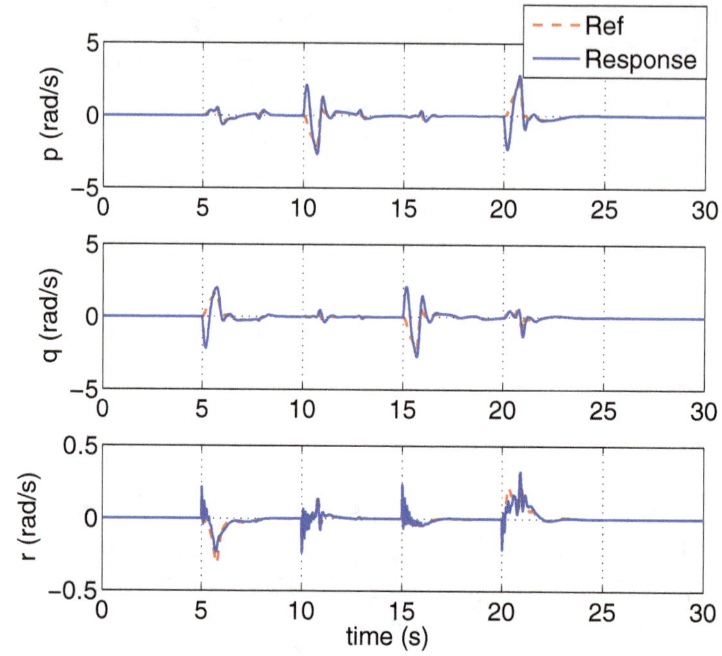

Fig. 13 Rate response using the FTC system in the presence of actuator faults

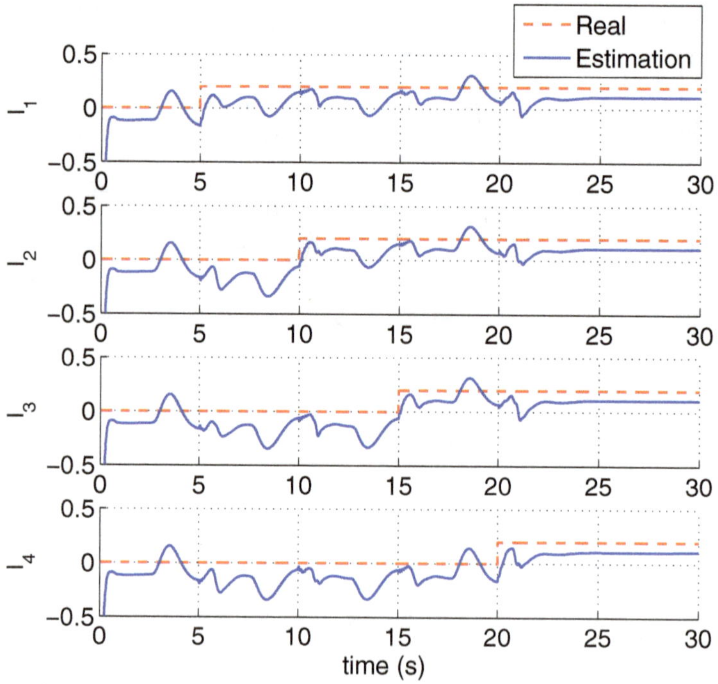

Fig. 14 LOE estimation using the FTC system in the presence of actuator faults

in the presence of actuator faults, the yaw torque is also influenced by the faults. Therefore, the yaw angle response will deviate from the reference as long as there is an actuator fault. The angular rates of the quadrotor response is shown in Fig. 13. Oscillations can be observed from the yaw rate response.

The key point of the FTC system is the LOE estimation. The estimation of the LOE during the faults can be found in Fig. 14. The LOE factor of the faults are denoted by the red dotted lines while the blue solid lines represent the estimation of the LOE factors using the FTC system. Note that the real LOE here means the LOE caused by the faults. The estimated LOE contains both the LOE caused by the faults and the maneuver of the quadrotor. Therefore, the estimated LOE does not coincide with the real LOE caused by the faults as seen in the figure.

6 Conclusions

This paper presents a new FTC system for the quadrotor which is capable of maintain the flight even in the presence of actuator faults. The system is composed of three subsystems which are the state estimation, the LOE estimation and the BS controller. All the subsystems play a vital role in the reconfiguration control. The state estimation can remove the biases in the sensors such as the IMU sensors which usually suffer from biases and drifts. The LOE estimation subsystem provides the real-time model of the quadrotor which is used by the BS controller to design adaptive control laws even in the presence of actuator faults.

The performance of the proposed FTC system is validated by two different simulations: with and without actuator faults. As for the situation when there are no actuator faults, the control performance of the FTC system shows more overshoot but its steady-state error is smaller than the BS controller without online model identification. In terms of actuator faults, the FTC system shows a satisfactory performance. It can maintain the flight even when all the rotors fail consecutively.

The proposed FTC system can be used by the quadrotor to enhance its control performance since it can achieve a better performance no matter whether there is an actuator fault or not. It can prevent the quadrotor from crashing when the quadrotor loses its effectiveness.

In the future, the present work should be implemented on a real quadrotor to enhance the safety and performance of the quadrotor. Another interesting topic would be to investigate the influence of the actuator dynamics. Since the actuator dynamics can also be influenced by the battery level, the actuator dynamics can change during the flight.

References

1. Amoozgar, M.H., Chamseddine, A., Zhang, Y.: Experimental Test of a Two-Stage Kalman Filter for Actuator Fault Detection and Diagnosis of an Unmanned Quadrotor Helicopter. Journal of Intelligent & Robotic Systems 70(1-4), 107–117 (2012)

2. Bouabdallah, S.: Design and control of quadrotors with application to autonomous flying. PhD thesis, Ecole Polytechnique Fedeale de lausanne (2007)
3. Bouabdallah, S., Siegwart, R.: Full control of a quadrotor. In: 2007 IEEE/RSJ International Conference on Intelligent Robots and Systems, oct 2007, pp. 153–158. IEEE Computer Society Press, Los Alamitos (2007)
4. Cen, Z., Noura, H., Susilo, T.B., Younes, Y.A.: Robust Fault Diagnosis for Quadrotor UAVs Using Adaptive Thau Observer. Journal of Intelligent & Robotic Systems 73(1-4), 573–588 (2013)
5. Fay, G.: Derivation of the Aerodynamic Forces for the Mesicopter Simulation, pp. 1–8 (2001)
6. Freddi, A., Longhi, S., Monteriù, A.: A Diagnostic Thau Observer for a Class of Unmanned Vehicles. Journal of Intelligent & Robotic Systems 67(1), 61–73 (2012)
7. Krstic, M., Kanellakopoulos, I., Kokotovic, P.: Nonlinear and Adaptive Control Design. John Wiley & Sons, Inc. (1995)
8. Levant, A.: Robust exact differentiation via sliding mode technique. Automatica 34(3), 379–384 (1998)
9. Lombaerts, T.J.J., Chu, Q.P., Mulder, J.a., Joosten, D.a.: Modular flight control reconfiguration design and simulation. Control Engineering Practice 19(6), 540–554 (2011)
10. Lu, P., Van Eykeren, L., Kampen, E.v., Chu, Q.P., Yu, B.: Adaptive Hybrid Unscented Kalman Filter for Aircraft Sensor Fault Detection, Isolation and Reconstruction. In: AIAA Guidance, Navigation, and Control Conference, National Harbor, Maryland, pp. 1–18 (2014)
11. Marconi, L., Naldi, R.: Robust full degree-of-freedom tracking control of a helicopter. Automatica 43(11), 1909–1920 (2007)
12. Mulder, J.A., Chu, Q.P., Sridhar, J.K., Breeman, J.H., Laban, M.: Non-linear aircraft flight path reconstruction review and new advances. Progress in Aerospace Sciences 35, 673–726 (1999)
13. Raffo, G.V., Ortega, M.G., Rubio, F.R.: An integral predictive/nonlinear control structure for a quadrotor helicopter. Automatica 46(1), 29–39 (2010)
14. Slegers, N., Kyle, J., Costello, M.: Nonlinear Model Predictive Control Technique for Unmanned Air Vehicles. Journal of Guidance, Control, and Dynamics 29(5), 1179–1188 (2006)
15. Van Eykerenand, L., Chu, Q.P.: Air Data Sensor Fault Detection using Kinematic Relations. In: Proceedings of the EuroGNC 2013, 2nd CEAS Special Conference on Guidance, Navigation & Control, pp. 414–428 (2013)
16. Zhang, X., Zhang, Y., Su, C.-y., Feng, Y.: Fault Tolerant Control for Quadrotor via Backstepping Approach. In: AIAA Aerospace Sciences Meeting, Orlandeo, Florida, pp. 1–12 (2010)
17. Zhang, Y., Chamseddine, A.: Fault Tolerant Flight Control Techniques with Application to a Quadrotor UAV Testbed (2011)

Comparison of L1 Adaptive Augmentation Strategies for a Differential PI Baseline Controller on a Longitudinal F16 Aircraft Model

Fabian Hellmundt, Andreas Wildschek, Rudolf Maier,
Robert Osterhuber, and Florian Holzapfel

Abstract. In this paper two different approaches are presented to design an adaptive augmentation using L1 Adaptive Control. In terms of reference dynamics, the first one takes the closed-loop aircraft with baseline controller into account. The second approach tries to maintain nominal open-loop aircraft dynamics, with the baseline controller wrapped around the adaptive augmentation. They are compared by application to a model of the longitudinal dynamics of a F16 aircraft. The aircraft model is equipped with a Differential PI (DPI) baseline controller. The design of the combination of baseline controller and augmentation takes saturation as well as rate limitation of the control signal directly into account. Simulation results show the nominal closed-loop behavior is not harmed by the adaptive augmentation and that an increase of performance in case of uncertainty can be achieved. As an exemplary uncertainty case a rapid shift of the center of gravity (CG) is examined. The robustness of the controller structure is assessed by the use of both linear and nonlinear methods to obtain the Time Delay Margin (TDM) and the Gain Margin (GM) of the closed loop system. It can be observed, the adaptive augmentation leads to a significant decrease of robustness w.r.t. the plant input channel. This drop in TDM and GM can be fully restored by the application of a hedging strategy.

Fabian Hellmundt · Andreas Wildschek · Rudolf Maier
Airbus Group Innovations, Ottobrunn, 81663, Germany
e-mail: {fabian.hellmundt,andreas.wildschek,
rudolf.maier}@airbus.com

Robert Osterhuber
Airbus Defence & Space, Manching, 85077, Germany
e-mail: robert.osterhuber@cassidian.com

Florian Holzapfel
Technische Universität München, Garching, 85748, Germany
e-mail: florian.holzapfel@tum.de

J. Bordeneuve-Guibé et al. (eds.), *Advances in Aerospace Guidance, Navigation and Control*, DOI: 10.1007/978-3-319-17518-8_7

1 Introduction

For many years Adaptive Control is investigated in order to be used within Flight Control Systems (FCS) to improve the performance in case of failure or uncertain aerodynamics. A common philosophy is to design a baseline controller, which ensures the desired performance of the nominal aircraft, by application of practical proven techniques (e.g. linear control [2, 11, 26], nonlinear dynamic inversion [7, 22], ...) That way, the existing know-how of the manufacturer in terms of controller design can be fully used. The baseline controller is then augmented by an adaptive controller, which significantly contributes to the total control signal only in case of an existing deficiency between a nominal reference model and the actual aircraft response. As an increase w.r.t. performance gained by Adaptive Control usually effects decreasing robustness properties, it is the challenge to find a suitable trade-off.

For the following application on a model of the F16 aircraft, L1 Adaptive Control [13] is used to augment the DPI baseline controller [21, 20, 4]. As stated in [28], L1 Adaptive Control offers guaranteed transient performance and robustness through its architecture. Moreover, estimation loop and control loop are decoupled by design, which allows application of fast adaptation. Furthermore, with the choice of Piecewise Constant Parameter Update Laws [27] the tuning effort can be reduced drastically, the adaptation process does not rely on persistency of excitation and the design results in a linear control law [3]. Thus, it is possible to use methods from linear control theory to assess the robustness. The speed of adaptation is theoretically only limited by the available CPU power (Sampling Time). It is also worth noting, that a standalone controller using L1 Adaptive Control was flight tested on the NASA AirSTAR, which is a turbine powered, subscaled model of a transport aircraft [8, 9, 10, 23].

2 Aircraft Model

For this paper a model of the nonlinear, rigid dynamics of a F16 aircraft is used. Aerodynamic data and inertia properties are taken from [25]. The model was originally introduced in [19]. Using this data the model was implemented by the *TUM Institute of Flight System Dynamics* (TUM-FSD) using MATLAB/Simulink. At this point of the study only longitudinal aircraft dynamics are considered.

The model is valid for low speed up to $Ma \approx 0.6$, since no dependence on Mach number is considered for the aerodynamic coefficients. [19] states a valid range for angle of attack as $-20^\circ \leq \alpha \leq 90^\circ$, whereas [25] claims that only $-10^\circ \leq \alpha \leq 45^\circ$ should be used. This is because on the one hand it can be avoided to dynamically model the post-stall region, on the other hand the aircraft can barely reach an angle of attack $\alpha \geq 25^\circ$ due to limitations of the control surfaces. In addition, the reference CG for the aircraft model is at $0.35\bar{c}$. This leads to static instability in the longitudinal motion. In order to be able to investigate the effect of large, off-nominal

CG-shifts, which destabilize the aircraft, the nominal CG is shifted forward to $0.3\bar{c}$, i.e. the considered nominal aircraft behavior is stable in the longitudinal motion.

The original aircraft uses mainly two control surfaces for the longitudinal motion, namely the elevator and the leading edge flap (LEF). The deflection of the LEF δ_{LEF} is determined as a function of α and Ma. As the Ma-dependence is very low for the considered valid envelope and neglecting the actuator dynamics, the control of the LEF was merged into the aerodynamic coefficients. This leaves the elevator deflection η as only aerodynamic control input for the longitudinal motion. The elevator is modeled as a well damped PT2 transfer function with a natural frequency $\omega_0 = 40\frac{rad}{s}$ and relative damping $\zeta = 0.71$. Its absolute deflection is limited to $|\eta| \leq 25°$ with a rate limit of $|\dot{\eta}| \leq 60\frac{°}{s}$.

The sensors of the aircraft states are uniformly modeled as a time delay of $T_D = 55ms$.

3 Baseline Controller

It is the task of the baseline controller to ensure the required Level 1 handling qualities for the nominal case, where no failure occurrs. In order to gain better comparability in terms of the basic FCS functionality with the Eurofighter Typhoon aircraft, the DPI approach was chosen for the design of the baseline controller [21, 20, 4]. It is very similar to the well-known PI controller, but distinguishes crucially with the introduction of differentiation operators at every input (i.e. sensor signals) and an integrator at the output. Thus, it is possible to limit both the absolute value and the rate of the control signal at the integrator, according to the properties of the actuator, without having to deal with integrator windup. Another advantage of this control structure is the possibility to use fast states e.g. α as scheduling parameters for the gains within the controller without generating hidden feedback. The integrator also has a smoothing effect on the control signal when using noisy measurements as scheduling parameters.

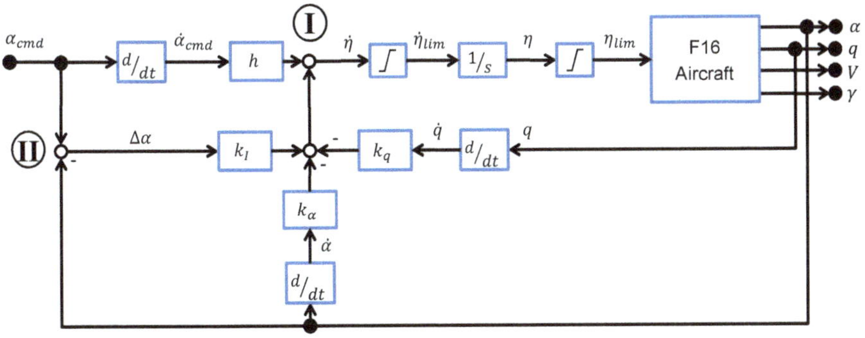

Fig. 1 Structure of the DPI Baseline Controller

The basic structure of the controller is depicted in Figure 1. For shaping of the closed-loop behavior w.r.t. the short-period, feedback of the states α and q is used. The commanded angle of attack α_{cmd} is fed in both the feed-forward path and $\Delta\alpha = \alpha_{cmd} - \alpha$. As it is not differentiated the $\Delta\alpha$-path leads to integral feedback of α, ensuring zero steady-state error. α_{cmd} is obtained by the stick input $\delta_{s,lon}$, where full stick deflection translates to $\alpha_{cmd,max}$ and $\alpha_{cmd,min}$, respectively. For $\delta_{s,lon} = 0$ an angle of attack α_0 is commanded, which trims the aircraft in the current flight state. α_0 is stored as a pre-calculated lookup-table.

Neglecting the limitations within the controller structure enables one to use the whole spectrum of linear control theory to design the gains h, k_I, k_α and k_q according to the desired handling qualities. Within this work, Eigenstructure Assignment is chosen for this task. Specifically in aircraft applications it is often used for the design of the lateral controller due to its capability to systematically and automatically calculate gains suiting the requirements on eigenvalues and to ensure decoupling of eigenmotions for MIMO systems. Examples can be found in [6, 15, 12]. In case of a SISO system presented here, Eigenstructure Assignment allows systematic pole placement with output feedback. This is required, because the actuator model and sensor delays, being modeled as Pade-approximations, are incorporated in the design process, for the sake of a more accurate representation of the actual plant within the design model of the controller. Actuator and sensor model contribute additional states to the design model, which are not available as outputs.

The DPI Controller for the F16 aircraft was designed with a CAP[1] of 1.3 and a relative damping of the short period $\zeta_{SP} = 0.95$. The chosen CAP leads to e.g. $\omega_{0,SP} = 5.98\frac{rad}{s}$ at the envelope point with the velocity $V = 177\frac{m}{s}$ and at the altitude $h = 5000m$. Since Eigenstructure Assignment without modifications inherits no robustness assessment, it has to be validated after, that robustness margins are ensured by the controller design.

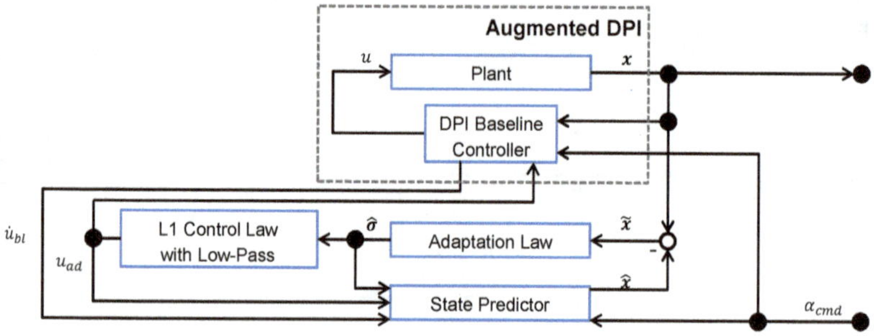

Fig. 2 Interconnection for L1 DPI Augmentation

[1] Control Anticipation Parameter: Ratio between the initial angular acceleration $\dot{q}_0$ and the resulting steady state load factor $n_{z,s}$ ($CAP = \frac{\dot{q}_0}{n_{z,s}}$), can be seen as measure for the longitudinal control sensitivity [4]

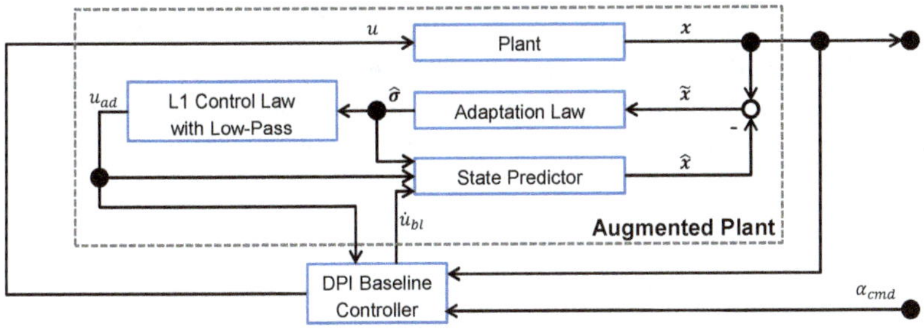

Fig. 3 Interconnection for L1 Plant Augmentation

4 L1 Adaptive Augmentation

In order to increase the controller performance and to ensure robustness in case of uncertainty, it is proposed, to augment the DPI baseline controller using L1 Adaptive Control with Piecewise Constant Parameter Update Laws. Simultaneously, the augmentation should neither degenerate the performance for the nominal case, nor should it significantly decrease the robustness properties of the closed-loop. As an additional requirement it should be ensured, that the control law

$$u = u_{bl} + u_{ad} \tag{1}$$

where u_{bl} is the control signal portion generated by the baseline controller and u_{ad} the one produced by the L1 Adaptive Controller, respectively, does not exceed given saturation and rate limits for the actuator, as they are stated above. Thus u_{ad} has to be fed in at summation ① of the baseline controller depicted in Figure 1, which is upstream the limiting structure of the DPI controller. In this paper, two different approaches to augment the interconnection of aircraft and DPI baseline controller and to generate the adaptive control signal u_{ad} are presented.

The first would be the most obvious choice and can be found quite often in literature, see e.g. [11, 16, 2]. The basic structure is illustrated in Figure 2. In that case the L1 Adaptive Controller with its elements *State Predictor*, *Adaptation Law* and *L1 Control Law with Low-Pass* tries to maintain the dynamics of the closed-loop aircraft with the baseline controller. For this reason the reference model to be used is the nominal closed-loop aircraft. This also requires the consideration of the baseline controller within the adaptive controller e.g. in the State Predictor. Thus it is called *DPI Augmentation* here.

An alternative augmentation method was developed at the *TUM Institute of Flight System Dynamics* (TUM-FSD) and is given in Figure 3. For this approach the baseline controller is wrapped around a unit consisting of aircraft and L1 Adaptive Controller. It is now the task of the adaptive augmentation to ensure, that the baseline controller always "sees" its design model. Thus the reference model of the L1 Adaptive Controller can be reduced to the open-loop aircraft model in this case. This approach is called *Plant Augmentation* in the context of this paper.

Although the approaches differ from each other from a structural point of view, the desired aircraft dynamics to be achieved by the combination of adaptive augmentation and baseline controller in terms of the transfer function $\alpha_{cmd} \rightarrow \alpha$ remains the same. The desired aircraft behavior is determined by the interconnection of aircraft and DPI baseline controller and can be conducted as

$$\begin{aligned} \dot{\mathbf{x}}_{ref}(t) &= \mathbf{A}_m \mathbf{x}_{ref}(t) + \mathbf{b}_m \cdot G_{Delay}(t) * G_{Actuator}(t) \\ &* sat\left\{ \int_{t_0}^{t} sat\left[\mathbf{k}_{DPI}^T \cdot \frac{d}{d\tau} \mathbf{x}_{ref}(\tau) + h \cdot \frac{d}{d\tau} \alpha_{cmd}(\tau) \right] \right\} d\tau \\ &+ \begin{bmatrix} 0 \\ 0 \\ 1 \end{bmatrix} \alpha_{cmd}(t) \end{aligned} \tag{2}$$

where $\mathbf{x}_{ref}(t)$ contains the states α_{ref}, q_{ref} and the integrated error $e_{I,ref} = \int (\alpha_{cmd} - \alpha_{ref})\, dt$. The nominal open-loop aircraft is described by the short-period approximation, which is enhanced by an integration of $-\alpha_{ref}$. Hence the system and input matrices are defined as

$$\mathbf{A}_m = \begin{bmatrix} Z_\alpha & Z_q + 1 & 0 \\ M_\alpha & M_q & 0 \\ -1 & 0 & 0 \end{bmatrix} \tag{3}$$

$$\mathbf{b}_m^T = \begin{bmatrix} Z_\eta & M_\eta & 0 \end{bmatrix}. \tag{4}$$

The dynamics w.r.t. the short-period and the error integrator are shaped by the feedback gains

$$\mathbf{k}_{DPI}^T = \begin{bmatrix} k_\alpha & k_q & -k_I \end{bmatrix}, \tag{5}$$

while the feed-forward contribution of α_{cmd} is scaled by the gain h. Actuator and sensors are considered by the transfer functions $G_{Actuator}(t)$ and $G_{Delay}(t)$, which is a 2nd-order Pade approximation. To reflect the combination of baseline controller and aircraft model as precise as possible w.r.t. the reference dynamics, it is important to also consider the specific structure of the DPI controller. For that matter the differentiations, the integrator and the saturation function [17]

$$sat(x) = \begin{cases} x & if\, |x| \leq x_{lim} \\ x_{lim} \cdot sgn(x) & if\, |x| > x_{lim} \end{cases} \tag{6}$$

are included in Equation (2). Especially the saturation function is the reason, why the reference dynamics cannot be summarized into one single $\mathbf{A}_m$ matrix and the feedback $\mathbf{k}_{DPI}^T$ has to be considered separately.

The reference dynamics are not determined for one single envelope point. Instead $\mathbf{A}_m$ and $\mathbf{b}_m$ as well as the gains of the modeled controller h and $\mathbf{k}_{DPI}$ are scheduled with the dynamic pressure $\bar{q}$, which is the same scheduling as applied to the baseline controller shown in Section 3.

4.1 DPI Augmentation

At first the elements of the *DPI Augmentation* are presented, namely the State Predictor, the Parameter Update Law and the Control Law. As it is stated above, the augmentation is based on L1 Adaptive Control with Piecewise Constant Parameter Update Laws [13].

State Predictor

As this approach requires modeling of aircraft and baseline controller within the adaptive controller, the state predictor can be constructed analogous to the reference model described in Equation (2) and is compiled as shown in Figure 4 and through

$$\begin{aligned}\dot{\hat{\mathbf{x}}}(t) &= \mathbf{A}_m\hat{\mathbf{x}}(t) + \Lambda_0 \cdot \mathbf{b}_m \cdot G_{Delay}(t) * G_{Actuator}(t) \\ &\quad * \; sat\left\{\int_{t_0}^{t} sat\left[\mathbf{k}_{DPI}^T \cdot \frac{d}{d\tau}\hat{\mathbf{x}}(\tau) + h \cdot \frac{d}{d\tau}\alpha_{cmd}(\tau) + \frac{d}{d\tau}u_{ad,1}(\tau)\right] d\tau\right\} \\ &\quad + \mathbf{b}_m\hat{\sigma}_m(t) + \mathbf{B}_{um}\hat{\sigma}_{um}(t) + \begin{bmatrix}0\\0\\1\end{bmatrix}\left[u_{ad,2}(t) + \alpha_{cmd}(t)\right] \\ &\quad + \mathbf{b}_m \cdot \Delta u(t) + \mathbf{K}_e\tilde{\mathbf{x}}(t)\end{aligned} \tag{7}$$

with estimations of the matched $\hat{\sigma}_m$ and unmatched $\hat{\sigma}_{um}$ uncertainties. Their estimation law is defined in Equation (10). The adaptive control law is subdivided into one portion $u_{ad,1}(t)$ to compensate matched uncertainties and another one $u_{ad,2}(t)$ to compensate unmatched uncertainties. The definition of the control laws can be found in equations (13) and (14). $u_{ad,2}(t)$ is directly inserted into the integrator of the DPI Controller in order to avoid additional integral error feedback [3]. Thus the insertion point for $u_{ad,2}(t)$ in Figure 1 is Ⓘ. $\mathbf{B}_{um}$ is a constant matrix chosen so that $\mathbf{b}_m^T\mathbf{B}_{um} = \mathbf{0}$ is satisfied [13]. Λ_0 is an initial guess on the control effectiveness and is set to $\Lambda_0 = 1$. Additional feedback of the estimation error

$$\tilde{\mathbf{x}}(t) = \hat{\mathbf{x}}(t) - \mathbf{x}(t) \tag{8}$$

is used to further tune the performance w.r.t. error dynamics by adjusting the gain matrix $\mathbf{K}_e$ =diag(k_1, k_2, k_3). To account for additional unmodeled dynamics at the plant input channel the hedging term

$$\begin{aligned}\Delta u &= \eta_{meas} - \Lambda_0 \cdot G_{Delay}(t) * G_{Actuator}(t) \\ &\quad * \; sat\left\{\int_{t_0}^{t} sat\left[\frac{d}{d\tau}u_{ad,1}(\tau) + \frac{d}{d\tau}u_{bl}(\tau)\right] d\tau\right\}\end{aligned} \tag{9}$$

is used. This term is constructed utilizing the idea of Pseudo-Control Hedging [14]. $\frac{d}{d\tau}u_{bl}(\tau)$ is the deflection rate, which is commanded by the baseline controller without any limitations applied. Thus, it is the output of summation point Ⓘ in Figure 1 without the contribution of $u_{ad,1}$. Δu compares the measured actuator deflection

η_{meas} with the expected control signal considering all elements, which are already implemented within the state predictor. As shown in Section 5.2, this term has a significant role in preserving the robustness properties of the baseline controller.

Just like the reference dynamics, the state predictor uses scheduled versions of $\mathbf{A}_m$, $\mathbf{b}_m$, $\mathbf{k}_{DPI}$ and h. This also leads to the introduction of scheduling within the Parameter Update Law and the Control Law shown below as well as for the matrix $\mathbf{B}_{um}$.

Parameter Update Law

The matched and unmatched uncertainties are gained through the piecewise constant update law [13]

$$\begin{bmatrix} \hat{\sigma}_m(t) \\ \hat{\sigma}_{um}(t) \end{bmatrix} = \mathbf{K}(T_s)\,\tilde{\mathbf{x}}(iT_s) \quad \forall t \in [iT_s, (i+1)T_s] \tag{10}$$

where

$$\mathbf{K}(T_s) = -\begin{bmatrix} \mathbf{b}_m & \mathbf{B}_{um} \end{bmatrix}^{-1} \left[\mathbf{A}_e^{-1}\left(e^{\mathbf{A}_e T_s} - \mathbf{I}_n\right)\right]^{-1} e^{\mathbf{A}_e T_s} \tag{11}$$

and with

$$\mathbf{A}_e = \mathbf{A}_{m,CL} + \mathbf{K}_e = \mathbf{A}_m + \mathbf{b}_m \hat{\mathbf{k}}_{DPI}^T + \mathbf{K}_e. \tag{12}$$

Since the sampling time T_s is constant, the adaptation gain $\mathbf{K}(T_s)$ can be pre-calculated for all $\mathbf{A}_m$ and $\mathbf{b}_m$. $\mathbf{A}_{m,CL}$ is constructed such, that it has the same eigenvalues as the closed-loop aircraft with the baseline controller in terms of short-period and error integration. As actuator and sensors are not considered for $\mathbf{A}_{m,CL}$ the set of gains $\hat{\mathbf{k}}_{DPI}^T$ differs from the one used within the baseline controller.

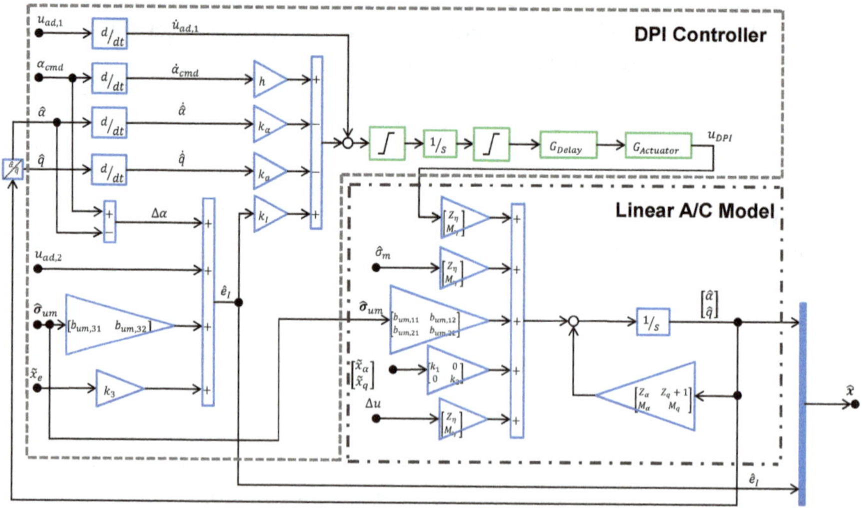

Fig. 4 DPI Augmentation - State Predictor

Control Law

The control law, which is based on [13], is subdivided in portions for matched and unmatched uncertainties, respectively. They are defined as

$$u_{ad,1} = C_1(s)\,\hat{\sigma}_m(s) \tag{13}$$

$$u_{ad,2} = \mathbf{H}_m^{-}1(s)\,\mathbf{H}_{um}(s)\begin{bmatrix} C_2(s) & 0 \\ 0 & C_3(s) \end{bmatrix}\underbrace{\begin{bmatrix} \hat{\sigma}_{um,1}(s) \\ \hat{\sigma}_{um,2}(s) \end{bmatrix}}_{\hat{\sigma}_{um}}. \tag{14}$$

Within the transfer functions

$$\mathbf{H}_m(s) = \mathbf{c}_m(s\mathbf{I}_n - \mathbf{A}_m)^{-1}\begin{bmatrix} 0 \\ 0 \\ 1 \end{bmatrix} \tag{15}$$

$$\mathbf{H}_{um}(s) = \mathbf{c}_m(s\mathbf{I}_n - \mathbf{A}_m)^{-1}\mathbf{B}_{um} \tag{16}$$

it is considered, that $u_{ad,2}$ is fed into the integrator of the baseline controller as mentioned above. For that reason, the input vector $\begin{bmatrix} 0 & 0 & 1 \end{bmatrix}^T$ is used instead of $\mathbf{b}_m$ for (15) [3]. The output vector is set as $\mathbf{c}_m = \begin{bmatrix} 1 & 0 & 0 \end{bmatrix}$. The Low-Pass Filters are chosen as PT1 transfer functions defined as

$$C_i(s) = \frac{\omega_{C,i}}{s + \omega_{C,i}} \quad \forall i \in \{1;2;3\}. \tag{17}$$

4.2 *Plant Augmentation*

Next the elements of the *Plant Augmentation* approach are conducted. For the sake of comparison only the differences to the *DPI Augmentation* are highlighted.

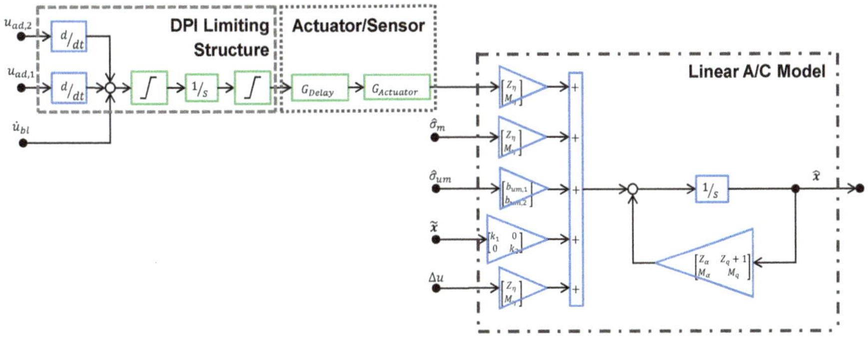

Fig. 5 Plant Augmentation - State Predictor

State Predictor

While designing the *Plant Augmentation*, modeling of the baseline controller within the adaptive controller can be neglected. Therefore $\mathbf{A}_m^\star$ and $\mathbf{b}_m^\star$ are reduced to

$$\mathbf{A}_m^\star = \begin{bmatrix} Z_\alpha & Z_q + 1 \\ M_\alpha & M_q \end{bmatrix} \tag{18}$$

$$\mathbf{b}_m^{\star T} = \begin{bmatrix} Z_\eta & M_\eta \end{bmatrix} \tag{19}$$

and the state predictor is constructed as

$$\begin{aligned} \dot{\hat{\mathbf{x}}}^\star(t) &= \mathbf{A}_m^\star \hat{\mathbf{x}}^\star(t) + \Lambda_0 \cdot \mathbf{b}_m^\star \cdot G_{Delay}(t) * G_{Actuator}(t) \\ &* \; sat\left\{\int_{t_0}^{t} sat\left[\frac{d}{d\tau}\left(u_{bl}(\tau) + u_{ad,1}^\star(\tau) + u_{ad,2}^\star(\tau)\right)\right] d\tau\right\} \\ &+ \mathbf{b}_m^\star \hat{\sigma}_m^\star(t) + \mathbf{b}_{um}\hat{\sigma}_{um}^\star(t) + \mathbf{b}_m^\star \cdot \Delta u^\star(t) + \mathbf{K}_e^\star \tilde{\mathbf{x}}^\star(t) \end{aligned} \tag{20}$$

with $\mathbf{b}_m^{\star T}\mathbf{b}_{um} = 0$. The structure is also depicted in Figure 5. With no integrator state available, $u_{ad,2}^\star$ does not generate an additional integral feedback and can thus be directly fed into the plant input. This is also why $u_{ad,2}^\star$ is added at summation point ① within the baseline controller shown in Figure 1. As $u_{ad,1}^\star$ is still fed in at the same position, the limiting structure of the DPI controller remains within the state predictor. According to the described changes, the hedging term is modified to

$$\begin{aligned} \Delta u^\star &= \eta_{meas} - \Lambda_0 \cdot G_{Delay}(t) * G_{Actuator}(t) \\ &* \; sat\left\{\int_{t_0}^{t} sat\left[\frac{d}{d\tau}u_{ad,1}^\star(\tau) + \frac{d}{d\tau}u_{ad,2}^\star(\tau) + \frac{d}{d\tau}u_{bl}(\tau)\right] d\tau\right\}. \end{aligned} \tag{21}$$

Parameter Update Law

The parameter update law [13] is defined as

$$\begin{bmatrix} \hat{\sigma}_m^\star(t) \\ \hat{\sigma}_{um}^\star(t) \end{bmatrix} = \mathbf{K}^\star(T_s)\tilde{\mathbf{x}}^\star(iT_s) \quad \forall t \in [iT_s, (i+1)T_s] \tag{22}$$

where

$$\mathbf{K}^\star(T_s) = -\begin{bmatrix} \mathbf{b}_m^\star & \mathbf{b}_{um} \end{bmatrix}^{-1} \left[\mathbf{A}_e^{\star -1}\left(e^{\mathbf{A}_e^\star T_s} - \mathbf{I}_n\right)\right]^{-1} e^{\mathbf{A}_e^\star T_s} \tag{23}$$

with $\mathbf{A}_e^\star = \mathbf{A}_m^\star + \mathbf{K}_e^\star$.

Control Law

At last the control law [13] can be described through

$$u_{ad,1}^\star = C_1^\star(s)\,\hat{\sigma}_m^\star(s) \tag{24}$$

$$u_{ad,2}^\star = C_2^\star(s)\,\mathbf{H}_m^{\star -1}(s)\,\mathbf{H}_{um}^\star(s)\,\hat{\sigma}_{um}^\star \tag{25}$$

where

$$\mathbf{H}_m^{\star}(s) = \mathbf{c}_m^{\star}\left(s\mathbf{I}_n - \mathbf{A}_m^{\star}\right)^{-1}\mathbf{b}_m^{\star} \tag{26}$$

$$\mathbf{H}_{um}^{\star}(s) = \mathbf{c}_m^{\star}\left(s\mathbf{I}_n - \mathbf{A}_m^{\star}\right)^{-1}\mathbf{b}_{um} \tag{27}$$

with the output vector $\mathbf{c}_m^{\star} = \left[\,1\ 0\,\right]$.

5 Simulation Results

In this section both the performance and the robustness properties of the proposed augmentation structures are shown. In order to not exceed the framework of this study, only one exemplary envelope point is considered for evaluation. The aircraft model described in Section 2 is trimmed at the velocity $V_0 = 177\frac{m}{s}$ in an altitude of $h_0 = 5000m$. Thus the simulation starts at $Ma \approx 0.55$.

The sample time is uniformly set to $T_s = 0.001s$ for the whole simulation environment, which includes the L1 Adaptive Controller. The Low-Pass Filters for the matched uncertainty is set to $\omega_{C,1} = 40\frac{rad}{s}$, which is the actuator bandwidth. The filters corresponding to the unmatched uncertainties are adjusted to $\omega_{C,2} = 1\frac{rad}{s}$ and $\omega_{C,3} = 0.01\frac{rad}{s}$ with the intention to tune the robustness properties w.r.t. the plant output channels. The additional error feedback gains are determined as $k_i = -230\ \forall i \in \{1;2;3\}$ after some manual tuning. It has to be noted that a reasonable choice of this gain is highly depended on the sample time of the controller. That means the smaller the sample time is set, the faster the error dynamics can be rendered.

5.1 *Performance*

In order to assess the performance, the controllers are interconnected with the nonlinear F16 aircraft model. The simulations are conducted for the nominal aircraft and for two different CG-shifts of 5% and 7% aft. The uncertainty is implemented as a step function, which is activated at $t = 10s$, in order to model a rapid change in system dynamics. Next to simulation plots, the performance metrics

$$M_{\mathcal{L}_2} = \left\|\alpha(t) - \alpha_{ref}(t)\right\|_{\mathcal{L}_2} = \sqrt{\int_{t_0}^{t_f}\left(\alpha(t) - \alpha_{ref}\right)^2 dt} \tag{28}$$

$$M_{\mathcal{L}_\infty} = \left\|\alpha(t) - \alpha_{ref}(t)\right\|_{\mathcal{L}_\infty} = \max_{t\in\left[t_0,t_f\right]}\left|\alpha(t) - \alpha_{ref}(t)\right| \tag{29}$$

are used to compare the different controller configurations [24]. $M_{\mathcal{L}_\infty}$ measures the maximum value of the control error, $M_{\mathcal{L}_2}$ is a measure for its energy. This means in both cases *the smaller the value, the better the performance*.

The simulation results for the nominal aircraft are depicted in Figure 6. The figure contains the history of angle of attack α for the configurations Baseline Controller only, *DPI Augmentation* and *Plant Augmentation*. α should follow the black dotted line for α_{ref} as close as possible. It can be seen, that the performance is not degraded by the application of adaptive augmentation for both cases. On the contrary, it even leads to slight improvements in terms of tracking, which is also confirmed by the performance metrics given in Table 1. Furthermore, the adaptive augmentation generates reasonable control signals without any oscillation. This is also illustrated in Figure 6 by both the plots of commanded actuator rate $\dot{\eta}_{cmd}$ and the integrated deflection η_{cmd} after application of rate limitation. Although $\dot{\eta}_{cmd}$ exceeds the limit several times, the performance of the adaptive controller is not perturbed.

Table 1 Comparison of performance metrics

	nominal		5% CG-Shift		7% CG-Shift	
Controller	$M_{\mathcal{L}_2}$ [-]	$M_{\mathcal{L}_\infty}$ [-]	$M_{\mathcal{L}_2}$ [-]	$M_{\mathcal{L}_\infty}$ [-]	$M_{\mathcal{L}_2}$ [-]	$M_{\mathcal{L}_\infty}$ [-]
Baseline	1.9957	0.0347	11.8138	0.1755	164.8498	2.1355
L1 DPI Augmentation	0.8759	0.017	1.6612	0.0358	2.8796	0.0516
L1 Plant Augmentation	0.8609	0.0144	1.7646	0.0375	3.3317	0.0583

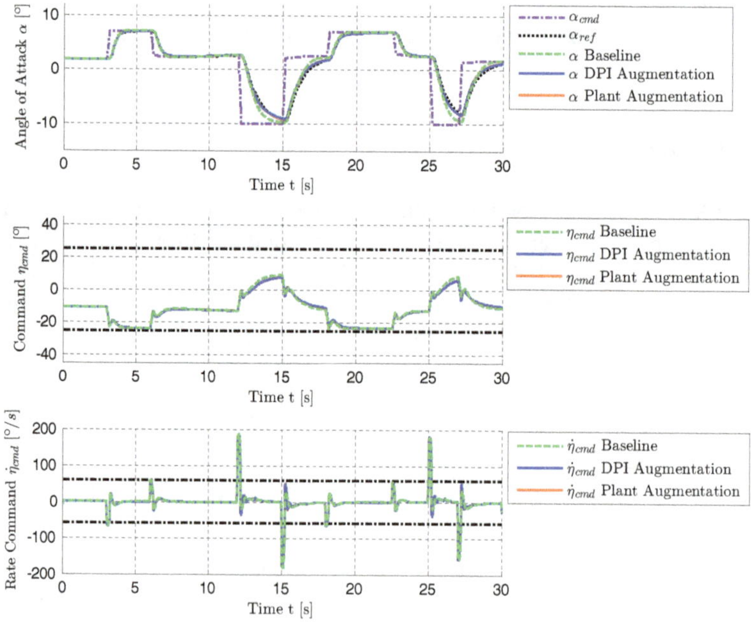

Fig. 6 Angle of attack and command: Nominal aircraft

The second case to be investigated is a CG-shift of 5% aft, which renders the open-loop aircraft unstable. The simulation results are outlined in Figure 7. The baseline controller still manages to stabilize the aircraft, albeit heavy overshoot is induced and α is oscillating, which is also reflected by an increase of $M_{\mathscr{L}_\infty}$ and $M_{\mathscr{L}_2}$, respectively. Both augmentation approaches are able to compensate for the uncertainty in reasonable time and restore good handling qualities. Differences between the two augmentation approaches are barely visible, only the performance metrics reveal a slight advantage of the *DPI Augmentation*.

At last a CG-shift of 7% aft is examined to exhaust the possibilities of the adaptive augmentation for the considered aircraft model. It has to be noted, that the capability of the actuator in terms of the possible control momentum and how fast the momentum can be build up sets this limit. This is not due to limitations of the adaptive augmentation. In Figure 9 it can be seen that activation of uncertainty makes the combination of baseline controller and aircraft unstable, whereas both augmentation approaches stabilize the aircraft and perform even when the deflection saturation is hit at $t \approx 15s$ and $t \approx 27s$ for a short period of time. Still *DPI Augmentation* and *Plant Augmentation* behave quite similar, whereas the latter exhibits slightly larger overshoot at $t \approx 18s$, $t \approx 23s$ and $t \approx 26s$. This also manifests in the performance metrics in Table 1. The history of the parameter estimations for this case is depicted in Figure 8. It is no surprise that $\hat{\sigma}_{um,2}$ remains close to zero, as it corresponds to the state e_I, which contains no uncertainty by design.

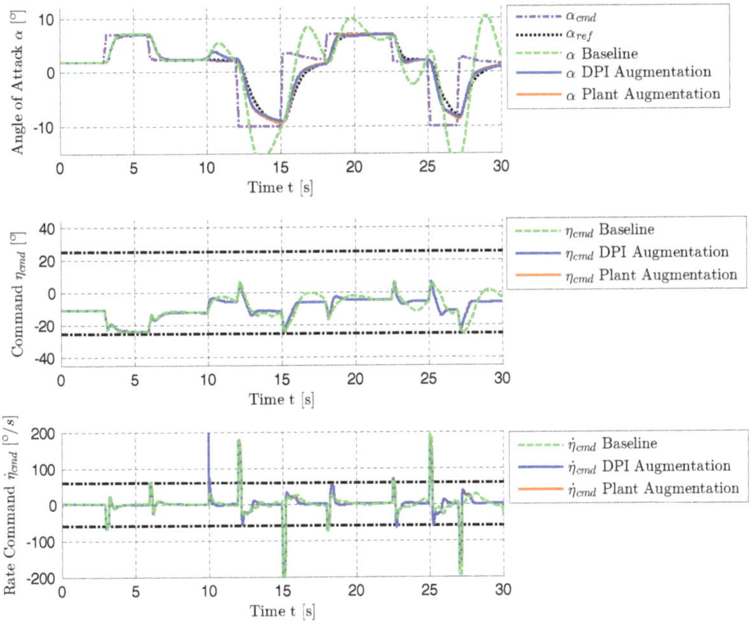

Fig. 7 Angle of attack and command: 5% CG-shift

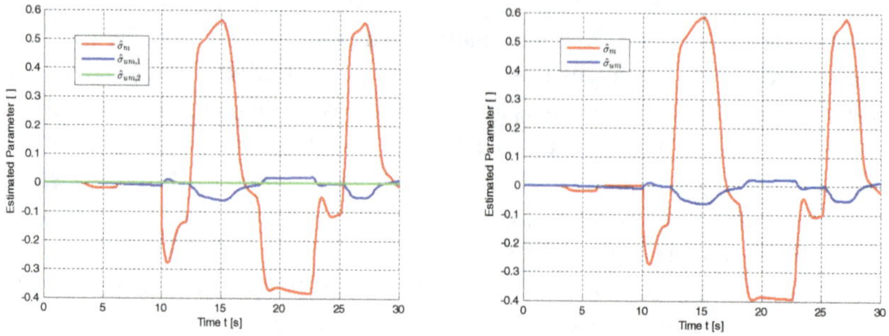

(a) DPI Augmentation (b) Plant Augmentation

Fig. 8 Parameter estimations of the L1 Adaptive Controller for 7% CG-shift

Additional experiments have shown that introducing the hedging term Δu has no effect on the overall performance of the controller.

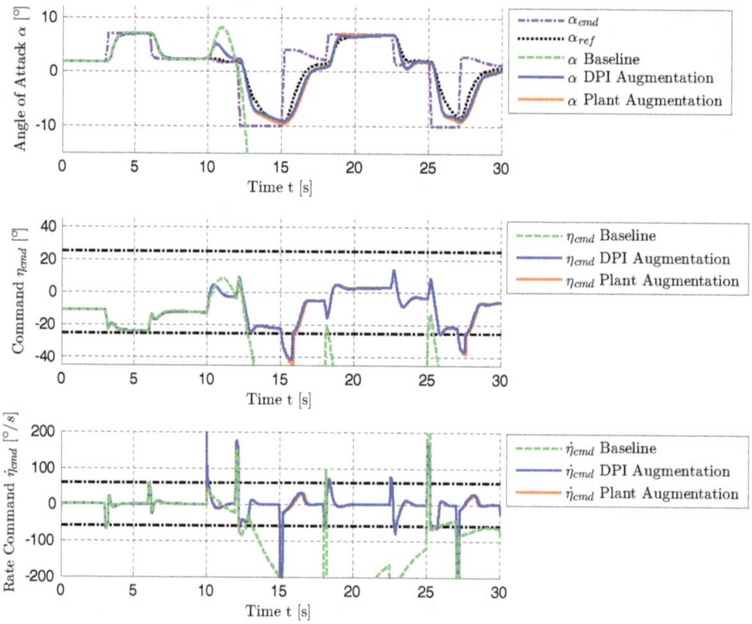

Fig. 9 Angle of attack and command: 7% CG-shift

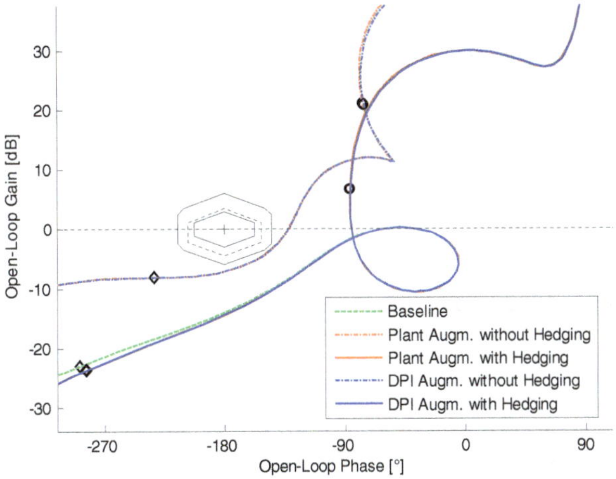

Fig. 10 Nichols plot - Open-Loop assessment at actuator input

5.2 *Robustness*

The assessment of the robustness properties of the proposed adaptive augmentation methods is performed in this section. These are measured by Time-Delay-Margin (TDM) and Gain-Margin (GM). Unlike for the resulting performance, the hedging term Δu decisively influences the robustness of the closed-loop system. Thus, it is always distinguished between controller configurations with and without hedging ($\Delta u = 0$) in the following statements.

Since the augmentation approaches use piecewise constant update laws, the closed-loop can be analyzed by the methods of linear control theory. This can also be an advantage in terms of certification issues, which are commonly associated with adaptive control laws. Nonlinear robustness measurements are performed additionally, to show comparability to the results gained by linear assessment.

The controllers are compared at the same envelope point as for the performance benchmark, which is $V_0 = 177\frac{m}{s}$ and $h_0 = 5000m$. The following evaluation considers the nominal CG position.

5.2.1 Linear Assessment

In order to be analyzed by linear methods, the closed-loops of the different controller configurations presented above and also considering activated and deactivated hedging are numerically linearized. The loop is at first cut at the actuator input η_{cmd} to measure the robustness at this location.

For the resulting systems Nichols plots are generated, which are shown in Figure 10. The required robustness margins are chosen according to [18] and are illustrated through the diamonds around the critical point $[-180^\circ \; 0dB]$. The relevant margin

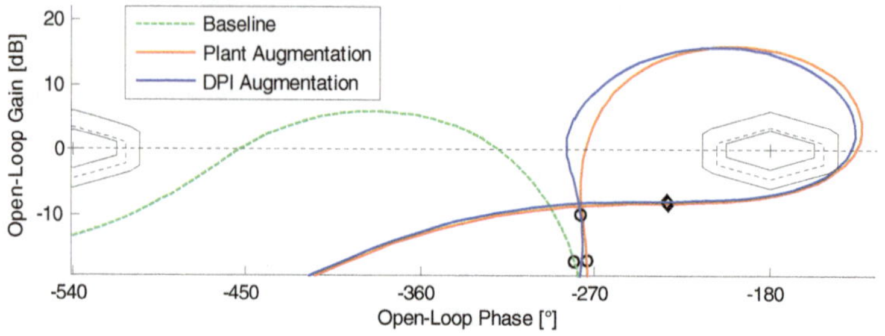

Fig. 11 Nichols plot - Open-Loop assessment at plant output q

for this assessment is marked by the outermost diamond, whereas the two smaller diamonds are applied e.g. in case of failure. The black circles show the frequency of the phugoid and the black diamonds are located at the estimated frequency of first wing-bending mode. First it can be seen quite clearly that the curves of the baseline controller and the adaptive augmentations with hedging barely differ in the relevant frequency range. Deactivation of hedging leads to crossing of the $0dB$ line far more left, which indicates degraded robustness. Although the minimal robustness requirements are still met, because the outermost diamond is not cut by the frequency response curve.

As the phase margin is not determinable in the nonlinear world, it is chosen to compare the TDM in addition to GM. The TDM is calculated from the phase margin Φ_m and the gain crossover frequency ω_{gc} by the following formula [1]:

$$TDM = \frac{\Phi_m}{\omega_{gc}} \tag{30}$$

The results are summarized in Table 2. It can be seen that both TDM and GM drop significantly without applied hedging for the adaptive augmentations. These are fully restored in comparison to the baseline controller by activation of hedging. As for the performance, there are only slight differences in the robustness properties in the two different augmentation methods.

As opposed to the plant input channel, the proposed hedging strategy has no effect on the robustness properties of the output channels, namely angle of attack α and pitch rate q. Especially for the pitch rate channel a significant drop in both TDM and GM can be observed, while the robustness requirements are still fulfilled. This can be seen in Figure 11. The baseline controller has a $TDM = 0.266s$, which drops to $TDM = 0.07s$ for the DPI Augmentation and to $TDM = 0.071s$ for the Plant Augmentation. The GM changes from $[13.47 \ -\infty]$ to $[6.88 \ -14.86]$ and $[7.12 \ -15.45]$, respectively.

Table 2 Linear robustness assessment: TDM and GM

	without Hedging		with Hedging	
Controller	TDM [s]	GM [dB]	TDM [s]	GM [dB]
Baseline	0.431	$[13.97\ -\infty]$	—	—
L1 DPI Augmentation	0.082	$[7.08\ -\infty]$	0.436	$[14.34\ -\infty]$
L1 Plant Augmentation	0.081	$[7.07\ -\infty]$	0.435	$[14.34\ -\infty]$

Table 3 Nonlinear robustness assessment: TDM and GM

	without Hedging		with Hedging	
Controller	TDM [s]	GM [dB]	TDM [s]	GM [dB]
Baseline	0.426	$[13.86\ -\infty]$	—	—
L1 DPI Augmentation	0.081	$[6.74\ -\infty]$	0.43	$[13.5\ -\infty]$
L1 Plant Augmentation	0.08	$[6.71\ -\infty]$	0.429	$[13.5\ -\infty]$

5.2.2 Nonlinear Assessment

Due to space limitations only the plant input channel is analyzed in this section. Yet, there is no generic way known to calculate the robustness properties TDM and GM mentioned above for a nonlinear system. Thus, they have to be determined using simulations. Therefor the actuator input is perturbed with artificial delay and gain. At the same time a slight deviation is applied on the trim deflection η_0. One can find TDM by systematically increasing the delay, analogous GM is found through increase of the gain. In both cases the margins are reached, if unstable oscillation in the system response can be detected. The drawback of this method is, that it obviously leads to a significant simulation effort for large systems.

The results gained under the same conditions used for the linear assessment are summarized in Table 3. The statement remains the same, the beneficial effect of hedging on the robustness can as well be detected at the nonlinear system. Furthermore the applied method shows good comparability to the results gained by linear methods in Table 2.

Finally, a method is presented to visualize the offset between the required and the actual robustness margins for nonlinear systems. Its aim is to give a similar representation of the margin between frequency response and appropriate diamond within Nichols plots for linear systems. The results are illustrated in Figure 12. Again the margins for the different controller configurations are determined by application of artificial gain and time delay at the plant input channel. In contrast to the assessment made above, the two parameters are now varied at the same time. Thus, the resulting points define the boundary between stability and instability in the whole space spanned by time delay and gain. The required margin is defined analog to the diamond in Figure 10 with the only difference, that the phase margin is replaced with a certain TDM limit. The minimum required TDM is determined according to [5] by

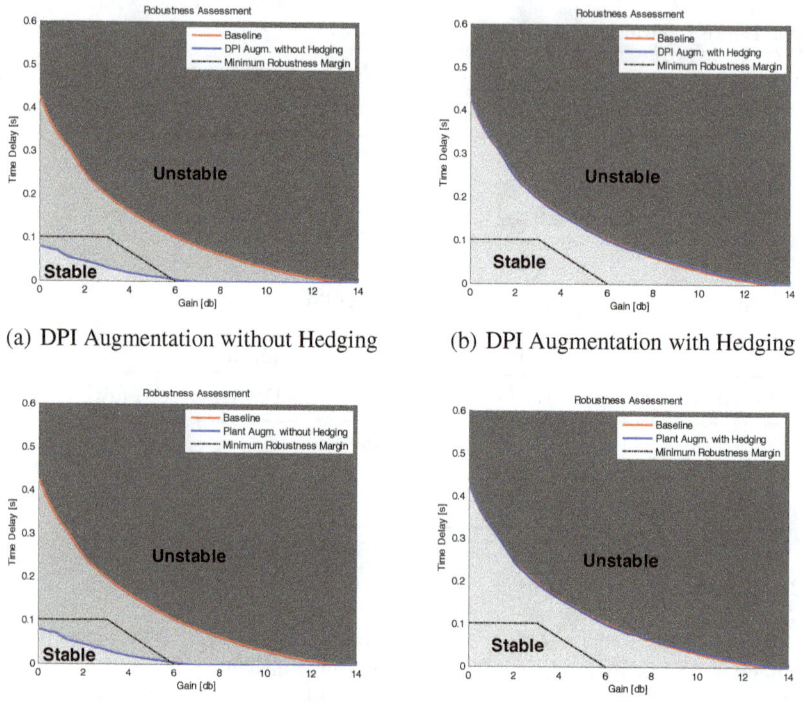

(a) DPI Augmentation without Hedging (b) DPI Augmentation with Hedging

(c) Plant Augmentation without Hedging (d) Plant Augmentation with Hedging

Fig. 12 TDM over GM for nonlinear assessment of robustness

demanding a phase margin of 35° at the system bandwidth, which is the frequency of the short-period. Then the TDM can be calculated using Equation (30), resulting in a $TDM \approx 0.1s$ for this case.

It can be clearly seen that there is sufficient distance between the minimal required margin and the margins for Baseline Controller and Augmentations with hedging applied. As for the linear assessment, hedging nearly leads to full match of the two curves, whereas for deactivated hedging the requirements are not met. Comparing this result to the linear assessment, which indicates sufficient margin, one could conclude for this case, the proposed nonlinear method leads to a more conservative statement on robustness margin fulfillment.

6 Conclusion

In this paper two different adaptive augmentations for a DPI baseline controller are presented using L1 Adaptive Control. It is their task to compensate uncertainties within the system. In this case rapid CG-shifts are exemplarily chosen. The approaches are both successfully applied on a nonlinear model of a F16 aircraft. Simulation results show only slight differences between the two approaches for the considered uncertainty in terms of performance. Thus, it is possible to simplify the design of the state predictor and also the parameter update law and the control law of the L1 Adaptive Controller choosing the *Plant Augmentation* approach. The controllers perform well within the physical limits of the aircraft, even for large CG-shifts and with the actuator hitting the saturation limit. One can state, the DPI baseline controller is well suited for adaptive augmentation, especially when actuator limitations play an important role for the application. The robustness properties are also examined by both linear and nonlinear methods. It is shown, that the robustness in the plant input channel can be restored by introduction of an additional hedging term, which contains feedback of the measured actuator deflection.

References

1. Bates, D., Postlethwaite, I.: Robust Multivariable Control of Aerospace Systems. DUP Science, Delft, Netherlands (2002)
2. Bichlmeier, M., Holzapfel, F., Xargay, E., Hovakimyan, N.: L1 Adaptive Augmentation of a Helicopter Baseline Controller. In: Guidance, Navigation, and Control and Co-located Conferences, American Institute of Aeronautics and Astronautics (August 2013)
3. Bierling, T.: Comparative Analysis of Adaptive Control Techniques for Improved Robust Performance. PhD thesis, TU München (2014)
4. Brockhaus, R., Alles, W., Luckner, R.: Flugregelung, 3rd edn. Springer, Berlin (2011)
5. Dorobantu, A., Seiler, P., Balas, G.J.: Time-Delay Margin Analysis for an Adaptive Controller. Journal of Guidance, Control, and Dynamics 35(5), 1418–1425 (2012)
6. Farineau, J.: Lateral Electric Flight Control Laws of a Civil Aircraft Based Upon Eigen Structure Assignment Technique. In: AIAA Guidance, Navigation and Control Conference, American Institute of Aeronautics and Astronautics (1989)
7. Geiser, M., Xargay, E., Hovakimyan, N., Bierling, T., Holzapfel, F.: L1 Adaptive Augmented Dynamic Inversion Controller for a High Agility UAV. In: Guidance, Navigation, and Control and Co-located Conferences, American Institute of Aeronautics and Astronautics (August 2011)
8. Gregory, I., Cao, C., Xargay, E., Hovakimyan, N., Zou, X.: L1 Adaptive Control Design for NASA AirSTAR Flight Test Vehicle. In: Guidance, Navigation, and Control and Co-located Conferences, American Institute of Aeronautics and Astronautics (August 2009)
9. Gregory, E., Xargay, C.: Flight Test of an L1 Adaptive Controller on the NASA AirSTAR Flight Test Vehicle. In: Guidance, Navigation, and Control and Co-located Conferences. American Institute of Aeronautics and Astronautics (August 2010)
10. Gregory, I., Xargay, E., Cao, C., Hovakimyan, N.: Flight Test of L1 Adaptive Control Law: Offset Landings and Large Flight Envelope Modeling Work. In: Guidance, Navigation, and Control and Co-located Conferences, American Institute of Aeronautics and Astronautics (August 2011)
11. Griffin, B., Burken, J., Xargay, E.: L1 Adaptive Control Augmentation System with Application to the X-29 Lateral/Directional Dynamics: A Multi-Input Multi-Output Approach. In: Guidance, Navigation, and Control and Co-located Conferences. American Institute of Aeronautics and Astronautics (2010)
12. Holzapfel, F., Heller, M., Weingartner, M., Sachs, G., da Costa, O.: Development of control laws for the simulation of a new transport aircraft. In: Proceedings of the Institution of Mechanical Engineers, Part G: Journal of Aerospace Engineering, vol. 223, pp. 141–156 (2009)

13. Hovakimyan, N., Chengyu, C.: L1 Adaptive Control Theory - Guaranteed Robustness with Fast Adaptation. Society for Industrial and Applied Mathematics (2010)
14. Johnson, E.N.: Limited Authority Adaptive Flight Control. PhD thesis, Georgia Institute of Technology (2000)
15. Lee, H.P., Youssef, H.M., Hanel, R.P.: Application of Eigenstructure Assignment to the Design of STOVL Flight Control System. In: Guidance, Navigation and Control Conference, American Institute of Aeronautics and Astronautics (1988)
16. Leman, T., Xargay, E., Dullerud, G., Hovakimyan, N., Wendel, T.: L1 Adaptive Control Augmentation System for the X-48B Aircraft. In: Guidance, Navigation, and Control and Co-located Conferences, American Institute of Aeronautics and Astronautics (August 2009)
17. Li, D., Hovakimyan, N., Cao, C.: L1 Adaptive Controller in the Presence of Input Saturation. In: Guidance, Navigation, and Control and Co-located Conferences. American Institute of Aeronautics and Astronautics (August 2009)
18. Moritz, N., Osterhuber, R.: Three-Stage Gradient-Based Optimization Scheme in Design of Feedback Gains within Eurofighter Primary Control Laws. In: Guidance, Navigation, and Control and Co-located Conferences, American Institute of Aeronautics and Astronautics (August 2006)
19. Nguyen, L.T., Ogburn, M.E., Gilbert, W.P., Kibler, K.S., Brown, P.W., Deal, P.L.: Simulator study of stall/post-stall characteristics of a fighter airplane with relaxed longitudinal static stability. Technical report, NASA Langley Research Center, Hampton, Virginia (1979)
20. Oelker, H.-C., Osterhuber, R., Hanel, M.: Experiences with Eurofighter Handling Qualities Testing. In: Guidance, Navigation, and Control and Co-located Conferences, American Institute of Aeronautics and Astronautics (August 2009)
21. Osterhuber, R., Hanel, M., Hammon, R.: Realization of the Eurofighter 2000 Primary Lateral/Directional Flight Control Laws with Differential PI-Algorithm. In: Guidance, Navigation, and Control and Co-located Conferences. American Institute of Aeronautics and Astronautics (August 2004)
22. Peter, F., Holzapfel, F., Xargay, E., Hovakimyan, N.: L1 Adaptive Augmentation of a Missile Autopilot. In: Guidance, Navigation, and Control and Co-located Conferences. American Institute of Aeronautics and Astronautics (August 2012)
23. Seiler, P., Dorobantu, A., Balas, G.: Robustness Analysis of an L1 Adaptive Controller. In: Guidance, Navigation, and Control and Co-located Conferences. American Institute of Aeronautics and Astronautics (August 2010)
24. Stepanyan, V., Krishnakumar, K., Nguyen, N., Eykeren, L.V.: Stability and Performance Metrics for Adaptive Flight Control. In: Guidance, Navigation, and Control and Co-located Conferences. American Institute of Aeronautics and Astronautics (August 2009)
25. Stevens, B., Lewis, F.: Aircraft Control and Simulation. John Wiley & Sons (2003)
26. Wise, K., Lavretsky, E., Hovakimyan, N., Cao, C., Wang, J.: Verifiable Adaptive Control: UCAV and Aerial Refueling. In: Guidance, Navigation, and Control and Co-located Conferences. American Institute of Aeronautics and Astronautics (August 2008)
27. Xargay, E., N., Hovakimyan, C.C.: L1 Adaptive Controller for Multi-Input Multi-Output Systems in the Presence of Nonlinear Unmatched Uncertainties. In: American Control Conference (ACC), pp. 874–879 (2010)
28. Xargay, E., Hovakimyan, N., Dobrokhodov, V., Statnikov, R., Kaminer, I., Cao, C., Gregory, I.: L1 Adaptive Flight Control System: Systematic Design and Verification and Validation of Control Metrics. In: Guidance, Navigation, and Control and Co-located Conferences, American Institute of Aeronautics and Astronautics (August 2010)

Flight Path Management System of EOLE UAV*

Frank Jouhaud

Abstract. Within the PERSEUS project of CNES (French National Space Centre), EOLE is a demonstrator for a reusable unmanned automatic carrier for airborne rocket launchers. EOLE will lift a small scale rocket from an airport up to 4000 m in altitude. A ground pilot controls EOLE for take off and landing manually via radio-controls at viewing distance. The guidance and autopilot system is then switched on for autonomous flight including climb, cruise flight, launch and separation maneuver, return flight and initial descent. This paper deals more particularly with the design of the guidance for the flight path management system, which was successfully used in flight tests.

Keywords: EOLE - UAV dynamics, control, guidance and navigation - Airborne Launcher

1 Introduction

PERSEUS is the European Academic and Scientific Space Student Research Project initiative of the CNES (French National Space Centre, see [2]). Within this context, EOLE is a UAV (unmanned automatic vehicle), demonstrator of a reusable carrier for an airborne rocket launchers. The development of EOLE is supervised by both CNES and the French Aerospace Lab Onera, see [5]. EOLE will lift a rocket from an airport up to 4000 m in altitude. A ground pilot controls EOLE take off and landing manually via radio-controls at viewing distance. The guidance and autopilot system is then switched on for autonomous flight including initial climb, cruise flight, launch and separation maneuver, return flight and final descent.

Frank Jouhaud
The French Aerospace Lab, ONERA/DCSD, Toulouse, France
e-mail: Frank.Jouhaud@onera.fr

* This study was supported by the French Space Agency CNES.

J. Bordeneuve-Guibé et al. (eds.), *Advances in Aerospace Guidance, Navigation and Control*, DOI: 10.1007/978-3-319-17518-8_8

At first a guidance system was defined (see [3]) for an automatic separation maneuver (see Fig. 1) which occurs at a high flight path angle of about 45 degrees ensuring a quick separation between the carrier Eole and the rocket such that Eole avoids the plume of the rocket.

As the separation maneuver must occur in a segregated airspace, out of view and with long outbound and return flights the guidance system was augmented in order to manage the monitoring of a flight plan with climbs and descents.

This article presents the details of the guidance law function which is divided in two parts (see Fig. 2):

- inner loops for the respect of short term objectives: regulation of altitude in the neighborhood of a target altitude (output is desired load factor n_{z_d}), regulation of ground track relatively (output is roll angle $\phi_\mathbf{d}$ desired value) to a prescribed ground track. If the altitude is too far from the target altitude, the altitude regulation law is switched to the respect of an ascent or descent profile with minimum and maximum air speed protection. The output is a desired flight path angle ($\gamma_\mathbf{d}$);
- flight plan monitoring: a sequencer computes, from the actual position and the current leg between two way points (defined by a their position with target values on altitude and air speed), the objectives for the inner loops. When the initial position is too far from any leg of the flight plan, then additional way points can be inserted. Between two legs, a circular turn is defined.

The outputs of the above guidance functions are either transmitted directly as inputs to a previously defined attitude control law (load factor $\mathbf{n}_{\mathbf{z_c}}$, side slip $\beta_\mathbf{c}$, roll angle $\phi_\mathbf{c}$ and equivalent air speed $\mathbf{V}_{\mathbf{a_c}}$ orders, see [6]) or either through a low level guidance loop linking the desired flight path angle to an order on load factor. This low level loop was also already defined for the separation maneuver (see [3]).

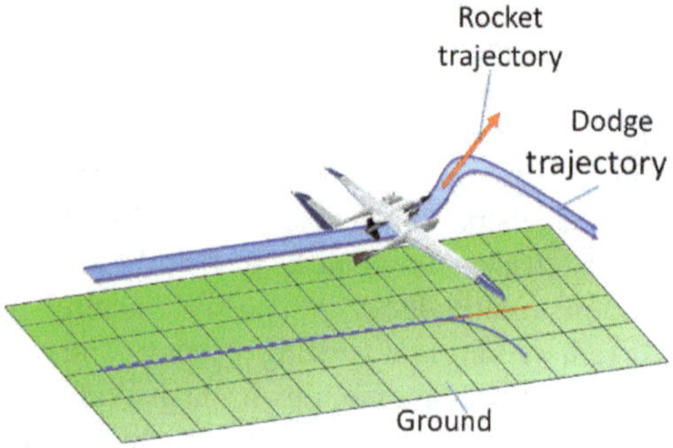

Fig. 1 Separation maneuver

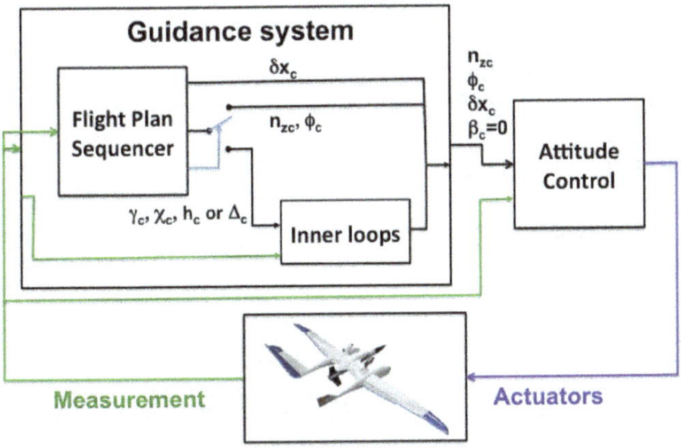

Fig. 2 Structure of Guidance System

List of Symbols

α	angle of attack
β	side slip
γ	air speed flight path
δ_l	aileron command
δ_m	elevator command
δ_n	rudder command
δ_x	reduced thrust command
λ	geocentric latitude
χ	air speed azimuth
μ	aerodynamic bank angle
ϕ	roll angle
ρ	air density
ρ_0	ground air density
θ	pitch angle
δ_T	engine pitch angle
C_D	drag coefficient, aerodynamic frame
C_L	lift coefficient, aerodynamic frame
g	ground gravity acceleration
G	longitude
h	altitude
n_x	total axial load factor, aerodynamic frame
n_z	normal load factor, aerodynamic frame
r	radius from Earth Center
V_a	true air speed
V_{eas}	equivalent air speed

x, y local horizontal North, East coordinates
T thrust
S_{ref} reference area of the vehicle

2 Modeling for Guidance

In order to shape the trajectory and tune the inner guidance laws, we used the following 3 degrees of freedom simplified model where attitude motion is supposed instantaneous (see [7, 4]):

$$\dot{r} = V_a \sin\gamma \, [= \dot{h}] \tag{1}$$
$$r\,\dot{\lambda} = V_a \cos\gamma \cos\chi \, [= \dot{x}] \tag{2}$$
$$r\,\dot{G} \cos\lambda = V_a \cos\gamma \sin\chi \, [= \dot{y}] \tag{3}$$

$$\dot{V_a} = g\, n_x - g \sin\gamma \tag{4}$$
$$V_a\,\dot{\gamma} = g\, n_z \cos\mu - g \cos\gamma \tag{5}$$
$$V_a\,\dot{\chi} = g\, n_z \frac{\sin\mu}{\cos\gamma} \tag{6}$$

where axial and normal load factors are respectively:

$$n_x = \frac{-D + \cos(\alpha + \delta_T)\, T}{m\, g} \simeq \frac{-D + T}{m\, g} \tag{7}$$
$$n_z = \frac{L + \sin(\alpha + \delta_T) T}{m\, g} \simeq \frac{L}{m\, g} \tag{8}$$

and:

$$\begin{aligned}
\text{Drag Force} \quad D &= p_{dyn}\, S_{ref}\, C_D \\
\text{Lift Force} \quad L &= p_{dyn}\, S_{ref}\, C_L \\
\text{Dyn. Pressure} \quad p_{dyn} &= \frac{1}{2}\,\rho\, V_a^2 = \frac{1}{2}\,\rho_0\, V_{eas}^2 \\
\text{Equivalent Air Speed} \quad V_{eas} &= \sqrt{\frac{\rho}{\rho_0}}\, V_a
\end{aligned}$$

This model is used only for the synthesis of guidance laws, for **validation**, the so synthesized guidance laws are used on the **complete** 6 degrees of freedom simulator.

The local horizontal North (x) and East (y) coordinates are defined using a chosen close reference point of latitude λ_{ref} and longitude G_{ref} and radius of Earth r_E:

$$x = r_E\, (\lambda - \lambda_{ref}) \tag{9}$$
$$y = r_E \cos\lambda_r\, (G - G_{ref}) \tag{10}$$

Given a desired flight path angle dynamic $\dot{\gamma}_d$, from (5), one can extract the desired value for n_{z_d} and similarly, given a desired azimuth dynamic $\dot{\chi}_d$, (6) yields the desired bank angle μ_d:

$$n_{z_d} = \frac{V_a \, \dot{\gamma}_d + g \, \cos\gamma}{g \, \cos\mu} \tag{11}$$

$$\sin\mu_d = \frac{V_a \, \cos\gamma}{g \, n_z} \, \dot{\chi}_d \tag{12}$$

3 Guidance Laws for Flight Path Monitoring

3.1 Longitudinal Guidance Laws

3.1.1 Altitude Keeping

Altitude keeping is obtained through a PID (Proportional, Integral and Derivative) controller toward a target altitude h_d:

$$n_{z_{ac}} = \left[1 + k_{h_p} \, (h - h_d) + k_{h_d} \, V_a \, \sin\gamma + k_{h_i} \int (h - h_d) \, dt\right] \frac{1}{\cos\mu_c} \tag{13}$$

Although this linear regulation shall in theory deal with any discrepancy in altitude, when the current altitude is too far from the target altitude, in practice, it is better to commute to an ascent or descent optimal trajectory, as presented in the next section.

3.1.2 Ascent and Descent Profiles

Several kinds of optimal trajectories can be defined, according to the chosen criteria:

1. maximum flight path angle: useful after take-off for obstacle avoidance;
2. minimum ascent or descent time. A slightly suboptimal solution corresponds to a trajectory with almost constant air speed and flight path angle and with an extremum vertical speed;
3. minimum time to reach both a target altitude and a target final air speed: optimal computation necessary.

In order to avoid delicate and costly optimal computation the extremum vertical speed solution is retained as it is compatible with the mission duration.

The mathematical representation of the problem is then:

Find $\mathbf{n_z}$ such that:

- $J = V_a \sin\gamma$ *is extremum*
- $\dot{V}_a = 0$ *constant speed*
- $\dot{\gamma} = 0$ *constant flight path angle*

The last two constraints, using (4) and (5), leads to a relation for speed independant of flight path angle:

$$\lambda^2\,(C_L^2+C_D^2)\,V_a^4-2\,\lambda\,C_D\,\bar{T}\,V_a^2+\bar{T}^2-1=0 \tag{14}$$

where: $\lambda(h)=\frac{1}{2}\,\frac{\rho(h)\,S}{m\,g}$, $\bar{T}=\frac{T_{max}}{mg}$ for ascent or $\frac{T_{min}}{mg}$ for descent (T_{max} is maximum continuous trust, ie 90% of maximum thrust available; T_{min} is the residual thrust at idle level).

The solution of this equation is given by:

$$V_a=\frac{\lambda\,C_D\,\bar{T}+\sqrt{\lambda^2\,C_D^2\,\bar{T}^2+(1-\bar{T}^2)\,\lambda^2\,(C_L^2+C_D^2)}}{\lambda^2\,(C_L^2+C_D^2)} \tag{15}$$

Hence the criterion J can be now rewritten as:

$$J=V_a\,(\bar{T}-\lambda\,C_D\,V_a^2) \tag{16}$$

The extremum of the criterion is then seeked numerically by iterations on C_L. The best value is called $C_{L\,ref}$.

The corresponding flight path angle is then:

$$\gamma_{ref}(h)=\arccos(\lambda\,C_{L\,ref}\,V_a^2) \tag{17}$$

All the parameters are depending on the altitude and so is the resulting flight path angle. Nevertheless, on Fig. 3, one can see that the variations are slow and compatible with the initial constraint of constant speed and path angle.

On this Fig. 3, one can see the effect of an arbitrary adjustment in order to reduce the minimum time to reach 4000 m in altitude and a maneuver speed of 80 m/s. The effect is a reduction of ascent time of about 50 s for an ascent of 7 min in nominal conditions. One can note also on Fig. 3 that the angle of ascent becomes negative just before an altitude of 6000 m : the trajectory is optimized for a climb up to 4000 m. For flying higher, it would be necessary to lower the speed by keeping a higher, positive, value of flight path angle.

The tabulated result is directly an input for the inner loop of the guidance law.

3.1.3 Air Speed Protection

An air speed protection is implemented because, for some combinations of input, the reference profiles computed above can lead to a too low speed in ascent (while at full throttle) or to a too high speed in descent (while at idle throttle). Furthermore, for the early flight tests, it is necessary to stay within viewing distance of the security pilot with a limited speed range.

As stated before, this protection is activated when the throttle are at their extrema, so the only remaining available parameter is the flight path angle : it has to be adjusted in order to regulate the speed toward the target value.

A Proportionnel and Integral (PI) regulation is chosen:

$$\gamma_c=\gamma_{ref}(h)+k_{v_p}\,(V_a-V_d)+k_{v_i}\int(V_a-V_d)\,dt \tag{18}$$

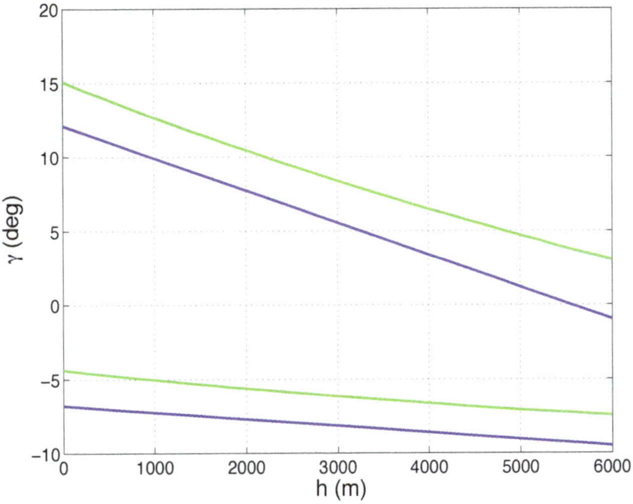

Fig. 3 Ascent (positive values) and descent profiles (negative values) *(dotted line: vertical speed optimum, solid line: adjusted optimum)*

When it is needed to respect a maximum speed v_{max} lower than the achievable maximum speed $\bar{v}_{max}$ of the vehicle, then:

- in descent: $\frac{v_{max}}{\bar{v}_{max}}\,\gamma_{ref}(h)$ is taken as new reference path angle;
- in ascent: $\gamma_{ref}(h)$ is kept but maximum value of throttle is limited at the following value: $\frac{v_{max}}{\bar{v}_{max}}\,\delta x_{max}$

In order to avoid oscillations different threshold are taken on speed for entering and leaving this mode.

3.2 Lateral Guidance Laws

Lateral guidance laws are elaborated in order to follow an assigned ground track made of several straight legs linked by circular arcs. The initial flight point when initiating flight plan monitoring does not need to be close to the beginning of the flight plan, see Fig. 4. The explicit definition of a link circle, see Fig. 5, instead of only a heading turn, allows to be more robust to unknown wind and to smooth the trajectory.

The ground track is followed using modulation of roll angle μ in order to regulate the cross distance Δ to the track (see Figs. 6 and 7). A proportional and derivative (PD) regulator is chosen:

$$\ddot{\Delta} = K_p\,\Delta + K_d\,\dot{\Delta} \tag{19}$$

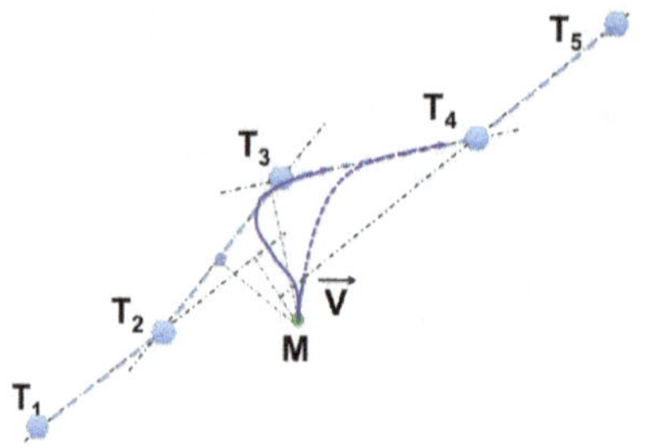

Fig. 4 Example of the ground track of a flight plan

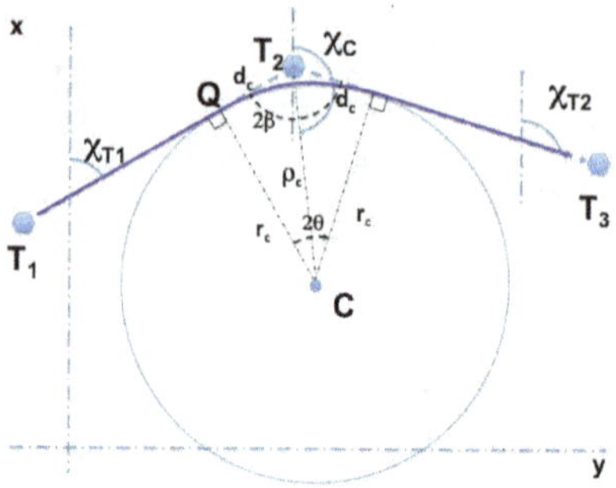

Fig. 5 Link turn

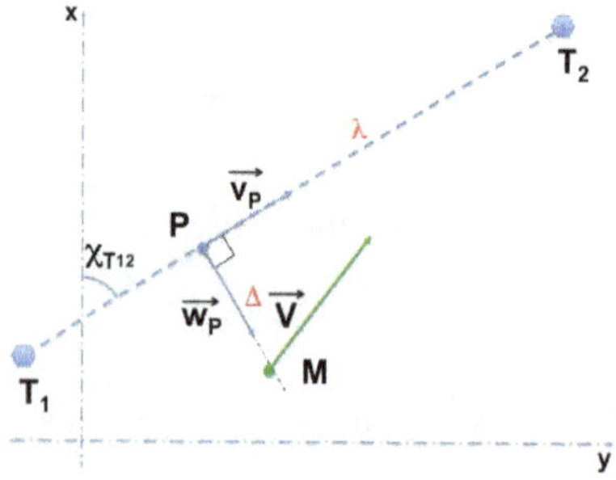

Fig. 6 Straight ground track

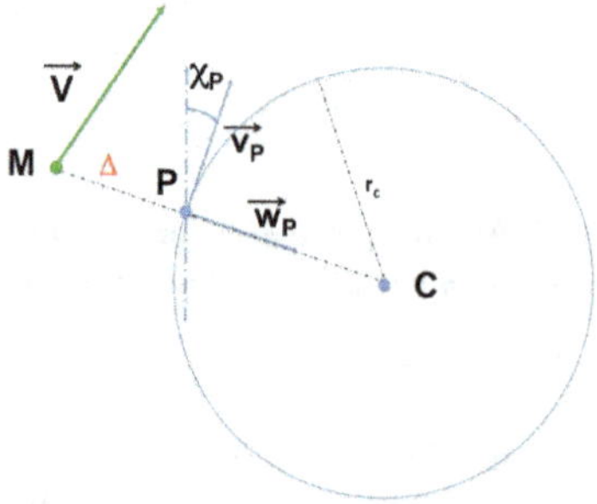

Fig. 7 Circular arc ground track

In effect, by derivation of Eqs. (2) and (3) for constant air speed and flight path angle, we see that the second derivative of position is directly linked to the derivative of azimuth, itself linked through (6) to the sine of the roll angle:

$$\ddot{\Delta} \simeq V_a \cos \gamma_a \, \dot{\chi}_a = g\, n_z \sin \mu \tag{20}$$

The approximation in (20) relies on $(\chi_{T_{12}} - \chi_a)$ being small, when it is not the case, a constant roll turn is initiated in a suitable direction as explained in the flight plan initialization section (see 3.3). This leads to the classical form for ground track monitoring:

$$\sin \mu_c = \sin \mu_d + \frac{K_p}{g\, n_z} \Delta + \frac{K_d}{g\, n_z} \dot{\Delta} \tag{21}$$

μ_d is 0 for a straight leg and is set to a chosen reference value for a circular leg, as detailed in the next sections.

3.2.1 Straight Leg

Using the notations of Fig. 6, we obtain:

$$\begin{aligned} \Delta &= \overrightarrow{\mathbf{wp}} \,.\, \overrightarrow{\mathbf{T_2M}} = [-(x - x_{T_2})\ \sin\chi_{T_{12}} + (y - y_{T_2})\ \cos\chi_{T_{12}}] \\ \dot{\Delta} &= \overrightarrow{\mathbf{wp}} \,.\, \overrightarrow{\mathbf{V}}_a = V_a\ \cos\gamma_a\ \sin(\chi_a - \chi_{T_{12}}) \end{aligned} \tag{22}$$

where $\overrightarrow{\mathbf{wp}}$ is the right cross unit vector:

$$\overrightarrow{\mathbf{wp}} = \begin{pmatrix} -\sin\chi_{T_{12}} \\ \cos\chi_{T_{12}} \end{pmatrix} \tag{23}$$

3.2.2 Circular Arc Ground Track

Using the notations of Figs. 5 and 7 , we have:

$$\begin{aligned} \Delta &= -\varepsilon(r - r_C) = -\varepsilon(\sqrt{(x - x_C)^2 + (y - y_C)^2} - r_C) \\ \dot{\Delta} &= \varepsilon^2\ \overrightarrow{\mathbf{wp}} \,.\, \overrightarrow{\mathbf{V}}_a = V_a\ \cos\gamma_a\ \sin(\chi_a - \chi_P) \end{aligned} \tag{24}$$

The sign of Δ is chosen such that it is positive when point M is on the right of point P according to the orientation of air speed.

The value of ε is +1 or -1 according to respectively right (increasing headings) or left turn. With such a convention we deduce:

$$\mu_d = \varepsilon\ \mu_{ref} \tag{25}$$

The radius r_C of the circle center C is linked to horizontal air speed and to reference rotation speed $\dot{\chi}_{ref}$ using following relation:

$$r_C = \frac{V_a\ cos\gamma_a}{\dot{\chi}_{ref}} \tag{26}$$

The expression of $\dot{\chi}_{ref}$ is a function of chosen reference roll angle μ_{ref} (from (6) and (5)):

$$\dot{\chi}_{ref} = \frac{g\ n_{z_{ref}}\ \sin\mu_{ref}}{V_a\ cos\gamma_a} = \frac{g\ \tan\mu_{ref}}{V_a} \tag{27}$$

The local coordinates x_C and y_C of the circle center C are based on those of the way point T_2 at distance ρ_C in direction χ_C according to:

$$x_C = x_{T_2} + \rho_C\ \cos\chi_C \tag{28}$$

$$y_C = y_{T_2} + \rho_C\ \sin\chi_C \tag{29}$$

The other characteristic values of Figs. 5 and 7 are then:

- half angle of turn, seen from center C: $\theta = \frac{\mathrm{diff}(\chi_{T_{23}} - \chi_{T_{12}})}{2}$
  (*function* diff *framed an angle between* $-\pi$ *and* π);
- half angle of turn, seen from summit T_2: $\beta = \frac{\pi}{2} - |\theta|$
- direction of C from T_2: $\chi_c = \chi_{T_{23}} + \mathrm{sign}(\theta)\ \beta$

- distance between C and T_2: $\rho_c = \frac{r_C}{\sin\beta}$
  this last formulae presents an indetermination when β tends toward 0: the flight plan must be tuned in order to avoid this reversal trajectory case.

Apart from last equation, all quantities are computed only once at flight plan initialization time. ρ_c is pre-computed using $r_C = 1$ and then actualized during flight.

One last quantity depends on actual position M during flight: P is the point of the circle located on the radius from C to M. The reference azimuth χ_P at this point is then:

$$\tan\chi_P = \frac{(x - x_C)}{-(y - y_C)} \tag{30}$$

Similarly to (23), unit cross vector $\overrightarrow{\mathbf{w_P}}$ is:

$$\overrightarrow{\mathbf{w_P}} = \begin{pmatrix} -\sin\chi_P \\ \cos\chi_P \end{pmatrix} \tag{31}$$

At the beginning and at the end of the circular leg, there is a transitory effect due to the fact that the roll angle is not instantaneous: a deviation occurs outside at the beginning and inside at the end. This last effect is attenuated by beginning the end of the turn for a heading deviation equal to the reference roll angle.

In order to reduce deviations a value of 45°is retained for the reference roll angle as 60°are allowed at maximum for corrections.

3.3 Flight Plan Initialization

The activation of the monitoring of the flight plan raised the issue of the choice of the first leg according to the current position.

A first simple mode consists in selecting the first leg, then a sequencer checks if the transverse projection of the current flight position lies within the boundaries of the leg. If it is not the case, then the test is done with the next leg and so on.

When the flight plan is almost in line (see Fig. 4), this process will find the correct leg. But when there are turns, the selected leg may be not the nearest from the current position.

So another mode is added in which an initial seek is led on all the eligible legs, choosing the nearest.

For a more extreme case, as the one displayed on Fig. 8, representative of an approach phase, the first selected leg might be inappropriate, whatever the mode: a short simulation is needed before any use of the flight plan.

In case of a too great cross distance at initial position, one or two supplementary legs are automatically inserted:

- from the initial flight position to the corresponding transverse point on the first leg;
- from the transverse point to the end point of the leg if the distance is lower than the turn radius, if not, the end point is selected instead of the transverse point.

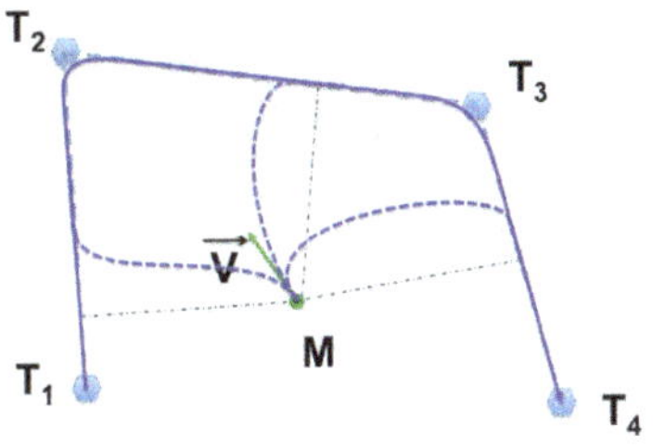

Fig. 8 Selection of first leg for complex flight plan

4 Simulations

The guidance laws were used along the attitude control laws (see [6]) in a full 6 degrees of freedom simulator, "MUSIC" (= MUlti Simulators in Combination, allowing several vehicles). The software is coded in Matlab® except for the attitude control and guidance laws that are implemented in C language, in order to be close to the onboard embedded software. MUSIC allowed in particular to generate many perturbation causes and to process, store and analyze Monte-Carlo simulations.

For the separation maneuver (about 30 s of flight) of the rocket from Eole, as presented in [3], 3000 trajectories were computed for Monte-Carlo analysis. For the flight plan monitoring, we restricted the computations to some worse cases detected for separation as the flight plan for complete ascent, acceleration and descent is about 16 min of flight long and as there is an infinite variety of ground tracks.

4.1 Perturbations Causes

Three main categories are considered for the perturbation causes:

- vehicle environment (moderate wind and turbulence):
  - static wind (up to 8 Kts, all directions)
  - Dryden turbulence (up to 3 Kts);
  - air density deviations (up to 4%).
- deviations on measurements of:
  - airspeed (up to 3% and 0.6 Kts bias);
  - clinometric values (up to 3% and 0.3° bias);
  - inertial measurement unit (up to 1°for atitude angles, 0.2 m/s^2 for accelerations);
  - GPS (up to 5 m for position, 1 m/s for velocities);
  - air surface position measurement deviations (up to 1% and 0.2° bias–used for re-initialisation of integral terms in attitude controller);

- modeling uncertainties on:
  - mass (up to 2 kg), centering (up to 3 cm), inertia (up to 10%);
  - thrust (up to 10%) with left/right deviation (up to 10%);
  - aerodynamic coefficients (up to 5% and 0.1% bias).

Most values are standard deviation for a Gaussian sampling. It implied that typical values could reach more than 3 standard deviations.

The level of wind and turbulence could be enforced by a no go for the experiment when weather conditions are not met.

4.2 Examples of Flight Plan Monitoring

First example of flight plan is constructed in order to test all the capacities of the guidance law: climb from take-off to 4000 m in altitude, acceleration to initial maneuver air speed,about 80 m/s of indicated air speed, then descent to 300 m in altitude with reduced air speed of 35 m/s. The ground track is wrapped in order to stay within a limited space and to test the turn capacities. The whole flight lasted about 16 min.

Figures 9 to 11 displayed successively:

- Fig. 9: ground track in local coordinates;
- Fig. 10/left: time evolutions of altitude h, air speed (true air speed in dotted line, equivalent air speed V_{eas} in solid line), flight path γ;
- Fig. 10/right: time evolutions of pitch angle θ, attack angle α and normal load factor n_z;
- Fig. 11/left: time evolutions of side slip β, aerodynamic bank μ, roll ϕ and azimuth χ angles;
- Fig. 11/right: time evolutions of air surface deflections dl, dm, dn and thrust command dx.

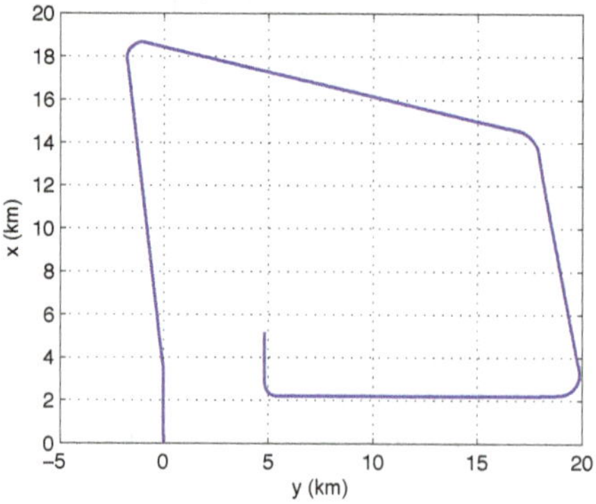

Fig. 9 Example of a flight plan in perturbed conditions (1/3)

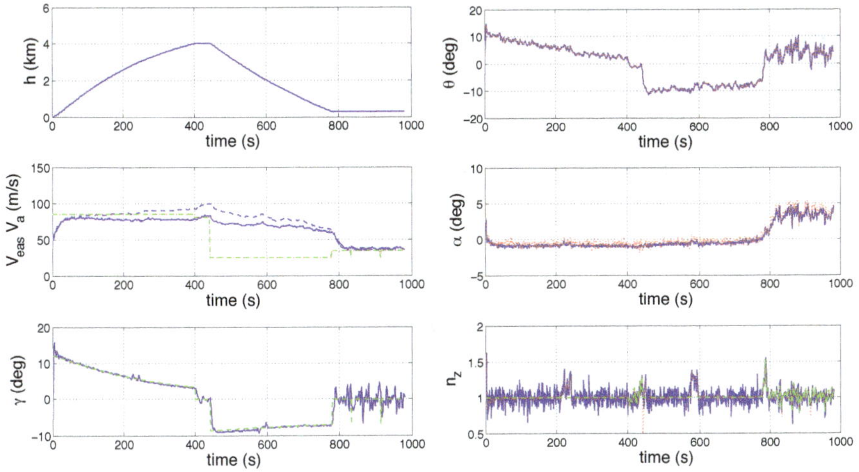

Fig. 10 Example of a flight plan in perturbed conditions (2/3)

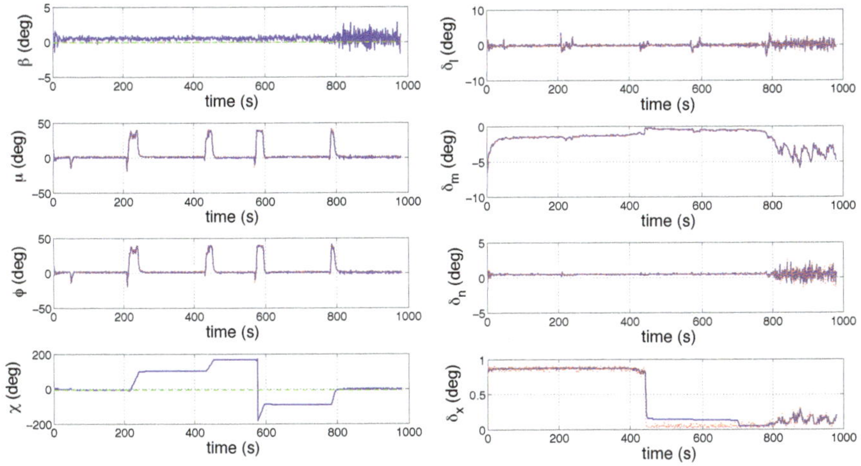

Fig. 11 Example of a flight plan in perturbed conditions (3/3)

The blue solid lines are the actual values, the green solid lines indicate desired or indicative reference values for objectives while the red solid lines represent either the reconstructed desired values for inputs that are not directly controlled (such as angle of attack and pitch angle) of orders for the actuators. In figure labels, the greek letters are written using their latin equivalent: *beta* for β, *theta* for θ ,... and so on.

On Fig. 10/left, the indicative reference value for air speed is at constant maximum value in ascent and constant minimum value in descent but the actual value respect the ascent then descent profiles at intermediate varying speed. On Fig. 11/left, the indicative reference value of azimuth is constantly zero as no good value is computed by ground track monitoring as it defines directly the roll angle from cross distance.

Transients can be seen at the beginning of the figures: initial condition is computed through equilibrium computed without turbulence. It takes then about 2 s for the integrators to counteract.

The obtained trajectory respect the aimed flight plan with the normal load factor between -0.5 and 1.5 as designed. The incidence is kept under 12°.

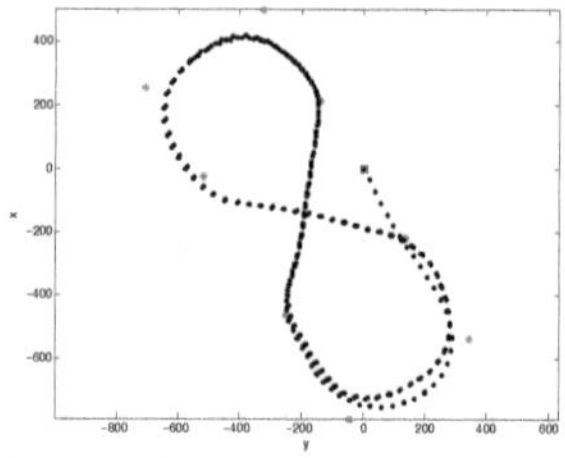

Fig. 12 Eight shape flight plan

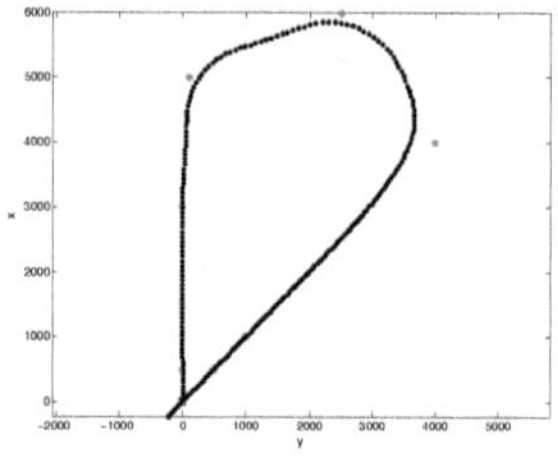

Fig. 13 Diamond like shape flight plan

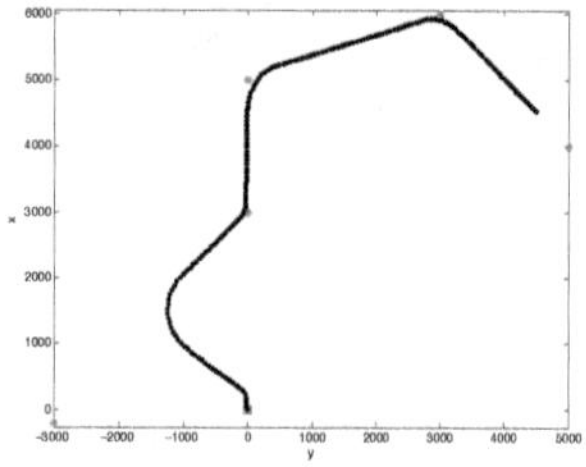

Fig. 14 Flight plan with replanning of initial point

One can see also on Fig. 11/left that the roll angle presents a little variation during turns as expected due to the latency in the following of the circle at beginning of turn.

On Fig. 11/right, we can see that the orders coming out of control law are much affected by turbulence. Instead of filtering these outputs, they are send directly to the actuators which dynamics act as a natural filter. More generally, turbulence effects can be seen on several variables after about 800 s of simulation at low altitude and low velocity.

Several other forms of flight plans were tested with different speed and altitude profiles, see Figs.12-14. On Fig. 12, we can see that two turns doesn't overlap: they are executed at different air speeds. Furthermore wind, supposed unknown, can distort the eight shape : up to 200 m for 10 m/s of wind for a low speed (about 35 m/s) flight plan.

On Fig. 14, the initial point is at some distance and with a different azimuth of the first leg (from point at lower left corner to point at about the middle of the figure): the flight plan is modified such that a new leg is inserted in order to reach first the initial cross point on former first leg.

5 Conclusion

The guidance laws for flight plan monitoring for the UAV launcher vehicle EOLE were presented. Simulations were conducted in presence of many perturbation causes (wind, turbulence, measurement errors, model uncertainties, ...).

These guidance laws were derived from a more generic ONERA Toolbox μFMS for integration in EOLE.

Test flight at low altitude and low air speed (in order to stay at visual range of a security human pilot) took place in October, 2014 at Saint-Yan airport (in the middle of France). The eight shaped flight plan monitoring was successfully achieved. On Fig. 15 (from [1]), the whole flight is represented : it has been verified that out of eight shaped track are sections under manuel control.

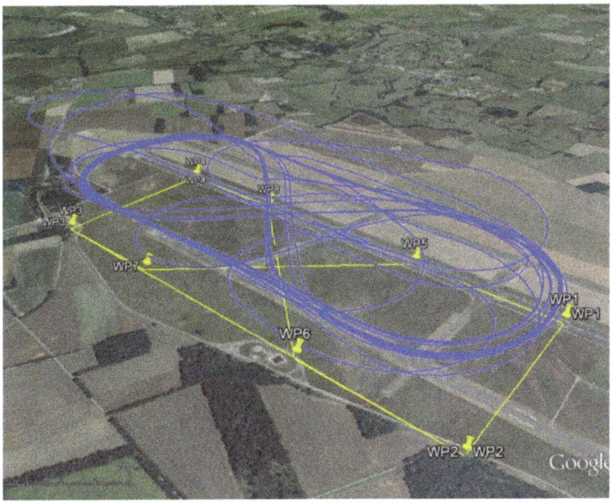

Fig. 15 October, 2014 flight recording – including manually controled sections

Acknowledgements. This study was supported by the French Space Agency CNES.

References

1. Hermetz, J.: Le lancement spatial aéroporté: du concept á la démonstration technologique. In: Conférence du groupe Ile-de-France 3AF. AAAF, Paris (2014)
2. Hingre, H., Bec, R., Oswald, J., Pernon, S.: European university and scientific space research program PERSEUS (projet étudiant de recherche spatiale européen universitaire et scientifique). In: 4th European Conference for Aerospace Sciences, EUCAS, Saint Petersburg (2011)
3. Jouhaud, F.: Guidance of UAV EOLE for launcher separation maneuver. In: ACA 2013, IFAC, Würzburg (2013)
4. Jouhaud, F., Ferreres, G., Vu, B.D., Döll, C.: Guidance and attitude control law for the final part of a reentry vehicle trajectory - Alternate laws for the experimental vehicle HSFD II. In: ACA 2007, IFAC, Toulouse (2007)
5. Ledogar, J., Hermetz, J., Sohier, H., Oswald, J.J., Rantet, E.: EOLE, an innovative flying scale demonstrator for air-launch-to-orbit automatic systems. In: International Council of the Aeronautical Sciences ICAS 2012, ICAS, Brisbane (2012)
6. Waitman, S., Minami, Y., Döll, C.: Autopilot design for the EOLE demonstrator. In: International Council of the Aeronautical Sciences ICAS 2012, ICAS, Brisbane (2012)
7. Wanner, J.C., Forestier, J., George, L., Vernet, J.F.: La Mécanique du vol: performances des avions et des engins – Cours professé à l'École nationale supérieure de l'aéronautique, 2nd edn. Dunod (1969)

Fault Tolerant $\mathcal{L}_1$ Adaptive Control Based on Degraded Models

Toufik Souanef and Walter Fichter

Abstract. This paper proposes a method for fault tolerant control. The design is based on an $\mathcal{L}_1$ adaptive controller with a nominal reference model and a set of degraded reference models. In a degraded model the criteria of stability margins and performance bounds are reduced. The simulation results, for the altitude hold control of a small fixed-wing UAV, show good compromise between performance and robustness against faults and large uncertainties.

1 Introduction

In flight control systems, fault tolerance is usually achieved using redundancy in sensors and actuators. This approach places additional supplementary constraints on the system design, maintenance, payload and costs. All these reasons make redundancy unsuitable for UAVs. A solution to this problem is the design of a fault tolerant control method [13, 15, 21].

Adaptive control is a very suitable method for fault tolerant control, due to its ability to automatically adapt to changes in the parameters of the system without a fault detection scheme [2, 18]. This leads to a relatively simple design and reduced computing effort, which make it suitable for low cost flight systems [6, 7, 11].

Among all the possible methods that can be used to design fault tolerant adaptive control, this paper focuses on $\mathcal{L}_1$ adaptive control [8]. This approach is beneficial because it offers fast and robust adaptation that leads to bounded transient performances for both input and output signals. Successful applications of $\mathcal{L}_1$ adaptive in fault tolerant control are reported in [9, 4, 16, 5, 12, 19].

Nonetheless, and despite the good performance of the $\mathcal{L}_1$ adaptive control, it is still true that when the uncertainties induced by disturbances, faults or failures

Toufik Souanef · Walter Fichter
Institute of Flight Mechanics and Control, Pfaffenwaldring 27, 70569 Stuttgart, Germany
e-mail: {toufik.souanef,fichter}@ifr.uni-stuttgart.de

J. Bordeneuve-Guibé et al. (eds.), *Advances in Aerospace Guidance, Navigation and Control*, DOI: 10.1007/978-3-319-17518-8_9

are too large, they may reduce the performance of the controller or even make the system unstable. Actually, if the uncertainties caused by a fault are larger than the system design, the $\mathcal{L}_1$ norm stability condition [8] will not be satisfied.

Furthermore, when a fault affects the system actuators, it reduces their capabilities and, if the nominal performance of the system is maintained, the actuators will work beyond their nominal set point, which might lead to severe failures that cannot be compensated by a fault tolerant controller.

Therefore, it is not reasonable to maintain the same desired performance of the system, because after a fault or a failure it is not possible to recover the nominal performance. This is especially true for non-redundant systems such as low cost UAVs.

An approach to this problem was presented in [10, 20] based on performance degradation of the nominal controller. The method considers a model that combines the control with a fault detection scheme. The design is made under the assumption that the model of the plant has no uncertainties, which is not realistic, especially for post-fault systems. Furthermore, only actuator faults are addressed while structural faults are not considered.

This paper proposes an approach to this problem that uses a multiple model $\mathcal{L}_1$ adaptive control. The key idea is to design an $\mathcal{L}_1$ adaptive controller with a nominal reference models and a set of degraded reference models. In a degraded model the criteria of stability margins and performance bounds are reduced.

The main contributions of this paper are:

- It introduces a new approach for fault tolerant control based on performance degradation of the nominal $\mathcal{L}_1$ adaptive controller.
- It proves the performance bounds of the architecture for both system input and output.
- It shows, through simulations, that this approach maintains system stability under large uncertainties.

2 Problem Formulation

Given a general class of systems with matched uncertainties represented as

$$\begin{aligned} \dot{x}(t) &= A_p x(t) + b\big(\omega u_p(t) + \sigma(t)\big), \\ y(t) &= c^\top x(t), \quad x(0) = x_0, \end{aligned} \tag{1}$$

where $A_p \in \mathbb{R}^{n\times n}$ is an unknown matrix, $b \in \mathbb{R}^n$ and $c \in \mathbb{R}^n$ are known constant vectors, $x(t) \in \mathbb{R}^n$ is the state vector, $u_p(t) \in \mathbb{R}$ is the control input, $y(t) \in \mathbb{R}$ is the output of the system, $\omega \in \mathbb{R}$ is an unknown constant with known sign and $\sigma(t) \in \mathbb{R}$ is a time-varying matched disturbance.

Consider the following control structure

$$u_p(t) = u_l(t) + u(t), \tag{2}$$

where $u_l = -K^\top x(t)$ is a linear control law that yields to the desired performance in the absence of uncertainties. The linear controller is designed such that $A_p = A_m + bK^\top$ where $A_m \in \mathbb{R}^{n\times n}$ is a known Hurwitz matrix that defines the desired dynamics of the system. The adaptive control law $u(t)$ will be defined later. Hence, the system in (1) an be written as

$$\begin{aligned} \dot{x}(t) &= A_m x(t) + b\big(\omega u(t) + \theta^\top x(t) + \sigma(t)\big), \\ y(t) &= c^\top x(t), \quad x(0) = x_0, \end{aligned} \tag{3}$$

where $A_m \in \mathbb{R}^{n\times n}$ is a known Hurwitz matrix that defines the desired dynamics of the system.

Assumption 1. The unknown parameters are bounded, i.e. $\theta \in \Theta$ where Θ is a known compact convex set, $0 < \omega_l \leq \omega \leq \omega_u$ and $|\sigma(t)| < \triangle$ where $\triangle \in \mathbb{R}^+$. We further assume that $\sigma(t)$ is continuously differentiable and its derivative is uniformly bounded, i.e. $|\dot{\sigma}(t)| < d_\sigma$ where $d_\sigma \in \mathbb{R}^+$.

Assumption 2. The pair (A_m, b) is controllable.

For the system in (3) the $\mathcal{L}_1$ adaptive controller is composed of a state predictor, an adaptation law and a control law [8].

The state predictor is defined by

$$\begin{aligned} \dot{\hat{x}}(t) &= A_m x(t) + b(\hat{\omega} u(t) + \hat{\theta}^\top(t) x(t) + \hat{\sigma}(t)), \\ \hat{y}(t) &= c^\top \hat{x}(t), \quad \hat{x}(0) = x_0, \end{aligned} \tag{4}$$

where $\hat{x}(t)$ is the state vector of the predictor and $\hat{\omega}(t)$, $\hat{\theta}(t)$ and $\hat{\sigma}(t)$ are respectively the estimations of the unknown parameters ω, θ and $\sigma(t)$ respectively, governed by the following adaptation law

$$\begin{aligned} \dot{\hat{\omega}}(t) &= -\Gamma Proj\big(\tilde{x}^\top(t),\, Pbu(t)\big), \\ \dot{\hat{\theta}}(t) &= -\Gamma Proj\big(\tilde{x}^\top(t),\, Pbx(t)\big), \\ \dot{\hat{\sigma}}(t) &= -\Gamma Proj\big(\tilde{x}^\top(t),\, Pb\big), \end{aligned} \tag{5}$$

where $\tilde{x}(t) = \hat{x}(t) - x(t)$ is the prediction error, $\Gamma \in \mathbb{R}^+$ is the adaptation rate, $P = P^\top > 0$ is the solution of the Lyapunov equation $A_m^\top P + PA_m = -Q$ with $Q = Q^\top > 0$ arbitrary and $Proj(\cdot,\cdot)$ is the projection operator [14].

The control law is given by

$$u(s) = kD(s)\big(k_g r(s) - \hat{\eta}(s)\big), \tag{6}$$

where $\hat{\eta}(s)$ is the Laplace transformation of the term $\hat{\omega}(t)u(t) + \hat{\theta}^\top(t)x(t) + \hat{\sigma}(t)$, $D(s)$ is a transfer function that leads to a strictly proper stable filter $C(s) = \omega kD(s)/(1 + \omega kD(s))$ with $C(0) = 1$ and the static gain is chosen as $k_g = -1/(c^\top A_m^{-1} b)$.

Furthermore, we define the following

$$L = \underbrace{max}_{\theta \in \Theta} \|\theta\|_1, H(s) = (sI - A_m)^{-1} b \text{ and } G(s) = \big(1 - C(s)\big) H(s).$$

The $\mathcal{L}_1$ adaptive controller defined by equations (4), (5) and (6) is subject to the following $\mathcal{L}_1$-norm condition

$$\|G(s)\|_{\mathcal{L}1} L < 1. \tag{7}$$

It is clear that if a fault or a failure occurs on the system the unknown parameters may go outside the predefined sets. As a consequence, the $\mathcal{L}_1$-norm condition in (7) may become not satisfied. Hence, it is necessary to maintain system stability and a minimum of good performance, this is done through the design of a set of degraded models which become effective when large uncertainties appear on the plant.

3 Multiple-Model $\mathcal{L}_1$ Adaptive Controller

Based on engineering insights, several models are built, including a nominal model and a set of degraded models. The desired performance of each degraded model is made through the design of the pair $(A_{m(i)}, b_i)$, for $i = 0...M$, where M is the number of degraded models and $i = 0$ is the nominal system.

Consequently, the system in (3) can be written as

$$\begin{aligned} \dot{x}(t) &= A_{m(i)} x(t) + b_i\big(\omega_i u(t) + \theta_i^\top x(t) + \sigma_i(t)\big), \\ y(t) &= c^\top x(t). \end{aligned} \tag{8}$$

Remark 1. The unknown parameters are within different known bounds, i.e. $\theta_i \in \Theta_i$ where Θ_i is a known compact convex set, $0 < \omega_{l(i)} \leq \omega_i \leq \omega_{u(i)}$ and $|\sigma_i(t)| < \triangle_i < +\infty$. Furthermore, $\sigma_i(t)$ are continuously differentiable and their derivatives are uniformly bounded, i.e. $|\dot{\sigma}_i(t)| < d_{\sigma(i)} < +\infty$.

3.1 Controller Design

For the system in (8) the proposed $\mathcal{L}_1$ adaptive controller is composed of a bank of state predictors, adaptation laws, a control law acting similarly on the plant and the state predictors and a switching logic.

The state predictors are defined by

$$\dot{\hat{x}}_i(t) = A_{m(i)} x(t) + b_i\big(\hat{\omega}_i u(t) + \hat{\theta}_i^\top(t) x_i(t) + \hat{\sigma}_i(t)\big), \qquad \hat{x}_i(0) = x_0, \tag{9}$$

with the adaptation laws

$$\begin{aligned}
\dot{\hat{\omega}}_i(t) &= -\Gamma_i Proj\big(\tilde{x}_i^\top(t)P_i bu(t)\big),\\
\dot{\hat{\theta}}_i(t) &= -\Gamma_i Proj\big(\tilde{x}_i^\top(t)P_i bx(t)\big),\\
\dot{\hat{\sigma}}_i(t) &= -\Gamma_i Proj\big(\tilde{x}_i^\top(t)P_i b\big),
\end{aligned} \tag{10}$$

where $\Gamma_i \in \mathbb{R}$ is the adaptation rate of each model, $P_i = P_i^\top > 0$ is the solution of the Lyapunov equation $A_{m(i)}^\top P_i + P_i A_{m(i)} = -Q_i$ with $Q_i = Q_i^\top > 0$ arbitrary.

3.2 The Switching Logic

The design of the $\mathcal{L}_1$ adaptive controller consists of a set of a candidate state predictors, adaptation laws and control outputs. At least one of these controllers is assumed to ensure system desired performance. The "best" controller is selected through the switching logic.

The quadratic criterion with forgetting factor is used for switching between controllers and it is given by

$$min\big(J_i(t)\big) = \int_0^t e^{-\beta(t-\tau)}\|\tilde{x}_i(\tau)\|^2 d\tau \tag{11}$$

where $\tilde{x}_i(t) = \hat{x}_i(t) - x(t)$ is the prediction error, and $\beta > 0$ is the forgetting factor.

The plant with the selected controller is written

$$\dot{x}(t) = A_{m(*)}x(t) + b_*\big(\omega_* u(t) + \theta^\top{}_*(t)x(t) + \sigma_*(t)\big). \tag{12}$$

The selected state predictor is given by

$$\dot{\hat{x}}_*(t) = A_{m(*)}x(t) + b_*\big(\hat{\omega}_* u(t) + \hat{\theta}_*^\top(t)x(t) + \hat{\sigma}_*(t)\big), \qquad \hat{x}_*(0) = x_0, \tag{13}$$

and the control law acting on the system is written

$$u(s) = k_* D_*(s)\big(k_{g(*)}r(s) - \hat{\eta}_*(s)\big), \tag{14}$$

where $\hat{\eta}_*(s)$ is the Laplace transformation of the term $\hat{\omega}_* u(t) + \hat{\theta}_*^\top(t)x(t) + \hat{\sigma}_*(t)$, $D_*(s)$, k_* and $k_{g(*)} = -1/(c^\top(A_{m(*)})^{-1}b_*)$ are the parameter of the selected model.

Define the following

$$L_* = \underbrace{max}_{\theta_* \in \Theta_*} \|\theta_*\|_1,\ H_*(s) = (sI - A_{m(*)})^{-1}b \text{ and } G_*(s) = \big(1 - C_*(s)\big)H_*(s).$$

The $\mathcal{L}_1$ adaptive controller defined by equations (4), (5) and (6) is subject to the following $\mathcal{L}_1$-norm condition

$$\|G_*(s)\|_{\mathcal{L}1} L_* < 1. \tag{15}$$

4 Controller Analysis

4.1 Analysis of the Multiple Predictors

The following lemma shows that the prediction errors $\tilde{x}_i(t) = \hat{x}_i(t) - x(t), i = 1...N$ of all the state predictors are bounded.

Lemma 1. The prediction error of each state predictor, $\tilde{x}_i(t)$ is bounded with respect to initial conditions and its bound is given by

$$\|\tilde{x}_i(t)\| \leq \rho_i = \sqrt{\frac{\theta_{m(i)}}{\lambda_{min}(P_i)\Gamma_i}}, \quad i = 0...M, \tag{16}$$

where $\theta_{m(i)} = 4\theta_{max(i)}{}^2 + 4\triangle_i^2 + \left(\omega_{u(i)} - \omega_{l(i)}\right)^2 and + 4d_\sigma \triangle \frac{\lambda_{max}(P_i)}{\lambda_{min}(Q_i)}$ and $\theta_{max(i)} = \max_{\theta_i \in \Theta_i} \|\theta_i\|$.

Proof. The prediction error dynamics of every state predictor can be written as

$$\dot{\tilde{x}}_i(t) = A_{m(i)}\tilde{x}_i(t) + b_i\left(\tilde{\omega}_i(t)u(t) + \tilde{\theta}_i^\top(t)x(t) + \tilde{\sigma}_i(t)\right), \tag{17}$$

where $\tilde{\theta}_i(t) = \hat{\theta}_i(t) - \theta$, $\tilde{\omega}_i(t) = \hat{\omega}_i(t) - \omega$ and $\tilde{\sigma}_i(t) = \hat{\sigma}_i(t) - \sigma(t)$ are, respectively, the estimation errors of the unknown parameters θ, ω and $\sigma(t)$.

We consider the set of Lyapunov function candidates

$$V_i(t) = \tilde{x}_i^\top(t)P_i\tilde{x}_i(t) + \frac{1}{\Gamma_i}\left(\tilde{\omega}_i^2(t) + \tilde{\theta}_i^\top(t)\tilde{\theta}_i(t) + \tilde{\sigma}_i^2(t)\right). \tag{18}$$

The derivatives of $V_i(t)$ are given by

$$\begin{aligned}\dot{V}_i(t) =& \tilde{x}_i^\top(t)\left(A_{m(i)}^\top P_i + P_iA_{m(i)}\right)\tilde{x}_i(t)\\ &+ 2\tilde{x}_i^\top(t)P_ib_i\tilde{\omega}_i(t)u(t) + 2\tilde{x}_i^\top(t)P_ib_i\tilde{\theta}_i^\top(t)x(t) + 2\tilde{x}_i^\top(t)P_ib_i\tilde{\sigma}_i(t)\\ &+ \frac{2}{\Gamma_i}\left(\tilde{\omega}_i(t)\dot{\hat{\omega}}_i(t) + \tilde{\theta}_i^\top(t)\dot{\hat{\theta}}_i(t) + \tilde{\sigma}_i(t)\dot{\hat{\sigma}}_i(t)\right) - \frac{2}{\Gamma_i}\tilde{\sigma}_i(t)\dot{\sigma}_i(t).\end{aligned} \tag{19}$$

Next, the adaptation laws (10) are included which yields

$$\begin{aligned}\dot{V}_i(t) =& -\tilde{x}_i{}^\top(t)Q_i\tilde{x}_i(t)\\ &+ 2\tilde{\omega}_i(t)\left(\tilde{x}_i^\top(t)P_ib_iu(t) + Proj(\hat{\omega}, -\tilde{x}_i^\top(t)P_ib_iu(t))\right)\\ &+ \tilde{\theta}_i^\top(t)\left(\tilde{x}_i^\top P_ib_ix(t)) + Proj(-\tilde{x}_i^\top(t)P_ib_ix_i(t))\right)\\ &+ \tilde{\sigma}_i(t)\left(\tilde{x}_i^\top P_ib_i + Proj(-\tilde{x}_i^\top(t)P_ib_i)\right) - \frac{2}{\Gamma_i}\tilde{\sigma}_i(t)\dot{\sigma}(t)\\ \leq& -\tilde{x}_i{}^\top(t)Q_i\tilde{x}_i(t) + \frac{2}{\Gamma_i}|\tilde{\sigma}_i\dot{\sigma}_i|.\end{aligned} \tag{20}$$

The projection operator also ensures that for all $t \geq 0$

$$\max_{t\geq 0}\Big(\tilde{\theta}_i^\top(t)\tilde{\theta}_i(t)+\tilde{\omega}_i^2(t)+\tilde{\sigma}_i^2(t)\Big). \leq 4\theta_{max}^2+4\triangle^2+(\omega_u-\omega_l)^2,$$

and

$$\tilde{\sigma}_i\dot{\sigma}_i \leq 2d_{\sigma(i)}\triangle_i.$$

If at any time $t_1 > 0$, one has $V_i(t_1) > \theta_{m(i)}/\Gamma_i$ the we have

$$\dot{V}_i(t_1) < 0, \tag{21}$$

and it follows that for all $t \geq 0$

$$\dot{V}_i(t) < \frac{\theta_{m(i)}}{\Gamma_i}. \tag{22}$$

Since $\lambda_{min}\|\tilde{x}_i(t)\|^2 \leq \tilde{x}_i^\top(t)P_i\tilde{x}_i(t) \leq V_i(t)$ it follows that

$$\|\tilde{x}_i(t)\|^2 \leq \frac{\theta_{m(i)}}{\lambda_{min}(P_i)\Gamma_i}, \tag{23}$$

and the proof is complete. □

Remark 2. It is clear that all the estimated states and parameters are bounded, independently from if one is selected by the switching logic or not.

Remark 3. Similar to $\mathcal{L}_1$ adaptive control [8], in multiple model $\mathcal{L}_1$ adaptive control, the prediction error of every model can be made arbitrarily large by selecting high adaptation rates.

4.2 Reference System Analysis

If a controller is selected by the switching logic, it becomes the reference model defined by

$$\dot{x}_*^{ref}(t) = A_{m(*)}x_*^{ref}(t) + b_*\big(\omega_* u^{ref}(t) + \theta_*{}^\top x_*^{ref} + \sigma_*(t)\big), \quad x_*^{ref}(0) = x_0 \tag{24}$$

where ω_*, θ_* and $\sigma_*(t)$ are the nominal parameters of the selected model.

The control law for the reference system is given by

$$u^{ref}(s) = \frac{C_*(s)}{\omega_*}\,(k_{g(*)}r(s) - \eta_*{}^{ref}(s)), \tag{25}$$

where $\eta_{ref}(s)$ is the Laplace transformation of the term $\omega_* u^{ref}(t) + \theta_*{}^\top x_*^{ref} + \sigma_*(t)$.

We define proper BIBO stable transfer functions $H_*(s) = (s\mathbb{I} - A_m(*))^{-1}b_*$, $G_*(s) = H_*(s)(1 - C_*(s))$, and we note $L_* = \max_{\theta_*\in\Theta_*}\|\theta_*\|_1$ and we demonstrate.

The stability of the closed loop reference system is stated through the following lemma.

Lemma 2. If the filter $C_*(s)$ is designed such that it verifies the $\mathcal{L}_1$ norm condition

$$\|G_*(s)\|_{\mathcal{L}_1} L_* < 1, \tag{26}$$

then the closed-loop reference system in (24) and (25) is BIBS stable with respect to the reference input and initial conditions.

Proof. The proof is similar to Lemma 2.2.1 in [8] and is omitted here.

4.3 Performance Bounds

The following Theorem shows that the performance bounds of closed loop system, generated by the selected model, are bounded.

Theorem 2. Given the system (12), the reference system (24), (25) and the selected $\mathcal{L}_1$ adaptive model (13) and (14) we have

$$\|x_*^{ref} - x\|_{\mathcal{L}_\infty} \le \frac{\gamma_{1*}}{\Gamma_*}, \tag{27}$$

and

$$\|u_*^{ref} - u\|_{\mathcal{L}_\infty} \le \frac{\gamma_{2*}}{\Gamma_*}, \tag{28}$$

where $\gamma_{1*} = \frac{\|C_*(s)\|_{\mathcal{L}_1}}{1-\|G_*(s)\|_{\mathcal{L}_1} L_*}\sqrt{\frac{\theta_{m*}}{\lambda_{min}(P_*)}}$, $\gamma_2 = \|\frac{C_*(s)}{\omega_*}\|_{\mathcal{L}_1} L_* \gamma_{1*} + \|\frac{H_{1*}(s)}{\omega_*}\|_{\mathcal{L}_1}\sqrt{\frac{\theta_{m*}}{\lambda_{min}(P_*)}}$ and $H_{1*}(s)$ is defined below.

Proof. The control law in (14) can be formulated as

$$u(s) = \frac{C_*(s)}{\omega}(k_{g*}r(s) - \theta_*{}^\top x(s) - \sigma_*(s) - \tilde{\eta}_*(s)), \tag{29}$$

where $\tilde{\eta}_*(s)$ is the Laplace transformation of the term $\tilde{\omega}_*(t)u(t) + \tilde{\theta}_*^\top(t)x(t) + \tilde{\sigma}_*(t)$. The Laplace transformation of the closed loop system (12) and (29) can be written

$$x(s) = H_*(s)C_*(s)k_{g*}r(s) + G_*(s)\left(\theta_*^\top x(s) + \sigma_*(s)\right) - H_*(s)C_*(s)\tilde{\eta}_*(s) + x_{in*}(s), \tag{30}$$

where $x_{in*}(s) = (sI - A_{m*})^{-1}x_0$.

On the other hand, the closed loop reference system (24) and (25) can be written

$$x_*^{ref}(s) = H_*(s)C_*(s)k_{g*}r(s) + G_*(s)\left(\theta_*^\top x_*^{ref}(s) + \sigma_*(s)\right) + x_{in*}(s). \tag{31}$$

Considering the difference of equations (31) and (30) it follows that

$$x_*^{ref}(s) - x(s) = G_*(s)\theta_*^\top \left(x_*^{ref}(s) - x(s)\right) + H_*(s)C_*(s)\tilde{\eta}_*(s). \tag{32}$$

From (17) the Laplace transformation of the prediction error of the selected system is written

$$\tilde{x}_*(s) = H(s)\tilde{\eta}_*(s). \tag{33}$$

Consequently

$$x_*^{ref}(s) - x(s) = G_*(s)\theta_*^\top \left(x_*^{ref}(s) - x(s)\right) + C_*(s)\tilde{x}_*(s). \tag{34}$$

This leads to the following bound

$$\|(x_*^{ref} - x)_\tau\|_{\mathcal{L}_\infty} = \|G_*(s)\|_{\mathcal{L}_1} L_* \|(x_*^{ref} - x)_\tau\|_{\mathcal{L}_\infty} + \|C_*(s)\|_{\mathcal{L}_1} \|(\tilde{x}_*)_\tau\|_{\mathcal{L}_\infty} \tag{35}$$

and the bound in (27) is straightforward.

To show the bound of the control law $u(t)$, from equations (25) and (29) one can write

$$u_*^{ref}(s) - u(s) = -\frac{C_*(s)}{\omega_*}\theta_*{}^\top \left(x_*^{ref}(s) - x(s)\right) + \frac{C_*(s)}{\omega_*}\tilde{\eta}_*(s). \tag{36}$$

From Lemma A.12.1 in [8] there exists a vector vector c_0 verifying

$$H_{1*}(s) = C_*(s)\frac{c_0^\top}{c_0^\top H_*(s)}. \tag{37}$$

Hence equation (36) can be written

$$u_*^{ref}(s) - u(s) = -\frac{C_*(s)}{\omega_*}\theta_*{}^\top \left(x_*^{ref}(s) - x(s)\right) + \frac{H_{1*}(s)}{\omega_*}\tilde{x}_*(s), \tag{38}$$

and leads to the bound in (28). This completes the proof. □

Remark 4. Similar to $\mathcal{L}_1$ adaptive control [8], in multiple model $\mathcal{L}_1$ adaptive control, the performance bounds of the selected model can be made arbitrarily large by taking high adaptation rates.

5 Application: Altitude Hold Control of a Small UAV

In this section, the proposed approach is demonstrated on a full nonlinear simulator of a BO-Munsun 209 small fixed wing UAV (Fig. 1). The numerical simulation model of the UAV runs on a standard desktop PC with Matlab/Simulink [1]. The objective is the design of an altitude hold controller.

It is proposed here an integrated design based on a full longitudinal model that includes the airspeed, the altitude and the attitude of the plant. The main advantage of the integrated approach, compared to a classical design with an outer-loop (guidance controller) and an inner-loop (attitude controller), is that it permits to solve the problem of computing feasible reference commands to the inner loop [3].

Fig. 1 Bo Monsun 209 model aircraft.

Assuming that the true airspeed is constant, the longitudinal model of a fixed wing airplane can be written in matrix form as [17]:

$$\underbrace{\begin{pmatrix} \dot{V} \\ \dot{\alpha} \\ \dot{q} \\ \dot{\theta} \\ \dot{h} \end{pmatrix}}_{\dot{x}} = \underbrace{\begin{bmatrix} X_V & X_\alpha & X_q & -gcos(\gamma_0) & 0 \\ \frac{Z_V}{V_0} & \frac{Z_\alpha}{V_0} & 1+\frac{Z_q}{V_0} & \frac{-gsin(\gamma_0)}{V_0} & 0 \\ M_V & M_\alpha & M_q & 0 & 0 \\ 0 & 0 & 1 & 0 & 0 \\ 0 & -V_0 & 0 & V_0 & 0 \end{bmatrix}}_{A_p} \underbrace{\begin{pmatrix} V \\ \alpha \\ q \\ \theta \\ h \end{pmatrix}}_{x} + \underbrace{\begin{pmatrix} X_\delta \\ \frac{Z_\delta}{V_0} \\ M_\delta \\ 0 \\ 0 \end{pmatrix}}_{b} \delta_e \tag{39}$$

where V is the airspeed, α is the angle of attack, q is the pitch angular velocity, θ is the pitch angle, h is the altitude, $(X_V, X_\alpha, X_q, X_\delta)$, $(Z_V, Z_\alpha, Z_q, Z_\delta)$ and $(M_V, M_\alpha, M_q, M_\delta)$ are the partial derivatives of the aerodynamic force Z and the pitching moment M, with respect to α, q, and δ_e respectively. V_0 is the trimmed airspeed and γ_0 is the trimmed flight path angle.

Numerical values for the considered UAV trimmed at $V_0 = 20\ m/s$, $\alpha_0 = 0.07$ rad, $\theta = 0$ rad, $q_0 = 0$ rad/s, and altitude $h_0 = 50$ m are given by

$$A_p = \begin{bmatrix} -0.12 & 13.07 & -0.64 & -9.77 & 0 \\ -0.03 & -11.04 & 0.94 & -0.02 & 0 \\ 0.42 & -248.8 & -14.47 & 0 & 0 \\ 0 & 0 & 1 & 0 & 0 \\ 0 & -20 & 0 & 20 & 0 \end{bmatrix}, \quad b = \begin{pmatrix} 0.09 \\ -0.2840 \\ -112.4126 \\ 0 \\ 0 \end{pmatrix}.$$

The elevator dynamics are assumed to be negligible.

The control approach is based on the augmentation of a baseline linear controller by the $\mathcal{L}_1$ adaptive controller as it is common in flight control systems. The total deflection of the elevator $\delta_e(t) = u_l(t) + u(t)$ is the sum of the commands from the baseline linear controller $u_l(t)$ and the nominal adaptive controller $u(t)$.

For the nominal controller, the desired dynamics matrix $A_{m(0)}$ is designed based on LQR approach. The model eigenvalues are $\lambda_{1,2} = -0.5 \pm 0.5j$, $\lambda_{3,4} = -14.5 \pm 10.8j$ and $\lambda_5 = -0.11$.

The system takes the form defined in equation (8) for $i = 0$

$$\begin{aligned} \dot{x}(t) &= A_{m(0)}x(t) + b_0\big(\omega_0 u(t) + \theta_0{}^\top x(t) + \sigma_0(t)\big) \\ y(t) &= c^\top x(t) \end{aligned} \tag{40}$$

For the implementation of the controller the compact sets are chosen $\Omega_0 = [0.5, 1.5]$, $\Theta_0 = \{\vartheta = [\vartheta_1, \vartheta_2] \in \mathbb{R}^2 : \vartheta_i \in [-0.5, 0.5], i = 1,..,4\}$, $\Delta_0 = 1$. The filter parameters are taken $D(s) = \frac{k_0}{s(s+2k_0)}$, and $k_0 = 350$. The adaptation rate is set as Γ_0=1000.

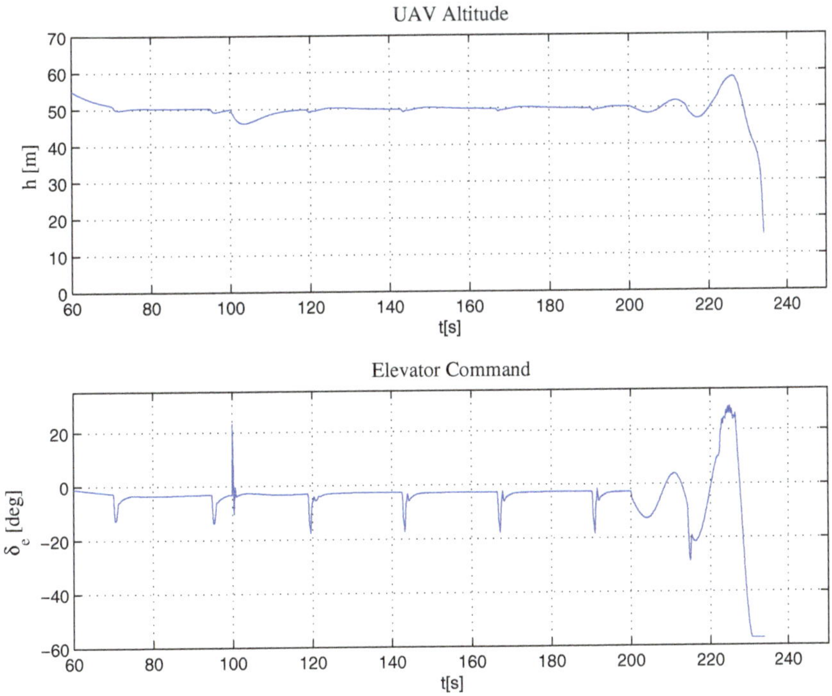

Fig. 2 Closed loop response of the system and control input with the nominal controller

The proposed scenario for the real time simulations, was to follow a path defined by four way points at a fixed altitude $h = 50m$. The adaptive controller was turned on at simulation time $t = 100s$.

To show the necessity of a multiple model design, three types of matched uncertainties were added to the system at simulation time $t = 200s$:

- linear-in-state uncertainty $\theta = [-0.1, 0.1, -0.1, -0.01, 0]^\top$
- A loss of actuator effectiveness ω=0.5
- A constant control bias $\sigma(t) = 0.25$.

It is shown in Fig. 2 that the system becomes unstable under this combination of uncertainties and faults. Note the presence of peaks due to the compensation of rolling motion of the UAV when turning.

The degraded model is chosen so that the desired dynamics of the system are slower than those of the nominal model. In this example the eigenvalues of the degraded model are chosen $\lambda_{1,2} = -0.35 \pm 0.35j$, $\lambda_{3,4} = -10.5 \pm 8.6j$ and $\lambda_5 = -0.08$. The compact sets are chosen $\Omega_1 = [0.25, 1.25]$, $\Delta_1 = 1$, $\Theta_1 = \{\vartheta = [\vartheta_i \in [-0.5, 0.5], i = 1, .., 4\}$. The filter parameters are taken $D(s) = \frac{k_1}{s(s+2k_1)}$, and $k_1 = 350$. The adaptation rate is set as Γ_1=1000.

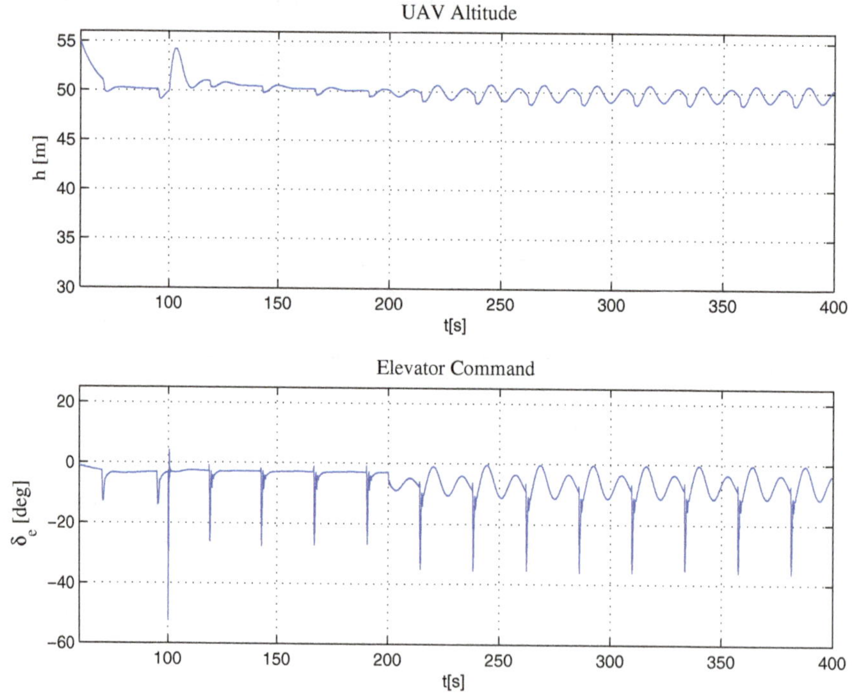

Fig. 3 Closed loop response of the system and control input with the degraded controller

In Fig. 3 it is shown that the degraded controller is robust against the disturbances. However, the performance of the disturbances-free system is relatively weak compared to the nominal controller. Particularly, it can be noted the presence of high frequency oscillations on the elevator command, due to the fact that degraded controller "steers" more than the nominal model to compensate for unknown parameters. Furthermore, it is noted the presence of low frequency oscillations in the control command, when faults and disturbances occur. This is caused by the compensation of the oscillations induced by the uncertainties, that have caused the instability of the nominal model.

Using the approach described above, based on the combination of the nominal model the degraded model, and the switching logic defined in equation (11), one can see that before the introduction of the large uncertainties, the system is controlled by the nominal controller and it switches to the degraded controller exactly when uncertainties appear on the system, as it is shown in figure Fig. 4.

It can be concluded that the design based on multiple models presents a better compromise between performance and robustness against faults and large uncertainties.

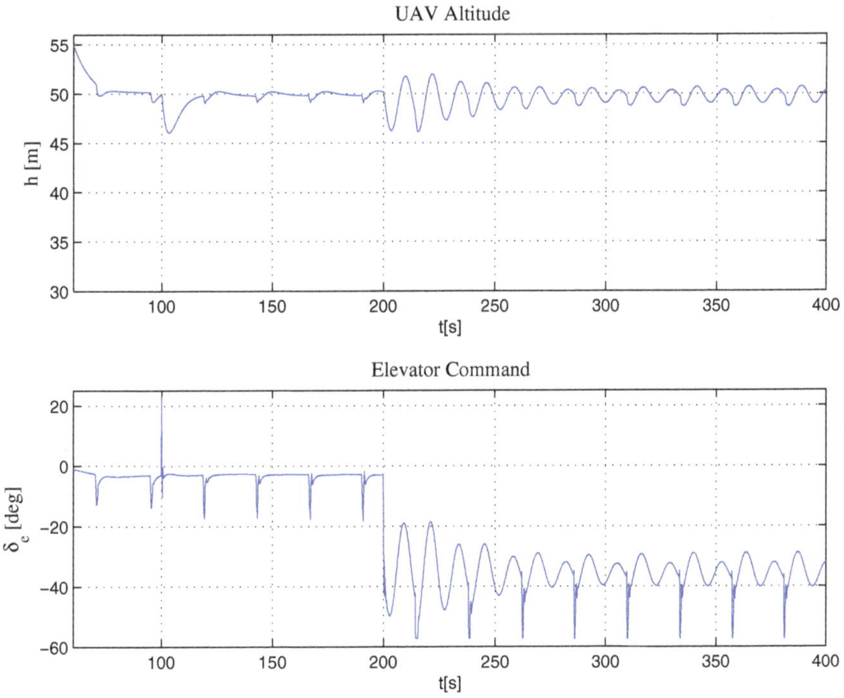

Fig. 4 Closed loop response of the system and control input with the multiple model controller

6 Conclusion

This paper presented an approach for $\mathcal{L}_1$ adaptive fault tolerant control. The design is based on a nominal model for a fault-free plant and a set of degraded models for the plant under failures or large disturbances. The switching between the models is based on a simple quadratic criterion.

The main advantage of this approach is that it allows a larger class of uncertainties and faults to be considered and can also achieve better accommodation. Furthermore, there is no need for a fault detection scheme, while using an adaptive fault tolerant control, which simplifies the design and the implementation.

References

[1] (2015), http://www.ifr.uni-stuttgart.de
[2] Bodson, M.: Reconfigurable nonlinear autopilot. Journal of guidance, control, and dynamics 26(5), 719–727 (2003)
[3] Chowdhary, G., Johnson, E.N., Chandramohan, R., Kimbrell, M.S., Calise, A.: Guidance and Control of Airplanes Under Actuator Failures and Severe Structural Damage. Journal of Guidance, Control, and Dynamics 36(4), 1093–1104 (2013)
[4] Dobrokhodov, V., Kaminer, I., Kitsios, I., Xargay, E., Hovakimyan, N., Cao, C., Gregory, I.M., Valavani, L.: Experimental Validation of L 1 Adaptive Control: The Rohrs Counterexample in Flight. Journal of Guidance, Control, and Dynamics 34(5), 1311–1328 (2011)
[5] Dobrokhodov, V., Xargay, E., Hovakimyan, N., Kaminer, I., Cao, C., Gregory, I.M.: Multicriteria analysis of an L1 adaptive flight control system. Proceedings of the Institution of Mechanical Engineers, Part I: Journal of Systems and Control Engineering 227(4), 413–427 (2013)
[6] Ducard, G.J.: Fault-tolerant flight control and guidance systems: Practical methods for small unmanned aerial vehicles. Springer (2009)
[7] Edwards, C., Lombaerts, T., Smaili, H.: Fault tolerant flight control: a benchmark challenge, vol. 399. Springer (2010)
[8] Hovakimyan, N., Cao, C.: L1 adaptive control theory: Guaranteed robustness with fast adaptation, vol. 21. Siam (2010)
[9] Hovakimyan, N., Xargay, E., Gregory, I.M.: Experience with the AirSTAR L1 Adaptive Flight Control System (2011)
[10] Jiang, J., Zhang, Y.: Accepting performance degradation in fault-tolerant control system design. IEEE Transactions on Control Systems Technology 14(2), 284–292 (2006)
[11] Kim, K.-S., Lee, K.-J., Kim, Y.: Reconfigurable flight control system design using direct adaptive method. Journal of Guidance, Control, and Dynamics 26(4), 543–550 (2003)
[12] Patel, V.V., Cao, C., Hovakimyan, N., Wise, K.A., Lavretsky, E.: $\mathcal{L}_1$ adaptive controller for tailless unstable aircraft in the presence of unknown actuator failures. International Journal of Control 82(4), 705–720 (2009)
[13] Patton, R.J., Hu, H.: Fault tolerant control: the 1997 situation (survey), pp. 1033–1054 (1997)
[14] Pomet, J.-B., Praly, L.: Adaptive nonlinear regulation: estimation from the lyapunov equation. IEEE Transactions on Automatic Control 37(6), 729–740 (1992)

[15] Sadeghzadeh, I., Zhang, Y.: A Review on Fault-Tolerant Control for Unmanned Aerial Vehicles (UAVs), pp. 1–12 (2011)
[16] Song, H., Zhang, T., Zhang, G., Lu, C.: Integrated Design of Interceptor Guidance and Control Based on L1 Adaptive Control. In: 2013 5th International Conference on Intelligent Human-Machine Systems and Cybernetics, pp. 525–528 (2013), http://ieeexplore.ieee.org/lpdocs/epic03/wrapper.htm?arnumber=6643943
[17] Stevens, B.L., Lewis, F.L.: Aircraft Control and Simulation. Wiley, New York (2003)
[18] Tao, G.: Adaptive control of systems with actuator failures. Springer (2004)
[19] Xargay, E., Dobrokhodov, V., Kaminer, I., Pascoal, A.M., Hovakimyan, N.: Time-Critical Cooperative Control of Multiple, pp. 49–73 (October 2012)
[20] Zhang, Y., Jiang, J.: Fault tolerant control system design with explicit consideration of performance degradation. IEEE Transactions on Aerospace and Electronic Systems 39(3), 838–848 (2003)
[21] Zhang, Y., Jiang, J.: Bibliographical review on reconfigurable fault-tolerant control systems. Annual Reviews in Control 32(2), 229–252 (2008)

$\mathcal{L}_1$ Adaptive Control of Systems with Disturbances of Unknown Bounds

Toufik Souanef, Ahsene Boubakir, and Walter Fichter

Abstract. This paper presents an $\mathcal{L}_1$ adaptive controller for systems with disturbances of unknown bounds. The estimation laws borrows insights from the sliding mode control to estimate the unknown bounds of disturbances. Simulation results are provided to show the benefits of the presented control scheme.

1 Introduction

$\mathcal{L}_1$ adaptive control was developed to improve the robustness of the model reference adaptive control in the presence of high frequency noise induced by fast adaptation [8]. An approach was also designed to deal with unmatched disturbances, where the problem of the inversion of the control matrix for unmatched disturbance compensation is reduced to the inversion of transfer functions involving the system's desired dynamics and the low pass filter [20]. The closed-loop $\mathcal{L}_1$ adaptive control system has uniform and quantifiable transient performance bounds. However, when the bounds of the external disturbances are not known, as it is in practice, the performance bounds cannot be clearly evaluated.

In this article, a disturbance estimator is presented based on the concept of the sliding surface, used in the sliding mode control [19]. We notice that adaptive schemes have been used with sliding mode control to estimate the unknown upper bounds of perturbations and achieve global asymptotic stability [12, 21, 9] and [7]. The main disadvantage of these approaches is that there is no systematic way to quantify and improve the transient performance. Moreover, fast adaptation may lead to poor robustness characteristics [10]. Furthermore, a major drawback of the sliding mode control is the lack of robustness against unmatched disturbances [6], that

Toufik Souanef · Walter Fichter
Institute of Flight Mechanics and Control, Pfaffenwaldring 27, 70569 Stuttgart, Germany
e-mail: {toufik.souanef,fichter}@ifr.uni-stuttgart.de

Ahsene Boubakir
University of Jijel, BP 98, Jijel 18000, Algeria
e-mail: ah_boubakir@yahoo.fr

J. Bordeneuve-Guibé et al. (eds.), *Advances in Aerospace Guidance, Navigation and Control*, DOI: 10.1007/978-3-319-17518-8_10

is, perturbations that do not lie in the range space of the input matrix. Undeniably, if a sliding mode controller is improperly designed, eliminating the matched disturbances could lead to amplification of the unmatched ones [17]. Several solutions to this problem were developed in [5, 16, 3, 1] and [2]. Their common issue is the need for a compromise between system performance and disturbance attenuation.

The main contributions of this work are the following:

- Improving the robustness of $\mathcal{L}_1$ adaptive control systems in the presence of disturbances with unknown bounds by using the sliding surface structure from sliding mode control;
- Proving stability and achieving quantifiable performance bounds both for the system output and the control signal.

In addition, the low pass filter used in the control law contributes to eliminate the high frequency oscillations that cause chattering and ensures continuity of the control signal [11].

In order to show the application potential of this approach, simulation results of a pitch rate control of a small fixed-wing Unmanned Aerial Vehicle (UAV) are presented, with large uncertainties in aerodynamic parameters.

2 Problem Formulation

Given a class of nonlinear single input single output systems defined by

$$\begin{aligned} \dot{x}(t) &= A_m x(t) + b\big(\omega\, u(t) + \theta^\top x(t)\big) + d\big(t, x(t)\big), \quad x(0) = x_0, \\ y(t) &= c^\top x(t), \end{aligned} \tag{1}$$

where $A_m \in \mathbb{R}^{n \times n}$ is a known Hurwitz matrix that defines the desired dynamics of the system; $b, c \in \mathbb{R}^n$ are known constant vectors; $x(t) \in \mathbb{R}^n$ is the state vector which is assumed available for measurement; $u(t) \in \mathbb{R}$ is the control input; $y(t) \in \mathbb{R}$ is the scalar performance output; $\omega \in \mathbb{R}$ is an unknown constant with known sign; $\theta \in \mathbb{R}^n$ is a vector of constant unknown parameters representing model uncertainties; and $d(t, x(t)) \in \mathbb{R}\text{x}\mathbb{R}^n \rightarrow \mathbb{R}^n$ is an unknown nonlinear map.

Assumption 1. The pair (A_m, b) is controllable.

Assumption 2. The nonlinear map $d(t, x(t))$ has matched and unmatched terms , i. e., there exist a nonlinear continuous functions $\eta_m(t, x) \in \mathbb{R}$ and $\eta_u(t, x) \in \mathbb{R}\text{x}\mathbb{R}^n \rightarrow \mathbb{R}^n$, such that system (1) is written

$$\dot{x}(t) = A_m x(t) + b\big(\omega u(t) + \theta^\top x(t) + \eta_m(t, x)\big) + \eta_u(t, x). \tag{2}$$

Assumption 3. The nonlinear functions $\eta_m(t, x)$ and $\eta_u(t, x)$ are uniformly bounded, i. e., there exists an unknown real constant $L_m > 0$ and $L_u > 0$, such that for all $t \geq 0$ the following bounds hold:

$$\|\eta_m(t,x)\| \leq L_m \text{ and } \|\eta_u(t,x)\| \leq L_u.$$

Assumption 4. The unknown model parameters are bounded, i.e., $\theta \in \Theta$, where Θ is a known compact convex set and $0 < \omega_l \leq \omega \leq \omega_u$.

The objective is to design a state-feedback controller to ensure that the output of the system tracks a given piecewise continuous bounded reference signal $r(t)$.

3 $\mathcal{L}_1$ Adaptive Control with Sliding Mode

We consider the architecture of $\mathcal{L}_1$ adaptive controller, as shown in Fig. 1, which is composed of the state predictor, the adaptation law and the control law. The main difference of it from prior versions is that we now have an estimation law of both the disturbance and its bound using a sliding surface, whereas the $\mathcal{L}_1$ adaptive control is based on the estimation of the disturbance within its known conservative bound.

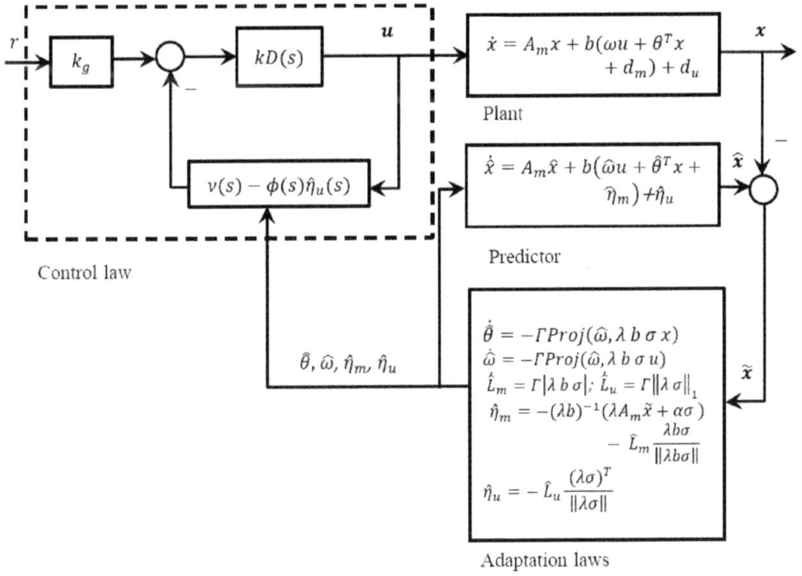

Fig. 1 Block diagram of the control architecture.

Given the state predictor

$$\dot{\hat{x}}(t) = A_m\hat{x}(t) + b\big(\hat{\theta}^\top(t)x(t) + \hat{\omega}(t)u(t) + \hat{\eta}_m(t)\big) + \hat{\eta}_u(t), \quad \hat{x}(0) = x_0, \tag{3}$$

where $\hat{x}(t)$ is the predicted state and, $\hat{\theta}(t)$, $\hat{\omega}(t)$, $\hat{\eta}_m(t)$, and $\eta_u(t)$ are the estimates of the unknown system parameters and disturbances.

Define the sliding surface by

$$\sigma(t) = \lambda \tilde{x}(t), \tag{4}$$

where $\tilde{x}(t) = \hat{x}(t) - x(t)$ is the state estimation error and $\lambda \in \mathbb{R}^{1\times n}$ is a constant arbitrary row vector, chosen such that $\lambda b \neq 0$.

The estimation of the matched disturbance $\hat{\eta}_m(t)$ is defined by

$$\hat{\eta}_m(t) = -(\lambda b)^{-1}\big(\lambda A_m \tilde{x}(t) + \alpha\sigma(t)\big) - \hat{L}_m(t)\frac{\lambda b \sigma(t)}{\|\lambda b \sigma(t)\|}, \tag{5}$$

where $\alpha > 0$ is arbitrary, and the estimated bound $\hat{L}_u(t)$ of the unmatched disturbance $d_m(t,x)$ is given by

$$\dot{\hat{L}}_m(t) = \Gamma|\lambda b \sigma(t)|, \, L_{m0} = \hat{L}_m(0), \tag{6}$$

where $\Gamma \in \mathbb{R}^+$ is the adaptation rate.

The estimation of the unmatched disturbance $\hat{\eta}_u(t)$ is defined by

$$\hat{\eta}_u(t) = -\hat{L}_u(t)\frac{\big(\lambda\sigma(t)\big)^\top}{\|\lambda\sigma(t)\|}, \tag{7}$$

where the estimated bound $\hat{L}_u(t)$ of the unmatched disturbance $d_u(t,x)$ is computed by

$$\dot{\hat{L}}_u(t) = \Gamma\|\lambda\sigma(t)\|_1, \, L_{u0} = \hat{L}_u(0), \tag{8}$$

where $\Gamma \in \mathbb{R}^+$ is the adaptation rate.

The estimation of the unknown parameter θ and the input gain ω are defined by the following equations:

$$\begin{aligned} \dot{\hat{\omega}}(t) &= -\Gamma \mathrm{Proj}\big(\hat{\omega}(t), \lambda\, b\, \sigma(t)\, u(t)\big), \\ \dot{\hat{\theta}}(t) &= -\Gamma \mathrm{Proj}\big(\hat{\theta}(t), \lambda\, b\, \sigma(t)\, x(t)\big), \end{aligned} \tag{9}$$

where $Proj(\cdot,\cdot)$ is the projection operator [15] defined, for two vectors $\xi, \varepsilon \in \mathbb{R}^n$ by

$$Proj(\xi,\varepsilon) = \begin{cases} \varepsilon - \frac{\nabla g(\xi)(\nabla g(\xi)^\top)}{\|\nabla g(\xi)\|^2}\varepsilon g(\xi) & \text{if } g(\xi) > 0 \wedge \nabla g^\top(\xi)\varepsilon > 0 \\ \varepsilon & \text{otherwise} \end{cases}$$

where $g : \mathbb{R}^n \to \mathbb{R}^n$ is an arbitrary convex function.

The projection operator ensures that the estimates of the unknown parameters remain within their sets.

The control law is given by

$$u(s) = kD(s)\Big(k_g\, r(s) - \hat{v}(s) - \phi(s)\hat{\eta}_u(s)\Big), \tag{10}$$

where $k > 0$ is arbitrary, $D(s)$ is a transfer function that leads to a strictly proper stable filter $C(s) = \omega k D(s)/(1+\omega k D(s))$ with $C(0) = 1$; the static gain is chosen as $k_g = -1/(c^\top A_m^{-1} b)$; $\hat{v}(s)$ is the Laplace transformation of the term $\hat{\theta}(t)x(t) + \hat{\omega}(t)u(t) + \hat{\eta}_m(t)$; $\phi(s) = c^\top(sI - A_m)^{-1}/c^\top(s\mathbb{I} - A_m)^{-1}b$; and $\hat{\eta}_u(s)$ is the Laplace transform of $\hat{\eta}_u(t)$.

4 Analysis of the Control Architecture

4.1 *Definitions and $\mathcal{L}_1$-norm Condition*

Let

$$L = \underbrace{\max}_{\theta \in \Theta} \|\theta\|_1,\ H(s) = (s\mathbb{I} - A_m)^{-1}b,\ G(s) = \big(1 - C(s)\big)H(s). \tag{11}$$

The $\mathcal{L}_1$ adaptive controller is defined via equations (4) to (10) and is subject to the following $\mathcal{L}_1$-norm condition

$$\|G(s)\|_{\mathcal{L}1} L < 1. \tag{12}$$

Moreover, the design of k and $D(s)$ needs to ensure that

$$G_u(s) = (s\mathbb{I} - A_m)^{-1} - C(s)H(s)\phi(s), \tag{13}$$

is proper and stable.

Furthermore, from Lemma A.12.1 in [8], there exists a vector c_0, such that

$$H_1(s) = C(s)\frac{1}{c_0 H(s)} c_0. \tag{14}$$

4.2 *Analysis of the State Predictor*

In this section, the dependence of the parameters on (t) is dropped unless it is not clear from the context.

From (2) and (3), the prediction error dynamics can be written

$$\dot{\tilde{x}} = A_m\tilde{x} + b\big(\tilde{\omega}u + \tilde{\theta}^\top x + \tilde{\eta}_m\big) + \tilde{\eta}_u, \tag{15}$$

where $\tilde{\theta} = \hat{\theta} - \theta$, $\tilde{\omega} = \hat{\omega} - \omega$, $\tilde{\eta}_m = \hat{\eta}_m - \eta_m$ and $\tilde{\eta}_u = \hat{\eta}_u - \eta_u$. We define also $\tilde{L}_m = \hat{L}_m - L_m$ and $\tilde{L}_u = \hat{L}_u - L_u$.

Lemma 1. The following uniform bound holds for the prediction error

$$\|\tilde{x}\|_{\mathcal{L}_\infty} \le \frac{\bar{\theta}}{\sqrt{\Gamma \|\lambda\|}}, \tag{16}$$

where $\bar{\theta} = \tilde{\theta}^\top(0)\tilde{\theta}(0) + \tilde{\omega}^2(0) + \tilde{L}_m^2(0) + \tilde{L}_u^2(0)$.

Proof. Consider the Lyapunov function candidate

$$V = \frac{1}{2}\sigma^2 + \frac{1}{2}\Gamma^{-1}\left(\tilde{\theta}^\top\tilde{\theta} + \tilde{\omega}^2 + \tilde{L}_m^2 + \tilde{L}_u^2\right). \tag{17}$$

The derivative of the Lyapunov function is given by

$$\dot{V} = \sigma\dot{\sigma} + \Gamma^{-1}\left(\tilde{\theta}^\top\dot{\tilde{\theta}} + \tilde{\omega}\dot{\tilde{\omega}} + \tilde{L}_m\dot{\tilde{L}}_m + \tilde{L}_u\dot{\tilde{L}}_u\right). \tag{18}$$

From (4), the derivative of the sliding surface can be written as

$$\dot{\sigma} = \lambda A_m\tilde{x} + \lambda b\left(\tilde{\theta}^\top x + \tilde{\omega}u + \tilde{\eta}_m\right) + \lambda\tilde{\eta}_u. \tag{19}$$

Replacing in (18), it follows that

$$\begin{aligned}\dot{V} =& \sigma\Big(\lambda A_m\tilde{x} + \lambda b\left(\tilde{\theta}^\top x + \tilde{\omega}u + (\hat{\eta}_m - \eta_m)\right) + \lambda(\hat{\eta}_u - \eta_u)\Big)\\ &+ \Gamma^{-1}\left(\tilde{\theta}^\top\dot{\hat{\theta}} + \tilde{\omega}\dot{\hat{\omega}} + \tilde{L}_m\dot{\hat{L}}_m + \tilde{L}_u\dot{\hat{L}}_u\right).\end{aligned} \tag{20}$$

Given $\hat{\eta}_m$ and $\hat{\eta}_u$, from (5) and (7) and the adaptation law (9) it can be written

$$\begin{aligned}\dot{V} =& -\alpha\sigma^2 - \lambda b\sigma\eta_m - \lambda\sigma\eta_u\\ &- |\lambda b\sigma|\hat{L}_m - \|\lambda\sigma\|\hat{L}_u + \Gamma^{-1}\left(\tilde{L}_m\dot{\hat{L}}_m + \tilde{L}_u\dot{\hat{L}}_u\right).\end{aligned} \tag{21}$$

Recalling that $\|\cdot\|_1 \geq \|\cdot\|$, the following upper bound can be derived

$$\begin{aligned}\dot{V} \leq& -\alpha\sigma^2 + |\lambda b\sigma|\|\eta_m\| + \|\lambda\sigma\|_1\|\eta_u\|\\ &- |\lambda b\sigma|\hat{L}_m - \|\lambda\sigma\|_1\hat{L}_u + \Gamma^{-1}\left(\tilde{L}_m\dot{\hat{L}}_m + \tilde{L}_u\dot{\hat{L}}_u\right).\end{aligned} \tag{22}$$

Using assumption 3, it can be written

$$\dot{V} \leq -\alpha\sigma^2 - |\lambda b\sigma|\tilde{L}_m - \|\lambda\sigma\|_1\tilde{L}_u + \Gamma^{-1}\left(\tilde{L}_m\dot{\hat{L}}_m + \tilde{L}_u\dot{\hat{L}}_u\right). \tag{23}$$

Considering the adaptation laws from (6) and (8), it follows that

$$\dot{V} \leq -\alpha\sigma^2. \tag{24}$$

it follows that the prediction error $\tilde{x}$, the estimation errors of $\tilde{\theta}$, $\tilde{\omega}$; and the disturbances errors $\tilde{L}_m$ and $\tilde{L}_u$ are uniformly bounded.

Furthermore, we have

$$\begin{aligned}\|\lambda\|\|\tilde{x}\|^2 &\leq V(t)\\ &\leq V(0)\\ &= \Gamma^{-1}\left(\tilde{\theta}^\top(0)\tilde{\theta}(0) + \tilde{\omega}^2(0) + \tilde{L}_m^2(0) + \tilde{L}_u^2(0)\right),\end{aligned} \tag{25}$$

and therefore

$$\|\tilde{x}\|^2 \leq \frac{\bar{\theta}}{\|\lambda\|\Gamma}. \tag{26}$$

Recalling, that $\|\cdot\|_\infty \leq \|\cdot\|$ this completes the proof. □

Remark 1. From (6) and (8) it follows that the estimation errors of the disturbances $\eta_m(t,x(t))$ and $\eta_u(t,x(t))$ are bounded.

Remark 2. Similar to $\mathcal{L}_1$ adaptive control [8], the bound of the prediction error can be made arbitrarily small by increasing the adaptation rates.

4.3 Reference System Analysis

The reference system in this case is the same as in all $\mathcal{L}_1$ adaptive control architectures; in fact, compensation of uncertainties is assumed within the bandwidth of the feedback channel. The reference system, i.e. the closed-loop system with nominal parameters, is considered, defined by

$$\begin{aligned}\dot{x}_r(t) =& A_m x_r(t) + b\big(\omega u_r(t) + \theta^\top x_r(t) + \eta_m(t,x_r)\big) + \eta_u(t,x_r), \qquad x_r(0) = x_0,\\ y_r(t) =& c^\top x_r(t).\end{aligned} \tag{27}$$

The control law of the reference system is given by

$$u_r(s) = \frac{C(s)}{\omega}\Big(k_g r(s) - v_r(s) - \phi(s)\eta_{ur}(s)\Big), \tag{28}$$

where $v_r(s)$ is the Laplace transformation of the term $\theta^\top x_r(t) + \eta_{mr}(t)$; $D(s)$, k_g and $\phi(s)$ are defined above; and $\eta_{mr}(s)$ and $\eta_{ur}(s)$ are the Laplace transform of the functions $\eta_m(t) = \eta_m(t,x_r)$ and $\eta_u(t) = \eta_u(t,x_r)$, respectively.

Lemma 2. If the filter $C(s)$ is designed such that it verifies the $\mathcal{L}_1$ norm condition in (17), then the closed-loop reference system in (27) and (28) is Bounded-Input Bounded-State (BIBS) stable with respect to the reference input and initial conditions.

Proof. The closed-loop reference system can be written

$$\begin{aligned}x_r(s) =& C(s)H(s)k_g r(s) + G(s)\theta^\top x_r(s)\\ &+ G(s)\eta_{mr}(s) + G_u(s)\eta_{ur}(s) + x_{in}(s),\end{aligned} \tag{29}$$

where $x_{in}(s) = (s\mathbb{I} - A_m)^{-1}x_0$.

Then, for all $t \in [0,\tau]$, we have

$$\begin{aligned}\|x_{r_\tau}\|_{\mathcal{L}_\infty} \leq& \|C(s)H(s)\|_{\mathcal{L}_1}k_g\|r_\tau\|_{\mathcal{L}_\infty} + \|G(s)\|_{\mathcal{L}_1}L\|x_{r_\tau}\|_{\mathcal{L}_\infty}\\ &+ \|G(s)\|_{\mathcal{L}_1}\|_{\mathcal{L}_1}\|\eta_{mr_\tau}\|_{\mathcal{L}_\infty} + \|G_u(s)\|_{\mathcal{L}_1}\|\eta_{ur_\tau}\|_{\mathcal{L}_\infty} + \|x_{in_\tau}\|_{\mathcal{L}_\infty}.\end{aligned} \tag{30}$$

Substituting the upper bounds of η_{mr} and η_{ur} and solving for $\|x_{r_\tau}\|_{\mathcal{L}_\infty}$ in the equation above to obtain the following bound

$$\begin{aligned}\|x_{r_\tau}\|_{\mathcal{L}_\infty} \leq & \frac{\|C(s)H(s)\|_{\mathcal{L}_1} k_g \|r_\tau\|_{\mathcal{L}_\infty} + \|G(s)\|_{\mathcal{L}_1} L_m}{1 - \|G(s)\|_{\mathcal{L}_1} L} \\ & + \frac{\|G_u(s)\|_{\mathcal{L}_1} L_u + \|x_{in}\|_{\mathcal{L}_\infty}}{1 - \|G(s)\|_{\mathcal{L}_1} L}.\end{aligned} \tag{31}$$

If the $\mathcal{L}_1$ norm condition in (11) is verified and $G_u(s)$ is proper and stable, then $\|x_{r_\tau}\|_{\mathcal{L}_\infty}$ is uniformly bounded for all $\tau > 0$, and the proof is complete. □

4.4 Performance Bounds

The following theorem shows the performance bounds of the $\mathcal{L}_1$ adaptive controller.

Theorem 1. Given the system (1), the reference system (27), (28) and the $\mathcal{L}_1$ adaptive controller (10), (3), (9), and (6), we have

$$\|x_r - x\|_{\mathcal{L}_\infty} \leq \gamma_1, \tag{32}$$

$$\|u_r - u\|_{\mathcal{L}_\infty} \leq \gamma_2, \tag{33}$$

where

$$\gamma_1 = \frac{\|C(s)\|_{\mathcal{L}_1}}{1 - \|G(s)\|_{\mathcal{L}_1} L}(\rho + 2L_m) + 2\frac{\|G_u\|_{\mathcal{L}_1}}{1 - \|G(s)\|_{\mathcal{L}_1} L} L_u,$$

and

$$\gamma_2 = \|\frac{C(s)}{\omega}\|_{\mathcal{L}_1}(L\gamma_1 + 2L_m) + \|\phi(s)\|_{\mathcal{L}_1} L_u + \|\frac{H_1}{\omega}\|_{\mathcal{L}_1}\rho.$$

Proof. The control law in (10) can be written as

$$u(s) = \frac{C(s)}{\omega}\left(k_g r(s) - \nu_1(s) - \phi(s)\eta_u(s) - \tilde{\nu}(s)\right), \tag{34}$$

where $\nu_1(s)$ is the Laplace transform of the function $\theta^\top x(t) + \eta_m(t)$, $\tilde{\nu}(s)$ is the Laplace transformation of the function $\tilde{\omega}u + \tilde{\theta}x(t) + \hat{\eta}_m(t) - \eta_m(t) + \tilde{\eta}_{u1}(t)$, where $\eta_{u1}(t)$ is the function with the Laplace transform $\tilde{\eta}_{u1}(s) = \phi(s)(\hat{\eta}_u(s) - \eta_u(s))$.

Hence, the Laplace transformation of the closed loop system (1) can be written

$$\begin{aligned}x(s) = & H(s)C(s)k_g r(s) + G(s)\theta^\top x(s) + G(s)\eta_m(s) \\ & + G_u(s)\eta_u(s) - C(s)H(s)\tilde{\nu}(s) + x_{in}(s).\end{aligned} \tag{35}$$

Taking the difference of (29) and (35), it follows that

$$\begin{aligned} x_r(s) - x(s) =& G(s)\theta^\top (x_r(s) - x(s)) + C(s)H(s)\tilde{v}(s) \\ &+ G(s)\big(\eta_m(s) - \eta_{mr}(s)\big) + G_u(s)\big(\eta_u(s) - \eta_{ur}(s)\big). \end{aligned} \tag{36}$$

From (15) the prediction error dynamics can be written

$$\dot{\tilde{x}}(t) = A_m \tilde{x}(t) + b\,\tilde{v}(t), \tag{37}$$

and its Laplace transform is

$$\tilde{x}(s) = H(s)\tilde{v}(s). \tag{38}$$

Substituting in (36) and solving for $x_r(s) - x(s)$, the following bound holds for $t \in [0, \tau]$

$$\begin{aligned} \|(x_r - x)_\tau\|_{\mathcal{L}_\infty} \le& \frac{\|C(s)\|_{\mathcal{L}_1}}{1 - \|G(s)\|_{\mathcal{L}_1} L} \|\tilde{x}_\tau + \eta_{m_\tau} - \eta_{mr_\tau}\|_{\mathcal{L}_\infty} \\ &+ \frac{\|G_u\|_{\mathcal{L}_1}}{1 - \|G(s)\|_{\mathcal{L}_1} L} \|(\eta_u - \eta_{ur})_\tau\|_{\mathcal{L}_\infty}. \end{aligned} \tag{39}$$

Given the upper bound of $\tilde{x}(t)$ from Lemma 1, and the disturbance bounds from Assumption 3, it follows that

$$\begin{aligned} \|(x_r - x)_\tau\|_{\mathcal{L}_\infty} \le& \frac{\|C(s)\|_{\mathcal{L}_1}}{1 - \|G(s)\|_{\mathcal{L}_1} L} (\rho + 2L_m) \\ &+ 2\frac{\|G_u\|_{\mathcal{L}_1}}{1 - \|G(s)\|_{\mathcal{L}_1} L} L_u, \end{aligned} \tag{40}$$

which holds uniformly for all $\tau \ge 0$ and leads to the bound in (32).

To show the second bound in (33), by taking the difference of (28) and (34), the control bound can be written

$$\begin{aligned} u_r(s) - u(s) =& -\frac{C(s)}{\omega}\big(\theta^\top x_r(s) - x(s)\big) + \frac{C(s)}{\omega}\tilde{\eta}(s) \\ &- \frac{C(s)}{\omega}\big(\eta_m(s) - \eta_{mr}(s)\big) - \frac{C(s)}{\omega}\phi(s)\big(\eta_u(s) - \eta_{ur}(s)\big). \end{aligned} \tag{41}$$

Thus

$$\frac{C(s)}{\omega}\tilde{\eta}(s) = \frac{1}{\omega} H_1(s)\tilde{x}(s). \tag{42}$$

From Lemma A.7.1 [8], it follows that (41) can be upper bounded as

$$\begin{aligned} \|(u_r - u)_\tau\|_{\mathcal{L}_\infty} \le& \left\|\frac{C(s)}{\omega}\right\|_{\mathcal{L}_1} \|(x_r - x)_\tau\|_{\mathcal{L}_\infty} + \left\|\frac{H_1}{\omega}\right\|_{\mathcal{L}_1} \|\tilde{x}_\tau\|_{\mathcal{L}_\infty} \\ &+ \left\|\frac{C(s)}{\omega}\right\|_{\mathcal{L}_1} (2L_m + \|\phi(s)\|_{\mathcal{L}_1} L_u), \end{aligned} \tag{43}$$

which holds uniformly for all $\tau \ge 0$, leading to the bound in (33). □

Remark 3. Similar to the classical $\mathcal{L}_1$ adaptive control, the performance of the developed sliding mode $\mathcal{L}_1$ adaptive controller is bounded for both input and output signals, and it can be arbitrarily small by increasing the adaptation rate.

Remark 4. From the results of Lemma 1 and Theorem 1 it can be shown that the prediction error $\tilde{x}(t)$ is asymptotically stable. Actually, boundedness of the control signal and the estimated parameters implies the boundedness of $\dot{\tilde{x}}(t)$ in (15), which results in uniform boundedness of the second derivative of the Lyapunov function in (17). Applying Barbalat's Lemma, it follows that $\lim_{t\to\infty}\tilde{x}(t)=0$.

5 Implementation Issues

In practice, the sliding surface $\sigma(t)$ does not go to zero. This results in a persistent increase of the estimated bounds of (6) and (8), and may lead to bound overestimation with negative effects [14] and [4]. A solution to this problem is the deadzone modification [13], which works by switching the estimator off when the prediction error gets below a certain threshold.

The adaptation laws in (6) and (8) are modified to be

$$\dot{\hat{L}}_m(t)=\begin{cases}\Gamma|\lambda b\sigma(t)| & \text{if}\,|\sigma(t)|>\varepsilon_m,\\ 0 & \text{if not,}\end{cases} \tag{44}$$

and

$$\dot{\hat{L}}_u(t)=\begin{cases}\Gamma\|\lambda\sigma(t)\|_1 & \text{if}\,|\sigma(t)|>\varepsilon_u,\\ 0 & \text{if not,}\end{cases} \tag{45}$$

where $\varepsilon_m>0$ and $\varepsilon_u>0$ are the arbitrary small thresholds of the dead-zone function.

Lemma 3. The same bound in (15) holds for prediction error when using the adaptation laws in (44) and (45).

Proof. Consider the Lyapunov function defined in (17) and following the same steps as in equations (18) to (23), it can be written

$$\begin{aligned}\dot{V}(t)\leq&-\alpha\sigma^2(t)+\tilde{L}_m(t)\big(\Gamma^{-1}\dot{\hat{L}}_m(t)-|\lambda b\sigma(t)|\big)\\&+\tilde{L}_u(t)\big(\Gamma^{-1}\dot{\hat{L}}_u(t)-\|\lambda\sigma(t)\|_1\big).\end{aligned} \tag{46}$$

By introducing the adaptation laws from (44) and (45), and if at any time $t_1>0$ one has $|\sigma(t_1)|>\delta=max\{\varepsilon_m,\varepsilon_u\}$, and it follows that

$$\dot{V}(t_1)\leq-\alpha\sigma^2(t_1), \tag{47}$$

Consequently, the prediction error $\tilde{x}(t)$, the estimation errors of $\tilde{\theta}(t)$, $\tilde{\omega}(t)$; and the disturbances errors $\tilde{L}_m(t)$ and $\tilde{L}_u(t)$ are uniformly bounded.

Similar to equations (25) and (26), it can be shown that the same bound in (15) holds and the proof is complete. □

6 Application: Pitch Control of a Small Unmanned Aerial Vehicle

The developed approach was applied to the pitch dynamics of a fixed wing UAV, to show its performances in the control of systems with unmodeled uncertainties, external disturbances and faults. The linearized short period dynamics of an aircraft, while assuming that an additive disturbance affects the plant, can be written in a general state space representation, from [18]

$$\dot{x}(t) = A\,x(t) + b\,\omega\,\delta_e(t) + d(t,x) \tag{48}$$

where $x = [\alpha,\ q]^\top$ is the state vector, α is the angle of attack, q is the pitch angular velocity, $A \in \mathbb{R}^2$ is the unknown matrix of the linearized short period dynamics of the UAV, $b \in \mathbb{R}^n$ is known and represents the input vector, $\omega \in \mathbb{R}$ is an unknown constant with known sign representing control vector uncertainty, and $d(t,x(t)) \in \mathbb{R}\text{x}\mathbb{R}^2 \to \mathbb{R}^2$ is the nonlinear map of external disturbances.

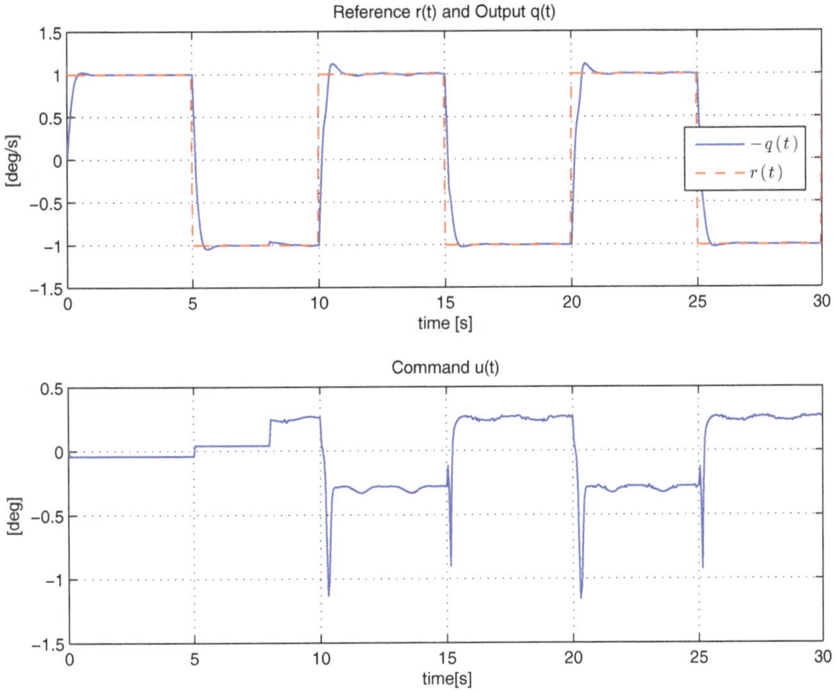

Fig. 2 Response of the closed loop system with uncertainties.

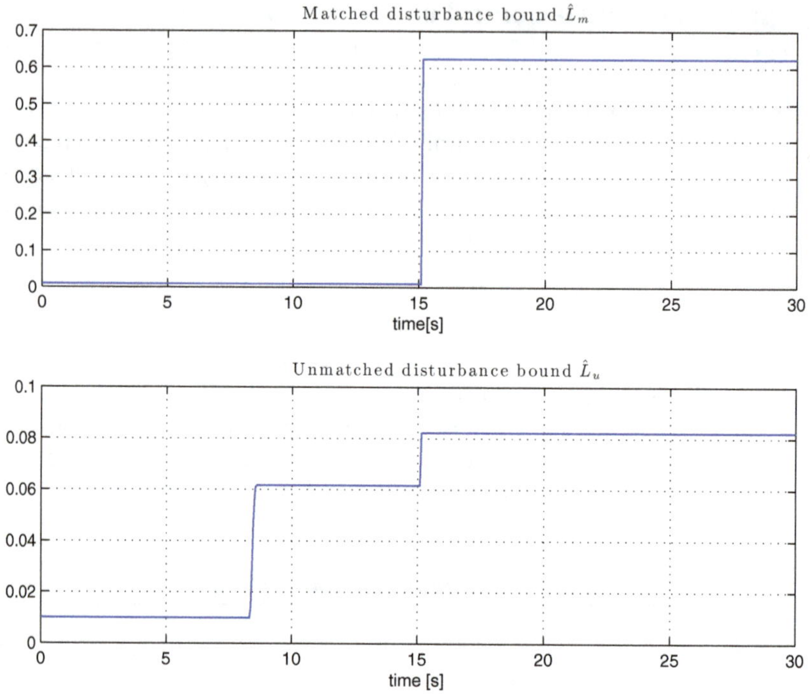

Fig. 3 Estimated disturbances bounds.

The control approach is based on the augmentation of a linear controller by the $\mathcal{L}_1$ adaptive controller, as it is common in flight control systems. The total deflection of the elevator $\delta_e(t) = u_b(t) + u(t)$ is the sum of the commands from the baseline linear controller $u_b(t)$ and the adaptive controller $u(t)$.

Consequently (48) takes the form defined in (2)

$$\dot{x}(t) = A_m x(t) + b\big(\omega u(t) + \theta^\top x(t) + \eta_m(t,x)\big) + \eta_u(t,x). \tag{49}$$

The controller is designed to be robust against model uncertainties within the compact sets $\omega = [0.4,\ 1.4]$, $\Theta = \{\vartheta = [\vartheta_1, \vartheta_2] \in \mathbb{R}^2 : \vartheta_i \in [-0.5,\ 1.5], i = 1,2\}$. The $\mathcal{L}_1$ adaptive controller parameters are set $\Gamma = 1000$, $D(s) = 1/s$ and $k = 500$. The controller is implemented with dead-zone thresholds $\varepsilon_m = 0.05$ and $\varepsilon_u = 0.05$.

To show system robustness, three types of matched uncertainties are introduced at t=8s:

1. Linear-in-state uncertainty representing 50% decrease in system damping and static stability;
2. Control efficiency constant uncertainty $\omega = 0.5$;

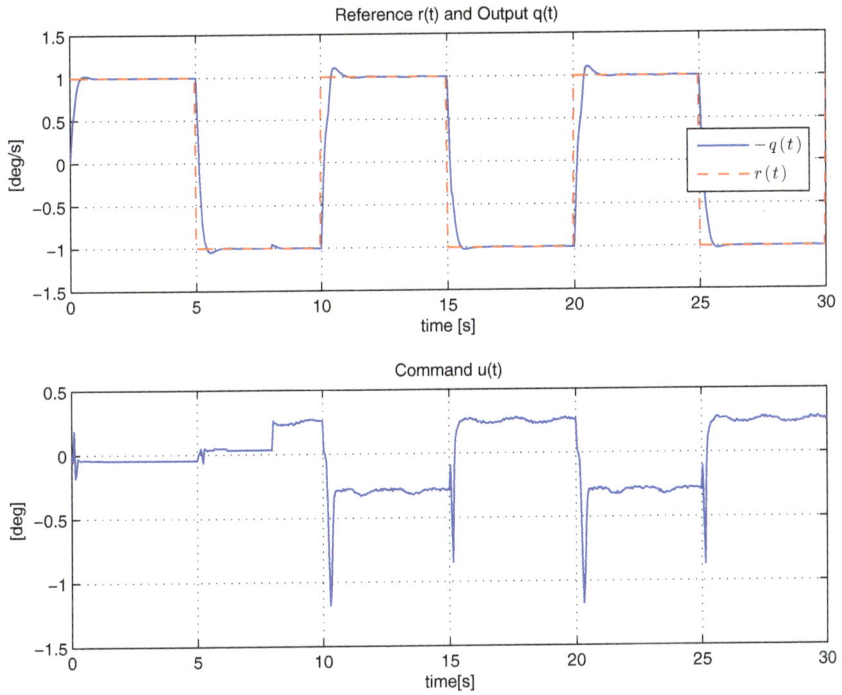

Fig. 4 Response of the closed loop system in the presence of unmodeled sensors and actuator dynamics.

3. A matched nonlinear uncertainty $\eta_m(x,t) = 0.5e^{[-(\alpha-\alpha_c)^2)/2\alpha_k^2]}$ with $\alpha_c = 2\pi/180$ and $\alpha_k = 0.0233$;
4. An unmatched component of the disturbance as a periodic signal $\eta_m(t) = 0.1 sin(\pi t)$.

This combination of uncertainties represents a situation where the vehicle becomes 50% more statically unstable, loses 50% of its pitch damping ability, and the UAV controllability decreases by 50%.

In Fig. 2, it can be seen that without any re-tuning, the controller copes with large parameter uncertainties and shows good performance. Moreover, the elevator command is within acceptable limits.

To underline the robustness of the controller against unmodeled system dynamics, the same plant given in (1) is considered with the actuator dynamics given by $\mu_a(s) = \frac{30}{s+30}$ and sensor dynamics $\mu_s(s) = \frac{1}{0.0001s^2+0.02s+1}$. Simulation results are given in Fig. 4. One can see that the controller is able to track the reference command, and no high frequency oscillations are noted in the control channel.

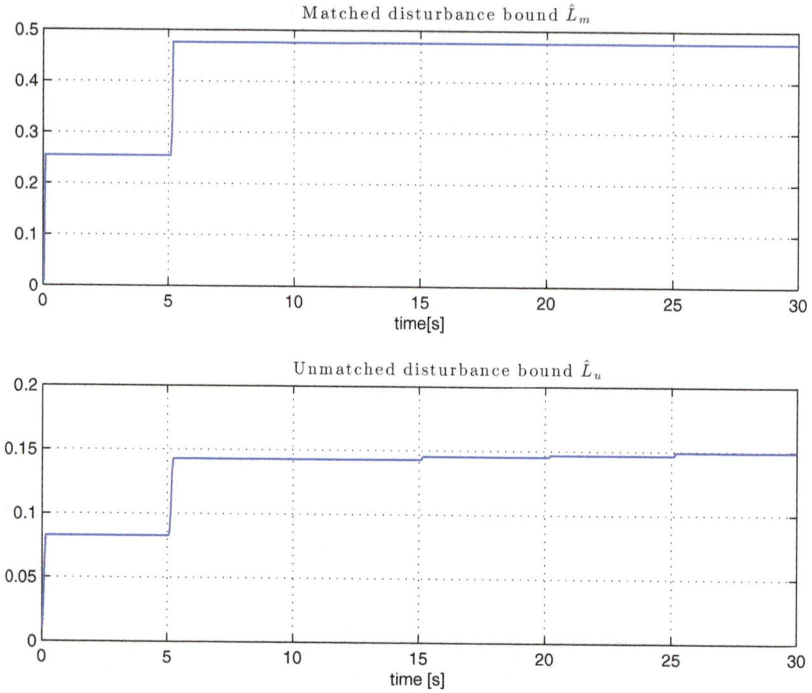

Fig. 5 Estimated disturbances bounds in the presence of unmodeled sensors and actuator dynamics.

7 Conclusion

An $\mathcal{L}_1$ adaptive controller for systems with matched and unmatched disturbances of unknown bounds is presented in this paper. The unknown bounds are estimated using a sliding surface. The proposed scheme guaranteed a fast transient response with bounded tracking. Simulation results show good performance for pitch angle control of a small fixed-wing UAV in the presence of large uncertainties.

Acknowledgements. The authors would like to acknowledge the valuable suggestions made by Pr. Naira Hovakimyan and Dr. Evgeny Kharisov, from the University of Illinois at Urbana-Champaign.

References

[1] Cao, W.J., Xu, J.X.: Nonlinear integral-type sliding surface for both matched and unmatched uncertain systems. IEEE Transactions on Automatic Control 49(8), 1355–1360 (2004)

[2] Castaños, F., Fridman, L.: Analysis and design of integral sliding manifolds for systems with unmatched perturbations. IEEE Transactions on Automatic Control 51(5), 853–858 (2006)
[3] Choi, H.H.: An lmi-based switching surface design method for a class of mismatched uncertain systems. IEEE Transactions on Automatic Control 48(9), 1634–1638 (2003)
[4] Cong, B., Chen, Z., Liu, X.: On adaptive sliding mode control without switching gain overestimation. International Journal of Robust and Nonlinear Control 24(3), 515–531 (2014)
[5] Davies, R., Spurgeon, S.: Robust implementation of sliding mode control schemes. International Journal of Systems Science 24(4), 733–743 (1993)
[6] Draženović, B.: The invariance conditions in variable structure systems. Automatica 5(3), 287–295 (1969)
[7] Fei, J.: Robust adaptive vibration tracking control for a micro-electro-mechanical systems vibratory gyroscope with bound estimation. IET Control Theory & Applications 4(6), 1019–1026 (2010), doi:10.1049/iet-cta.2008.0199
[8] Hovakimyan, N., Cao, C.: L1 adaptive control theory: Guaranteed robustness with fast adaptation, vol. 21. SIAM (2010)
[9] Huang, Y.J., Kuo, T.C., Chang, S.H.: Adaptive sliding-mode control for nonlinear systems with uncertain parameters. IEEE Transactions on Systems, Man, and Cybernetics, Part B: Cybernetics 38(2), 534–539 (2008)
[10] Ioannou, P.A., Sun, J.: Robust adaptive control. Courier Dover Publications (2012)
[11] Kharisov, E., Hovakimyan, N.: Generalization of $\mathcal{L}_1$ adaptive control architecture for switching estimation laws. In: 2012 American Control Conference (ACC), pp. 1907–1912 (2012), doi:10.1109/ACC.2012.6315651
[12] Perruquetti, W., Barbot, J.P.: Sliding mode control in engineering. CRC Press (2002)
[13] Peterson, B., Narendra, K.S.: Bounded error adaptive control. IEEE Transactions on Automatic Control 27(6), 1161–1168 (1982)
[14] Plestan, F., Shtessel, Y., Bregeault, V., Poznyak, A.: New methodologies for adaptive sliding mode control. International Journal of Control 83(9), 1907–1919 (2010)
[15] Pomet, J.B., Praly, L.: Adaptive nonlinear regulation: estimation from the lyapunov equation. IEEE Transactions on Automatic Control 37(6), 729–740 (1992)
[16] Poznyak, A., Fridman, L., Bejarano, F.J.: Mini-max integral sliding-mode control for multimodel linear uncertain systems. IEEE Transactions on Automatic Control 49(1), 97–102 (2004)
[17] Spurgeon, S., Davies, R.: A nonlinear control strategy for robust sliding mode performance in the presence of unmatched uncertainty. International Journal of Control 57(5), 1107–1123 (1993)
[18] Stevens, B.L., Lewis, F.L.: Aircraft Control and Simulation. Wiley, New York (2003)
[19] Utkin, V.: Survey paper variable structure systems with sliding modes. IEEE Transactions on Automatic Control 22(2) (1977)
[20] Xargay, E., Hovakimyan, N., Gregory, I.M.: L 1 Adaptive Flight Control System: Systematic Design and Verification and Validation of Control Metrics, pp. 1–34 (August 2010)
[21] Yoo, D.S., Chung, M.J.: A variable structure control with simple adaptation laws for upper bounds on the norm of the uncertainties. IEEE Transactions on Automatic Control 37(6), 860–865 (1992)

Differential Games Based Autonomous Rendezvous for Aerial Refueling

Ezra Tal and Tal Shima

Abstract. An integrated guidance law and auto-pilot for autonomous rendezvous towards aerial refueling using the probe-and-drogue system is presented. For the derivation the rendezvous problem is considered as a differential game in which the trailing aircraft's objective is to capture the drogue. A linear quadratic cost formulation is utilized in order to develop an optimal control expression for pursuer aircraft elevator, ailerons, and rudder control, as well as optimal evasive action. Optimal evasive action herein represents the worst-case drogue movement. Results of numerical simulations using a longitudinal lateral-directional flight dynamic model of a realistic aircraft are presented.

1 Introduction

Rendezvous for aerial refueling refers to the connecting of the probe on the receiver aircraft, and the drogue at the end of the hose on the tanker aircraft, before commencement of the actual refueling. For a successful rendezvous both position and velocity must be equal, which is complicated by the movement of the drogue. Due to aerodynamic coupling between tanker aircraft, drogue, and receiver aircraft the rendezvous is one of the most challenging maneuvers to perform for pilots.

Ezra Tal
Delft University of Technology, Faculty of Aerospace Engineering,
Delft, 2629 HS, Netherlands
e-mail: e.a.tal@student.tudelft.nl

Tal Shima
Technion - Israel Institute of Technology, Faculty of Aerospace Engineering,
Haifa, 32000, Israel
e-mail: tal.shima@technion.ac.il

J. Bordeneuve-Guibé et al. (eds.), *Advances in Aerospace Guidance, Navigation and Control*, DOI: 10.1007/978-3-319-17518-8_11

The emergence of autonomous unmanned aerial vehicles calls for the development of guidance and control laws that enable autonomous execution of the rendezvous for aerial refueling. The problem has been approached using various methods for robust and adaptive controller design [Wang et al.(2008), Pachter et al.(1997)], and incorporating sensor systems [Tandale et al.(2006)].

Owing to similarities to the problem of missile guidance, control methods based on the concepts of proportional navigation and line of sight angle control have been proposed [Ochi and Kominami(2005)]. Also originating from missile guidance is the concept of zero-sum pursuit-evasion differential game-based guidance laws [Bryson and Ho(1969)]. The method presented in this paper is based on the formulation of the rendezvous for aerial refueling as such a zero-sum pursuit-evasion differential game. Within this formulation the drogue acts as evader and the recipient aircraft as pursuer. The cost of the zero-sum game consists of the final position and velocity differences between drogue and recipient aircraft. It has been shown that for the case of unpredictable evasive action differential game guidance laws surpass optimal control based guidance laws in terms of performance [Anderson(1981)].

In order to account for control boundaries, weighed quadratic terms are added to the cost function. These terms act as penalty functions on the otherwise unbounded control inputs of both pursuer and evader, leading to a linear quadratic (LQ) differential game formulation [Bryson and Ho(1969)].

Within the zero-sum differential game it is the objective of the pursuer to minimize the cost, whereas it is the aim of the evader to maximize this cost. The solution of the game consists of optimal control actions for both pursuer and evader. The former serves as the autonomous aerial refueling guidance law, while the latter can be considered a worst-case movement of the drogue. Hence, the optimal evasive action can be used to give a measure of the robustness of the guidance law with respect to any possible movement by the drogue.

The recipient aircraft is represented by a model involving both dynamic and kinematic states leading to the derivation of an integrated guidance law and autopilot. In previous work the development of an integrated controller for missile autopilot guidance motivated by a differential game formulation has been presented [Shima et al.(2006)]. Application to elevator control for autonomous rendezvous using a longitudinal flight dynamic model has been shown as well [Tal and Shima(2015)].

In the current work, the integrated controller commands elevator, ailerons, and rudder. The concept of a differential games multi-control guidance law has been presented in the context of a dual-control guidance law for interception using missiles with multiple control surfaces [Shima and Golan(2007)].

2 Dynamics Modeling

A linearized rendezvous model considering movement perpendicular to the fixed relative horizontal closing speed V_c with which the recipient aircraft is approaching the drogue is used. The fixed relative horizontal closing speed leads to a known time of interception t_f. An overview of the rendezvous geometry is given in Figure 1.

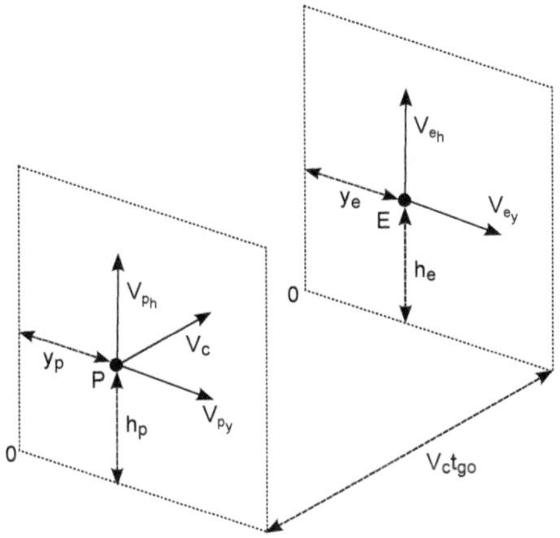

Fig. 1 Rendezvous geometry

The drogue, i.e. evader, has direct control of its acceleration in both the horizontal and vertical plane. The position is given by double integration of the corresponding acceleration input. For vertical position

$$\ddot{h}_e = a_e^{max} v_1 \tag{1}$$

and equivalently for lateral position

$$\ddot{y}_e = a_e^{max} v_2 \tag{2}$$

where a_e^{max} is the maximum magnitude of acceleration per direction, which is set equal for both directions, with $a_e^{max} > 0$, and v_1 and v_2 are normalized control variables with $|v_i| \leq 1$ for $i \in \{1, 2\}$.

By introducing the vertical evader speed V_{e_h} and the lateral evader speed V_{e_y}, the evader kinematics can be formulated as a system of first order differential equations

$$\begin{cases} \dot{h}_e = V_{e_h} \\ \dot{y}_e = V_{e_y} \\ \dot{V}_{e_h} = a_e^{max} v_1 \\ \dot{V}_{e_y} = a_e^{max} v_2 \end{cases} \tag{3}$$

The aircraft, i.e. pursuer, kinematic and dynamic equations are linearized around a horizontal symmetric steady flight condition. The state variables correspond to the state's deviation from this trim condition.

The aircraft vertical position is given by integration of its vertical speed

$$\dot{h}_p = V_{p_h} = V_0(\theta - \alpha) \tag{4}$$

where V_0 is the total airspeed in trimmed condition, θ is the pitch angle, and α is the angle of attack. Assuming that V_0 is constant, the vertical acceleration is

$$\dot{V}_{p_h} = V_0(\dot{\theta} - \dot{\alpha}) = V_0(q - \dot{\alpha}) \tag{5}$$

where q is the pitch rate. The lateral speed is given by

$$\dot{y}_p = V_{p_y} = V_0(\psi + \beta - \alpha_0\phi) \tag{6}$$

where α_0 and θ_0 correspond to the trimmed angle of attack and pitch angle, respectively. The sideslip, heading, and roll angles are indicated by β, ψ, and ϕ, respectively. The lateral acceleration is then given by

$$\dot{V}_{p_y} = V_0(\dot{\psi} + \dot{\beta} - \alpha_0\dot{\phi}) = V_0(r + \dot{\beta} - \alpha_0\dot{\phi}) \tag{7}$$

where r is the yaw rate.

The force and moment contributions of $\dot{\alpha}$ and $\dot{\beta}$ are neglected. It is assumed that the aircraft and flight condition are perfectly symmetrical and that there is no longitudinal lateral-directional coupling due to engine rotor angular momentum. Consequently the linear system of kinematic and dynamic equations is uncoupled. The differential equations for the aircraft dynamics are formulated using dimensional stability and control derivatives. For the control surface dynamics first order lags with time constant τ_δ are assumed. The resulting system of first order equations is given by

$$\begin{cases} \dot{h}_p = V_{p_h} \\ \dot{y}_p = V_{p_y} \\ \dot{V}_{p_h} = -Z_\alpha\alpha - Z_q q - Z_{\delta_{ele}}\delta_{ele} \\ \dot{V}_{p_y} = g\phi + Y_\beta\beta + (Y_p - V_0\alpha_0)p + Y_r r + Y_{\delta_{ail}}\delta_{ail} + Y_{\delta_{rud}}\delta_{rud} \\ \dot{\phi} = p + \theta_0 r \\ V_0\dot{\alpha} = Z_\alpha\alpha + (V_0 + Z_q)q + Z_{\delta_{ele}}\delta_{ele} \\ V_0\dot{\beta} = g\phi + Y_\beta\beta + Y_p p + (Y_r - V_0)r + Y_{\delta_{ail}}\delta_{ail} + Y_{\delta_{rud}}\delta_{rud} \\ \dot{p} = L_\beta\beta + L_p p + L_r r + L_{\delta_{ail}}\delta_{ail} + L_{\delta_{rud}}\delta_{rud} \\ \dot{q} = M_\alpha\alpha + M_q q + M_{\delta_{ele}}\delta_{ele} \\ \dot{r} = N_\beta\beta + N_p p + N_r r + N_{\delta_{ail}}\delta_{ail} + N_{\delta_{rud}}\delta_{rud} \\ \dot{\delta}_{ele} = -\frac{1}{\tau_\delta}\delta_{ele} + \frac{1}{\tau_\delta}\delta_{ele}^{max}u_1 \\ \dot{\delta}_{ail} = -\frac{1}{\tau_\delta}\delta_{ail} + \frac{1}{\tau_\delta}\delta_{ail}^{max}u_2 \\ \dot{\delta}_{rud} = -\frac{1}{\tau_\delta}\delta_{rud} + \frac{1}{\tau_\delta}\delta_{rud}^{max}u_3 \end{cases} \tag{8}$$

where p is the roll rate, and δ_{ele}^{max}, δ_{ail}^{max}, and δ_{rud}^{max} are the maximum elevator, aileron, and rudder deflection magnitudes, respectively. The inputs u_1, u_2, and u_3 are normalized control variables with $|u_i| \leq 1$ for $i \in \{1, 2, 3\}$.

Several observations can be made regarding the resulting system of differential equations. The model is considerably more complicated than the idealized missile dynamics models that are often used for the application of differential game-based guidance [Shinar and Shima(2012)], as it is composed of both dynamical and kinematical states leading to an integrated guidance law and autopilot design. The system is not fully controllable by $\mathbf{u}$, as is typical for linear aircraft models involving both longitudinal and lateral-directional dynamic and kinematic states.

An interesting feature is the fact that the aircraft vertical speed is non-minimum phase with regard to elevator control: The additional lift of a positive elevator deflection will initially result in a positive vertical acceleration, after which it leads to pitch down and negative vertical acceleration. Similar non-minimum phase behavior is exhibited by the response of the lateral speed to aileron and rudder inputs, due to the direct side-force and subsequent respective roll and yaw rates that these inputs cause. These phenomena are relevant, as they lead to control reversal during the final part of rendezvous.

A state-space system incorporating both the receiver aircraft (pursuer) and drogue (evader) dynamics can now be defined. Since the aircraft model is already formulated including vertical and lateral positions and speeds as states, the composite system can straightforwardly be formulated as

$$\dot{\mathbf{x}} = \mathbf{A}\mathbf{x} + \mathbf{B}\mathbf{u} + \mathbf{C}\mathbf{v} \tag{9}$$

where

$$\mathbf{x} = \begin{bmatrix} h_p - h_e \\ y_p - y_e \\ V_{p_h} - V_{e_h} \\ V_{p_y} - V_{e_y} \\ \phi \\ \alpha \\ \beta \\ p \\ q \\ r \\ \delta_{ele} \\ \delta_{ail} \\ \delta_{rud} \end{bmatrix}, \tag{10}$$

$$
\mathbf{A}=\begin{bmatrix}
\mathbf{0}_{2\times2} & \mathbf{I}_{2\times2} & & & \mathbf{0}_{2\times6} & & & & \mathbf{0}_{2\times3} & \\
 & & 0 & -Z_\alpha & 0 & 0 & -Z_q & 0 & -Z_{\delta_{ele}} & 0 & 0 \\
 & & g & 0 & Y_\beta & Y_p - V_0\alpha_0 & 0 & Y_r & 0 & Y_{\delta_{ail}} & Y_{\delta_{rud}} \\
 & & 0 & 0 & 0 & 1 & 0 & \theta_0 & 0 & 0 & 0 \\
\mathbf{0}_{8\times2} & \mathbf{0}_{8\times2} & 0 & \frac{Z_\alpha}{V_0} & 0 & 0 & 1+\frac{Z_q}{V_0} & 0 & \frac{Z_{\delta_{ele}}}{V_0} & 0 & 0 \\
 & & \frac{g}{V_0} & 0 & \frac{Y_\beta}{V_0} & \frac{Y_p}{V_0} & 0 & \frac{Y_r}{V_0}-1 & 0 & \frac{Y_{\delta_{ail}}}{V_0} & \frac{Y_{\delta_{rud}}}{V_0} \\
 & & 0 & 0 & L_\beta & L_p & 0 & L_r & 0 & L_{\delta_{ail}} & L_{\delta_{rud}} \\
 & & 0 & M_\alpha & 0 & 0 & M_q & 0 & M_{\delta_{ele}} & 0 & 0 \\
 & & 0 & 0 & N_\beta & N_p & 0 & N_r & 0 & N_{\delta_{ail}} & N_{\delta_{rud}} \\
\mathbf{0}_{3\times2} & \mathbf{0}_{3\times2} & & & \mathbf{0}_{3\times6} & & & & -\frac{1}{\tau_\delta}\mathbf{I}_{3\times3} &
\end{bmatrix}, \tag{11}
$$

$$
\mathbf{B}=\begin{bmatrix} & \mathbf{0}_{10\times3} & \\ \frac{\delta_{ele}^{max}}{\tau_\delta} & 0 & 0 \\ 0 & \frac{\delta_{ail}^{max}}{\tau_\delta} & 0 \\ 0 & 0 & \frac{\delta_{rud}^{max}}{\tau_\delta} \end{bmatrix}, \quad \mathbf{C}=\begin{bmatrix} \mathbf{0}_{2\times2} \\ -a_e^{max}\mathbf{I}_{2\times2} \\ \mathbf{0}_{9\times2} \end{bmatrix} \tag{12}
$$

For a successful rendezvous the first four states, vertical and lateral position and speed difference, must be nulled. Hence the pursuer is minimizing these values, whereas the evader is maximizing these values. Both are assumed to have perfect knowledge of the entire state.

The admissible gamespace is limited by several constraints. The states ϕ, α, β, p, q, and r are limited by operational limits during the rendezvous. The aircraft control surface states are limited by their respective maximum deflection magnitudes.

3 Guidance Law

Three elements are taken into account in the formulation of the linear-quadratic cost function that is used for guidance law optimization: Terminal relative position and relative speed contributions, the cost of pursuer control, and the cost of evader control. The relative position R is the Euclidean distance between the evader and the pursuer in the vertical-lateral plane and the relative speed V_r is the Euclidean norm of the difference between their velocities in the aforementioned plane. The cost function is given by

$$
J = \frac{1}{2}R^2(t_f) + \frac{Q_V}{2}{V_r}^2(t_f) + \frac{1}{2}\int_{t_0}^{t_f} \left(\alpha_1{u_1}^2(t) + \alpha_2{u_2}^2(t) + \alpha_3{u_3}^2(t) - \beta_1{v_1}^2(t) - \beta_2{v_2}^2(t)\right) dt \tag{13}
$$

where Q_V, α_1, α_2, α_3, β_1, and β_2 are all non-negative weights. Note that if the weights approach zero an interception guidance law is found. Equivalently, setting $Q_V \to \infty$ leads to a nulling of relative speed.

3.1 Order Reduction

In order to reduce the order of the problem the terminal cost $\mathbf{Z}(t_f)$ is introduced. In order to express the terminal cost the zero-effort vector is used. Making use of the terminal projection transformation [Bryson and Ho(1969)] the zero-effort vector is defined as

$$\mathbf{Z}(t) = \mathbf{D}\boldsymbol{\Phi}(t_f, t)\mathbf{x}(t) \tag{14}$$

where

$$\mathbf{D} = \begin{bmatrix} \mathbf{I}_{4\times 4} & \mathbf{0}_{9\times 4} \end{bmatrix} \tag{15}$$

and $\boldsymbol{\Phi}(t_f, t)$ is the transition matrix associated with Eq. (9)

$$\boldsymbol{\Phi}(t_f, t) = e^{\mathbf{A}t_{go}} \tag{16}$$

with

$$t_{go} = t_f - t \tag{17}$$

Given the time-derivative of the transition matrix

$$\dot{\boldsymbol{\Phi}}(t_f, t) = -\boldsymbol{\Phi}(t_f, t)\mathbf{A} \tag{18}$$

the time-derivative of the zero-effort vector is

$$\dot{\mathbf{Z}}(t) = \tilde{\mathbf{B}}(t_f, t)\mathbf{u}(t) + \tilde{\mathbf{C}}(t_f, t)\mathbf{v}(t) \tag{19}$$

where

$$\tilde{\mathbf{B}}(t_f, t) = \mathbf{D}\boldsymbol{\Phi}(t_f, t)\mathbf{B}, \; \tilde{\mathbf{C}}(t_f, t) = \mathbf{D}\boldsymbol{\Phi}(t_f, t)\mathbf{C} \tag{20}$$

The zero-effort vector includes contributions of both the vertical and lateral position and speed differences. It represents the vertical and lateral position and speed differences that would be created by the time t_f if none of the parties were to apply any control from time t onward. $Z_1(t)$ is referred to as the vertical zero-effort miss distance (VZEM), $Z_2(t)$ is referred to as the lateral zero-effort miss distance (LZEM), $Z_3(t)$ is referred to as the vertical zero-effort speed difference (VZES), and $Z_4(t)$ is referred to as the lateral zero-effort speed difference (LZES). The zero-effort vector at t_f is given by

$$\mathbf{Z}(t_f) = \mathbf{D}\mathbf{x}(t_f) = \begin{bmatrix} x_1(t_f) \\ x_2(t_f) \\ x_3(t_f) \\ x_4(t_f) \end{bmatrix} \tag{21}$$

Using the zero-effort vector and the definition of the Euclidean norm the LQ cost function J can now be reformulated as

$$J = \frac{1}{2}\mathbf{Z}^T(t_f)\mathbf{Q}\mathbf{Z}(t_f) + \frac{1}{2}\int_{t_0}^{t_f} \left(\mathbf{u}^T(t)\boldsymbol{\alpha}\mathbf{u}(t) - \mathbf{v}^T(t)\boldsymbol{\beta}\mathbf{v}(t)\right) \mathrm{d}t \tag{22}$$

where

$$\mathbf{Q} = \begin{bmatrix} \mathbf{I}_{2\times2} & \mathbf{0}_{2\times2} \\ \mathbf{0}_{2\times2} & Q_V\mathbf{I}_{2\times2} \end{bmatrix}, \ \boldsymbol{\alpha} = \begin{bmatrix} \alpha_1 & 0 & 0 \\ 0 & \alpha_2 & 0 \\ 0 & 0 & \alpha_3 \end{bmatrix}, \ \boldsymbol{\beta} = \begin{bmatrix} \beta_1 & 0 \\ 0 & \beta_2 \end{bmatrix} \tag{23}$$

It should be noted that no explicit constraints are defined; the conditions $|u_i(t)| \leq 1 \ \forall t$ for $i \in \{1,2,3\}$ and $|v_i(t)| \leq 1 \ \forall t$ for $i \in \{1,2\}$ as well as the operational constraints on ϕ, α, β, p, q, and r are to be satisfied by appropriate selection of $\mathbf{Q}$, $\boldsymbol{\alpha}$ and $\boldsymbol{\beta}$.

3.2 Optimal Control

Using the reduced-order LQ cost function the differential game can now be solved. The Hamiltonian is given by (time indices are omitted for brevity)

$$H = L + \boldsymbol{\lambda}_{\mathbf{Z}}^T\mathbf{f} \tag{24}$$

where

$$L = \frac{1}{2}\left(\mathbf{u}^T(t)\boldsymbol{\alpha}\mathbf{u}(t) - \mathbf{v}^T(t)\boldsymbol{\beta}\mathbf{v}(t)\right),\ \mathbf{f} = \dot{\mathbf{Z}} = \tilde{\mathbf{B}}\mathbf{u} + \tilde{\mathbf{C}}\mathbf{v} \tag{25}$$

The adjoint equation is then given by

$$\dot{\boldsymbol{\lambda}}_{\mathbf{Z}}^T = -\left(\frac{\partial L}{\partial \mathbf{Z}} + \boldsymbol{\lambda}_{\mathbf{Z}}^T\frac{\partial \mathbf{f}}{\partial \mathbf{Z}}\right) = -\left(\mathbf{0}_{1\times4} + \boldsymbol{\lambda}_{\mathbf{Z}}^T\mathbf{0}_{4\times4}\right) = \mathbf{0}_{1\times4} \tag{26}$$

and

$$\boldsymbol{\lambda}_{\mathbf{Z}}^T(t_f) = \frac{\partial\left(\frac{1}{2}\mathbf{Z}^T(t_f)\mathbf{Q}\mathbf{Z}(t_f)\right)}{\partial \mathbf{Z}} = \mathbf{Z}^T(t_f)\mathbf{Q} \tag{27}$$

which results in

$$\boldsymbol{\lambda}_{\mathbf{Z}}(t) = \mathbf{Q}\mathbf{Z}(t_f) \tag{28}$$

In order to find the optimal control law the derivative of the Hamiltonian with regard to the control input is set to zero

$$\frac{\partial H}{\partial \mathbf{u}} = \mathbf{u}^T \boldsymbol{\alpha} + \mathbf{Z}^T(t_f)\mathbf{Q}\tilde{\mathbf{B}} = 0 \tag{29}$$

This results in the optimal pursuer control

$$\mathbf{u}^* = -\boldsymbol{\alpha}^{-1}\tilde{\mathbf{B}}^T \mathbf{Q}\mathbf{Z}(t_f) \tag{30}$$

Similarly it can be shown that the optimal evader control is

$$\mathbf{v}^* = \boldsymbol{\beta}^{-1}\tilde{\mathbf{C}}^T \mathbf{Q}\mathbf{Z}(t_f) \tag{31}$$

Integration after the substitution of the expressions for the optimal control actions into Eq. (19) gives

$$\mathbf{Z}(t_f) = \mathbf{Z}(t) + \mathbf{F}_{\alpha\beta}(t_f, t)\mathbf{Z}(t_f) \tag{32}$$

where

$$\mathbf{F}_{\alpha\beta}(t_f, t) = \int_t^{t_f} -\tilde{\mathbf{B}}(t_f, \xi)\boldsymbol{\alpha}^{-1}\tilde{\mathbf{B}}^T(t_f, \xi)\mathbf{Q} + \tilde{\mathbf{C}}(t_f, \xi)\boldsymbol{\beta}^{-1}\tilde{\mathbf{C}}^T(t_f, \xi)\mathbf{Q}\mathrm{d}\xi \tag{33}$$

Now the control laws can be written as functions of the current zero-effort vector as

$$\mathbf{u}^*(t) = -\boldsymbol{\alpha}^{-1}\tilde{\mathbf{B}}^T(t_f, t)\mathbf{Q}\left(\mathbf{I}_{4\times 4} - \mathbf{F}_{\alpha\beta}(t_f, t)\right)^{-1}\mathbf{Z}(t) \tag{34}$$

$$\mathbf{v}^*(t) = \boldsymbol{\beta}^{-1}\tilde{\mathbf{C}}^T(t_f, t)\mathbf{Q}\left(\mathbf{I}_{4\times 4} - \mathbf{F}_{\alpha\beta}(t_f, t)\right)^{-1}\mathbf{Z}(t) \tag{35}$$

3.3 Conjugate Point Analysis

Eq. (34) and Eq. (35) signify that a conjugate point exists if the matrix $(\mathbf{I} - \mathbf{F}_{\alpha\beta}(t_f, t))$ is singular. In this case the trajectory may not be optimal going back in time beyond the conjugate point.

Considering that $\mathbf{F}_{\alpha\beta} \to \mathbf{0}$ as $t_{go} \to 0$, $(\mathbf{I} - \mathbf{F}_{\alpha\beta}(t_f, t)) \to \mathbf{I}$ and consequently $\Delta \to 1$, where $\Delta = \det(\mathbf{I} - \mathbf{F}_{\alpha\beta}(t_f, t))$. Since $\mathbf{F}_{\alpha\beta}$ is a continuous function of t, the optimal LQ differential game solution exists if

$$\Delta = \det(\mathbf{I} - \mathbf{F}_{\alpha\beta}(t_f, t)) > 0 \quad \forall t \tag{36}$$

or equivalently using the analytic expression of the determinant

$$\Delta = (\kappa_1 + \kappa_2 - \kappa_1\kappa_2 + \kappa_3\kappa_4 - 1)(\kappa_5 + \kappa_6 - \kappa_5\kappa_6 + \kappa_7\kappa_8 - 1) > 0 \quad \forall t \tag{37}$$

where

$$\kappa_1 = \int_t^{t_f} \frac{a_e^{max\,2}}{\beta_1}{\phi_{1,3}}^2(t_f,\xi) - \frac{\delta_{ele}^{max\,2}}{\alpha_1 {\tau_\delta}^2}{\phi_{1,11}}^2(t_f,\xi)\mathrm{d}\xi \tag{38}$$

$$\kappa_2 = \int_t^{t_f} \frac{Q_V a_e^{max\,2}}{\beta_1}{\phi_{3,3}}^2(t_f,\xi) - \frac{Q_V \delta_{ele}^{max\,2}}{\alpha_1 {\tau_\delta}^2}{\phi_{3,11}}^2(t_f,\xi)\mathrm{d}\xi \tag{39}$$

$$\kappa_3 = \int_t^{t_f} \frac{Q_V a_e^{max\,2}}{\beta_1}\phi_{3,3}(t_f,\xi)\phi_{1,3}(t_f,\xi) - \frac{Q_V \delta_{ele}^{max\,2}}{\alpha_1 {\tau_\delta}^2}\phi_{3,11}(t_f,\xi)\phi_{1,11}(t_f,\xi)\mathrm{d}\xi \tag{40}$$

$$\kappa_4 = \int_t^{t_f} \frac{a_e^{max\,2}}{\beta_1}\phi_{3,3}(t_f,\xi)\phi_{1,3}(t_f,\xi) - \frac{\delta_{ele}^{max\,2}}{\alpha_1 {\tau_\delta}^2}\phi_{3,11}(t_f,\xi)\phi_{1,11}(t_f,\xi)\mathrm{d}\xi \tag{41}$$

$$\kappa_5 = \int_t^{t_f} \frac{a_e^{max\,2}}{\beta_2}{\phi_{2,4}}^2(t_f,\xi) - \frac{\delta_{ail}^{max\,2}}{\alpha_2 {\tau_\delta}^2}{\phi_{2,12}}^2(t_f,\xi) - \frac{\delta_{rud}^{max\,2}}{\alpha_3 {\tau_\delta}^2}{\phi_{2,13}}^2(t_f,\xi)\mathrm{d}\xi \tag{42}$$

$$\kappa_6 = \int_t^{t_f} \frac{Q_V a_e^{max\,2}}{\beta_2}{\phi_{4,4}}^2(t_f,\xi) - \frac{Q_V \delta_{ail}^{max\,2}}{\alpha_2 {\tau_\delta}^2}{\phi_{4,12}}^2(t_f,\xi) \tag{43}$$

$$- \frac{Q_V \delta_{rud}^{max\,2}}{\alpha_3 {\tau_\delta}^2}{\phi_{4,13}}^2(t_f,\xi)\mathrm{d}\xi$$

$$\kappa_7 = \int_t^{t_f} \frac{Q_V a_e^{max\,2}}{\beta_2}\phi_{2,4}(t_f,\xi)\phi_{4,4}(t_f,\xi) - \frac{Q_V \delta_{ail}^{max\,2}}{\alpha_2 {\tau_\delta}^2}\phi_{2,12}(t_f,\xi)\phi_{4,12}(t_f,\xi) \tag{44}$$

$$- \frac{Q_V \delta_{rud}^{max\,2}}{\alpha_3 {\tau_\delta}^2}\phi_{2,13}(t_f,\xi)\phi_{4,13}(t_f,\xi)\mathrm{d}\xi$$

$$\kappa_8 = \int_t^{t_f} \frac{a_e^{max\,2}}{\beta_2}\phi_{2,4}(t_f,\xi)\phi_{4,4}(t_f,\xi) - \frac{\delta_{ail}^{max\,2}}{\alpha_2 {\tau_\delta}^2}\phi_{2,12}(t_f,\xi)\phi_{4,12}(t_f,\xi) \tag{45}$$

$$- \frac{\delta_{rud}^{max\,2}}{\alpha_3 {\tau_\delta}^2}\phi_{2,13}(t_f,\xi)\phi_{4,13}(t_f,\xi)\mathrm{d}\xi$$

with $\phi_{i,j}(t_f,t)$ the element on the i-th row of the j-th column of $\mathbf{\Phi}(t_f,t)$. An analytic expression for $\mathbf{\Phi}(t_f,t)$ can be found by performing an inverse Laplace transform of $(s\mathbf{I}-\mathbf{A})^{-1}$, but is too lengthy to be shown here.

Examination of Eq. (37) - Eq. (45) shows that $\Delta \to 1\ \forall t$ as $\mu \to \infty$, where $\mu = \alpha_1 = \alpha_2 = \beta_1 = \beta_2 = \beta_3$. The maximum value of Δ on the closed interval $\{t_{go} \in \mathbf{R} | 0 \leq t_{go} \leq 10\}$ was calculated as a function of μ in order to assess the existence of an LQ differential game solution on the aforementioned interval for the case with $Q_V = 0.5$, and $a_e^{max} = 2$ ft/s^2 using parameters corresponding to an F-16 in trimmed wings-level flight at 15,000 ft and 315 KTAS (532 ft/s) as given in Table 1 [Stevens and Lewis(1992)]. The value for Q_V was selected such that the rendezvous is completed with minimal distance error, whereas some speed error is permissible. Results are shown in Figures 2 and 3. The numerical results shown in the figures conform to the statements made above regarding the condition $\mu \to \infty$. It can be seen that a conjugate point exists for $\mu < 0.015$.

A similar analysis was performed for the interception problem with $Q_V = 0$, and $a_e^{max} = 2$ ft/s^2. It was found that in this case a conjugate point exists only for $\mu < 1 \cdot 10^{-5}$.

Table 1 Parameter values for F-16 longitudinal dynamics model using ft-sec-rad units

M_α	-1.773	Y_β	-114.5	L_β	-24.69	N_β	6.771	τ_δ	0.0495
M_q	-0.932	Y_p	35.17	L_p	-2.416	N_p	-0.035	δ_{ele}^{max}	$25\frac{\pi}{180}$
$M_{\delta_{ele}}$	-7.381	Y_r	-1060	L_r	0.537	N_r	-0.334	δ_{ail}^{max}	$21.5\frac{\pi}{180}$
Z_α	-362.2	$Y_{\delta_{ail}}$	5.614	$L_{\delta_{ail}}$	-29.77	$N_{\delta_{ail}}$	-1.604	δ_{rud}^{max}	$30\frac{\pi}{180}$
Z_q	-32.72	$Y_{\delta_{rud}}$	16.39	$L_{\delta_{rud}}$	3.896	$N_{\delta_{rud}}$	-3.036		
$Z_{\delta_{ele}}$	-43.95	θ_0	0.066	V_0	532	g	32.17		

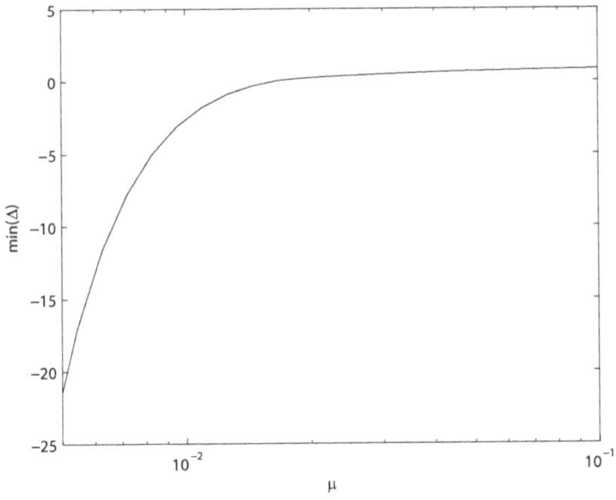

Fig. 2 Minimum Δ-values for $Q_V = 0.5$, $a_e^{max} = 2$ ft/s^2

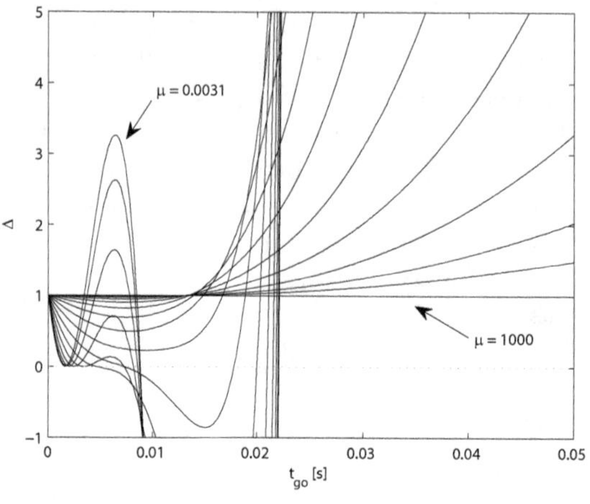

Fig. 3 Δ-values for $Q_V = 0.5$, $a_e^{max} = 2$ ft/s^2, $3.1 \cdot 10^{-3} \leq \mu \leq 1.0 \cdot 10^3$

3.4 Navigation Gains

By rewriting Eq. (34) the control law can be formulated as a function of the navigational gains $\mathbf{N}_{VZEM}$, $\mathbf{N}_{LZEM}$, $\mathbf{N}_{VZES}$, and $\mathbf{N}_{LZES}$ as

$$\mathbf{u}^*(t) = \frac{\mathbf{N}_{VZEM}}{\tau_\delta V_0} Z_1(t) + \frac{\mathbf{N}_{LZEM}}{\tau_\delta V_0} Z_2(t) + \frac{\mathbf{N}_{VZES}}{V_0} Z_3(t) + \frac{\mathbf{N}_{LZES}}{V_0} Z_4(t) \quad (46)$$

where the navigational gains are defined by (time indices are omitted for brevity)

$$\mathbf{N}_{VZEM} = V_0 \begin{bmatrix} \delta_{ele}^{max} \frac{\phi_{1,11} - \phi_{1,11}\kappa_2 + Q_V \kappa_4 \phi_{3,11}}{\alpha_1(\kappa_1 + \kappa_2 - \kappa_1\kappa_2 + \kappa_3\kappa_4 - 1)} \\ \mathbf{0}_{2\times 1} \end{bmatrix} \quad (47)$$

$$\mathbf{N}_{LZEM} = V_0 \begin{bmatrix} 0 \\ \delta_{ail}^{max} \frac{\phi_{2,12} - \phi_{2,12}\kappa_6 + Q_V \kappa_8 \phi_{4,12}}{\alpha_2(\kappa_5 + \kappa_6 - \kappa_5\kappa_6 + \kappa_7\kappa_8 - 1)} \\ \delta_{rud}^{max} \frac{\phi_{2,13} - \phi_{2,13}\kappa_6 + Q_V \kappa_8 \phi_{4,13}}{\alpha_3(\kappa_5 + \kappa_6 - \kappa_5\kappa_6 + \kappa_7\kappa_8 - 1)} \end{bmatrix} \quad (48)$$

$$\mathbf{N}_{VZES} = V_0 \begin{bmatrix} \delta_{ele}^{max} \frac{Q_V \phi_{3,11} + \phi_{1,11}\kappa_3 - Q_V \kappa_1 \phi_{3,11}}{\alpha_1 \tau_\delta(\kappa_1 + \kappa_2 - \kappa_1\kappa_2 + \kappa_3\kappa_4 - 1)} \\ \mathbf{0}_{2\times 1} \end{bmatrix} \quad (49)$$

$$\mathbf{N}_{LZES} = V_0 \begin{bmatrix} 0 \\ \delta_{ail}^{max} \frac{Q_V \phi_{4,12} + \phi_{2,12}\kappa_7 - Q_V \kappa_5 \phi_{4,12}}{\alpha_2 \tau_\delta(\kappa_5 + \kappa_6 - \kappa_5\kappa_6 + \kappa_7\kappa_8 - 1)} \\ \delta_{rud}^{max} \frac{Q_V \phi_{3,13} + \phi_{2,13}\kappa_7 - Q_V \kappa_5 \phi_{4,13}}{\alpha_3 \tau_\delta(\kappa_5 + \kappa_6 - \kappa_5\kappa_6 + \kappa_7\kappa_8 - 1)} \end{bmatrix} \quad (50)$$

The control input $\mathbf{u}^*$ is linear in the zero-effort vector and is bounded if all navigational gains are bounded, which is the case if and only if no conjugate point exists. The denominator of every non-zero element in the navigational gain matrices contains one of the factors of Δ as defined in Eq. (37). Since $\Delta = 0$ at a conjugate point the navigational gains will become unbounded at such a point. In case a conjugate point exists the unbounded navigational gains will asymptotically approach $\pm\infty$ at the conjugate point and change sign if the sign of Δ changes.

Sign changes due to the non-minimum phase dynamics of the aircraft with regard to elevator control can be observed for all of the elements of the transition matrix that appear directly in the gains in Eq. (47) - Eq. (50). The sign changes occur at the point where the contribution by direct lift due to control surface deflection becomes larger than the contribution of the resulting angle of attack or side-slip angle. This point occurs at small values for t_{go} where the value of κ_i with $i \in \{1, 2, ..., 8\}$ is approaching zero. Hence the nominators of the navigational gains are dominated by the terms that are not multiplied by these values. Consequently, the navigational gains exhibit a change in sign at approximately the same time as the element of the transition matrix that occurs in the first term of their nominator.

If Q_V is set to zero (the interception problem), $\mathbf{N}_{VZES}$ and $\mathbf{N}_{LZES}$ are equal to zero for all t_{go}.

3.5 Optimal Trajectories

If Q_V, $\boldsymbol{\alpha}$, and $\boldsymbol{\beta}$ are set such that no conjugate point exists, then optimal trajectories can be calculated by integration of Eq. (19) with $\mathbf{u}(t) = \mathbf{u}^*(t)$ and $\mathbf{v}(t) = \mathbf{v}^*(t)$ as defined in Eq. (34) and Eq. (35), as long as $|u_i^*(t)| \leq 1 \; \forall t$ for $i \in \{1, 2, 3\}$ and $|v_i^*(t)| \leq 1 \; \forall t$ for $i \in \{1, 2\}$. Since $\dot{\mathbf{Z}}(t)$ is a linear function of $u(t)$ and $v(t)$, and $u^*(t)$ and $v^*(t)$ are both linear functions of $\mathbf{Z}(t)$; $\dot{\mathbf{Z}}(t)$ is also a linear function of $\mathbf{Z}(t)$. Consequently, any optimal trajectory $\mathbf{Z}(t)$ obtained using some Q_V, $\boldsymbol{\alpha}$, and $\boldsymbol{\beta}$ forms a linearly dependent set with any other optimal trajectory that is obtained using the same Q_V, $\boldsymbol{\alpha}$, and $\boldsymbol{\beta}$, if their initial conditions $\mathbf{Z}(0)$ are linearly dependent, no conjugate point exists, and the control saturation constraints are not violated.

Due to the fact that a linearized aircraft model is used, the pursuer's maximum acceleration is very large. If a negative unit step input is given to the elevator, an equilibrium for the vertical acceleration $\dot{V}_{p_h}$ is obtained at $\dot{V}_{p_h} \approx 500$ ft/s^2. However, in practice the maximum acceleration is never obtained by a control law based on a cost function with well tuned weights, due to the inclusion of $u(t)$ in Eq. (13).

Despite its larger maximum acceleration, the pursuer cannot guarantee a zero miss vector. This is due to the fact that the pursuer has second order dynamics, whereas the evader has ideal dynamics and thus can instantly apply an acceleration. As t_{go} nears zero the evader applies a large control action. Due to its higher order dynamics the pursuer is incapable of immediately responding to the evader acceleration, resulting in an increase in miss distance and miss speed.

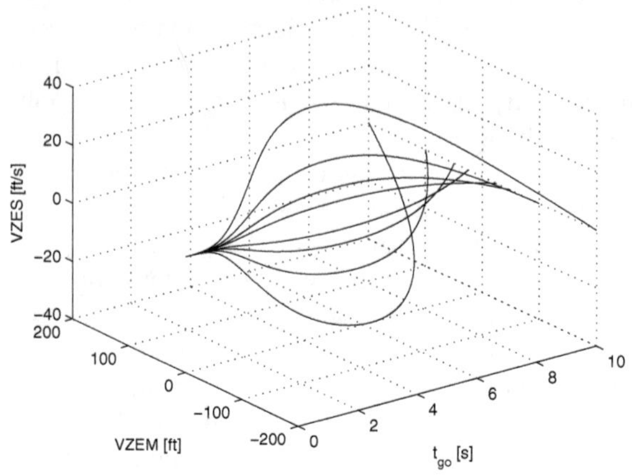

Fig. 4 Vertical optimal trajectories for $Q_V = 0.5$, $(\alpha_1, \alpha_2, \alpha_3) = (1, 2, 1)$, $(\beta_1, \beta_2) = (0.75, 0.75)$, and $a_e^{max} = 5$ ft/s^2

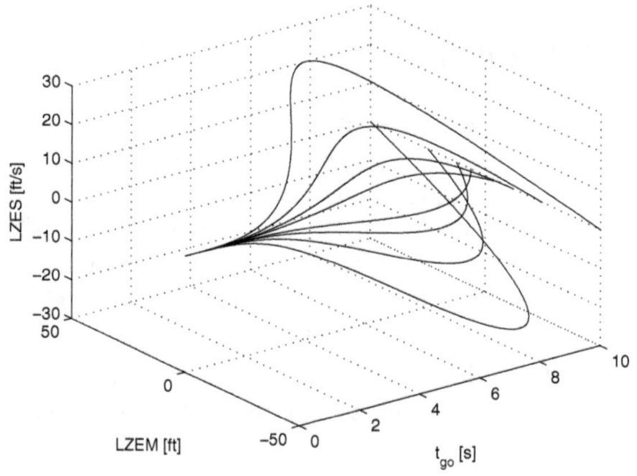

Fig. 5 Lateral optimal trajectories for $Q_V = 0.5$, $(\alpha_1, \alpha_2, \alpha_3) = (1, 2, 1)$, $(\beta_1, \beta_2) = (0.75, 0.75)$, and $a_e^{max} = 5$ ft/s^2

Figures 4 and 5 show optimal trajectories for $Q_V = 0.5$, $(\alpha_1, \alpha_2, \alpha_3) = (1, 2, 1)$, $(\beta_1, \beta_2) = (0.75, 0.75)$, and $a_e^{max} = 5$ ft/s^2. Again Q_V is selected such that rendezvous is achieved with minimum miss distance, while some speed difference is acceptable. The weight on aileron control action is increased in order to keep the roll angle within reasonable limits. Finally, $\boldsymbol{\beta}$ and a_e^{max} are chosen such that no conjugate point occurs, so that the obtained trajectory is optimal.

It should be noted that the vertical optimal trajectories in Figure 4 cannot be directly related to the lateral optimal trajectories in Figure 5. Due to decoupling of longitudinal and lateral-directional dynamics, there is no longitudinal control action due to lateral-directional states and vice versa. This is evidenced by the zero-valued elements of the navigation gains, Eq. (47) - Eq. (50), and results in uncoupled vertical and lateral optimal trajectories.

4 Simulation Results

In order to assess the performance of the guidance law two types of simulations were performed: The first type involves the LQ differential game guidance laws for both evader and pursuer, whereas the second type involves a burst noise signal as evader control.

4.1 Optimal Evader Guidance

Figures 6 and 7 show trajectory simulations corresponding to Figures 4 and 5. The simulation is initialized with the drogue located 100 ft above and 25 ft to the right of the pursuer aircraft. In order to prevent the occurrence of a conjugate point, the control weights $\boldsymbol{\alpha}$ and $\boldsymbol{\beta}$ cannot be decreased indefinitely, as described in Section 3.3. This causes both the evader and pursuer control inputs to remain small. The pursuer is however able to achieve small miss distance and speed. The terminal vertical and lateral distance are respectively 0.22 ft and -0.0093 ft, and the terminal vertical and lateral speed are 0.13 ft/s and 0.17 ft/s. The same values can also be found at $t_{go} = 0$ on the lines corresponding to $VZEM(t_{go} = 10) = -100$ ft in Figure 4 and $LZEM(t_{go} = 10) = -25$ ft in Figure 5. Vertical and lateral distance are reduced smoothly, while the angle of attack, sideslip angle, roll angle, and rotation rates remain well within operational limits.

4.2 Random Evader Guidance

For the simulation using burst noise evader control to mimic accelerations due to atmospheric turbulence, the evader control signal is divided into blocks of 0.4 s. Each block consists of a Gaussian white noise signal with $\sigma = 0.15$ and μ randomly selected to be either -0.5 or 0.5 separately for vertical and lateral acceleration.

An example of a simulation run using $Q_V = 0.5$, $(\alpha_1, \alpha_2, \alpha_3) = (1, 2, 1)$, $(\beta_1, \beta_2) = (0.75, 0.75)$, and $a_e^{max} = 5$ ft/s^2 and starting at -100 ft vertical distance and -25 ft lateral distance is shown in Figures 8 and 9. It can clearly be seen that the evader control action is much larger than for the optimal evasion guidance. Consequently, a larger control action is also applied by the pursuer. The condition $|u_i| \leq 1$ for $i \in \{1, 2, 3\}$ is still satisfied though.

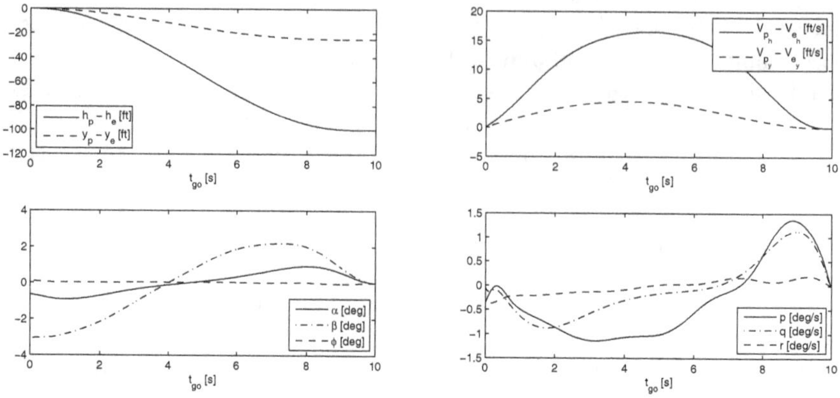

Fig. 6 Optimal evader and pursuer guidance trajectory for $Q_V = 0.5$, $(\alpha_1, \alpha_2, \alpha_3) = (1, 2, 1)$, $(\beta_1, \beta_2) = (0.75, 0.75)$, and $a_e^{max} = 5$ ft/s^2

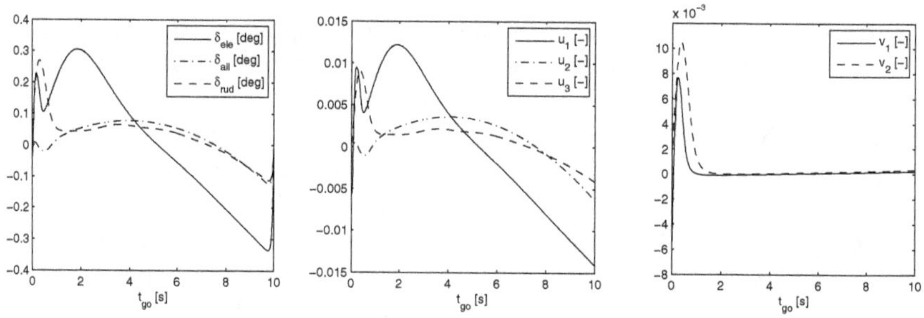

Fig. 7 Optimal evader control surface deflection and control inputs, and pursuer control inputs for $Q_V = 0.5$, $(\alpha_1, \alpha_2, \alpha_3) = (1, 2, 1)$, $(\beta_1, \beta_2) = (0.75, 0.75)$, and $a_e^{max} = 5$ ft/s^2

Larger angle of attack, sideslip angle, and roll angle are reached, due to the increase in control action. Also the roll rate reaches rather large values, but does reduce during the final part of the trajectory. However, for runs in which the evader control had several successive blocks with equal μ values towards the end of the trajectory larger terminal rotation rates and roll angles were observed.

A 10,000 run Monte Carlo simulation was performed using the burst noise evader control signal and $Q_V = 0.5$, $(\alpha_1, \alpha_2, \alpha_3) = (1, 2, 1)$, $(\beta_1, \beta_2) = (0.75, 0.75)$, and $a_e^{max} = 5$ ft/s^2. Accumulated terminal total distance and total relative speed are shown in Figure 10. These values are defined by the Euclidean norm of their respective vertical and lateral components. In all of the runs the magnitude of the terminal distance is below 0.50 ft, and the magnitude of the terminal speed is below 0.30 ft/s. The mean terminal distance is 0.22 ft, and the mean terminal speed is 0.20 ft/s. The terminal distance and speed results are negatively correlated with a Kendall's τ value of -0.23 ($p < 1 \cdot 10^{-100}$).

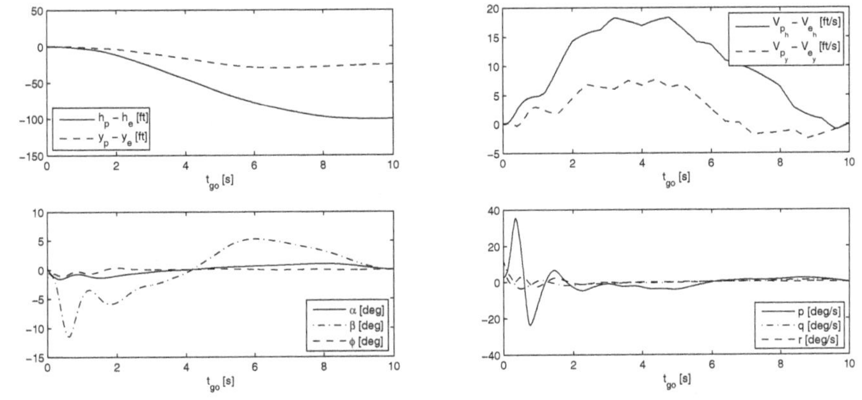

Fig. 8 Random evader and optimal pursuer guidance trajectory for $Q_V = 0.5$, $(\alpha_1, \alpha_2, \alpha_3) = (1, 2, 1)$, $(\beta_1, \beta_2) = (0.75, 0.75)$, and $a_e^{max} = 5$ ft/s^2

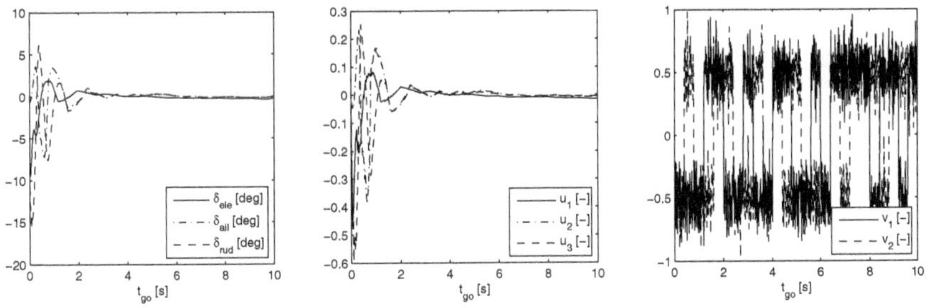

Fig. 9 Random evader control surface deflection and control inputs, and pursuer control inputs for $Q_V = 0.5$, $(\alpha_1, \alpha_2, \alpha_3) = (1, 2, 1)$, $(\beta_1, \beta_2) = (0.75, 0.75)$, and $a_e^{max} = 5$ ft/s^2

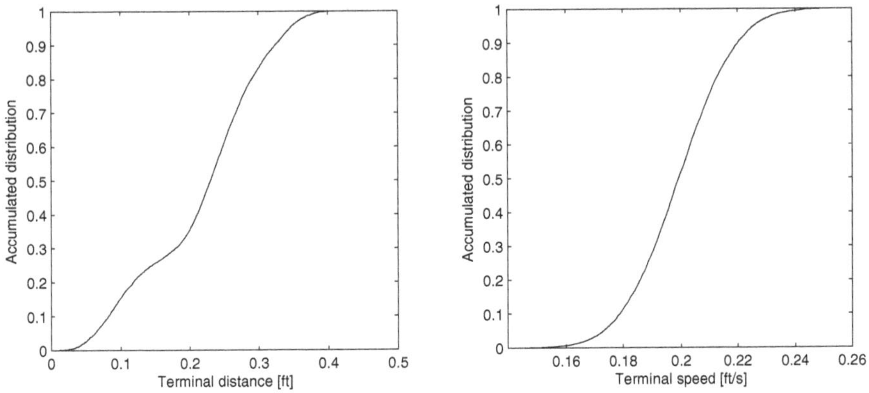

Fig. 10 Accumulated properties of Monte Carlo simulation results using $Q_V = 0.5$, $(\alpha_1, \alpha_2, \alpha_3) = (1, 2, 1)$, $(\beta_1, \beta_2) = (0.75, 0.75)$, and $a_e^{max} = 5$ ft/s^2, and burst noise evader control

5 Concluding Remarks

In this article LQ differential game-based guidance was presented to be a viable option for control of an aircraft during refueling rendezvous. The control method is capable of dealing with the complexity of the dynamics. It is however not able to guarantee zero miss for the rendezvous problem.

The results of a Monte Carlo simulation using burst noise evader control were presented. Results show promise, although there were issues with large aircraft rotations and rotation rates. Inclusion of these state variables in the cost function could potentially be a way to mitigate these issues.

This article showed an initial study along with practical results. A more elaborate dynamics model will result in more realistic results. A non-linear 6 degree of freedom flight dynamic model could be used to study the effects of longitudinal lateral-directional coupling. During this research the probe position was assumed to coincide with the aircraft center of gravity, thus neglecting the influence of aircraft rotation on the miss.

Further research into the movement of the drogue is also recommended. The application of realistic turbulence models and the addition of a coupling that simulates the effect of the wake of the drogue on the receiver aircraft are recommended for the purpose of obtaining more accurate simulation results.

Acknowledgements. This research was partially supported by the German-Israeli Foundation for Scientific Research and Development.

References

Anderson(1981). Anderson, G.: Comparison of optimal control and differential game intercept missile guidance laws. Journal of Guidance, Control, and Dynamics 4(2), 115–190 (1981)

Bryson and Ho(1969). Bryson, A., Ho, Y.: Applied Optimal Control. Blaisdell Publishing Company, New York (1969)

Ochi and Kominami(2005). Ochi, Y., Kominami, T.: Flight control for automatic aerial refueling via png and los angle control. In: AIAA Guidance, Navigation, and Control Conference and Exhibit (2005)

Pachter et al.(1997). Pachter, M., Houpis, C., Trosen, D.: Design of an air-to-air automatic refueling flight control system using quantitative feedback theory. International Journal of Robust and Nonlinear Control 7, 561–580 (1997)

Shima and Golan(2007). Shima, T., Golan, O.: Linear quadratic differential games guidance law for dual controlled missiles. IEEE Transactions on Aerospace and Electronic Systems 43(3), 834–842 (2007)

Shima et al.(2006). Shima, T., Idan, M., Golan, O.: Sliding-mode control for integrated missile autopilot guidance. Journal of Guidance, Control, and Dynamics 29(2), 250–260 (2006)

Shinar and Shima(2012). Shinar, J., Shima, T.: Differential game-based interceptor missile guidance. In: Balakrishnan, S., Tsourdos, A., White, B. (eds.) Advances in Missile Guidance, Control, and Estimation. Automation and Control Series, ch. 9, pp. 307–342. CRC Press, Boca Raton (2012)

Stevens and Lewis(1992). Stevens, B., Lewis, F.: Aircraft Control and Simulation. Wiley Inter-Science, New York (1992)

Tal and Shima(2015). Tal, E., Shima, T.: Linear quadratic differential games guidance law for autonomous aerial refueling. In: Israel Annual Conference on Aerospace Sciences (2015)

Tandale et al.(2006). Tandale, M., Bowers, R., Valasek, J.: Trajectory tracking controller for vision-based probe and drogue autonomous aerial refueling. Journal of Guidance, Control, and Dynamics 29(4), 846–857 (2006)

Wang et al.(2008). Wang, J., Patel, V., Cao, C., Hovakimyan, N., Lavretsky, E.: Novel $\mathcal{L}_1$ adaptive control methodology for aerial refueling with guaranteed transient performance. Journal of Guidance, Control, and Dynamics 31(1), 182–193 (2008)

Nonlinear and Fault-Tolerant Flight Control Using Multivariate Splines

H.J. Tol, C.C. de Visser, E. van Kampen, and Qiping P. Chu

Abstract. This paper presents a study on fault tolerant flight control of a high performance aircraft using multivariate splines. The controller is implemented by making use of spline model based adaptive nonlinear dynamic inversion (NDI). This method, indicated as SANDI, combines NDI control with nonlinear control allocation based on an onboard aerodynamic spline model and a real-time identification routine. The controller is tested for an aileron hardover failure and structural damages which change the global aerodynamic properties of the aircraft. It is shown that the controller can quickly tune itself in failure conditions without the need of failure detection and monitoring algorithms. Instead, self-tuning innovation based forgetting is applied to reconfigure the onboard aerodynamic model. The controller is able to tune itself each time a model error is detected and does not require any external triggers for re-identification. Multivariate splines have a high local approximation power and are able to accurately model nonlinear aerodynamics over the entire flight envelope of an aircraft. As a result the identification routine gives a robust adaption of the aerodynamic model in case of a failure.

H.J. Tol · C.C. de Visser · E. van Kampen · Qiping P. Chu
Delft University of Technology, 2600 GB Delft
e-mail: {h.j.tol,c.c.devisser,E.vanKampen,q.p.chu}@tudelft.nl

J. Bordeneuve-Guibé et al. (eds.), *Advances in Aerospace Guidance, Navigation and Control*, DOI: 10.1007/978-3-319-17518-8_12

Glossary

A_x, A_y, A_x specific forces along the body X/Y/Z axis $[m/s^2]$

C_l, C_m, C_n aerodynamic moment coefficients around the body X/Y/Z axis

I inertia matrix

S wing area $[m^2]$

V airspeed [m/s]

b wing span [m]

$\bar{c}$ mean aerodynamic chord [m]

p, q, r roll, pitch and yaw rate around the body X/Y/Z axis [rad/s]

p_s static pressure [Pa]

$\bar{q}$ dynamic pressure [Pa]

u, v, w velocity components along the body X/Y/Z axis [m/s]

$\mathbf{u}$ input vector

$\mathbf{x}$ state vector

α, β angle of attack and sideslip angle [rad]

ϕ, θ, ψ roll, pitch and yaw angle [rad]

$\delta_e, \delta_a, \delta_r, \delta_{lef}$ elevator, aileron, rudder and leading edge flap control surface deflection [rad]

ε model error

λ forgetting factor

ρ air density $[kg/m^3]$

τ virtual input

ν virtual input

Σ_0 filter information content

σ time constant

1 Introduction

This paper presents a study on fault tolerant flight control using multivariate splines. Self-tuning adaptive control is applied in which a real time identification routine is used to supply updated information to the control system in case of a failure. The controller is based on the nonlinear dynamic inversion principle in which the aerodynamic model is used to linearize the system after which single linear controllers can be used to control the system [10, 16, 20]. Currently, most adaptive NDI controllers use either polynomial structures [11, 12, 14] or neural networks [1, 8] for their internal model.

In this study multivariate simplex splines are used for the internal model. A multivariate spline is a piecewise defined polynomial function in which each basis polynomial is defined on a simplex [9, pp. 18-25]. Any number of basis polynomials can be combined with predefined continuity by combining simplices into a geometric structure called a triangulation. The approximation power of simplex spline functions is therefore is not only proportional to the polynomial degree, but also to the number and density of the polynomial pieces. Multivariate splines are linear in the parameters and are easily integrated into standard and recursive parameter estimation routines [3, 5].

Recently a new framework for spline model based NDI flight control is introduced in [19]. It is shown that the use of splines significantly improves the performance of NDI based control systems compared with ordinary polynomial based NDI. The controller from [19] is now augmented with a real time identification and applied for fault-tolerant flight control. First the aircraft model and the failure scenarios are discussed in Sec. 2. In Sec. 3 the fault tolerant control method is presented and in Sec. 4 the controller is evaluated.

2 Aircraft Model and Failure Cases

The aircraft to be controlled in this study is a model of the F-16 fighter aircraft from NASA, which is based on a set of data tables based on wind-tunnel measurements [15]. The model has the traditional aerodynamic control surfaces: elevator, ailerons and rudder for pitch, roll and yaw control. In addition, the leading edge flap is scheduled with angle of attack and $\frac{\bar{q}}{p_s}$ to optimize performance [15]. Models for the actuators are included in the form of first order lags:

$$\dot{u} = \frac{1}{\sigma}(u_{com} - u) \tag{1}$$

In which the commanded input is bounded by $u_{min} \leq u_{com} \leq u_{max}$ and the deflection rate is bounded by $|\dot{u}| \leq \dot{u}_{lim}$. The time constants σ and actuator limits are listed in table 1 which are taken from [15] and [18, pp.633-664]. For simulating the response and for flight control design the flat earth, body axis six degree of freedom equations of motion are used [18, pp. 107-116]. All simulations are performed in a deterministic environment. No external disturbances like wind gusts are added to the models and the sensor information is considered to be noise free. Stochastic adaptive control is reserved for future studies. Two failure cases are considered:

Table 1 Actuator model

	deflection limit	rate limit	time constant
Elevator	$\pm 25.0^o$	60°/s	0.0495 s lag
Ailerons	$\pm 21.5^o$	80°/s	0.0495 s lag
Rudder	$\pm 30.0^o$	120°/s	0.0495 s lag
leading edge flap	$0^o - 25^o$	25°/s	0.136 s lag

Aileron Hardover

The first failure scenario is a aileron hardover in which the right aileron gets stuck in an offset position. The failure scenario is described by:

$$\delta_{a_{actual}} = \delta_{a_{left}} + \delta_{a_{right}} = \frac{1}{2}\delta_{a_{command}} + \delta_{a_{offset}} \tag{2}$$

The offset position is chosen as $\delta_{a_{offset}} = 7^o$. Due to the hardover failure the aileron channel will become less effective. Furthermore, the aircraft will get a constant nominal rolling moment which needs to be compensated by the left aileron.

Structural Damage

The second failure case considered is a structural damage. It is assumed the structural damage changes the global aerodynamic properties of the aircraft. Two types of aerodynamic changes are considered based on earlier work by Choi et al [2]:

1. Magnitude scaling: In this case the coefficients in failure conditions are obtained by scaling the magnitude of the nominal aerodynamic coefficients from the look-up table:
$$C_{act}(\mathbf{x}) = [1 + a_{mag}]C_{nom}(\mathbf{x}) \tag{3}$$
2. Variable scaling: In this case the coefficients in failure conditions are obtained by scaling the independent variable of the aerodynamic coefficient from the look-up table:
$$C_{unc}(\mathbf{x}) = C_{act}([1 + a_{var}]\mathbf{x}) \tag{4}$$

A combined magnitude scaling and variable scaling with $a_{mag} = a_{var} = -0.3$ applied to all coefficients from the lookup table. In addition the center of gravity is shifted 10% rearward to make the actual aircraft model less stable. Fig. 1 visualizes the uncertainties for the pitch moment coefficients. It can be observed that these uncertainties have a nonlinear effect on the model especially in the high angle of attack operating region.

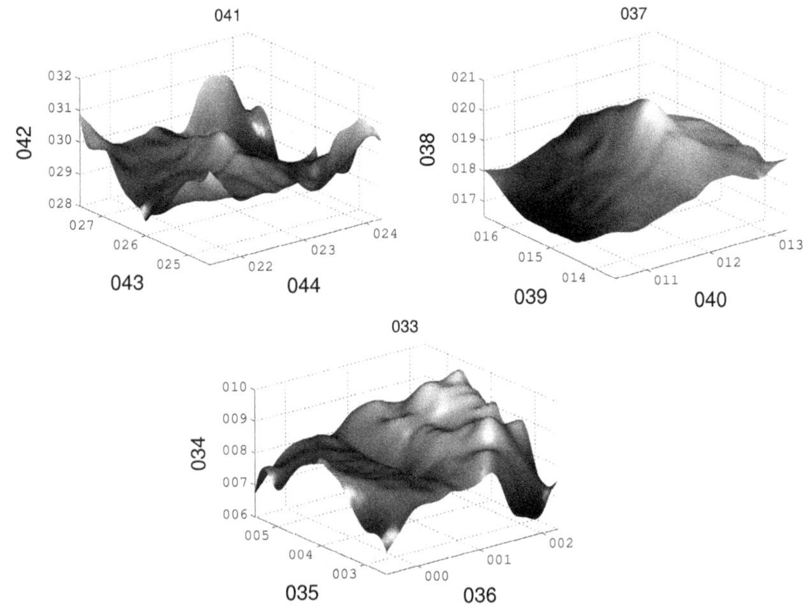

Fig. 1 C_m surface plots generated from data tables without uncertainties (top left) and with uncertainties (top right) ($\tilde{q} = 0.0015,\ \delta_e = 5^o,\ \delta_{lef} = 10^o$)

3 Fault Tolerant Flight Control Method

The fault tolerant control method used in this study is spline model based adaptive NDI (SANDI). A structure for spline based NDI control was first introduced in [9] which combines NDI control with nonlinear control allocation based on the onboard spline model. This structure is now augmented with a self tuning online estimator for simplex splines and applied to control the F-16 aircraft in failure conditions. The control diagram is shown in Fig. 2. The architecture has three major assemblies: 1) NDI control augmented with 2) control allocation based on the onboard aerodynamic spline model and 3) a real-time model identification routine. The aim of this setup is to update an a-priory aerodynamic model online in case of a failure with adaptive simplex splines to eliminate model errors within the NDI control system. In [19] the F-16 aerodynamic model is identified with multivariate splines using flight testing based system identification techniques from [3, 4, 6]. This is an accurate global model and is used as the a-priori nominal onboard aerodynamic model in this study. Refer to [19] for the complete structure, the accuracy and error bounds of the model.

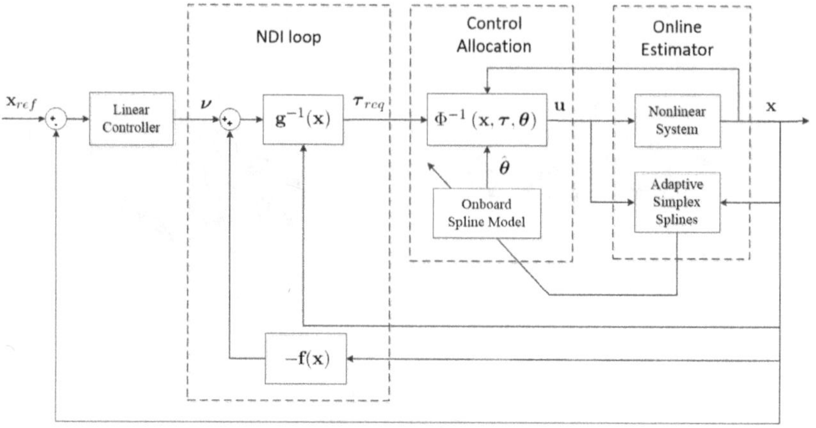

Fig. 2 Overview of the spline based adaptive control setup

Consider the aircraft state equations in the input affine form:

$$\dot{\mathbf{x}} = \mathbf{f}(\mathbf{x}) + g(\mathbf{x})\tau \tag{5}$$

$$\tau = \Phi(\mathbf{x}, \mathbf{u}, \Theta). \tag{6}$$

with $\mathbf{x} \in \mathbb{R}^n$ the state vector, $\mathbf{u} \in \mathbb{R}^m$ the control input vector, $\tau \in \mathbb{R}^l$ representing the aerodynamics assumed to be a nonlinear function of the aircraft state and control input, and with $\Theta \in \mathbb{R}^r$ the parameter vector to be updated in case of a failure. With NDI the system is linearized by solving for the input τ by introducing an outer loop control input ν:

$$\tau_{req} = g^{-1}(\mathbf{x})(\nu - \mathbf{f}(\mathbf{x})) \tag{7}$$

Which results in a closed-loop system with a decoupled linear input-output relation:

$$\dot{\mathbf{x}} = \nu \tag{8}$$

Eq. (6) represents the aerodynamic model and maps the physical control inputs to the virtual controls, i.e. to the forces and moments of the aircraft:

$$\tau = \Phi(\mathbf{x}, \mathbf{u}, \Theta) : \mathbb{R}^m \rightarrow \mathbb{R}^l \tag{9}$$

The spline model for τ is non-affine in the controls. Furthermore, the basis polynomials of the simplex splines are defined locally on each simplex in terms of barycentric coordinates instead of globally in terms of Cartesian coordinates [9, pp.18-25]. Therefore the NDI control structure is augmented with a separate control allocation module that determines the physical control inputs for a required demand τ_{req} from the NDI control law Eq. (7). For flight control purposes the required moment coefficients have to be translated into control surface deflections based on the onboard spline model. Note that by separating the NDI loop from the control allocation task the need to reconfigure the NDI control laws to accommodate for the failure is avoided. Instead, reconfiguration is completely encapsulated within the control allocation module. The control allocation problem is solved by minimizing the least square errors with respect to the control input $\mathbf{u}$:

$$\min_{\underline{\mathbf{u}} \leq \mathbf{u} \leq \overline{\mathbf{u}}} \mathscr{J} = \|s(\mathbf{x},\mathbf{u},\Theta) - \tau_{req}\|_2^2 = \sum_{i=1}^{N} \left(s_i(\mathbf{x},\mathbf{u},\Theta) - \tau_{req_i}\right)^2 \tag{10}$$

The derivation of the solution for Eq. (10) is not trivial, refer to [19] for the analytic derivation and solution for this optimization problem.

3.1 Adaptive Simplex Splines

A real time identification algorithm is used to reconfigure the aerodynamic model in failure conditions. The algorithm combines a recursive squares least estimator for multivariate splines [5] with an innovation based self-tuning forgetting factor [7], see Fig. 3.

A disadvantage of applying real time identification routines for fault-tolerant control is that the parameter vector converges quickly after which new data has no or little impact on the parameter estimate; the estimator has saturated. However, a fault-tolerant controller must be able to identify time-varying models in failure conditions. A frequently used approach to ensure adaptivity of the identification routine is to incorporate a trigger for re-identification of the aerodynamic model, see e.g [12, 14]. A disadvantage of this approach is that it requires monitoring algorithms to identify and isolate the failure and it requires significant off-line analysis in order to define robust reset triggers.

In this study adaptivity is ensured by applying a variable forgetting factor proposed by Fortescue in [7] in the RLS algorithm which is tuned based on the RLS innovation:

$$\lambda(t) = 1 - [1 - X(t)L(t)]\,\varepsilon^2(t)/\Sigma_0 \tag{11}$$

with X the row regression matrix for a new observation, L the adaptation gain of the filter and, ε the innovation and with Σ_0 a measure of the filter's information content. By keeping the information content constant it is ensured that the estimation is always based on the same amount of information. The amount of forgetting at each time step corresponds to the amount of new information in the latest measurement. In nominal flight conditions the estimation error will be small. Either the system has not been excited, there has been an excitation with the correct set of parameter values, or the estimator is sensitive enough to reduce the estimation error. In all these cases the forgetting factor is close to one. However in failure conditions when the error becomes larger the sensitivity of the estimator is increased and the estimator starts discounting past data in favor of more recent data. Σ_0 controls the speed of the adaptation and determines the sensitivity of the system. A small value for Σ_0

gives a large covariance matrix and a sensitive system; a larger value will give a less sensitive system and slower adaptation. However choosing a larger Σ_0 increases the robustness of the parameter estimate.

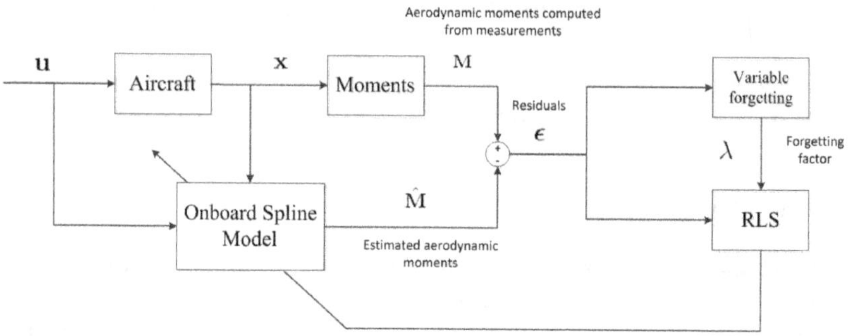

Fig. 3 Adaptive splines

3.2 *NDI Flight Control System*

An NDI flight control system is implemented using a two loop time-scale separated design [16], an inner rate control loop and an outer angle of sideslip control loop. This control setup is frequently used for manual flight control [12, 13, 17, 21] and is shown in Fig. 4. First order lag prefilters ($H_{pf} = \frac{1}{\sigma s+1}$) are added to ensure comfortable aircraft response and to avoid unachievable commands due to the actuator rate limits. Only proportional control is used for the sideslip feedback and pitch rate feedback.

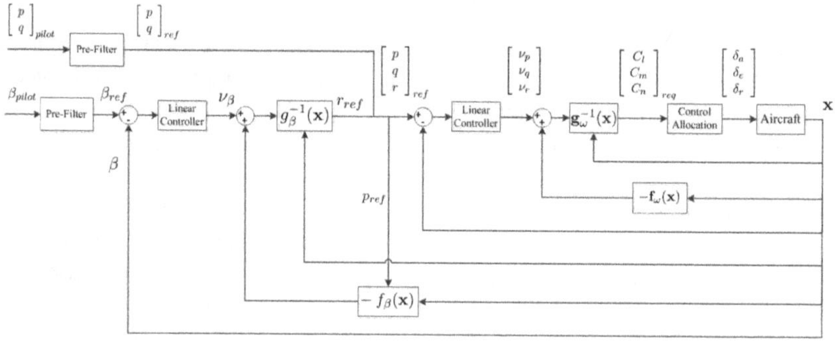

Fig. 4 Control setup. An angular rate inner NDI loop combined with a sideslip angle outer NDI loop

In the inner loop the system is influenced by commanding the moments coefficients of the aircraft. The inner loop quantities are the body angular rates:

$$\begin{bmatrix}\dot{p}\\ \dot{q}\\ \dot{r}\end{bmatrix} = -I^{-1}\begin{bmatrix}p\\ q\\ r\end{bmatrix}\times I\begin{bmatrix}p\\ q\\ r\end{bmatrix} + I^{-1}\frac{1}{2}\rho V^2 S\begin{bmatrix}b & 0 & 0\\ 0 & \bar{c} & 0\\ 0 & 0 & b\end{bmatrix}\begin{bmatrix}C_l\\ C_m\\ C_n\end{bmatrix} \tag{12}$$

With I the moment of inertia, b the wingspan and $\bar{c}$ the mean aerodynamic chord. Rewriting for the moment coefficients into the form of Eq. (7) by introducing a virtual control input ν_ω gives:

$$\begin{bmatrix}C_l\\ C_m\\ C_n\end{bmatrix}_{req} = \frac{I}{\frac{1}{2}\rho V^2 S}\begin{bmatrix}b & 0 & 0\\ 0 & \bar{c} & 0\\ 0 & 0 & b\end{bmatrix}^{-1}\left\{\begin{bmatrix}\nu_p\\ \nu_q\\ \nu_r\end{bmatrix} + I^{-1}\left(\begin{bmatrix}p\\ q\\ r\end{bmatrix}\times I\begin{bmatrix}p\\ q\\ r\end{bmatrix}\right)\right\} \tag{13}$$

$$= g_\omega^{-1}(\mathbf{x})\left[\nu_\omega - \mathbf{f}_\omega(\mathbf{x})\right] \tag{14}$$

An outer sideslip control loop is implemented for the yaw channel using time scale separation with the yaw rate r as control input. The sideslip angle can be written as:

$$\beta = \arcsin\frac{v}{V} \tag{15}$$

with V the total true airspeed:

$$V = \sqrt{u^2 + v^2 + w^2} \tag{16}$$

Taking the time derivative of Eq. (15) gives:

$$\dot{\beta} = \frac{\dot{v}V - v\dot{V}}{V\sqrt{u^2+w^2}} = \frac{\dot{v}}{\sqrt{u^2+w^2}} - \frac{v(u\dot{u}+v\dot{v}+w\dot{w})}{(u^2+v^2+w^2)\sqrt{u^2+w^2}} \tag{17}$$

Substituting the following body axis equations for $\dot{u}$, $\dot{v}$ and $\dot{w}$ in Eq. (17):

$$\dot{u} = A_x - g\sin\theta + rv - qw \tag{18}$$

$$\dot{v} = A_y + g\sin\phi\sin\cos\theta - ru + pw \tag{19}$$

$$\dot{w} = A_z + g\cos\theta\cos\phi + qu - pv \tag{20}$$

and writing as an affine function of the yaw rate gives:

$$\dot{\beta} = \frac{1}{\sqrt{u^2+v^2}}\left[\frac{-uv}{V^2}(A_x - g\sin\theta) + \left(1 - \frac{v}{V^2}\right)(A_y + g\sin\phi\cos\theta) - \frac{vw}{V^2}(A_z + g\cos\phi\cos\theta) + wp_{ref}\right] + \frac{-u}{\sqrt{u^2+w^2}}r_{ref} \tag{21}$$

Solving for the input r by introducing the second virtual control input ν_β gives:

$$r_{ref} = \left(\frac{-u}{\sqrt{u^2+w^2}}\right)^{-1}\left(\nu_\beta - \frac{1}{\sqrt{u^2+v^2}}\left[\frac{-uv}{V^2}(A_x - g\sin\theta) + \left(1-\frac{v}{V^2}\right)\right.\right. \quad (22)$$

$$\left.\left. \times (A_y + g\sin\phi\cos\theta) - \frac{vw}{V^2}(A_z + g\cos\phi\cos\theta) + wp_{ref}\right]\right) \quad (23)$$

$$= g_\beta^{-1}(\mathbf{x})\left[\nu_\beta - f_\beta(\mathbf{x})\right] \quad (24)$$

Parallel to the control structure in Figure 4 an airspeed NDI loop is added for thrust control. In the velocity axis reference frame the dynamics of the true airspeed is given by:

$$\dot{V} = \left(A_x + \frac{T}{m}\right)\cos\alpha\cos\beta + A_y\sin\beta + A_z\sin\alpha\cos\beta \quad (25)$$

Rewriting for the thrust lever input T by introducing a third virtual control input ν_V gives:

$$T = \left(\frac{1}{m}\cos\alpha\cos\beta\right)^{-1}\left[\nu_V - (A_x\cos\alpha\cos\beta + A_y\sin\beta + A_z\sin\alpha\cos\beta)\right] \quad (26)$$

$$= g_V^{-1}(\mathbf{x})\left[\nu_V - f_V(\mathbf{x})\right] \quad (27)$$

4 Performance Evaluation

A high performance maneuver is used to evaluate the fault-tolerant controller. The maneuver is a combined pitch $q_{ref} = 8^o/s$ and turn command $p_{ref} = \pm 15^o/s$ at zero side-slip and constant airspeed $V_{trim} = 83.5m/s$. The aircraft is trimmed at $\alpha_{trim} = 10^o$. The maneuver is performed in three flight scenarios: 1. The unfailed scenario (Fig. 5), 2. aileron hardover (Fig. 6) and 3. the structural damage scenario (Fig. 7). In both failure cases, the failure is activated after two seconds. The unfailed scenario serves as a comparison basis for the two failure cases.

The results for each scenario are presented through three sub figures. Subfigure (a) shows the response of the tracking quantities and aircraft states, subfigure (b) shows the control inputs and subfigure (c) shows the performance quantities which are the forgetting factor (Eq. (11)), the RLS innovation which is the error between the actual and the estimated moment coefficients:

$$\varepsilon C_M(t) = C_M(t) - \hat{C}_M(t)$$

and the control allocation error. The control allocation error is the error between the required moment coefficient delivered by the NDI controller and the actual moment coefficient delivered by the control allocator:

$$\Delta C_M(t) = C_{M_{req}}(t) - C_M(t)$$

The action of the identification algorithm is illustrated in Fig 8. This figure shows the onboard spline models for C_l, C_m and C_n at two time intervals during the maneuver. The models are parameterized as a function of the angle attack. The flight trajectories for the three scenarios are shown in Fig. 9. The performance assessment is based on the RMS values of the model errors, allocation errors and forgetting factor which are listed in Table 2.

Nominal Unfailed

In the unfailed scenario tracking is satisfactory and minimal couplings exist in the three control channels as can be seen from Fig. 5a. From the forgetting factors in Fig. 5c and the updated models in Fig. 8 it can be observed that the controller is able to make small scale local modifications to the onboard spline model to reduce the errors within the control system. This property is a direct result of the high local approximation power of multivariate splines.

Aileron Hardover

From Fig. 6 it can be observed that this failure does not represent a major challenge for the control system. The controller immediately compensates for the disturbing moment generated by the stuck aileron. There is no significant deterioration in the performance of the control system. There is only a slight increase in RLS innovations and control allocation errors as compared with the unfailed scenario, see Table 2, and the flight trajectories are nearly identical as can be seen from Fig. 9 .

Structural Damage

This failure does also not represent a major challenge for the control system as can be seen from Fig. 7. Although the global aerodynamic properties of the aircraft have changed significantly the controller is still able to perform the requested maneuver. Again the control allocation and innovation RMS values have not increased significantly compared to the unfailed case. The higher value for RMS ΔC_m is also due to saturation of the elevator actuator as can be seen from Fig. 7b and Fig. 7c. From the innovations in Fig. 7c it can be seen that during excitation the controller detects the error and reduces it to zero mean quickly. The flight trajectory has changed significantly compared to the unfailed case which is to be expected since the aerodynamic properties have changed.

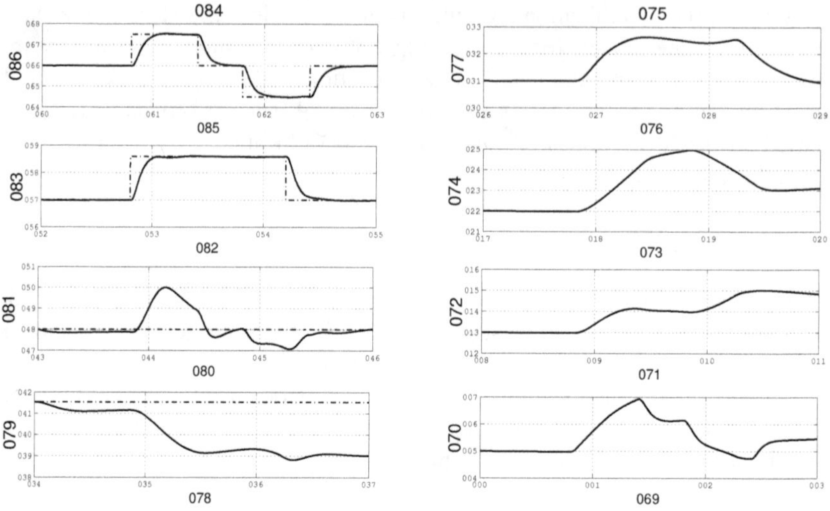

(a) Tracking response and states

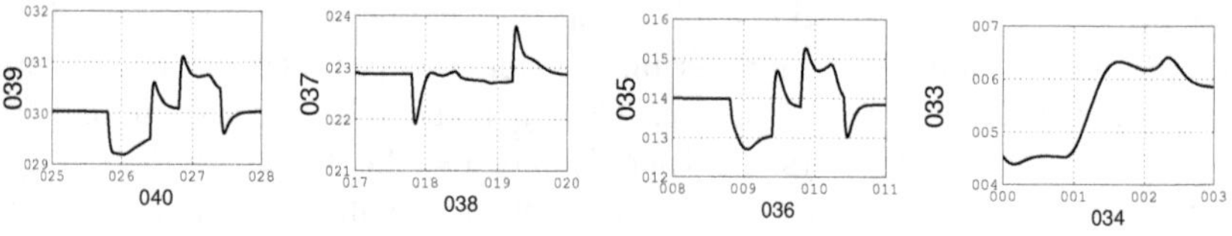

(b) Control deflections

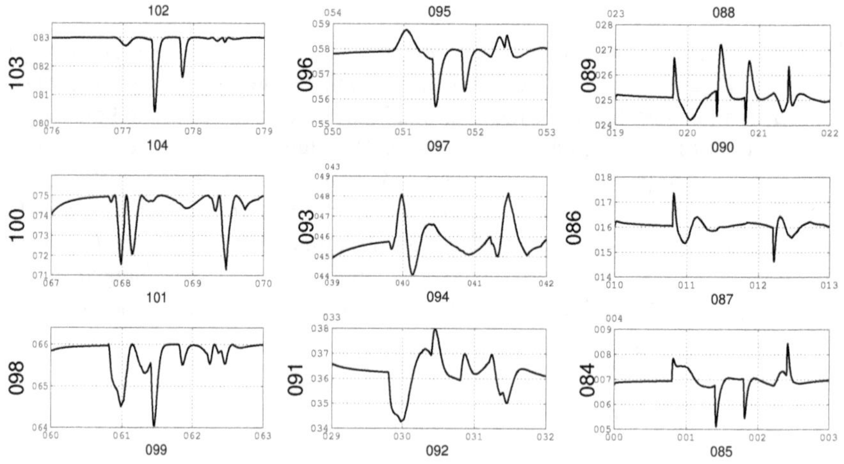

(c) Performance parameters: Forgetting factor, innovation and CA error

Fig. 5 Tracking quantities, states, control deflections and performance parameters for the unfailed scenario.

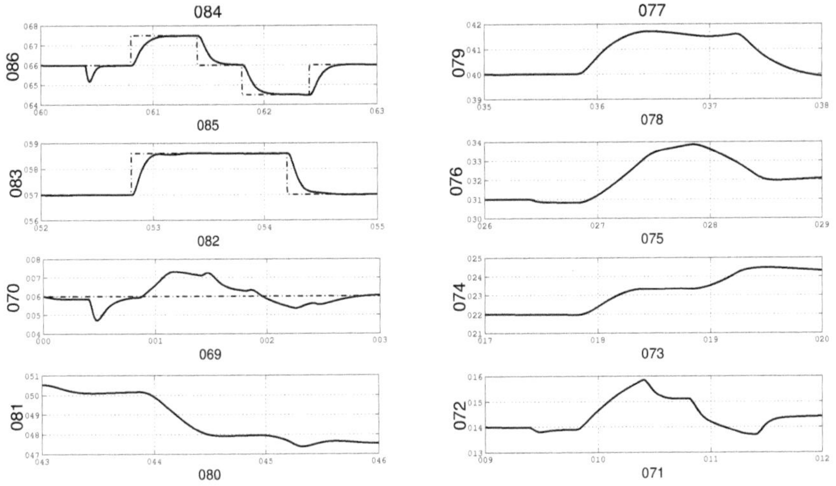

(a) Tracking response and states

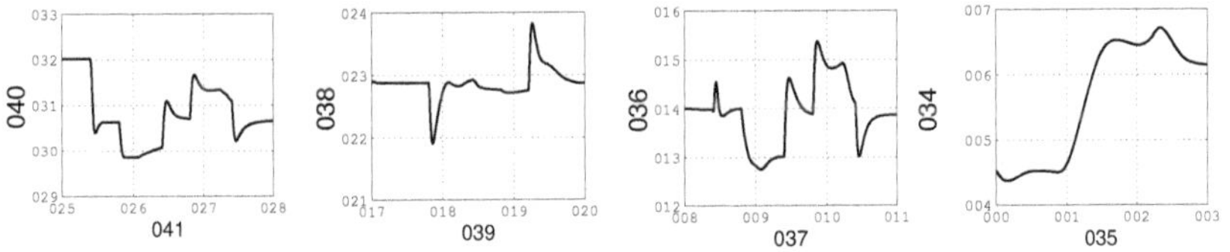

(b) Control deflections

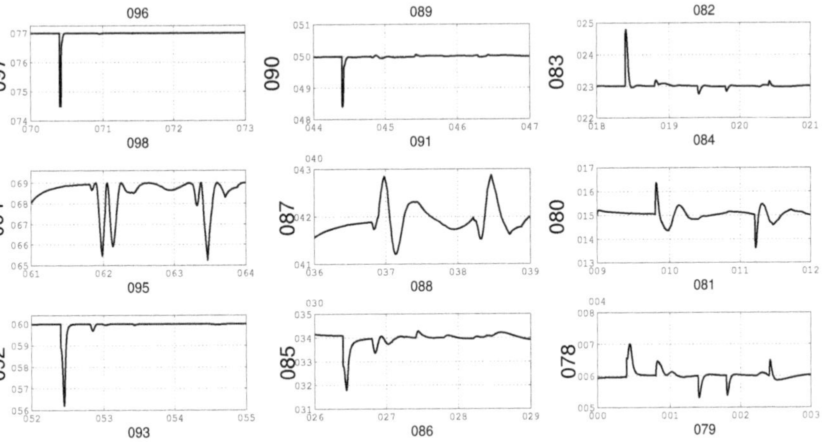

(c) Performance parameters: Forgetting factor, innovation and CA error

Fig. 6 Tracking quantities, states, control deflections and performance parameters for the aileron hard-over scenario. The failure is activated after 2 seconds.

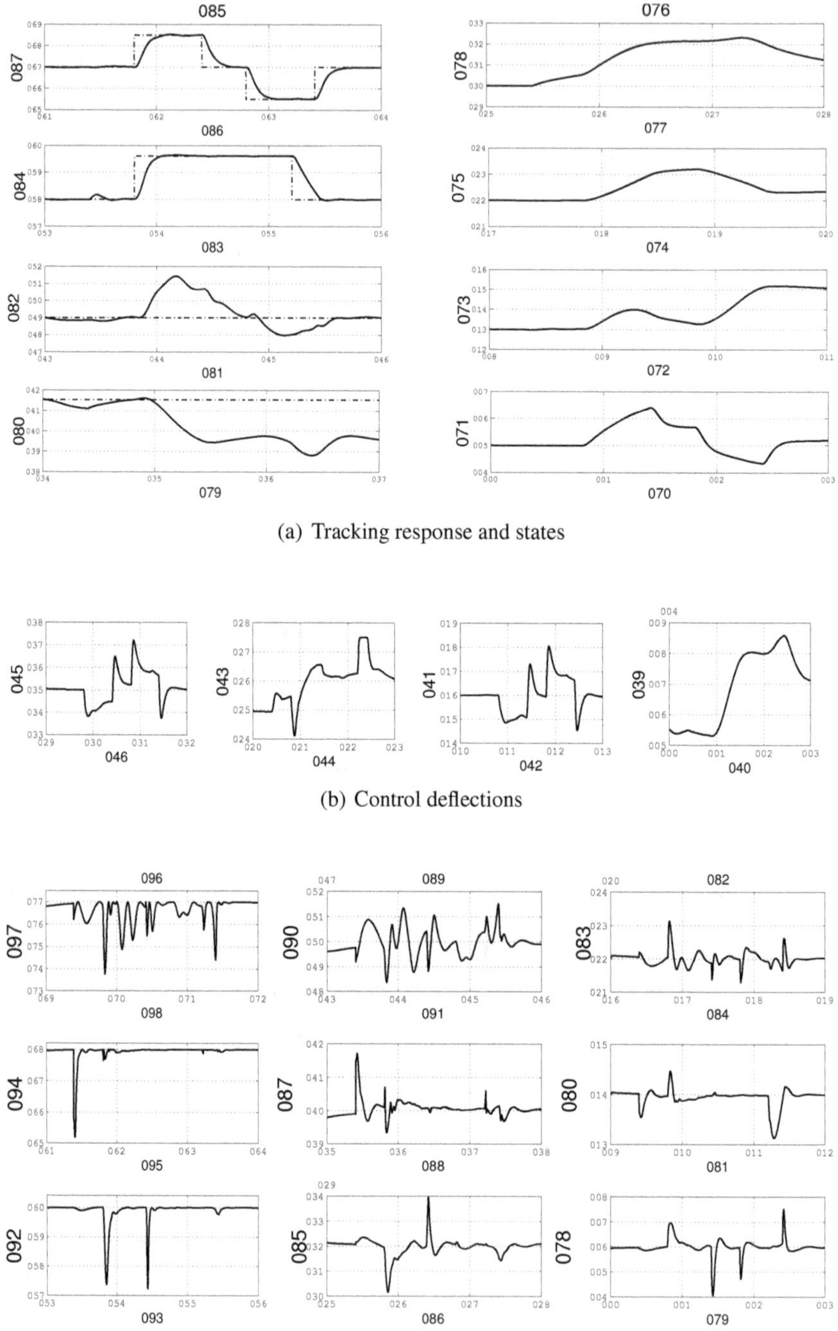

(a) Tracking response and states

(b) Control deflections

(c) Performance parameters: Forgetting factor, innovation and CA error

Fig. 7 Tracking quantities, states, control deflections and performance parameters for the structural failure scenario. The failure is activated after 2 seconds.

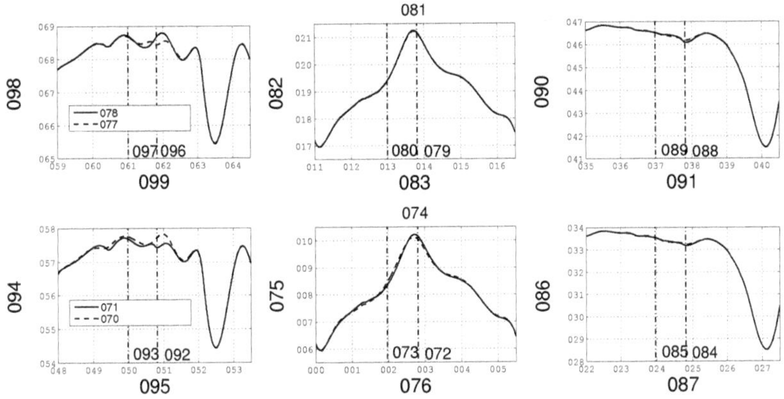

(a) Unfailed scenario

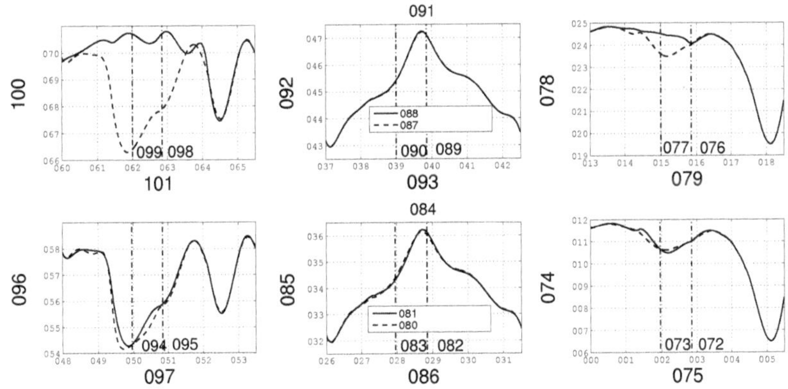

(b) Aileron hard-over scenario

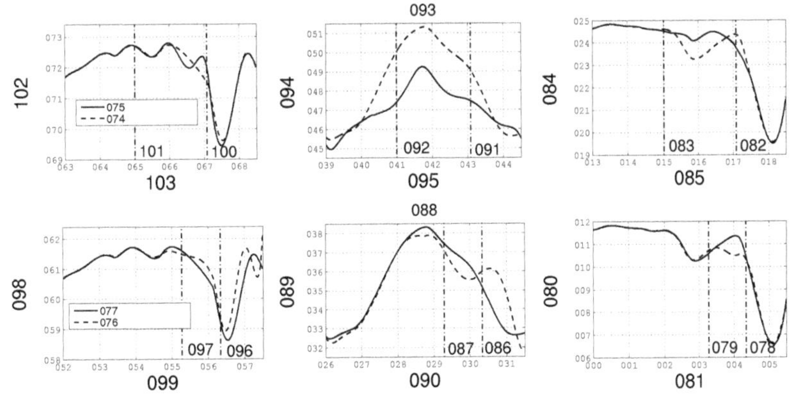

(c) Structural damage

Fig. 8 Updated models within two time intervals with: $\underline{\alpha} \leq \alpha \leq \overline{\alpha}$ the operating region within each interval. ($\beta = 0^o$, $\tilde{p} = 0.0049$, $\tilde{q} = 0.0018$, $\tilde{r} = 0.0049$, $\delta_a = -5^o$, $\delta_e = 5^o$, $\delta_r = -5^o$, $\delta_{lef} = 15^o$)

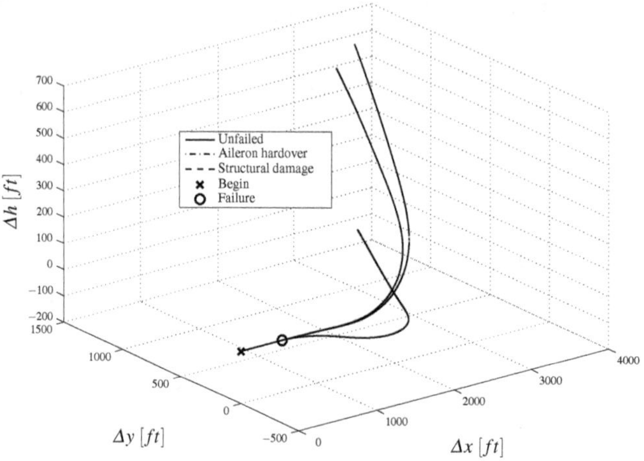

Fig. 9 Flight trajectories for the three scenarios

Table 2 Performance parameters for the three scenarios

	nominal	aileron hardover	structural damage
RMS εC_l	0.0002	0.0012	0.0003
RMS εC_m	0.0016	0.0016	0.0025
RMS εC_n	0.0004	0.0006	0.0007
RMS ΔC_l	0.0003	0.0017	0.0004
RMS ΔC_m	0.0026	0.0026	0.0094
RMS ΔC_n	0.0007	0.0009	0.0016
Mean λ_l	0.9951	0.9942	0.9932
Mean λ_m	0.9975	0.9975	0.9942
Mean λ_n	0.9987	0.9963	0.9950

5 Conclusions

This study shows that self-tuning adaptive control using multivariate splines is an effective and robust approach for fault-tolerant flight control. It is shown that the use of innovation based variable forgetting can circumvent the need for fault-detection and monitoring algorithms. Furthermore, the use of multivariate splines for real time model identification gives a robust adaptation of the internal model in case of a failure.

References

1. Calise, A.J., Lee, S., Sharme, M.: Nonlinear adaptive flight control using neural networks. IEEE Control Systems 18, 14–25 (1998)
2. Choi, J.Y., Chwa, D., Kim, M.: Adaptive control for feedback-linearized missiles with uncertainties. IEEE Transactions on Aerospace and Electronic Systems 36(2), 467–481 (2000)
3. de Visser, C.C., Chu, Q.P., Mulder, J.A.: A new approach to linear regression with multivariate splines. Automatica 45(12), 2903–2909 (2009)
4. de Visser, C.C., Chu, Q.P., Mulder, J.A.: A multidimensional spline based global nonlinear aerodynamic model for the cessna citation ii. In: AIAA Atmospheric Flight Mechanics Conference, number AIAA-2010-7950 (2010)
5. Visser, C.C.d., Chu, Q.P., Mulder, J.A.: Differential constraints for bounded recursive identification with multivariate splines. Automatica 47(9), 2059–2066 (2011)
6. de Visser, C.C., Chu, Q.P., Mulder, J.A.: Validating the multidimensional spline based global aerodynamic model for the cessna citation II. In: AIAA Atmospheric Flight Mechanics Conference, number AIAA-2011-6356 (2011)
7. Fortescue, T.R., Kershenbaum, L.S., Ydstie, B.E.: Implementation of self-tuning regulators with variable forgetting factors. Automatica 17(6), 831–835 (1981)
8. Kim, B.S., Calise, A.J.: Nonlinear flight control using neural networks. AIAA Journal of Guidance, Control and Dynamics 20(1), 26–33 (1997)
9. Lai, M.J., Schumaker, L.L.: Spline Functions on Triangulations. Cambridge University Press (2007)
10. Lane, S.H., Stengel, R.F.: Flight control design using non-linear inverse dynamics. Automatica 24(4), 471–483 (1988)
11. Lombaerts, T.J.J., Chu, Q.P., Mulder, J.A., Joosten, D.A.: Modular flight control reconfiguration design and simulation. Control Engineering Practice 19(6), 540–554 (2011)
12. Lombaerts, T.J.J., Huisman, H.O., Chu, Q.P., Mulder, J.A., Joosten, D.A.: Nonlinear reconfiguring flight control based on online physical model identification. AIAA Journal of Guidance, Control, and Dynamics 32, 727–748 (2009)
13. Lombaerts, T.J.J., Smali, M.H., Stroosma, O., Chu, Q.P., Mulder, J.A., Joosten, D.A.: Piloted simulator evaluation results of new fault-tolerant flight control algorithm. AIAA Journal of Guidance, Navigation, and Control 32(6), 1747–1765 (2009)
14. Lombaerts, T.J.J., Van Oort, E.R., Chu, Q.P., Mulder, J.A., Joosten, D.A.: Online aerodynamic model structure selection and parameter estimation for fault-tolerant control. AIAA Journal of Guidance, Control, and Dynamics 33(3), 707–723 (2010)
15. Nguyen, L.T., Ogburn, M.E., Gilbert, W.P., Kibler, K.S., Brown, P.W., Deal, P.L.: Simulator study of stall/post-stall characteristics of a fighter airplane with relaxed longitudinal static stability. Technical Report 1538, NASA (1979)
16. Reiner, J., Balas, G.J., Garrard, W.L.: Flight control design using robust dynamic inversion and time-scale separation. Automatica 32(11), 1493–1504 (1996)
17. Sieberling, S.: Robust flight control using incremental nonlinear dynamic inverstion and angular acceleration prediction. AIAA Journal of Guidance, Control, and Dynamics 33, 1732–1742 (2010)
18. Stevens, B.L., Lewis, F.L.: Aircraft Control and Simulation, 2nd edn. John Wiley & Sons, Hoboken (2003)

19. Tol, H.J., de Visser, C.C., van Kampen, E., Chu, Q.P.: Nonlinear multivariate spline-based control allocation for high-performance aircraft. AIAA Journal of Guidance, Control, and Dynamics 36(6), 1840–1862 (2014)
20. Walker, G.P., Allen, D.A.: X-35b stovl flight control law design and flying qualities. In: Biennial International Powered Lift Conference and Exhibit (2002)
21. Wang, Q., Stengel, R.F.: Robust nonlinear flight control of a high-performance aircraft. IEEE Transactions on Systems Technology 13(1), 15–26 (2005)

Rotor State Feedback in the Design of Rotorcraft Attitude Control Laws

Simone Panza and Marco Lovera

Abstract. Helicopter flight control law design including rotor state feedback (RSF) is considered. A mathematical model suitable for analysis and design of RSF control systems is obtained and a structured H_∞ approach to the problem is proposed, capable of guaranteeing stability and performance robustness. The framework also encompasses fault tolerance with respect to failures of the rotor state sensors. Simulation results comparing the proposed approach to results obtained using conventional attitude control laws are presented and discussed.

1 Introduction

Future rotorcraft will be required to operate in degraded visual environments and in poor weather conditions, to achieve high levels of agility and maneuverability, and at the same time to relieve the pilot workload when tackling demanding mission task elements. In such a context, the capability to provide a fast and accurate attitude response while rejecting external disturbances, such as wind gusts, is of fundamental importance, both in a military and a civil framework. This can be summarized in the notion of disturbance rejection bandwidth [1]: the higher the bandwidth, the faster the rejection of disturbances will be, resulting in a more "crisp" vehicle response and helping the pilot in the task of governing the vehicle, allowing him to focus on higher-level tasks. Of course other factors come into play into defining the desirable vehicle behaviour: the efforts of the helicopter community led in the 90's to the definition of the notion of *handling qualities*, and to the formulation of the ADS-33 specification [2], which defines the handling quality requirements for US army military rotorcraft; this specification has gained widespread acceptance and became the standard reference document in the community.

Simone Panza · Marco Lovera
Dipartimento di Scienze e Tecnologie Aerospaziali, Politecnico di Milano
e-mail: {simone.panza,marco.lovera}@polimi.it

J. Bordeneuve-Guibé et al. (eds.), *Advances in Aerospace Guidance, Navigation and Control*, DOI: 10.1007/978-3-319-17518-8_13

Modern helicopters are equipped with a flight control system (FCS); it usually has the capability to stabilize the vehicle and to assist the pilot. The control system acquires measurements related to fuselage motion (usually attitude, angular rates and linear accelerations) and feeds them back to the control inputs. If the FCS operates on the basis of fuselage measurements only, control law gains can be tuned so as to achieve a trade-off between disturbance rejection bandwidth and damping of oscillations in the fuselage response. This trade-off between disturbance rejection bandwidth and damping ratio became of interest in recent years ([3–5]) and it is of fundamental importance to understand its mechanism in order to assess the closed-loop performance boundaries the control law can achieve.

At this point, the notion of *rotor state feedback* (RSF) comes into play. In fact, it can be shown that introducing rotor state measurements in the feedback control law, in addition to classical fuselage measurements, has the potential to overcome the trade-off between bandwidth and damping: higher disturbance rejection bandwidth can be achieved, while keeping sufficiently high levels of damping. The main motivation of this performance increment lies in the fact that the fuselage attitude dynamics are coupled to rotor dynamics in the range of frequency in which the attitude control loop is closed: thus, adding rotor-state measurements to the feedback control law introduces information about rotor dynamics in the loop and drives in turn to better performance; the issue of fuselage-rotor coupling and its importance in the synthesis of high-bandwidth attitude control laws was studied by [6–9]. The interested reader can refer to [4, 6, 10, 11] for further details about RSF.

However, the available literature offers few studies about the role of attitude control law in determining the closed-loop properties of the system, both in the case traditional feedback measurements are used (fuselage angles and rates) and in the case the novel rotor-state measurements are introduced; on the other hand, both the literature state-of-the-art and the industry know-how are in agreement on the fact that the control law structure should be kept simple: in this way, the feedback control law can be given a physical interpretation. As far as the attitude control law design is concerned, the most common structure of feedback control law consists in static gains on the attitude angles and rates [12]. Thus, a better understanding of how the control law gains affect the closed-loop performance and, more in general, the handling qualities, would certainly help in the stage of attitude control law design.

In addition, the issue of closed-loop robustness with respect to model uncertainty is often neglected, nor the fault tolerance to rotor-sensor faults: the latter issue is of particular importance, as stated in [11].

This work proposes itself as a continuation and extension of the methodology defined in [11]; while [11] constituted a preliminary study intended to show the potential of RSF on a simple, second order, uncoupled helicopter model (an articulated rotor UH-60 helicopter), in this paper the same methodology is applied to a high order, fully coupled, articulated rotor helicopter model provided by the industry. The same procedure performed in the cited paper, which consists in model reduction, study of the closed-loop performance and control law synthesis, was undertaken; in particular, it turns out that a second order approximation of the coupled fuselage-rotor dynamics is valid to represent the attitude dynamics, as was shown in [6, 11].

Results similar to those presented in the cited papers were obtained. In addition, this work accomplishes also a closed-loop robustness analysis with respect to the model uncertainty due to physical parameters variations; also, a study of the closed-loop attitude performance limitations is performed, with the twofold objective of assessing the performance limits achievable with a traditional attitude control law (resulting in a trade-off between bandwidth and damping ratio), and evidencing the benefits deriving from RSF, which show up in the form of improved performance as a consequence of the relaxation of the bandwidth-damping trade-off; in alternative, better noise rejection properties can be achieved by means of using RSF. Also, the methodology proposed in [11] allows to address fault tolerance requirements.

2 Control-Oriented Attitude Model

For the purpose of this study, a sample linearized FLIGHTLAB model has been provided by AgustaWestland. The model, corresponding to 58 states, linearized in hover, takes into account both fuselage and rotor dynamics; in particular, body dynamics (Euler angles, angular rates, translational speed), main rotor dynamics (flap and lag angles in Multi-Blade Coordinates (MBC) and their derivatives), main rotor inflow model (10 states), main rotor wake model (3 states) and tail rotor dynamics (collective inflow and coning angle) are modeled. The model takes into account the four classical helicopter control inputs: main rotor collective, lateral and longitudinal cyclic, and tail rotor collective.

Actuator and sensor dynamics have been cascaded upstream and downstream to the FLIGHTLAB model. Focus was put on the lateral and longitudinal axes. Models for the fly-by-wire actuators of the lateral and longitudinal cyclic inputs have been provided by the manufacturer, according to which they can be modeled as third-order systems with 8Hz bandwidth. As far as sensors are concerned, roll and pitch rate gyros can be modeled as second-order systems, with 8Hz bandwidth and 0.7 damping ratio. Moreover, an equivalent time delay of 20ms was cascaded to the model, in order to take into account delays related to signal processing and ZOH. A third-order Padé approximation of the pure time delay was used.

For the sake of attitude control law synthesis, the full-order augmented model was reduced. Starting from the 58 states model, a 25 states reduced model has been obtained by truncating the less relevant states, with the aim of obtaining a model which describes the lateral and longitudinal attitude dynamics, in order to achieve an accurate model in the frequency range in which the attitude control loop bandwidth lies. The truncation approach retains the faster dynamics associated to attitude and rotor states, while the low frequency behaviour associated to the states of linear velocities is lost. Also, states related to other axes than the lateral and longitudinal ones, *e.g.*, the vertical and directional axes, were truncated too, because they were not relevant to the pitch and roll attitude control loop synthesis.

The only states retained in the 25 states model were:

- roll and pitch Euler angles and angular rates;
- main rotor inflow and wake model;

- lateral and longitudinal cyclic flap and lag angles, and their derivatives.

The 25 states model was then further reduced, in order to obtain a model which retained only the dominant attitude dynamics and, as a consequence, to shed light over the most relevant closed-loop dynamics of the system. For the same reasons stated in [6,11], such a reduced-order model needs to take into account not only the fuselage attitude dynamics, but also rotor dynamics: indeed, when the attitude loop is closed at high bandwidth, the rotor dynamics get coupled to fuselage dynamics; in particular, as far as the lateral axis is concerned, it turns out the regressive flap mode (which is associated to two real-valued eigenvalues) plays a fundamental role in this coupling, while the longitudinal attitude dynamics are dominated by pitch and roll subsidence modes.

Starting from the 25 states model, a second order model which approximates the lateral attitude response was hence obtained; the modal decomposition of the frequency response of the transfer function from lateral cyclic to roll rate (Figure 1) showed that, as far as the magnitude is concerned, the regressive flap modal components (in this case, regressive flap is given by the sum of two real-valued modal components, associated respectively with -3.7 and -5.6 rad/s eigenvalues) is the most relevant in the frequency response, at least up to 10 rad/s (Figure 2).

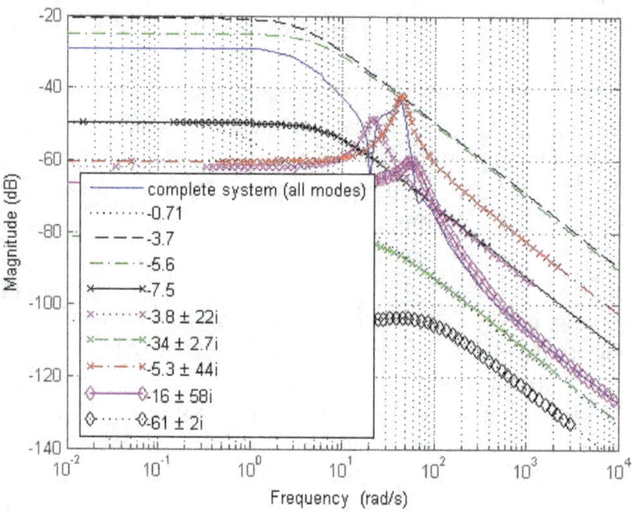

Fig. 1. Modal decomposition, 25 states lateral cyclic to roll rate frequency response.

In order to take into account the phase delay due to neglected higher order rotor modes, and to actuator and sensor dynamics, a pure time delay of 71ms has been introduced in the second order model; this was computed as the time delay necessary for the second order model in order to fit the phase of the 58 states augmented model at 10 rad/s. Figure 2 shows the second order model compared to the 58 states augmented model, and shows the phase contribution given by the equivalent time delay.

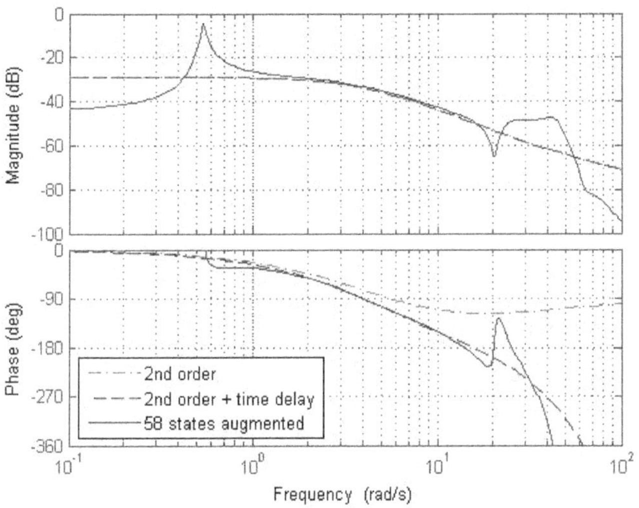

Fig. 2. Lateral cyclic to roll rate frequency response, 58 states augmented model vs 2nd order model, with and without 71ms time delay.

The transfer function from lateral cyclic to roll angle was obtained by multiplying the roll rate transfer function by an integrator (1/s), as suggested by [6].

Finally, the transfer function from lateral cyclic to lateral flap was obtained in a similar way: again, the most relevant modal component in the frequency response magnitude turned out to be the regressive flap mode; a fitting routine was run, which had the objective of minimizing the difference between the lateral flap response of the 25 states augmented model and a second order transfer function where the 2 poles coincided with the regressive flap poles, the static gain was fixed to the 25 states one, and the only free parameters were a zero, which turned out to be non-minimum phase, and a time delay, in order to take into account the unmodeled dynamics (high frequency rotor modes, actuator, sensor, transport delay) phase lag. This resulted in a 23ms equivalent time delay.

The longitudinal axis was addressed in a similar fashion, thus obtaining a second order approximation.

As a conclusion, Table 1 shows the parameters of the 2nd order models for the pitch and roll axes, defined

$$G(s) = g\frac{(1+s/z)}{s^2+a_1s+a_2}a_2e^{-\tau s} \tag{1}$$

as the transfer function from the proper input to attitude rate (in which case g, z, τ are referred to with the $\dot{f}$ subscript, thus becoming respectively $g_{\dot{f}}$, $z_{\dot{f}}$, $\tau_{\dot{f}}$) and rotor flap (r subscript: g_r, z_r, τ_r); poles (defined by a_1, a_2) are the same for both

Table 1. Second order approximated model parameters (pitch and roll dynamics)

	a_1	a_2	g_f	$z_f[rad/s]$	$\tau_f[ms]$	g_r	$z_r[rad/s]$	$\tau_r[ms]$
Roll	9.267	20.58	0.0362	27.12	71	-9.793×10^{-4}	-0.797	23
Pitch	8.173	5.322	0.0508	22.52	71	-0.0011	-0.159	11

fuselage and rotor responses; attitude response can be obtained by multiplying the rate response by $1/s$.

3 Closed-Loop Analysis

In this section the closed loop behaviour of the helicopter will be studied. An ACAH (attitude command, attitude hold) command configuration will be taken into consideration in the following (see [2] for details). Focus will be put on the attitude control loop (inner loop), which is in charge of tracking the attitude commanded by pilot; this is the control loop which can benefit from the most relevant improvements deriving from RSF, and is closed at high frequency. The velocity (outer) loop, which is closed at lower frequency and is not influenced by RSF, is neglected. The attitude control loop is closed for each axis separately. No decoupling between axes is taken into consideration, for the time being.

3.1 Baseline Attitude Control Law

The baseline control law is a PD controller on the attitude; as far as the roll axis is concerned, the required measurements are attitude angle φ and attitude rate p. The lateral cyclic control input is then computed as a linear combination of these measurements.

The same holds true for the pitch axis, where the measurements are respectively θ, q and the control input is the longitudinal cyclic.

Focus on the roll axis, and let $G(s)$ be the transfer function from lateral cyclic to roll rate as defined by Equation 1. Closing the loop with the baseline PD control law ($K = [K_p,\, K_\varphi]$) and breaking the loop in the actuator, the (scalar) loop transfer function $L(s)$ can be computed as[1]

$$L(s) = g\frac{(1+s/z)}{s^2+a_1 s+a_2}a_2 e^{-\tau s}\frac{\left(\frac{K_p}{K_\varphi}s+1\right)}{s}K_\varphi. \tag{2}$$

$L(s)$ takes the classical form of a loop transfer function with an integrator (its magnitude is infinite at 0 rad/s, then it decreases in magnitude as the frequency

[1] Consistently with the notation adopted previously in this work, the model parameters are referred to fuselage and can be found in Table 1 under the f subscript, here not reported for simplicity.

increases); the effect of the two control law gains $K_\varphi \frac{[\%]}{[rad]}$ and $K_p \frac{[\%]}{[rad/s]}$ can be interpreted in this way:

- K_φ moves vertically upwards or downwards the whole magnitude plot of the frequency response of $L(s)$; thus, it determines the position of the point where the magnitude plot crosses the 0 dB axis (which is, approximately, the closed-loop bandwidth location): the higher K_φ, the higher the bandwidth (since the 0 dB cut point moves rightwards);
- the ratio $\frac{K_p}{K_\varphi}$ determines the location of the $L(s)$ zero introduced by the control law; this is a real-valued zero, and its interpretation is less immediate than K_φ.

Figure 3 shows bandwidth and damping ratio for different lateral axis control laws; bandwidth is computed based on the sensitivity function (the closed-loop transfer function from disturbance on φ to φ), which is also referred to as the disturbance rejection bandwidth, DRB[2]; the damping ratio is computed based on the closed-loop regressive flap poles location, since the regressive flap mode dominates the roll attitude dynamics.

The graph is meant to be read along two directions:

- each curve represents a different value of K_φ. Starting from the top-left corner of the figure, and proceeding towards the bottom-right corner, K_φ increases (from 50 to 290); the curves - which have a "hill" shape - move to the right, and their peak reduces. The shift to the right is due to the fact that, when K_φ increases, then also bandwidth increases; the peak reduction is related to a smaller damping; so, the trend is: the higher the bandwidth, the lower the damping;
- focusing on one curve at a time, each point of the curve represents a different K_p/K_φ ratio value. Starting from the right end of the curve, and proceeding to the left, climbing the hill and descending, the ratio K_p/K_φ increases, from 2% to 100%. Each curve reaches a peak, which is the maximum damping ratio achievable, having K_φ fixed.

As Figure 3 clearly shows, there exists a limit, represented by the envelope of all the points, beyond which no point can stand: in other words, there is a trade-off between bandwidth and damping ratio, a maximum performance achievable by the baseline control law; the best combinations of bandwidth/damping ratio lie on the envelope of the graph, and that is the limit performance which can be obtained with a baseline control law. The bandwidth cannot be increased arbitrarily, because at a certain point, damping would become too low to be tolerated; points that lie under the 0 damping ratio horizontal line represent unstable systems, and the rightmost point in the graph which is at the same time stable, lies at a bandwidth of about 3 rad/s.

---

[2] Sensitivity bandwidth is computed as the highest frequency at which the sensitivity function cuts -3 dB axis.

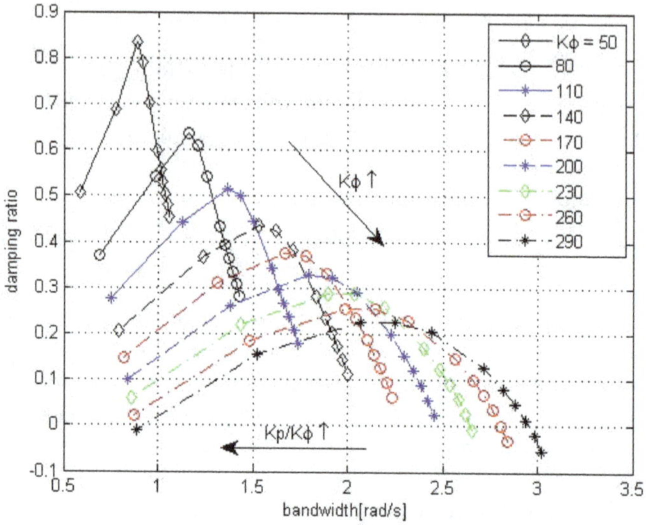

Fig. 3. Baseline control law: trade-off between bandwidth and damping ratio, lateral axis. $K_p/K_\varphi = [0.02, 0.04, 0.06, 0.08, 0.1, 0.13, 0.2, 0.25, 0.3, 0.5, 1]$.

3.2 RSF vs Baseline Performance

Up to now, only a baseline control law has been considered, which exploits fuselage measurements.

Let now rotor state measurements be introduced. The MBC transformation is applied to flap and lag angles; in particular:

- the lateral axis is most influenced by lateral flap angle β_{1s} (which is related to the lateral thrust component, which in turn causes a roll moment) and by lateral lag angle ζ_{1c} (which can be interpreted as the rotor blades center of mass offset in the lateral direction, thus again related to a roll moment), so these two measurements are eligible to be introduced in the roll axis feedback control law;
- on the other hand, the longitudinal axis is most influenced by longitudinal flap β_{1c} and longitudinal lag ζ_{1s}, for reasons analogous as above.

In particular, as far as the roll attitude is concerned, a control law which exploits feedback measurement of lateral flap β_{1s} will be considered, so that the control law is a static gain matrix

$$K = \left[K_{\beta_{1s}} \; K_p \; K_\varphi \right].$$

Figure 4 shows the effect of increasing the $K_{\beta_{1s}}$ gain only, starting from a baseline control law (K_φ, K_p gains fixed). It can be seen that the regressive flap ($R\beta$) root locus is well described by the second order approximate model (at least for realistic values of $K_{\beta_{1s}}$); the introduction of lateral flap measurement in the control law has the power of moving the regressive flap pole into regions of the complex plane where the damping ratio is much higher with respect to a simple baseline control law ($K_{\beta_{1s}} = 0$), up to a certain point beyond which the damping ratio no longer increases. Also, the progressive flap ($P\beta$) pole is subject to a damping ratio increment, even though less remarkable with respect to the regressive flap damping increment. As far as the lag (regressive, $R\zeta$ and progressive, $P\zeta$) poles are concerned, their displacement in the complex plane is significantly less sensitive to $K_{\beta_{1s}}$ variations, with respect to flap modes.

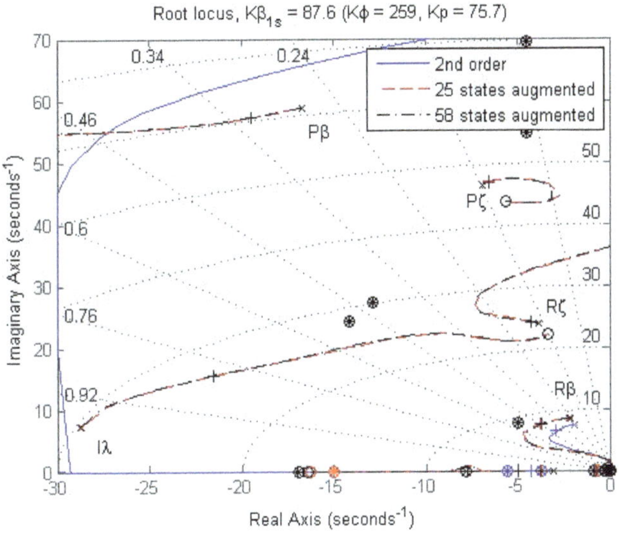

Fig. 4. Root locus varying lateral flap gain $K_{\beta_{1s}}$.

As far as the loop closure on the lateral flap measurement is concerned, the interested reader can refer to [6, 11] for a more detailed analysis about the RSF and baseline control law gains and their effect on the closed-loop system. One difference with respect to the cited studies is that, while in the problem formulation proposed by [6] all the three output channels were given the same time delay, in the work here presented it was deemed more proper to define for the flap feedback channel a different (smaller) time delay with respect to fuselage attitude measurements (see Table 1).

The same procedure undertaken when assessing the bandwidth-damping trade-off for the baseline case was repeated for the RSF case, for different values of $K_{\beta_{1s}}$.

In Figure 5, which represents the roll axis, a cloud of red points shows a set of the possible control law configurations for the baseline case (which basically correspond to the same points of Figure 3, in which $K_{\beta_{1s}} = 0$), while blue points are referred to the RSF case (different combinations for the three gains $K_{\beta_{1s}}$ K_p K_φ). The effect of introducing rotor state measurements in the control law is that the cloud of points expands in the top-right direction with respect to the baseline red cloud, and this enables to reach higher bandwidth at equal damping, or vice versa. RSF thus makes it possible to overcome the trade-off between bandwidth and damping which emerged in the baseline control law.

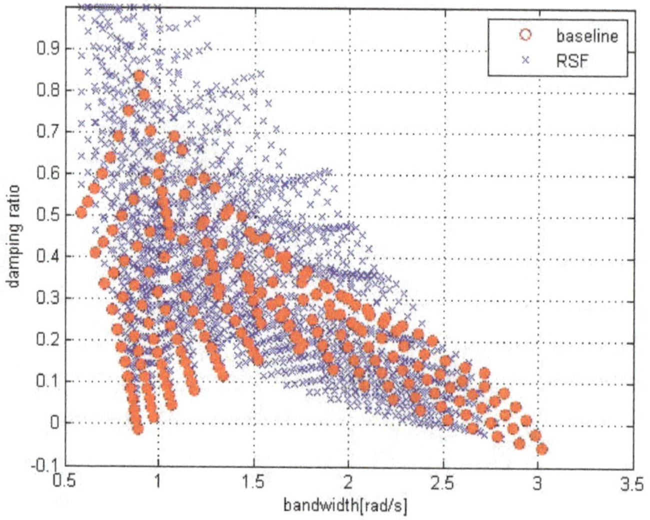

Fig. 5. Baseline vs RSF: trade-off between bandwidth and damping ratio, lateral axis.

It is important to remark that introducing RSF does not allow to raise the *maximum* reachable bandwidth (in fact, there are no blue points at the right end of the graph, beyond red points): the bandwidth is essentially determined by K_φ; however, at equal bandwidth, RSF allows to achieve a higher damping ratio, that is, RSF makes it possible to cover complex plane regions that were not reachable in the baseline case, due to inadequate damping ratio. In this sense, RSF makes it possible to *raise bandwidth*.

4 Model Uncertainty

In this section, the issue of model uncertainty with respect to physical parameters variations is addressed. The physical parameters which have been taken into account are mass, altitude and center of mass offset in the three directions. A set of linearized model was provided by AgustaWestland, taking into account different combinations of these parameters (details are omitted for confidentiality reasons). Notice that the model used throughout this work up to now is linearized in the nominal parameters values and will be referred to as the nominal model. The nominal model will be the basis on which the model uncertainty description will be defined. The other models in the set will be referred to as the "perturbed", or "off-nominal" models.

An analysis of sensitivity to the parameter variations, performed on the frequency responses on the main axes, highlighted that the most relevant effects of mass, altitude, *CG* offset variations show up at low frequency, on the translational dynamics; as far as the rotor modes are concerned, the lateral axis frequency response does not seem to be particularly influenced by these variations, at least in the frequency range interested by fuselage attitude dynamics. In conclusion, since little variability is evidenced on the fuselage attitude dynamics, parameter variations are expected not to influence in a significant way the closed-loop robustness of the attitude loop.

In order to keep into account model uncertainty during the control synthesis phase, it is necessary to obtain a representation of the model uncertainty in the frequency domain. Since this work is a natural extension and application of the methodology proposed in [11], a multiplicative uncertainty representation is to be obtained; such uncertainty representation will then be introduced in the control synthesis process, in order to satisfy the requirement of robustness with respect to model uncertainty.

As an example, the roll rate frequency response was considered; Figure 6 shows the frequency response of the nominal and the perturbed models. A multiplicative uncertainty description was computed, both for the 58 and 25 states model (Figure 7). While the two of them are very similar at high frequency (> 2 rad/s), the same does not hold true at low frequency; in particular, as far as the 58 states model is concerned, there is a significant amount of uncertainty below 0.1 rad/s (almost 0 dB), and a very remarkable peak of uncertainty in the neighborhood of the lateral oscillation mode. This is due to the fact the 58 states model takes into account low frequency dynamics related to translational modes, which are very sensitive to parameter variations, and which are neglected by the 25 states model. However, uncertainty seems to be of small entity in the 1-10 rad/s range, which is the one interested by the attitude dynamics (as confirmed by the roll rate frequency response for the perturbed models in Figure 6); this suggests uncertainty should not be a concern in the control law synthesis.

5 Inner Loop Control Synthesis

Once the structure of the control law has been chosen as seen in Section 3 and the control requirements have been defined, a control law synthesis approach is needed, such that the closed-loop system achieves the control requirements. Such synthesis approach shall provide some means to formulate the control requirements, and must be able to tune the control law parameters, given an arbitrary control law architecture. Following the methodology proposed in [11], the structured H_∞ approach was chosen.

Not unlike the classical H_∞ approach, in the structured H_∞ case the plant is augmented with performance inputs and outputs, which are defined as signals that need in some sense to be minimized; an optimal regulator is computed such that the H_∞ norm of the system closed in loop with the regulator, from the performance inputs to the performance outputs, is minimized [13]. In this framework, performance inputs and outputs can be chosen such that the norm to be minimized is a function of

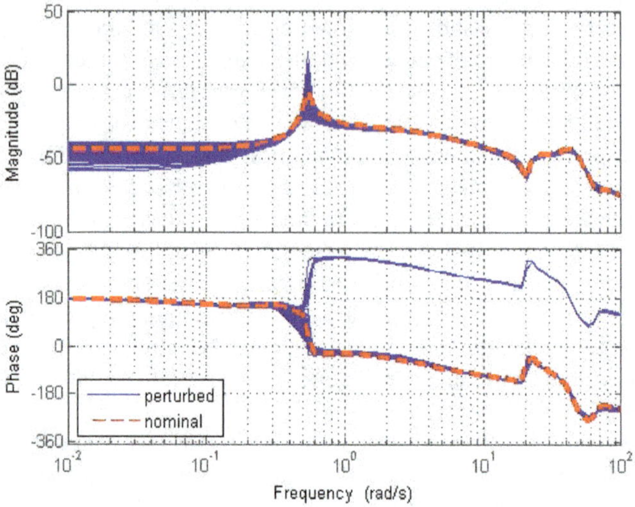

Fig. 6. Nominal vs perturbed models, 58 states, roll rate frequency response.

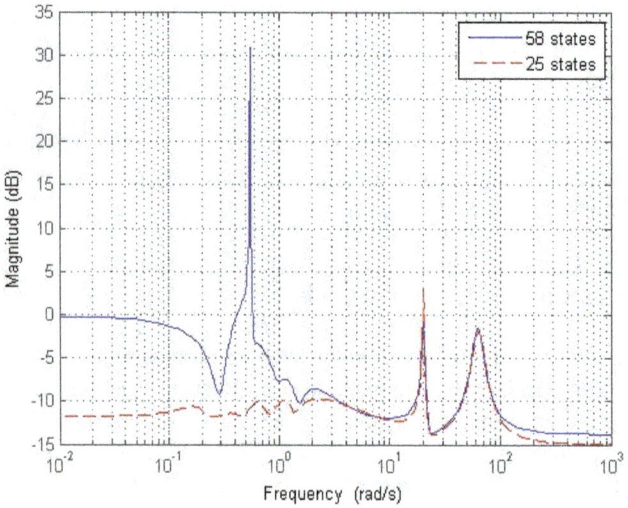

Fig. 7. Multiplicative uncertainty on the roll rate frequency response, 58 vs 25 states models.

the closed-loop sensitivity functions (sensitivity, control sensitivity, complementary sensitivity), weighted with proper frequency weights; control requirements are thus encoded as the frequency weights. In the classical H_∞ approach the optimization problem is solved without constraints on the structure and order of the controller, which implies that the obtained regulator is high order (the sum of the order of the plant and of the frequency weights) and it is a full matrix from all the measurable outputs to all the control inputs of the plant; in the case of rotorcraft flight control system, both these issues represent a limitation to the feasibility of the classical H_∞ approach, since the control law architecture is often fixed and simple, and the computational power on board is limited. The structured H_∞ approach, on the other hand, can cope with both these issues and overcome the limitations of the classical H_∞ approach: the main difference with respect to the latter is that the structure of the control law can be defined *a priori*; the optimization algorithm tunes the parameters of the control law such that the H_∞ norm is minimized. This requires non-smooth optimization techniques to be employed [14]; such techniques allow new constraints to be added to the optimization process, which would not be available by means of classical H_∞, hence control requirements of different nature can be introduced in the control synthesis problem [15]. It should be remarked that by means of structured H_∞, the methodology presented herein can be applied to already existing flight control law structures, provided that the control law parameters are not fixed a priori and can be tuned. On the other hand, of course the so-obtained solution is sub-optimal with respect to the solution of the classical H_∞ problem, since the number of degrees of freedom is significantly lower.

The control law synthesis methodology presented in [11] was applied to the 25 states augmented model here obtained; control requirements were formulated as follows:

- performance requirements were formulated as weights on the attitude angle sensitivity functions (for φ and θ);
- control moderation requirements were formulated as a weight matrix on control sensitivity function: these requirements can also be interpreted from the point of view of measurement noise rejection requirements;
- the requirement of robustness with respect to uncertainty is formulated as a multiplicative uncertainty description.

In order to highlight the benefits of RSF with respect to a baseline control law, two sets of performance requirements for the feedback system have been defined: a set of *soft* requirements, with low bandwidth; a set of *hard* requirements, with higher bandwidth; the parameters of the weighting functions on roll sensitivity are summarized in Table 2. Pitch requirements were chosen in the same fashion and are not shown here, for simplicity.

Similarly, two sets of requirements concerning the rejection of measurement noise have been defined: a *soft* and a *hard* set of requirements, the latter emphasizing improved performance in terms of high frequency noise attenuation. In selecting the control sensitivity weight, the high frequency pole was chosen keeping into account the actuators' bandwidth, assuming a value of 50 rad/s, in order to

Table 2. Performance requirements formulated as weight on the roll attitude sensitivity function.

	soft	*hard*
Desired bandwidth [rad/s]	1.5	2
DC gain K_{DC}	500	500
High frequency gain K_{HF}	0.9	0.5

Table 3. Control action moderation requirements formulated as weight on roll control sensitivity function.

	soft	*hard*
High frequency gain	$3.5 \cdot 10^{-3}$	$12 \cdot 10^{-3}$
Pole frequency [rad/s]	50	50
Ratio K_{HF}/K_{DC}	$5 \cdot 10^5$	$5 \cdot 10^5$

reduce out-of-bandwidth control action. Again, parameters for the weights on the roll control sensitivity functions are summarized in Table 3.

As far as the requirement of robustness to uncertainty is concerned, since the results of the uncertainty analysis shown in Section 4 evidence that uncertainty is negligible in the frequency range interested by attitude dynamics, it was deemed sufficient to perform just a robustness analysis *a posteriori* on the perturbed models closed in loop with the regulator computed on the nominal model.

6 Simulation Results

Following the methodology proposed in [11], in order to show the benefits of RSF over a traditional attitude control law, four different laws were designed with the H_∞ approach: a *baseline* (B) law, a *RSF hard* (RSF_H) law, a *soft* (RSF_S) law, and a *fault robust* (RSF_{FR}) law; for further details about the rationale that drove this choice, refer to [11]. For the sake of simplicity, only results related to the roll axis will be presented in the following.

Table 4 shows the values of the four control law gains for the lateral axis, tuned by H_∞ optimization (based on the 25 states model), along with the DRB bandwidth ω_B and phase margin[3]. In particular, the latter was computed both in nominal conditions and in case of fault of the lateral flap sensor, in which case the β_{1s} measurement is simply put to zero.

It has to be remarked that, in order to tune control laws for the lateral axis and avoid interferences due to the coupling with longitudinal axis, it was necessary to synthesize the lateral axis control law with the longitudinal attitude loop closed;

[3] Bandwidth was computed based on the 58 states augmented model; phase margin was computed breaking the loop in the actuator.

the gains for the (*baseline*) longitudinal attitude loop were chosen according to *soft* longitudinal performance requirements.

Some remarks follow:

- K_φ is related to bandwidth: as this gain increases, so bandwidth does;
- when switching from the nominal case to the faulty one, phase margin drops.
- the *baseline* control law has been set $K_{\beta_{1s}} = 0$, since it does not consider lateral flap measurement; for the same reason, the fault condition for such a control law is not defined.
- the RSF_S controller appears to be the least sensitive to rotor sensor fault, since its bandwidth is lower with respect to the other two RSF control laws (the PM drop in RSF_S is negligible if compared to the others); this is in some way intuitive, since $K_{\beta_{1s}}$ is very low and the β_{1s} measurement is expected to have a less relevant role in the feedback loop, with respect to other laws in which $K_{\beta_{1s}}$ is higher.

Table 4. Comparison between controller gains for the four control laws obtained using the H_∞ approach, lateral axis.

	$K_{\beta_{1s}}\left[\frac{\%}{rad}\right]$	$K_p\left[\frac{\%}{rad/s}\right]$	$K_\varphi\left[\frac{\%}{rad}\right]$	$\omega_B[rad/s]$	PM[°]	PM fault
RSF hard	88	76	259	2.15	37.8	23.5
RSF soft	12	45	91	1.21	56.8	54.3
RSF fault robust	231	63	158	1.49	76.2	36.8
Baseline	0	65	119	1.27	42.5	-

Some comparisons between the four control laws for the lateral axis follow; the comparison is based on the 58 states augmented model.

6.1 Step Responses

Comparing the φ attitude responses of B, RSF_S, RSF_H to a step reference φ_0 in Figure 8, it appears evident the RSF_H response is much faster than the other two, which have basically the same response. Moreover, RSF_H shows some oscillations in the first second of the response. The drift in the response that shows up after a few seconds is due to the fact the 58 states model describes also the low frequency translational dynamics, which includes the unstable longitudinal phugoid mode; the velocity (outer) loop is in charge of stabilizing this low frequency modes, exploiting fuselage speed measurements, since it is apparently not possible to stabilize the longitudinal phugoid mode with attitude measurements (θ, q) only.

Figure 9 shows the lateral flap response to a 10° step change in φ reference; again, the RSF_H flap response is more remarkable than other control laws, due to the higher bandwidth; RSF_S lateral flap reaches a lower peak than B.

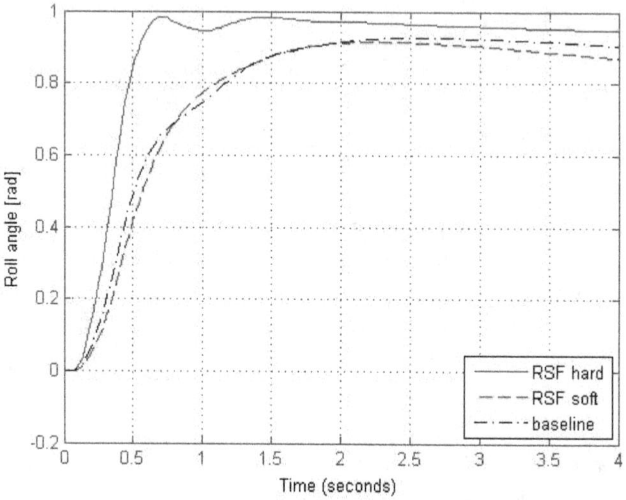

Fig. 8. Step response, φ: baseline vs RSF_H vs RSF_S.

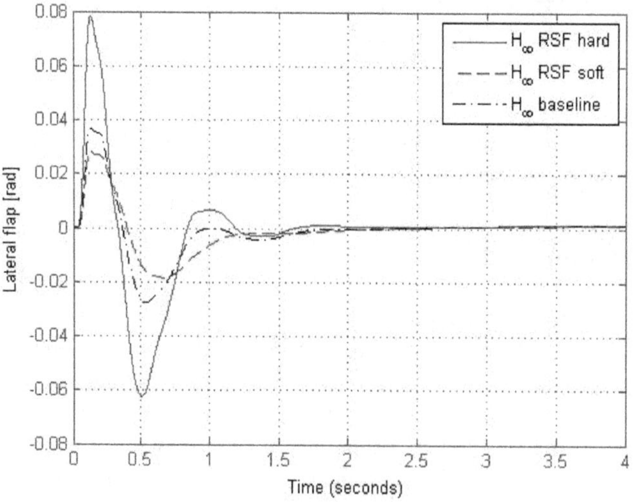

Fig. 9. Lateral flap response to a 10° step change in φ: baseline vs RSF_H vs RSF_S.

6.2 Control Sensitivity: Baseline vs RSF_S

Figure 10 shows the magnitude of the control sensitivity function on the three channels (from β_{1s}, p, φ measurements to lateral cyclic input), both for the B and RSF_S laws. This is the closed loop transfer function from measurement noise to control input, but it can also be interpreted as the function from disturbances on the output (such as wind gusts) to control input; in both cases, it is desirable to have a low control sensitivity, in order to reduce control action due to measurement noise in the former case (especially at high frequencies), and to reduce control action as a response to disturbances in the latter (especially in the range of frequencies in which the disturbance acts).

From the inspection of the values of the gains reported in Table 4 one can see that the gains K_p, K_φ of the RSF_S controller have a smaller magnitude with respect to the ones for the B controller, so one expects a smaller control sensitivity in this case. As can be seen from Figure 10, the RSF_S controller leads to a smaller control sensitivity at high frequencies, which implies a reduced sensitivity of the control action to measurement noise; also, peaks in magnitude are lower.

At the same time, while showing such improvements in terms of control sensitivity reduction, RSF_S achieves similar results to B in terms of performance: the closed loop bandwidth is basically the same, and also step responses are similar (Figure 8).

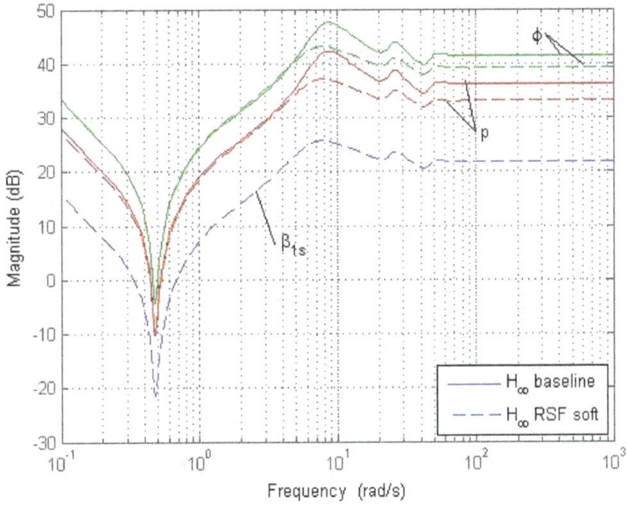

Fig. 10. Control sensitivity function on the three channels, magnitude: baseline vs RSF_S.

6.3 Fault Robustness

The last point of interest in the analysis of the performance of the considered control laws is the occurrence of faults leading to the loss of the measurement of β_{1s}, *e.g.*, $\beta_{1s} = 0$. In this respect one wishes the closed-loop system to fall back to a performance level comparable to the one for the *baseline* controller. Figure 11 shows a comparison between the closed-loop responses to a step change in the set-point for φ for the fault robust RSF_{FR} controller and the baseline B one. As can be seen, the response of the fault robust controller remains faster than the baseline one even in faulty conditions, a result which is consistent with the values for the bandwidth presented in Table 4.

Similarly, in Figure 12 the closed-loop step responses corresponding to the RSF_{FR} and RSF_H controllers in fault condition are compared. Again, the simulation results are consistent with the analysis, as they show that the RSF_{FR} controller in the faulty case leads to a non-oscillatory step response, unlike the RSF_H control law, which has not been designed to be fault robust and shows remarkable oscillations in the response; the price to pay, however, is that the (nominal) bandwidth achieved by RSF_{FR} is lower than the *hard* control law, although RSF_{FR} was intended to achieve the same *hard* performance requirements as RSF_H.

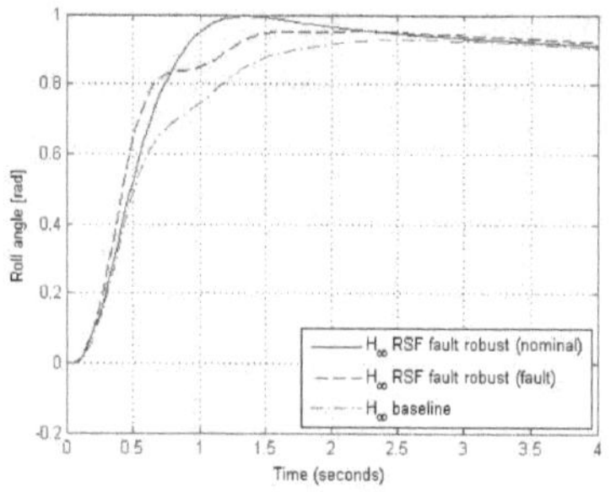

Fig. 11. Step response, φ: baseline vs RSF_{FR} (nominal) vs RSF_{FR} (fault).

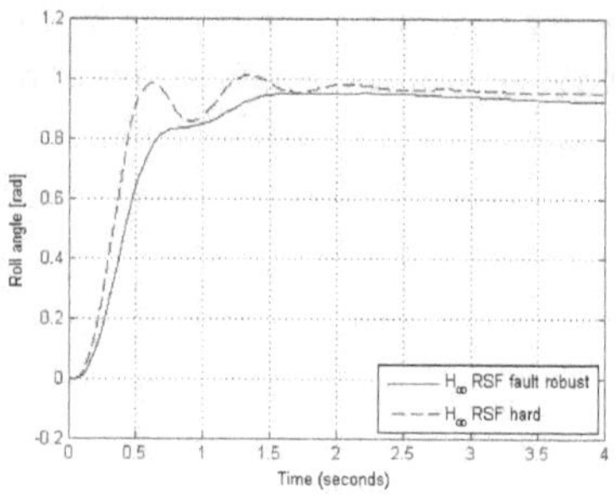

Fig. 12. Step response, φ: RSF_H (fault) vs RSF_{FR} (fault).

6.4 A Posteriori *Robustness Analysis*

A robustness analysis *a posteriori* on the closed-loop system, based on the control law gains obtained with the H_∞ approach and shown in Table 4, has been performed. Notice that these control laws were synthesized based on the nominal model, without taking into account the uncertainty description obtained in Section 4.

Since the multiplicative uncertainty description may be interpreted as a weight on the complementary sensitivity in order to show the amount of uncertainty as a function of frequency, the complementary sensitivity function was computed for the nominal model closed in loop with the B, RSF_H, RSF_S control laws, and then weighted with the multiplicative uncertainty description, in order to show possible closed-loop robustness issues. The complementary sensitivity was computed based on the model outputs β_{1s}, p, φ and is a $[3 \times 3]$ transfer matrix; it can be interpreted as the closed-loop transfer matrix from set-points on these variables to the outputs.

In order to analyze the closed-loop robustness to multiplicative uncertainty, it is useful to inspect the singular values plot of the weighted complementary sensitivity function: indeed, the largest singular value of the weighted complementary sensitivity represents an indication of the maximum magnitude the frequency response of the $[3 \times 3]$ MIMO transfer matrix can reach, as a function of frequency; the maximum singular value magnitude over frequency is defined as the H_∞ norm of the transfer matrix. From such inspection, it is evident that the B and RSF_S control laws always remain under the 0 dB threshold, thus indicating the robustness requirement is accomplished; RSF_H exceeds the 0 dB bound in the 3-10 rad/s range of frequencies, although the peak is low (3 dB); this indicates the RSF_H control law anyway has good robustness properties.

Figure 13 shows the φ step responses of a set of perturbed (off-nominal) models closed in loop with the RSF_H control law. This is just an example to show that parameter variations do not significantly affect the attitude response when the system is closed in loop with the attitude control law computed on the nominal model; the variation in the peak height between 0.5 and 1s is probably due to mass variation; the overall response shape remains the same even in presence of parameter variations.

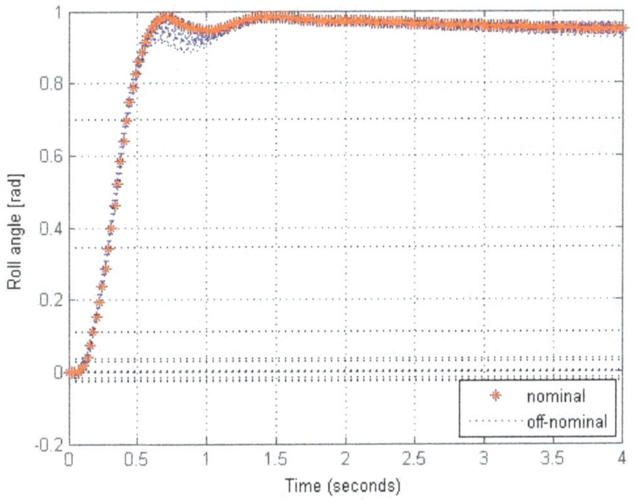

Fig. 13. RSF_H, φ step responses, off-nominal models.

7 Concluding Remarks

A linearized helicopter model in hover was obtained; the model was reduced up to a second order approximation of roll dynamics, keeping into account the dynamic coupling between rotor and fuselage; in this way, it was possible to assess the role of the gains of a traditional attitude control law in determining the dominant closed-loop properties of the system. As a main result, the presence of a trade-off was evidenced between bandwidth and damping ratio, which constitutes a limit on the achievable performance in the case a traditional control law is used; RSF has the potential to overcome this limitation, allowing to achieve higher bandwidth being the damping ratio equal, or the other way round; as an alternative, RSF allows to decrease the gains at equal bandwidth, resulting in better noise rejection properties. In addition, a robustness analysis to parameters variations was performed, suggesting that the uncertainty stemming from such variations does not constitute a limitation in achieving the desired performance while maintaining adequate levels of robustness. As a final remark, it has to be pointed out that this study is based on the assumption that ideal rotor state measurements are available; future work is going to keep into account also realistic rotor measurements, and on the basis of the results herein obtained, to assess if the same benefits achieved by RSF in the case of ideal measurements can be obtained even in the case of realistic ones.

Acknowledgements. This research was supported by the CleanSky GRC5 MANOEUVRES project, grant agreement no. 620068.

References

1. Blanken, C.L., Hoh, R.H., Mitchell, D.G., Key, D.L.: Test guide for ADS-33E-PRF. Tech. Rep. AMR-AF-08-07, AMRDEC (2008)
2. ADS-33E-PRF, Aeronautical Design Standard, Performance Specification. Handling Qualities Requirements for Military Rotorcraft (March 2000)
3. Mansur, M.H., Lusardi, J.A., Tischler, M.B., Berger, T.: Achieving the best compromise between stability margins and disturbance rejection performance. In: 65th Annual Forum of the American Helicopter Society, Grapevine, USA (2009)
4. Horn, J.F., Guo, W., Ozdemir, G.T.: Use of rotor state feedback to improve closed-loop stability and handling qualities. Journal of the American Helicopter Society 57(2), 1–10 (2012)
5. Blanken, C.L., Lusardi, J.A., Ivler, C.M., Tischler, M.B., Hofinger, M.T., Decker, W.A., Malpica, C.A., Berger, T., Tucker, G.E.: An investigation of rotorcraft stability-phase margin requirements in hover. In: 65th Annual Forum of the American Helicopter Society, Grapevine, USA (2009)
6. Takahashi, M.D.: Rotor-state feedback in the design of flight control laws for a hovering helicopter. Journal of the American Helicopter Society 39(1), 50–62 (1994)
7. Tischler, M.B.: System identification requirements for high-bandwidth rotorcraft flight control system design. In: 1991 American Control Conference, Boston, USA (1991)
8. Jr. Hall, W.E., Jr. Bryson, A.E.: Inclusion of rotor dynamics in controller design for helicopters. Journal of Aircraft 10(4), 200–206 (1973)
9. Ingle, S.J., Celi, R.: Effects of higher order dynamics on helicopter flight control law design. Journal of the American Helicopter Society 39(3), 12–23 (1994)
10. Ivler, C.M.: Development and comparison of explicit and implicit rotor-state feedback control systems for a fly-by-wire UH-60. In: AHS Rotorcraft Handling Qualities Specialists Meeting, Huntsville, USA (2014)

11. Panza, S., Lovera, M.: Rotor state feedback in helicopter flight control: robustness and fault tolerance. In: IEEE Multi-Conference on Systems and Control, Antibes-Nice, France (2014)
12. Padfield, G.D.: Helicopter flight dynamics: the theory and application of flying qualities and simulation modelling. Blackwell (2007)
13. Skogestad, S., Postlethwaite, I.: Multivariable feedback control: analysis and design. J. Wiley and Sons (1996)
14. Apkarian, P., Noll, D.: Nonsmooth H_∞ synthesis. IEEE Transactions on Automatic Control 51(1), 71–86 (2006)
15. Apkarian, P.: Tuning controllers against multiple design requirements. In: American Control Conference, Washington DC, USA (2013)

Cooperative 2-On-1 Bounded-Control Linear Differential Games

Shmuel Y. Hayoun and Tal Shima

Abstract. A linearized 2-on-1 engagement is considered, in which the players' controls are bounded and have first order dynamics and equal terminal instants. A capturability analysis is performed, presenting necessary and sufficient conditions for the feasibility of exact capture against any target maneuver and for arbitrary control dynamics. Wishing to formulate the engagement as a zero-sum differential game, a suitable cost function is proposed and validated, and the resulting optimization problem and its solution are presented. Construction and analysis of the game space for the case of strong pursuers is shown, and the players' closed form optimal controls are derived.

1 Introduction

Cooperative strategies are becoming more and more popular with the continuing evolution and advancement in decision making capabilities of autonomous vehicles. Utilizing multiple agents to perform a given task can be beneficial even in cases when the goal is achievable by a single agent. With regard to interception engagements with a single target a single superior missile could be replaced by a group of inferior ones in order to maintain or improve capturability. Through shared information and coordinated actions the capability requirements and/or the number of required agents may be relaxed and reduced respectively, resulting in an overall reduced engagement cost.

Since Isaacs's [10] pioneering work on pursuit-evasion games, both linear and non-linear two-player differential games have been extensively studied. Differential game based solutions to 1-on-1 interception engagements, in cases where the target is likely to execute unknown evasive maneuvers, provide the optimal strategies for

Shmuel Y. Hayoun · Tal Shima
Department of Aerospace Engineering, Technion-Israel Institute of Technology, Haifa, 32000 Israel
e-mail: shmuli@tx.technion.com, tal.shima@technion.ac.il

J. Bordeneuve-Guibé et al. (eds.), *Advances in Aerospace Guidance, Navigation and Control*, DOI: 10.1007/978-3-319-17518-8_14

both the pursuing missile and the evading target. Since the interests of the adversaries are typically diametrically opposed, such engagements are typically formulated as zero-sum differential games [21]. The nature of such games promises that in case of a deviation from its optimal solution any player can only cause a decrease in its own payoff, and thereby an increase (of the same magnitude) in its opponent's payoff. Hence the optimal pursuit-evasion strategies couplet was dubbed a "saddle point". This property implies that the optimal differential game based strategy of each player holds regardless of its opponent's actual behavior.

Apart from the optimal strategies the game solution also includes a mapping of optimal trajectories depicting the guaranteed outcome of the game, known as the game space. An interesting phenomenon in games of pursuit is the existence of singular surfaces. In [10] the initial analysis and terminology of the different surface types was presented. The theory regarding the methods of solving problems with singular surfaces was later given in [13], and was briefly discussed in [12] for a more general case with more than two players. Usually such surfaces emerge within voids or between intersecting fields of regular optimal trajectories. In finite horizon and some infinite horizon engagements the existence of voids is also possible. Voids are bounded by optimal trajectories emanating from same-cost terminal states, and constitute singular zones in which one or more players' optimal strategy is indeterminate. In both cases of singularity the behaviour of the players is not trivial, and must be analyzed with care. Generally the solution finally fills the void with so called "tributaries" emanating from the singular surface, in a manner defined by the singular surface's type.

In many cases the motion of each of the adversaries during the endgame can be linearized about its corresponding initial collision course. The two main approaches to the solution of such linear games of pursuit include the linear quadratic cost & unbounded controls formulation [9, 1] and the norm cost & bounded controls formulation [8, 7, 20, 19]. The closed form optimal controls derived for the latter are typically bang-bang controls, whereas the optimal controls of the former can generally be represented in linear form by the product of a time-varying gain and the expected miss distance.

Solutions to linearized two-player zero-sum differential games of both types also include variations on the order of the players' control dynamics. In [8] the solution to a simplified linear pursuit-evasion game in which both players have bounded controls and ideal control dynamics (DGL/I) is presented. A general solution for an arbitrary set of linear system dynamics was later presented in [7], which also included a demonstration for the specific case of a pursuer with first order control dynamics intercepting an ideal evader (DGL/0). This was followed by [20], in which a more general case of both players having first order control dynamics (DGL/1) was solved. In [19] five types of possible game structures of the DGL/1 solution were presented, dependant on the dynamics capabilities of the pursuer relative to those of the evader: maneuverability (μ) and agility ($\mu\varepsilon$).

There have also been studies regarding multi-player linear pursuit-evasion games. Solutions to a linear quadratic three-player nonzero-sum pursuit-evasion game, in which two pursuers attempt to intercept a single evader, were presented in [5] and in [14]. The cost functions to be minimized by each of the pursuers included the corresponding pursuer's quadratic miss and control effort, and the evader's payoff, to be maximized, included the minimum of the pursuers' quadratic misses and its own control effort. In [15] a study of a two-person zero-sum linear quadratic game was given, in which a pursuing missile strives to intercept a maneuvering target while avoiding capture by a defending missile (the target-defender team is regarded as a single player with two controls). The common cost function included a sum of the quadratic target-missile and missile-defender misses and of the adversaries' control efforts. Solutions to a bounded controls 2-on-1 pursuit evasion game were presented in [11] and in [6]. In the former an optimal evasion strategy in the framework of DGL/1 against two *strong* pursuers ($\mu \geq 1, \mu\varepsilon \geq 1$) was proposed, for cases in which the evader initiates and terminates between the two pursuers. The latter presented an extensive numerical study of value function level sets, and proposed a switching lines based method for constructing optimal feedback controls. In both studies the engagement was formulated as a two-person zero-sum game (the pursuing team regarded in both cases as a single player with two controls), where the performance index was defined as the minimum of both pursuers' misses.

The examined previous studies in which the two-person approach was used, demonstrated two acceptable forms of the shared cost function: the weighted sum of the misses and the minimum of the misses. The former is typical of the linear quadratic formulation, and the latter of the bounded controls formulation. Of these two forms the first will produce an overall convergence of the pursuers toward the target, but does not accurately represent the evader's interest. The second better represents the evader's goal, however in many cases the outcome of the game is determined by the actions of only a few or even just one of the pursuers. This crucial aspect of the definition of the performance index coupled with the existing approaches begs the investigation of alternative definitions.

A great point of interest when analyzing games of pursuit is whether or not the evader's capture can be guaranteed. The first work to specifically address the issue of capturability was presented in [4]. In it the case of a 1-on-1 planar engagement was considered, in which both adversaries have constant speeds, bounded lateral accelerations and ideal dynamics. It was proven that the pursuer can capture (defined as position coincidence) the evader from any initial state if and only if it has a speed advantage and is at least as maneuverable as the evader. This theory was later extended in [16] to address motion in three dimensional space. It was shown that a sufficient condition for capture of the target is the pursuer's superiority both in speed and in maneuverability. In [2] an analogous study to that in [4] was presented. It was proven that in the 1-on-1 planar engagement the evader can avoid capture for any initial conditions if and only if: a) it has a speed advantage and is at least as maneuverable as the pursuer, b) its speed is equal to the pursuer's and it has a maneuverability advantage. This work was extended to a planar case of n pursuers vs. a single evader in [17]. It was shown that if the evader has a speed advantage and is at least as maneuverable as the pursuer it can avoid capture indefinitely. In a later publication [19] it was shown that in the framework of DGL/1 capture is possible only if the pursuer does not have a maneuverability or an agility disadvantage ($\mu \geq 1, \mu\varepsilon \geq 1$) or is only more agile ($\mu < 1, \mu\varepsilon > 1$). This work was also extended to include biproper control dynamics of the adversaries in [18].

These previous capturability analyses have yielded important conclusions with regard to the necessary and sufficient requirements of pursuers in interception engagements. As opposed to the non-linear case studies, the linear analysis included an additional factor which is the adversaries controls dynamics. Following [19] it will be interesting to examine the necessary and sufficient conditions for capture in the more general case of an n-on-1 engagement, the results of which have important implications on the merits of utilizing a multiplicity of pursuers.

This paper presents an initial analytical study of a perfect information linear 2-on-1 pursuit-evasion game in the framework of DGL/1, in a scenario where the pursuers are launched simultaneously from a single platform. Conditions for the feasibility of exact capture are derived for arbitrary control dynamics, and compared to those obtained for the 1-on-1 game in the case of first order control dynamics. The engagement is treated as a two-person zero-sum differential game, for which a new cost function is proposed. Solution to the game is presented, including construction and analysis of the game space and the derivation of closed form controls for both the pursuers and the evader.

The remainder of this paper is organized as follows: The next section includes the 2-on-1 engagement formulation and its mathematical model in the framework of DGL/1. It is followed by a general capturability analysis, which provides the necessary and sufficient conditions for the feasibility of the evading target's exact capture. In the next section the problem is regarded as a two-person zero-sum differential game, a new cost function is proposed and utilized and the game solution is presented. The game space construction and analysis is shown in the following section, for the case of two *strong* pursuers.

2 2-On-1 Engagement Formulation

Consider the endgame geometry of a planar 2-on-1 pursuit-evasion game as shown in Fig. 1, where the X axis is along the initial line of sight (LOS) of pursuer P_1 and the Y axis is perpendicular to it. V and $a^{\perp}$ denote the velocity and lateral acceleration respectively, and γ and λ denote the path and LOS angles respectively from the X axis. Subscript P_i denotes the i-th pursuer and E the evader. We will denote the group of entities $G = \{P_1, P_2, E\}$ and the pursuer tags $N = \{1, 2\}$. The non-linear kinematics are given by

$$\begin{aligned} \dot{r}_{P_i} &= -V_E \cos(\gamma_E + \lambda_{P_i}) - V_{P_i} \cos(\gamma_{P_i} - \lambda_{P_i}) \\ \dot{\lambda}_{P_i} &= \frac{V_E \sin(\gamma_E + \lambda_{P_i}) - V_{P_i} \sin(\gamma_{P_i} - \lambda_{P_i})}{r_{P_i}} \end{aligned} \quad , \quad i \in N \tag{1}$$

where r is the range between a pursuer and the evader.

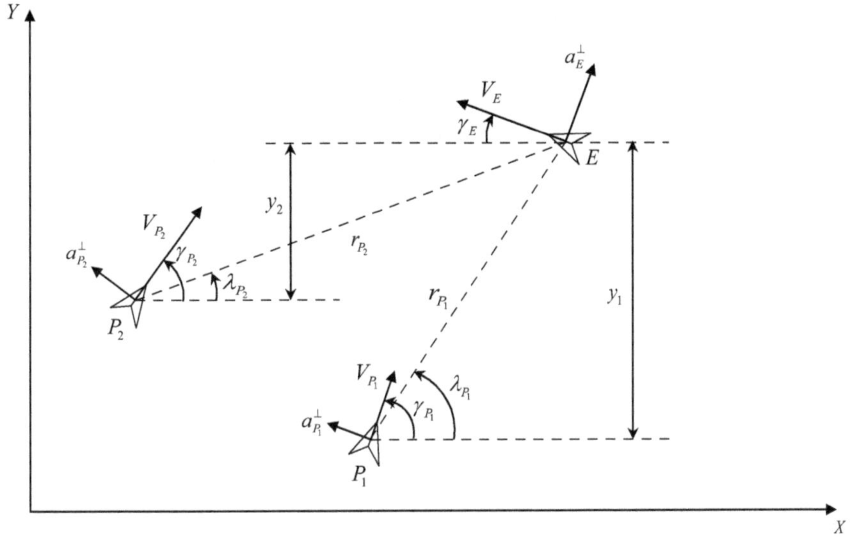

Fig. 1 2-on-1 Engagement Scheme

The evolution of the path angles is according to

$$\dot{\gamma}_j = \frac{a_j^\perp}{V_j}, \quad j \in G. \tag{2}$$

Assuming near-collision course flight the engagement can be linearized around a single fixed reference line, of angular slope λ_o. For simplicity we choose $\lambda_o = 0$ $(LOS_o = LOS_{oP_1})$.

2.1 Linearized System Kinematics

The linearized kinematics of the intercept geometry are given by

$$\begin{aligned} \dot{y}_i &= V_E \sin(\gamma_{oE}) - V_{P_i} \sin(\gamma_{oP_i}) \\ \ddot{y}_i &= a_E - a_{P_i} \end{aligned} \quad , \quad i \in N \tag{3}$$

where y_i is the relative displacement between the evader E and pursuer P_i normal to X, and where a_j are the lateral accelerations of the adversaries normal to X. γ_{oj} denotes player j's initial path angle.

$$\begin{aligned} a_{P_i} &= a_{P_i}^\perp \cos(\gamma_{oP_i}) \\ a_E &= a_E^\perp \cos(\gamma_{oE}) \end{aligned} \quad , \quad i \in N \tag{4}$$

Assuming first order pursuer and evader control dynamics

$$\frac{a_j(s)}{a_j^c(s)} = \frac{1}{1+s\tau_j}, \quad j \in G \tag{5}$$

where a_j^c is the commanded lateral accelerations of player j.
Assuming constant speeds of the adversaries during the engagement the interception time of the i-th pursuer can be approximated by

$$t_{f_i} = \frac{r_{oP_i}}{V_{c_i}}, \quad i \in N \tag{6}$$

where r_{oP_i} is the initial i-th pursuer-evader range. V_{c_i} is the closing speed, which is approximately constant for each pursuer, and equal to

$$V_{c_i} = V_E \cos(\gamma_{oE} + \lambda_{oP_i}) + V_{P_i} \cos(\gamma_{oP_i} - \lambda_{oP_i}), \quad i \in N \tag{7}$$

with λ_{oP_i} denoting pursuer i's initial LOS angle. Assuming equal terminal instants

$$t_{f_1} = t_{f_2} = t_f \tag{8}$$

Subsequently, the time-to-go is defined by

$$t_{go} = t_f - t \tag{9}$$

Defining the state vector as follows

$$\mathbf{x}(t) = [y_1(t) \;\; \dot{y}_1(t) \;\; a_{P_1}(t) \;\; y_2(t) \;\; \dot{y}_2(t) \;\; a_{P_2}(t) \;\; a_E(t)]^T \tag{10}$$

and substituting the commanded lateral accelerations with the following expression

$$a_j^c = a_j^{max} u_j, \quad j \in G \tag{11}$$

results in the following linear system

$$\dot{\mathbf{x}} = \mathbf{A}\mathbf{x} + \mathbf{B}_1 u_{P_1} + \mathbf{B}_2 u_{P_2} + \mathbf{C} u_E, \quad x(t_o) = x_o, \quad |u_j| \le 1, \quad j \in G \tag{12}$$

where

$$\mathbf{A} = \begin{bmatrix} 0 & 1 & 0 & 0 & 0 & 0 & 0 \\ 0 & 0 & -1 & 0 & 0 & 0 & 1 \\ 0 & 0 & -\frac{1}{\tau_{P_1}} & 0 & 0 & 0 & 0 \\ 0 & 0 & 0 & 0 & 1 & 0 & 0 \\ 0 & 0 & 0 & 0 & 0 & -1 & 1 \\ 0 & 0 & 0 & 0 & 0 & -\frac{1}{\tau_{P_2}} & 0 \\ 0 & 0 & 0 & 0 & 0 & 0 & -\frac{1}{\tau_E} \end{bmatrix}, \mathbf{B}_1 = \begin{bmatrix} 0 \\ 0 \\ \frac{a_{P_1}^{max}}{\tau_{P_1}} \\ 0 \\ 0 \\ 0 \\ 0 \end{bmatrix}, \mathbf{B}_2 = \begin{bmatrix} 0 \\ 0 \\ 0 \\ 0 \\ 0 \\ \frac{a_{P_2}^{max}}{\tau_{P_2}} \\ 0 \end{bmatrix}, \mathbf{C} = \begin{bmatrix} 0 \\ 0 \\ 0 \\ 0 \\ 0 \\ 0 \\ \frac{a_E^{max}}{\tau_E} \end{bmatrix} \tag{13}$$

The corresponding transition matrix, satisfying

$$\dot{\Phi}(t,t_o) = \mathbf{A}\Phi(t,t_o); \;\; \Phi(t_o,t_o) = \mathbf{I}$$

is

$$\Phi(t_f,t)=\begin{bmatrix}1 & t_{go} & -\tau_{P_1}^2\psi(\theta\varepsilon_1) & 0 & 0 & 0 & \tau_E^2\psi(\theta)\\ 0 & 1 & -\tau_{P_1}\left(1-e^{-\theta\varepsilon_1}\right) & 0 & 0 & 0 & \tau_E\left(1-e^{-\theta}\right)\\ 0 & 0 & e^{-\theta\varepsilon_1} & 0 & 0 & 0 & 0\\ 0 & 0 & 0 & 1 & t_{go} & -\tau_{P_2}^2\psi(\theta\varepsilon_2) & \tau_E^2\psi(\theta)\\ 0 & 0 & 0 & 0 & 1 & -\tau_{P_2}\left(1-e^{-\theta\varepsilon_2}\right) & \tau_E\left(1-e^{-\theta}\right)\\ 0 & 0 & 0 & 0 & 0 & e^{-\theta\varepsilon_2} & 0\\ 0 & 0 & 0 & 0 & 0 & 0 & e^{-\theta}\end{bmatrix} \tag{14}$$

where

$$\begin{aligned}\psi(\alpha) &= e^{-\alpha}+\alpha-1\\ \theta &= \frac{t_{go}}{\tau_E}\\ \varepsilon_i &= \frac{\tau_E}{\tau_{P_i}}, \quad i\in N\end{aligned} \tag{15}$$

2.2 Order Reduction

By means of the terminal projection transformation introduced in [3] we are able to reduce the order of the problem to two state variables, each known as the zero-effort miss (ZEM)

$$z_i(t)=\mathbf{D}_i\Phi(t_f,t)\mathbf{x}(t), \quad i\in N$$

where

$$\mathbf{D}_i=\left[[0]_{1\times 3(i-1)}\ 1\ [0]_{1\times 3(3-i)}\right], \quad i\in N.$$

resulting in the normalized reduced order dynamic equations

$$\begin{aligned}\dot z_i(t) &= \mathbf{D}_i\left[\dot\Phi(t_f,t)\mathbf{x}+\Phi(t_f,t)\dot{\mathbf{x}}\right]=\mathbf{D}_i\Phi(t_f,t)\left[\mathbf{B}_i\mathbf{u}_{P_i}+\mathbf{C}u_E\right]\\ z_i(t_o) &= \mathbf{D}_i\Phi(t_f,t_o)\mathbf{x}(t_o)=z_{oi}\end{aligned}, \quad i\in N.$$

Substituting the system matrices

$$\dot z_i(t)=a_E^{max}\tau_E\psi(\theta)u_E-a_{P_i}^{max}\tau_{P_i}\psi(\theta\varepsilon_1)u_{P_i}, \quad i\in N, \quad z_i(t_o)=z_{oi}. \tag{16}$$

We define a non-dimensional state vector as the normalized ZEM vector

$$\mathbf{Z}(\theta)\triangleq\frac{\mathbf{z}(t)}{\tau_E^2 a_E^{max}} \tag{17}$$

yielding the normalized dynamic system

$$\frac{dZ_i}{d\theta}=\frac{\mu_i}{\varepsilon_i}\psi(\theta\varepsilon_i)u_{P_i}-\psi(\theta)u_E, \quad i\in N, \quad Z_i(\theta_o)=Z_{oi} \tag{18}$$

where

$$\mu_i = \frac{a_{P_i}^{max}}{a_E^{max}}, \quad i \in N \tag{19}$$

3 Capturability Analysis

The following section deals with the derivation of necessary and sufficient conditions for the feasibility of the evader's exact capture by the pursuing team in the case of bounded adversaries' controls. Meaning we are interested in finding the conditions under which there exists a set of initial conditions for which point capture is possible, regardless of the evader's behavior, i.e. the existence of a capture zone.

Definition 1. Reachable Set. Given its state the reachable set of any player at instant t is the set of achievable vertical distances from the fixed reference line at instant t_f.

Lemma 1. *The necessary and sufficient condition for the existence of a capture zone in case of equal terminal instants is*

$$\exists t_o < t_f : \; S_E(\xi_E, t_{go}) \subseteq \bigcup_{i \in N} S_{P_i}(\xi_{P_i}, t_{go}) \quad \forall t \in [t_o, t_f] \tag{20}$$

where the state of each player is

$$\xi_j(t) = [y_j(t) \;\; \dot{y}_j(t) \;\; a_j(t)]^T$$

and where $S_j : \xi_j(t) \twoheadrightarrow [\underline{y}_j(t_f), \overline{y}_j(t_f)]$ is the reachable set of player j at a given instant t.
y_j denotes player j's displacement perpendicular to the reference line LOS_o, and $\underline{y}_j(t_f)$ and $\overline{y}_j(t_f)$ represent the lower and upper limits of $y_j(t_f)$'s value range respectively.

Proof. Assuming this condition is not satisfied, meaning at any instant there exists a subset of the evader's reachable set which is not in the union of the pursuers' reachable sets ($\forall \tilde{t} < t_f \; \exists \tilde{S}_E(\xi_E, \tilde{t}_{go}) \subseteq S_E(\xi_E, \tilde{t}_{go}) : \; \tilde{S}_E(\xi_E, \tilde{t}_{go}) \nsubseteq \bigcup_{i \in N} S_{P_i}(\xi_{P_i}, \tilde{t}_{go})$), it is clear that at any instant $\tilde{t} < t_f$ the evader can avoid capture by maneuvering towards any point in $\tilde{S}_E(\xi_E, \tilde{t}_{go})$, thereby proving necessity.
In order to prove sufficiency we merely consider the limit value $\lim_{t \to t_f} S_j(\xi_j, t_{go})$ of each player j, which, under physical constraints, converges to a single point $y_j(t_f)$. By definition, if (20) is satisfied it is satisfied particularly for the aforementioned limit values, meaning necessarily that there exists a pursuer P_i that coincides with the evader at the terminal instant ($\exists i \in N : y_{P_i}(t_f) = y_E(t_f)$).

□

Corollary 1. *Condition (20) is equivalent to the following sub-conditions:*

1. *The upper (lower) bound of the union of the pursuers reachable sets must be greater (lesser) than the upper (lower) bound of the evaders reachable set*

$$S_E(\xi_E,t_{go}) \subseteq \left[min \bigcup_{i\in N} S_{P_i}(\xi_{P_i},t_{go}), max \bigcup_{i\in N} S_{P_i}(\xi_{P_i},t_{go}) \right] \quad \forall t \in [t_o,t_f] \quad (21)$$

2. *If* $S_{P_i}(\xi_{P_i},t_{go}) \bigcap S_E(\xi_E,t_{go}) \neq S_E(\xi_E,t_{go})\ \forall i \in N$ *then the intersection of the pursuers reachable sets must not be empty*

$$\bigcap_{i\in N} S_{P_i}(\xi_{P_i},t_{go}) \neq \emptyset \quad \forall t \in [t_o,t_f] \quad (22)$$

Clearly if both sub-conditions hold then the existence of a capture zone is assured.

Consider a more general case of (16)

$$\dot{z}_i(t) = a_E^{max} f_E(t_f,t)u_E - a_{P_i}^{max} f_{P_i}(t_f,t)u_{P_i}, \quad i \in N. \quad (23)$$

$f_j(t_f,t)$ represents player j's control dynamics, and is assumed to be a bounded function of t throughout the engagement ($-\infty < f_j(t_f,t) < \infty\ \forall t \in [t_o,t_f]$).
By integrating from t to t_f we obtain

$$z_i(t_f) = z_i(t) + a_E^{max} \int_t^{t_f} \left[f_E(t_f,\tau)u_E - \mu_i f_{P_i}(t_f,\tau)u_{P_i} \right] d\tau, \quad i \in N \quad (24)$$

Remark 1. Each player j's reachable set is a closed interval, the upper (lower) bound of which is obtained for $u_j = sign\left[f_j(t_f,t)\right]$ ($u_j = -sign\left[f_j(t_f,t)\right]$) $\forall t \leq t_f$.

Theorem 1. *The necessary and sufficient condition for the existence of a capture zone in a 2-on-1 engagement for the case of equal terminal instants is*

$$\exists i \in N:\ \mu_i \frac{F_{P_i}(t_f,t)}{F_E(t_f,t)} \geq 1 \quad \forall t \in [t_o,t_f] \quad (25)$$

where

$$F_j(t_f,t) = \int_t^{t_f} \left| f_j(t_f,t) \right| d\tau, \quad j \in G \quad (26)$$

Proof. Necessity is proven regarding two separate cases.

Case A: $sign\left[z_1(t)\right] = sign\left[z_2(t)\right]$

Without loss of generality assume

$$z_1(t) > 0, \quad z_2(t) > 0$$

In this case in order to satisfy (21) there must exist a pursuer capable of reaching or surpassing the evader on either side, particularly for the upper bound of the evader's reachable set $\left(u_E = sign\left[f_E(t_f,t)\right]\ \forall t \le t_f\right)$

$$\exists i \in N:\ z_i(t_f)\big|_{u_E=sign\left[f_E(t_f,t)\right]} \le 0 \tag{27}$$

Substituting (24) with $u_E = sign\left[f_E(t_f,t)\right]\ \forall t \le t_f$ and pursuer i's best response $u_{P_i} = sign\left[f_{P_i}(t_f,t)\right]\ \forall t \le t_f$ yields the necessary condition

$$\exists i \in N:\ z_i(t) \le a_{P_i}^{max} F_{P_i}(t_f,t) - a_E^{max} F_E(t_f,t) \tag{28}$$

Even for the most lenient case of $z_i(t) \to 0$ we therefore require

$$\exists i \in N:\ \mu_i \frac{F_{P_i}(t_f,t)}{F_E(t_f,t)} \ge 1 \tag{29}$$

Case B: $sign\left[z_1(t)\right] \neq sign\left[z_2(t)\right]$

Without loss of generality assume

$$z_1(t) > 0,\quad z_2(t) < 0$$

and that $S_{P_i}(\xi_{P_i},t_{go}) \bigcap S_E(\xi_E,t_{go}) \neq S_E(\xi_E,t_{go})\ \forall i \in N$.

In order to satisfy (21) the most lenient requirement is that each pursuer must be capable of maintaining its ZEM sign for any target maneuver

$$\begin{aligned} z_1(t_f)\big|_{u_E=-sign\left[f_E(t_f,t)\right]} &\ge 0 \\ z_2(t_f)\big|_{u_E=sign\left[f_E(t_f,t)\right]} &\le 0 \end{aligned} \tag{30}$$

As in the previous case the best response of each of the pursuers is

$$u_{P_1} = -sign\left[f_{P_1}(t_f,t)\right];\quad u_{P_2} = sign\left[f_{P_2}(t_f,t)\right].$$

In order to satisfy (22) the pursuers' reachable sets must overlap

$$z_1(t_f)\big|_{u_{P_1}=sign\left[f_{P_1}(t_f,t)\right]} - z_2(t_f)\big|_{u_{P_2}=-sign\left[f_{P_2}(t_f,t)\right]} \le 0 \tag{31}$$

Define

$$\Delta z_{12}(t) \triangleq z_1(t) - z_2(t) \tag{32}$$

By substituting (24) we derive the following necessary conditions on the relative ZEM displacement of the pursuers at any instant t

$$\Delta z_{12}(t) \ge 2a_E^{max} F_E(t_f,t) - a_{P_1}^{max} F_{P_1}(t_f,t) - a_{P_2}^{max} F_{P_2}(t_f,t) \tag{33a}$$

$$\Delta z_{12}(t) \le a_{P_1}^{max} F_{P_1}(t_f,t) + a_{P_2}^{max} F_{P_2}(t_f,t) \tag{33b}$$

where (33a) assures that (30) holds, and (33b) assures that (31) holds.
Recalling that all players' control dynamics functions are bounded during the engagement

$$\lim_{t \to t_f} F_j(t_f, t) = \lim_{t \to t_f} \int_t^{t_f} \left| f_j(t_f, \tau) \right| d\tau = 0, \quad j \in G \tag{34}$$

Therefore, looking at the limit case $t \to t_f$

$$\lim_{t \to t_f} \Delta z_{12}(t) = 0 \tag{35}$$

meaning simultaneous capture is required. For simultaneous capture to be possible for *any* target maneuver each pursuer's reachable set must include the intersection of the reachable sets of the second pursuer and the evader ($S_{P_j}(\xi_{P_j}, t_{go}) \bigcap S_E(\xi_E, t_{go}) \subseteq S_{P_i}(\xi_{P_i}, t_{go}), \; i, j \in N, \; i \neq j$). This implies that each pursuer's reachable set must include the entire evader's reachable set, which in terms of the ZEM translates into

$$\begin{array}{r} z_i(t_f)\big|_{u_E = -sign\left[f_E(t_f, t)\right]} \geq 0 \\ z_i(t_f)\big|_{u_E = sign\left[f_E(t_f, t)\right]} \leq 0 \end{array}, \quad i \in N \tag{36}$$

Substituting (24) and the respective best responses $u_{P_i} = -sign\left[f_{P_i}(t_f, t)\right]$ and $u_{P_i} = sign\left[f_{P_i}(t_f, t)\right]$ yields the condition

$$\mu_i \frac{F_{P_i}(t_f, t)}{F_E(t_f, t)} \geq 1, \quad i \in N \tag{37}$$

which in particular satisfies (25).
In conclusion (25) is a necessary condition for exact capture.
Sufficiency is easily proved by regarding the 2-on-1 engagement as a 1-on-1 pursuit-evasion game between pursuer *i*, for which (25) holds, and the evader. The same initial conditions which assure exact capture of the evader by pursuer *i* in the 1-on-1 scenario necessarily assure exact capture in the 2-on-1 scenario.

□

Corollary 2. *It can be readily shown that (25) is also necessary and sufficient for the existence of a capture zone for the 1-on-1 case.*

4 Optimization Problem and Solution

Due to the desired element of cooperation between the interceptors and the opposing nature of the pursuing team and evader, the two-person zero-sum game formulation is chosen, defined by

$$\min_{u_{P_1}; u_{P_2}} \max_{u_E} J \tag{38}$$

4.1 Cost Function

Formulation of the cost function is a key part in any optimization problem. The payoff must be suitable to the desired problem formulation, and more importantly, it must reflect the desired interests (or goals) of each of the players.

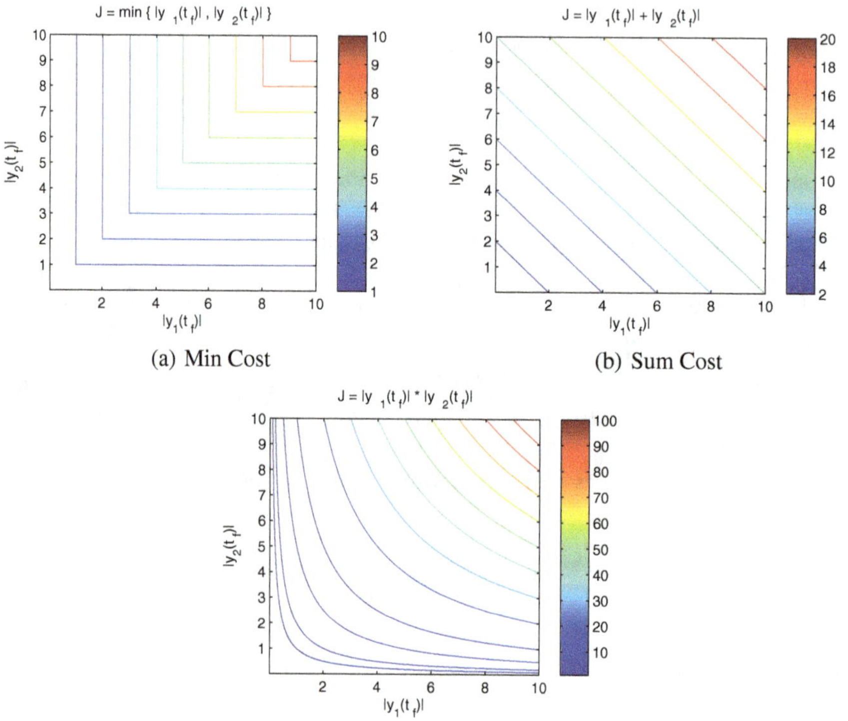

(a) Min Cost

(b) Sum Cost

(c) Product Cost

Fig. 2 Level Sets of Different Type Cost Functions

For a 2-on-1 engagement a logical choice for the cost function is the minimum of both pursuers' miss distances

$$J_{min} = \min\{|y_1(t_f)|, |y_2(t_f)|\} \tag{39}$$

It can also be represented by a piecewise differentiable function as follows

$$J_{min} = \frac{1}{2}\left[|y_1(t_f)| + |y_2(t_f)| - \left|\,|y_1(t_f)| - |y_2(t_f)|\,\right|\right] \tag{40}$$

Although this cost function suitably reflects the players goals in the 2-on-1 game several issues arise, of which the main two are:

- *Expansion to $n \geq 3$ pursuers.* For a large number of pursuers the complexity of the resulting cost's algebraic form makes the derivation of analytical results and their analysis problematic
- *Individual performance.* In this form of the cost the individual performance of each of the pursuers is not reflected (see Fig. 2 (a)). This becomes evident in cases when one of the pursuers is only "slightly more threatening" than the other (one pursuers is capable of terminating slightly closer to the evader than the other). In such cases the second, less threatening pursuer, is initially not provided with a single optimal strategy.

A rather trivial choice for a cost function which addresses these issues is the sum of both pursuers' miss distances

$$J_{sum} = |y_1(t_f)| + |y_2(t_f)| \tag{41}$$

However, it is quite plain to see that this cost does not reflect the significance of the minimum of the miss distances (see Fig. 2 (b)), and therefore does not suitably reflect any of the players' true goals.

After considering the desired behavior of the players in several basic cases, the product of the miss distances was proposed as a suitable cost function

$$J = |y_1(t_f)| \cdot |y_2(t_f)| \tag{42}$$

Comparing the behavior of this cost function (Fig. 2 (c)) with the behavior of the *min* and *sum* cost functions we realize the following:

- When $|y_1(t_f)| \approx |y_2(t_f)|$ (nearly equally threatening pursuers) the proposed cost function tends towards the *sum* cost function.
- When $|y_1(t_f)| \gg |y_2(t_f)|$ or $|y_1(t_f)| \ll |y_2(t_f)|$ (single threatening pursuer) the proposed cost function tends towards the *min* cost function.

This implies that the proposed cost function encompasses the desired qualities of both the *min* and the *sum* cost functions. Using the ZEM variables

$$J = |z_1(t_f)| \cdot |z_2(t_f)| \tag{43}$$

4.2 Optimization Solution

Using the normalized ZEM variables

$$J = \left(\tau_E^2 a_E^{max}\right)^2 |Z_1(0)| \cdot |Z_2(0)| \tag{44}$$

Since $\frac{d}{dt} = -\frac{1}{\tau_E}\frac{d}{d\theta}$ the Hamiltonian of the normalized reduced order game is

$$H = -\frac{1}{\tau_E}\lambda^T \frac{dZ}{d\theta} = -\frac{1}{\tau_E}\lambda_1 \frac{dZ_1}{d\theta} - \frac{1}{\tau_E}\lambda_2 \frac{dZ_2}{d\theta} \tag{45}$$

Substituting (18) and rearranging yields

$$H = (\lambda_1 + \lambda_2)\frac{\psi(\theta)}{\tau_E}u_E - \lambda_1\frac{\mu_1}{\varepsilon_1}\frac{\psi(\theta\varepsilon_1)}{\tau_E}u_{P_1} - \lambda_2\frac{\mu_2}{\varepsilon_2}\frac{\psi(\theta\varepsilon_2)}{\tau_E}u_{P_2} \tag{46}$$

The adjoint equations are obtained from

$$\frac{d\lambda_i}{d\theta} = \tau_E\frac{\partial H}{\partial Z_1}; \quad \lambda_i(0) = \frac{\partial J}{\partial Z_1(0)}, \quad i \in N \tag{47}$$

The optimal controllers are such that

$$\begin{aligned} \left(u_{P_1}^*, u_{P_2}^*\right) &= arg \min_{u_{P_1}, u_{P_2}} H \\ u_E^* &= arg \max_{u_E} H \end{aligned} \tag{48}$$

therefore

$$\begin{aligned} u_{P_i}^* &= sign\left[\lambda_i(\theta)\psi(\theta\varepsilon_i)\right], \quad i \in N \\ u_E^* &= sign\left[(\lambda_1(\theta) + \lambda_2(\theta))\,\psi(\theta)\right] \end{aligned} \tag{49}$$

It can be readily proven that $\psi(\alpha) > 0\ \forall\alpha > 0$, by which it follows

$$\begin{aligned} u_{P_i}^* &= sign\left[\lambda_i(\theta)\right], \quad i \in N \\ u_E^* &= sign\left[\lambda_1(\theta) + \lambda_2(\theta)\right] \end{aligned} \tag{50}$$

Substituting (44) and (46) in (47) yields

$$\begin{aligned} \frac{d\lambda_i}{d\theta} &= 0; \quad \lambda_1(0) = \left(\tau_E^2 a_E^{max}\right)^2 sign\left[Z_1(0)\right]\cdot|Z_2(0)| \\ \frac{d\lambda_2}{d\theta} &= 0; \quad \lambda_2(0) = \left(\tau_E^2 a_E^{max}\right)^2 sign\left[Z_2(0)\right]\cdot|Z_1(0)| \end{aligned} \tag{51}$$

therefore

$$\begin{aligned} \lambda_1(\theta) &= \left(\tau_E^2 a_E^{max}\right)^2 sign\left[Z_1(0)\right]\cdot|Z_2(0)| \\ \lambda_2(\theta) &= \left(\tau_E^2 a_E^{max}\right)^2 sign\left[Z_2(0)\right]\cdot|Z_1(0)| \end{aligned} \tag{52}$$

Substituting into (50) yields

$$\begin{aligned} u_{P_i}^* &= sign\left[Z_i(0)\right], \quad i \in N \\ u_E^* &= sign\left\{|Z_2(0)|sign\left[Z_1(0)\right] + |Z_1(0)|sign\left[Z_2(0)\right]\right\} \end{aligned} \tag{53}$$

A simple investigation shows that these strategies lead to the players' expected/desired behavior.

4.3 Candidate Optimal Trajectories

Substituting (53) into (18) yields the following dynamic equations

$$\begin{aligned}\frac{dZ_1^*}{d\theta} &= \frac{\mu_1}{\varepsilon_1}\psi(\theta\varepsilon_1)\, sign\,[Z_1(0)] - \psi(\theta)\, sign\{|Z_2(0)|sign\,[Z_1(0)] + |Z_1(0)|sign\,[Z_2(0)]\}\\ \frac{dZ_2^*}{d\theta} &= \frac{\mu_2}{\varepsilon_2}\psi(\theta\varepsilon_2)\, sign\,[Z_2(0)] - \psi(\theta)\, sign\{|Z_2(0)|sign\,[Z_1(0)] + |Z_1(0)|sign\,[Z_2(0)]\}\end{aligned} \tag{54}$$

Integration from θ to 0 (from t to t_f) yields the candidate optimal trajectories

$$\begin{aligned}Z_1^*(\theta) &= Z_1(0) - sign\,[Z_1(0)]\left[\frac{\mu_1}{\varepsilon_1}W_1(\theta) - sign\{|Z_2(0)| + |Z_1(0)|sign\,[Z_1(0)Z_2(0)]\}W_0(\theta)\right]\\ Z_2^*(\theta) &= Z_2(0) - sign\,[Z_2(0)]\left[\frac{\mu_2}{\varepsilon_2}W_2(\theta) - sign\{|Z_2(0)|sign\,[Z_1(0)Z_2(0)] + |Z_1(0)|\}W_0(\theta)\right]\end{aligned} \tag{55}$$

where

$$\begin{aligned}W_0(\theta) &= \int_\theta^0 \psi(\tau)\,d\tau = \psi(\theta) - \frac{\theta^2}{2}\\ W_i(\theta) &= \int_\theta^0 \psi(\tau\varepsilon_i)\,d\tau = \frac{1}{\varepsilon_i}\left[\psi(\theta\varepsilon_i) - \frac{(\theta\varepsilon_i)^2}{2}\right], \quad i \in N\end{aligned} \tag{56}$$

5 Case of Two Strong Pursuers

We now present an analysis of the solution for the case that $\mu_i > 1,\ \mu_i\varepsilon_i > 1\ \forall i \in N$.

5.1 Game Space Construction

5.1.1 Regular Zones

We begin by finding the boundary trajectories $Z_i^{\pm}(\theta) = \lim_{Z_i(0)\to 0^{\pm}} Z_i(\theta),\ i \in N$ of the regular zones

Case 1: $sign\,[Z_1(0)] = sign\,[Z_2(0)]$

$$Z_i^{\pm}(\theta) = \pm\left[W_0(\theta) - \frac{\mu_i}{\varepsilon_i}W_i(\theta)\right] \tag{57}$$

Case 2: $sign[Z_1(0)] \neq sign[Z_2(0)]$

Case 2.1: $|Z_i(0)| > |Z_j(0)|, \ i,j \in N, \ i \neq j$

$$Z_i^{\pm}(\theta) = \mp\left[W_0(\theta) + \frac{\mu_i}{\varepsilon_i}W_i(\theta)\right] \tag{58}$$

Case 2.2: $|Z_i(0)| < |Z_j(0)|, \ i,j \in N, \ i \neq j$

$$Z_i^{\pm}(\theta) = \pm\left[W_0(\theta) - \frac{\mu_i}{\varepsilon_i}W_i(\theta)\right] \tag{59}$$

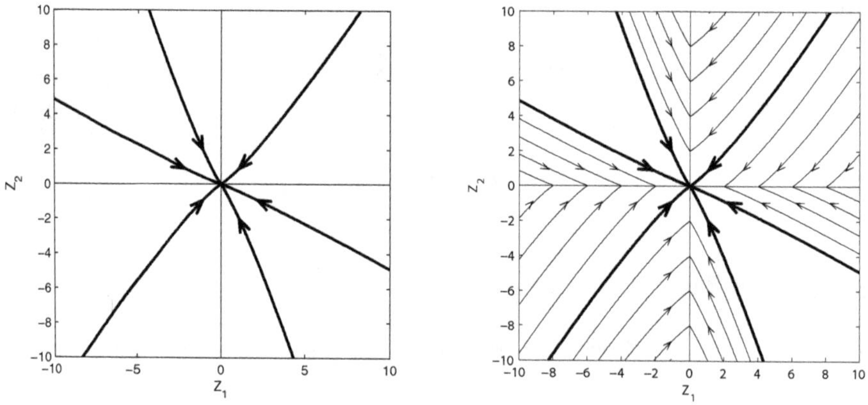

(a) Boundary Trajectories

(b) Boundary Trajectories

Fig. 3 Regular Zone Boundaries

As depicted in Fig. 3 (a), appropriate combinations between the two pursuers' trajectories yields the following three boundary lines

$$Z_2 = \frac{W_0(\theta) - \frac{\mu_2}{\varepsilon_2}W_2(\theta)}{W_0(\theta) - \frac{\mu_1}{\varepsilon_1}W_1(\theta)}Z_1;\ Z_2 = \frac{W_0(\theta) - \frac{\mu_2}{\varepsilon_2}W_2(\theta)}{W_0(\theta) + \frac{\mu_1}{\varepsilon_1}W_1(\theta)}Z_1;\ Z_2 = \frac{W_0(\theta) + \frac{\mu_2}{\varepsilon_2}W_2(\theta)}{W_0(\theta) - \frac{\mu_1}{\varepsilon_1}W_1(\theta)}Z_1 \tag{60}$$

Regular trajectories, emanating from the vertical and horizontal axes, are parallel to each of these lines according to the zone in question (Fig. 3 (b)). These boundary trajectories lead to the target's capture by either one or both of the pursuers. Complete the regular zones by constructing the trajectories emanating from all points on the $\theta = 0$ plane. It can be easily verified that no intersections exist between the resultant curves.

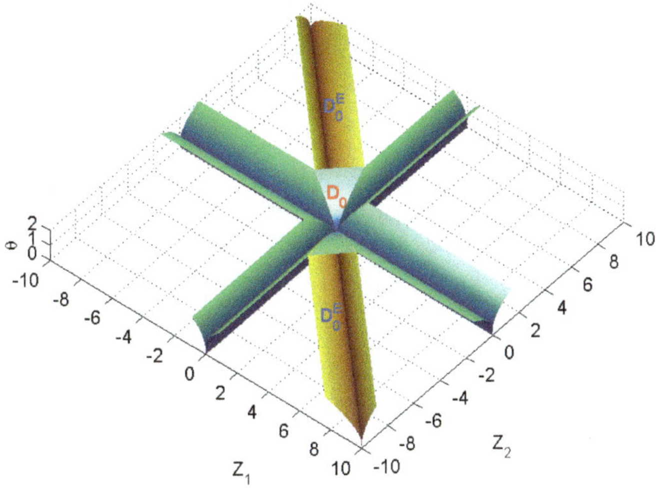

Fig. 4 Singular Zones

5.1.2 Singular Zones

Upon completion of the regular trajectories field's construction the existence of three voids becomes apparent (Fig. 4). The central void, representing the *capture zone*, is denoted as D_0. In it $J = 0$ and all players' optimal controls are not uniquely defined. Additionally, by comparison to the two 1-on-1 games of P_i vs. E, $i \in N$, the zones defined by

$$Z_1(\theta)Z_2(\theta) < 0 \tag{61a}$$

$$|Z_i(\theta)| > -\frac{\mu_i}{\varepsilon_i}W_i(\theta) + W_0(\theta), \quad i \in N \tag{61b}$$

$$|Z_1(\theta) - Z_2(\theta)| \leq -\frac{\mu_1}{\varepsilon_1}W_1(\theta) - \frac{\mu_2}{\varepsilon_2}W_2(\theta) \tag{61c}$$

constitute an enlargement of the two 1-on-1 games' combined capture zone, obtained simply by the presence of an additional pursuer.

The two remaining voids, representing the *no escape zones*, are denoted by D_0^E. In them the pursuers' optimal controls are equal to those in the relevant quadrant, but the evader's optimal control is not uniquely defined. Interestingly, these zones' interior is defined by

$$|Z_1(\theta) - Z_2(\theta)| = -\frac{\mu_1}{\varepsilon_1}W_1(\theta) - \frac{\mu_2}{\varepsilon_2}W_2(\theta) + |Z_1(0) - Z_2(0)| \quad \forall Z_1(0) = -Z_2(0) \tag{62}$$

relating to the required capture condition (33b), subject to equal pursuers' misses ($|Z_1(0)| = |Z_2(0)|$). In effect these zones are made up of curvilinear slices in which J is constant and equal to the cost at the point of origin.

5.2 *Closed Form Controls*

In order to obtain the closed-loop form optimal controls we examine (55) in two cases: $sign[Z_1(0)] = sign[Z_2(0)]$ and $sign[Z_1(0)] \neq sign[Z_2(0)]$. We are assisted by the readily proven characteristics of $\psi(\alpha)$ and its integral.

Case 1: $sign[Z_1(0)] = sign[Z_2(0)]$

$$\begin{aligned} u_{P_i}^* &= sign[Z_i(\theta)], \quad i \in N \\ u_E^* &= sign[Z_1(\theta)] = sign[Z_2(\theta)] \end{aligned} \tag{63}$$

Case 2: $sign[Z_1(0)] \neq sign[Z_2(0)]$

$$\begin{aligned} u_{P_i}^* &= sign[Z_i(\theta)], \quad i \in N \\ u_E^* &= -sign\left[(Z_1(\theta)+Z_2(\theta)) + \left(\frac{\mu_1}{\varepsilon_1}W_1(\theta) - \frac{\mu_2}{\varepsilon_2}W_2(\theta)\right) sign[Z_1(\theta)]\right] \end{aligned} \tag{64}$$

Finally, we deduce that the evader's control can be represented by a single term, yielding the following closed form controls

$$\begin{aligned} u_{P_i}^* &= sign[Z_i(\theta)], \quad i \in N \\ u_E^* &= sign[Z_1(\theta)Z_2(\theta)]\, sign\left[(Z_1(\theta)+Z_2(\theta)) + \left(\frac{\mu_1}{\varepsilon_1}W_1(\theta) - \frac{\mu_2}{\varepsilon_2}W_2(\theta)\right) sign[Z_1(\theta)]\right] \end{aligned} \tag{65}$$

6 Conclusions

A linear bounded control 2-on-1 pursuit-evasion game has been presented. A study of the feasibility of exact capture yielded the necessary and sufficient conditions, which evidently are unchanged in comparison to the conditions in a 1-on-1 scenario.

The engagement was treated as a two sided optimization problem. A new product-form cost was presented and utilized in the game solution. It was verified that the proposed cost yields the expected and/or desired behavior of the players compared to the *min* and *sum* cost functions.

Solution of the case of two strong pursuers was presented in full. The game space construction was shown, revealing the existence of three singular zones: a central *capture zone*, and two additional *no escape zones* in which the evader's optimal strategy is not uniquely defined. Within these zones, assuming optimal pursuit, the outcome is guaranteed. Comparison of the obtained game space with that of a 1-on-1 engagements showed that the presence of an additional pursuer enlarges the capture zone. Further analysis of the obtained solution yielded optimal closed form controls for all players: the optimal pursuit strategies were identical to those obtained in a 1-on-1 scenario, whereas the optimal evasion was found to be different.

References

1. Ben-Asher, J., Yaesh, I.: Advances in Missile Guidance Theory (Progress in Astronautics and Aeronautics), vol. 180, pp. 89–126. AIAA, Washington D.C (1998)
2. Borowko, P., Rzymowski, W.: On the Game of Two Cars. Journal of Optimization Theory and Applications 44(3), 381–396 (1984)
3. Bryson, A.E., Ho, Y.C.: Applied Optimal Control. Ann. Mat. Pura. Appl. Blaisdell Publishing Company, Waltham, Massachusetts (1969)
4. Cockayne, E.: Plane Pursuit with Curvature Constraints. SIAM Journal on Applied Mathematics 15(6), 1511–1516 (1967)
5. Foley, M.H., Schmitendorf, W.E.: A Class of Differential Games with Two Pursuers Versus One Evader. IEEE Transactions on Automatic Control 19(3), 239–243 (1974)
6. Ganebny, S.A., Kumkov, S.S., Le Mènec, S., Patsko, V.S.: Model Problem in a Line with Two pursuers and One Evader. Dynamic Games and Applications 2(2), 228–257 (2012)
7. Gutman, S.: On Optimal Guidance for Homing Missiles. Journal of Guidance and Control 3(4), 296–300 (1979)
8. Gutman, S., Leitmann, G.: Optimal Strategies in the Neighborhood of a Collision Course. AIAA Journal 14(9), 1210–1212 (1976)
9. Ho, Y.C., Bryson Jr, A.E., Baron, S.: Differential Games and Optimal Pursuit-Evasion Strategies. IEEE Transactions on Automatic Control AC-10(4), 385–389 (1965)
10. Isaacs, R.: Differential Games. John Wiley, New York (1965)
11. Le Mènec, S.: Linear Differential Game with Two Pursuers and One Evader. Advances in Dynamic Games 25(6), 209–226 (2011)
12. Leatham, A.L.: Some Theoretical Aspects of Nonzero Sum Differential Games and Applications to Combat Problems. Wright-Patterson AFB OH School of Engineering 10(AFIT/DS/MC/71-3) (1971)
13. Lewin, J.: Differential games: Theory and Methods for Solving Game Problems with Singular Surfaces. Springer, New York (1994)
14. Liu, Y., Qi, N., Tang, Z.: Linear Quadratic Differential Game Strategies with Two-Pursuit Versus Single-Evader. Chinese Journal of Aeronautics 25(6), 896–905 (2012)
15. Perelman, A., Shima, T., Rusnak, I.: Cooperative Differential Games Strategies for Active Aircraft Protection from a Homing Missile. AIAA Journal of Guidance, Control, and Dynamics 34(3), 761–773 (2011)
16. Rublein, G.T.: On Pursuit with Curvature Constraints. SIAM Journal on Control 10(1), 37–39 (1972)
17. Rzymowski, W.: On The Game of 1+n Cars. Journal of Mathematical Analysis and Applications 99(1), 109–122 (1984)
18. Shima, T.: Capture Conditions in a Pursuit-Evasion Game Between Players with Biproper Dynamics. Journal of Optimization Theory and Applications 126(3), 503–528 (2005)
19. Shima, T., Shinar, J.: Time-varying linear pursuit-evasion game models with bounded controls. AIAA Journal of Guidance, Control, and Dynamics 25(3), 425–432 (2002)
20. Shinar, J., Shinar, J.: Solution Techniques for Realistic Pursuit-Evasion Games. Advances in Control and Dynamic Systems 17, 63–124 (1981)
21. Von Neumann, J., Morgenstern, O.: Theory of Games and Economic Behavior. Princeton University Press, Princeton (1944)

A New Impact Angle Control Guidance Law to Reduce Sensitivity on Initial Errors

Hyo-Sang Shin, Jin-Ik Lee, and Antonios Tsourdos

Abstract. This paper proposes a new homing guidance which reduces sensitivity on initial zero effort miss and error in flight path angle to the desired angle while achieving the terminal angle constraint on the impact. To develop such a guidance law, the guidance problem is formulated as an optimal guidance problem with a new performance index and constraints including the terminal angle one. The main idea enabling this approach is introduction of distribution functions to the input weighting. The distribution functions in the new performance index allows distributing of the relative input weighting over the entire homing phase. Then, a homing guidance law is derived by applying the optimal control theory with the new performance index. Consequently, the proposed guidance law generates a time-varying gain to ease the sensitivity at the initial homing phase. The characteristics of the proposed guidance algorithm are investigated and its performance is demonstrated through numerical simulations.

1 Introduction

The satisfaction of desired terminal angle reference, which is often called the impact angle, is an important requirement for anti-ship and anti-tank missile systems. The guidance laws meeting the terminal angle requirement can significantly improve the performance and effectiveness of the missile. For example, such a guidance law enables missiles to attack a weak spot on a target and enhance the warhead effects by improving penetration. Furthermore, this type of guidance could improve

Hyo-Sang Shin · Antonios Tsourdos
Cranfield University, College Road, Cranfield, MK43 0AL UK
e-mail: {h.shin,a.tsourdos}@cranfield.ac.uk

Jin-Ik Lee
Agency for Defence Development (ADD), Daejeon 305-600 Rep. of Korea
e-mail: jinjjangu@gmail.com

J. Bordeneuve-Guibé et al. (eds.), *Advances in Aerospace Guidance, Navigation and Control*, DOI: 10.1007/978-3-319-17518-8_15

survivability of the missile since it could allow the missile to approach to the target with a terminal angle for which the enemy's counter measures are difficult to effectively strike the missile. For this reason, over the past decades, there have been extensive studies on the impact angle control guidance laws, based on optimal control theory [1, 2, 3, 4, 5]and biased proportional navigation guidance (BPNG) [6, 7, 8].

For medium or long range missiles, mid course guidance is required as effectiveness ranges of missile seekers are limited. In practice, mid course guidance could be different from terminal homing guidance, especially when impact angle control guidance is applied at the homing phase. For instance, command guidance such as beam riding guidance, inertial navigation guidance or PN guidance could be used in the mid course phase. Since each guidance strategy produces each own collision course, having different guidance strategies likely generates different collision course geometry for the two phases. Thus, zero effort miss, *ZEM*, generated by the midcourse and homing guidance algorithms might be different from each other at the handover. Moreover, an error in the flight path angle to the desired terminal orientation angle is likely present. Consequently, achieving zero *ZEM* at the mid course phase can still result in initial errors at the beginning of the homing phase.

Most of impact angle control problems are solved by optimal control theory in which the command profiles and the flight trajectories are determined by a given initial geometry. In impact angle control guidance, non-zero ZEM and the flight path angle error are two dominant sources generating the acceleration command at the homing phase. When homing guidance is sensitive to the initial ZEM and flight path angle error, a small ZEM and angle error could result in an abrupt change of the missile acceleration at the initial homing phase. This sudden change in acceleration also causes an abrupt transition manoeuvre which can be regarded as external disturbance to the seeker. This phenomenon is undesirable for the terminal homing guidance. It cannot be decoupled completely from the guidance system even though it can be stabilised by the internal stabilisation loop. The disturbance makes the seeker provide undesirable and incorrect measurements to the guidance system. Since consecutive incorrect information may prevent the seeker from tracking its target, the guidance system may not be able to effectively intercept and destroy the target. Although transition filters such as low pass filters are often introduced to mitigate this problem in the guidance loop, it is difficult to successfully cope with this issue.

The key idea of this paper is that the sensitivity issue could be mitigated by having a small navigation gains at the initial homing phase: if with the same ZEM and flight path angle error, the smaller the navigation gains are, the smaller acceleration commands will be generated. However, in order to guarantee interception and improve robustness, it is desired to have a navigation gain larger than a certain value. If a navigation gain varies over time, these issues could be naturally resolved.

Therefore, this study aims to develop a new homing guidance law which has time-varying navigation gains. In our previous studies, it is shown that PN-type homing guidance with varying navigation gains can be obtained by introducing new weighing functions to an optimal guidance problem and applying the optimal control theory [9, 10]. Therefore, introducing an appropriate weighing function to the conventional Linear Quadratic (LQ) energy optimal problem should enable development of a new homing guidance law with varying gains.

There have been a few weighted performance indexes introduced not only to the LQ based PN guidance problem, but also to the optimal guidance problems with terminal angle constraints [11, 12, 4, 13, 14, 15]. Ben-Asher *et al.* [12] proposed an exponentially weighted performance index to the optimal guidance problem to relax manoeuvrability as air density decreases for aerodynamically manoeuvring ground-to-air missiles. This resulted in a new PN law with time-varying navigation gain where larger navigation gains are obtained at the early engagement and they are decreasing over engagement. Ryoo *et al.* [11] introduced a weighted performance index, where the weighting is a function of air density and missile varying velocity, to the optimal guidance problem. The optimal guidance law obtained minimises the drag for subsonic or supersonic missiles depending on design parameters values.

Relative weighting with respect to the minimum weighting in the LQ optimal problem represents the relative importance on minimising the corresponding acceleration command. If it is possible to distribute this relative weighting over the entire engagement, then shaping acceleration commands will be possible: increasing relative weighting at the initial homing phase will reduce the initial acceleration commands compared with having the identical relative weighting over the engagement. The linear quadratic optimal problem providing conventional impact angle control guidance with constant gains have either identical or monotonically increasing weighting on the cost function [4, 16]. Introducing a second order time-to-go weighted polynomial to the weighting function of the optimal guidance problem could generate a non-monotonic weighting function which possibly has a stationary point. One can readily distribute the relative weighting over the entire engagement, by determining the characteristics of the second order polynomial function including minimum and maximum values, their positions, ratio of initial and final values to the minimum value, etc.

Our previous studies showed that a PN guidance law optimally obtained from this weighting function can reduce sensitivity on the initial heading errors [9]. For this reason, in the proposed weighting function, a second order time-to-go weighted polynomial is introduced into a inverse weighted the time-to-go function. This way, the proposed impact angle control guidance law will be able to alleviate the sensitivity on the initial *ZEM* and flight path angle error at the homing phase, as well as handle other design requirements.

In the next section, the effect of *ZEM* and flight path angle to the impact angle control guidance and its sensitivity issue at the initial homing phase are reviewed. Section 3 first introduces the main idea of this paper and then derives the proposed guidance law. In Section 4, characteristics of the new impact angle control guidance law are analysed. Then, Section 6 concludes this study after demonstrating the performance of the new homing guidance via some numerical simulations in Section 5.

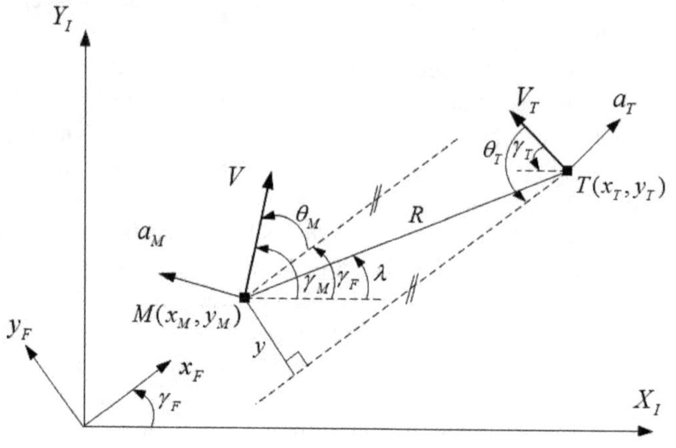

Fig. 1 Planar engagement geometry

2 Preliminary

This paper considers a 2D engagement geometry shown in Fig. 1. As shown in the geometry, the inertial reference frame is denoted as $(X_I, , Y_I)$. For convenience, a new coordinate system, called the guidance frame, is defined and expressed as $(x_F, , y_F)$. Variables with subscripts of M and T represent those of the missile and target, respectively. The notations of r, λ, and γ_F are the relative range, line-of-sight, and desired impact angle. γ and θ denote the flight path angle defined in the inertial reference frame and in the guidance frame, respectively, so that:

$$\begin{aligned} \theta_M &= \gamma_M - \gamma_F, \\ \theta_T &= \gamma_T - \gamma_F. \end{aligned} \tag{1}$$

The velocity and lateral acceleration are denoted as V and a. The lateral distance of the missile perpendicular to the impact course, expressed by y, is given by:

$$y = y_T - y_M \tag{2}$$

The conventional impact angle control guidance problem is to find guidance commands which minimises the following performance index

$$J = \frac{1}{2}\int_{t_0}^{t_F} a_M^2 dt, \quad t_F : \textit{given} \tag{3}$$

subject to the missile kinematics and the terminal constraints such that:

$$\begin{aligned} y(t_F) &= y_F \\ \dot{y}(t_F) &= \dot{y}_F \end{aligned} \tag{4}$$

Under the assumption of the lag-free missile system, the optimal solution is given in of the form [3, 16]:

$$a_{IACG} = \frac{N_1 y_{go} + N_2 \dot{y}_{go} t_{go} + N_3 \dot{y}_{go}(t_F) t_{go}}{t_{go}^2} \tag{5}$$

where $(N_1, N_2, N_3) = (6, 4, 2)$, $y_{go} = y_F - y(t)$ and $t_{go}(=: t_F - t)$ denotes time-to-go. Note that this guidance law is well known as OGL (Optimal Guidance Law) or trajectory shaping guidance law.

Zero effort miss, *ZEM*, can be approximated as [16]:

$$y_{go} + \dot{y}_{go} t_{go} \approx ZEM \tag{6}$$

With small angle approximation, we have:

$$\begin{aligned} \dot{y}_{go} &= V_C \theta_M(t) \\ \dot{y}_{go}(t_F) &= V_C \theta_{MF} \end{aligned} \tag{7}$$

where V_C denotes the closing velocity. Substituting Eqns. (6) and (7) into Eqn. (5) yields:

$$a_{IACG} = N_{ZEM} \frac{ZEM(t)}{t_{go}^2} + N_\theta V_C \frac{\theta_{MF} - \theta_M(t)}{t_{go}} \tag{8}$$

where $N_{ZEM} = 6$ and $N_\theta = 2$.

If impact angle control guidance is applied at the homing phase, as shown in Eqn. (8), there are two main components generating the missile commanded acceleration at the hand-over: *ZEM* and the error in the flight path angle from the desired one. The midcourse guidance law plays an important role in leading the missile into the so-called "lock-on basket" for the seeker to track the target. If the mid course guidance provides hand-over conditions where the two main contributors to the commanded acceleration are very small or even close to zero, the initial commanded acceleration remains small. However, in general, the energy minimised manoeuvring is accomplished to follow the optimal trajectory or command to line-of-sight guidance is applied in the mid course phase. The potential discrepancy in the mid course and homing guidance results in non-zero *ZEM* and flight path angle error Eqn. (8) at the hand-over. Furthermore, zero *ZEM* and flight path angle error are hard to be fulfilled in practice with measurement noise and uncertainties on the future target acceleration.

From Eqn. (8), the initial commanded acceleration of the missile at the hand-over, i.e., $t = 0$, is given by:

$$a_M(0) = N_{ZEM} \frac{ZEM(0)}{t_F^2} + N_\theta V_C \frac{\theta_{MF} - \theta_M(0)}{t_F} \tag{9}$$

Therefore, the initial commanded acceleration has some value unless the initial *ZEM* and flight path angle error are zero. Since an initial condition is unlikely to satisfy the zero *ZEM* and flight path angle error condition at the initial homing phase, initial errors on these two may exist. These errors trigger an abrupt transition of the acceleration command at the initial homing phase.

The commanded acceleration is proportional to the effective navigation constants N_{ZEM} and N_θ. Adjusting N_{ZEM} and N_θ, rather than applying a navigation constant, could enable mitigation of an abrupt acceleration transition caused by initial errors on *ZEM* and flight path angle. Thus, the following section develops a new homing guidance having time varying navigation gains to reduce the sensitivity on the initial errors.

3 New Impact Angle Control Guidance Law

From the engagement geometry shown in Figure 1, let us define relative position of the missile w.r.t. the target in y_F axis as a new state z_1 and its first time derivative as z_2:

$$\begin{aligned} z_1 &:= y = y_T - y_M \\ z_2 &:= \dot{y} = \dot{y}_T - \dot{y}_M \end{aligned} \tag{10}$$

Under the assumption that the target flies on a straight course, the state equation of the homing guidance problem is given by:

$$\dot{\mathbf{z}} = A\mathbf{z} + Bu, \qquad z_0 = z(t_0) \tag{11}$$

where $\mathbf{z} := [z_1,\, z_2]^T$ and

$$A = \begin{bmatrix} 0 & 1 \\ 0 & 0 \end{bmatrix}, \quad B = \begin{bmatrix} 0 \\ 1 \end{bmatrix}. \tag{12}$$

Now, let us consider following optimal control problem: find optimal control command $u(t)$ which minimizes the performance index defined by:

$$J = \int_{t_0}^{t_f} \frac{1}{2} u(t)^T R(t) u(t)\, dt, \quad t_f : \textit{given} \tag{13}$$

subject to Eqn. (11) with terminal constraints given by

$$D\mathbf{z}(t_F) = E, \tag{14}$$

where

$$D = \begin{bmatrix} 1 & 0 \\ 0 & 1 \end{bmatrix}, \qquad E = \begin{bmatrix} 0 \\ z_{2F} \end{bmatrix}. \tag{15}$$

In typical homing guidance such as PN, the terminal constraint is given only to guarantee the interception of the target, that is, z_1 should converge to zero. However, in terminal impact angle control guidance, one more constraint is imposed to the relative velocity along the y_F axis to meet the terminal impact angle constraint as shown in Eqn. (15). If the target is non-manoeuvring and has a constant speed, the optimal solution minimising the performance index (13) is identical to (8) with $N_{ZEM} = 6$ and $N_\theta = 2$ for a constant weighting function, $R(t) = 1$.

The key idea of this study is to shape the guidance command as required by changing the weighting function, $R(t)$. In order to enable this idea, this paper introduces a second order polynomial function to the weighting function:

$$R(t) = \mu_1^2 (t_{go} - \mu_2)^2 + 1 \tag{16}$$

where $\mu_1 \in \mathbb{R}_+$ and $\mu_2 \in \mathbb{R}_+$ are the distribution parameters to be designed. Note that $\mathbb{R}_+$ represents the set of non-negative real numbers. This relative weighting function with respect to the minimum weighting can be interpreted as relative importance on minimising the corresponding acceleration command. If the relative weighting is more significant at some engagement instance than any other points, the solution of the linear quadratic optimal problem puts more efforts to reduce the guidance commands on this instance.

Unlike the conventional impact angle control guidance having either identical or monotonically increasing weighting on the cost function, the proposed approach can generate non-monotonic weighting functions which could have a stationary point. This means that the proposed approach in this paper enables guidance designers or engineers to shape the relative weighting on the guidance command profile over the entire homing phase by tuning the twi design parameters μ_1 and μ_2. The weighting function has the minimum value, 1, at $t = t_f - \mu_2$ and the maximum value either at the initial homing phase ($t = 0$) or the terminal homing phase ($t = t_f$).

Figure 2 shows an example illustrating the difference on the weighting functions between the conventional impact angle control guidance and the proposed guidance law. The weighting function for the conventional impact angle control guidance is assumed to be 1 and that for the proposed guidance is $\left[0.3 * (t_{go} - 4)^2 + 1\right]$. Whereas the weighting in the impact control guidance law remains constant, the proposed guidance has its minimum value at $t_{go} = 4 (= \mu_2)$ and its maximum value at the interception, that is when $t_{go} = 0$. Therefore, relative weighting, which is relative importance, during the entire homing time window with respect to the minimum weighting and its position can be completely determined by the values of the design parameters. This allows the proposed guidance law to shape the acceleration commands over the entire homing phase. For example, increasing the relative weighting at the initial homing phase will generate smaller acceleration commands than those of the equivalent impact angle control guidance law. Therefore, introducing the performance index Eqn. (16) will enable the homing guidance not only to reduce sensitivity to *ZEM* and the initial errors in the flight path angle at the beginning of the homing phase, but also to consider other design requirement such as overall robustness.

The optimal guidance problem described in Equations (13) and (14) is a linear quadratic problem. The optimal solution of this quadratic problem can be derived by using the sweep method considering the normality and the convexity conditions [1]:

$$u^*(t) = K(F^T \mathbf{z} - E) \tag{17}$$

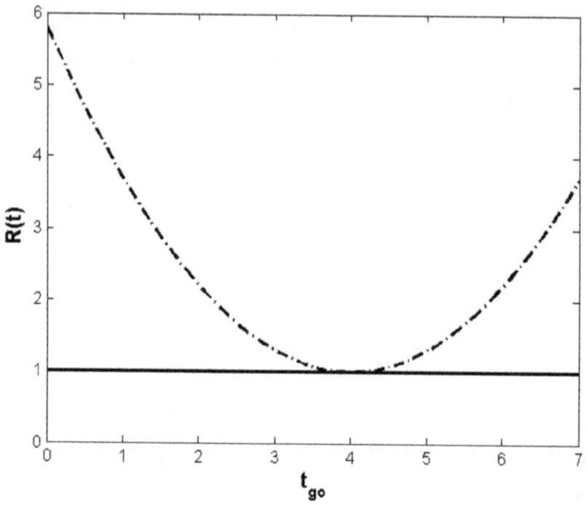

Fig. 2 An example of difference on the weighting function between a PN guidance law and the proposed guidance law: The weighting function for the PN guidance law is $R(t) = 1$ and that for the proposed guidance is $R(t) = 0.3 * (t_{go} - 4)^2 + 1$.

where

$$K = R^{-1} B^T F G^{-1} \tag{18}$$

$$\begin{cases} \dot{F} = -A^T F \\ \dot{G} = F^T B R^{-1} B^T F \end{cases}, \tag{19}$$

with terminal conditions:

$$F(t_f) = D^T,\; G(t_f) = \mathbf{0} \tag{20}$$

From Eqn. (19) and (20), $\dot{F}$ is given by:

$$\dot{F} = -\begin{bmatrix} 0 & 0 \\ 1 & 0 \end{bmatrix} \begin{bmatrix} f_{11} & f_{12} \\ f_{21} & f_{22} \end{bmatrix} = -\begin{bmatrix} 0 & 0 \\ f_{11} & f_{12} \end{bmatrix}; \quad F(t_F) = \begin{bmatrix} 1 & 0 \\ 0 & 1 \end{bmatrix}^T. \tag{21}$$

Integrating Eqn. (21) yields,

$$F = \begin{bmatrix} 1 & 0 \\ t_{go} & 1 \end{bmatrix} \tag{22}$$

Substituting $F(t)$ derived and the new weighting $R(t)$ into $\dot{G}(t)$ in Equation (19) yields:

$$\begin{aligned} \dot{G} &= F^T B R^{-1} B^T F \\ &= \frac{1}{\mu_1^2 (t_{go} - \mu_2)^2 + 1} \begin{bmatrix} t_{go}^2 & t_{go} \\ t_{go} & 1 \end{bmatrix}; \quad G(t_F) = \mathbf{0}_{2\times 2}; \end{aligned} \tag{23}$$

From this equation, we have:

$$G = \begin{bmatrix} g_{11} & g_{12} \\ g_{21} & g_{22} \end{bmatrix} = -\int_t^{t_f} \frac{1}{\mu_1^2(t_{go}-\mu_2)^2+1} \begin{bmatrix} t_{go}^2 & t_{go} \\ t_{go} & 1 \end{bmatrix} dt \tag{24}$$

Solving Eqn. (24) yields:

$$\begin{aligned} g_{11} &= -\tfrac{1}{\mu_1^3}\left\{\mu_1 t_{go} + \mu_1\mu_2 \ln \tfrac{W(t_{go})}{W(0)} + (W(0)-2)\tan^{-1} g(t_{go})\right\}, \\ g_{12} = g_{21} &= -\tfrac{1}{2\mu_1^2}\left\{\ln \tfrac{W(t_{go})}{W(0)} + 2\mu_1\mu_2 \tan^{-1} g(t_{go})\right\} \\ g_{22} &= -\tfrac{1}{\mu_1}\tan^{-1} g(t_{go}) \end{aligned} \tag{25}$$

where

$$W(t_{go}) = \mu_1^2(t_{go}-\mu_2)^2+1; \tag{26}$$

$$g(t_{go}) = \frac{\mu_1 t_{go}}{1-\mu_1^2\mu_2(t_{go}-\mu_2)}. \tag{27}$$

After some algebra, the close loop solution is given by:

$$u = -\frac{K^T\eta}{\Lambda}, \tag{28}$$

where the initial states η is defined as

$$\eta = \begin{bmatrix} z_{10} & z_{20} & z_{2F} \end{bmatrix}^T \tag{29}$$

and the control gains are given by:

$$K = \begin{bmatrix} k_1 \\ k_2 \\ k_3 \end{bmatrix} = \begin{bmatrix} 0 & 2\alpha^2 & 4\alpha^3(t_{go}-\beta) \\ 4t_{go}\alpha^2 & 4\alpha^2(t_{go}-\beta) & 4\alpha(R-2) \\ 4t_{go}\alpha^2 & 2\alpha^2(t_{go}-2\beta) & 4\alpha(R_p-2) \end{bmatrix} \begin{bmatrix} 1 \\ \xi_1 \\ \xi_2 \end{bmatrix} \tag{30}$$

where

$$\begin{aligned} &\alpha = \mu_1 t_F, \qquad \beta = 1-\tfrac{\mu_2}{t_F}, \\ &\Lambda = R(\xi_1^2 + 4\xi_2^2 - 4t_{go}R\alpha\xi_2), \\ &\xi_1 = \ln\tfrac{R_F}{R}, \qquad \xi_2 = \tan^{-1}\tfrac{\mu_1 t_{go}}{R_p}, \\ &R_F =: R(t_F) = \mu_1^2\mu_2^2+1, \quad R_p =: \mu_1^2\mu_2(t_{go}-\mu_2). \end{aligned} \tag{31}$$

Alternative form of Eqn. (28) is obtained as:

$$u = \frac{N_1 z_{10} + N_2 t_{go} z_{20} + N_3 t_{go} z_{2F}}{t_{go}^2}, \tag{32}$$

where

$$N = \begin{bmatrix} N_1 \\ N_2 \\ N_3 \end{bmatrix} = \Lambda^{-1} t_{go}^2 \begin{bmatrix} k_1 \\ k_2/t_{go} \\ k_3/t_{go} \end{bmatrix}. \tag{33}$$

4 Characteristics of the New Impact Angle Control Guidance Law

Note that the alternative form in Eqn. (32) is the same form as in Eqn. (5). Unlike the conventional impact angle control guidance algorithms, as can be deduced from Eqns (30) and (33), the navigation gains of the proposed guidance algorithm could vary over the entire engagement. This enables the guidance algorithm developed in this paper to shape the guidance command over the flight time depending on the two design parameters, μ_1 and μ_2. Therefore, the proposed algorithm can handle more design requirements. Consequently, by simply tuning the three design parameters, the proposed algorithm is able to reduce the sensitivity to the errors at the initial homing phase and handle other design requirements.

It is obvious that the parameters, N_1, N_2, and N_3, are directly related to the design parameters, μ_1 and μ_2. For the simplicity of comparison, some variables are defined as:

$$\rho =: \mu_1 t_{go} \tag{34}$$

$$t_\rho =: \tan^{-1} \rho \tag{35}$$

$$l_\rho =: \ln\left(\rho^2 + 1\right) \tag{36}$$

Let us investigate the characteristics of guidance law in Eqn. (32) with respect to the design parameters. If μ_1 goes to zero which means that the weighting function is constant, $R = 1$, the commanded acceleration is given by

$$u = \frac{6z_{10} + 4z_{20}t_{go} + 2z_{2F}t_{go}}{t_{go}^2} \tag{37}$$

Since $z_{10} = y_{go}$, $z_{20} = \dot{y}_{go}$, and $z_{2F} = \dot{y}_{go}(t_F)$, the commanded acceleration for $\mu_1 = 0$ is equivalent to that of the OGL which is represented in in Eqn. (5).

If μ_2 becomes zero and μ_1 equals to unity which mean the weighting function is proportional to the square of the time-to-go, $R = t_{go}^2 + 1$, the commanded acceleration is obtained as:

$$u = \frac{2(l_1 - 2t_{go}t_1)z_{10} + 4\left\{t_{go}(l_1 - 1) - (t_{go}^2 - 1)t_1\right\}z_{20} + 2\left\{t_{go}(2 + l_1) - 2t_\alpha\right\}z_{2F}}{(t_{go}^2 + 1)(l_1^2 + 4t_1^2 - 4t_{go}t_1)} \tag{38}$$

5 Numerical Simulations

In this section, the performance of the proposed method is investigated via numerical simulations for several several values of $\boldsymbol{\mu} = (\mu_1, \mu_2)$. The initial conditions for numerical simulations are represented in Table 1. This paper examines the performance of the proposed guidance laws for the following sets of design parameters, $\boldsymbol{\mu}$, combinations, (0.1, 0), (0.3, 3), and (0.5, 6).

Table 1 Initial simulation conditions

Initial separation	Closing velocity	Initial Heading error	Flight time t_F
6 *km*	1,000 m/s	-5 *deg*	6.0 *s*

The proposed guidance law with all given $\boldsymbol{\mu}$ combinations succeeded to intercept the target. The simulation results for the navigation gains and acceleration commands are shown in Figs. 3- 6. The proposed guidance laws are compared with the equivalent the OGL. As designed, navigation gains of the proposed guidance law vary over the engagement. The navigation gains of the proposed approach are smaller than the equivalent navigation gains of the OGL at the beginning of the homing phase, but become bigger than the equivalent gains. It is also shown that the gains of the proposed guidance laws always converge to the equivalent navigation gains of the OGL from Figs. 3, 4 and 5. As illustrated in Fig. 6, varying navigation gains influences the guidance command profiles compared with those of the equivalent OGL.

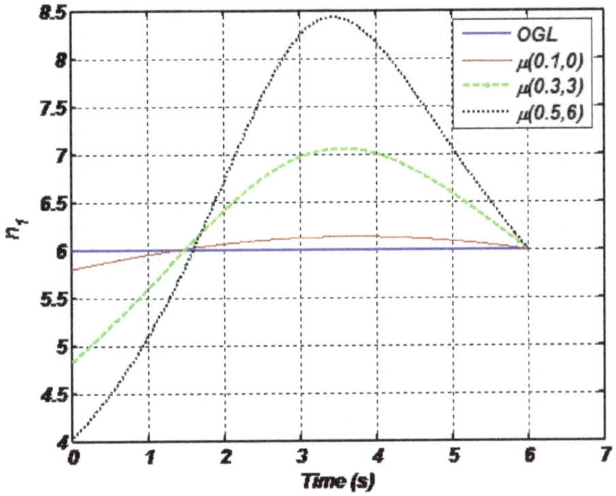

Fig. 3 Time history of the navigation gain N_1

From all the numerical simulation results, it is shown that the navigation gains are varying over the homing phase and their characteristics change depending on the values of design parameters, μ_1, and μ_2. This enables the new homing guidance to shape the acceleration commands and consequently missile trajectories. In all of the design parameters combinations, the acceleration commands of the proposed guidance law are smaller than the OGL commands at the early homing phase. As smaller acceleration commands generates less manoeuvres, the proposed approach

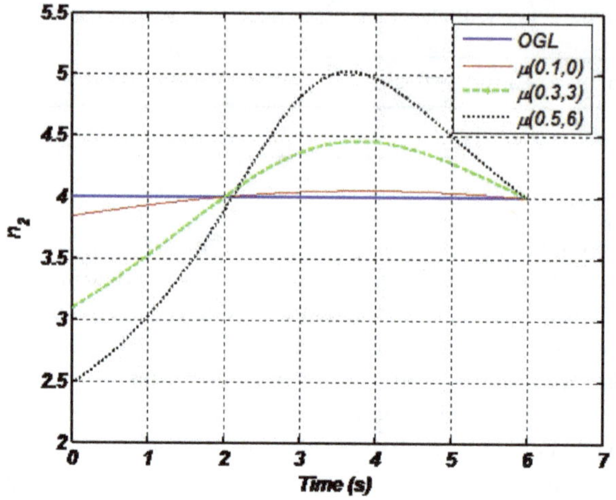

Fig. 4 Time history of the navigation gain N_2

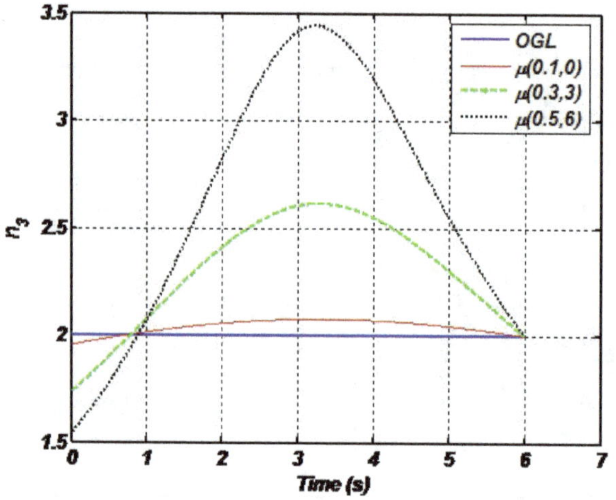

Fig. 5 Time history of the navigation gain N_3

can reduce heading error sensitivity at the beginning of the homing phase. However, one should carefully select appropriate design parameters, μ_1 and μ_2, to meet design requirements.

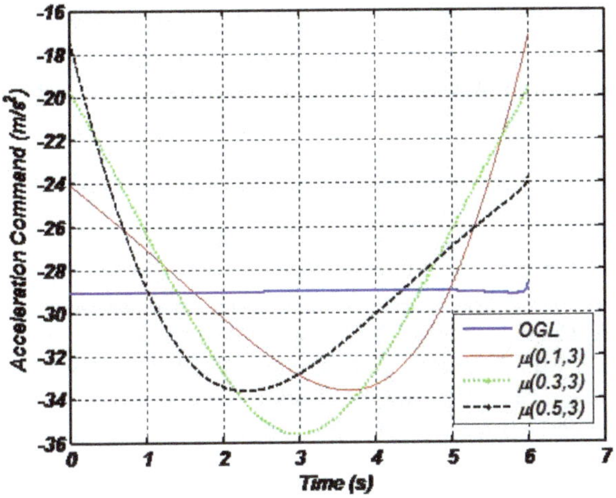

Fig. 6 Time history of acceleration commands

6 Conclusion

In this paper, a new impact angle control guidance law is developed using the linear quadratic optimal control theory with a new weighting function. The proposed weighting function is formulated with a second order polynomial of time-to-go to reduce sensitivity on zero effort miss and flight path angle error at the initial homing phase. The characteristics of the proposed guidance law are investigated and its performance is demonstrated using engagement simulations. From the analysis and simulations, it is shown that the proposed guidance law can reduce the sensitivity on the initial errors. It is also shown that the guidance law can readily shape the acceleration command profile over the entire engagement by distributing the relative weighting with respect to the minimum weighting. Consequently, the proposed guidance law can simultaneously achieve design requirements on miss distance, sensitivity on initial errors, and/or trajectory modulation by selecting an appropriate set of the design parameters, μ_1, and μ_2.

References

1. Bryson, A.E., Ho, Y.-C.: Applied optimal control: optimization, estimation, and control. Taylor & Francis (1975)
2. Ohlmeyer, E.J., Phillips, C.A.: Generalized vector explicit guidance. Journal of Guidance, Control, and Dynamics 29(2), 261–268 (2006)
3. Ryoo, C.K., Cho, H., Tahk, M.J.: Optimal guidance laws with terminal impact angle constraintvol., no. 4, 2005, pp.. Journal of Guidance, Control, and Dynamics 28(4), 724–732 (2005)

4. Ryoo, C.-K., Cho, H., Tahk, M.-J.: Time-to-go weighted optimal guidance with impact angle constraints. IEEE Transactions on Control Systems Technology 14(3), 483–492 (2006)
5. Shaferman, V., Shima, T.: Linear quadratic guidance laws for imposing a terminal interception angle. Journal of Guidance, Control, and Dynamics 31(5), 1400–1412 (2008)
6. Erer, K.S., Merttopcuoglu, O.: ndirect impact-angle-control against stationary targets using biased pure proportional navigation. Journal of Guidance, Control, and Dynamics 35(2), 700–703 (2012)
7. Kim, B.S., Lee, J.G., Han, H.S.: Biased png law for impact with angular constraint. IEEE Transactions on Aerospace and Electronic Systems 29(2), 261–268 (2006)
8. Ratnoo, A., Ghose, D.: Impact angle constrained interception of stationary targets. Journal of Guidance, Control, and Dynamics 31(6), 1816–1821 (2008)
9. Shin, H.-S., Lee, J.-I., Tsourdos, A., Tahk, M.-J.: Homing guidance law for reducing sensitivity on heading error. In: AIAA Guidance, Navigation, and Control Conference, Minneapolis, Minnesota. AIAA 2012-4473 (August 2012)
10. Lee, J.-I.: New homing guidance law for alleviating sensitivity to initial heading errors via input weight shaping. In: Korea Institute of Military Science and Technology, Daejeon Korea (2007)
11. Cho, H., Ryoo, C.K.: Implementation of optimal guidance laws using predicted missile velocity profiles. Journal of Guidance, Control, and Dynamics 22(4), 579–588 (1999)
12. Joseph, Z.: Ben-Asher and Sergei Levinson. New proportional navigation law for ground-to-air systems. Journal of Guidance, Control, and Dynamics 26(5), 822–825 (2003)
13. Lee, Y.-I., Kim, S.-H., Tahk, M.-J.: Optimality of linear time-varying guidance for impact angle control. IEEE Transactions on Aerospace and Electronic Systems 48(4), 2802–2817 (2012)
14. Harrison, G.A.: Hybrid guidance law for approach angle and time-of-arrival control. Journal of Guidance, Control, and Dynamics 35(4), 1104–1114 (2012)
15. Lee, C.H., Tahk, M.J., Lee, J.I.: Generalized formulation of weighted optimal guidance laws with impact angle constraint. IEEE Transactions on Aerospace and Electronic Systems 49(2), 1317–1322 (2013)
16. Zarchan, P.: Tactical and strategic missile guidance, progress in astronautics and aeronautics. AIAA, Reston (2002)

Part II
Estimation and Navigation

On-Line Safe Flight Envelope Determination for Impaired Aircraft

Thomas Lombaerts, Stefan Schuet, Diana Acosta, and John Kaneshige

Abstract. The design and simulation of an on-line algorithm which estimates the safe maneuvering envelope of aircraft is discussed in this paper. The trim envelope is estimated using probabilistic methods and efficient high-fidelity model based computations of attainable equilibrium sets. From this trim envelope, a robust reachability analysis provides the maneuverability limitations of the aircraft through an optimal control formulation. Both envelope limits are presented to the flight crew on the primary flight display. In the results section, scenarios are considered where this adaptive algorithm is capable of computing online changes to the maneuvering envelope due to impairment. Furthermore, corresponding updates to display features on the primary flight display are provided to potentially inform the flight crew of safety critical envelope alterations caused by the impairment.

1 Introduction

In all transportation systems, but especially in civil aviation, safety is of paramount importance. Many developments focus on improving safety levels and reducing the risks of life threatening failures. In a recent study by the Commercial Aviation Safety Team (CAST) and the International Civil Aviation Organization (ICAO), it can be observed that loss of control in flight (LOC-I) is the most frequent primary accident cause. This study is based on a statistical analysis of aircraft accidents

Thomas Lombaerts
German Aerospace Center (DLR), Robotics and Mechatronics Center,
Institute of System Dynamics and Control, Oberpfaffenhofen, D-82234 Weßling, Germany
e-mail: Thomas.Lombaerts@dlr.de

Stefan Schuet · Diana Acosta · John Kaneshige
NASA Ames Research Center, Intelligent Systems Division, CA 94035 Moffett Field, USA
e-mail: {Stefan.R.Schuet,Diana.M.Acosta,
John.T.Kaneshige}@nasa.gov

J. Bordeneuve-Guibé et al. (eds.), *Advances in Aerospace Guidance, Navigation and Control*, DOI: 10.1007/978-3-319-17518-8_16

between 2002 and 2011, and indicates that this category accounts for as much as 23% of all fatal aircraft accidents and involves most fatalities[1]. LOC-I can have various causes, occurring individually or in combination, such as a system malfunction, atmospheric disturbances (e.g. turbulence or icing), and loss of situational awareness by the crew. An important aspect during operations under system malfunction or atmospheric disturbance is that the crew needs to maintain awareness through possibly time changing degradation of aircraft performance characteristics. The CAST established a specialized international safety analysis team to study the apparent growing trend in loss of Airplane State Awareness (ASA) by the flight crew. This team has produced recommended safety enhancements that include research to develop and implement technologies for enhancing flight crew awareness of airplane energy state (SE 207)[2]. The approach developed in this paper focuses on increasing awareness of the boundaries of the safe flight envelope over which the pilot can maneuver without losing control over the plane. These boundaries can be updated based on possibly time changing flow of information regarding the aircraft state.

A variety of methods conforming to this concept have been investigated in previous studies. The most straightforward methods include wind tunnel testing, flight test experiments and high-fidelity model-based computation of attainable equilibrium sets or achievable trim points[23], possibly with bifurcation analysis [7] or a vortex lattice algorithm combined with an extended Kalman filter[16]. More complex methods include formulating flight envelope estimation as a reachability problem and solving this with level set methods and Hamilton-Jacobi equations [15], possibly with time scale separation [5] or semi-Lagrangian level sets [19]. Alternative methods rely on linearization and region of attraction analysis [20], determining controllability/maneuverability limits in a quaternion-based control architecture[3] or robustness analysis for determination of reliable flight regimes [22]. An approach suggested by Boeing uses Control-Centric Modeling, dynamic flexible structure and load models [24]. In the frequency domain, stability margins can be estimated in real time via nonparametric system identification [8]. More focused techniques inspired by flight dynamics exist as well, such as determining the minimum lateral control speed[6]. In this approach, the trim envelope is estimated through efficient high-fidelity model-based computations of attainable equilibrium sets based on aerodynamic coefficient identification from air data, inertial and GPS measurements. The corresponding maneuverability limitations of the aircraft are then determined through a robust reachability analysis (relative to the trim envelope) through an optimal control formulation and based on the principle of time scale separation. The theoretical underpinnings covering the overall approach are available in previous publications[12, 21].

The structure of this paper is as follows. In Sec. 2, the estimation of the envelope boundaries is discussed. The method for presenting this information to the crew is elaborated in Sec. 3. Sec. 4 discusses some relevant application example scenarios. Conclusions and recommendations can be found in Sec. 5. Further related research based on the results presented here is described in Sec. 6.

2 Estimation of the Envelope Boundaries

The safe maneuvering envelope is a fundamental property of the aircraft's design and overall current state of health. By definition it determines the overall capability of the aircraft. If actively monitored in time, it may function as an early warning system as well as provide anticipatory guidance to help avoid loss of control. For example, automated planning tools may use it to help pilots land safely under emergency landing conditions[17], or when combined with a display it may result in overall better pilot awareness of the state of the aircraft. This can be particularly useful when an automation system switches off. Additionally, the physics-based maneuverability envelope can be analyzed separately from the control strategy, and knowledge of the envelope may for example unmask control limitations clouded by adaptive controllers, and even lower barriers to the introduction of more advanced unconventional control strategies[9]. For these reasons, improved methods for tracking aircraft maneuverability in real-time may effectively help pilots avoid inappropriate crew response and further prevent or recover aircraft from upset conditions. Generally, the maneuvering envelope is the set of safe aircraft state and control inputs. Unfortunately, because of the underlying nonlinear aircraft dynamics, it is challenging to calculate this set of states accurately and rapidly enough to provide the pilot or automation system with reliable information in a diverse and rapidly changing environment.

2.1 Mathematical Model Postulation and Parameter Identification

The aircraft model used for this simulator study is the nonlinear RCAM (Research Civil Aircraft Model) simulation model, which is representative of a large two-engine jet transport with general characteristics of a wide-body, conventional tail and low wing airplane configuration with twin turbofan engines located under the wings[14]. The physical dimensions are similar to an Airbus A300 aircraft, with flight characteristics representative of a large jet transport in landing configuration (flaps at 32.5 deg and gear down) at sea level. A selection of RCAM model data is provided in Table 1.

Table 1 Selection of parameter values for RCAM model

Constants	Aerodynamic Coefficients		Input Bounds
	$C_{D_0} = 0.1599$		
$S = 260\ \text{m}^2$	$C_{D\alpha} = 0.5035$	$C_{m_0} = -0.015$	$T \in [20546, 410920]$ N
$m = 120 \times 10^3$ kg	$C_{D_{\alpha^2}} = 2.1175$	$C_{m\alpha} = -2.15$	$\alpha \in [0, 14.5]$ deg.
$g = 9.81\ \text{m/s}^2$	$C_{L_0} = 1.0656$	$C_{m_q} = -14.0$	$\delta_e \in [-30.0, 20.0]$ deg.
$\rho = 1.225\ \text{kg/m}^3$	$C_{L\alpha} = 6.0723$	$C_{m\delta_e} = -2.87$	
	$C_{Y_\beta} = -1$		

Nonlinear aircraft dynamics with higher dimensions can be simplified by considering the principle of time scale separation[5]. The structure of time scale separation is analogous as applied for the fault tolerant control algorithm developed earlier[9]. The overview can be found in Fig. 1, which illustrates that a nine dimensional nonlinear problem is decoupled in three consecutive three dimensional optimization problems.

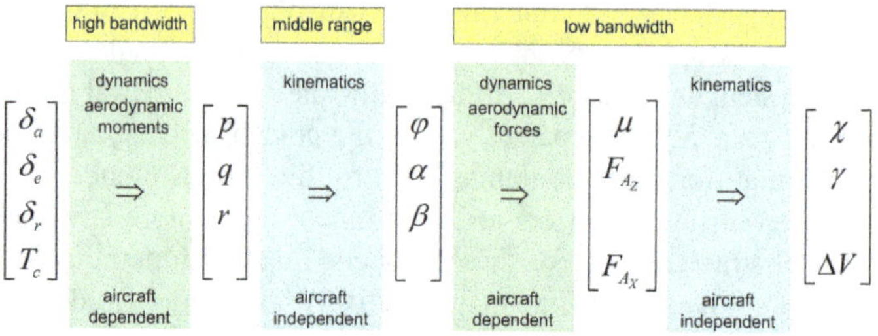

Fig. 1 Separation of dynamics over high bandwidth, middle range and low bandwidth

2.1.1 Model Postulation

A nonlinear 3D aircraft example is considered. Main focus of this research is on the slow aircraft dynamics as specified in Fig. 1, involving the aerodynamic forces lift L, drag D and sideforce Y_{aero}. Also the pitching moment M, as part of the fast dynamics, has been included in the model but this moment can be considered separately from the aerodynamic forces, thanks the the principle of time scale separation, see Fig. 1. The acting forces on the aircraft are illustrated in Fig. 2 for a symmetric flight condition.

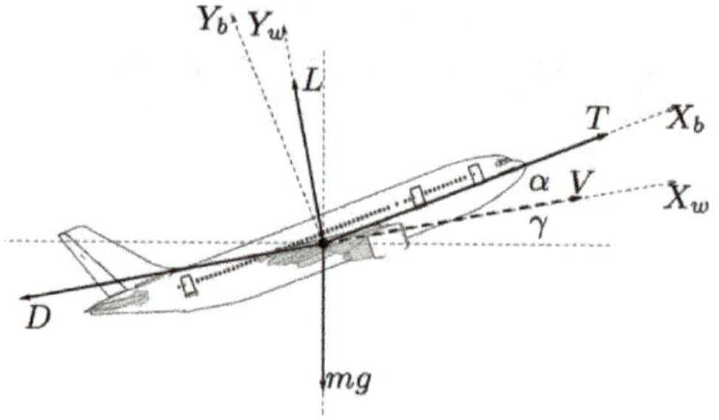

Fig. 2 Acting forces on the aircraft model, source: Lygeros[15]

For the complete 3D situation, the equations of motion are written as follows[9]:

$$F_{A_X} - W\sin\gamma = m\dot{V} \tag{1}$$

$$F_{A_Z}\cos\varphi + F_{A_Y}\sin\varphi + W\cos\gamma = -mV\dot{\gamma} \tag{2}$$

Where the aerodynamic forces can be simplified assuming small aerodynamic angles α and β:

$$F_{A_X} = T\cos\beta\cos\alpha - D(V,\alpha) \approx T - D(V,\alpha) \quad (3)$$
$$F_{A_Z} = -T\sin\alpha - L(V,\alpha) \approx -L(V,\alpha) \quad (4)$$
$$F_{A_Y} = -T\sin\beta\cos\alpha + Y_{\text{aero}}(V,\beta) \approx Y_{\text{aero}}(V,\beta) \quad (5)$$

with the following expansions for lift L, drag D and sideforce Y_{aero}:

$$D(V,\alpha) = \bar{q}S\left(C_{D_0} + C_{D\alpha}\alpha + C_{D_{\alpha^2}}\alpha^2\right) \quad (6)$$
$$L(V,\alpha) = \bar{q}S\left(C_{L_0} + C_{L\alpha}\alpha\right) \quad (7)$$
$$Y_{\text{aero}}(V,\beta) = \bar{q}S\left(C_{Y_\beta}\beta\right) \quad (8)$$

where the dynamic pressure $\bar{q} = 1/2\rho V^2$.

The aircraft dynamics are combined in the following matrix structure:

$$\begin{bmatrix}\dot{V}\\ \dot{\gamma}\end{bmatrix} \approx \begin{bmatrix}-\frac{\rho S}{2m}V^2 C_{D_0} - g\sin\gamma\\ -\frac{g}{V}\cos\gamma\end{bmatrix} + \begin{bmatrix}1\\0\end{bmatrix}\frac{T}{m} + \begin{bmatrix}-\frac{\rho S}{2m}V^2\left(C_{D\alpha}\alpha + C_{D_{\alpha^2}}\alpha^2\right)\\ \frac{\rho S}{2m}V\left(C_{L_0} + C_{L\alpha}\alpha\right)\cos\varphi\end{bmatrix} + \begin{bmatrix}0\\ -\frac{\rho S}{2m}VC_{Y_\beta}\beta\sin\varphi\end{bmatrix} \quad (9)$$

where thrust T and angle of attack α are treated as virtual inputs, based on the time scale separation principle as illustrated in Fig. 1. An affine matrix structure in the inputs is obtained by simplifying for small angles of attack α and sideslip β.

The pitching moment structure typically varies for every aircraft type. For this specific model, a model structure selection analysis[10, 9] has provided the following structure:

$$C_m = C_{m_0} + C_{m\alpha}\alpha + C_{m\frac{q\bar{c}}{V}}\frac{q\bar{c}}{V} + C_{m_{\delta_e}}\delta_e + C_{m_{i_h}}i_h + C_{m_T}\frac{T}{\bar{q}d_{\text{eng}}^2} \quad (10)$$

2.1.2 Parameter Identification

In order to characterize the flight envelope of a particular aircraft, the as of yet undetermined aerodynamic coefficients need to be reconstructed using available sensor measurements. This is a necessary step, that essentially fits the above described dynamics model to any particular aircraft, or aircraft simulation of higher fidelity. An inertial sensor package, consisting of rate gyros and accelerometers, provides angular rates and accelerations as well as specific forces. It is assumed that a state estimation algorithm[13] is available, taking into account sensor disturbances (biases and/or noise), and compensating for them.

The forces in the body fixed reference frame are reconstructed from:

$$X_{\text{tot}} = mA_x \qquad Y_{\text{tot}} = mA_y \qquad Z_{\text{tot}} = mA_z \quad (11)$$
$$C_X = \frac{X}{\bar{q}S} = \frac{T - X_{\text{tot}}}{\bar{q}S} \qquad C_Y = \frac{Y}{\bar{q}S} \qquad C_Z = \frac{Z}{\bar{q}S} \quad (12)$$

where A_x, A_y and A_z are the specific forces as measured by the accelerometers. The force coefficients in the body fixed reference frame are then transformed to the aerodynamic reference frame using

$$\begin{pmatrix} -C_D \\ C_{Y_{\text{aero}}} \\ -C_L \end{pmatrix} = \begin{pmatrix} \cos\alpha & 0 & \sin\alpha \\ 0 & 1 & 0 \\ -\sin\alpha & 0 & \cos\alpha \end{pmatrix} \cdot \begin{pmatrix} \cos\beta & \sin\beta & 0 \\ -\sin\beta & \cos\beta & 0 \\ 0 & 0 & 1 \end{pmatrix} \cdot \begin{pmatrix} C_X \\ C_Y \\ C_Z \end{pmatrix} . \tag{13}$$

Finally, the pitching moment coefficient is reconstructed from the corresponding moment equation

$$C_m = \frac{M}{\bar{q}S\bar{c}} = \frac{\dot{q}I_{yy} + pr\,(I_{xx} - I_{zz}) + \left(p^2 - r^2\right) I_{xz}}{\bar{q}S\bar{c}}, \tag{14}$$

where I_{xx}, I_{yy}, I_{zz}, and I_{xz} are the components of the inertia matrix in the body axis, and where p, q, and r are respectively the roll, pitch, and yaw rate components of the body-axis angular velocity.

The above defined flight dynamics model provides the means through which one can assess the flight performance capability of the aircraft. However, it depends on a set of effective aerodynamic coefficients, which should be estimated from the combination of available sensor data and the known physical relationships between quantities specified by the model. The aerodynamic coefficient vector to be identified for the forces is defined as:

$$\mathbf{c} = \left[C_{D_0}, C_{D\alpha}, C_{D_{\alpha^2}}, C_{L_0}, C_{L\alpha}\right]^T,$$

and the moment coefficient vector is:

$$\mathbf{m} = \left[C_{m_0}, C_{m_\alpha}, C_{m_{\frac{q\bar{c}}{V}}}, C_{m_{\delta_e}}, C_{m_{i_h}}, C_{m_T}\right]^T.$$

A specialized Bayesian probabilistic approach was developed to infer the unknown aerodynamic coefficients from the noisy sensor measurements and to quantify the estimation uncertainty; an overview of which is provided next. More information about the identification procedure, including results, can be found in Schuet et. al. [21].

A state measurement process $\mathbf{x}(k) = [V(k), \gamma(k)]$ is defined, which proceeds, for $k = 1, 2, \ldots, m-1$, according to a midpoint-Euler discrete approximation:

$$\mathbf{x}(k+1) = \mathbf{x}(k) + h\mathbf{f}(\tilde{\mathbf{x}}, \mathbf{u}; \mathbf{c}) + \boldsymbol{\tau}(k), \tag{15}$$

where $\mathbf{f}(\mathbf{x}, \mathbf{u}; \mathbf{c})$ represents the continuous flight dynamics model (9), $\mathbf{u}$ is the vector of known virtual inputs at time instance k,

$$\mathbf{u}(k) = [T, \alpha, \beta]^T,$$

h is the time-sample resolution, $\tilde{\mathbf{x}} = [\mathbf{x}(k) + \mathbf{x}(k+1)]/2$ is the midpoint, and $\boldsymbol{\tau}(k) \sim \mathcal{N}(0, h\mathbf{S}^{-1})$ is the additive independent and identically distributed Gaussian measurement noise process. While the maneuverability model is linear in $\mathbf{c}$, it is still nonlinear in $\mathbf{x}$ and $\mathbf{u}$, and the midpoint-Euler approximation enables better representation of the true nonlinear system, for negligible additional computational cost. The inverse covariance matrix $\mathbf{S}$ is an important statistical parameter that characterizes the process noise $\boldsymbol{\tau}(k)$, which will also be estimated from the observed data.

In similar fashion, one can also define processes for the accelerometer measurements

$$\mathbf{x}_{\mathrm{a}}(k) = \mathbf{A}_k\mathbf{c} + \boldsymbol{\tau}_{\mathrm{a}}(k),$$

where $\mathbf{A}_k$ is a matrix, depending on $\mathbf{u}(k)$, that implements the combination of (13) and (6)–(8). Finally, the moment measurement process is defined as

$$x_{\mathrm{m}}(k) = \mathbf{M}_k\mathbf{m} + \tau_{\mathrm{m}}(k),$$

where $\mathbf{M}_k$ is a row vector implementing (10), that depends on the moment model input variables. Also, $\boldsymbol{\tau}_{\mathrm{a}}$ and τ_{m} represent independent additive Gaussian measurement noise terms for the accelerometer and moment measurements, with associated inverse covariance matrix $\mathbf{S}_{\mathrm{a}}$ and scalar inverse variance S_{m}, respectively. Furthermore, the moment measurement process is independent of the dynamics and accelerometer measurement processes, and can therefore be treated by a separate inference process, that follows the same approach used to infer the aerodynamic force coefficients summarized next.

Focusing on the estimation of the force coefficient vector $\mathbf{c}$, the above measurement model equations and Gaussian noise distribution assumptions enable one to specify the multivariate probability density function (pdf) for the measured data given the model parameters

$$p(\mathbf{X}, \mathbf{X}_{\mathrm{a}} | \mathbf{c}, \mathbf{S}, \mathbf{S}_{\mathrm{a}}, \mathbf{U}), \tag{16}$$

where $\mathbf{X}$, $\mathbf{X}_{\mathrm{a}}$, and $\mathbf{U}$ represent matrices containing the entire history of state, accelerometer, and input data measurements, respectively. Through Bayes' Theorem, the *likelihood* pdf (16) can then be combined with appropriate prior information to determine the *posterior* pdf

$$p(\mathbf{c}, \mathbf{S}, \mathbf{S}_{\mathrm{a}} | \mathbf{X}, \mathbf{X}_{\mathrm{a}}, \mathbf{U}) \tag{17}$$

that specifies the probability density of the unknown parameter values $\mathbf{c}$ and inverse measurement noise covariance matrices, given all of the observed data and prior information. The optimal parameter estimate is then found by solving

$$\text{maximize } p(\mathbf{c}, \mathbf{S}, \mathbf{S}_{\mathrm{a}} | \mathbf{X}, \mathbf{X}_{\mathrm{a}}, \mathbf{U}) \tag{18}$$

with respect to the unknown aerodynamics coefficients $\mathbf{c}$ *and* inverse covariance matrix terms $\mathbf{S}$ and $\mathbf{S}_{\mathrm{a}}$.

With the modeling and additive noise assumptions used in the derivation, the posterior pdf (17) is log-concave in any one of the variables $\mathbf{c}$, $\mathbf{S}$, or $\mathbf{S}_\mathrm{a}$ when the other two variables are held fixed. This enables the efficient computation of a local optimal estimate using a block coordinate descent method to break the solution process into steps that require solving simple convex optimization problems. Furthermore, the coefficient parameter estimation uncertainty and Bayesian model evidence can also be estimated using closed form expressions. See [21] for details.

2.2 *Estimation of the Trim Envelope*

One important feature of the maneuverability model (9) is that it enables a rapid numerical analysis of the trimmable states. To see how this works first note the definition of the set of trimmable states is

$$\{\mathbf{x}\,|\,\mathbf{f}(\mathbf{x},\mathbf{u};\mathbf{c}) = 0,\ (\mathbf{x},\mathbf{u}) \in \mathscr{B}\}, \tag{19}$$

where $\mathscr{B}$ represents the set of overall allowable states and virtual inputs. This set is important because it represents an a-priori safe maneuverability envelope.

For the maneuverability model, and a given aerodynamic parameter vector $\mathbf{c}$, characterizing the set of trimmable points then involves setting the top and bottom equations on the right hand side of (9) equal to zero. The bottom equation is solved for angle of attack α in terms of the other variables in that equation, which do not include thrust T. The top equation is then solved for T into which the previous solution for α is substituted. The result is a closed form solution for the required thrust and alpha needed to achieve trim for any given trim state and other virtual inputs. This enables a fast numerical sweep to determine the non-convex trim envelope as follows:

1. Setup a grid of state values in $\mathscr{B}$. For most practical applications a coarse resolution is sufficient.
2. Fix values for roll angle ϕ and side-slip angle β, and make them equal to the current values.
3. For each point in the state grid solve for the thrust T and angle of attack α needed to achieve trim.
4. Return only those points for which T and α are within $\mathscr{B}$.

A crude visualization of the (V,γ)-trim envelope is then obtained by simply plotting the trimmable points from the above calculation. In addition, one should check whether the achieved trim points are stable. This involves checking the eigenvalues associated with the local linear approximation to (9) at each trim point, and as before a closed form expression can be found. The entire computation is fast enough on a modern PC to enable dynamic re-computation as aircraft conditions change, or to compute extended envelopes by sweeping over values for ϕ and β.

The actual maneuverability calculation for a grid covering 5029 points is completed in 21 ms on a 2.6 GHz MacBook Pro. The calculation includes a check for

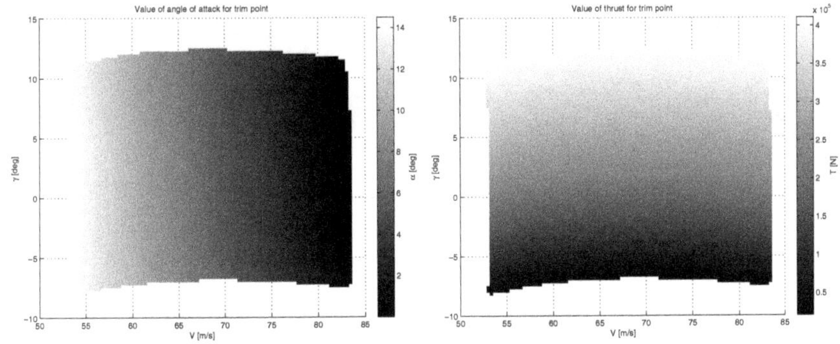

(a) Values of angle of attack for trim points (b) Values of thrust for trim points

Fig. 3 Trim envelope for the RCAM model and required inputs for trim points

stability[21]. An example trim envelope is shown in Fig. 3, and is based on the RCAM simulation model.

In Fig. 3, it can be seen that the trim envelope boundaries are imposed by the input saturation limits. The upper boundary corresponds to maximum thrust $T_{\max} =$ 410920N, the lower boundary is imposed by minimum thrust $T_{\min} = 20546$N. The range for the angle of attack results in the boundaries left ($\alpha_{\max} = 14.5°$, prior to stall) and right ($\alpha_{\min} = 0°$). It can also be observed that more thrust is needed for larger flight path angles, because the thrust force has to counteract the exponentially increasing drag force, since a larger angle of attack occurs for slower speeds. Analyzing the range of thrust values over airspeed for constant flight path angle, shows that more thrust is needed for a further speed decrease below 69m/s. This region corresponds to the range of the angle of attack $\alpha > 4.5°$. Analysis of the Lift-Drag Polar and the power required curve, shown in Fig. 4, confirms that this region is the back side of the power curve, and that $V = 69\text{m/s}$ is the minimum drag airspeed.

2.3 *Estimation of the Maneuvering Envelope*

In this context, the preferred interpretation of the safe maneuvering envelope considers reachability from the trim envelope. The stable and controllable trim envelope is considered an a-priori safe set. The backwards reachable set is defined as the set of states from where (at least one point in) the trim envelope can be reached. The forwards reachable set is defined as the set of states which can be reached from (at least one point in) the trim envelope. Then the safe maneuvering flight envelope is the cross section between the forwards and backwards reachable sets. This interpretation is illustrated in Fig. 5. In addition to the safe envelope, the backwards reachable set is considered as the survivable flight envelope. After an upset due to damage, turbulence, a wake encounter etc., it is possible to bring the aircraft back to a safe trim condition as long as the current flight condition is situated inside the backwards reachable set.

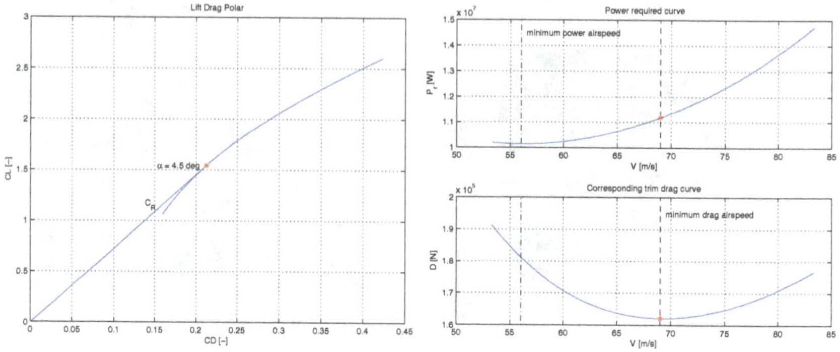

(a) Lift-Drag Polar (b) Power required curve

Fig. 4 Lift-Drag Polar and power required curve of the RCAM model confirm that minimum drag airspeed is $V = 69\mathrm{m/s}$, and the region to the left is the back side of the power curve

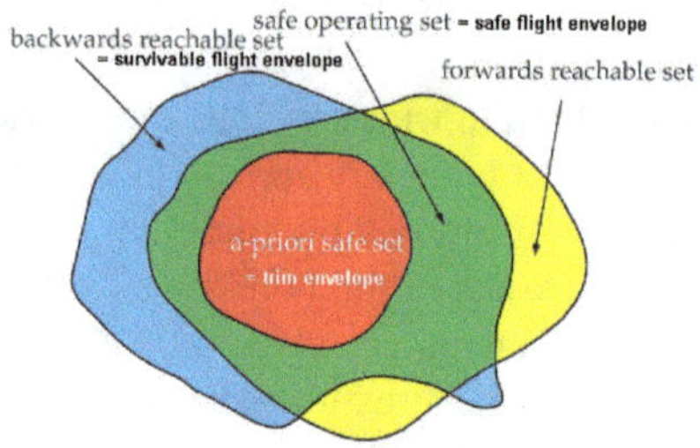

Fig. 5 Safe maneuvering envelope as intersection between forwards and backwards reachability, modified from source: van Oort[19]

The aim is to perform a combined forward and backward reachability analysis from the trim envelope as efficiently as possible, for on-line implementations. Based on previous research[4], level set methods are an excellent candidate. Finally, robustness is an important aspect to be considered in this context as well.

It has been shown in the literature that maneuvering envelope estimation through reachability can be reformulated in the optimal control framework[15]. Consider a continuous time control system:

$$\dot{\mathbf{x}} = \mathbf{f}(\mathbf{x}, \mathbf{u}, \boldsymbol{\Delta}) \tag{20}$$

with $\mathbf{x} \in \mathbb{R}^n$, $\mathbf{u} \in U \subseteq \mathbb{R}^m$, $\boldsymbol{\Delta} \in D \subseteq \mathbb{R}^k$, $\mathbf{f}(\cdot,\cdot) : \mathbb{R}^n \times U \to \mathbb{R}^n$, a function:

$$l(\cdot) : \mathbb{R}^n \to \mathbb{R} \tag{21}$$

and an arbitrary time horizon $T \geq 0$. Let $\mathscr{U}_{[t,t']}$denote the set of Lebesgue and bounded measurable functions from the interval $[t, t']$ to U. Define $\boldsymbol{\phi}(\tau, t, \mathbf{x}, \mathbf{u}(\cdot), \boldsymbol{\Delta})$as the state trajectory. $\boldsymbol{\Delta}$ are defined as parameter uncertainties.

Given a set of states $K \subseteq \mathbb{R}^n$, the reachability question can be naturally formulated regarding the relation between the set K and the state trajectories $\boldsymbol{\phi}$ of Eq. (20) over the horizon T. Problem of interest is the following:

Robust reachability: Does there exist a $\mathbf{u} \in \mathscr{U}_{[0,T]}$ and a $t \in [0,T]$ such that the trajectory $\boldsymbol{\phi}$ of the state $\mathbf{x}$ satisfies $\mathbf{x} \in K$, irrespective of $\boldsymbol{\Delta}$?

The optimization problem can be formulated as a pursuit evasion game over the horizon $T \geq 0$ with target set $K \subseteq \mathbb{R}^n$ [5]. It is assumed that $\mathbf{u}$ is trying to bring or keep the state in the set K, whereas $\boldsymbol{\Delta}$ is trying to drive it out of K. To ensure the game is well-posed, $\mathbf{u}$ is restricted to play non-anticipative strategies with respect to the unknown uncertainties $\boldsymbol{\Delta}$.

For the types of safety problems considered here, a set of initial states has to be established such that $\mathbf{u}$ can win the game, in other words the set Reach can be characterized as follows:

$$\begin{aligned}\text{Reach}_{\text{robust}}(t,K) = \{&\mathbf{x} \in \mathbb{R}^n | \forall \boldsymbol{\Delta} \in D, \exists \mathbf{u} \in \mathscr{U}_{[t,T]}, \\ &\exists \tau \in [t,T], \boldsymbol{\phi}(\tau,t,\mathbf{x},\mathbf{u}(\cdot),\boldsymbol{\Delta}) \in K\}\end{aligned}$$

As done elsewhere in the literature[15], the characterization of this set can be done according to the principle of duality:

$$\text{Reach}(t,K) = (\text{Inv}(t,K^c))^c \tag{22}$$

Through this principle, it can be characterized as an INFMIN problem[15]. The crux is to include the Δ's as disturbances in the optimization function, they oppose the optimization over $\mathbf{u}$. Consider a closed set K, that can be written as the level set of a continuous function $l : \mathbb{R}^n \to \mathbb{R}$, i.e. $K = \{\mathbf{x} \in \mathbb{R}^n | l(\mathbf{x}) \geqslant 0\}$. As a consequence, the Invariance optimization formulation becomes[5]:

$$\text{Inv}(t,K) = \{\mathbf{x} \in \mathbb{R}^n | V_2(\mathbf{x},t) \geqslant 0\} \tag{23}$$

with:

$$V_2(\mathbf{x},t) = \inf_{\mathbf{u}(\cdot)\in\mathscr{U}_{[t,T]}} \sup_{\boldsymbol{\Delta}\in D} \min_{\tau\in[t,T]} l(\boldsymbol{\phi}(\tau,t,\mathbf{x},\mathbf{u}(\cdot),\boldsymbol{\Delta})) \tag{24}$$

This can be reformulated into an Hamilton-Jacobi-Bellmann Partial Differential Equation[15, 5]:

$$\frac{\partial V_2}{\partial t}(\mathbf{x},t) + \min_{\tau\in[t,T]} \left\{ \inf_{\mathbf{u}(\cdot)\in\mathscr{U}_{[t,T]}} \sup_{\boldsymbol{\Delta}\in D} \frac{\partial V_2}{\partial \mathbf{x}}(\mathbf{x},t)\mathbf{f}(\mathbf{x},\mathbf{u},\boldsymbol{\Delta}) \right\} = 0 \tag{25}$$

where $V_2(\mathbf{x},T) = l(\mathbf{x})$ holds for backward integration and $V_2(\mathbf{x},t) = l(\mathbf{x})$ applies to forward integration. These HJB PDE's can be solved by level sets, for which a toolbox is available in Matlab®[18].

A more elaborate discussion of the optimization strategy, and the detailed approach followed, is discussed by Lombaerts et al[12].

2.4 Maximum Bank Angle Calculation

The maximum bank angle ϕ can be calculated as the bank angle at which the vertical component of the maximum attainable lift vector balances the weight vector of the aircraft. This means that banking the aircraft at the current airspeed beyond this maximum angle will result in a stall. This concept is illustrated in Fig. 6.

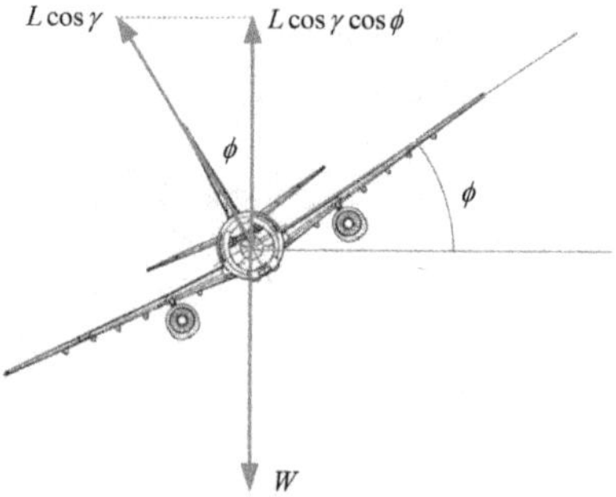

Fig. 6 Force balance between lift and weight in a turn

Equilibrium of forces in the vertical plane requires that:

$$L \cos\gamma \cos\phi = W \tag{26}$$

Rewriting for bank angle ϕ:

$$\cos\phi = \frac{W}{L\cos\gamma} = \frac{W}{C_L {}^1/_2 \rho V^2 S \cos\gamma}, \tag{27}$$

For extreme bank angles the following relationship can be derived:

$$\cos\phi_{\max} = \frac{W}{L_{\max}\cos\gamma} = \frac{W}{C_{L_{\max}} {}^1/_2 \rho V^2 S \cos\gamma}, \quad C_{L_{\max}} = C_{L_0} + C_{L_\alpha}\alpha_{\max} \tag{28}$$

Therefore, it can be stated that:

$$\pm\phi_{\max} = \pm\min\left\{\arccos\left(\frac{W}{C_{L_{\max}} {}^1/_2 \rho V^2 S \cos\gamma}\right), 35^\circ\right\} \tag{29}$$

where $C_{L_{\max}} = C_{L_0} + C_{L_\alpha}\alpha_{\max}$. In this calculation the current values for airspeed V and flight path angle γ are used, the up-to-date values for C_{L_0} and C_{L_α} as provided by the identification algorithm, as well as the maximum angle of attack $\alpha_{\max}$. For normal maneuvers of a conventional civil airliner, the maximum bank angle is not expected to exceed 35°.

Reducing speed will restrict the available bank range to lower values of $\pm\phi_{max}$. At stall speed, no bank authority will be left. Some typical values for the nominal RCAM model are given in Table 2.

Table 2 Some typical values of maximum bank angle at different airspeeds for the RCAM model

V [m/s]	α[°]	ϕ_{max}[°]
75	3	$\pm$ 60
59	9	$\pm$ 35
53	14.5	$\approx$ 0

3 Additional Information Provided to the Pilot over the Cockpit Displays

The information obtained from the envelope estimation algorithm can then be presented to the pilot. This is done in the primary flight display (PFD). The speed and flight path angle boundaries, which apply for the current bank angle and sideslip angle, are shown on the relevant parts of the PFD. Also the bank angle limits are displayed on the PFD.

The calculated true airspeed boundaries (TAS) are converted to indicated airspeed (IAS) by taking into account altitude varying air density and presented on the speedtape at the left hand side of the artificial horizon in the PFD. The flight path angle information γ is translated into vertical speed $\dot{h}$ and presented on the vertical speed tape at the right hand side of the artificial horizon in the PFD.

Safe envelope information presented on the vertical speed tape, as illustrated in Fig. 7, is absent in current PFDs. In this new setup, the amber lines mark the vertical speed ranges where no equilibrium can be established. In practice, this means that speed will increase in the lower amber region, even for idle thrust, and that speed will decrease in the upper amber region, even with full thrust. The red barber poles mark the ranges where the maneuverability envelope is left. In practice, this means that it is impossible to reach this range and return to trim within 5 s for each direction. For the bank angle limits, the amber region is static on the 35° marker. The red barber regions indicate where stall will occur and are determined by the maximum bank angle as calculated in Sec. 2.4.

4 Application Examples

Two main application examples have been considered in this research, both scenarios have an impact on the speed boundaries of the safe flight envelope. The first one is a generic simplified icing scenario, the latter is an unscheduled stabilizer deflection. The nature of both scenarios is elaborated in detail, after which the calculated effects on the envelope are analysed. Finally the observations are verified from a flight dynamics point of view.

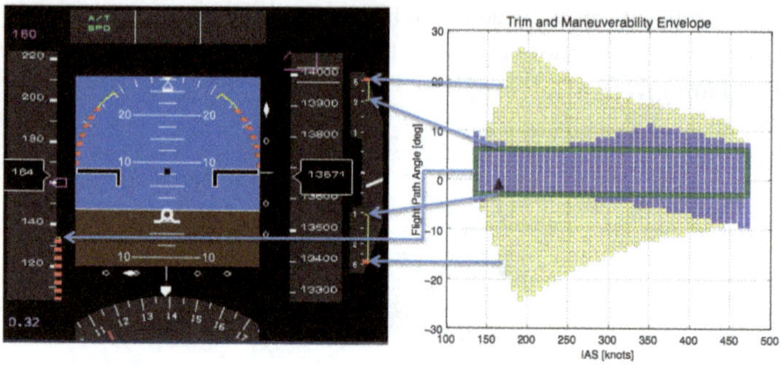

Fig. 7 The envelope limits as displayed on the primary flight display, and how they are driven by the envelope calculation

4.1 *Icing Scenario*

In specific atmospheric conditions, ice accretion is possible on the aircraft wings. In this example, it is assumed that the regular thermal anti-ice systems, which are conventional equipment on all civil airliners, are operating but are not successful in removing all and/or preventing any ice being built up on the wings. This example will focus on the impact of ice accretion on the flight performance envelope of the aircraft.

4.1.1 Impact on the Simulation Model

The influence of icing can be injected into the maneuverability model through two primary mechanisms. First, changes in the flight dynamics are represented through the aerodynamic parameters. Typically, icing will decrease lift C_L and increase drag C_D. For this simplified example, all lift coefficients (C_{L_0} and $C_{L\alpha}$) have been scaled down by 20%, and all drag coefficients (C_{D_0}, $C_{D\alpha}$ and $C_{D_{\alpha^2}}$) up by 20%. The second mechanism is modeling diminished control authority through the set $\mathscr{B}$ of overall allowable virtual inputs. For example, the maximum permissible angle of attack may diminish due to icing. In this generic example, the stall angle of attack $\alpha_{\max}$ is reduced from 14.5° to 8°, in addition to the coefficient changes, and its impact on the safe flight envelope has been analysed.

4.1.2 Effect on the Safe Flight Envelope

Fig. 8(a) shows that a 20% decrease in lift combined with a 20% increase in drag results in a shift of the trim envelope towards higher airspeeds and lower flight path angles. The following force equilibrium equations explain this physically:

$$C_L(\alpha)\frac{1}{2}\rho V^2 S = W\cos\gamma \approx W \tag{30}$$

$$T - C_D(\alpha)\frac{1}{2}\rho V^2 S = W\sin\gamma \approx W\gamma \tag{31}$$

Due to the reduced lift capability, it can be seen in Eq. 30 that a higher airspeed is needed to compensate for the aircraft weight. On the other hand, an increase in drag means that less net excess thrust is available for climb in Eq. 31. If the maximum angle of attack limit α_{max} is reduced, then there is a further reduction in maximum lift capability, which is translated into a shift to the right of the left limit of the trim envelope, as shown in Fig. 8(b). Note that a decrease in α_{max} leads to the loss of any possibility to establish trim below approximately $V = 70\text{m/s}$.

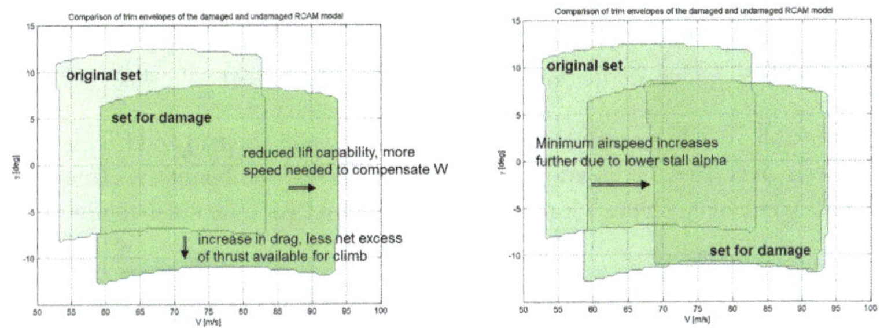

(a) icing effect (b) icing effect and reduced α_{max}

Fig. 8 Comparison of trim envelopes of the damaged and undamaged RCAM model

Next the V,γ maneuvering envelope is calculated for bank angle $\varphi = 0$, with and without generic icing scenario involving 20% decrease in lift force and increase in drag force, and with an additional envelope degradation scenario caused by a reduced maximum angle of attack. This example builds further on the results for the trim envelope, as presented in Fig. 8. The intersection of forwards and backwards reachability is defined as the safe maneuvering envelope and is shown in Fig. 9. The time horizon in this example is set at $T = 5\text{s}$. This choice is based on the time constants of the considered relevant dynamics. The blue rectangular contour corresponds to the rectangle with the largest area which can be drawn in the trim envelope of the nominal aircraft as depicted in Fig. 3. Similarly the red contours correspond to the rectangles with the largest area which can be spanned in the trim envelopes of both damage scenarios in Fig. 8(b).

Comparing undamaged and damaged envelope boundaries shows the influence of the damage characteristics. In Fig. 9(a), it can be seen that the safe maneuvering envelope shift is similar as for the trim envelope. A higher speed range is needed to compensate for the loss in the lift force coefficient, and not as large positive flight path angles can be reached due to the smaller excess of net thrust due to the increased drag. In Fig. 9(b), it can be seen that the restriction on maximum angle

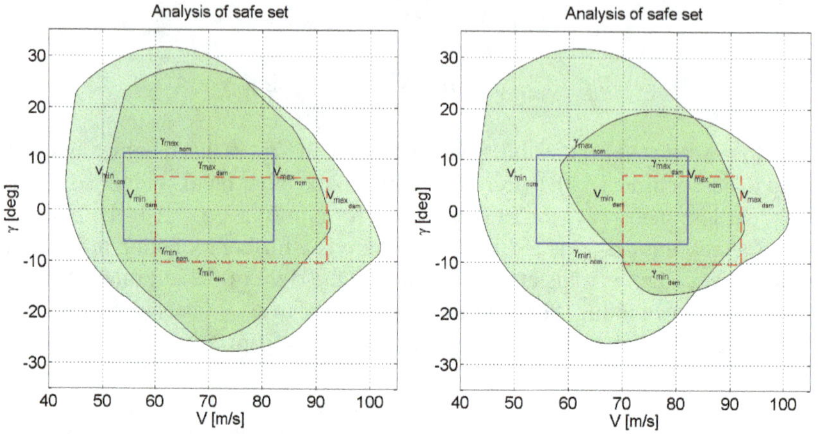

(a) Effect of 20% decrease in lift coefficient and 20% increase in drag coefficient

(b) Additional envelope degradation caused by reduction of maximum angle of attack

Fig. 9 Calculation of safe maneuvering envelope sets through forward and backward reachability analysis over $T = 5$s, based on calculated trim envelope boundaries and identified aerodynamic parameters. Upper left envelope areas correspond to nominal configuration.

of attack leads to a further shrinking of the maneuvering envelope, especially for slower speeds since stall will occur for higher airspeeds.

Extensive Monte Carlo analyses have been performed in order to verify the accuracy of the boundaries of the estimated maneuvering envelopes. These analyses have been based on the non-simplified aircraft model, ignoring the assumption that the aerodynamic angles α and β should be small. All these Monte Carlo analyses have confirmed that the results provided here are accurate and that the simplifications hold for the current ranges of the aerodynamic angles, namely $\alpha \in [0°; 14.5°]$ (no icing) and $\beta \in [-5°; +5°]$. This is an important conclusion which makes a relevant on-line safe maneuvering envelope estimation tool much more feasible.

4.1.3 Maximum Bank Angle in Icing Conditions

Since the total maximum lift capability $C_{L_{max}}$ of the aircraft characteristics will decrease in an icing scenario, while the total weight W will remain unchanged, it can be expected that this will have a significant impact on the maximum bank angle. As a consequence $\cos\phi_{max}$ will increase, which corresponds to a decrease in ϕ_{max}. The effect and comparison without icing is shown in table 3.

4.2 *Stabilizer Misalignment*

Another scenario considered in this study, was a nose down stab trim misalignment, which generates a pitch up moment. The disturbing moment must then be

Table 3 Some typical values of maximum bank angle at different airspeeds for the RCAM model with and without icing

V [m/s]	84	75	66	59	53
$\phi_{\max}$ [°] with ice	± 60	± 50	± 35	≈ 0	∅
$\phi_{\max}$ [°] no ice	± 66	± 60	± 49	± 35	≈ 0

counteracted by the elevators, which come close to their saturation limits. This results in significantly stricter upper envelope limits on speed and climb rate. For practical purposes, it is important to achieve zero angular rates around all body axes, in order to establish a stabilized final approach. Relying on Newton's second law for rotations: $\mathbf{M}_a = \mathbf{I}\dot{\boldsymbol{\omega}} + \boldsymbol{\omega} \times \mathbf{I}\boldsymbol{\omega}$, the requirement for $\boldsymbol{\omega} = 0$ as well as $\dot{\boldsymbol{\omega}} = 0$, results in the requirement that the total aerodynamic moment $\mathbf{M}_a$ (the pitching moment M around the Y-axis in the specific case considered here) should be equal to zero.

Considering Eq. (10) for this specific case:

$$C_m = 0 = C_{m_0} + C_{m_\alpha}\alpha + \underbrace{C_{m_q}\frac{q\bar{c}}{V}}_{q=0} + C_{m_{\delta_e}}\delta_e + \underbrace{C_{m_{i_h}} i_h}_{\text{disturbance}} + C_{m_T}\frac{T}{\bar{q}d_{\text{eng}}^2} \tag{32}$$

for the combination of the required pitch rate $q = 0$ and the disturbance i_h caused by the misaligned stab, together with the given trim values for angle of attack α and thrust T calculated earlier, a specific value for elevator deflection δ_e is needed in order to balance the total Eq. (32), while taking into account the upper saturation limit of the elevator $\delta_{e_{\max}}$. Eq. (32) is an additional trim requirement, besides Eq. (9). In most conventional flight conditions without stab misalignment, Eq. (32) has no effect on the shape of the trim envelope. However, the impact on the trim envelope is significant for large angle stabilizer misalignment, as can be seen in Fig. 10.

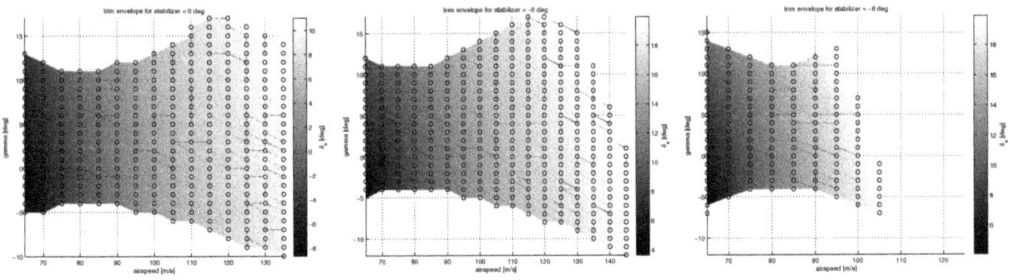

(a) $i_h = 0°$ (b) $i_h = -6°$ (c) $i_h = -8°$

Fig. 10 V,γ trim envelopes for different stabilizer misalignment scenarios for bank angle $\varphi = 0$ and sideslip angle $\beta = 0$. Elevator saturation limits are $-30°$ and $+20°$.

Fig. 10(a) shows that the required elevator deflections for a stabilizer stuck at 0° vary between $-8°$ and 15° over the entire trim envelope, which is well between the elevator saturation limits $-30°$ and $+20°$. However, Fig. 10(b) and 10(c) show a totally different situation for a stabilizer misalignment at 6° or 8° nose down respectively. The pitch up moment which is generated requires significant elevator deflection and they hit their maximum limit at $\delta_{e_{\max}} = 20°$ within the trim envelope calculated based on lift an drag. This additional input constraint results in the fact that the envelope is significantly smaller. For level flight $\gamma = 0°$, the conventional maximum airspeed of $V = 135\text{m/s}$ is reduced to $V = 100\text{m/s}$ for $i_h = -8°$ to prevent elevator saturation. This is relevant information, which increases the crew's situational awareness significantly in this type of scenario.

5 Conclusions and Recommendations

The methods presented in this paper have three distinguishing features. First, instead of linearizing the full aircraft dynamics model, a representative compact nonlinear model has been determined. This enabled rapid numerical computations, which are more representative of the aircraft performance than would be obtained through linearized models. Second, the integrated modular approach involving system identification – trim envelope calculation – maneuvering envelope calculation leverages the same dynamics model to estimate the safe flight envelope from currently available flight data as provided by current sensor packages on-board civil airliners, while also establishing confidence regions biased towards greater uncertainty when insufficient input excitation is available. Furthermore, the system identification process makes the global approach adaptive for (unintentional) configuration changes and damage, which has been shown in the application examples. Third, the safe maneuvering envelope maximizes the options pilots or automation systems have to recover the aircraft in damage scenarios, while remaining robust to the uncertainty in the system identification process. The net result is an increased flexibility for developing advanced aircraft diagnostics that provide the bottom line maneuverability of the aircraft as an output, and this is expected to have important applications to flight planning, trajectory generation, guidance algorithms, and pilot displays.

6 Further Research

The display features presented here, providing up to date safe flight envelope information to the flight crew, have been evaluated in the Advanced Concepts Flight Simulator at NASA Ames Research Center to investigate the impact on aircraft energy state awareness of the crew. Commercial airline crews have flown multiple challenging approach and landing scenarios in a relevant environment simulating the airspace around Memphis International Airport. More details and extensive results of this specific simulator study can be found in Ref. [11].

Acknowledgements. This work is supported by NASA's Aviation Safety Program, through the Vehicle Systems Safety Technologies (VSST) project, and by a Marie Curie International Outgoing Fellowship (IOF) within the 7th European Community Framework Program.

References

1. Statistical summary of commercial jet airplane accidents – worldwide operations 1959 - 2011. Technical report, Boeing Commercial Aircraft (July 2012), http://www.boeing.com/news/techissues/pdf/statsum.pdf
2. Cast safety enhancement plan. Technical Report SE207, Commercial Aviation Safety Team (2013), http://www.skybrary.aero/index.php/Portal:CAST_SE_Plan
3. Bacon, B.J.: Quaternion based control architecture for determining controllability/maneuverability limits. In: Guidance, Navigation, and Control and Co-located Conferences, AIAA 2012-5028. American Institute of Aeronautics and Astronautics (2012)
4. Nithin Govindarajan. An optimal control approach for estimating aircraft command margin – with applications to loss-of-control prevention. Masters thesis, Delft University of Technology (October 2012), http://repository.tudelft.nl/view/ir/uuid%3A48dbad61-728a-4c7e-ba3e-cf8382ef1cef/
5. Kitsios, I., Lygeros, J.: Launch pad abort flight envelope computation for a personnel launch vehicle using reachability. In: AIAA Guidance, Control and Navigation, AIAA 2005-6150 (August 2005)
6. Koolstra, H.J., Damveld, H.J., Mulder, J.A.: Envelope determination of damaged aircraft. In: Guidance, Navigation, and Control and Co-located Conferences, AIAA 2012-4699. American Institute of Aeronautics and Astronautics (August 2012)
7. Kwatny, H.G., Allen, R.C.: Safe set maneuverability of impaired aircraft. In: Guidance, Navigation, and Control and Co-located Conferences, AIAA 2012-4405. American Institute of Aeronautics and Astronautics (August 2012)
8. Lichter, M.D., Bateman, A.J., Balas, G.: Flight test evaluation of a rum-time stability margin estimation tool. In: AIAA Guidance, Navigation and Control Conference, AIAA 2009-6257 (August 2009)
9. Lombaerts, T.: Fault Tolerant Flight Control – A Physical Model Approach. Phd dissertation, Delft University of Technology (May 2010), http://repository.tudelft.nl/view/ir/uuid%3A538b0174-fe84-43af-954d-02f256b2ec50/
10. Lombaerts, T., Van Oort, E., Chu, Q., Mulder, J.A., Joosten, D.: On-line aerodynamic model structure selection and parameter estimation for fault tolerant control. Journal of Guidance, Control and Dynamics 33(3), 707–723 (2010)
11. Lombaerts, T., Schuet, S., Acosta, D., Kaneshige, J., Shish, K., Martin, L.: Piloted simulator evaluation of maneuvering envelope information for flight crew awareness. In: AIAA GNC Conference, AIAA-2015-1546 (January 2015)
12. Lombaerts, T., Schuet, S., Wheeler, K., Acosta, D., Kaneshige, J.: Safe maneuvering envelope estimation based on a physical approach. In: AIAA Guidance, Navigation and Control (GNC) Conference, AIAA-2013-4618 (August 2013)
13. Lombaerts, T.J.J., Chu, Q.P., Mulder, J.A., Joosten, D.A.: Real time damaged aircraft model identification for reconfiguring control. In: Proceedings of the AIAA AFM Conference and Exhibit, AIAA-2007-6717 (August 2007)
14. Looye, G., Bennani, S.: Description and analysis of the research civil aircraft model (rcam). Technical report TP-088-27, Group for Aeronautical Research and Technology in Europe GARTEUR (1997)
15. Lygeros, J.: On reachability and minimum cost optimal control. Automatica 40, 917–927 (2004)

16. Menon, P., Kim, J., Sengupta, P., Vaddi, V., Yang, B.-J., Kwan, J.: Onboard estimation of impaired aircraft performance envelope. In: Guidance, Navigation, and Control and Co-located Conferences, AIAA 2011-6688, American Institute of Aeronautics and Astronautics (August 2011)
17. Meuleau, N., Neukom, C., Plaunt, C., Smith, D., Smith, T.: The emergency landing planner experiment. In: ICAPS-11 Scheduling and Planning Applications Workshop (2011)
18. Mitchell, I.M.: The flexible, extensible and efficient toolbox of level set methods. Journal of Scientific Computing 35, 300–329 (2008)
19. Oort, E.V., Chu, P., Mulder, J.A.: Advances in Aerospace Guidance, Navigation and Control. In: Holzapfel, F., Theil, S. (eds.) Maneuvering Envelope Determination through Reachability Analysis, pp. 91–102. Springer, Heidelberg (2011)
20. Pandita, R., Chakraborty, A., Seiler, P., Balas, G.: Reachability and region of attraction analysis applied to gtm dynamic flight envelope assessment. In: AIAA Guidance, Navigation and Control Conference, AIAA 2009-6258 (August 2009)
21. Schuet, S., Lombaerts, T., Acosta, D., Wheeler, K., Kaneshige, J.: An adaptive nonlinear aircraft maneuvering envelope estimation approach for online applications. In: AIAA SciTech Guidance, Navigation and Control (GNC) Conference, AIAA 2014-0268 (January 2014)
22. Shin, J.-Y., Belcastro, C.: Robustness analysis and reliable flight regime estimation of an integrated resilient control system for a transport aircraft. In: AIAA Guidance, Navigation and Control Conference, AIAA 2008-6656 (August 2008)
23. Tang, L., Roemer, M., Bharadwaj, S., Belcastro, C.: An integrated health assessment and fault contingency management system for aircraft. In: AIAA Guidance, Navigation and Control Conference, AIAA 2008-6505 (August 2008)
24. Urnes, J.M., Reichenbach, E.Y., Smith, T.A.: Dynamic flight envelope assessment and prediction. In: AIAA Guidance Navigaton and Control Conference, AIAA 2008-6983 (August 2008)

A Sigma-Point Kalman Filter for Remote Sensing of Updrafts in Autonomous Soaring

Martin Stolle, Yoko Watanabe, and Carsten Döll

Abstract. Autonomous soaring is a promising approach to augment the endurance of small UAVs. Most of the existing work on this field relies on accelerometers and/or GPS receivers to sense thermals in the proximity of the vehicle. However, thermal updrafts are often visually indicated by cumulus clouds that are well characterized by their sharp baselines. This paper focuses on a cloud mapping algorithm which estimates the 3D position of cumulus clouds. Using the meteorological fact of a uniform cloud base altitude a state-constrained sigma-point Kalman filter (SC-SPKF) is developed. A method of using the resulting cloud map and its uncertainty in the path planning task to realize a soaring flight to a given wayoint is presented as a perspective of this work.

1 Introduction

Accelerated by a breakthrough in micro electromecanical systems (MEMS), small UAVs and the role they play in our society, be it military or civil, have grown in importance in the near past. However, their utility is still restricted due to small payload capacities as well as poor endurance and small operational ranges. One existing idea to overcome these still predominant drawbacks, is to apply flight control and guidance algorithms for soaring flight [1, 2, 3]. The soaring flight makes use of updrafts to lift the UAV and hence to reduce the transported mass dedicated to energy (battery or fuel). Moreover, soaring UAVs operate silently which clearly is a benefit for military purposes.

Martin Stolle · Yoko Watanabe · Carsten Döll
Department of Systems Control and Flight Dynamics (DCSD),
The French Aerospace Laboratory (ONERA),
2 avenue Edouard Belin, 31055 Toulouse Cedex 4, France
e-mail: {martin.stolle,yoko.watanabe,carsten.doell}@onera.fr

J. Bordeneuve-Guibé et al. (eds.), *Advances in Aerospace Guidance, Navigation and Control*, DOI: 10.1007/978-3-319-17518-8_17

Generally speaking, soaring flight combines all kind of techniques to keep an unpowered aircraft airborne. Dynamic soaring for instance is a technique where the vehicle harnesses energy from horizontal wind gradients. In thermal soaring energy is gained by relying on uprising currents of air. These buoyant plumes of rising air result from gradients in the earth's surface heating and can reach heights of up to 4000 m above ground according to [4]. In cross-country soaring, gliders fly beyond the gliding distance from the initial take-off point performing waypoint navigation.

Amongst existing approaches to automatic cross-country soaring, the work of Edwards et al. [5] is the only one which includes flight testing. His work lead to the participation in a cross-country soaring challenge for remotely piloted gliders and the performance of a fully autonomous soaring flight over a distance of 50 km. With no a priori information about thermal locations in the far environment, the flight path was defined as the direct line between two consecutive waypoints. The aircraft flight control mode was set to thermal centering mode, when encountering strong enough thermals on the path - detected by Inertial Measurement Unit (IMU) and GPS measurements. With this suboptimal flight path, the UAV could only benefit from a subset of possible updrafts - more precisely those that were directly located on the line of sight to a given waypoint. Evidently this approach is limited to conditions where a strong density of thermals is provided along the direct path and by consequence carries a significant risk of mission failure.

The author of [6] considered autonomous cross-country soaring from a top down approach and proposed path planning algorithms assuming that a perfect map with pinpoint thermal locations is at hand which raises doubts about its applicability beyond the synthetic case of computer simulations.

Human glider pilot mostly rely on their vision to locate thermal updrafts indicated by cumulus clouds. Doing so they can fly distances of up to 3000 km. Inspired by these performances, the paper on hand describes the development of an algorithm for remotely sensing thermal updrafts by locating cumulus clouds. An increased uncertainty of a thermal position estimate can significantly augment the time the UAV will spend on hitting the thermal and thus impacts the cross-country soaring performance. Therefore, the filter was designed to not only provide fast convergence but also a confident estimation of the thermal position uncertainty. Finally, a perspective is presented on how to take into account the uncertainty of estimated thermal positions in the cost of a cross-country path planning algorithm.

2 Cumulus Clouds and Thermal Updrafts

Consider a UAV flying in a sky that is partly covered by cumulus clouds. Depending on their stage, these clouds are the most important visual indicator for thermals that glider pilots rely on during thermal soaring.

2.1 Thermals and Their Visible Features

Vision-based object recognition algorithms detect objects in the real world from an image of the real world based on models. Since algorithmic description of this task still remains difficult, especially when dealing with objects such as clouds, varying in shape, color and texture, most simple and informative features are to be used in order to augment the recognition performance.

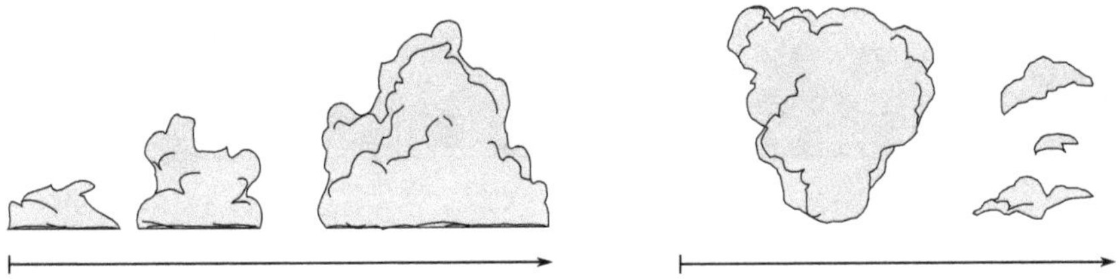

(a) Growth (b) Decay

Fig. 1. Growth and decay of a cumulus cloud

Clouds that are based on thermals, in general undergo a certain decay and rebirth process consisting of two different stages. As long as a thermal source on the ground feeds the cloud, it will continue growing and remains in the first building stage (fig. 1a). In case the thermal source vanishes, the cloud will start dying out (fig. 1b). The stages of a cumulus cloud are indicated by a variety of visible signs. For a growing cumulus, these features include sharp outlines as well as a dark and flat baseline. In contrast, the shape of a dying cloud is poorly defined and its baseline is rather frayed. In addition to these contrast and shape indicators, the color of a cumulus cloud varies as well during its cycle. While a growing cumulus cloud will tend to be gray or white, a dying cloud appears to be off color since its moisture particles evaporate which results in a change of its reflectivity.

Regarding these facts, the most simple-to-detect feature of a far away growing or mature thermal is the baseline of its related cumulus. In this paper, it is supposed that an image processing algorithm capable of extracting the baseline center of a cumulus cloud, as illustrated in fig. 2 is available. A simple but yet efficient algorithm for edge detection could therefore be used as presented by the author of [7].

Fig. 2. Baseline detection of a cumulus cloud

2.2 *Dynamics of Cumulus Clouds*

Solar radiation causes heating of the earth's surface. Variations in the heating of the ground result in rising parcels of air. On a day with typical soaring weather conditions, these parcels will first rise under dry adiabatic conditions i.e. without condensation. Once reaching the Cumulus Condensation Level (CCL) which is the point where the relative humidity of the parcel attains 100 %, water vapor starts condensing and cumulus cloud droplets are formed. As shown in [8] and illustrated in fig. 3, the cumulus cloud base altitude z_c is the line intersection of the Dry Adiabatic Lapse Rate (DALR) and the Dew Point (DP).

$$z_c = \frac{T_0 - T_{d,0}}{\frac{\partial T}{\partial z} - \frac{\partial T_d}{\partial z}} \approx 125(T_0 - T_{d,0})[\frac{m}{K}] \quad (1)$$

with T as the air temperature and T_d the dew point temperature. Note that the factor 125 corresponds to a temperature of $20°C$ and barely varies with the temperature. The index 0 stands for the temperature on ground. The index e (as used in fig. 3) is used for a local North East Down (NED) frame.

When flying over a region with constant geological and orographical structure, only tiny variations of the temperature difference between air and DP can be found [9]. Consequently, the cloud base is nearly uniform as depicted in fig. 3 which linearly constraints the altitude of each individual cloud base. The following section describes a cloud mapping algorithm that incorporates this constraint.

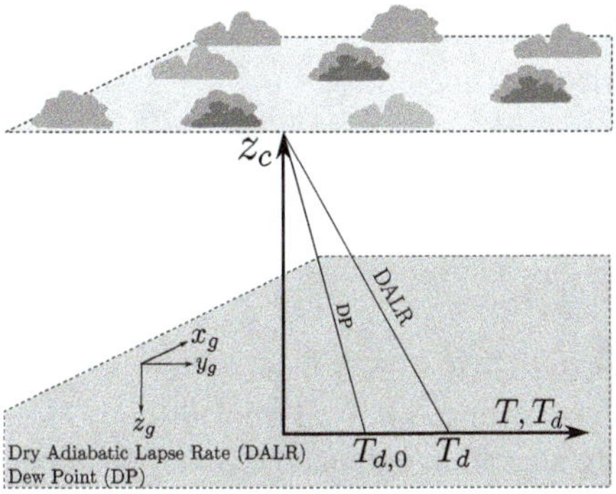

(a) Cloud base definition

(b) Cumulus clouds with uniform cloud base altitude

Fig. 3. Dynamics of cumulus clouds

3 Cloud Mapping Algorithm

Combining the UAV's state estimates with the output of the image processing algorithm, it becomes possible to estimate the 3D position of clouds in the inertial reference frame (index g). This problem is referred to as bearings-only target localization.

3.1 State Definition and Process Model

With the cloud map containing the individual positions of all n clouds that are encountered during a flight, the $3 \times n$ dimensional state vector $\mathbf{x}$ is defined as

$$\mathbf{x} = \left[\mathbf{x}_1^T \ \mathbf{x}_2^T \ \dots \ \mathbf{x_n}^T \ \right]^T \tag{2}$$

where $\mathbf{x}_i$ represents the cloud position of a single cloud in a local NED frame. The index g is not further carried for the sake of better readability. In general, the wind velocity has an effect on the drift of cumulus clouds, even if due to the inertia of the thermal air (note that the mass of a cumulus cloud can measure into the thousands of tons), cumulus clouds drift much slower than the surrounding air. Therefore, we consider a scenario with no cloud drift corresponding to weather conditions with only little or no horizontal wind. This assumption is legitimate, since algorithms will be tested on a small UAV glider whose flight envelope restricts operation to low wind conditions. In that case, the state transition equation for a single cloud $\mathbf{x}_i$ can be modeled as

$$\mathbf{x}_{i,k} = f(\mathbf{x}_{i,k-1}) + \mathbf{w}_{i,k-1} = \mathbf{x}_{i,k-1} + \mathbf{w}_{k-1} \tag{3}$$

where $\mathbf{w}$ is the white and Gaussian process noise with covariance $\mathbf{Q}$ i.e.

$$\mathbf{w} \sim \mathbf{w}(0, \mathbf{Q}) \tag{4}$$

3.2 Measurement Model

3.2.1 Pixel Coordinates of Cloud Baseline's Center

A forward looking camera is mounted on the UAV with a fixed offset from the vehicles center of gravity (CG) as well as a known angular offset from the body axis. At each time instance, the image processing algorithm outputs the center positions of the m cumulus cloud baselines in the image resulting in a $2 \times m$ dimensional measurement vector

$$\mathbf{y}_{ip} = \left[\mathbf{y}_{ip,1}^T \ \mathbf{y}_{ip,2}^T \ \cdots \ \mathbf{y}_{ip,m}^T\right]^T \tag{5}$$

Note that the index ip is used for vectors in the image frame in pixels and the index im denotes vectors in the image frame in meters as described in fig. 4. To project vectors from the camera frame (index c) onto the image plane, a pinhole camera model as shown in fig. 4 is used.

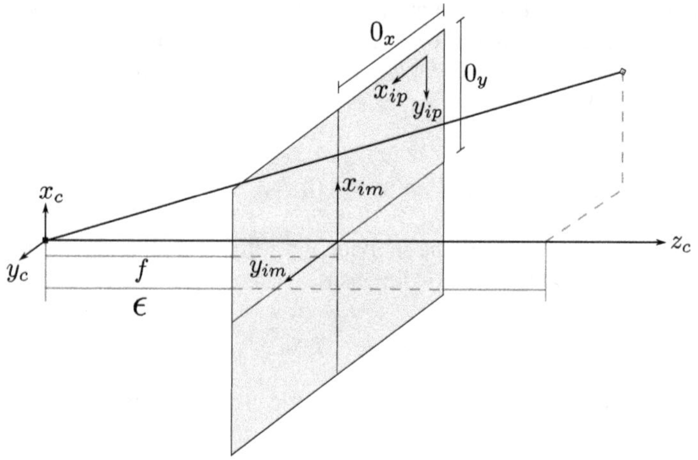

Fig. 4. Pinhole camera projection model

A transformation from the inertial to the image frame in pixels is given by

$$\mathbf{y}_{ip} = \frac{1}{\varepsilon} \mathbf{T}_{ipg} \mathbf{x} \tag{6}$$

where ε is the image depth. The transformation matrix $\mathbf{T}_{ipg}$ includes the translation from the target to the vehicle $\mathbf{T}_{vg}$, the rotation from the vehicle to the body frame $\mathbf{T}_{bv}$, the combined translation and rotation from the body to the camera frame $\mathbf{T}_{cb}$

as well as the combined projection and unit conversion (m to px) $\mathbf{C}$ from the camera frame onto the image plane (see fig. 4)

$$\mathbf{T}_{ipg} = \mathbf{C}\mathbf{T}_{cb}\mathbf{T}_{bv}\mathbf{T}_{vg} \tag{7}$$

with the camera calibration matrix $\mathbf{C}$ being defined as

$$\mathbf{C} = \begin{bmatrix} 0 & f_x & 0_x & 0 \\ -f_y & 0 & 0_y & 0 \end{bmatrix} \tag{8}$$

The two quantities f_x and f_y are function of the focal length f and the unit conversion factors S_x and S_y

$$f_x = \frac{f}{S_x} \text{ and } f_y = \frac{f}{S_y}$$

where the unit conversion is given by

$$S_x = \frac{y_{im}}{x_{ip} - 0_x} \text{ and } S_y = \frac{x_{im}}{-y_{ip} - 0_y}$$

Note that the parameters 0_x and 0_y are the offsets to the center of the image from the upper left hand corner.

Adding Gaussian white measurement noise $\mathbf{v}$, with zero mean and covariance $\mathbf{R}$, the discrete measurement equation is stated as

$$\mathbf{y}_k = \mathbf{h}(\mathbf{x}_k) + \mathbf{v}_k = \frac{1}{\varepsilon}\mathbf{T}_{ipg,k}\mathbf{x}_k + \mathbf{v}_k = \mathbf{H}\mathbf{x}_k + \mathbf{v}_k \tag{9}$$

Note that from here on the index of the measurement vector (ip) is not further carried for the sake of better readability.

3.2.2 Pseudo Measurement for the Altitude Constraint

Significant filter performance augmentations can be reached when including the dynamic relation between cloud base altitudes as a state constraint in the estimation process. Applied to the path planning, a faster convergence of position estimates and covariances will as well invoke a faster reduction of the uncertainty ellipses. Consequently, the cross- country speed of the UAV glider increases, since less time will be spent on encountering thermals.

However, this benefit comes with a price. The assumption of a uniform cloudbase is an approximation of the reality, and thus only a soft constraint where it is hard to detect constraints violations during estimation.

To comprise the uniform cloud base state constraint, the $2 \times m$ dimensional measurement vector eq. (9) is augmented with the pseudo measurement $\mathbf{d}$

$$\mathbf{y}_a = \begin{bmatrix} \mathbf{y} \\ \mathbf{d} \end{bmatrix} = \begin{bmatrix} \mathbf{H} \\ \mathbf{D} \end{bmatrix} \mathbf{x} + \begin{bmatrix} \mathbf{v} \\ \mathbf{v}_1 \end{bmatrix} \tag{10}$$

where $\mathbf{d}$ is a null vector of dimension n and $\mathbf{D}$ is the $n \times 3n$-dimensional constraint matrix with diagonal elements $\mathbf{D}_1 = \left[0\ 0\ (1 - \frac{1}{n})\right]$ and off-diagonal elements $\mathbf{D}_0 = \left[0\ 0\ \frac{-1}{n}\right]$ carrying the geometrical state restriction that the individual cloud bases z_i equals to the mean cloud base $\bar{z}$. $\mathbf{v}_1$ is the white Gaussian noise of the state constraint with covariance $\mathbf{R}_1$.

3.3 Estimation Algorithm

The bearings-only target localization is a highly nonlinear estimation problem. A variety of nonlinear filters has been proposed to solve this problem. What is common to nearly all of these methods, is the idea of providing a least squares estimate of the process's state. The standard approach for nonlinear estimation is the Extended Kalman Filter (EKF) that however comes with two significant drawbacks. Not only that the computation of the Jacobians is usually cumbersome, but if the linearization is poor, the estimated state covariance will tend to be inconsistent and in the worst case overconfident as discussed in [10]. Projected to the problem of autonomous cross-country soaring, this will erroneously tighten the error ellipsoid associated to the estimated position of a cloud and potentially results in a thermal search within an area of sinking air.

A common way to cope with this known weakness of the EKF is to artificially magnify the state covariance after each update or simply to drop certain observations. This is however an unfortunate and iterative procedure, since it discards information that is potentially useful.

A main challenge of the bearings-only target localization is caused by its lack of depth-observability. With the trajectory having a significant impact on the observability, there have been attempts [11] to design trajectories that optimize the target observability. However, in the case of autonomous cross-country soaring where the clouds are far away from the observing vehicle and the UAV aims to minimize energy consumption, favoring target observability in the trajectory design is inefficient. More recently, a group of algorithms [12, 13, 14, 15] has been published to address the issues of the EKF by using deterministic sampling approaches circumventing both laborious linearization and suboptimal performance due to poor linearizations. These algorithms referred to as sigma-point Kalman filters (SPKF) as well follow the prediction-correction procedure of the Kalman filter. But rather than linearizing the nonlinear system equations, they use the intuition that it is easier to approximate a probability distribution than it is to approximate an arbitrary nonlinear function or transformation. This is done by first propagating a weighted set of samples called sigma points $\boldsymbol{\mathcal{X}}$ through a nonlinear function. Then, the statistic properties of the propagated state are recaptured. The principle behind this probability distribution approximation is called Unscented Transform (UT) and was first presented in [12]. SPKF show a certain resemblance to Particle Filters (PF) in the way that the probability distribution is approximated by a set of points. However, they operate with a much lower number of points reducing computational effort which renders them more appropriate for real time implementation. This has led to the SPKF of [14]

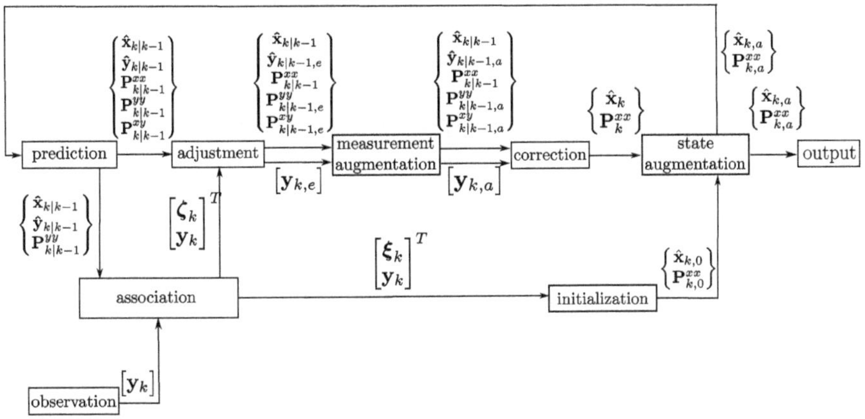

Fig. 5. Data flow during estimation

where the authors present a filter, capable of operating with a minimum set of sigma points that contains the most important information of the state's probability distribution.

A state-constrained version of this filter has been developed to estimate the position of clouds. The data flow during the estimation process is depicted in fig. 5. In the following, the individual tasks during estimation are discussed.

3.3.1 Prediction

Before conducting the actual prediction using the UT, a set of weight vectors $\boldsymbol{\eta}$ is computed. These vectors depend on both the scaling parameters α, β, κ and the state dimension $L = 3 \times n$.

$$\begin{aligned}
\lambda &= \alpha^2 + (L+\kappa) - L \\
\eta_0^m &= \lambda/(L+\lambda) \\
\eta_0^c &= \lambda/(L+\lambda) + 1 - \alpha^2 - \beta \\
\eta_i^m &= \eta_i^c = 1/[2(L+\lambda)], i = 1, ..., 2L
\end{aligned} \tag{11}$$

According to this notation, the indices m and c in eq. (11) stand for measurement and covariance respectively. With the state estimate prior to k, the sigma points follow as

$$\boldsymbol{\mathcal{X}}_{k-1} = \left[\hat{\mathbf{x}}_{k-1} \; \hat{\mathbf{x}}_{k-1} + \sqrt{L+\lambda}\sqrt{\mathbf{P}^{xx}_{k-1}} \; \hat{\mathbf{x}}_{k-1} - \sqrt{L+\lambda}\sqrt{\mathbf{P}^{xx}_{k-1}}\right] \tag{12}$$

Note that there are different approaches to compute the square root of a matrix. As suggested in [16], the lower Cholesky decomposition method is applied i.e. $\sqrt{\mathbf{P}} = chol(\mathbf{P})$. Each of the sigma points $\boldsymbol{\mathcal{X}}^{(i)}$ is then propagated through the state transition function eq. (3) yielding the propagated state

$$\mathcal{X}_{k|k-1}^{(i)} = \mathbf{f}(\mathcal{X}_{k-1}^{(i)}) \text{ for } i = 1, ..., 2L+1 \tag{13}$$

In this notation, the index $k|k-1$ stands for the state at time k incorporating knowledge prior to k and the parenthesized superscript stands for the index of the sigma-point. Also note that (eq. (13)) is only mentioned for the purpose of completeness, since the propagation does not impact the state as can be seen in eq. (3).

With the weight vector $\boldsymbol{\eta}_i^m$ the mean of the propagated state is

$$\hat{\mathbf{x}}_{k|k-1} = \sum_{i=1}^{2L+1} \eta_i^m \mathcal{X}_{k|k-1}^{(i)} \tag{14}$$

Given the process noise covariance $\mathbf{Q} = E[\mathbf{w}\mathbf{w}^T]$, the propagated state covariance matrix yields

$$\mathbf{P}_{k|k-1}^{xx} = \mathbf{Q} + \sum_{i=1}^{2L+1} \eta_i^c (\mathcal{X}_{k|k-1}^{(i)} - \hat{\mathbf{x}}_{k|k-1})(\mathcal{X}_{k|k-1}^{(i)} - \hat{\mathbf{x}}_{k|k-1})^T \tag{15}$$

Each of the sigma points is then processed through the nonlinear measurement equation, leading to a set of $2L+1$ predicted observations

$$\mathcal{Y}_{k|k-1}^{(i)} = \mathbf{h}(\mathcal{X}_{k|k-1}^{(i)}) \text{ for } i = 1, ..., 2L+1 \tag{16}$$

This yields the mean of the predicted measurement

$$\hat{\mathbf{y}}_{k|k-1} = \sum_{i=1}^{2L+1} \eta_i^m \mathcal{Y}_{k|k-1}^{(i)} \tag{17}$$

Summing the measurement covariance $\mathbf{R}$ and the covariance of the transformed state, the predicted measurement covariance is

$$\mathbf{P}_{k|k-1}^{yy} = \mathbf{R} + \sum_{i=1}^{2L+1} \eta_i^c (\mathcal{Y}_{k|k-1}^{(i)} - \hat{\mathbf{y}}_{k|k-1})(\mathcal{Y}_{k|k-1}^{(i)} - \hat{\mathbf{y}}_{k|k-1})^T \tag{18}$$

The prediction step is accomplished with the computation of the cross covariance matrix

$$\mathbf{P}_{k|k-1}^{xy} = \sum_{i=1}^{2L+1} \eta_i^c (\mathcal{X}_{k|k-1}^{(i)} - \hat{\mathbf{x}}_{k|k-1})(\mathcal{Y}_{k|k-1}^{(i)} - \hat{\mathbf{y}}_{k|k-1})^T \tag{19}$$

3.3.2 Data Association, Measurement Augmentation and Adjustment

Depending on meteorological conditions, the density of the thermals in an area can significantly vary [4]. Assuming that each thermal is visible through convection i.e. brings out a cumulus cloud, multiple clouds will simultaneously lie in the camera's field of vision. Therefore, precise matching between incoming measurements and already registered estimates is required to avoid filter divergence. Also, measurements from newly detected clouds have to be distinguished from those belonging to already initialized ones.

Data association

In this work, we apply a gated nearest neighbor approach based on the Mahalaboni distance. Where the underlying idea is to compute the probability that a predicted measurement corresponds to an incoming measurement. This technique has proven to work reliably [17, 18], provided that the uncertainty of the predicted measurements $\mathbf{P}^{yy}_{k|k-1}$ is sufficiently small.

At each time instance with incoming measurements for m detected clouds, the measurement vector is defined by eq. (9). A score r is defined and computed for the $m \times n$ combinations between predicted measurements and incoming measurements

$$r_k^{(ij)} = (\mathbf{y}_k^{(j)} - \mathcal{Y}_{k|k-1}^{(i)})\mathbf{P}_{k|k-1}^{yy,i}(\mathbf{y}_k^{(j)} - \mathcal{Y}_{k|k-1}^{(i)})^T \tag{20}$$

An estimate with index i is updated with a measurement with index j if their common score r_{ij} is the minimum score of all the scores belonging to the measurement j and is smaller than some fixed threshold known as gate g. This procedure leads to a $2 \times l$ dimensional vector $\boldsymbol{\zeta}_k = \left[\boldsymbol{\zeta}_{k,\mathbf{y}}\ \boldsymbol{\zeta}_{k,\hat{\mathbf{y}}}\right]$ containing the indices of the l associated pairs of estimates i and measurements j.

If not all measurements have been related to an initialized estimate, those measurements that have erroneously not been related and the ones that arise from a newly detected cloud have to be distinguished. Therefore, for all of the measurements that have not been associated, it is checked if they attain the minimum score to any of the n estimates. In case this statement is false, measurement j is considered as a newly detected cloud and used to initializue a new cloud state. Otherwise, it is rejected. The indices of measurements that are used to initialize new clouds are stored in a vector $\boldsymbol{\xi}_k$.

Adjustment

As illustrated in fig. 5, the predicted quantities $\hat{\mathbf{y}}_k, \mathbf{P}^{yy}_{k|k-1}, \mathbf{P}^{xy}_{k|k-1}$ and the measurement vector $\mathbf{y}_k$ are adjusted by selecting the relevant elements (index vector $\boldsymbol{\zeta}$) which have been related to a measurement. Where the index e stands for effect (see fig. 5).

$$\begin{aligned}
\mathbf{y}_{k,e} &= \hat{\mathbf{y}}_k(\boldsymbol{\zeta}_{k,\mathbf{y}}) \text{ with } \boldsymbol{\zeta}_{k,\mathbf{y}} = \left[j_1, \ldots, j_l\right]^T \text{ and } 0 \leq l \leq m \\
\hat{\mathbf{y}}_{k|k-1,e} &= \hat{\mathbf{y}}_{k|k-1}(\boldsymbol{\zeta}_{k,\hat{\mathbf{y}}}) \text{ with } \boldsymbol{\zeta}_{k,\hat{\mathbf{y}}} = \left[i_1, \ldots, i_l\right]^T \\
\mathbf{P}^{yy}_{k|k-1,e} &= \mathbf{P}^{yy}_{k|k-1}(\boldsymbol{\zeta}_{k,\hat{\mathbf{y}}}, \boldsymbol{\zeta}_{k,\hat{\mathbf{y}}}) \\
\mathcal{Y}_{k|k-1,e} &= \mathcal{Y}_{k|k-1}(\boldsymbol{\zeta}_{k,\mathbf{y}}, \boldsymbol{\alpha}) \text{ with } \boldsymbol{\alpha} = \left[1, \ldots\ n\right] \\
\mathbf{P}^{xy}_{k|k-1,e} &= \mathbf{P}^{xy}_{k|k-1}(\boldsymbol{\alpha}, \boldsymbol{\zeta}_{k,\hat{\mathbf{y}}})
\end{aligned} \tag{21}$$

Measurement augmentation

According to eq. (10), both the adjusted measurement $\mathbf{y}_{k,e}$ as well as the adjusted and predicted quantities $\hat{\mathbf{y}}_{k,e}, \mathcal{Y}_{k|k-1,e}$ vector are augmented using the state constraint on the uniform cloud base yielding

$$\mathbf{y}_{k,a} = \begin{bmatrix} \mathbf{y}_{k,e} \\ \mathbf{d}_k \end{bmatrix} \text{ and } \hat{\mathbf{y}}_{k,a} = \begin{bmatrix} \hat{\mathbf{y}}_{k|k-1,e} \\ \hat{\mathbf{d}}_k \end{bmatrix} \tag{22}$$

where $\hat{\mathbf{d}}$ containts the n variations from the individual cloud base z_i to the mean cloud base $\bar{z}$

$$d_i = z_i - z_c$$

Also, the measurement noise is augmented such that $\mathbf{R}_a = diag(\mathbf{R}, \mathbf{R}_1)$.

The prediction steps eqs. (17) to (19) are then recomputed for the augmented quantities yielding $\hat{\mathbf{y}}_{k,a}, \mathbf{P}^{yy}_{k|k-1,a}$ and $\mathbf{P}^{xy}_{k|k-1,a}$.

3.3.3 Correction

Using the adjusted predicted quantities as well as the adjusted measurement vector, the classical Kalman correction step is accomplished following

$$\begin{aligned} \mathbf{K}_k &= \mathbf{P}^{xx}_{k|k-1} (\mathbf{P}^{yy}_{k|k-1,a})^{-1} \\ \hat{\mathbf{x}}_k &= \hat{\mathbf{x}}_{k|k-1} + \mathbf{K}_k (\mathbf{y}_{k,a} - \hat{\mathbf{y}}_{k|k-1,a}) \\ \mathbf{P}^{xx}_k &= \mathbf{P}^{xx}_{k|k-1} - \mathbf{K}_k \mathbf{P}^{yy}_{k|k-1,a} \mathbf{K}^T_k \end{aligned} \tag{23}$$

3.3.4 Cloud Initialization and State Augmentation

Each time a new cloud is detected, both its initial estimate $\mathbf{x}$ its error covariance $\mathbf{P}^{xx}$ have to be computed from only one measurement. The state initialization causes potential difficulties, because the data association is prone to errors in the covariance of the predicted estimate.

Clouds are assumed to have approximately the same base altitude. Therefore, it is straightforward to compute the initial state estimate by calculating the plane-line intersection between the cloud base plane and the line-of-sight from the current vehicle position $\mathbf{p}$ along the bearing $\mathbf{b}$ to the new cloud, if some knowledge of the cloud base z_0 is given. For the very first cloud, an a-priori estimate of the cloud base z_0 is used. Subsequent clouds are initialized based on the actual estimated altitude. The initial cloud position ($\begin{bmatrix} x \ y \ z_0 \end{bmatrix}^T$) can be obtained by the function $\mathbf{s}$ with an input vector $\mathbf{m} = \begin{bmatrix} \mathbf{p} \ \mathbf{q} \ z_0 \end{bmatrix}^T$ as shown in eq. (24)

$$\mathbf{x} = \begin{bmatrix} x \\ y \end{bmatrix} = \mathbf{s}(\mathbf{m}) = \begin{bmatrix} p_x \\ p_y \end{bmatrix} + \mu \begin{bmatrix} b_x \\ b_y \end{bmatrix} \tag{24}$$

Where the bearing $\mathbf{b}$ and the magnitude μ are defined as:

$$\mathbf{b} = \mathbf{T}_{ce}^{-1} \begin{bmatrix} y_x \\ y_y \\ 1 \\ 1 \end{bmatrix} \qquad \text{and} \qquad \mu = \frac{z_0 - p_z}{b_z} \tag{25}$$

Unscented transformation proves again to be a convenient method to convert the measurement uncertainty $\mathbf{P}^m$ into an initial state covariance

$$\mathbf{P}^m = \begin{bmatrix} \sigma_{y_x}^2 & 0 & 0 \\ 0 & \sigma_{y_y}^2 & 0 \\ 0 & 0 & \sigma_{z_0}^2 \end{bmatrix} = \begin{bmatrix} \mathbf{R} & 0 \\ 0 & \sigma_{z_0}^2 \end{bmatrix} \tag{26}$$

where σ_{z_0} is the standard deviation of the a-priori knowledge on z_0. Defining the incoming measurement vector $\mathcal{M}_0$, the related $2L_i + 1$ sigma points result as

$$\mathcal{M} = \left[\mathcal{M}_0 \quad \mathcal{M}_0 + \sqrt{L+\lambda}\sqrt{\mathbf{P}^m} \quad \mathcal{M}_0 - \sqrt{L+\lambda}\sqrt{\mathbf{P}^m}\right] \tag{27}$$

Where the state dimension is $L_i = 3$ when dealing with a single cloud. Each of the $2L + 1$ sigma points is instantiated through the initialization function $\mathbf{s}(\mathbf{m})$ which yields the matrix $\mathcal{O}$ containing seven 3D positions of the cloud

$$\mathcal{O}_j = s(\mathcal{M}_j) \tag{28}$$

Both the initial state estimate $\mathbf{x}_0$ and the state covariance $\mathbf{P}_0^{xx}$ are then obtained

$$\mathbf{x} = \bar{\mathbf{o}} = \sum_{j=0}^{2L} \eta_i^c \mathbf{o}_j \qquad \mathbf{P}^{xx} = \sum_{j=0}^{2L} \eta_j^c (\mathcal{O}_j - \bar{\mathbf{o}})(\mathcal{O}_j - \bar{\mathbf{o}})^T \tag{29}$$

As illustrated in fig. 5, the corrected state and covariance estimate are augmented with the state and covariance of the initialized clouds.

4 Simulation Results

3DOF simulations were conducted for the following two purposes: First, to demonstrate that the SPKF is able to provide a convergent and confident estimation of cumulus cloud positions that can reliably be used for path planning algorithms. Second, that the filter formulation with a soft state constraint based on the assumption of a uniform cloud base leads to faster and still confident convergence of both state and covariance estimation - even in case of strongly varying cumulus cloud bases.

4.1 Simulation Scenario and Settings

A forward looking camera was moved along a circular and climbing trajectory (as shown in fig. 6) for an observation duration of 300 s to simulate the estimation pro-

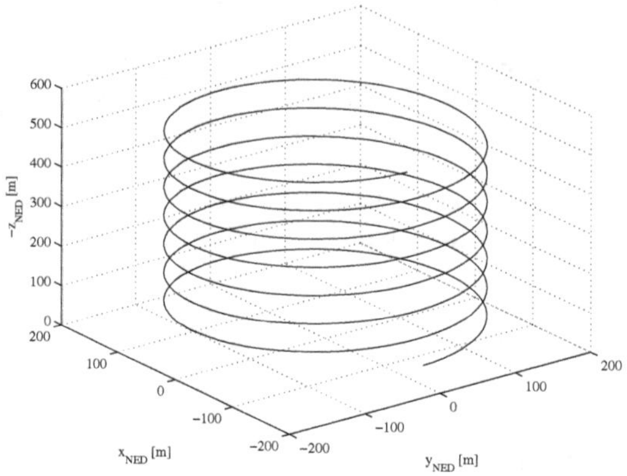

Fig. 6. Camera trajectory

cess during a standard thermaling flight where the clouds repeatedly appear and disappear on the image sensor due to the circular trajectory.

Cumulus clouds were located around the center of the trajectory for two scenarios as depicted in fig. 7.Note that the camera's trajectory is the circle around the origin. It appears small due to the small scale of the map.

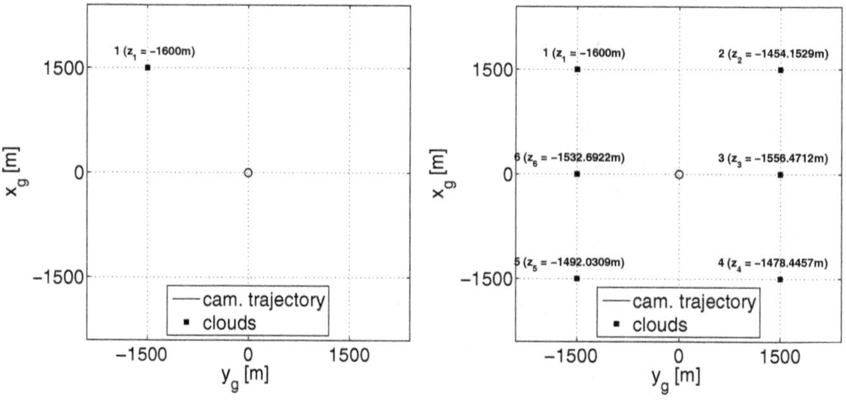

(a) First scenario (b) Second scenario

Fig. 7. Cloud position estimation scenarios

The first scenario (fig. 7a) was chosen to demonstrate the performance of an unconstrained filter formulation by observing only one cloud which suppresses the last n rows of eq. (10). In the second scenario (fig. 7b), six clouds were positioned around the center of the trajectory. In order to obtain realistic estimation results, the six cloud bases were normally distributed around a mean $\bar{z}_c = -1500\,m$ with a standard deviation of $\sigma_{z_c} = 25\,\mathrm{m}$. Subsequently, the cloud base altitude of cloud number one was biased by $100\,m$ to simulate strong variations. Note that cloud number one has precisely the same horizontal position as the cloud in the first scenario.

The parameters of the camera model eq. (8) $\mathbf{C}$, i.e. the focal length, field of view and image size were selected to represent the performance of a small low-cost camera. A constant and unbiased measurement noise $\mathbf{v}$ with a standard deviation of 5 % of the image size was used to simulate uncertainties evoked by the image detection algorithm. Where the primary scaling parameter α which determines the spread of the sigma points was set to 1, the secondary scaling parameter β was set to 2 which according to [16] is the optimum value for Gaussian distributions. The tertiary parameter κ was set to a commonly-used value of 0. The process noise covariance $\mathbf{Q}$ was chosen to be $0.01\,(m/s)^2$. The initial cloud base z_0 can for example be obtained as a cloud ceiling provided in METAR/TAF information. In this simulation it was set to a value of $z_0 = -1200\,\mathrm{m}$ with a significant uncertainty $\sigma_{z_0} = 600\,\mathrm{m}$. The gate g in the data association was set to 9.

As discussed in section 2, the soft state constraint enables faster convergence of cloud position estimates. This is because, even if a cloud is out of the field of view, its position estimate can be corrected based on the state constraint measurement (last n rows in eq. (10)). However, the position estimation will be biased and/or the error interval will be predicted too tight, if the assumption of a uniform cloud base does not hold for a particular cloud, i.e. the state constraint is biased. A too small covariance $\mathbf{R}_1$ will reduce the slackness of the constraint and cause a fast overconfidence during estimation. Therfore, the parameter $\mathbf{R}_1$ has to be selected carefully depending on cloud base variations that can be encountered in the real world. That being said, $\mathbf{R}_1$ was tuned with the second scenario such that the filter ensures estimation confidence for clouds with an altitude variation of up to 100 m. This value roughly corresponds to the maximum the main author has observed during various cross-country soaring flights and has been confirmed by a meteorologist. The procedure lead to a parameter value of $\mathbf{R}_1 = 750000\,m^2$.

4.2 Estimation Performance

The position estimate of the first cloud is considered to compare the filter's performance in terms of confidence and convergence for the unconstrained (noted *unc.* in fig. 8) and *c.* for the constrained formulation as depicted in fig. 8. Recall that this corresponds to the estimation performances of the two defined scenarios. While the dark envelope and the dashed line denote the 3σ envelope and the position error for the unconstrained case, the light envelope and the solid line correspond to the 3σ envelope and the estimation error when considering the state constraint.

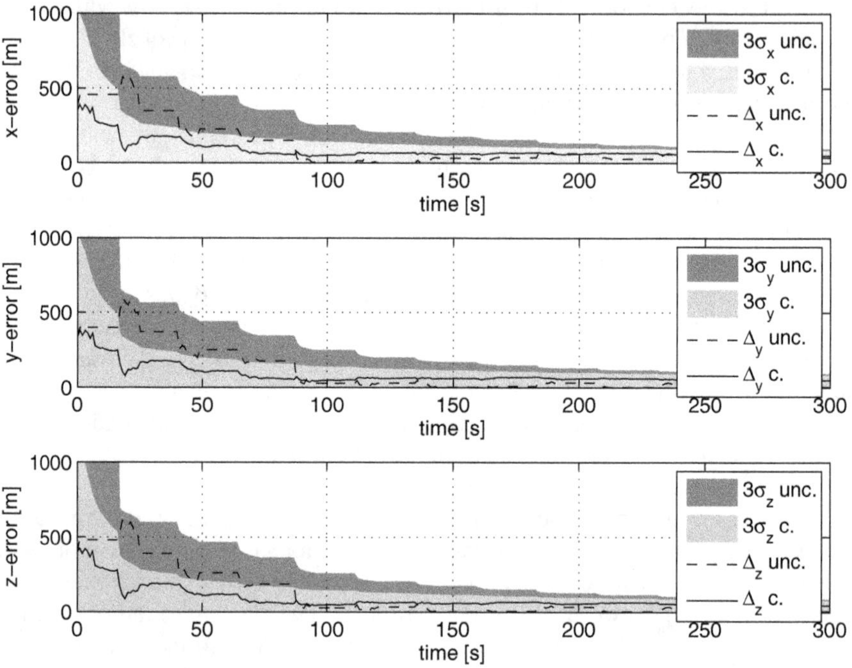

Fig. 8. Estimation performance

As expected, both regarding rate of convergence and the error, the SCSPKF outperforms the standard SPKF. In all three cases, the upper bound of the position error is reliably predicted. The huge uncertainty of the initial cloud base impacts the filter's transient behavior which can be seen in terms of estimation overshoots in the beginning of the estimation process. Also, the settling time for the unconstrained filter process is extended since each cloud is visible only for approximately 35 % of the estimation duration. Periods with no measurements can be seen at the long horizontal segments within the graph. In contrast, for the constrained formulation a significant reduction of the state's settling time is obtained due to the measurement augmentation.

Estimation degradation is expected whenever the vertical position of a cloud strongly varies from the mean and its position update is only performed using augmented measurements. This is particularly the case for clouds that lie behind the camera's field of view when flying towards the next waypoint in cross-country soaring. However, this degradation is not predominant, since clouds that lie behind the vehicle have no impact on the future path.

5 Perspectives

This paper presented a method providing functionality to the remote sensing of thermal updrafts. The information of both state and covariance estimation should be taken into account in path planning algorithms to enable more efficient autonomous cross-country soaring.

In general, the problem of autonomous cross-country soaring can be stated as a waypoint navigation (from A to B) as illustrated in fig. 9.

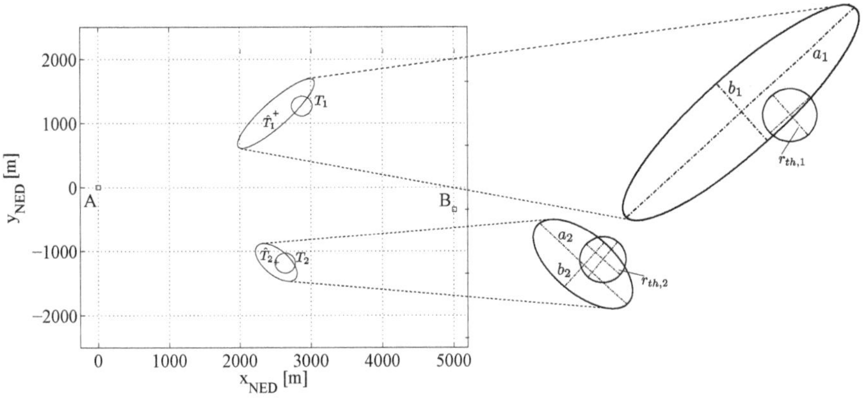

Fig. 9. Path planning problem

In this example, an unpowered UAV glider has to fly from waypoint A to waypoint B given position estimates for the two thermal updrafts T_1 and T_2, where the true thermal centers are supposed to lie somewhere within the 2D error ellipses of the estimates. The mission starts at waypoint A where the UAV is scanning the sky for thermals while climbing in a thermal before planning the path to the next intermediate or global target B. The ability of the glider to perform this mission in minimum time depends on three factors. Firstly, the vehicle's performance in terms of its glide ratio i.e. its capacity to transform potential energy into travelled distance. Secondly, the flight control's performance to center around a given thermal. Thirdly and most importantly meteorological conditions and the pilot's capacity to read them i.e. to locate far away thermals in order to plan the most efficient path.

If the direct path from A to B is not feasible due to the vehicle's limited glide ratio, it has to fly a detour via one of the two thermals to regain altitude. The total flight time for the two path options is given by

$$
\begin{aligned}
(A \rightarrow T_1 \rightarrow B): \quad & t_{AB} = t_{AT_1} + \underbrace{t_{en,T_1} + t_{th,T_1}}_{\text{time spent at } T_1} + t_{T_1B} \\
(A \rightarrow T_2 \rightarrow B): \quad & t_{AB} = t_{AT_2} + \underbrace{t_{en,T_2} + t_{th,T_2}}_{\text{time spent at } T_2} + t_{T_2B}
\end{aligned}
\tag{30}
$$

where the encounter time t_{en} is the time to hit the thermal while searching within the error ellipse, and the thermal time t_{th} is the time spent in the thermal updraft during climb. The latter depends on the initial altitude at which the vehicle enters the thermal and the strength of the updraft as well as the cloud base. The vehicle is supposed to leave the thermal once the cloud base is reached. Assuming equal thermal strength, the two routes seem to be on par regarding the flight time. However, the larger position uncertainty of T_1 might require more time to encounter the thermal using some search pattern whose size is determined by the error ellipse. That being said, the uncertainty of the thermal position impacts the flight time. According to Allen's research on modeling thermal updrafts for autonomous UAV soaring [4], the thermal radius r_{th} as illustrated in fig. 9 can be predicted when knowing z_c and the altitude z at which the vehicle reaches the thermal

$$r_{th} = 0.5\left[0.203\left(\frac{z}{z_c}\right)^{\frac{1}{3}}\left(1 - 0.25\frac{z}{z_c}\right)z_c\right] \tag{31}$$

Whenever the thermal radius r_{th} is larger than the half of the semi minor b belonging to the 2D error ellipse (see $\hat{T}_1$ in fig. 9), the maximum time to encounter the thermal can be predicted by the speed V (which is considered to be constant during operation) and the semi major a

$$t_{en} = \frac{a}{V} \tag{32}$$

Otherwise, a systematic search pattern has to be flown within the error ellipse. Regardless of the pattern's shape, the maximum time to encounter the thermal is

$$t_{en} = \frac{l_p}{V} \text{ where } l_p = l_p(\mathbf{P}_T, p) \tag{33}$$

where the pattern length l_p depends on the uncertainty $\mathbf{P}_T$ as well as on the shape of the pattern p.

These upper bounds on t_e render it possible to incorporate the uncertainty of thermal position estimates into the cost function thus reducing the total flight time.

Future work will concentrate on two fields. First, the design of path planning algorithms for autonomous cross-country soaring including crucial meteorological aspects as thermal updrafts and wind. Second, the design of image processing algorithms capable of deducing information about thermals given images of clouds. This includes for instance thermal strength prediction based on the color and contrast of the related cumulus as well as cloud size and shape. If those visible features can be detected, and thermal strength can fairly be predicted, even more efficient path planning becomes possible by taking into account this additional information.

6 Conclusion

In this paper, a SCSPKF was developed for remotely sensing thermal updrafts indicated by cumulus clouds in autonomous soaring. Two design efforts were focused on:

- Including the state constraint of a uniform cumulus cloud base for faster convergence
- Maintaining estimation consistency and providing a confident estimate of the uncertainty

Simulation results clearly demonstrate the benefits of the constrained filter formulation in terms of convergence rate. The filter still provides consistent estimation for strong model deviations with biases in the cumulus cloud base of up to 100 m.

Finally, a perspective for a new path planning approach for autonomous cross-country soaring was presented considering the uncertainty of thermal position estimates to augment the efficiency of future UAV operations.

References

[1] Allen, M.J., Lin, V.: Guidance and Control of an Autonomous Soaring UAV with Flight TTest Results. In: 45th AIAA Aerospace Sciences and Meeting and Exhibit (January 2007)
[2] Andersson, K., Kaminer, I., Dobrokhodov, V., Cichella, V.: Thermal Centering Control for Autonomous Soaring; Stability Analysis and Flight Test Results. Journal of Guidance Navigation and Control 35, 963–975 (2012)
[3] Akhtar, N., Whidborne, J.F., Cooke, A.K.: Real-time trajectory generation technique for dynamic soaring UAVs. In: Proceedings of the UKACC International Conference on Control (2008)
[4] Allen, M.J.: Updraft model for development of autonomous soaring uninhabited air vehicles. In: 44th AIAA Aerospace Sciences Meeting and Exhibit, American Institute for Aeronautics and Astronautics, AIAA (January 2006)
[5] Edwards, D.J., Silberberg, L.M.: Autonomous Soaring: The Montague Cross-Country Challenge. AIAA Journal of Aircraft 47, 1763–1769 (2010)
[6] Kahveci, N.: Robust Adaptive Control For Unmanned Aerial Vehicles. PhD thesis, University of Southern California - Faculty of the Graduate School, USA (2007)
[7] Canny, J.: A Computational Approach to Edge Detection. IEEE Transactions on Pattern Analysis and Machine Intelligence PAMI-8, 679–698 (1986)
[8] Kleinschmidt, E.: Handbuch der Meteorologischen Instrumente und ihrer Auswertung. Verlag von Julius Springer (1935)
[9] Pagen, D.: Understanding the sky. Dennis Pagen (February 1992)
[10] Yaakov, T.K., Bar-Shalom, X., Li, R.: Estimation with Applications To Tracking and Navigation. John Wiley and Sons, Inc. (2001)

[11] Ponda, S.S.: Trajectory Optimization for Target Localization Using Small Unmanned Aerial Vehicles. Master's thesis, Massachusetts Institute of Technology (September 2008)
[12] Julier, S.J.: A skewed approach to filtering. In: SPIE Conference on Signal and Data Processing of Small Targets, vol. 3373, pp. 271–282. SPIE (1998)
[13] Julier, S.J., Uhlman, J.K.: A New Extension of the Kalman Filter to Nonlinear systems. In: Proc. SPIE 3068, Signal Processing, Sensor Fusion, and Target Recognition VI (April 1997)
[14] Van der Merwe, R.: The square-root unscented Kalman filter for state and parameter estimation. In: 2001 IEEE Acoustics, Speech, and Signal Processing, Proceedings (ICASSP 2001), vol. 6, pp. 3461–3464 (May 2001)
[15] Van der Merwe, R., Wan, E.A.: The unscented Kalman filter for nonlinear estimation. In: The IEEE 2000 Adaptive Systems for Signal Processing, Communications, and Control Symposium 2000. AS-SPCC, pp. 153–185. IEEE (October 2000)
[16] Rhudy, M., Gu, Y.: Understanding Nonlinear Kalman Filters, Fart II: An Implementation Guide. Interactive Robotics Letters (2013)
[17] Langelaan, J.W.: State estimation for autonomous flight in cluttered environments. PhD thesis, Stanford University (March 2006)
[18] Montemerlo, M., Thrun, S.: FastSLAM: A Scalable Method For The Simulations Localization and Mapping Problem in Robotics. Springer (2007)

Multiple-Model Adaptive Estimation of Time-Varying Residual Magnetic Moment for Small Satellites

Halil Ersin Soken and Shin-ichiro Sakai

Abstract. As the satellite size gets smaller, the residual magnetic moment (RMM) becomes the dominant attitude disturbance for the low Earth orbit satellites. Especially for advanced space missions such as astronomical observation, the RMM must be in-orbit estimated and compensated to increase the attitude pointing accuracy. Classical estimators can estimate the RMM terms accurately as long as the terms are constant. However, if there is unmodeled changes in the RMM parameters, as experienced for small satellite missions, then the estimations may deteriorate for a long time until the estimator catch the new values. In such cases the designer must sacrifice either the accuracy or the tracking capability of the estimator. In this paper, we propose a Multiple-Model Adaptive Estimation (MMAE) technique for the RMM estimation. By the use of a newly defined likelihood function both the steady state accuracy and tracking agility are secured for the estimator.

1 Introduction

The main attitude disturbance sources for Earth orbiting satellites are usually categorized in four as the gravity gradient, sun pressure, aerodynamic drag and residual magnetic moment (RMM). Specifically for the low Earth orbit (LEO) small satellites, the RMM becomes dominant amongst these disturbance sources and effects of the rest decrease as the satellite size gets smaller [1, 2]. However, the magnetic disturbance is mainly caused by the onboard electric current loop, small permanent magnet in some devices or some special material on the satellite, and does not strongly depend on the satellite size [3]. Moreover, other disturbance sources can

---

Halil Ersin Soken · Shin-ichiro Sakai
Japan Aerospace Exploration Agency, Institute of Space and Astronautical Science, Yoshinodai 3-1-1 Sagamihara, Kanagawa, Japan
e-mail: ersin_soken@ac.jaxa.jp, sakai@isas.jaxa.jp

J. Bordeneuve-Guibé et al. (eds.), *Advances in Aerospace Guidance, Navigation and Control*, DOI: 10.1007/978-3-319-17518-8_18

be modeled accurately with onground simulations and further minimized during the design process. On contrary, pre-launch testing for onboard RMM characteristics is not easy because of the difficulties for providing a magnetically clean environment and more importantly, these characteristics differ after the launch [4].

The effects of the RMM on the attitude determination and control accuracy and the necessities for the RMM compensation are well discussed in [1-3]. In an early study for the RMM estimation Steyn and Hashida use the extended Kalman filter (EKF) for estimation and show the clear improvement in the attitude stabilization when the disturbance magnetic moment is cancelled [5]. In [1] an observer is designed for the RMM estimation and then the disturbance effect is cancelled out using a feed-forward technique. It is stated that the feed-forward cancellation is preferred over the feedback controller because of being more efficient. Inamori *et al.* propose a Kalman filter (KF) based method and compare the RMM estimation accuracy of the EKF and unscented Kalman filter (UKF) [2]. In common, these studies prove that the RMM must be estimated and compensated for increasing the attitude pointing accuracy of small satellites. However, they mostly assume the RMM components are constant in time. Such assumption is not correct in practice and the RMM parameters may change with sudden shifts because of the instantaneous variations in the onboard electrical current. Instantaneous variations in the current may be caused by switching on/off of the onboard electronic devices or going into/out of eclipse. Among the mentioned studies only Hosonuma *et al.* refer to the possibility of the sudden changes in the RMM terms but the authors state these changes are estimated with an off-line method and they do not give a clear description for the estimation process [3]. Should there be sudden changes in the estimated parameters, it becomes difficult to design the estimator and the estimator cannot catch the new values quickly if it is designed for high steady state accuracy. In general this situation brings about a dilemma for the designer; either the steady state accuracy or the tracking agility of the filter should be sacrificed unless an adaptive approach is used.

Estimation of constant parameters with sudden changes (or piecewise constant parameters) has been addressed in many researches in various different disciplines. The most widespread approach is to use an adaptive Kalman filter and tune the filter's covariance matrices either intuitively or by an optimization method. The adaptation might be performed during the whole estimation process [6], but the widely accepted technique is to integrate the adaptation procedure with a change detector and adapt the filter only when the change is detected [7-9]. These methods that emphasize on adapting the filter covariances use the joint estimation procedure. In other words, they treat the estimated parameters together with the states of the dynamical system. There are also dual estimation methods that decompose the estimation problem into two parts as state and parameter estimations. In this concept both the states and parameters can be estimated by parallel running two KFs [10]. Moreover, the parameters can be estimated using an optimization algorithm such as the maximum likelihood (ML) while the states are still the estimates of a KF [11].

There are many difficulties of using optimization methods for parameter estimation. First of all, especially if the parameters are directly estimated as in dual estimation, the initial values for the algorithm should be carefully selected to obtain the global optimal solution. Otherwise the algorithm may converge to local minimums and that will lead the overall estimation algorithm to deterioration. Besides, the computational load of the optimization algorithm may be too high for on-line estimation – e.g. if the ML algorithm in [11] is run for a long window, *N*. In such cases the algorithm can be only used as a batch estimator. On contrary, if the optimization performed for a short window then the window size should be carefully selected since it affects the accuracy and agility of the parameter estimation. As discussed in [12, § 2.7], short window size for the optimization algorithm makes the estimator more agile but decreases the steady state accuracy and vice versa. Nevertheless, the methods that rely on adapting the KF, particularly after the change is detected with a statistical method, have their own disadvantages. Change detection is a difficult task always with a chance of false detection or undetected changes and the KF adaptation is very sensitive to the parameters used for tuning the covariances [13].

In previous researches we proposed an adaptive UKF algorithm that tunes the filter's covariance matrix regarding the magnitude of the change once the changes are detected by a low pass filter [14, 15]. The algorithm gives considerably good results for the RMM estimation. However, the parameters for the change detector and adaptation are yet to be selected carefully. That encouraged us to search for a more autonomous adaptive algorithm, in particular without any change detection scheme. In this study, we propose a Multiple-Model Adaptive Estimation (MMAE) technique for the RMM estimation. Several EKFs run under different models with different process noise covariance matrices and the estimates of each filter are integrated by the use of a likelihood function. Initial results were presented in [16]. In this paper, we propose an improved likelihood function that significantly increases the steady state accuracy. Newly defined likelihood function is composed of two measures, one for weighting during the steady state and the other for weighing in case of change in the estimated RMM terms. In this sense, the function is different than the classical Bayesian likelihood function [17, 18] used for the MMAE. Rather than assuming only one model is true, each EKF contributes to the estimation depending on the assigned weights. Performance of the algorithm is investigated for different cases and number of models.

The rest of the paper is organized as follow: In section two, firstly the dynamics model of the satellite, which will be used as the model for the RMM estimation, is given. Subsequently, the RMM estimation procedure is presented and the basic Kalman filter algorithm is briefly discussed. The third section introduces the novel MMAE algorithm for estimating the RMM should there be sudden changes. Likelihood function design is discussed in detail in this section. In the final section, the proposed method is tested for different cases and the results are discussed by comparing with the estimates of each EKF contributing to the MMAE algorithm.

2 The RMM Estimation

The RMM estimation model is based on the satellite dynamics. Because of inherent nonlinearity in the dynamics model, the filter that is used for the RMM estimation must be a nonlinear version of the KF, such as the EKF or the UKF. In [2, 14] it is shown both of them can be used with this purpose. We preferred using the EKF in this study mainly because of two reasons: 1) Nonlinearity in the dynamics is low and the UKF becomes more advantageous only in case of high nonlinearity in the system equations; 2) Computational load of the EKF is a bit lesser than the UKF, so it is more suitable if multiple filters run under different models.

In this section, we first present the satellite dynamics model. Then, the KF algorithm for the RMM estimation is given with a brief reasoning on why we chose the EKF as a part of our algorithm.

2.1 Satellite Dynamics Model

The dynamic equations of the satellite can be easily derived using the Euler's equations describing the rigid body motion [19];

$$\frac{d\boldsymbol{\omega}_{BI}}{dt} = J^{-1}\left[\boldsymbol{M}\times\boldsymbol{B} - \boldsymbol{\omega}_{BI}\times\left(J\boldsymbol{\omega}_{BI}\right)\right]. \tag{1}$$

Here, J is the inertia matrix formed of the principal moments of inertia as $J = diag\left(J_x, J_y, J_z\right)$, $\boldsymbol{\omega}_{BI}$ is the body angular rate vector with respect to the inertial axis frame that can be represented as $\boldsymbol{\omega}_{BI} = \left[\omega_x \ \ \omega_y \ \ \omega_z\right]^T$, $\boldsymbol{M} = \left[M_x \ \ M_y \ \ M_z\right]^T$ is the RMM vector and $\boldsymbol{B}$ is the Earth's magnetic field vector measured in the body frame. Note that the magnetometers that are carried onboard directly supply the $\boldsymbol{B}$ information, and we assume that they are calibrated using one of the existing techniques [20, 21]. Besides the body angular rates with respect to the inertial frame, $\boldsymbol{\omega}_{BI}$, are measured using the onboard gyros, which are also in-flight calibrated with a KF algorithm other than the one used for the RMM estimation [22].

In this study, the RMM terms are modeled as constant but with sudden changes as discussed in the introduction. Assuming the RMM as piecewise constant is valid in general since the high frequency time-varying components of the RMM are negligibly small compared to the constant components and magnitude of the sudden changes. Moreover, the middle frequency time-variation in the RMM may be suppressed in the design process of the satellite as described in [3]. Hence, the hypothesis for the RMM model is

$$\boldsymbol{M}(t) = \begin{cases} \boldsymbol{M}_0 & t_0 \le t < t_1 \\ \boldsymbol{M}_1 & t_1 \le t < t_2 \\ \vdots & \vdots \\ \boldsymbol{M}_\alpha & t_\alpha \le t < t_{orb} \end{cases}; \tag{2}$$

where

$$\dot{\boldsymbol{M}}_j = 0. \tag{3}$$

Here, t_j are the unknown time instances that a change occurs within one orbit period t_{orb}, $\boldsymbol{M}_j$ are constant RMM vectors after the changes at t_j and $\Delta M_j = \left\|\boldsymbol{M}_j - \boldsymbol{M}_{j-1}\right\|$ are the magnitude of the changes at t_j for $j = 1...\alpha$. $\boldsymbol{M}_0$ is assumed to be the initial value of the RMM.

2.2 Kalman Filter for the RMM Estimation

In this study, we use a joint estimation procedure so the RMM parameters are estimated together with the states of the dynamic system. The estimated state vector is composed of the body angular rates with respect to the inertial frame and RMM terms as given with

$$\boldsymbol{x} = \begin{bmatrix} \boldsymbol{\omega}_{BI} \\ \boldsymbol{M} \end{bmatrix} . \tag{4}$$

The filtering system model is given in discrete-time by;

$$\boldsymbol{x}_{k+1} = f(\boldsymbol{x}_k, k) + \boldsymbol{w}_k , \tag{5a}$$

$$\tilde{\boldsymbol{y}}_k = h(\boldsymbol{x}_k, k) + \boldsymbol{v}_k . \tag{5b}$$

Here, $\boldsymbol{x}_k$ is the state vector and $\tilde{\boldsymbol{y}}_k$ is the measurement vector; $f(\cdot)$ is the non linear process function and $h(\cdot)$ is the nonlinear measurement function. Moreover, $\boldsymbol{w}_k$ and $\boldsymbol{v}_k$ are the process and measurement error noises, which are assumed to be Gaussian white noise processes with the covariances of Q_k and R_k respectively. The nonlinear process model is obtained by discrete-time integration of (1, 3). Nevertheless, since the onboard gyros directly supply $\boldsymbol{\omega}_{BI}$ information, the measurement model is represented with a linear equation as

$$\tilde{\boldsymbol{y}}_k = \begin{bmatrix} I_{3\times3} & 0_{3\times3} \end{bmatrix} \boldsymbol{x}_k + \boldsymbol{v}_k , \tag{6}$$

where $I_{3\times3}$ and $0_{3\times3}$ are 3×3 identity and null matrices, respectively.

The EKF and UKF are two well known nonlinear versions of the KF that can be used for the RMM estimation. As it is extensively discussed in the literature the UKF's estimation accuracy is better than the EKF principally because of having no need for any linearization process and using more accurate approximation for solving the multidimensional integrals [23]. Yet we believe these two filters should be compared on application basis since there are many factors affecting the filters' performance such as the complexity of the problem. We tested both the Kalman filtering methods for the RMM estimation before deciding which one to use as a part of the MMAE procedure.

The simulations are done for a hypothetical small satellite. The inertia matrix of the satellite model is;

$$J = \begin{bmatrix} 310kg.m^2 & 0 & 0 \\ 0 & 180kg.m^2 & 0 \\ 0 & 0 & 180kg.m^2 \end{bmatrix}. \tag{7}$$

For the magnetometer measurements, which provide $\boldsymbol{B}$ information for dynamic modeling, the sensor noise is characterized by zero mean Gaussian white noise with a standard deviation of $\sigma_m = 300nT$; whereas, the gyro random error is assumed as $\sigma_v = 1\times10^{-3}[\deg/\sqrt{h}]$. In this scenario, where the RMM terms are constant and there is no abrupt change in time, the real RMM terms are $M = (0.1, \quad 0.02, \quad -0.05)\,Am^2$. The process noise covariance matrix of the filters is selected as

$$Q = \begin{bmatrix} (1E-20)\,I_{3x3} & 0_{3x3} \\ 0_{3x3} & (1E-10)I_{3x3} \end{bmatrix}. \tag{8}$$

Simulations are performed for 20000 sec. In Fig.1 the estimation result for the RMM in the x axis is presented for the initial 500 sec. of the simulations. As seen both filters can accurately estimate the RMM parameters. Yet the UKF converges more quickly and has slightly better estimation accuracy. We can more clearly understand this fact from the Root Mean Squared Error (RMSE) for the RMM terms of the state vector (x_k^j such that $j = 4...6$). The RMSE is calculated in between the 1000th and 1500th seconds (for 5000 samples since $\Delta t = 0.1\sec$ - see eq.9) and tabulated in Table 1.

$$RMSE^j = \sqrt{\frac{1}{5000}\sum_{k=10001}^{15000}\left[x_k^j - \hat{x}_k^j\right]^2} \qquad j = 4...6\,, \tag{9}$$

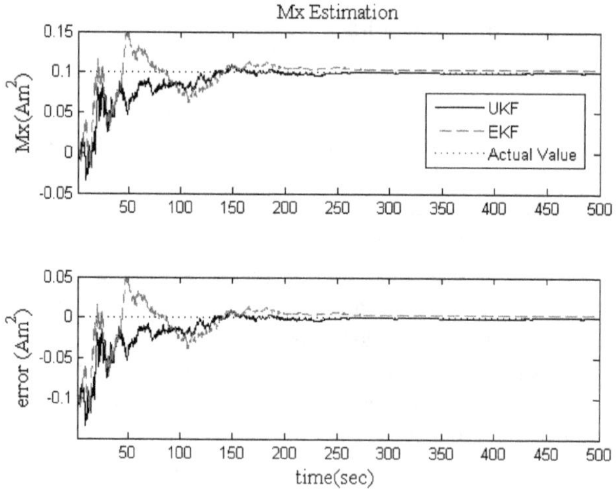

Fig. 1 Estimation of the RMM in x axis. Top plot compares the UKF and EKF estimation results with the actual value and lower plot gives the estimation errors.

Table 1 RMSE and computation time for the simulations by the UKF and EKF. Given values are the means for Monte Carlo runs.

		UKF	*EKF*
RMSE	M_x	3.814×10^{-4}	4.269×10^{-4}
(Am^2)	M_y	2.084×10^{-4}	3.146×10^{-4}
	M_z	4.386×10^{-4}	6.456×10^{-4}
Computation time* (sec)		8.048	5.798

*For the complete simulation of 2000 sec.

Although the UKF is better in terms of accuracy, the difference in the accuracy of two filters is not so significant and the EKF estimates are also good enough for the RMM compensation scheme (we require an estimation error less than $0.001Am^2$). The nonlinearity in the dynamics equations is not high and the measurement equation for the filters is linear; this is why we get a limited accuracy improvement over the EKF with the UKF. Moreover, when we use the EKF the computation time is reduced more than 25% (Table 1). Regarding that multiple filters run simultaneously for the MMAE method, 25% is a considerable decrease in the computational load specifically if we would like to use the algorithm for on-line estimation. Considering these facts we decided the EKF is more suitable for the RMM estimation in this study.

3 The MMAE Algorithm for the RMM Estimation

As discussed, in case there is a change in the RMM parameters the estimation performance of the KF mainly depends on the selection of the process noise covariance matrix, Q. When the Q is composed of small values – e.g. as given in (8) –the filter can accurately estimate the parameters but it does not respond to the changes quickly and long time passes till the filter catch new values. In contrast, if the process noise covariance is high, the filter's tracking ability improves but the estimation accuracy deteriorates because of the imposed noise. For brevity, here, we do not present the detailed analysis for the RMM estimation with different Q values and the reader may refer to [15].

The MMAE is one of the adaptive estimation techniques where multiple KFs run under different models and the estimates coming from all filters are merged depending on their likelihoods for being the true estimate [18]. In our case, multiple EKFs run with different Q matrices and their RMM estimations are combined. Main purpose is to obtain an overall algorithm that is accurate in steady state regime and can quickly respond and catch the new values when there are sudden changes in the RMM parameters.

In this section, first we give the classical MMAE algorithm and briefly discuss why it is not suitable for solving our problem. Then we propose the MMAE algorithm for the RMM estimation with a description on the newly proposed likelihood function.

3.1 Classical Approach for the MMAE

The MMAE method is based on implementing a bank of KFs with different models and merging the estimations of all filters by using each model's probabilities for being the true one. The state estimate might be either the output of the model with the highest probability or a weighted sum of all the outputs. This is a highly appreciated method especially for fault detection and isolation. For example for fault tolerant aircraft control, each model in the bank represents a different sensor/actuator failure condition and the fault can be detected when the probability for any model increases above an alarm threshold [24]. The MMAE scheme is given in Fig.2. In figure $\hat{x}_{k/k}^{(i)}$, $e_k^{(i)}$, $q_k^{(i)}$ represent the estimated state vector, likelihood function input and likelihood for $i = 1 \ldots \Omega$ KFs, respectively and $\hat{x}_{k/k}$ is the state estimation of the MMAE algorithm.

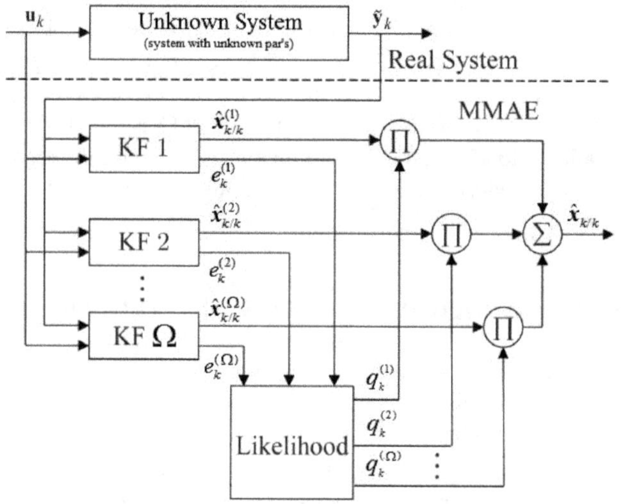

Fig. 2 The MMAE scheme.

The classical MMAE algorithm has a drawback of assuming only one of the models in the bank is true and this is due to the employed likelihood function. In the classical approach, the Bayesian likelihood for each KF is [18],

$$q_k^{(i)} = \frac{\text{pdf}_k^{(i)} q_{k-1}^{(i)}}{\sum_{j=1}^{\Omega} \text{pdf}_k^{(j)} q_{k-1}^{(j)}}, \tag{10}$$

where,

$$\text{pdf}_k^{(i)} = \frac{1}{(2\pi)^{m/2} \left| S_k^{(i)} \right|^{1/2}} \exp\left(\frac{-\left[\tilde{y}_k - \hat{y}_k^{(i)}\right]^T \left(S_k^{(i)}\right)^{-1} \left[\tilde{y}_k - \hat{y}_k^{(i)}\right]}{2} \right), \tag{11}$$

and

$$S_k^{(i)} = H_k P_{k/k-1}^{(i)} H_k^T + R_k \,. \tag{12}$$

Here, $\hat{y}_k^{(i)}$ is the predicted measurement, $S_k^{(i)}$ is the innovation covariance and $P_{k/k-1}^{(i)}$ is the predicted covariance for KFs. Besides, H_k is the measurement matrix and m is the number of measurements.

When we use the classical Bayesian likelihood in the MMAE algorithm the likelihood for one of the models, which is assumed to be the true model with the lowest residual, goes to 1 as the other likelihoods diminish and become 0 eventually. In essence, this is as a result of including the memory of previous $q_{k-1}^{(i)}$ values in the calculation of the current likelihood, $q_k^{(i)}$. As an example, suppose that we apply the classical MMAE algorithm for the RMM estimation problem. Two EKFs run in parallel, one with low and the other with high Q values for the RMM parameters as given with (13):

$$Q = \begin{bmatrix} (1E-20)I_{3x3} & 0_{3x3} \\ 0_{3x3} & \Lambda \times (1E-10)I_{3x3} \end{bmatrix} \quad \begin{cases} \Lambda = 1 \quad \text{for} \quad Q_{low} \\ \Lambda = 10000 \quad \text{for} \quad Q_{high} \end{cases} . \tag{13}$$

In Fig. 3 the trends for the likelihood of both EKFs are given. The MMAE algorithm assumes that the model with the high process noise is true and approximately after 250th sec. the likelihood for the EKF with $\Lambda = 10000$ becomes 1 while the likelihood for the EKF with $\Lambda = 1$ vanishes. In other words, the MMAE estimations become exactly same as the estimations of the EKF with $\Lambda = 10000$ after 250th sec. Once the likelihood values are settled as shown, they do not change throughout the whole estimation procedure, even when the estimated RMM parameters change.

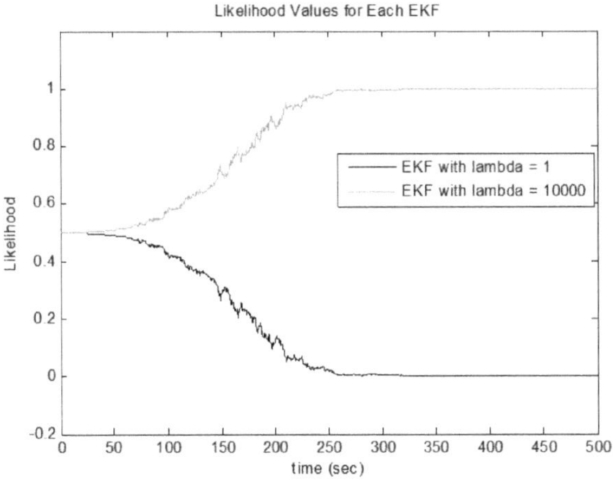

Fig. 3 Likelihood values for each EKF in the MMAE algorithm for the RMM estimation. Likelihoods are calculated using the classical Bayesian approach.

Plainly, this example shows that the MMAE with the classical Bayesian likelihood is a method suitable only for identifying the true parameters as also discussed by Karasolo and Hu [6]. In contrast, we want to use the MMAE algorithm for a system with unknown dynamics, specifically a system for which the parameters are suddenly changing. In this case, rather than operating the MMAE with the assumption that there is one true model, each model should contribute on the estimates depending on their likelihood. In fact, if we consider the RMM estimation problem, the EKF with low noise should be highly weighted during the steady state regime to get more accurate estimation results and the EKF with high noise should be more likely only when there is change in the RMM parameters to catch the new parameter values quickly. This leads us to a search for a MMAE algorithm with more appropriate likelihood function for systems with sudden parameter changes. In the next section, we present the MMAE algorithm with the newly proposed likelihood function.

3.2 *The MMAE for the RMM Estimation in Case of Sudden Changes*

Main idea behind the new likelihood function for the MMAE is to make the algorithm suitable for switching between the modes such that it is both accurate when the system parameters are in steady state and highly agile when the parameters change. As stated the essence is to use multiple KFs with different Q values (different lambdas, Λ, for the RMM estimation problem). The estimations of the KFs with lower Q values, which are more accurate, should have more weight than the others during the steady state regime and the estimations of the KFs with higher Q values, which are more agile, should be more likely when the parameters change.

There are two components of the proposed likelihood function, one as a measure for the steady state error and the other as a measure for the agility of the filter. Together, these measures form the likelihood function as

$$q_k^{(i)} = \frac{\exp\left(-q_{1,k}^{(i)} - q_{2,k}^{(i)}\right)}{\sum_{j=1}^{\Omega} \exp\left(-q_{1,k}^{(j)} - q_{2,k}^{(j)}\right)}. \tag{14}$$

Here, $q_{1,k}^{(i)}$ is the measure of steady state error and $q_{2,k}^{(i)}$ is the measure of filter agility. These measures are defined in the following subsections.

3.2.1 Measure of Steady State Error

Assume that we are estimating piecewise constant parameters (e.g. RMM) and the vector of estimated parameters is shown as $\hat{\boldsymbol{M}}_k^{(i)}$. When we average the parameter estimations within a moving window we smooth the results for each KF. For a sufficiently large window size the averaged estimation results are similar for every KF since the noise imposed on the results by the different values of Q is attenuated.

$$\hat{\boldsymbol{M}}_{mean,k}^{(i)} = \frac{1}{\mu} \sum_{l=k-\mu+1}^{k} \hat{\boldsymbol{M}}_{l}^{(i)}, \tag{15}$$

where, μ is the size of the moving window and $\hat{\boldsymbol{M}}_{mean,k}^{(i)}$ are the averaged estimation results. If the current value of the estimated parameters is subtracted from this average value such that,

$$\mathrm{Z}_{k}^{(i)} = \left[\hat{\boldsymbol{M}}_{mean,k}^{(i)} - \hat{\boldsymbol{M}}_{k}^{(i)}\right]^{T} \left[\hat{\boldsymbol{M}}_{mean,k}^{(i)} - \hat{\boldsymbol{M}}_{k}^{(i)}\right], \tag{16}$$

then $\mathrm{Z}_{k}^{(i)}$ can be used as a measure of noise in the estimation results.

When $\mathrm{Z}_{k}^{(i)}$ is directly used as the input for the likelihood function there are two issues. First, $\mathrm{Z}_{k}^{(i)}$ may have abrupt changes since the noise in $\hat{\boldsymbol{M}}_{k}^{(i)}$ is random. Suppose that at *k*, $\left|\hat{\boldsymbol{M}}_{mean,k}^{(i)} - \hat{\boldsymbol{M}}_{k}^{(i)}\right| \approx 0$ while at $k+1$, $\left|\hat{\boldsymbol{M}}_{mean,k+1}^{(i)} - \hat{\boldsymbol{M}}_{k+1}^{(i)}\right| \gg 0$. In this case abrupt changes will be also observed in the weights. So that we prevent this and obtain a smoother weighting scheme, $\mathrm{Z}_{k}^{(i)}$ should be filtered with a low-pass filter,

$$g_{1,k}^{(i)} = \lambda_1 g_{1,k-1}^{(i)} + (1-\lambda_1) Z_{k}^{(i)}, \tag{17}$$

where λ_1 is the tuning parameter for the low-pass filter and $g_{1,k}^{(i)}$ are the outputs.

Secondly, input values for the exponential function should be scaled (or normalized) for each KF such that they fall into the desired range $[b,\ a]$,

$$q_{1,k}^{(i)} = b + \frac{(a-b)\left[g_{1,k}^{(i)} - \min(g_{1,k})\right]}{\max(g_{1,k}) - \min(g_{1,k})} \tag{18}$$

where *b* and *a* are lower and upper normalization bounds. The lower bound may be taken simply as $b = 0$ for the likelihood function.

As a result, we obtain the first component of the proposed likelihood function which is a measure for the steady state error. $q_{1,k}^{(i)}$ takes greater values (closer to the upper bound *a*) for a KF with high noise and this KF's likelihood decreases since $q_{k}^{(i)} \propto \exp\left(-q_{1,k}^{(i)}\right)$. In contrast, when the Q is small for a specific KF, $q_{1,k}^{(i)}$ takes smaller values (closer to the lower bound *b*) and this KF's estimations become more likely.

3.2.2 Measure of Filter Agility

The normalized innovation sequence of the KF is a commonly used parameter for detecting the changes in the estimated parameters. For vector valued measurements (or innovation sequence) the sum of the normalized innovation sequence is used [7, 15],

$$\mathrm{En}_k^{(i)} = \frac{1}{\sqrt{m}} \mathbf{1}_m^T \left\{ H P_{k-1/k}^{(i)} H^T + R_k \right\}^{-1/2} \left\{ \tilde{y}_k - \hat{y}_k^{(i)} \right\}. \tag{19}$$

Here, $\mathbf{1}_m$ is unit vector with m unit elements, by which the sum of the elements of the normalized innovation sequence is taken. In general a low-pass filter is applied to $\mathrm{En}_k^{(i)}$ values before using as an indicator for the change,

$$g_{2,k}^{(i)} = \lambda_2 g_{2,k-1}^{(i)} + (1 - \lambda_2) \mathrm{En}_k^{(i)}, \tag{20}$$

where λ_2 is the tuning parameter for the low-pass effect similarly with λ_1.

Low-pass filtered values of the normalized innovation sequence, which has zero-mean white noise trend in normal case, increases/decreases when there is a sudden change in the estimated parameters. The increment/decrement level is related with the agility of the filter. If the filter has high Q values and agilely catches the new parameters then the increment/decrement level is relatively low and if the filter is agile enough such variation in $g_{2,k}^{(i)}$ will not be even observed. Hence, $g_{2,k}^{(i)}$ can be also used as a measure of the filter agility. There should be change in the estimated parameters, $g_{2,k}^{(i)}$ will take greater values for the KFs with low Q values and smaller values for the KFs with high Q values. Nonetheless, it should be scaled before using as an input to the exponential function,

$$q_{2,k}^{(i)} = \left| g_{2,k}^{(i)} \right| / \beta, \tag{21}$$

where, β is the scaling parameter that should be selected depending on the behavior of $g_{2,k}^{(i)}$ in case of change in the estimated parameters.

The $q_{2,k}^{(i)}$ value that we obtain using (21) constitutes the second component of the likelihood function $q_k^{(i)}$. Since $q_k^{(i)} \propto \exp\left(-q_{2,k}^{(i)}\right)$, the likelihood $q_k^{(i)}$ for estimates of the KFs with low process noise covariance will decrease ($q_{2,k}^{(i)}$ will increase) when the estimated parameters change. On the other hand, $q_{2,k}^{(i)}$ will not vary significantly ($g_{2,k}^{(i)} \approx 0$) for the KFs with high process noise covariance and these KFs' estimates will become more likely.

In the next section, we apply the proposed MMAE algorithm to the RMM estimation problem.

4 Numerical Results

In this section, the MMAE algorithm with the newly proposed likelihood function is used for estimating the RMM parameters. Simulations are run for a hypothetical small satellite, the inertia and sensor characteristics for which are described in Section 2.2. This time, we assume the estimated RMM parameters change at 2000th sec. as,

$$\boldsymbol{M} = \begin{cases} [0.1 \quad 0.02 \quad -0.05]^T \; Am^2 & t < 2000 \sec \\ [0.25 \quad 0.1 \quad -0.15]^T \; Am^2 & t \geq 2000 \sec \end{cases}. \tag{22}$$

We use the EKF for the estimations of each model as discussed before. Specific parameters that we chose for calculating the measures of steady state error and filter agility are tabulated in Table 2.

Table 2 Parameters for calculating the proposed likelihood function.

	For measure of steady state error	*For measure of filter agility*
Parameters	$\mu = 150$	$\lambda_2 = 0.997$
	$\lambda_1 = 0.9$	$\beta = 0.2$
	$a = 5$	
	$b = 0$	

Simulations are performed for 40,000 seconds. We tested the MMAE algorithm with 2 and 3 EKFs. Regarding the Q matrix given as,

$$Q = \begin{bmatrix} (1E-20)\, I_{3x3} & 0_{3x3} \\ 0_{3x3} & \Lambda \times (1E-10) I_{3x3} \end{bmatrix}, \tag{23}$$

each EKF has different Λ values, which makes them run under different levels of process noise covariance for the RMM terms.

In the first case, there are two EKFs running parallel. Λ values are selected as $\Lambda_1 = 1$ for the EKF1 and $\Lambda_2 = 10000$ for the EKF2. In Fig. 4 M_y estimation results for the MMAE algorithm is given together with the estimations of two EKFs. As seen the MMAE algorithm with the newly proposed likelihood function gives accurate results during the steady state regime and quickly catches the new values when the RMM parameters change. This is clearer in the subfigures; Fig.4-*a* zooms a section from the steady state and Fig.4-*b* presents the tracking of the filters immediately after the parameters change. In principle, the newly proposed likelihood function makes the EKF1 estimations more likely when the parameters are in steady state. Hence, the MMAE estimations are almost same with the EKF1 estimations during that period. On contrary, when the RMM parameters change, likelihood of the EKF1 decreases and the EKF2 estimations become more likely. As a result, the MMAE estimations coincide with the EKF 2 estimations for a period after the change (approximately 600 sec.) and the algorithm catches the new parameter values quickly. Variation of the likelihoods for the EKF1 and EKF2 can be seen in Fig. 5.

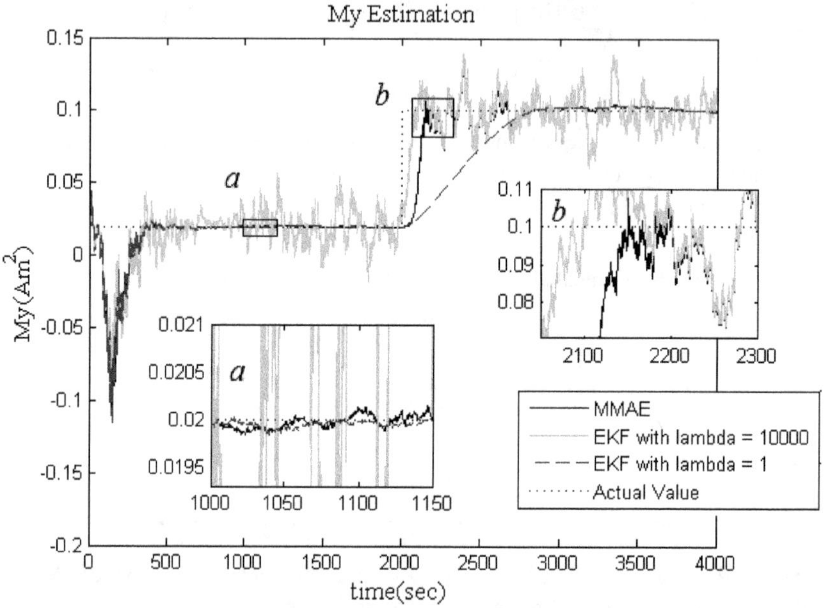

Fig. 4 Estimation of the RMM in *y* axis using the proposed MMAE algorithm with two EKFs. The MMAE estimation results are plotted together with the estimations of two EKFs used as a part of the algorithm. Subfigures *a* and *b* zoom to the indicated sections in the main figure.

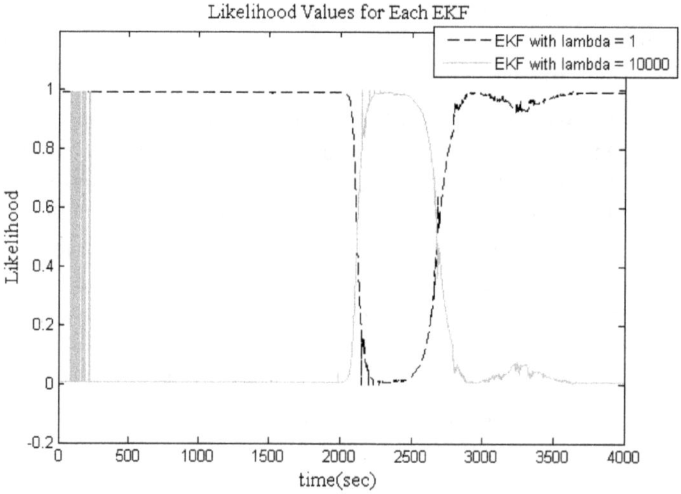

Fig. 5 Likelihood values for each EKF in the first case where two EKFs run in parallel. Values are calculated using the proposed method.

During the steady state estimation period the measure of filter agility for both filters, $q_{2,k}^{(1)}$ and $q_{2,k}^{(2)}$ are almost equal since there is no change in the parameters. On the other hand, measure of steady state error for the EKF2 is higher than the EKF1's ($q_{1,k}^{(2)} > q_{1,k}^{(1)}$) and so the likelihood of the EKF2 is lower. As a result EKF1's likelihood is close to 1 and EKF2's likelihood is close to 0 until 2000th sec. When the parameters change the increase in the measure of filter agility for the EKF1 reduces this filter's likelihood and the EKF2 likelihood increases. Fig. 6 gives a deeper look to the variation of the measures throughout the estimation procedure. We see that change in the parameters also affects the measure of steady state for the EKF2 (similarly with the initial convergence period for the filters) but the dominant factor for determining the likelihood is the measure of filter agility which increases drastically for the EKF1 after the change but remains same for the EKF2.

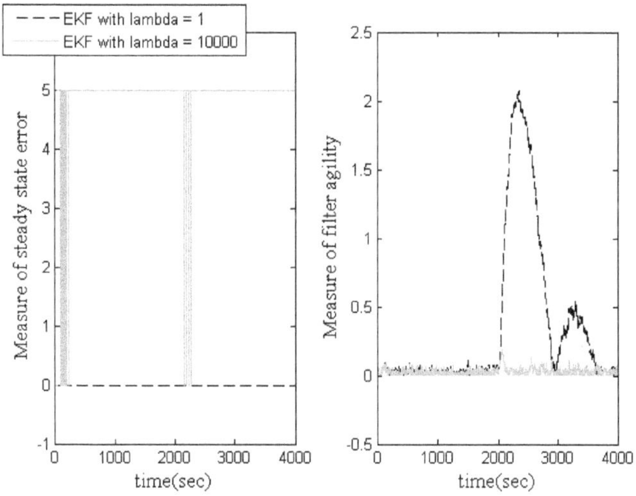

Fig. 6 Variations of the measures of steady state error and filter agility.

In the second case another EKF is added to the bank and the MMAE is operated with three EKFs. The third EKF, which is named as the EKF3 straightforwardly, has a process noise covariance in between the other two's as $\Lambda_3 = 100$. In Fig.7 M_y estimation results are given. As can be seen this time the MMAE algorithm has a smoother tracking performance when the RMM parameters change (see also Fig. 8 where the MMAE estimations for two cases are compared). This is a clear contribution of the EKF3 with a noise level in between. Once the parameters change, the EKF2 likelihood increases for a shorter period compared to the first case with only two EKFs (see Fig. 9 for the variation of the likelihood functions). Nonetheless, the EKF3 likelihood increases and remains close to 1 for almost 600 sec. until the EKF1 also catches the new values. In other words, after the change in the RMM terms, the EKF2 becomes likely only for a short duration and then the main contribution for the agility comes from the EKF3.

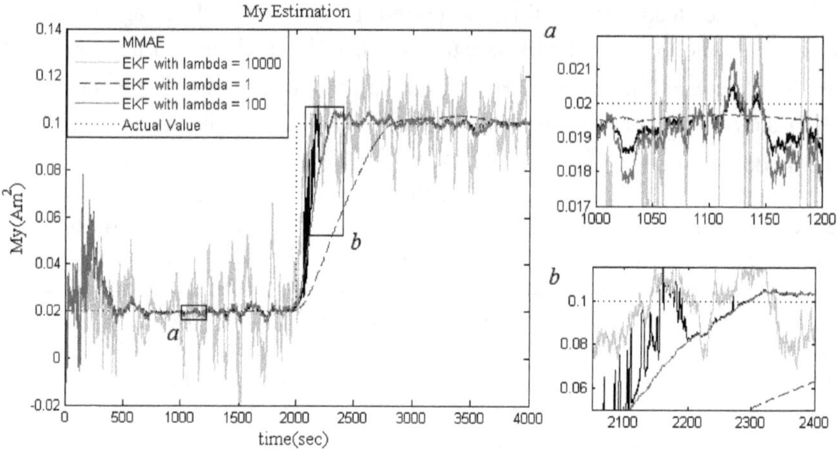

Fig. 7 Estimation of the RMM in y axis using the proposed MMAE algorithm with three EKFs. The MMAE estimation results are plotted together with the estimations of three EKFs used as a part of the algorithm. Subfigures a and b zoom to the indicated sections in the main figure.

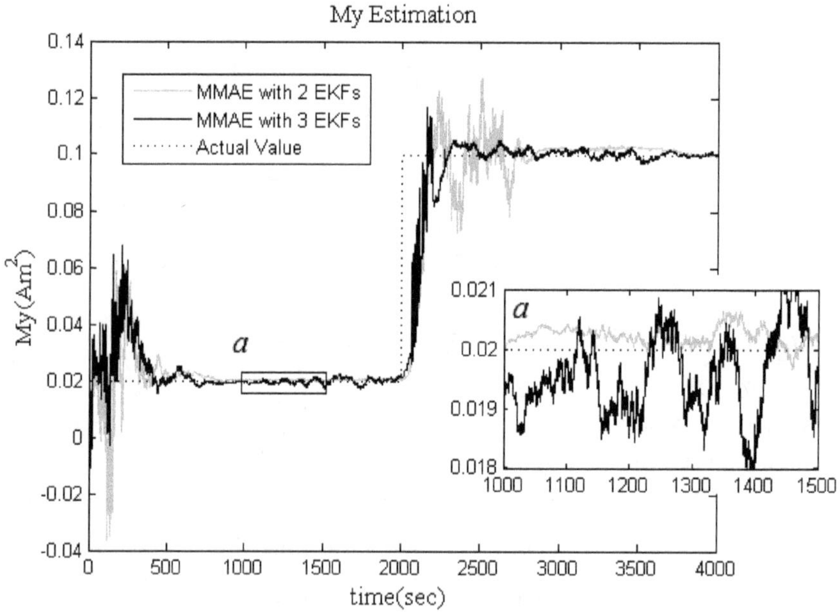

Fig. 8 Comparison of the estimations for the MMAE algorithms with two and three EKFs. Subfigure a zooms to the indicated section in the main figure.

In the second case, in contrast with the relative improvement in tracking smoothness, the steady state estimation accuracy decreases. This is a well known disadvantage of the MMAE method. When the EKF1 gets competition from the other EKFs like the EKF3 the MMAE estimation accuracy decreases in the steady state regime. This is explicitly seen both in Fig.7-*a* and Fig.8-*a*. The likelihood of the EKF3 that is about 0.48 in the steady state supports this statement (Fig.9).

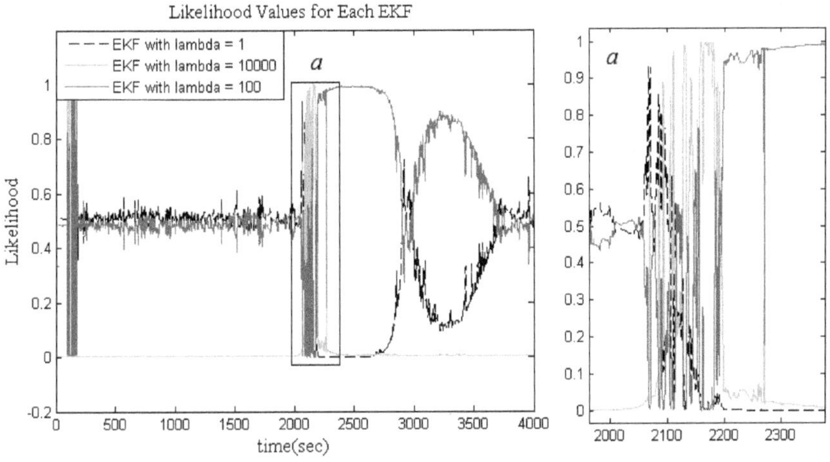

Fig. 9 Likelihood values for each EKF in the second case where three EKFs run in parallel. Values are calculated using the proposed method. Subfigure *a* zooms to the indicated section in the main figure.

In summary, the tracking performance of the proposed MMAE method can be improved by including EKFs with different models to the algorithm especially if we consider that changes with different magnitudes is possible in the RMM parameters. However adding other filters also means sacrificing the steady state accuracy as the most accurate filter (EKF1 in our case) gets more competition. Besides, inevitably, using more filters increases the computational load of the algorithm. The performance of the algorithm might be further tested by implementing with EKFs with different noise levels. Moreover, the effects of the tuning parameters for the measures should be investigated.

5 Conclusions

In this study we proposed a Multiple-Model Adaptive Estimation (MMAE) method for estimating the time-varying Residual Magnetic Moment (RMM), which is a dominant disturbance source for the small satellites. Several EKFs run with different process noise covariance matrices and the estimates of each filter are integrated by the use of a newly defined likelihood function. Newly defined likelihood function is composed of two measures: measure for the steady state accuracy and measure for the filter agility. In this sense, the likelihood function is different than the classical Bayesian likelihood function used for the MMAE and we show this new function is more appropriate for solving our problem. The performance of the algorithm is investigated for different cases. The results prove that the MMAE algorithm with the newly proposed likelihood function gives accurate results during the steady state regime and quickly catches the new values when the RMM parameters change.

References

1. Sakai, S., Fukushima, Y., Saito, H.: Design and on-orbit evaluation of magnetic attitude control system for the "REIMEI" microsatellite. In: Proc. 10th IEEE International Workshop on Advanced Motion Control, Trento, Italy (2008), doi:10.1109/AMC.2008.4516132
2. Inamori, T., Nakasuka, S., Sako, N.: In-orbit magnetic disturbance estimation and compensation using UKF in nano-satellite mission. In: Proc. AIAA Guidance, Navigation, and Control Conference, Chicago, USA (2009), doi:10.2514/6.2009-5905
3. Hosonuma, T., Inamori, T., Nakasuka, S.: A precise attitude determination and control strategy for small astrometry satellite "Nano-Jasmine". In: Proc. 26th AIAA/USU Conference on Small Satellites, Logan, USA (2012)
4. Burton, R., Rock, S., Springman, J., Cutler, J.: Dual attitude and parameter estimation of passively magnetically stabilized nano satellites. Acta Astronautica 94, 145–158 (2014), doi:10.1016/j.actaastro.2013.08.017
5. Steyn, W.H., Hashida, Y.: In-orbit attitude performance of the 3-axis stabilised SNAP-1 nanosatellite. In: Proc. 15th AIAA/USU Conference on Small Satellites, Logan, USA (2001)
6. Karasolo, M., Hu, X.: An optimization approach to adaptive Kalman filtering. Automatica 47, 1785–1793 (2011), doi:10.1016/j.automatica.2011.04.004
7. Gustafsson, F.: Adaptive filtering and change detection. Wiley, New York (2001)
8. Basseville, M.: Nikiforov IV Detection of abrupt changes: Theory and application. Prentice Hall, New Jersey (1993)
9. Bisht, S.S., Singh, M.P.: An adaptive unscented Kalman filter for tracking sudden stiffness changes. Mechanical Systems and Signal Processing 49, 181–195 (2014), doi:10.1016/j.ymssp.2014.04.009
10. Bavdekar, V.A., Prakash, J., Shah, S.L., Gopaluni, R.B.: Constrained dual ensemble Kalman filter for state and parameter estimation. In: Proc. American Control Conference, Washington DC, USA (2013), doi:10.1109/ACC.2013.6580306
11. Sun, Z., Yang, Z.: Joint parametric fault diagnosis and state estimation using KF-ML method. In: Proc. 19th IFAC World Congress, Cape Town, South Africa (2014)
12. Sun, Z.: Nonlinear system identification and its applications in fault detection and diagnosis, PhD. Thesis. Aalborg University, Aalborg, Denmark (2013)
13. Hartikainen, E., Ekelin, S.: Enhanced network-state estimation using change detection. In: Proc. 31st IEEE Conference on Local Computer Networks, Florida, USA (2006), doi:10.1109/LCN.2006.322178
14. Soken, H.E., Sakai, S.: In-orbit estimation of time-varying residual magnetic moment for small satellite applications. In: Proc. AIAA Guidance, Navigation and Control Conference, Boston, USA (2013)

15. Soken, H.E., Sakai, S., Wisniewski, R.: In-orbit estimation of time-varying residual magnetic moment. IEEE Transactions on Aerospace and Electronic Systems 50(4), 3126–3136 (2014), doi:10.1109/TAES.2014.130225
16. Soken, H.E., Sakai, S.: Investigation of estimation methods for time-varying residual magnetic moment. In: Proc. 6th Asia-Pacific International Symposium on Aerospace Technology, Shanghai, China (2014)
17. Maybeck, P.S.: Stochastic models, estimation, and control, vol. 2. Academic Press, New York (1982)
18. Simon, D.: Optimal state estimation. Wiley-Interscience, New Jersey (2006)
19. Wertz, J.R.: Spacecraft attitude determination and control. Kluwer Academic Publishers, Dordrecht (1988)
20. Soken, H.E.: UKF adaptation and filter integration for attitude determination and control of nanosatellites with magnetic sensors and actuators, PhD. Thesis. The Graduate University for Advanced Studies (Sokendai), Sagamihara, Japan (2013)
21. Springmann, J.C., Cutler, J.W.: Attitude-independent magnetometer calibration with time-varying bias. Journal of Guidance, Control, and Dynamics 35(4), 1080–1088 (2012), doi:10.2514/1.56726
22. Crassidis, J.L., Markley, F.L.: Unscented filtering for spacecraft attitude estimation. Journal of Guidance, Control and Dynamics 26(4), 536–542 (2003), doi:10.2514/2.5102
23. Daum, F.: Nonlinear filters: Beyond the Kalman filter. IEEE Aerospace and Electronic Systems Magazine 20(8), 57–69 (2005), doi:10.1109/MAES.2005.1499276
24. Ducard, G.J.J.: Fault-tolerant flight control and guidance systems. Springer, New York (2009)

Sliding Mode Observers for Fault Estimation in Multisensor Avionics Systems

Jérome Cieslak, Alejandra Ferreira de Loza, David Henry, Jorge Dávila, and Ali Zolghadri

Abstract. The paper addresses the problem of sensor fault estimation in avionics multisensor systems. Under the assumption of system strong observability, sliding mode observers are designed to estimate the faults in finite time and in the presence of bounded disturbances. It is shown that the fault estimation error is bounded in the L_∞-norm sense, and an upper bound is theoretically derived. The method is applied to the problem of sensor fault estimation of a large transport aircraft. Simulation results as well as a pilot experiment are presented to demonstrate the potential of the proposed method.

1 Introduction

The aircraft state is measured by a set of sensors delivering e.g. anemometric and inertial measurements that characterize the aircraft attitude, speed and altitude. The data are acquired using an acquisition system comprising several dedicated redundant units. For instance, a typical commercial aircrafts navigation sensing system can contain triple-redundant inertial references plus triple-redundant air data sensors.

Jérome Cieslak · David Henry · Ali Zolghadri
Université de Bordeaux, CNRS, IMS-lab, 351 cours de la Libération,
33405 Talence cedex, France
e-mail: {jerome.cieslak,david.henry,
ali.zolghadri}@ims-bordeaux.fr

Alejandra Ferreira de Loza
Instituto Politecnico Nacional, Avenida del Parque 1310, Mesa de Otay, Tijuana, Mexico
e-mail: da_ferreira@yahoo.com

Jorge Dávila
National Polytechnic Institute, Section of Graduate Studies and Research, ESIME-UPT, C.P. 07340, México D.F., México
e-mail: jadavila@ipn.mx

J. Bordeneuve-Guibé et al. (eds.), *Advances in Aerospace Guidance, Navigation and Control*, DOI: 10.1007/978-3-319-17518-8_19

A voting system monitors and checks the state of each sensor and detects any abnormal behaviour (see for instance [1, 5, 4]). Several common sensor faults in aerospace applications are discussed in [6]. Usual failures include bias, drift, scaling (abnormal gain variation), hard failures (frozen output, loss of signal, ...), oscillation and intermittent faults which are difficult to track because they can appear and disappear several times and the frequency of such faults is often random. A promising method to estimate sensor faults consists in reformulating the problem as an actuator fault reconstruction problem. Here, unknown input observers can be exploited to achieve the fault identification and reconstruction. In [8] sensor fault reconstruction was addressed using sliding mode observers based on the output injection concept. However, the principal drawback of this approach is related to its implementation due to the discontinuous nature of the sliding mode techniques. Indeed, in the presence of sampling time, it is necessary to apply low-pass filters to obtain the equivalent output injection [9]. This disadvantage has been overcome with higher order sliding mode techniques [10] which have been applied to the observation of a wide class of systems [11, 13], but also to sensor fault tolerant control [14]. Finite time exact observers based on high order sliding modes (HOSM) are proposed in [15, 16, 17]. These kind of observers provide theoretically exact observation and the best possible accuracy order with respect to the sampling step and bounded deterministic noises [18]. Applications of HOSM methodologies to fault diagnosis can be found in [12, 19]. In [20] a non-homogeneous high order sliding mode differentiator which provides faster convergence compared to the standard homogeneous differentiator was introduced. This characteristic becomes propitious for the intermittent fault case.

The work presented in this paper deals with sensor fault reconstruction. First, a non-homogeneous HOSM based observation approach is proposed to estimate the fault, theoretically in finite time. Next, the effect of noise on the fault reconstruction accuracy is analysed. It is shown that the fault estimation error is bounded in the L_∞-norm sense, and an upper bound is theoretically derived, leading to a viable candidate solution in a noisy environment. The proposed scheme is tested using an aircraft simulator.

The paper is organized as follows. Section II is devoted to the problem formulation and sections 3 and 4 address the proposed solution. Theoretical backgrounds are given and it is shown how the sensor fault reconstruction problem is solved. Section 4 addresses the measurement noise effects on the fault reconstruction. Section 5 is dedicated to an aircraft application taken from the European GARTEUR[1] action FM-AG16 [21]. A simulation campaign in the case of faults appearing on the angle-of-attack sensor with different time profiles, together with a pilot experiment, demonstrate the efficiency of the proposed strategy. Finally some concluding remarks are given in Section 6.

Notations:

The following notations are used all along the paper. The set of real numbers is denoted R. A square matrix $X \in R^{n\times n}$ is called Hurwitz if all eigenvalues have strictly

[1] Group for Aeronautical Research and Technology in Europe.

negative real parts. $\lambda_i(X)$, $i = 1,\dots,n$ is used to denote an eigenvalue of the matrix X. For a matrix $X \in R^{n\times m}$ with $rank(X) = r$, the matrix $X^{\perp} \in R^{n-r\times n}$ with $rank(X^{\perp}) = n - r$ is defined such that $X^{\perp}X = 0$. If $X \in R^{n\times m}$ with $rank(X) = m$, the matrix $X^{+} = \left(X^T X\right)^{-1} X^T$ is the Moore-Penrose pseudo inverse of X. For any $x \in R$ the symbol $|x|$ denotes its absolute value. If $x \in R^n$ then $\|x\|$ is the Euclidean norm. If $X \in R^{n\times m}$ is a matrix, the symbol $\|X\|$ denotes the induced norm (or spectral norm) $\|X\| = \max_{1\le i\le m}\sqrt{\lambda_i(X^T X)}$. For a measurable function $d : R_+ \to R^d$, the L_∞ norm is defined as $\|d\|_{[0,T)} = ess\sup_{0\le t<T}|d(t)| \Rightarrow \|d\|_{[0,\infty)} = \|d\|_\infty$.

2 Problem Formulation

Consider a linear time invariant model affected by unknown inputs and subjected to sensor faults, i.e.

$$\dot{x}(t) = Ax(t) + Bu(t) + Ev(t) \tag{1}$$

$$y(t) = Cx(t) + Df(t) \tag{2}$$

$$\check{y}(t) = y(t) + \eta(t) \tag{3}$$

where $x \in R^n$, $y \in R^p$, $f \in R^q$, $u \in R^m$, $v \in R^{q_v}$ are the state, output, fault signal, known input and unknown input vectors respectively with $q < p \le n$, $\check{y}(t)$ is the available signal from online measurements affected by a locally bounded Lebesgue-measurable noise $\eta(t)$. The matrices A, B, C, D and E are known and they are of appropriate dimensions. The components of the vector $f(t)$ are the faults to be estimated. In Eq. (2), the sensor faults are modeled as an additive disturbance. Here, it is assumed that only "q" sensors are potentially faulty out of the "p" output measurements. This is a valid assumption since some sensors may be more vulnerable and predisposed to fail. From the practical point of view, this is a realistic issue considering that some sensors may be more delicate in terms of sensing mechanism and construction than others [22]. For instance, some sensors are exposed to the external environment and may be spoiled by blockages caused by ice or dirt [7].

Based on this assumption and without loss of generality, the fault distribution matrix can be rearranged as $D = \begin{bmatrix} 0 & I_q \end{bmatrix}^T$ leading to the following canonical form:

$$y(t) = \begin{bmatrix} y_h(t) \\ y_f(t) \end{bmatrix} = \underbrace{\begin{bmatrix} C_h \\ C_f \end{bmatrix}}_{C} x(t) + \underbrace{\begin{bmatrix} 0 \\ I_q \end{bmatrix}}_{D} f(t) \tag{4}$$

The outputs subset $y_h \in R^{p-q}$ represents the healthy sensors while $y_f \in R^q$ is the set of sensors prone to fault. The matrices C_h and C_f are of appropriate dimensions. The components of the vector $f(t)$ are the faults to be estimated.

Furthermore, the following assumptions are made:

A1. $\|v(t)\|_\infty < v^+$ where $v^+ > 0$.
A2. The triplet (A, C_h, E) is strongly observable.

Let us recall that the triple (A, C_h, E) is called strongly observable if the triple (A, C_h, E) does not contain invariant zeros, i.e. $rank(\mathbf{R}(s_0)) = n + rank(E)$, $\forall s_0 \in \mathbb{C}$, where the invariant zeros of the triple (A, C_h, E) are the values of $s_0 \in \mathbb{C}$ for which

$$rank(\mathbf{R}(s_0)) < n + rank(E) \tag{5}$$

$\mathbf{R}$ refers to the Rosenbrock matrix of the triple (A, C_h, E) which is defined by:

$$\mathbf{R}(s) = \begin{bmatrix} sI - A & -E \\ C_h & 0 \end{bmatrix} \tag{6}$$

It follows that the sensor fault estimation problem (1)-(2) can be formulated as a state estimation problem where unknown inputs occur in the state equation.

The goal is now to design an observer to reconstruct, in finite time, the fault signal $f(t)$ in the presence of $v(t)$. To this aim, the algebraic observer, first introduced in [23], is exploited in order to estimate the state x from a linear combination of the output and its successive derivatives from which the effects of the unknown inputs $f(t)$ have been decoupled. The ideal output (2) will be exploited to describe the observer and, in Section 4, the noise effects in (3) will be discussed.

Remark 1. The strategy presented in this article allows the estimation of the fault signal $f(t)$. Now, suppose that a priori information about a fault model is available, e.g.:

$$\begin{aligned} \dot{z}_f(t) &= A_f z_f(t) + w(t) \\ f(t) &= z_f(t) \end{aligned}$$

where $z_f \in R^q$ is an internal state, $A_f \in R^{q \times q}$ is a matrix used to model the dynamic of faults and $w(t)$ is an unknown function introduced to consider modeling errors for $f(t)$. The subsequent techniques can be adapted in order to consider the fault dynamics as a part of an extended vector $\bar{x} = [x^T \; z_f^T]^T$ and the following extended system

$$\begin{aligned} \dot{\bar{x}} &= \underbrace{\begin{bmatrix} A & 0 \\ 0 & A_f \end{bmatrix}}_{\bar{A}} \bar{x} + \underbrace{\begin{bmatrix} B \\ 0 \end{bmatrix}}_{\bar{B}} u(t) + \underbrace{\begin{bmatrix} E & 0 \\ 0 & I_q \end{bmatrix}}_{\bar{E}} \underbrace{\begin{bmatrix} v(t) \\ w(t) \end{bmatrix}}_{\bar{w}} \\ y(t) &= \underbrace{\begin{bmatrix} C & D \end{bmatrix}}_{\bar{C}} \bar{x} \end{aligned}$$

with $\|\bar{w}(t)\|_\infty < \bar{w}^+$, $\bar{w}^+ > 0$. The great advantage is when dealing with anticipated fault situations, i.e. when dedicated monitoring are needed for specific faults (drift, freezing, intermittent faults, oscillatory faults among others). Although the availability of fault models are rather common when dealing with flight systems, the availability of a fault model is not assumed in the following.

3 Finite-Time Convergent Observer

Recalling that A2 is satisfied and following the developments reported in [24], the design of the observer involves two stages. First, the boundedness of the estimation error is achieved by means of a Luenberger auxiliary system. Then, the state vector is reconstructed via a differentiation scheme. Here, a non-homogeneous HOSM differentiator will be considered [20]:

Stage 1: A dynamic auxiliary system is proposed to bound the observation error, i.e.

$$\dot{\tilde{x}}(t) = A\tilde{x}(t) + Bu(t) + L(y_h(t) - \tilde{y}(t)) \tag{7}$$

$$\tilde{y}(t) = C_h\tilde{x}(t) \tag{8}$$

where $\tilde{x} \in R^n$ and the gain matrix L is designed such that $\tilde{A} := A - LC_h$ is Hurwitz. Let $e(t) = x(t) - \tilde{x}(t)$ whose dynamics follows from (1), (4) and (7), (8), i.e.

$$\dot{e}(t) = \tilde{A}e(t) + Ev(t) \tag{9}$$

Stage 2: Let the output of the auxiliary estimation error be defined according to

$$y_e(t) = y_h(t) - \tilde{y}(t) \tag{10}$$

The error vector will be represented as an algebraic expression of the output and its derivatives. To this aim, the effects of the unknown input $v(t)$ must be decoupled from the output $y_e(t)$ and its derivatives. Hence, a decoupling algorithm is involved [25]. It follows:

1. Define $M_1 := C_h$.
2. Derive a linear combination of the output $y_e(t)$ which is not affected by the unknown inputs, i.e. $(M_1E)^{\perp}\dot{y}_e(t) = (M_1E)^{\perp}M_1e(t)$ and from (10) and (9) it yields to

   $$(M_1E)^{\perp}\dot{y}_e(t) = (M_1E)^{\perp}M_1\tilde{A}e(t)$$

   thus, taking (10) and forming an extended vector

   $$\begin{bmatrix} \frac{d}{dt}(M_1E)^{\perp}y_e(t) \\ y_e(t) \end{bmatrix} = \underbrace{\begin{bmatrix} (M_1E)^{\perp}M_1\tilde{A} \\ M_1 \end{bmatrix}}_{M_2} e(t)$$

   taking out the derivative operator, the above equation can be rewritten after as

   $$\frac{d}{dt}\begin{bmatrix} J_1 & 0 \\ 0 & I_p \end{bmatrix}\begin{bmatrix} y_e(t) \\ \int y_e(t) \end{bmatrix} = M_2e(t)$$

   where $J_1 = (M_1E)^{\perp}$ and $I_p \in R^p$ is an identity matrix.

3. Derive a linear combination of $M_2e(t)$ unaffected by the unknown inputs, it is $d/dt(M_2E)^{\perp}M_2e(t) = (M_2E)^{\perp}M_2\tilde{A}\ e(t)$ afterward, take the output $y_e(t) = M_1e(t)$ and construct the extended vector $M_3e(t) = \left[\left((M_2E)^{\perp}M_2\tilde{A}e(t)\right)^T (M_1e(t))^T\right]^T$ which can be represented as a second order derivative of the $y_e(t)$ it is

$$\frac{d^2}{dt^2}\begin{bmatrix} J_2 & 0 \\ 0 & I_p \end{bmatrix}\begin{bmatrix} y_e(t) \\ \vdots \\ \int\int y_e(t) \end{bmatrix} = M_3e(t)$$

where $J_2 = (M_2E)^{\perp}\begin{bmatrix} J_1 & 0 \\ 0 & I_p \end{bmatrix}$.

j. The $j-th$ step of the decoupling algorithm for $j = 4,..,k$ is summarized as follows: derive a linear combination of the vector $M_{j-1}e(t)$ such that it is not affected by the unknown inputs, i.e. $d/dt\left((M_{j-1}E)^{\perp}M_{j-1}e(t)\right)$, then form the identity

$$\begin{bmatrix} \frac{d}{dt}(M_{j-1}E)^{\perp}M_{j-1}e(t) \\ y_e(t) \end{bmatrix} = \underbrace{\begin{bmatrix} (M_{j-1}E)^{\perp}M_{j-1}\tilde{A} \\ M_1 \end{bmatrix}}_{M_j} e(t)$$

rearranging the terms, the next expression is obtained

$$\frac{d^{j-1}}{dt^{j-1}}\begin{bmatrix} J_{j-1} & 0 \\ 0 & I_p \end{bmatrix}\begin{bmatrix} y_e(t) \\ y_e^{[1]}(t) \\ \vdots \\ y_e^{[j-1]}(t) \end{bmatrix} = M_je(t) \tag{11}$$

where $J_{j-1} = (M_{j-1}E)^{\perp}\begin{bmatrix} J_{j-2} & 0 \\ 0 & I_p \end{bmatrix}$ and $y_e^{[i]}(t)$ represents the $i-th$ anti-differentiator of $y_e(t)$, that is, $y_e^{[i]}(t) = \int_0^t \cdots \int_0^{\tau_i} y_e(\tau_i)\, d\tau_i \ldots dt$.

Due to $A2$, there exists a unique positive integer $k \leq n$ such that the matrix M_k generated recursively by the application of step j and in particular of (11) satisfies the condition $rank(M_k) = n$ (see [25]).

Therefore, from the following algebraic expression

$$M_ke(t) = \underbrace{\frac{d^{k-1}}{dt^{k-1}}\begin{bmatrix} J_{k-1} & 0 \\ 0 & I_p \end{bmatrix}\begin{bmatrix} y_e(t) \\ \vdots \\ y_e^{[k-1]}(t) \end{bmatrix}}_{Y(t)} \tag{12}$$

a solution for $e(t)$ exists, i.e.

$$e(t) = M_k^{+}\frac{d^{k-1}}{dt^{k-1}}Y(t) \tag{13}$$

where $M_k^+ = \left(M_k^T M_k\right)^{-1} M_k^T$. Indeed, $e(t)$ is the vector of minimal norm among those which minimize $\left\|M_k e(t) - \frac{d}{dt}Y(t)\right\|_\infty$. Consequently, from the output $y_e(t)$ and its successive$(k-1)-th$ derivatives, $e(t)$ is recovered leading to the estimation of $x(t)$ and thereafter the fault $f(t)$ in noise-free environment. To this aim, the real time HOSM non-homogeneous differentiator introduced in [20] will be used to provide, in the absence of noise, exact and finite time differentiation of $Y(t)$.

The non-homogeneous HOSM differentiator is given by

$$\begin{aligned} \dot{z}_{i,0} &= g_{i,0}\left(z_{i,0} - Y_i(t)\right) + z_{i,1} \\ \dot{z}_{i,1} &= g_{i,1}\left(z_{i,1} - \dot{z}_{i,0}\right) + z_{i,2} \\ &\vdots \\ \dot{z}_{i,j} &= g_{i,j}\left(z_{i,j} - \dot{z}_{i,j-1}\right) + z_{i,k-1} \\ \dot{z}_{i,k-1} &= g_{i,k-1}\left(z_{i,k-1} - \dot{z}_{i,k-2}\right) \end{aligned} \tag{14}$$

where $i = 1,\dots,n$ are the components of $Y(t)$, $z_j = \left[z_{1,j} \ \cdots \ z_{n,j}\right]^T$. and the functions $g_{i,j}$ have the following form

$$g_{i,j}(\cdot) = \lambda_{k-1-j}\Gamma^{\frac{1}{k-i}} |\cdot|^{\frac{k-j-1}{k-j}} sign(\cdot) - \mu_{k-1-j} * (\cdot) \tag{15}$$

with $\mu_{k-1-j} > 0,\ j = 1,\dots,k-1$, and positive sequence of λ_j, μ_j can be selected as in [20]. The gain Γ is a Lipschitz constant of $Y^{(k-1)}(t)$, which in this case can be calculated from (9) as $\Gamma \geq \left\|\tilde{A}\right\| e^+ + \|E\| v^+$ where $\|e(t)\|_\infty < e^+$, $\|v(t)\|_\infty < v^+$, see [16].

Consequently, the state is estimated by the equation:

$$\hat{x}(t) := M_k^+ z_{k-1}(t) + \tilde{x}(t) \tag{16}$$

where $\hat{x} \in R^{\bar{n}}$ is the estimated value of x. Therefore, in noise-free case, $\hat{x}(t)$ represents the estimate of $x(t)$, for all $t > T$. Therefore, as a result it follows that $\hat{f}(t)$ is the optimal estimate of $f(t)$ for all $t > T$.

To derive a constructive design, the following proposition is first given:

Proposition 1: Under the assumption that A1-A2 are satisfied, the sensor faults of the system (1), (2) are estimated by means of the state observer (7), (8), (16) and the following relation:

$$\hat{f} = y_f - C_f \hat{x} \tag{17}$$

in the presence of bounded disturbances $v(t)$.

Proof. The proof can be divided into three parts. First, the boundedness of the estimation error, corresponding to the application of the auxiliary system (7), (8), is proven. The finite-time convergence of the state observer (16) is proven in the second part of the proof. Finally the sensor fault reconstruction is proven in the last part of the proof.

1. Notice that given that $A1$ holds, $v(t)$ is a bounded unknown function which satisfies $||v||_\infty \leq v^+$. Under the assumption $A2$, it is well known that $e(t)$ will have a bounded norm, i.e. there exist some positive constants $\beta_1, \beta_2, \beta_3$ such that

$$\|e(t)\|_\infty < \beta_1 \exp(-\beta_2 t) \|e(0)\|_\infty + \|E\| v^+ + \beta_3$$

Hence, from the above inequality, it follows that

$$\|e(t)\|_\infty < e^+, e^+ > 0 \text{ for all } t \geq t_e \quad (18)$$

$$t_e > \frac{1}{\beta_2}(\ln(\beta_1 ||e(0)||_\infty) - \ln(e^+ - |||E||v^+ - \beta_3)) \quad (19)$$

2. The convergence of the state estimation can be proven from previous results. In [20] was shown that with the proper choice of the constants λ_j, μ_j and Γ for all $j = 0, \ldots, k-1$ there is a finite time $t > T + t_e$ such that the next identity is fulfilled

$$z_j(t) = \frac{d^j}{dt^j} Y(t) \quad \text{for all } \; j = 0, ..., k-1 \quad (20)$$

with $z_j(t) = \left[z_{1,j}(t) \ldots z_{\bar{n},j}(t) \right]^T$. Hence, the vector $e(t)$ in (13) is recovered from the $(k-1)-th$ sliding dynamics:

$$e(t) = M_k^+ z_{k-1}(t), \quad \forall t > T + t_e.$$

Now, given the definition of $e(t) = x(t) - \tilde{x}(t)$, the state reconstruction (16) is straightforward by algebraic manipulation.

3. Equation (4) defines an algebraic relation between the measurable signal y_f, the state $\hat{x}$ and the unknown sensor fault $f(t)$. The application of the observer (16) allows us to consider that $\hat{x} = x$ after a finite-time transient of duration T. The equation (17) is directly obtained from (4). Then, the sensor faults of (2) can be estimated after a finite time transient of duration $T + t_e$.

4 Fault Reconstruction Error

In this section, the effect of noise on the accuracy of the estimation error $x - \hat{x}$ is analysed. It is shown that $x - \hat{x}$ is bounded in the L_∞-norm sense and an upper bound is theoretically derived.

Let the real available output signal (3) be considered. Taking into account (3) instead of (2) in (10), it follows that

$$\check{y}_e(t) = y_e(t) + \eta(t) \quad (21)$$

where $\check{y}_e(t)$ is the actual differentiator input affected by a locally bounded Lebesgue-measurable noise $\eta(t)$. It is supposed that the noise is bounded, i.e.

$$\|\eta(t)\|_\infty < \eta^+ \tag{22}$$

where $\eta^+ > 0$ is unknown.

Since the addition of a bounded noise term does not affect the convergence of the error signal to a bounded region around the origin, it follows that (9) yields to $\dot{e}(t) = \tilde{A}e(t) + Ev(t) - L\eta(t)$. Thus, since $\eta(t)$ is bounded, for $\beta_3 > \|L\|\,\eta^+$, equation (19) is still valid. The input of the differentiator is thus given by $\check{Y}(t)$. Now, let the effect of noise on the accuracy of the derivatives and fault reconstruction be studied. To this aim, first consider a preliminary result [20].

Proposition 2: ([20]) Let the positive sequences λ_i and μ_i be chosen properly. Then, there exist positive constants γ_i, depending on differentiator gain Γ, such that for $t > T$, the following differentiation accuracy is obtained in (14):

$$\left\|z_{i,j} - Y_i^{(j)}(t)\right\|_\infty \leq \gamma_j \left(\eta^+\right)^{\frac{k-j}{k}} \qquad j = 0, \ldots, k-1 \tag{23}$$

□

For the fault reconstruction accuracy specification, the observation error from (13) and (16) leads to

$$\begin{aligned}\|x(t) - \hat{x}(t)\|_\infty &= \left\|Y^{(k-1)}(t) - z_{k-1}(t)\right\|_\infty \\ &\leq \gamma_k \left(\eta^+\right)^{\frac{1}{k}}.\end{aligned} \tag{24}$$

Finally, based on (23) and (24), the fault estimation error can be upper bounded by the following constant in the presence of bounded noise:

$$\left\|\hat{f}(t) - f(t)\right\|_\infty = \|C(x(t) - \hat{x}(t)) + \eta(t)\|_\infty \leq \|C\|_\infty \gamma_k \left(\eta^+\right)^{1/k} + \eta^+ \tag{25}$$

Remark 2. Note that if the extended system given in remark 1 is considered, the fault estimation error can be upper bounded by $\left\|\hat{f}(t) - f(t)\right\|_\infty \leq \gamma_k \left(\eta^+\right)^{1/k}$.

5 Boeing 747-100/200 Aircraft Example

The fault estimation algorithm described in the above sections is now considered for validation using both the GARTEUR FM AG-16 simulation benchmark and a pilot experiment from the SIMONA flight simulator [21]. Both simulators are based on a Boeing 747-100/200 large transport aircraft with 26 control surfaces and 4 jet engines, with its standard autopilot system [26]. For faulty situations, different fault time profiles of the angle-of-attack sensor have been considered [27, 3].

5.1 Modelling

The dynamic behaviour of the Boeing 747-100/200 aircraft is described by a non-linear state representation that can be written in the following general form (the interested reader can refer to [26] for a complete description of the model)

$$\dot{x}_{NL}(t) = f(x_{NL}(t), u_{NL}(t), v(t))$$
$$y_{NL}(t) = g\left(x_{NL(t)}, u_{NL(t)}\right) + \eta(t)$$

where x_{NL}, u_{NL} and y_{NL} are the state, input and output vectors, respectively, of the full aircraft nonlinear model. $v(t)$ denotes the wind and atmospheric turbulences and $\eta(t)$ is the measurement noise. Once the trim condition is established for the non-linear aircraft model, a linear model is generated to capture the dynamics around an equilibrium point [28]. Simplified models for the longitudinal and lateral modes can then be derived. Here, the trim conditions are fixed to $263,000kg$ for the mass, $133.8m/s$ for the true airspeed and an altitude fixed to $980m$ with 25% of maximum thrust at 1° of flap position. To assess the proposed approach, the longitudinal model given in [29] is used. This model has been validated through several longitudinal simulations [29]. It is a suitable linear model for the approach described in Sections III and IV. The state vector is composed of the pitch rate $q[rad]$, true airspeed $V_{TAS}[m/s]$, angle of attack $\alpha[rad]$, pitch angle $\theta[rad]$, altitude $h[m]$, that is $x = \left[q\ V_{TAS}\ \alpha\ \theta\ h\right]^T$. The input vector corresponds to the four elevator deflections (right inner, left inner, right outer and left outer elevators) and the trimmable horizontal stabilizer surface. The following state space matrix definition is finally obtained:

$$A = \begin{bmatrix} -6.7e^{-1} & -1.4e^{-6} & -8.8e^{-1} & 0 & -3.4e^{-6} \\ -1.6e^{-1} & -7.5e^{-3} & 4.9 & -9.8 & 4.5e^{-5} \\ 10.0e^{-1} & -10.0e^{-4} & -6.7e^{-1} & 0 & 5.9e^{-6} \\ 1 & 0 & 0 & 0 & 0 \\ 0 & 0 & -1.3e^{2} & 1.3e^{2} & 0 \end{bmatrix}$$

$$B = \begin{bmatrix} -4.9e-3 & -4.9e-3 & -4.7e-3 & -4.7e-3 & -4.5e-2 \\ 0 & 0 & 0 & 0 & 0 \\ -1.8e-4 & -1.8e-4 & -1.9e-4 & -1.9e-4 & -1.9e-3 \\ 0 & 0 & 0 & 0 & 0 \\ 0 & 0 & 0 & 0 & 0 \end{bmatrix}$$

The outputs are selected from the available measurements given in [21] so that

$$y = \left[q\ V_{TAS}\ \theta\ h\ \alpha\right]^T \tag{26}$$

Since the case of angle-of-attack sensor fault is investigated, the fault distribution matrix D is:

$$D = \left[0\ 0\ 0\ 0\ 1\right]^T \tag{27}$$

It follows that $n=5, p=5, q=1, m=5$. The subset $y_h \in R^4$ that represents the healthy sensors is thus given by $y_h = \begin{bmatrix} q \; V_{TAS} \; \theta \; h \end{bmatrix}^T$. Note that the strong observability of the triplet (A, C_h, E) leads to strongly observable quadruple (A, E, C, D). The fault can then be reconstructed theoretically in finite time.

5.2 Design

The proposed procedure described in sections III and IV is applied step by step:

- First, the auxiliary system (7) is considered. Here, an empirical tuning procedure has been used to choose the eigenvalue of $\tilde{A}$. It follows that poles are fixed to -20, -40, -60, -80, -100 and -120. Thus, the decoupling algorithm given in Section 3 can be used.
- Following Section 3, it can be verified that the index k is equal to $k=2$. Thus, from (26), it yields to:

$$J_1 = \begin{bmatrix} 0 & 1 & 0 & 0 & 0 \\ 0 & 0 & 1 & 0 & 0 \\ 0 & 0 & 0 & 1 & 0 \\ -1 & 0 & 0 & 0 & 0 \end{bmatrix}$$

$$M_2 = \begin{bmatrix} 0.92 & -105.59 & 2.51 & 0 & 6.06 & -2.48 \\ 0 & 0 & -2.3e-1 & -20 & 1.2e-3 & -2.3e-1 \\ 7.23 & 42.4 & -67.23 & 1.2e-3 & -174.16 & 66.56 \\ 79.91 & -9.9e-1 & 4.4e-1 & 0 & 1.08e-1 & -4.4e-1 \\ 1 & 0 & 0 & 0 & 0 & 0 \\ 0 & 1 & 0 & 0 & 0 & 0 \\ 0 & 0 & 0 & 1 & 0 & 0 \\ 0 & 0 & 0 & 0 & 1 & 0 \\ 0 & 0 & 1 & 0 & 0 & 1 \end{bmatrix}$$

- From (13), it follows:

$$e(t) = M_2^+ \frac{d}{dt} \underbrace{\begin{bmatrix} J_1 & 0 \\ 0 & I_p \end{bmatrix} \begin{bmatrix} y_e(t) \\ y_e^{[1]}(t) \end{bmatrix}}_{Y(t)}$$

- The differentiator (14)-(15) becomes

$$\begin{aligned} \dot{z}_{i,0} &= -1.5\Gamma^{\frac{1}{2}} \left| z_{i,1} - Y_i(t) \right|^{\frac{1}{2}} sign\left(z_{i,1} - Y(t)\right) \\ &\quad -5\left(z_{i,1} - Y_i(t)\right) + z_{i,1} \\ \dot{z}_{i,1} &= -1.1\Gamma sign\left(z_{i,1} - \dot{z}_{i,0}\right) - 4\left(z_{i,1} - \dot{z}_{i,0}\right) \end{aligned} \tag{28}$$

  with $\Gamma = 5 \geq \left\|\tilde{A}\right\| e^+ + \|E\| v^+$ for $i = 0, ..., 5$.
- Thus, the estimated state vector (16) is given by $\hat{x} = \tilde{x} + M_2^+ z_1$.
- From the estimated state $\hat{x}$, the reconstructed fault $\hat{f}$ is directly obtained.

Note that the effect of the terms $\mu_j * (\cdot)$ given in (15) becomes obvious for large errors, while in a sufficiently small vicinity of zero their effects are negligible. Hence, for a fixed set of λ_j the trade-off is the following: the larger the parameter Γ is, the faster the convergence is with a higher noise sensitivity. Similarly, the larger the parameters μ_j is, the faster the convergence is with a stronger peaking effect.

5.3 Simulation Results

Simulations are performed for different fault time profiles in the α sensor. The noise level of q, V_{TAS}, θ, h and α sensors are set to $1.73 \times 10^{-4} rd/s$, $0.1 m/s$, $1.73 \times 10^{-3} rd$, $1.73 \times 10^{-4} rd$ and $0.1m$ respectively [26]. The fault estimate $\hat{f}$ is first filtered by a low pass filter of order 1 with cutting frequency $\omega_c = 20 rd/s$. Next, the GLR (Generalized Likelihood Ratio) test [30] is applied to $\hat{f}_f$ in order to make a decision about the faulty situation. An alarm is triggered (or disabled) when the GLR test boolean variable is greater (or lower) than a pre-specified threshold during a *confirmation time* t_{conf} [27]. For rising and falling edge of the GLR test boolean variable, t_{conf} is equal to 0.3 and 2 seconds respectively. Here, $GLR(\hat{f}_f(t))$ refers to the application of the GLR test to the signal $\hat{f}_f(t)$. This reduces to a boolean signal denoted $alarm(t)$ equals to "1" if and only if a fault has been detected (and confirmed, thanks to t_{conf}) by the fault detection unit.

The investigated scenarios correspond to four main cases that correspond to:
i) no fault occurs;
ii) an oscillatory fault occurs at $t = 4.85s$;
iii) an intermittent bias fault with different "time ON" values and
iv) a drift-like fault type in the time interval $[5, 15]$ seconds.

To appreciate the considered flight maneuver, the longitudinal states for $h = 980m$, $V_{TAS} = 145 m/s$, $m = 263000 kg$ and the X-component of center of gravity equal to 0.25 in fault-free case (case *i*)) is shown in [35]. All these scenarios are related to the terminal phase of a landing trajectory, i.e. after the glideslope capture, when the longitudinal model is valid (see [29]). To make a better assessment of the proposed scheme, the aircraft operation range considered in the simulations corresponds to the combination of the following parameters:
- altitude: $h = [980\ 1200\ 1500\ 3500]m$
- True air-speed: $V_{TAS} = [125\ 133.8\ 145] m/s$;
- mass: $m = [214000\ 263000\ 317000] kg$;
- X-component of center of gravity: $[0.17\ 0.25\ 0.31]\%/100$

As a consequence, each considered scenario involve 108 runs. Fig. 1, Fig. 2, Fig. 3 and Fig. 4 give the results for the case *i*), *ii*), *iii*) and *iv*) respectively. It can be seen that the proposed approach succeeds to reconstruct the fault signal despite the parameter variations. In particular, the use of the non homogeneous HOSM differentiator makes quick intermittent fault reconstruction possible thanks to its property

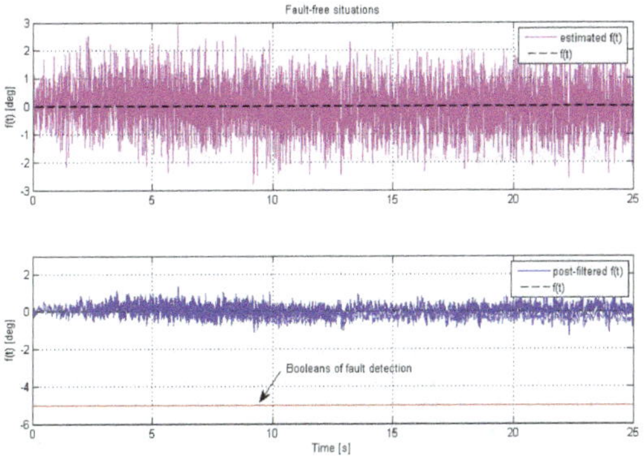

Fig. 1 Behaviour of $f(t), \hat{f}(t), \hat{f}_f(t)$ and the decision test for 108 runs - fault-free case

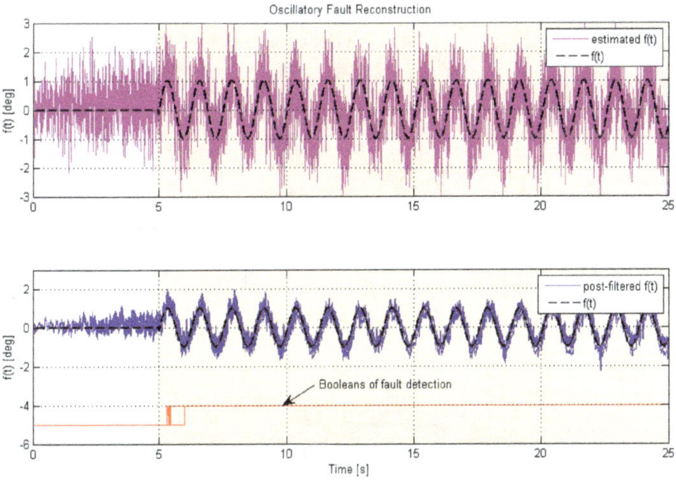

Fig. 2 Behaviour of $f(t), \hat{f}(t), \hat{f}_f(t)$ and the decision test for 108 runs - OFC

given in [20]. This highlights an interesting contribution of the proposed work, i.e. early fault reconstruction can be achieved.

Furthermore, and as expected, the estimation error depends on the measurement noise bound. This has been explained in Section 4 where it is stated that an upper bound of the fault estimation error is given in the L_∞ norm sense, by $\left\|\hat{f}(t) - f(t)\right\|_\infty \leq \gamma_2 (\eta^+)^{1/2} + \eta^+ \approx 3^\circ$ (an investigation of the data reveals a bound η^+ equal to $\eta^+ = 1.86$ thus $\gamma_2 \approx 0.84$). Note that the 'on' and 'off' characteristic of fault detection boolean in Fig. 3 is a feature of great importance in the aircraft structural design optimization. Indeed, the availability of viable measurement is improved, especially for the case of intermittent faults.

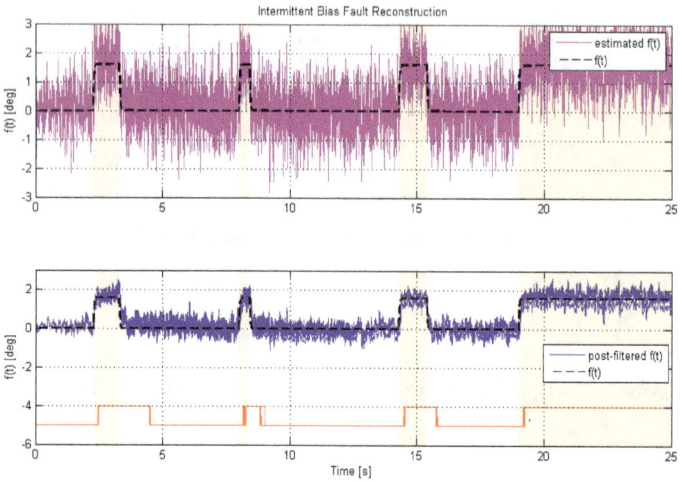

Fig. 3 Behaviour of $f(t), \hat{f}(t), \hat{f}_f(t)$ and the decision test for 108 runs - intermittent faults

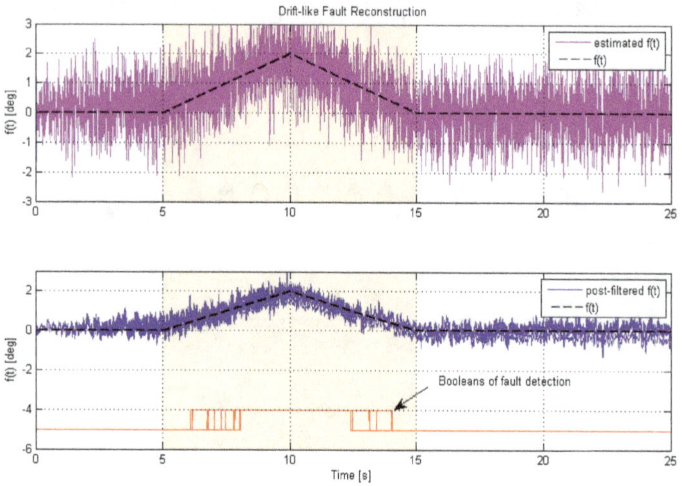

Fig. 4 Behaviour of $f(t), \hat{f}(t), \hat{f}_f(t)$ and the decision test for 108 runs - drift-like faults

Finally, fault detection performance statistics are computed to evaluate the performance of the fault detection scheme. They are summarized in table 1. The detection time performance (DTP) index is used to quantify the detection time. From table 1, it follows that all faults are successfully detected despite the parameter variations and the time profile of the faults, i.e. the 0% of false alarms ($FA\%$) and 100% of true detection ($TD\%$) for all considered parameter variations and fault types. Furthermore, the DTP index reveals reasonable detection delays.

Table 1 Statistics for the four considered faulty time profile

	$FA\%$	$TD\%$	DTP Min	DTP Max
no fault	0%	–	–	–
oscillatory fault	–	100%	$0.46s$	$1.14s$
intermittent bias fault	–	100%	$0.15s$	$0.27s$
drift-like fault	–	100%	1.09	3.06

5.4 A Pilot Experiment

The HOSM differentiator–based fault detection unit is finally used for a SIMONA–based pilot evaluation. The goal is to evaluate the robustness performance of the proposed fault detection unit in a more realistic environment, especially in the presence of a human pilot and when both longitudinal and lateral motions are performed.

The SIMONA flight simulator is located at the Delft University of Technology, Netherlands, see Fig. 5 for an illustration. It is a six degrees of freedom hydraulic motion system. It was designed to give realistic inertial motion cues. Specifically it was developed for human-machine interface and handling qualities research. The simulator's flexible architecture [31] allows for a high-fidelity integration of the B747-100/200 model (with a realistic aerodynamic coefficients database) by using the Matlab/Simulink Real-Time Workshop code generation. The inputs and outputs of SIMONA simulator have been standarised such that the actuators are driven by a dSPACE/SIMULINK architecture. The interested reader can refer to [32, 31, 33, 34] for more details about the visual system, simulator cab and flight desk of the SIMONA flight simulator.

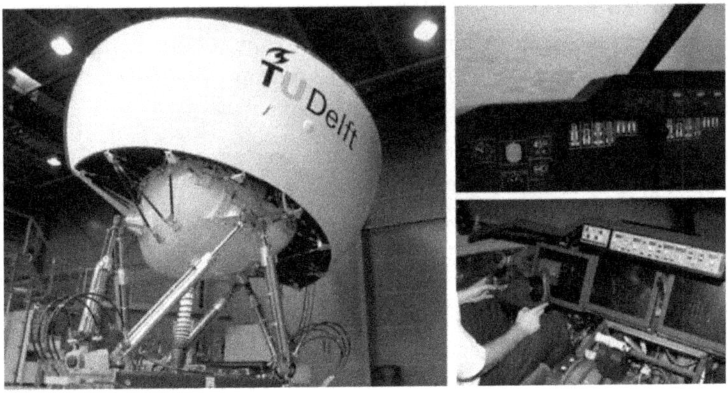

Fig. 5 SIMONA flight simulator

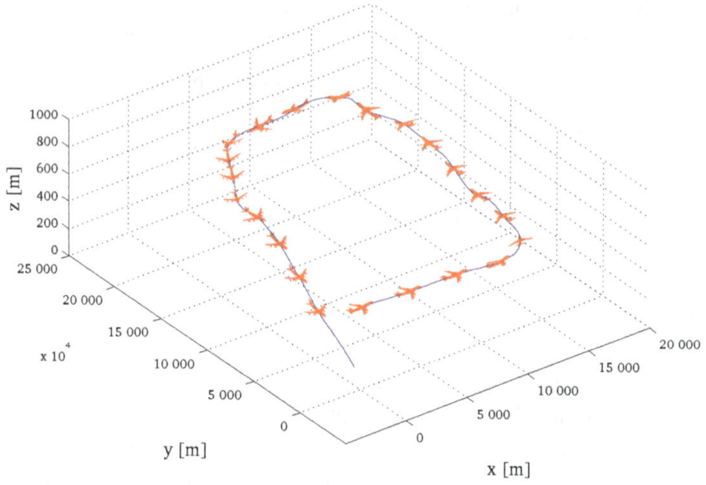

Fig. 6 The flight trajectory

The flight trajectory and attitude of the aircraft done by the pilot are illustrated in Fig. 6. The trajectory corresponds to a landing trajectory so that:

- the flight starts at the trimming conditions;
- then, the pilot performs a manual left turn;
- the altitude reference is changed to $1000m$ using the MCP (Mode Control Panel), a device interface that is used by the FCS defined in [21] to automatically generate trajectory reference (altitude in this case);
- then, another altitude change is done followed by a manual left turn;
- the next phase is concerned with the landing phase, i.e. when nearing the localizer, a final left turn is done and the localizer signal is captured. In the meantime, flaps are deployed. After the glideslope capture, the altitude decreases at low speed. The experiment ends close to the runway (the final altitude is $\approx 18m$) with the proper speed and alignment for landing.

Of course, this is a fault-free flight since it is a human experience. Again, the goal is there to evaluate the robustness performance of the proposed fault detection unit in a realistic flight, especially when both longitudinal and lateral motions are performed.

The two plots in the top of Fig. 7 illustrates the behaviour of the measure $\alpha(t)$, the estimate $\hat{\alpha}(t)$, the fault $f(t)$ and its estimate $\hat{f}(t)$ and the fault indicating signal $alarm(t)$ for the flight trajectory given in Fig. 6. As it be seen, the accuracy of the fault estimation error depends on:
- the measurement noise bound: Again, it can be identified a bound of the estimation error given approximatively by $\left\|\hat{f}(t)-f(t)\right\|_\infty \leq \gamma_2(\eta^+)^{1/2}+\eta^+ \approx 3°$
- and (obviously) the aircraft model. So, it can easily be noticed that the estimation error increases when the aircraft is performing "strong" lateral motions (i.e. left turns with important roll motion that correspond to $150s \leq t \leq 250s$ and $320s \leq t \leq 380s$).

Remark 3. To avoid false alarm during the "strong" lateral maneuveur, the GLR threshold is increased when the pilot command relative lateral motion is higher than a fixed threshold. This lead to a time-varying decision rule that can be obvious conservative in lateral maneuvers. In order to reduce the conservativeness of the proposed scheme, longitudinal model should be abandoned and the full aircraft model should be considered.

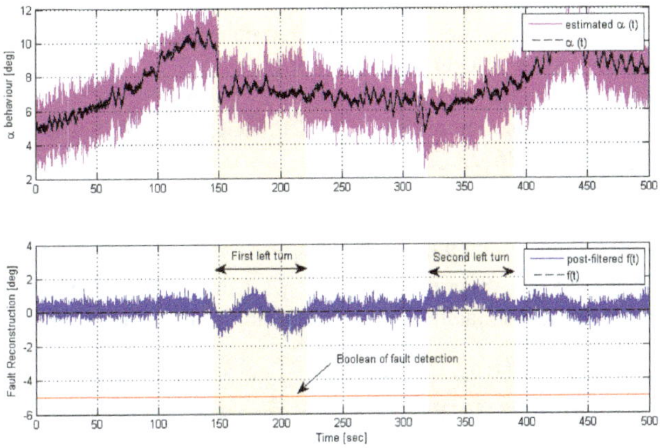

Fig. 7 Behaviour of $\alpha(t), \hat{\alpha}(t), f(t), \hat{f}_f(t), alarm(t)$ for the first 500 seconds of the overall flight trajectory - fault free situation

6 Concluding Remarks

In this work, a sensor fault reconstruction methodology dedicated to avionics systems has been developed. The early fault reconstruction is achieved by means of a non-homogeneous high order sliding mode differentiator. Theoretically the exact reconstruction of the sensor faults are carried in finite time even in the presence of bounded perturbations on the system. Nevertheless, experiments shown that in the presence of bounded measurement noises, the fault reconstruction is attained with an accuracy related to the noise L_∞ norm and the differentiation order. This scheme has been validated using a simulator of a large commercial aircraft. Further investigations will be necessary to integrate the developed fault identification strategy into a general fault tolerant flight control system.

Acknowledgements. Authors gratefully acknowledge the financial support from CONACyT(Mexico) under grant 151855, from Fondo Institucional del CONACYT under grant 209247, SIP-IPN under grant 20130983, from CDA-IPN and from the CNRS, France, under grant 218671. We would like to thank the TuDelft (NL) staff involved in the GARTEUR RECOVER project for making the SIMONA simulator results available to us.

References

1. Berdjag, D., Cieslak, J., Zolghadri, A.: Fault diagnosis and monitoring of oscillatory failure case in aircraft inertial system. Control Engineering Practice 20, 1410–1425 (2012)
2. Allerton, D., Jia, H.: Distributed data fusion algorithms for inertial network systems. IET Radar, Sonar and Navigation 2, 51–62 (2008)
3. Henry, D., Cieslak, J., Zolghadri, A., Efimov, D.: A non-conservative H_-/H_∞ solution for early and robust fault diagnosis in aircraft control surface servo-loops. Control Engineering Practice 31, 183–199 (2014)
4. Oosterom, M., Babuska, R., Verbruggen, H.: Soft computing applications in aircraft sensor management and flight control law reconfiguration. IEEE Transactions on Systems Man, and Cybernetics, Part C (Applications and Reviews) 32, 125–139 (2008)
5. Hegg, J.: Enhanced space integrated GPS/INS (SIGI). IEEE Aerospace and Electronic Systems Magazine 17, 26–33 (2002)
6. Zolghadri, A., Henry, D., Cieslak, J., Efimov, D., Goupil, P.: Fault Diagnosis and Fault-Tolerant Control and Guidance for Aerospace Vehicles: From theory to application. Series: Advances in Industrial Control. Springer (2014) ISBN 978-1-4471-5312-2
7. Balaban, E., Saxena, A., Bansal, P., Goebel, K., Curran, S.: Modeling, detection, and disambiguation of sensor faults for aerospace applications. IEEE Sensors Journal 9, 1907–1917 (2009)
8. Tan, C.P., Edwards, C.: Sliding mode observers for detection and reconstruction of sensor faults. Automatica 38, 1815 (2002)
9. Utkin, V.I.: Sliding modes in control and optimization. Springer, Berlin (1992)
10. Levant, A.: Higher-order sliding modes, differentiation and output-feedback control. International Journal of Control 76, 924–941 (2003)
11. Ferrara, A., Pisu, P.: Minimum sensor second-order sliding mode longitudinal control of passenger vehicles. IEEE Transactions on Intelligent Transportation 5, 20–32 (2004)
12. Cieslak, J., Efimov, D., Zolghadri, A., Henry, D., Goupil, P.: Design of a non-homogeneous differentiator for actuator oscillatory failure case reconstruction in noisy environment. Proc. IMechE Part I: J. Systems and Control Engineering (2014), doi:10.1177/0959651814561091
13. Imine, H., Madani, T., Srairi, S.: High order sliding mode observer to estimate vertical forces: experimental results. In: 11th International IEEE Conf. on Intelligent Transportation Systems, pp. 523–527 (2008)
14. Edwards, C., Tan, C.P.: Sensor fault tolerant control using sliding mode observers. Control Engineering Practice 14, 897–908 (2006)
15. Fridman, L., Levant, A., Davila, J.: Observation of linear systems with unknown inputs via high-order sliding-modes. International Journal of Systems Science 38, 773–791 (2007)
16. Bejarano, F., Fridman, L.: Output integral sliding mode control based on algebraic hierarchical observer. International Journal of Control 9, 1920–1929 (2010)
17. Fridman, L., Davila, J., Levant, A.: High-order sliding-mode observation for linear systems with unknown inputs. Nonlinear Analysis: Hybrid Systems 5(2), 189–205 (2011)
18. Kolmogorov, A.N.: On inequalities between upper bounds of consecutive derivatives of an arbitrary function defined on an infinite interval. American Mathematical Society Translations, 233–242 (1962)
19. Bejarano, F., Figueroa, M., Pacheco, J., Rubio, J.D.J.: Robust fault diagnosis of disturbed linear systems via a sliding mode high order differentiator. International Journal of Control 85, 648–659 (2012)

20. Levant, A.: Non-homogeneous finite-time convergent differentiator. In: Proc. of the Conference on Decision and Control, pp. 8399–8404 (2009)
21. Edwards, C., Lombaerts, T., Smaili, H.: Fault Tolerant Flight Control: A Benchmark Challenge. Springer (2010)
22. Alwi, H., Edwards, C., Tan, C.: Sliding mode observers for detection and reconstruction of sensor faults. Automatica 45, 1679–1685 (2009)
23. Fridman, L., Levant, A., Davila, J.: Observation of linear systems with unknown inputs via high-order sliding-modes. Int. J. System Science 38(10), 773–791 (2007)
24. Bejarano, F., Fridman, L., Poznyak, A.: Exact state estimation for linear systems with unknown inputs based on hierarchical super-twisting algorithm. Intern. J. of Robust and Nonlinear Control 17, 1734–1753 (2007)
25. Molinari, B.P.: A strong controllability and observability in linear multivariable control. IEEE Transactions on Automatic Control 21, 761–764 (1976)
26. Van Der Linden, C.: DASMAT-Delft university aircraft simulation model and analysis tool. Report LR-781 Technical University Delft (1996)
27. Goupil, P., Marcos, A.: The european addsafe project: Industrial and academic efforts towards advanced fault diagnosis. Control Engineering Practice (2014)
28. Cieslak, J., Henry, D., Zolghadri, A.: Fault tolerant flight control: from theory to piloted flight simulator experiments. IET Control Theory and Applications 4, 1451–1464 (2010)
29. Cieslak, J., Henry, D., Zolghadri, A., Goupil, P.: Development of an active fault-tolerant flight control strategy. Journal of Guidance, Control and Dynamics 31, 135–147 (2008)
30. Basseville, M., Nikiforov, I.: Detection of abrupt changes. Theory and application. Prentice Hall Information and System Sciences Series (1993)
31. Stroosma, O., Van Paassen, M., Mulder, M.: Using the simona research simulator for humanmachine interaction research. In: AIAA Modeling and Simulation Technologies Conf. (2003)
32. Koekebakker, S.: Model based control of a flight simulator motion base, Delft, Netherlands (2001)
33. Berkouwer, W., Stroosma, O., Van Paassen, M., Mulder, M., Mulder, J.: Measuring the performance of the simona research simulators motion system. In: AIAA Modeling and Simulation Conf. (2005)
34. Stroosma, O., Smaili, H., Mulder, J.: Pilot-in-the-loop evaluation of fault-tolerant flight control systems. In: SAFEPROCESS 2009 (2009)
35. Ferreira de Loza, A., Cieslak, J., Henry, D., Dávila, J., Zolghadri, A.: Sensor fault diagnosis using a non-homogeneous high-order sliding mode observer with application to a transport aircraft. IET Control Theory and Applications (2015), doi:10.1049/iet-cta.2014.0226

CubeSat Attitude Estimation via AUKF Using Magnetometer Measurements and MRPs

Francesco Sanfedino, Marco Scardino, Jérémie Chaix, and Stéphanie Lizy-Destrez

Abstract. In this article the Attitude and Control system of a CubeSat is presented. The attitude estimation design approach used is based on Adaptative Unscented Kalman Filter (AUKF) using three-axis magnetometer measurements. A set of modified Rodrigues Parameters (MRPs) is used to evaluate the attitude. Finally in order to have an complete ADCS system two control laws are introduced (Bdot and Sliding Mode) to best simulate a real CubeSat mission. The first one allows the spacecraft the control during the detumbling phase (phase at high angular rates) and in case of reaction wheels saturation and the second one is used for the nominal control (phase at low angular rates).

1 Introduction

During the last decades there has been a great development of cheap and small satellites, especially in CubeSat projects.

The CubeSat concept created in 1999 by Jordi Puig-Suari of California Polytechnic State University and Bob Twiggs of Stanford University in order to allow students to achieve all the skills which a complex satellite project needs [6].

CubeSats are picosatellites of standardised dimensions (cubes of $10\,cm$ per side with a maximum mass of $1\,kg$). The standard $10\text{x}10\text{x}10\,cm$ basic CubeSat is often called one unit or 1U CubeSat. CubeSats are scalable along only one axis by 1U increments. As in all satellites also in CubeSats the Attitude and Determination Control System (ADCS) plays a key role in their lives, because it guarantees the right pointing for the communication with the ground station.

Francesco Sanfedino · Marco Scardino · Jérémie Chaix · Stéphanie Lizy-Destrez
Institut Supérieur de l'Aéronautique et de l'Espace, 10,
Avenue Edouard-Belin, 31055 Toulouse
e-mail: {francesco.sanfedino,marco.scardino}@supaero.isae.fr,
{jeremie.chaix,stephanie.lizy-destrez}@isae.fr

J. Bordeneuve-Guibé et al. (eds.), *Advances in Aerospace Guidance, Navigation and Control*, DOI: 10.1007/978-3-319-17518-8_20

In a CubeSat application the ADCS is based above all on three-axis magnetometer employment. This sensor has several advantages such as relative low cost, low required power and continuous availability. In fact the major part of CubeSat orbits are Low Earth Orbit (LEO) and for these ones, during solar eclipse, only the Earth's magnetic field observations are available. Besides magnetometers can also serve as backup attitude estimators [11].

An example of CubeSat which exploits this idea is the JumpSat. The JumpSat is a 3U CubeSat mission proposed by ISAE Supaero in collaboration with TELECOM Bretagne, Massachusetts Institute of Technology (MIT), Centre National dEtude Spatial (CNES) and ONERA.

The goals of this mission are [9]

- Technological verification and Space qualification of a star tracker, which is currently under development by ISAE Supaero for future use in small satellite systems.
- Mapping of the properties of the Earth radiation belt with emphasis of the South Atlantic Anomaly using a directional radiation sensor under development by ONERA.
- Technological verification and Space qualification of the three-axis attitude control system of the Jumpsat space segment.

The ADCS is the system which has the role of satellite attitude control in each phase of its life-cycle. The information taken from some sensors is exploited by actuators in order to produce correction torques. The way of control is based on the specific operational phase.

In the JumpSat mission several phases have been identified.

The most influent ones are two [9]:

- *Rotational Rate Reduction Mode* (Angular rates higher than $5\,^{\circ}/s$): an operational mode to eliminate the rotational energy of the system after separation from the lunch vehicle or after idle times of the system. It is based on the B-dot control law [5] and utilizes the Magnetometer and Magnetorquer only.
- *Attitude Acquisition Mode* (Angular rates smaller than $5\,^{\circ}/s$): the main operation mode of the ACS system, based on all available sensors and actuators. It allows pointing of the satellite in any direction in any of the reference frame.

The attention has been focused on the *Rotational Rate Reduction Mode*, which guarantees the satellite mission survival. Thus the model of a real magnetometer has been made by adding a noise and a bias term to the magnetic field. Thanks to this model and a Kalman Filter employment an estimation of the system states (attitude and angular rate) can be obtained.

For this non-linear application it is necessary to use the Unscented Kalman Filter (UKF) algorithm, based on the Unscented Transformation (UT) [7]. The UT uses a set of sigma points in order to compute the statistics behaviour (propagation of means and covariance) of variables undergone to a nonlinear transformation. Sigma points can be selected according to the symmetric and spherical simplex sigma point [8].

In the next sections the implemented methods and the simulation results will be presented.

2 Methods

2.1 Unscented Kalman Filter

2.1.1 Unscented Transformation

The Unscented Transformation is a method for calculating the statistics of random variable, which undergoes under a non linear transformation. In order to do this, it uses a set of sigma points that guarantees the propagation of means and covariance through the non linear equations. Supposing a random variable $x \in R^n$ has mean $\bar{x}$ and covariance P_x, and x is propagated trough a non linear function, $y = g(x)$. In order to calculate the statistics of y, the sigma points can be selected according to the symmetric and spherical simplex sigma points. For the symmetric sigma points it is necessary to have $2n$ sigma points to represent the mean and covariance, while for the spherical simplex sigma points it is necessary to have $n+2$ points. Generally, the computational cost of Unscented Transformation are proportional to the number of sigma points. For this reason the spherical sigma points approach is chosen. In order to evaluate the spherical silmplex sigma points, the algorithm shows [8] is used.

2.1.2 State Estimation

The Unscented Transformation and the sigma points permit the estimation of a non linear dynamic system state vector.

An example of this system in discrete time is:

$$\begin{cases} \mathbf{x}_k = \mathbf{F}(\mathbf{x}_{k-1}, \mathbf{w}_{k-1}) \\ \mathbf{y}_k = \mathbf{H}(\mathbf{x}_k, \mathbf{v}_k) \end{cases} \tag{1}$$

where $\mathbf{x}_k$ represents the states of the system, $\mathbf{y}_k$ is the measurement of the system, $\mathbf{w}_k$ and $\mathbf{v}_k$ are noise, respectively, of the system and of the measurement. The non-linear dynamic equation system considered in this work is:

$$\begin{cases} \mathbf{x}_k = \mathbf{f}(\mathbf{x}_{k-1}) + \mathbf{w}_{k-1} \\ \mathbf{y}_k = \mathbf{h}(\mathbf{x}_k) + \mathbf{v}_k \end{cases} \tag{2}$$

For this particular non-linear system the Unscented Kalman Filter is called *Additive Unscented Kalman Filter*. The *Additive Unscented Kalman Filter*, respect to classical Unscented Kalman Filter used for the non-linear system 1, provides a greater estimation error, but it is more difficult to tune. The formulation for *Additive Unscented Kalman Filter* is given as follows.

Firstly, the filter is initialized as:

$$\hat{\mathbf{x}}_0 = E[\mathbf{x}_o] = [\mathbf{x}_o]^T \tag{3}$$

$$P_0 = E[(\mathbf{x}_o - \hat{\mathbf{x}}_0)(\mathbf{x_o} - \hat{\mathbf{x}}_0)^T] = diag(P_o) \tag{4}$$

Then the predicted state mean and covariance are computed using Unscented Transformation:

$$\chi_{i,k|k-1} = \mathbf{f}(\chi_{i,k-1}) \quad i = 0, ..., n+1 \tag{5}$$

$$\hat{\mathbf{x}}_k^- = \sum_{i=0}^{n+1} w_i^m \chi_{i,k|k-1} \tag{6}$$

$$P_{x_{k+1}}^- = \sum_{i=0}^{n+1} w_i^c (\chi_{i,k|k-1} - \hat{\mathbf{x}}_k^-)(\chi_{i,k|k-1} - \hat{\mathbf{x}}_k^-)^T + Q_k \tag{7}$$

where χ represents the matrix of sphercial simplex sigma points. The mean and covariance observations are found by:

$$\mathbf{Y}_{i,k} = h(\chi_{i,k|k-1}) \quad i = 0, ..., n+1 \tag{8}$$

$$\hat{\mathbf{y}}_k^- = \sum_{i=0}^{n+1} w_i^m \mathbf{Y}_{i,k} \tag{9}$$

$$P_{y_k} = \sum_{i=0}^{n+1} w_i^c (\mathbf{Y}_{i|k} - \hat{\mathbf{y}}_k^-)(\mathbf{Y}_{i|k} - \hat{\mathbf{y}}_k^-)^T + R_k \tag{10}$$

where Q_k and R_k are the covariance matrix noise, respectively, of the state and measurement. The cross correlation covariance is calculated using:

$$P_{x_k y_k} = \sum_{i=0}^{n+1} w_i^c (\chi_{i,k} - \hat{\mathbf{x}}_k^-)(\mathbf{Y}_{i,k} - \hat{\mathbf{y}}_k^-)^T \tag{11}$$

Finally, the correction stage is defined as follows:

$$K_k = P_{x_k y_k} P_{y_k}^{-1} \tag{12}$$

$$\hat{\mathbf{x}}_k = \hat{\mathbf{x}}_k^- + K_k(\mathbf{y}_k - \hat{\mathbf{y}}_k) \tag{13}$$

$$P_{x_k} = P_{x_k}^- - K_k P_{y_k} K_k^T \tag{14}$$

2.1.3 The Adaptive Tuning of the Q Matrix

The estimation error of UKF depends on the initial choice of the covariance matrix noise Q and of the measurements covariance matrix. For this reason it is necessary to present the Adaptive Unscented Kalman Filter (AUKF), that is a method based on UKF, but the value of Q changes at each step time, in order to reduce the estimation error. An each time step the observation of Q can be written as [13]:

$$Q^* = \Delta \mathbf{x_{k+1}} \Delta \mathbf{x_{k+1}^T} + P_k^- - P_k + Q_k \tag{15}$$

where Q_k is the current covariance matrix noise and $\Delta \mathbf{x_{k+1}}$ is the difference between the estimated and the predicted state.

$$\Delta \mathbf{x_{k+1}} = \hat{\mathbf{x}}_\mathbf{k} - \hat{\mathbf{x}}_k^- \tag{16}$$

So the estimation for the covariance matrix noise is

$$Q_{k+1} = Q_k + \frac{1}{\gamma}[Q^* - Q_k] \tag{17}$$

where γ represents the window size that sets the level of expected change in the noise covariance.

2.2 *Attitude Dynamics and Sensor Models*

This section provides a brief review of spacecraft attitude dynamics. The attitude parameters here introduced are the Modified Rorigues Parameters. A quaternion system is generally applied for spacecraft pointing and regulation thanks to the absence of singularities in its kinematic equations. However, the use of quaternions requires an extra parameter which leads to a non-minimal parametrization. The Rodrigues parameters provide a minimal (i.e., three dimensional) parametrization. However, a singularity exists for 180° rotations, which hinders this parametrization for extremely large angle rotations. The compromise between the two models it is the modified Rodrigues parameters application, whose singularity at 360° can be solved by a method explained in the section 2.2.1. Moreover they answer to the minimal parametrization need.

As in [2] this parametrization is derived by employing a stereographic projection of the quaternions. The quaternion representation is given by:

$$\mathbf{q} \equiv \begin{bmatrix} \mathbf{q}_{13} \\ q_4 \end{bmatrix} \tag{18}$$

with

$$\mathbf{q}_{13} \equiv \begin{bmatrix} q_1 \\ q_2 \\ q_3 \end{bmatrix} = \hat{\mathbf{n}} \sin\left(\frac{\theta}{2}\right) \tag{19}$$

$$q_4 = \cos\left(\frac{\theta}{2}\right) \tag{20}$$

where $\hat{\mathbf{n}}$ is the unit vector corresponding to the axis of rotation and θ is the angle of rotation.

The Modified Rodrigues Parameters are defined by Equation (21):

$$\mathbf{p} = \frac{\mathbf{q}_{13}}{1+q_4} = \hat{\mathbf{n}} \tan\left(\frac{\theta}{4}\right) \tag{21}$$

where $\mathbf{p}$ is a 3×1 vector. The kinematic equations of motion are derived by using the spacecraft's angular velocity $\mathbf{w}$, given by:

$$\dot{\mathbf{p}} = \frac{1}{2}\left\{\frac{1}{2}\left(1-\mathbf{p}^T\mathbf{p}\right) I_{3\times3} + [\mathbf{p}\times] + \mathbf{p}\mathbf{p}^T\right\}\mathbf{w} \tag{22}$$

The non-linear three-axis rotational dynamics of the rigid spacecraft with momentum wheel may be expressed as:

$$\dot{\mathbf{w}} = J^{-1}\left\{\mathbf{T}_c - \dot{\mathbf{H}}_i - [\mathbf{w}\times]\mathbf{H}_i - [\mathbf{w}\times]J\mathbf{w} + \Delta\mathbf{T}\right\} \tag{23}$$

where, J is the moment of the inertia matrix, $\mathbf{T}_c$ is the magnetorquer control torque, $\mathbf{H}_i$ is the angular momentum vector and , $\dot{\mathbf{H}}_i$ is the wheel control torque and $\Delta\mathbf{T}$ is the disturbance torque.

For convenience, defining the state vector $\mathbf{x} \in \mathbb{R}^{7\times1}$ in the attitude estimator as $\mathbf{x} = \begin{bmatrix}\mathbf{p}^T & \mathbf{w}^T & t\end{bmatrix}^T$, where the time t is added as estimation variable for a simpler implementation.

The non-linear dynamics equation for propagating $\mathbf{x}$ is rewritten as

$$\dot{\mathbf{x}} = F\left(\mathbf{x}, \mathbf{w}_T\right) = \mathbf{f}(\mathbf{x}) + \mathbf{w}_T \tag{24}$$

where

$$\mathbf{f}(\mathbf{x}) = \begin{bmatrix} \frac{1}{2}\left\{\frac{1}{2}\left(1-\mathbf{p}^T\mathbf{p}\right) I_{3\times3} + [\mathbf{p}\times] + \mathbf{p}\mathbf{p}^T\right\}\mathbf{w} \\ J^{-1}\left\{\mathbf{T}_c - \dot{\mathbf{H}}_i - [\mathbf{w}\times]\mathbf{H}_i - [\mathbf{w}\times]J\mathbf{w} + \Delta\mathbf{T}\right\} \\ 1 \end{bmatrix} \tag{25}$$

and the process noise $\mathbf{w}_T$ is zero-mean white noise described by the process noise matrix Q. The attitude measurement model for a single sensor is given by:

$$\mathbf{y}_k = \begin{bmatrix} \mathbf{B}_{\mathbf{body}\,k} \\ t_k \end{bmatrix} + \mathbf{v}_k = \begin{bmatrix} A(\mathbf{p}_k) \\ 1 \end{bmatrix} \begin{bmatrix} \mathbf{B}_k \\ t_{OBC} \end{bmatrix} + \mathbf{v}_k \tag{26}$$

where $\mathbf{y}_k$ is the kth measurement vector, t_{OBC} is the time provided by the on-board computer, and $\mathbf{v}_k$ is measurement zeros-mean white noise. So the measurements of the Earth magnetic field in the orbital frame are translated into the spacecraft body frame by using the matrix $A(\mathbf{p})$

$$A(\mathbf{p}) = I_{3\times3} - \frac{4\left(1-\mathbf{p}^T\mathbf{p}\right)}{\left(1+\mathbf{p}^T\mathbf{p}\right)^2}[\mathbf{p}\times] + \frac{8}{\left(1+\mathbf{p}^T\mathbf{p}\right)^2}[\mathbf{p}\times]^2 \tag{27}$$

2.2.1 How to Avoid the Singularity of Modified Rodrigues Parameters

In order to avoid the singularity of Modified Rodrigues Parameters, it is possible to use the similar parameters, which are called *shadow* Modified Rodrigues Parameters [12]:

$$p_i^S = -\frac{q_i}{1-q_4} = -\frac{p_i}{\mathbf{p}^T\mathbf{p}} \tag{28}$$

The shadow parameters $\mathbf{p}^S$ have some interesting properties. They have a singularity at the zero rotation and they go to zero at $\pm 360^\circ$ of principal rotation. This is the exact opposite of $\mathbf{p}$. For this reason they can be used when the Modified Rodrigues Parameters go to singular. So with these shadow points it is possible describe any rotation of satellite without singularity, but one discontinuity is present at the switching point. In terms of $\mathbf{p}$ the cosine matrix and the kinematic equation are exactly the same as 22 and 27.

In order to switch between these parameters it is possible to use the following relationship:

$$|\mathbf{p}||\mathbf{p}^s| = 1 \tag{29}$$

When using $\mathbf{p}$ to represent the attitude, there is switch from $\mathbf{p}$ to $\mathbf{p}^s$ if $|\mathbf{p}| > 1$ and thus:

$$\mathbf{p}^s = -\frac{\mathbf{p}}{|\mathbf{p}|^2} \tag{30}$$

While using $\mathbf{p}^s$ to represent the attitude, there is switch from $\mathbf{p}^s$ to $\mathbf{p}$ if $|\mathbf{p}^s| > 1$, and thus:

$$\mathbf{p} = -\frac{\mathbf{p}^s}{|\mathbf{p}^s|^2} \tag{31}$$

So with this definition the magnitude of $\mathbf{p}$ and $\mathbf{p}^s$ will never exceed 1, which results in avoiding the singularity.

2.3 *Control Law*

This section will present the two control laws implemented in this project: the B-Dot for the *Rate-Reduction Mode* and the Sliding Mode for the *Attitude-Acquisition Mode*.

2.3.1 B-Dot Control Law

This section describes the B-Dot control law for the *Rate-Reduction Mode*. The principle on which the B-Dot is based [5] is the minimization of the derivative of the Earth's magnetic field vector $\mathbf{B}$ measured by a magnetometer. The rate of its change depends on the spacecraft rotation rate. Thus the minimization of this derivative determines a decrease of the satellite angular rate that corresponds to a reduction of the rotational kinetic energy. This is defined as:

$$\dot{\mathbf{E}}_{\mathbf{rot}} = \frac{d}{dt}\left(\frac{1}{2}\mathbf{w_{body}}^T \cdot \mathbf{I_{sat}} \cdot \mathbf{w_{body}}\right) \tag{32}$$

This means that the scalar product of the angular rate of satellite body and the control torque must be smaller than zero:

$$\mathbf{w_{body}}^T \cdot \mathbf{T_c} < 0 \tag{33}$$

The control torque $\mathbf{T_c}$ is the result of the interaction of the Earth magnetic field vector $\mathbf{B}$ and the magnetorquers magnetic momentum $\mathbf{M_{torquer}}$:

$$\mathbf{T_c} = \mathbf{M_{torquer}} \times \mathbf{B_{Earth}} \tag{34}$$

Thus:

$$\mathbf{w_{body}}^T \cdot (\mathbf{M_{torquer}} \times \mathbf{B_{Earth}}) < 0 \tag{35}$$

After rearranging, Equation (35) becomes:

$$\mathbf{M_{torquer}} \cdot \left(\mathbf{w_{body}}^T \times \mathbf{B_{Earth}}\right) < 0 \tag{36}$$

From this inequality it can be deduced that the unique negative parameter has to be $\mathbf{M_{torquer}}$. Thus a negative control scalar gain C_{b-dot} is introduced in order to minimise the rotation kinetic energy.

The commanded control torque becomes:

$$\mathbf{M_{torquer}} = -C_{b-dot} \cdot (\mathbf{w_{body}} \times \mathbf{B_{Earth}}) \tag{37}$$

In Equation 37 the cross product between the angular body rate $\mathbf{w_{body}}$ and the Earth's magnetic field vector is equal to the time derivative of the Earth's magnetic field vector $\dot{\mathbf{B}}_{\mathbf{Earth}}$:

$$\dot{\mathbf{B}}_{\mathbf{Earth}} = \mathbf{w_{body}} \times \mathbf{B_{Earth}} \tag{38}$$

The control law finally becomes:

$$\mathbf{M_{torquer}} = -C_{b-dot} \cdot \dot{\mathbf{B}}_{\mathbf{Earth}} \tag{39}$$

2.3.2 Synthesis of Sliding Mode Control

Considering the following non-linear system:

$$\dot{\mathbf{x}} = \mathbf{f}(\mathbf{x}) + \mathbf{g}(\mathbf{x})\mathbf{U} \tag{40}$$

where $\mathbf{x}$ is the state vector of the system and $\mathbf{U}$ represents the command that stabilizes $\mathbf{x}$. For this system the goal is to find the value of $\mathbf{U}$. In order to do this, one approach is to choose a surface in the state space, called sliding surface $\mathbf{S}(\mathbf{x})$, where the command objective is:

- If $\mathbf{S}(\mathbf{x}) = 0$, the value of state vector has to be zero ($\mathbf{x} = 0$)
- To bring the state vector from an arbitrary position to the sliding surface.

Besides, the command of sliding mode can be characterized by the principle of attractiveness ($\mathbf{S}(\mathbf{x})\dot{\mathbf{S}}(\mathbf{x}) < 0$) and of invariance ($\dot{\mathbf{S}}(\mathbf{x})) = 0$ for $\mathbf{S}(\mathbf{x})) = 0$). With these principles it is possible to compute the value of $\mathbf{U}$ [1]:

$$\mathbf{U} = -\left(\left(\frac{\delta\mathbf{S}}{\delta\mathbf{x}}\right)^T \mathbf{g}(\mathbf{x})\right)^{-1} \left(\left(\frac{\delta\mathbf{S}}{\delta\mathbf{x}}\right)^T \mathbf{f}(\mathbf{x})\right) - Ksat(\mathbf{S}(\mathbf{x},\varepsilon)) \tag{41}$$

where K is a diagonal matrix that permits $\mathbf{U}$ to assure the condition of attractiveness and $sat(\mathbf{S}(\mathbf{x}))$ is the saturation function, that generally is equal to:

$$sat(\mathbf{S}(\mathbf{x},\varepsilon)) = \begin{cases} -1 \ if & \mathbf{S}(\mathbf{x}) < -\varepsilon \\ \frac{\mathbf{S}(\mathbf{x})}{\varepsilon} \ if & |\mathbf{S}(\mathbf{x})| < \varepsilon \\ 1 \ if & \mathbf{S}(\mathbf{x}) > \varepsilon \end{cases} \tag{42}$$

Sliding Surface using Modified Rodrigues and reaction wheels

For the *operational mode*, only the reaction wheels represent the way to stabilize the JumpSat. For this reason the control torque of magnetotorques is equal to zero and the Equation (23) of dynamics becomes:

$$J\dot{\mathbf{w}} + [\mathbf{w}\times]J\mathbf{w} = -[\mathbf{w}\times]\mathbf{H}_i - \dot{\mathbf{H}}_i + \Delta\mathbf{T} \tag{43}$$

Finally the linear model for spacecraft motion is:

$$\dot{\mathbf{p}} = \mathbf{F}(\mathbf{p})\mathbf{w} \tag{44}$$

$$\dot{\mathbf{w}} = \mathbf{f}(\mathbf{x}) - J^{-1}\mathbf{U} - J^{-1}\Delta\mathbf{T} \tag{45}$$

where,

$$\mathbf{F}(\mathbf{p}) = \frac{1}{2}\left\{\frac{1}{2}\left(1-\mathbf{p}^T\mathbf{p}\right)I_{3\times3} + [\mathbf{p}\times] + \mathbf{p}\mathbf{p}^T\right\} \tag{46}$$

$$\mathbf{f}(\mathbf{p}) = -J^{-1}[\mathbf{w}\times]J\mathbf{w} \tag{47}$$

$$\mathbf{U} = -[\mathbf{w}\times]\mathbf{H}_i - \dot{\mathbf{H}}_\mathbf{i} \tag{48}$$

For this dynamics the sliding surface is[3]:

$$\mathbf{S}(\mathbf{p}) = \mathbf{w} - \mathbf{m}(\mathbf{p}) \tag{49}$$

The value of $\mathbf{m}(\mathbf{p})$ is calculated from a desired vector of Modified Rodrigues Parameters and the kinematics equation:

$$\mathbf{m}(\mathbf{p}) = \mathbf{F}^{-1}(\mathbf{p})\mathbf{d}(\mathbf{p}) \tag{50}$$

where,

$$\mathbf{F}^{-1}(\mathbf{p}) = 4(1+\mathbf{p}^T\mathbf{p})^{-2} \\ \{(1+\mathbf{p}^T\mathbf{p})I_{3\times3} - 2[\mathbf{p}\times] + 2\mathbf{p}\mathbf{p}^T\}$$

and

$$\mathbf{d}(\mathbf{p}) = \Lambda(\mathbf{p} - \mathbf{p}_d) \tag{51}$$

where $\mathbf{P}_d$ is the desired reference of Modified Rodrigues Parameters and Λ is a diagonal matrix with negative elements. So with these elements the value of command is:

$$\mathbf{U} = -J\left\{\mathbf{f}(\mathbf{w}) - \frac{\delta \mathbf{m}}{\delta \mathbf{p}}[\mathbf{F}(\mathbf{p})\mathbf{m}(\mathbf{p}) + \mathbf{F}(\mathbf{p})\mathbf{S}(\mathbf{p})]\right\} \\ - JKsat(\mathbf{S}(\mathbf{p}), \varepsilon)$$

Supposing that the matrix Λ is given by a scalar λ times the identity matrix, the quantity $\mathbf{m}(\mathbf{p})$ becomes:

$$\mathbf{m}(\mathbf{p}) = 4\lambda(1+\mathbf{p}^T\mathbf{p})^{-1} - 4\lambda(1+\mathbf{p}^T\mathbf{p})^{-2} \\ \{(1-\mathbf{p}^T\mathbf{p})I_{3\times3} - 2[\mathbf{p}\times] + 2\mathbf{p}\mathbf{p}^T\}\mathbf{p}_d$$

ans its derivative respect $\mathbf{p}$ is equal to:

$$\frac{\delta \mathbf{m}(\mathbf{p})}{\delta \mathbf{p}} = 4\lambda(1+\mathbf{p}^T\mathbf{p})^{-1}\left\{I_{3\times3} - 2(1+\mathbf{p}^T\mathbf{p})^{-1}\mathbf{p}\mathbf{p}^T\right\} \\ - 8\lambda(1+\mathbf{p}^T\mathbf{p})^{-2}\{\mathbf{p}\mathbf{p}_d^T - \mathbf{p}_d\mathbf{p}^T + [\mathbf{p}_d\times] + (\mathbf{p}_d^T\mathbf{p})I_{3\times3}\} \\ + 16\lambda(1+\mathbf{p}^T\mathbf{p})^{-3}\{(1-\mathbf{p}^T\mathbf{p})I_{3\times3} - 2[\mathbf{p}\times] + 2\mathbf{p}\mathbf{p}^T\}\mathbf{p}_d\mathbf{p}^T$$

Finally in order to eliminate the effects of external disturbances, it is necessary to add to the quantity $\mathbf{U}$ another torque $\mathbf{U}_{dist}$, so the command to stabilise the system becomes:

$$\mathbf{U}_{tot} = \mathbf{U} + \mathbf{U}_{dist} \tag{52}$$

where

$$\mathbf{U}_{dist} = -\Delta\hat{\mathbf{T}} \tag{53}$$

$\Delta\hat{\mathbf{T}}$ is generally computed by an observer, but in our case it is a value computed by the *Unscented Kalman Filter*.

3 Results and Simulation

In this section two implemented models will be presented. Firstly the efficiency of the UKF estimation without the introduction of the control law will be analysed in order to give confidence to the measurements of the three-axis magnetometers. Then the results of the controlled system will be presented. For the simulations a model of the Earth magnetic field has been introduced according to the specification of IGRF11 available until 2015 [4]. The magnetic field vector has been computed at each $0.1\,s$ interval. In the reality only the measurements of the sensor will be used in order to evaluate the spacecraft attitude. The model of the magnetometer consists in a sensor with a bias and a white Gaussian noise to best simulate the errors accumulated by a real instrument. According to the verification of UKF algorithm all the control part has been removed. For the JumpSat mission a system of reaction wheels and magnetorquers is used to actuate the control. The elimination of the control system has as consequence the suppression of the wheel control torque $\dot{\mathbf{H}}_i$ and of the angular momentum vector $\mathbf{H}_i$, which have an important role in the dynamics Equation (23), but also of $\mathbf{T_c}$.

The complete system scheme is presented in Figure 1, where also the control law block is considered.

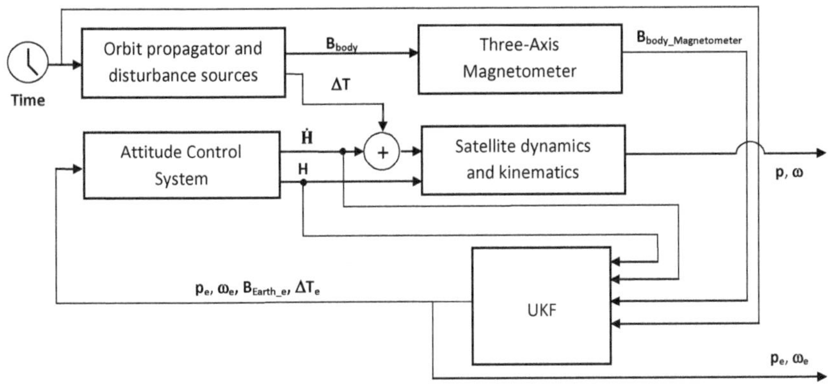

Fig. 1 Simulation scheme

For the test of estimation the blocks architecture except for the UKF are heritage of Christoph Pierls work [10], but all the equations has been translated into the modified Rodrigues Parameters $\mathbf{p}$ from the quaternion system and the control part has been adapted to the current case.

The *Orbit propagator and disturbance sources* block takes the time to calculate the evolution of the spacecraft trajectory in its orbit. Thanks to the orbit parameters evaluation it is possible to obtain the Earth magnetic field in the body frame $\mathbf{B_{body}}$ (here calculated by the IGRF11 model for a better precision) and all the considerable disturbing torques $\Delta\mathbf{T}$, i.e. the magnetic and the gravity gradient ones. The block *Satellite dynamics and kinematics* comprehends the Equations (22) and (23). The *Attitude Control System* takes into account the control law and the reaction wheel model in order to provide the control torque $\dot{\mathbf{H}}_i$ and the angular momentum $\mathbf{H}_i$.

For all the simulations two important hypothesis have been established:

- The Earth magnetic field model is the same in the block *Orbit propagator and disturbance sources* and in the filter model. In the reality a simpler algorithm than the IGRF11 is used in the embedded code for computational reasons. Thus it introduces other uncertainties not considered here.
- The model of the orbit is not estimated but taken from the environment modelled in the block *Orbit propagator and disturbance sources*. Thus a development of this work would be the implementation of an algorithm which predicts the satellite position in its orbit.

A set of parameters has been chosen to best produce the estimation. They are listed in Table 1.

Table 1 Parameters used in the estimation model. $\mathbf{B_{bias}}$ is the bias on the three-axis magnetometer, $\mathbf{B_{res\,dip}}$ is the residual dipole momentum of the satellite. $\mathbf{B_{pds}}$ is the deviation standard of the magnetic noise.

Parameter	*Value*
J	$diag[0.045, 0.045, 0.005]\,kg\,m^2$
$\mathbf{B_{bias}}$	$[25, -25, 25]^T\,nT$
$\mathbf{B_{res\,dip}}$	$[5\cdot10^{-8}, 5\cdot10^{-8}, 5\cdot10^{-6}]\,Am^2$
$\mathbf{B_{pds}}$	$[2\cdot10^{-9}; 2\cdot10^{-9}; 2\cdot10^{-9}]\,T$
W_0	0.5
P_0	$diag[0.087; 0.087; 0.87; 1; 1; 1; 0.1]\cdot10^{-1}$
Q_0	$diag[1.563; 1.43; 1.984; 0; 0; 0; 0.3]\cdot10^{-14}$
R_0	$diag[25; 25; 25; 100]\cdot10^{-15}$
$\mathbf{p_0}$	$[0; 0.414; 0]$
$\mathbf{w_0}$	$[0; 0; 0]\,rad/s$
γ	50
λ	−0.015
ε	0.01
k	0.0015
C_{b-dot}	−20000

In the following graphs different simulation of the implemented methods are presented. Particularly the variation of Euler's angles with the time will present . These angles represent the rotation needed to bring the body frame to the orbital frame:

- The angle ψ is the rotation around the axis Z_{body}, which brings the body frame to an intermediate frame identified with $X1_{body}$, $Y1_{body}$ and $Z1_{body} = Z_{body}$. This rotation is considered positive if it is an anti-clockwise rotation.
- The angle θ is the rotation around $Y1_{body}$, which brings the $F1_{body}$ to another intermediate frame, $X2_{body}$, $Y2_{body} = Y1_{body}$ and $Z2_{body}$. This rotation is considered positive if it is an anti-clockwise rotation.
- The angle ϕ is the rotation around $X2_{body} = X_{orbital}$, which aligns the two frames. It is positive if it is an anti-clockwise rotation.

An algorithm permit us to compute these angles from the values of the Modified Rodrigues Parameters. The first step of this algorithm is the evaluation of the quaternions:

$$q_4 = \frac{1 - \mathbf{p}^T \mathbf{p}}{1 + \mathbf{p}^T \mathbf{p}} \quad q_i = \frac{2p_i}{1 + \mathbf{p}^T \mathbf{p}} \quad for \quad i = 1,2,3 \tag{54}$$

With the values of the quaternions the Euler's angles are equals to:

$$\psi = atan\left(\frac{2(q_4 q_3 + q_1 + q_2)}{1 - 2(q_2^2 + q_3^2)}\right) \tag{55}$$

$$\theta = asin\,(2(q_4 q_2 - q_3 q_1)) \tag{56}$$

$$\phi = atan\left(\frac{2(q_4 q_1 + q_2 q_3)}{1 - 2(q_1^2 + q_2^2)}\right) \tag{57}$$

The limitations of this algorithm implemented on Matlab is that the values of ϕ and ψ are comprised between $-180\,^\circ$ and $180\,^\circ$ and for θ the values are comprised between $-90\,^\circ$ and $90\,^\circ$.

3.1 Performances of Modified Rodrigues Parameters

In this section a simulations will be presented in order to evaluate the filter efficiency.

Tacking as initial condition,

$$(\psi, \theta, \phi) = (140^\circ, 90^\circ, 160^\circ) \tag{58}$$

$$(r, q, p) = (1.3^\circ/s, 1^\circ/s, 0.85^\circ/s) \tag{59}$$

the estimation of UKF with Modified Rodrigues Parameters is presented in Figure 2, Figure 3 and Figure 4 for a simulation time of $1000\,s$ and using the IGRF11 model for the Earth's magnetic field.

3.2 Interest of the Adaptive Method

This section shows as the adaptive method for the noise covariance matrix guarantees better results in terms of errors between the real system and the estimated one. Tacking as initial condition,

$$(\psi, \theta, \phi) = (140^\circ, 90^\circ, 160^\circ) \tag{60}$$

$$(r, q, p) = (1.3^\circ/s, 1^\circ/s, 0.85^\circ/s) \tag{61}$$

a comparison between normal UKF and its adaptive version is shown in Figure 5.

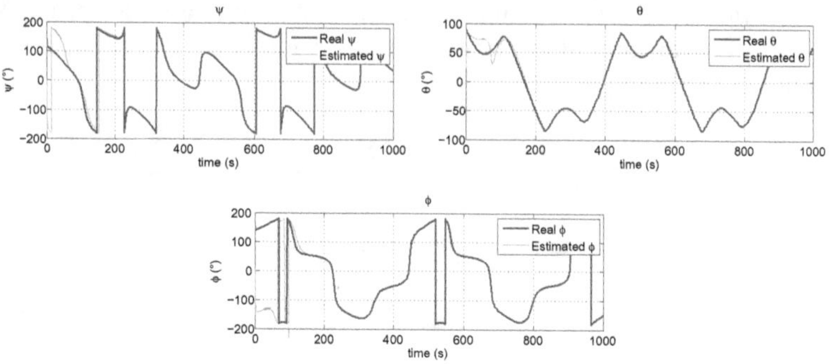

Fig. 2 Comparison between the estimate Euler angles and real value.

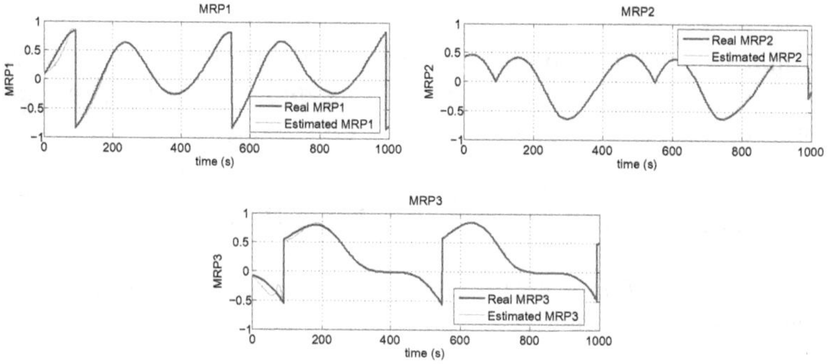

Fig. 3 Comparison between the estimate Modified Rodrigues Parameters value and real value.

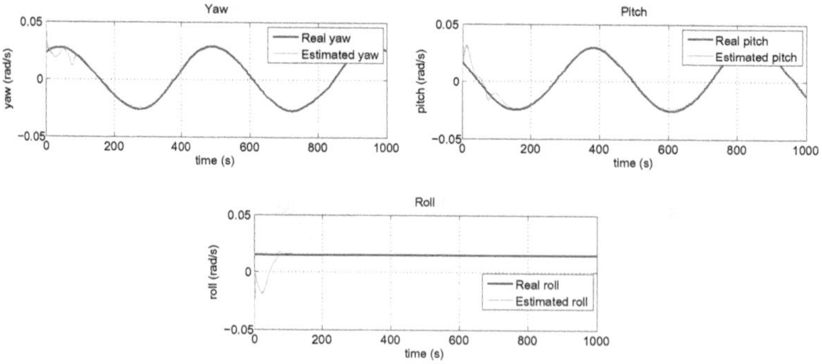

Fig. 4 Comparison between the estimate Angular rates value and real value.

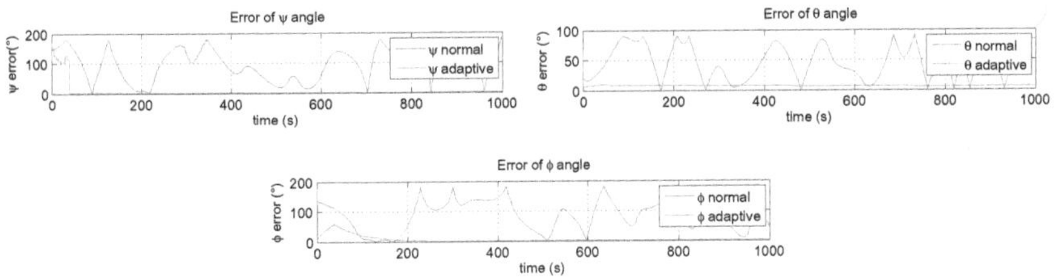

Fig. 5 Error between classical and adaptive UKF

3.3 Attitude Control

This section provides the complete control attitude simulation. Both B-Dot and Sliding Mode are applied: the first one in order to reduce the angular rates during the detumbling phase and in case of the reaction wheels saturation (*Rate-Reduction-Mode*). The second one is then used to reach the need attitude (*Attitude-Acquisition-Mode*), which corresponds to the condition 0° for all three Euler's angles.

For the simulation the magnetorquers provide a nominal magnetic momentum of $0.2 Am^2$ and the reaction wheel have a saturation torque of $0.635\, mNm$. The switch from a control law to the other one consists in the condition suggested in [10]: intervention of B-Dot if all the angular rates are bigger than $5^\circ/s$. In case of switch a little retard ($0.05\, s$) is introduced to permit the system a fluent passage from a mode to the other one without discontinuities.

Tacking as initial conditions:

$$(\psi, \theta, \phi) = (-100^\circ, 30^\circ, -100^\circ) \tag{62}$$

$$(r, q, p) = (8.6^\circ/s, -8.6^\circ/s, 14.3^\circ/s) \tag{63}$$

In Figure 6 the Euler's angles evolution is presented for a simulation time of $9500\, s$.

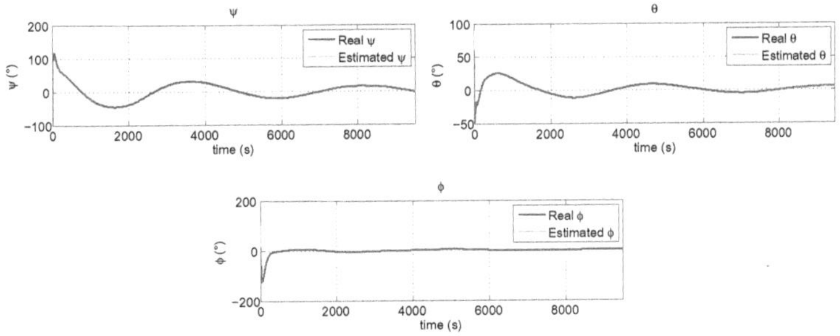

Fig. 6 Euler angles

3.4 Monte Carlo Simulation

Monte Carlo simulations are computational algorithms that rely on repeated random sampling to obtain numerical results, typically one runs simulations many times over in order to obtain the distribution of an unknown entity.

The verification of our procedure, based on UKF algorithms and Modified Rodrigues Parameters, will be made with Monte Carlo Simulations of the SIMULINK model with random sets of parameters. The goal is:

- to verify or dismiss the current estimation filter for actual implementation to the space segment.
- to obtain an estimation on the performance of the algorithm.

The assumptions for the Monte Carlo methods are:

- Each simulation has a duration of $3500\,s$;
- The system is considered converged, if the filter provides an estimation of the space segment attitude with an error of less than 5° around all three axes of the local orbital frame in less than $2400\,s$ and maintains this limit for $900\,s$

The error ε of Monte Carlo approach is defined as $\varepsilon = \frac{1}{\sqrt{N}}$, where N is the total number of simulations. In order to provide a result with a 10% error, a number of 100 simulations are necessary.

For simulating the truth model of satellite a set of random parameters were chosen, particularly the standard deviation value of the sensor σ_B and the bias value of the sensor B_{bias} as the set of random parameters.

But the initial value of the angular rate and satellite attitude were also chosen as random parameters . Particularly the bounds of these two parameters are:

- The angular rate w_z and w_y are comprised between $-8.6^\circ/s$ and $8.6^\circ/s$, while the w_x between $-14.3^\circ/s$ and $14.3^\circ/s$.
- The attitude value of ψ and ϕ are comprised between -180° and 180°, while the angle θ is comprised between -90° and 90°.

3.4.1 Results of Monte Carlo Simulation

The Monte Carlo simulation was run with the highest number of iterations possible with the given limitation on time, 153 times. Among these 153 simulations, 144 lead to success so:

$$P_{convergence} = \frac{144}{153} = 94.12\% \tag{64}$$

The average time needed for the system to reach convergence is equal to:

$$t_{average_{conv}} = \frac{\sum_{i=0}^{N_{convergence}} t_{i_{conv}}}{N_{convergence}} = 131.2\,s \tag{65}$$

In the Figure 7 and 8, it is possible to see the variation of the convergence time with different values of ψ_{ini}, ϕ_{ini}, θ_{ini}, the norm of angular rate($\sqrt{w_x^2 + w_y^2 + w_z^2}$),norm of the sensor bias ($\sqrt{B_{bias_x}^2 + B_{bias_y}^2 + B_{bias_z}^2}$) and norm of the sensor deviation standard ($\sqrt{B_{psd_x}^2 + B_{psd_y}^2 + B_{psd_z}^2}$). These simulations prove that the Adaptive Unscented Kalman filter, using Modified Rodrigues Parameters, is stable for various initial states.

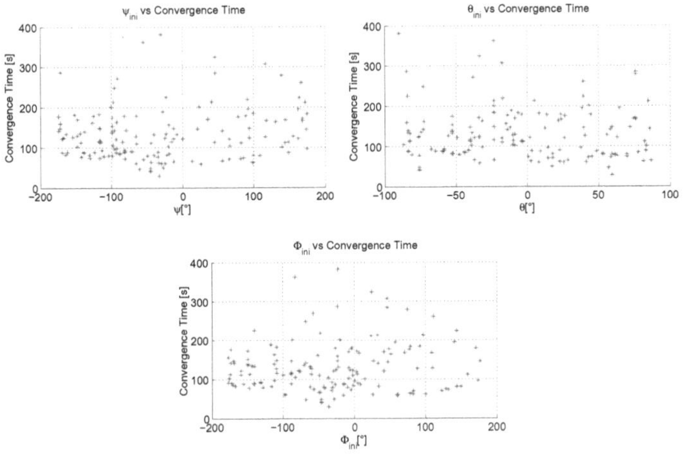

Fig. 7 Resulte of Monte Carlo simulation for Euler's angles.

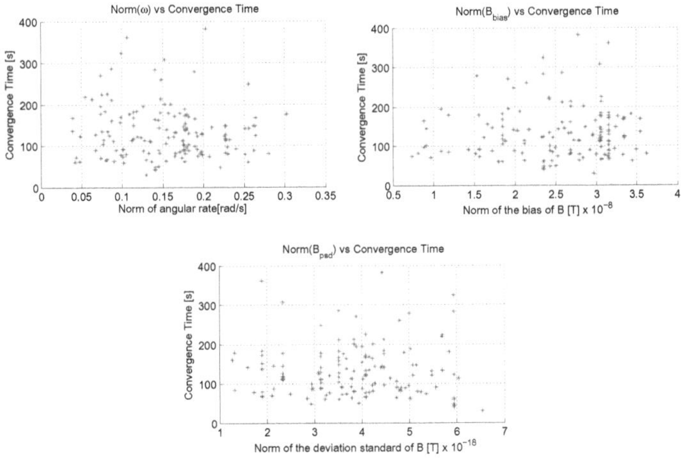

Fig. 8 Results of Monte Carlo simulation for the angular rate norm, the Earth's magnetic bias norm and the Earth's magnetic deviation standard norm.

4 Conclusions

In this paper the UKF advantage in spacecraft attitude estimation were presented and discussed . It was explored the possibility to apply the Modified Rodrigues Parameters and it was verified the efficiency of the estimation algorithm. In order to verify the reliability for future embedded applications a Monte Carlo simulation has been made by changing all the unknown parameters. A large set of initial condition after the orbit injection has been speculated in order to manage to control the satellite in each situation.

At this step of project development some hypothesis have been taken into account to simplify the problem. Future extensions consist as said before in developing also an estimation for the orbit propagator and testing two types of Earth's magnetic field: one more precise (e.g. IGRF11) for the real model and a simplified version for the estimator. This operation results compulsory because of the impossibility to charge the on-board computer with a complex model. Another thing here neglected is the influence of Aerodynamic disturbance torque. Its effect has to be considerate significant especially in detumbling phase, when spacecraft surfaces are largely exposed to the density of the low atmosphere in their rotation. Thus the implementation of a model for this disturbance torque has to be carried out.

References

1. Bellar, A., Seba, B., Mohammed, A., Sweeting, M.N.: Tree axis attitude control using sliding mode for leo microsatellite
2. Crassidis, J.L., Markiey, F.L.: Attitude estimation using modified rodrigues parameters (1996)
3. Crassidis, J.L., Markley, F.L.: Sliding mode control using modified rodrigues parameters. Journal of Guidance, Control, and Dynamics 19(6), 1381–1383 (1996)
4. Finlay, C.C., Maus, S., Beggan, C.D., Bondar, T.N., Chambodut, A., Chernova, T.A., Chulliat, A., Golovkov, V.P., Hamilton, B., Hamoudi, M., et al.: International geomagnetic reference field: the eleventh generation. Geophysical Journal International 183(3), 1216–1230 (2010)
5. Flatley, T., Morgenstern, W., Reth, A., Bauer, F.: A b-dot acquisition controller for the radarsat spacecraft. In: NASA Conference Publication, pp. 79–90. NASA (1997)
6. Heidt, H., Puig-Suari, J., Moore, A., Nakasuka, S., Twiggs, R.: Cubesat: A new generation of picosatellite for education and industry low-cost space experimentation (2000)
7. Julier, S.J., Uhlmann, J.K.: Unscented filtering and nonlinear estimation. Proceedings of the IEEE 92(3), 401–422 (2004)
8. Lozano, J.G.C., Carrillo, L.G., Dzul, A., Lozano, R.: Spherical simplex sigma-point kalman filters: a comparison in the inertial navigation of a terrestrial vehicle. In: American Control Conference, pp. 3536–3541. IEEE (2008)
9. Lucas, W., Rouanne-Labe, A., Grave, J., Peille, P., Lizy-Destrez, S.: Jumpsat: Qualifying three equipments in one cubesat mission (2013)
10. Pierl, C.: Preliminary design of the attitude control system and operational mode transition of the jumpsat project. Final report (2013)

11. Psiaki, M.L., Martel, F., Pal, P.K.: Three-axis attitude determination via kalman filtering of magnetometer data. Journal of Guidance, Control, and Dynamics 13(3), 506–514 (1990)
12. Schaub, H., Junkins, J.L.: Stereographic orientation parameters for attitude dynamics: A generalization of the rodrigues parameters. Journal of the Astronautical Sciences 44(1), 1–19 (1996)
13. Soken, H.E., Sakai, S.-I.: Adaptive unscented kalman filter for small satellite attitude estimation (2012)

Sensor Fault Detection and Estimation for Quadrotors Using Kinematic Equations

Peng Lu, Laurens Van Eykeren, Erik-Jan van Kampen, and Qiping P. Chu

Abstract. This paper proposes a new method for detecting and estimating the faults in the sensors of quadrotor Unmanned Aerial Vehicles. The model used for the fault detection is the kinematic model of the quadrotors, which reduces the influence of model uncertainties. The faults in the sensors are modelled by a random walk process. The state vector of the Unscented Kalman Filter is augmented with the faults, which allows the faults to be estimated. The proposed approach is validated by two scenarios: in the presence and absence of sensor faults. Simulation result shows that the Augmented Unscented Kalman Filter can estimate both the state and faults well, which enables the quadrotor to maintain the flight even in the presence of sensor faults.

1 Introduction

Recently, autonomous Unmanned Aerial Vehicles (UAV)s have attracted considerable attention due to their strong autonomy and ability to fulfill complex tasks without human intervention. The quadrotor is one type of UAV system which is easy to build and fly. It is able to take of and land vertically, and hover at a fixed point. The purpose of these quadrotors are various, ranging from scientific exploration and data collection, to provision of commercial services, military reconnaissance and intelligence gathering [2, 1]. Since these vehicles operate in an environment subjected to a high degree of uncertainties and disturbances, the problem of precise and accurate control and state estimation of these vehicles is difficult and requires advanced control and estimation techniques [2]. Furthermore, due to the increasing requirement for control systems to be more safe and reliable, Fault Detection and

Peng Lu · Laurens Van Eykeren · Erik-Jan van Kampen · Qiping P. Chu
Delft University of Technology, Delft, 2600 GB, The Netherlands
e-mail: {P.Lu-1,L.VanEykeren,E.vanKampen,q.p.chu}@tudelft.nl

J. Bordeneuve-Guibé et al. (eds.), *Advances in Aerospace Guidance, Navigation and Control*, DOI: 10.1007/978-3-319-17518-8_21

Diagnosis (FDD) and Fault Tolerant Control (FTC) technologies are becoming more and more critical and significant [2].

This paper is concerned with the FDD of the quadrotor sensors. Few papers [2] deal with the sensor FDD of quadrotors. However, sensor FDD is important for quadrotors. For one thing, most quadrotors are equipped with low-price, low-precision sensors which are easy to suffer from bias and drift. E.g., the Inertial Measurement Unit (IMU) sensors usually contains bias which is varying with temperature or even accelerations. For another, since the UAV has limited room for sensors, hardware redundancy is not an optimal solution. Therefore, a viable alternative is the development of an analytical based FDD system which is capable of handling sensor faults without using multiple sensors.

Freddi et al. [2] used a Thau observer to detect the faults in pitch angle and roll angle sensors. However, their approach has no ability to estimate the faults. In the present paper, both detection and diagnosis of the sensor faults are considered. In order to cope with model uncertainties, the kinematic equations of the quadrotor are used instead of the dynamics equations. If the dynamic equations are used, there are a number of uncertainties which could degrade the performance of the FDD system. First of all, the inertia of the quadrotor has to be accurate. Second, the thrust coefficient and torque coefficient of the quadrotor is changing according to the rotational speed of the rotor. If constant coefficients are used, it causes uncertainties. Finally, it is cumbersome to accurately determine the external forces and moments of the rotors such as hub forces and hub moments.

This paper proposes to use the kinematic model of the quadrotor, which could eliminate the uncertainties when using the dynamic equations. Sensor faults are modeled as a random walk process, which allows drift faults to be detected and estimated. The Unscented Kalman Filter (UKF) is used for the state estimation and its state vector is augmented with the faults to estimate the faults. Two different simulations are used to validate the proposed approach. The simulation results demonstrate that the proposed approach could estimate the state and faults accurately in the presence or absence of sensor faults. This enables the quadrotor to maintain the flight even in the presence of sensor faults, which enhances the safety of the quadrotor.

The structure of the paper is as follows: Sect. 2 presents the process model and the measurement model of the quadrotor which are used for the detection and diagnosis of sensor faults. Sensor faults are also included in the measurement model. Sect. 3 introduces the method which is used to estimate the state and fault of the system. Two different simulations are conducted, in Sect. 4, to test the performance of the proposed approach. The simulation demonstrates the effectiveness of the approach. Finally, Sect. 5 concludes the paper.

2 Kinematic Model of the Quadrotor

This section first introduces the process model of the quadrotor, which includes the translational and rotational dynamics. The assumptions which are required are also presented. Then, the measurement model is given which contains sensor faults.

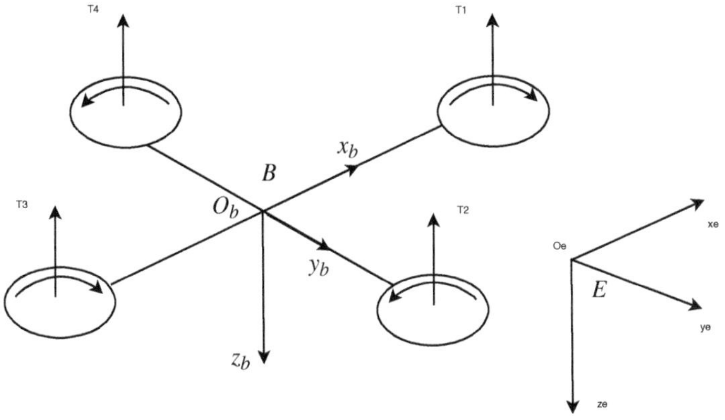

Fig. 1 Quadrotor and the reference frame

2.1 Process Model of the Quadrotor

Define a earth frame $\{\Sigma_E\}(O_e, x_e, y_e, z_e)$ and a body frame $\{\Sigma_B\}(O_b, x_b, y_b, z_b)$ in which O_b is fixed to the quadrotor (see Fig. 1). The earth frame is the North East Down (NED) frame in which the z_e axis points down. The z_b axis of the body frame also points down. The rotation of the body frame with respect to the earth frame is denoted by the following rotation matrix **R**:

$$\mathbf{R} = \begin{bmatrix} \cos\theta\cos\psi & \sin\phi\sin\theta\cos\psi - \cos\phi\sin\psi & \cos\phi\sin\theta\cos\psi + \sin\phi\sin\psi \\ \cos\theta\sin\psi & \sin\phi\sin\theta\sin\psi + \cos\phi\cos\psi & \cos\phi\sin\theta\sin\psi - \sin\phi\cos\psi \\ -\sin\theta & \sin\phi\cos\theta & \cos\phi\cos\theta \end{bmatrix} \tag{1}$$

where ϕ, θ and ψ are the Euler angles.

Before introducing the equations of motion of the quadrotor, the following assumptions have to be made:

1. The structure is supposed rigid.
2. The structure is supposed symmetrical.
3. The center of Gravity and the origin of the body fixed frame are assumed to coincide.
4. The propellers are supposed rigid.
5. The thrust and the drug are proportional to the square of the propeller's rotational speed.

Let η denote the Euler angles and ω denote the angular rates of the quadrotor:

$$\eta = [\phi, \theta, \psi]^T \tag{2}$$
$$\omega = [p, q, r]^T \tag{3}$$

ϕ and θ are assumed to satisfy the following conditions in this paper:

$$\phi \in [-\pi/2, \pi/2] \tag{4}$$
$$\theta \in (-\pi/2, \pi/2) \tag{5}$$

The position of the quadrotor is denoted as $d = (x, y, z)$ while the velocity of the quadrotor expressed in the body frame is denoted as $V = (u, v, w)$.

The quadrotor system model using the kinematic equations is described as follows:

$$\dot{x} = u \cos\psi \cos\theta \tag{6}$$
$$+ v(\cos\psi \sin\theta \sin\phi - \sin\psi \cos\phi) \tag{7}$$
$$+ w(\cos\psi \sin\theta \cos\phi + \sin\phi \sin\psi) \tag{8}$$
$$\dot{y} = u \sin\psi \cos\theta \tag{9}$$
$$+ v(\sin\psi \sin\theta \sin\phi + \cos\phi \cos\psi) \tag{10}$$
$$+ w(\sin\psi \sin\theta \cos\phi - \cos\psi \sin\phi) \tag{11}$$
$$\dot{z} = -u \sin\theta + v \cos\theta \sin\phi + w \cos\theta \cos\phi \tag{12}$$
$$\dot{u} = A_x - g \sin\theta + rv - qw \tag{13}$$
$$\dot{v} = A_y + g \cos\theta \sin\phi + pw - ru \tag{14}$$
$$\dot{w} = A_z + g \cos\theta \cos\phi + qu - pv \tag{15}$$
$$\dot{\phi} = p + q \sin\phi \tan\theta + r \cos\phi \tan\theta \tag{16}$$
$$\dot{\theta} = q \cos\phi - r \sin\phi \tag{17}$$
$$\dot{\psi} = q\frac{\sin\phi}{\cos\theta} + r\frac{\cos\phi}{\cos\theta} \tag{18}$$

Note that the dynamics of the angular rates of the quadrotor are not given here since they are not used in the system model. It should be noted that this kinematic equations have been used by some researchers to cope with the fixed-wing aircraft. Mulder et al. [9] and Lombaerts [5] used Extended Kalman Filter (EKF) and Iterated Extended Kalman Filter (IEKF) to cope with flight path reconstruction. Van Eykeren et al. [11, 12] and Lu et al. [7, 6] used this model for sensor fault detection. More recently, Lu et al. [8] dealt with the sensor fault detection of real flight test data of a fixed-wing aircraft using a Double-Model Adaptive Estimation approach.

Remarks:

1. This model is different from the dynamic model of the quadrotor which is commonly used. This kinematic model is insensitive to model uncertainties such as the mass and the inertia of the quadrotors. This is advantageous because the mass uncertainty and asymmetry of the propellers does not affect the model accuracy.
2. The sensor FDD is not influenced by the actuator faults of the quadrotor. The reason is that the actuator output of the quadrotor is not used in the system model. This is also an advantage since the FDD of the sensors and actuator are separated.

2.2 *Measurement Model of the Quadrotor*

The position and velocity of the quadrotor is assumed to be measured by a optic track system. It is assumed that there are no biases in the measurement of the position and velocity. The Euler angles are measured by the attitude heading and reference system. The angular rates and linear accelerations are measured by the IMU. The faults in the IMU are considered. Therefore, the complete measurement model is as follows:

$$x_m = x + \nu_x \tag{19}$$
$$y_m = y + \nu_y \tag{20}$$
$$z_m = z + \nu_z \tag{21}$$
$$v_{xm} = v_x + \nu_{vx} \tag{22}$$
$$v_{ym} = v_y + \nu_{vy} \tag{23}$$
$$v_{zm} = v_z + \nu_{vz} \tag{24}$$
$$\phi_m = \phi + \nu_\phi \tag{25}$$
$$\theta_m = \theta + \nu_\theta \tag{26}$$
$$\psi_m = \psi + \nu_\psi \tag{27}$$

The IMU measurement model is as follows:

$$p_m = p + \nu_p + f_p \tag{28}$$
$$q_m = q + \nu_q + f_q \tag{29}$$
$$r_m = r + \nu_r + f_r \tag{30}$$
$$A_{xm} = A_x + \nu_{A_x} + f_x \tag{31}$$
$$A_{ym} = A_y + \nu_{A_y} + f_y \tag{32}$$
$$A_{zm} = A_z + \nu_{A_z} + f_z \tag{33}$$

where f_x, f_y, f_z, f_p, f_q and f_r are the faults in the IMU measurement. It is common that there are biases and drifts in the IMU sensors, especially for the IMU sensors which are used by small UAVs. The objective of this paper is to estimate these faults. In this way, an improved estimation can be provided to enhance the controller performance of the quadrotor.

3 Sensor Fault Estimation

The objective is to estimate the biases and drifts in the IMU sensors. Since angular rate measurement is used by the controller, it is necessary to obtain an accurate angular rate information. This section introduces the method which is able to estimate the faults in the IMU sensors and obtain an accurate estimation of the angular rates.

3.1 Unscented Kalman Filter

This section presents the general UKF framework[3], which is used for the state estimation. It contains the following steps[10]:

Step 1 Sigma Points Calculation

$$\mathscr{X}_{0,k-1} = \hat{x}_{k-1} \tag{34a}$$

$$\mathscr{X}_{i,k-1} = \hat{x}_{k-1} - (\sqrt{(L+\gamma)P_{k-1}})_i, \quad i = 1,2,...,L \tag{34b}$$

$$\mathscr{X}_{i,k-1} = \hat{x}_{k-1} + (\sqrt{(L+\gamma)P_{k-1}})_i, \quad i = L+1, L+2,...,2L \tag{34c}$$

$$w_0^{(m)} = \gamma/(L+\gamma) \tag{35a}$$

$$w_0^{(c)} = \gamma/(L+\gamma) + (1-\alpha^2+\beta) \tag{35b}$$

$$w_i^{(m)} = w_i^{(c)} = 1/\{2(L+\gamma)\}, \quad i = 1,2,...,2L \tag{35c}$$

where L is the dimension of the state vector, $\gamma = \alpha^2(L+\mathscr{K})$ is a scaling factor, α determines the spread of the sigma points around $\bar{x}$, $\mathscr{K}$ is a secondary scaling factor, β is used to incorporate the prior knowledge of the distribution of x.

Step 2 Time Update

After the creation of the sigma points through the nonlinear transformation, the predicted mean and covariance are computed as follows

$$\mathscr{X}_{i,k|k-1} = f(\mathscr{X}_{i,k-1}) \tag{36}$$

$$\hat{x}_{k|k-1} = \sum_{i=0}^{2L} w_i^{(m)} \mathscr{X}_{i,k,k-1} \tag{37}$$

$$P_k^- = \sum_{i=0}^{2L} w_i^{(c)} [\mathscr{X}_{i,k|k-1} - \hat{x}_{k|k-1}][\mathscr{X}_{i,k|k-1} - \hat{x}_{k|k-1}]^T + Q \tag{38}$$

$$\mathscr{X}^*_{i,k|k-1} = [\mathscr{X}_{0:2L,k|k-1} \quad \mathscr{X}_{0,k|k-1} - \nu\sqrt{Q} \quad \mathscr{X}_{0,k|k-1} + \nu\sqrt{Q}]_i \tag{39}$$

$$\mathscr{Y}^*_{i,k|k-1} = h(\mathscr{X}^*_{i,k|k-1}) \tag{40}$$

$$\hat{y}_k = \sum_{i=0}^{2L^a} w_i^{*(m)} \mathscr{Y}^*_{i,k|k-1} \tag{41}$$

$$P_{xy,k} = \sum_{i=0}^{2L} w_i^{*(c)} [\mathscr{X}_{i,k|k-1} - \hat{x}_{k|k-1}][\mathscr{Y}_{i,k|k-1} - \hat{y}_k]^T \tag{42}$$

$$P_{yy,k} = \sum_{i=0}^{2L} w_i^{*(c)} [\mathscr{Y}_{i,k|k-1} - \hat{y}_k][\mathscr{Y}_{i,k|k-1} - \hat{y}_k]^T + R \tag{43}$$

where $L^a = 2L$, $\nu = \sqrt{L+\gamma}$, $w_i^{*(m)}$ and $w_i^{*(m)}$ are calculated in the same way as Eq. 35 with the only change from L to L^a.

Step 3 Measurement Update
This step is to calculate the Kalman gain and the updated states and covariance.

$$K_k = P_{xy,k} P_{yy,k}^{-1} \tag{44}$$

$$\hat{x}_{k|k} = \hat{x}_{k|k-1} + K_k(y_k - \hat{y}_k) \tag{45}$$

$$P_k = P_k^- - K_k P_{yy,k} K_k^T \tag{46}$$

The convergence of the UKF is not sensitive to the initial states and can obtain a high order of accuracy for nonlinear systems. Therefore, the UKF is selected for the state estimation.

3.2 *Modelling of the Sensor Faults*

The UKF is able to provide an unbiased state estimation when there are no faults in the system. If there is a bias in the system input, the UKF can still obtain a good performance of state estimation by augmenting the bias as an additional state. This approach is called Augmented Unscented Kalman Filter (AUKF) [7].

Suppose there is a bias in each input signal.

$$b = [b_1, b_2, ..., b_m]^T \tag{47}$$

Then the system states are denoted as

$$x' = [x_1, x_2, ..., x_n, b_1, b_2, ..., b_m]^T \tag{48}$$

The dynamics of the bias is

$$\dot{b} = 0 \tag{49}$$

It should be noted that the AUKF is able to cope with the bias in the system input since the bias is treated as a constant which means that the dynamics of the bias is relatively slow. However, if there is a time-varying fault in the input signal, the normal AUKF is not able to track the real states.

In order to cope with time-varying faults, the faults can be modelled as a random walk process as follows:

$$\dot{x}'(t) = \begin{bmatrix} \dot{x}(t) \\ \dot{f}_{i,k} \end{bmatrix} = \begin{bmatrix} f(x(t), u(t), t) \\ 0 \end{bmatrix} + \begin{bmatrix} w(t) \\ w_k'(t) \end{bmatrix} \tag{50}$$

where w_k' is assumed to be white noise. Other modelling methods of the faults can also can used [9]. However, it is without the scope of this paper.

4 Simulation Results

This section validates the performance of the proposed approach to detect and estimate the faults in the IMU sensors. The simulation consists of two scenarios: with and without sensor faults.

The controller of the quadrotor is a Backstepping (BS)-based [4] approach. The control objective is to control the altitude and position of the quadrotor. The controller design is omitted here since it is not the focus of the paper.

The command for the quadrotor is to climb up 1 meter, hover for some time and then go back to the initial height and hover. Then, the command is to descend one meter, hover for a period of time and then go back to the initial height.

4.1 Simulation in the Absence of Sensor Faults

This section tests the performance of the proposed approach during the fault-free scenario, to evaluate the occurrence of possible false alarms.

The state estimation of the AUKF is given in Fig. 2. The red dotted lines represent the real state of the system while the blue solid lines denote the estimation using the AUKF. As can be seem from the figure, these two lines coincide with each other.

The estimation error of the position, velocities and attitude angles using the AUKF are given in Figs. 3, 4 and 5, respectively. It can be seen that all of the estimation errors are close to zero-mean. However, the estimation error of the position appears

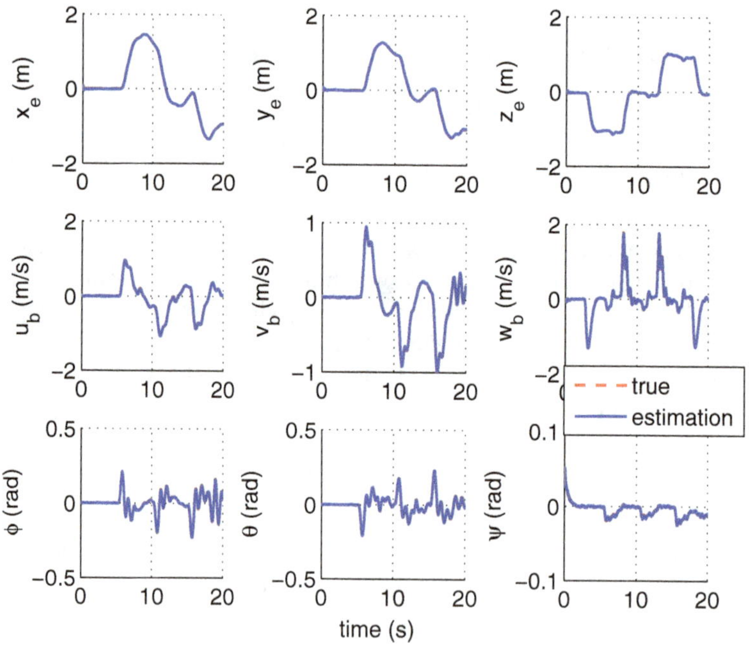

Fig. 2 State estimation using the AUKF and the real states. There are no faults in the sensors. The red dotted lines denote the real state while the blue solid lines represent the estimated states using the AUKF.

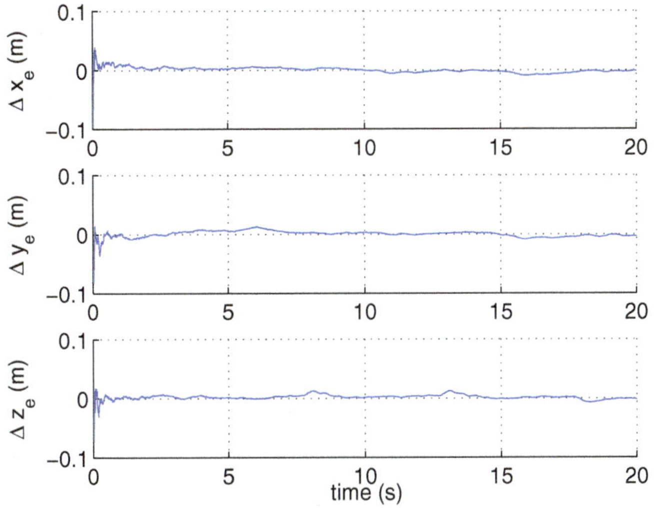

Fig. 3 Estimation error of the position using the AUKF approach when there are no sensor faults

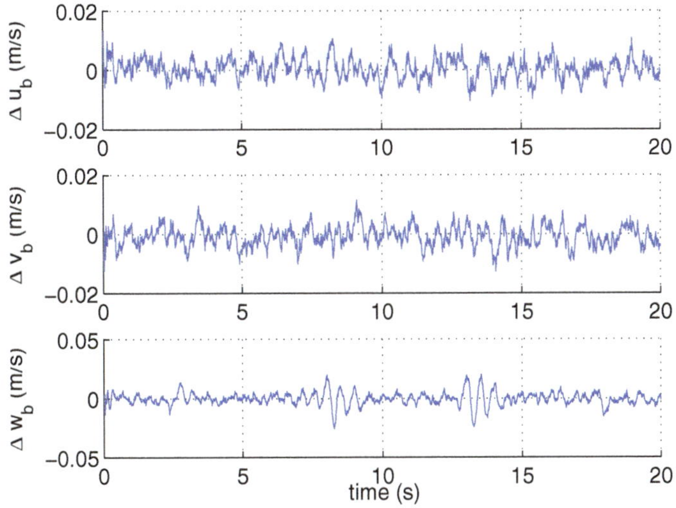

Fig. 4 Estimation error of the velocity using the AUKF approach when there are no sensor faults

to be more smooth than that of the velocity and attitude angles. The reason is that the position is measured by the camera system which is composed of a number of cameras. Some oscillations in the estimation error of the velocity and attitude angles

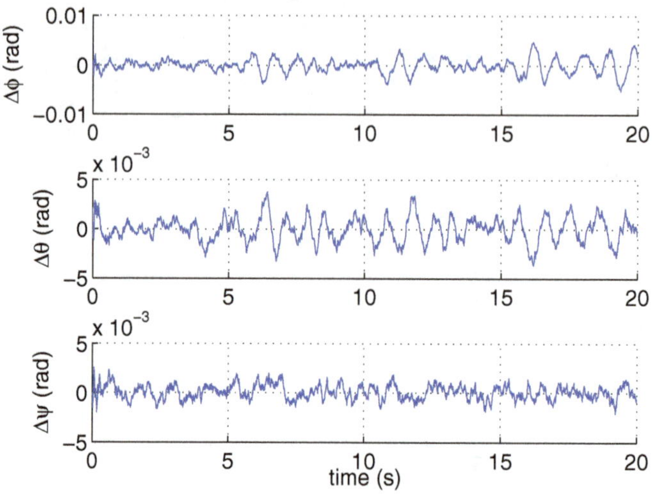

Fig. 5 Estimation error of the Euler angles using the AUKF approach when there are no sensor faults

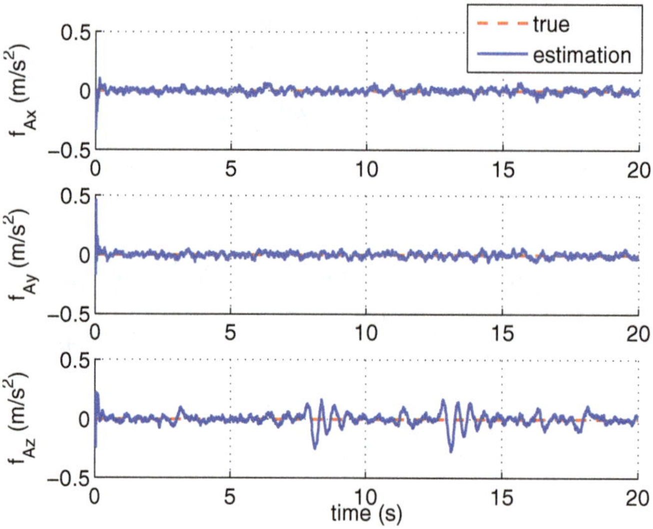

Fig. 6 Fault estimation of the accelerometer sensors using the AUKF and the real faults. There are no faults in the sensors. The red dotted lines denote the real accelerometer sensor faults while the blue solid lines represent the estimated ones using the AUKF.

can be observed. This is due to the maneuvers the quadrotor performs. However, the magnitude of the oscillations is small and can be neglected.

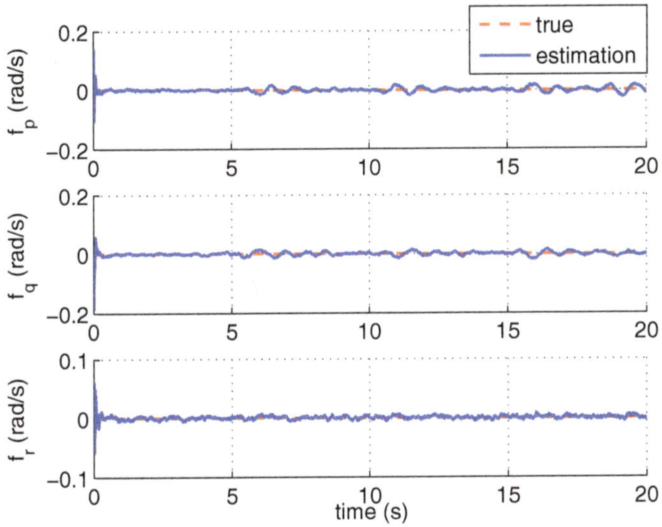

Fig. 7 Fault estimation of the rate gyro sensors using the AUKF and the real faults. There are no faults in the sensors. The red dotted lines denote the rate gyro sensor faults while the blue solid lines represent the estimated ones using the AUKF.

The fault estimation of the IMU sensors is presented in Figs. 6 and 7 respectively. The fault estimation of A_x and A_y sensor is zero-mean. However, the estimation of the A_z sensor fault shows some oscillations. This is due to the strong maneuver performed by the quadrotor. The vertical speed is around 2 m/s, which is twice as much as the speed along the x and y axis which can be seen in Fig. 2. The fault estimation of the angular rate sensors are all zero-mean, despite some small oscillations.

4.2 Simulation in the Presence of Sensor Faults

This section tests the performance of the AUKF when some faults are injected to the IMU sensors. The fault scenario is given in Table 1.

As can be seen from the table, there are three fault types including bias, drift and oscillatory. The faults of the sensors occur simultaneously, which is a challenging situation. A scenario where all six sensors fail is less likely to happen than the occurrence of a single fault. The purpose of this scenario is to demonstrate the performance of the proposed approach under extreme situations.

The real and estimated states are shown in Fig. 8. The red dotted lines show the real state while the blue solid lines represent the estimated states. Due to the strong maneuver, some oscillations are observed from the states such as the velocity and attitude angles.

The estimation error using the AUKF is also presented. The estimation error of the position of the quadrotor is shown in Fig. 9. It can be seen that before the oc-

Table 1 IMU sensor fault scenario

Faulty sensor	Fault time	Fault type	Fault magnitude	Fault units
A_x	$t = 10$ s	Bias	0.5	m/s^2
A_y	$t = 10$ s	Drift	-0.1	m/s^3
A_z	$t = 10$ s	Bias	0.5	m/s^2
p	$t = 10$ s	Bias	0.05	rad/s
q	$t = 10$ s	Oscillatory	$0.1\sin(0.005\pi t)$	rad/s
r	$t = 10$ s	Bias	-0.05	rad/s

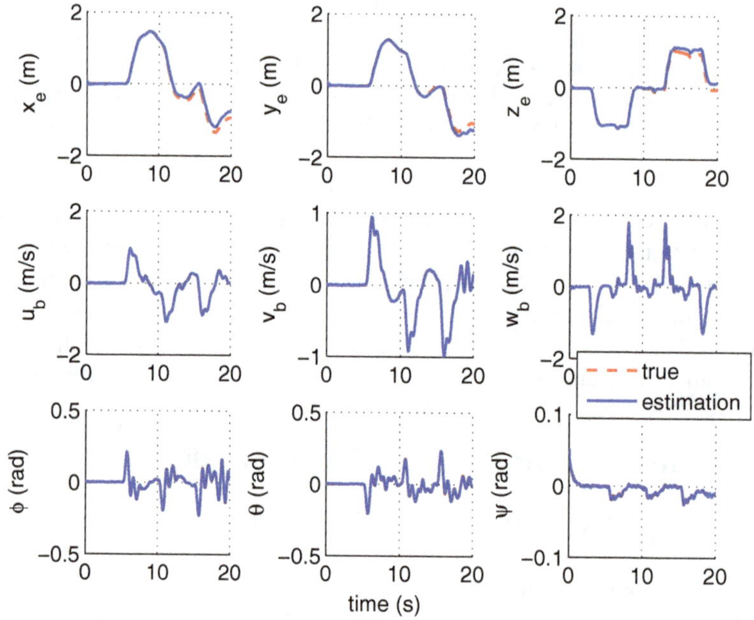

Fig. 8 State estimation using the AUKF and the real states. The scenario is with IMU sensor faults. The red dotted lines denote the real state while the blue solid lines represent the estimated states using the AUKF.

currence of the faults, the estimation error is minimum. Once the faults occur, the estimation error starts to increase. However, the error is small considering the fact that all the IMU sensors fail simultaneously.

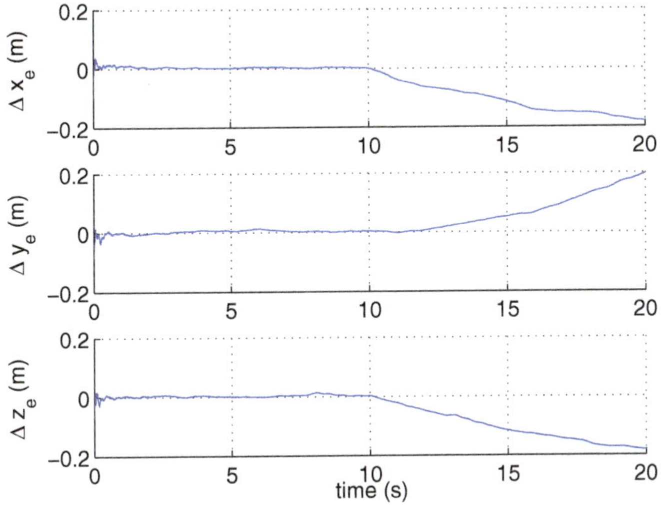

Fig. 9 Estimation error of the position using the AUKF approach when there are IMU sensor faults

The estimation error of the velocities and the attitude angles, as shown in Figs. 10 and 11 respectively, do not show a big difference from the situation where there are no sensor faults. This also shows the estimation performance of the AUKF in the presence of sensor faults. However, the estimation error of the pitch angle θ increases after the occurrence of the faults.

Finally, the estimation of the faults using the AUKF is shown. The accelerometer fault estimation is shown in Fig. 12. The red dotted lines denote the real faults which are injected to the sensors while the blue lines represent the estimated faults using the AUKF. It can be seen that the estimated faults can follow the real faults well, even in the presence of drift fault. However, it is noticed that the estimation of the fault in the A_x sensor shows some oscillations while the faults injected to the A_x sensor is bias fault. The reason is that the estimation of the A_x sensor fault is related to the estimation of the pitch angle, which is influenced by the fault in the p sensor. The fault estimation of the roll rate p sensor can track the oscillatory fault well. However, there is still a small delay which affects the estimation of the pitch angle. This is also the reason why the estimation of the pitch angle shows a larger error (as shown in Fig. 9) compared to the roll angle and yaw angle. The fault estimation of the roll rate sensor and yaw rate sensor is satisfactory, as can be seen from Fig. 13.

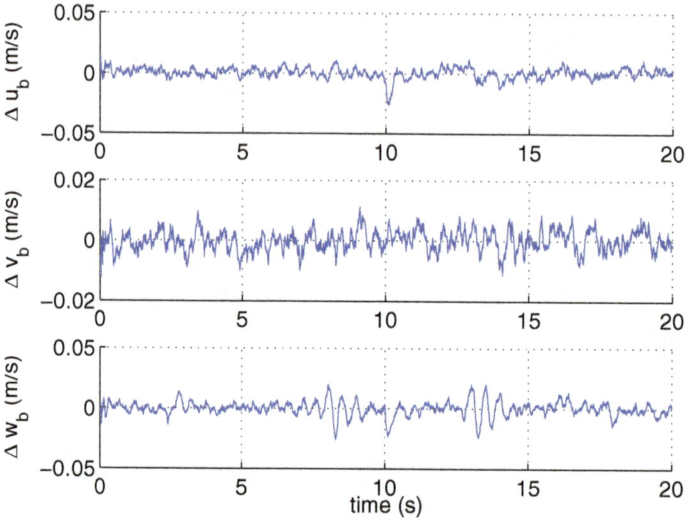

Fig. 10 Estimation error of the velocity using the AUKF approach when there are IMU sensor faults

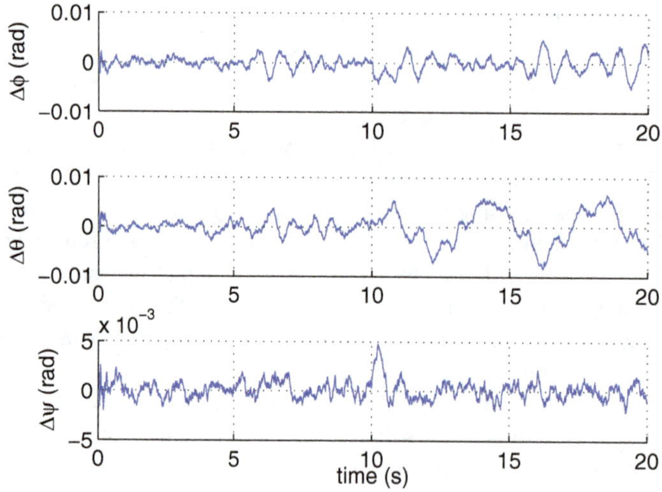

Fig. 11 Estimation error of the Euler angles using the AUKF approach when there are IMU sensor faults

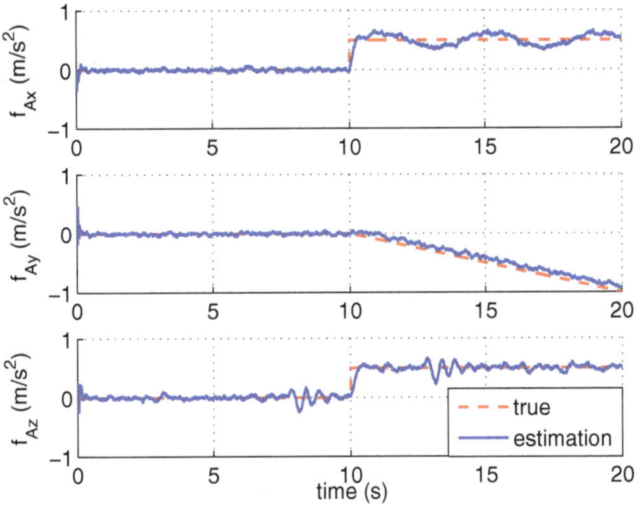

Fig. 12 Fault estimation of the accelerometer sensors using the AUKF and the real faults. There are IMU sensor faults in the system. The red dotted lines denote the real accelerometer sensor faults while the blue solid lines represent the estimated ones using the AUKF.

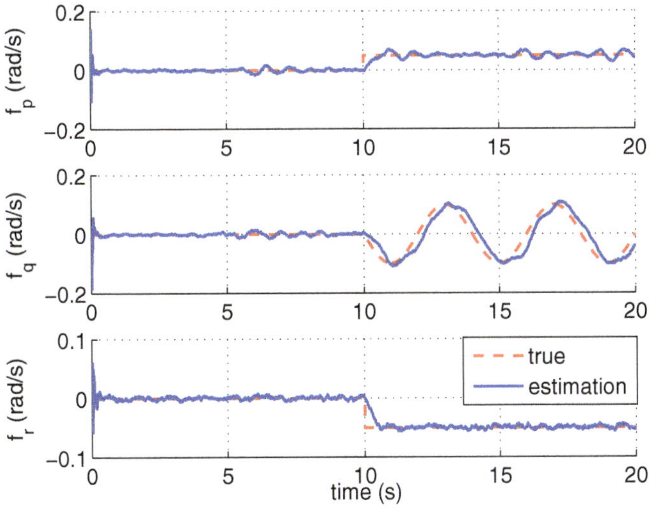

Fig. 13 Fault estimation of the rate gyro sensors using the AUKF and the real faults. There are IMU sensor faults in the system. The red dotted lines denote the rate guro sensor faults while the blue solid lines represent the estimated ones using the AUKF.

5 Conclusions

This paper proposes a new approach for the sensor fault detection and diagnosis of quadrotors. In order to cope with model uncertainties, this paper proposes to use the kinematic model of the quadrotor instead of the dynamic equations. The AUKF is used to estimate the state and the faults. Simulation results demonstrates the effectiveness of the proposed approach both in the presence and absence of sensor faults.

The proposed approach could provide correct information of the system states and sensor conditions, which enables the controller to realize fault tolerant control even in the presence of sensor faults. In the future, this approach will be implemented on real quadrotors to test the performance of the approach. It can also be implemented to enhance the safety and fault tolerance of the quadrotors.

References

1. Amoozgar, M.H., Chamseddine, A., Zhang, Y.: Experimental Test of a Two-Stage Kalman Filter for Actuator Fault Detection and Diagnosis of an Unmanned Quadrotor Helicopter. Journal of Intelligent & Robotic Systems 70(1-4), 107–117 (2012)
2. Freddi, A., Longhi, S., Monteriù, A.: A Diagnostic Thau Observer for a Class of Unmanned Vehicles. Journal of Intelligent & Robotic Systems 67(1), 61–73 (2012)
3. Julier, S.J., Uhlmann, J.K.: A new extension of the Kalman filter to nonlinear systems. In: Proc. of AeroSense: The 11th Int. Symp. on Aerospace/Defence Sensing, Simulation and Control (1997)
4. Krstic, M., Kanellakopoulos, I., Kokotovic, P.: Nonlinear and Adaptive Control Design. John Wiley & Sons, Inc. (1995)
5. Lombaerts, T.J.J., Chu, Q.P., Mulder, J.A., Joosten, D.A.: Modular flight control reconfiguration design and simulation. Control Engineering Practice 19(6), 540–554 (2011)
6. Lu, P., Van Kampen, E.: Aircraft Inertial Measurement Unit Fault Identification with Application to Real Flight Data. In: AIAA Guidance, Navigation and Control Conference, number 0859, Kissimmee, Florida, pp. 1–20 (2015)
7. Lu, P., Van Eykeren, L., van Kampen, E., Chu, Q., Yu, B.: Adaptive Hybrid Unscented Kalman Filter for Aircraft Sensor Fault Detection, Isolation and Reconstruction. In: AIAA Guidance, Navigation, and Control Conference, number 1145, National Harbor, Maryland, pp. 1–18 (2014)
8. Lu, P., Van Eykeren, L., van Kampen, E., de Visser, C., Chu, Q.: Double-model adaptive fault detection and diagnosis applied to real flight data. Control Engineering Practice 36, 39–57 (2015)
9. Mulder, J.A., Chu, Q.P., Sridhar, J.K., Breeman, J.H., Laban, M.: Non-linear aircraft flight path reconstruction review and new advances. Progress in Aerospace Sciences 35, 673–726 (1999)
10. Van Der Merwe, R., Wan, E.A.: The Square-root Unscented Kalman Filter for State and Parameter-estimation. In: IEEE International Conference on Acoustics, Speech, and Signal Processing, pp. 3461–3464 (2001)

11. Van Eykeren, L., Chu, Q., Mulder, J.A.: Sensor Fault Detection and Isolation using Adaptive Extended Kalman Filter. In: the 8th Symposium on Fault Detection, Supervision and Safety of Technical Processes, number 1969, Mexico City, Mexico, pp. 1155–1160 (2012)
12. Van Eykeren, L., Chu, Q.P.: Air Data Sensor Fault Detection using Kinematic Relations. In: Proceedings of the EuroGNC 2013, 2nd CEAS Special Conference on Guidance, Navigation & Control, pp. 414–428 (2013)

An Image Processing Algorithm for Ground Navigation of Aircraft

Kevin Theuma and David Zammit Mangion

Abstract. Aircraft taxiing can be challenging in periods of bad weather and pilots tend to face a considerable increase in workload. In order to minimise the impact of adverse weather such as low visibility conditions, we propose a solution that can automatically navigate aircraft by using image processing techniques to determine its position relative to the taxiway. The output position is intended to provide feedback to a dedicated controller. The task of the image processing algorithm is to identify the taxiway centreline markings by extracting features from the image and processing this information. Afterwards, the detected centreline markings are modelled through curve fitting techniques. The cross-track and heading errors of the aircraft are measured from these curves and these define its position. Results show that the developed algorithm provides the position of the aircraft with centimetric accuracy. The algorithm performs well in various weather conditions including clear, stormy and foggy weather. It also works well during day and night time conditions.

1 Introduction

As air traffic keeps on growing, so does the interest in automating all phases of flight. The aim is to shift tasks from the pilot to an autonomous system so that operations can be carried out quicker and in a safer way. Statistics published by PlaneCrashInfo.com [1] show that the majority of aircraft accidents are a result of

Kevin Theuma
University of Malta, Msida MSD 2080, Malta
e-mail: kevin.theuma@um.edu.mt

David Zammit Mangion
University of Malta, Msida MSD 2080, Malta and Cranfield University, Cranfield, Bedfordshire, MK43 0AL, UK
e-mail: david.zammit-mangion@um.edu.mt

J. Bordeneuve-Guibé et al. (eds.), *Advances in Aerospace Guidance, Navigation and Control*, DOI: 10.1007/978-3-319-17518-8_22

human error, possibly due to fatigue or distraction. Aircraft taxi, together with take-off has still not been automated while other phases such as cruise and landing have already been automated.

An autonomous system is particularly helpful in low visibility conditions. In this kind of weather, pilots encounter more challenges than usual because they face a considerable increase in workload. Also, there is an increase in hazards which could lead to possible incidents and accidents, and a reduction in the efficiency of the operation of the National Airspace System. In low visibility conditions, it is common for flights to be delayed, rerouted or cancelled. This has a negative effect on the industry's economics as it adds to their costs and lowers their revenue [2].

Today, the most popular systems for position fixing are satellite navigation systems. The traditional Global Positioning System (GPS) can pinpoint the position of the receiver with inaccuracies of up to 10m [3]. Unfortunately GPS is susceptible to various errors including atmospheric effects, multipath effects, relativistic effects, and ephemeris and clock errors [4]. Additionally, it is disrupted by stormy weather and solar flares [5]. Due to these problems, GPS fails to attain centimetric accuracy without additional hardware and reference ground stations. Therefore, on its own, GPS is not reliable enough for ground navigation of aircraft because an error of a few metres can result in a collision or the aircraft going off track. The use of advanced satellite navigation systems that make use of reference ground stations, such as Differential GPS (DGPS) [6] and Wide Area Augmentation System (WAAS) [7], is not feasible because it would require the airport to be equipped with the reference ground stations, which is not always the case.

An alternate solution is to use vision-based systems for determining the position of the aircraft relative to the taxiway or runway. The multinational corporation Google has already managed to construct a driverless car by equipping it with a Light Detection And Ranging (LIDAR) system together with an image processing system. Unlike satellite navigation systems, image processing systems use relative position fixing rather than absolute position fixing. Therefore errors related to absolute position fixing such as the path definition error are eliminated. These kind of systems are promising for achieving position fixing with centimetric accuracy in all weather conditions. With specialised equipment and image processing techniques, the system can be made robust to bad weather conditions, even low visibility. This work focused on image processing techniques that can be used to guide aircraft on the ground in real-time. The algorithm was developed for all weather conditions and for all times of the day.

By analysing and processing images, the algorithm calculates the cross-track error (i.e. the position of the aircraft relative to the centreline) and the heading error (i.e. the orientation of the aircraft relative to the centreline). The cross-track error ρ and the heading error θ are illustrated in Figure 1. Centimetric accuracy was desired for use in aircraft guidance on the ground. After finding the cross-track and heading errors of the aircraft, these can be used to provide feedback to the controller developed by Zammit *et al.* [8].

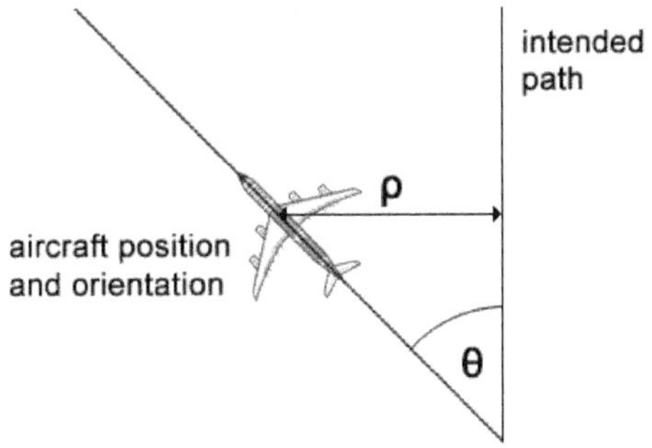

Fig. 1 An illustration of the cross-track error ρ and the heading error θ

2 Literature Review

Due to the lack of literature concerning the identification of taxiway markings through image processing techniques, the developed system was based on concepts and ideas taken from lane detection and tracking systems intended for road vehicles. These techniques are relevant to this work because road markings are similar to the ones present in the taxiway. In a road with two lanes, lane markings are identical to centrelines and markings of road boundaries are identical to taxiway side-markings. Therefore, due to these similarities, techniques from lane detection and tracking can be readily applied to systems that identify the taxiway markings.

The present research focused on work that takes a feature extraction approach in order to accurately identify the edges of the taxiway centreline and use this information to measure the cross-track and heading errors. Notable work on lane detection and tracking includes those by Bertozzi *et al.* [9], Wang *et al.* [10], Yu *et al.* [11] and Hota *et al.* [12]. Unfortunately, these systems suffer from certain shortcomings that make them unsuitable for the intended application (of detecting and tracking taxiway centreline markings) without any adjustments and modifications. The system proposed by Bertozzi *et al.* requires the centreline and both side-markings to be present in the input image. Also, the intensities of the taxiway markings have to be considerably higher than the rest of the ground. The Canny / Hough Estimation of Vanishing Point (CHEVP) algorithm proposed by Wang *et al* .did not work correctly when tested on video sequences having parallel lines. Since the CHEVP algorithm divides the input images into five sections, the lines detected in each section did not necessarily belong to the same edge, thus causing erroneous results. The method proposed by Yu *et al.* does not use a sufficiently flexible curve model for representing the centreline edges because the parabolic curves do not superimpose the centreline edges at the bottom of the image. The cross-track and heading errors are measured at the bottom of the image so, in that part of the image, the curves of the centreline edges have to accurately represent the edges by superimposing them. The line clustering approach proposed by Hota *et al.* assumes that edges are straight or almost straight. However, this work was expected to handle scenarios in which the centreline is curved so such assumptions could not be taken. Considering that there was no work in the reviewed literature that provided the desired results, the concepts and ideas from these systems were analysed, tested and compared in order to choose the best components for integrating them into a system tailor-made for the desired application.

3 Design of the Algorithm

The algorithm was designed and developed in Matlab environment, which includes an image processing toolbox that facilitates the integration of image processing techniques and provides an interface for debugging and checking the performance of the individual components. The complete system was intended to handle images from an external VGA CCD camera, so the algorithm was designed to handle images with a resolution of 640×480. Colour information is not used throughout the algorithm. Consequently, the algorithm was designed to handle grayscale images in order to simplify processing and minimise execution time. Colour images are therefore converted into grayscale prior to inputting them into the algorithm.

The flowchart of the system proposed is presented in Figure 2. Once the algorithm reads the image, it increases its contrast using a novel technique referred to as the Contrast-Limited Local Histogram Equalization (CLLHE). Then a Sobel edge detector identifies the left and right edges of the image and produces two separate binary images. These binary images are thinned using the morphological thinning technique [13]. The Hough transform [14] identifies the most dominant lines in the binary images and after finding the peaks, the line segments are reconstructed in the Cartesian space. From the line segments, a pair of line segments which best identifies the centreline edges is chosen. The line segments are mapped from the image plane to the ground plane by using the Homographic Transform [15]. Other line segments which appear to belong to the same edges described by the chosen pair of line segments are found and clustered. Points are sampled from the clustered line segments and inputting in the Weighted Least Squares Fit (WLSF) [16] technique which attempts to fit the best curves through these points. At this point, the algorithm will have generated two curves that represent the left and right edges of the centreline. The curve describing the actual centreline is derived from these curves by calculating the coefficients of the curve that lies in the middle of these two curves. The curve describing the actual centreline is used to measure the cross-track and heading errors. These values are tracked using the Kalman Filter [17] in order to minimise noise. The individual components are explained in detail in the following sections.

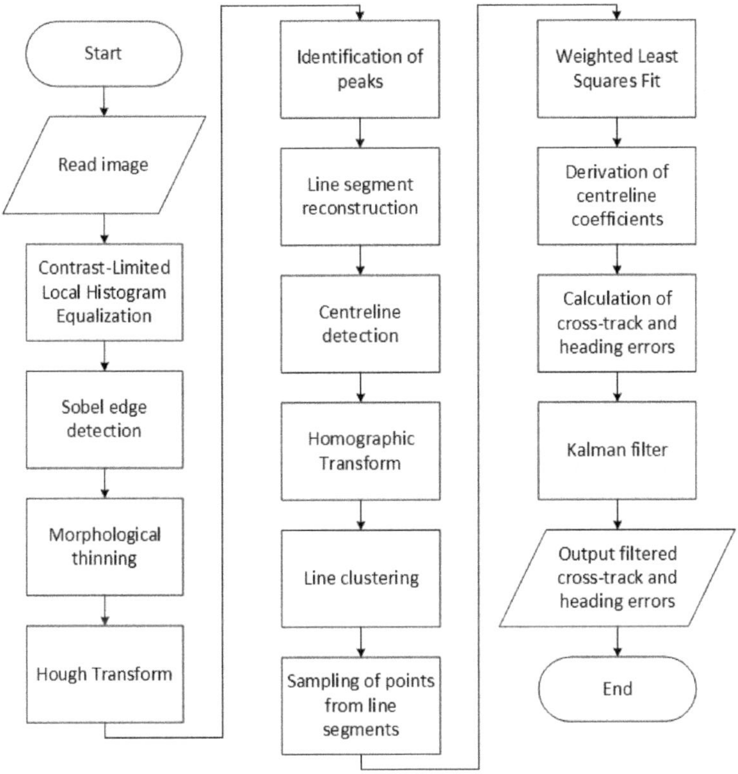

Fig. 2 The flowchart of the developed image processing algorithm

3.1 Contrast Enhancement

The contrast of the input images is enhanced in order to bring out details that might be obscured due to bad weather or poor illumination. This ensures that the desired features are extracted in various weather conditions and at different times of the day, thus providing a solution that works even in scenarios of low visibility and at night. The technique used for enhancing the contrast is referred to as the Contrast-Limited Local Histogram Equalisation which is based on the Contrast-Limited Adaptive Histogram Equalisation (CLAHE) [18] but is adjusted to execute faster than the CLAHE so that it can be used for real-time applications on embedded devices.

The CLLHE starts by diving the image into tiles with dimensions 80×80. The Histogram Equalisation technique is used individually on each tile. This consists of constructing the histogram by counting the occurrence of each grey level in the tile. Next, the cumulative histogram is constructed by cumulatively adding the number of occurrences in each bin. This process can be summarised by the cumulative distribution function (cdf) shown in Equation 1.

$$cdf_x(i) = \sum_{j=0}^{i} p_x(j) \tag{1}$$

where $p_x(j)$ is the occurrence of grey level j and $cdf_x(i)$ is the cumulative distribution function of grey level i.

Contrast enhancement is restricted so that in cases where the number of different grey levels is small, the contrast in the image is not enhanced excessively. This helps limit the amount of noise that can be produced by the Histogram Equalisation technique. The contrast is limited by spreading bin counts that exceed a specified limit. The limit per bin was chosen as 32 counts. Whenever a bin exceeds this limit, the excess is distributed by dividing it by the total number of bins (i.e. 256) and adding the result to each bin. This means that bins that exceed the limit still receive a share of the excess, but since they do not receive the full amount, the contrast enhancement is restricted. The reason the excess is not redistributed one by one as in the CLAHE technique is to minimise execution time. Furthermore, a division by 256 can be translated to 8 logical shifts to the right.

The mapped grey levels are found by using Equation 2:

$$h(k) = \frac{L-1}{MN} \times cdf(k) \tag{2}$$

where L is the total number of grey levels than can appear in the image, M and N are the width and height of the tiles respectively, cdf(k) is the cumulative distribution function of grey level k and h(k) is the mapped value of grey level k. Finally, the pixels in the tile are mapped to the new values by using this transformation.

An example of the output produced by the CLLHE process after being applied on the image in Figure 3 is presented in Figure 4. Since the CLLHE algorithm works on individual tiles, it produces a block-effect. However, this visual effect is not an issue because the output is not meant to be pleasing to the human eye but is meant to be processed further by the algorithm. The downside is that the edge detector tends to detect the boundaries of these blocks as edges. This is unwanted, so the edges at these locations are suppressed and therefore edge information is lost. The contrast can be increased even further by increasing the value of the limit but this also increases the noise in the image so a compromise between the two has to be reached.

3.2 Edge Detection and Morphological Thinning

After enhancing the contrast of the input images, the algorithm identifies the edges by using the Sobel edge detector [19]. When compared to other edge detection techniques such as the Roberts Cross and the Canny edge detectors [20], the Sobel filter appears to provide the best trade-off between sensitivity to noise and execution time. For example, when compared to the Roberts Cross technique, it takes

Fig. 3 An image of a taxiway in low visibility conditions used for testing the CLLHE algorithm

Fig. 4 An image showing the result of the CLLHE algorithm on Figure 3, demonstrating the block-effect

longer to execute but is less sensitive to noise. On the other hand, when compared to the Canny edge detector, the Sobel edge detector executes faster but is more sensitive to noise.

The Sobel edge detector works by sliding a 3x3 window over the entire image. For every set of pixels in the window, the elements in the window are convoluted with the vertical and horizontal Sobel masks, thus producing the vertical and horizontal gradients denoted by G_y and G_x respectively. This operation is summarized by Equation 3 and Equation 4:

$$G_x = \begin{bmatrix} -1 & 0 & 1 \\ -2 & 0 & 2 \\ -1 & 0 & 1 \end{bmatrix} \times W \tag{3}$$

$$G_y = \begin{bmatrix} -1 & -2 & -1 \\ 0 & 0 & 0 \\ 1 & 2 & 1 \end{bmatrix} \times W \tag{4}$$

where W is the 3x3 sliding window. These gradients are used to calculate the gradient magnitude G and the direction of the edges Θ as shown in Equation 5 and Equation 6:

$$G = \sqrt{G_x^2 + G_y^2} \tag{5}$$

$$\Theta = \tan^{-1}\left(\frac{G_y}{G_x}\right) \tag{6}$$

In order to distinguish between left and right edges, the thresholding method is replaced by another technique that apart from checking that the gradient magnitude exceeds the threshold, it also checks the sign of G_x. The value of G_x is negative when the intensity along the horizontal axis increases and positive when the intensity along the horizontal axis decreases. The centreline markings are assumed to be lighter than the rest of the ground and consequently a negative value of G_x indicates a left edge while a positive value of G_x indicates a right edge. The thresholding process is represented by Equation 7 and Equation 8, where T is the threshold and S_L and S_R are the binary outputs indicating the left and right edges respectively. The value of T is set to 77.

$$S_L = \begin{cases} 1, & \mathrm{G} \geq \mathrm{T} \text{ and } \mathrm{G_x} < 0 \\ 0, & \text{elsewhere} \end{cases} \tag{7}$$

$$S_R = \begin{cases} 1, & \mathrm{G} \geq \mathrm{T} \text{ and } \mathrm{G_x} > 0 \\ 0, & \text{elsewhere} \end{cases} \tag{8}$$

The binary outputs are used to construct two new binary images indicating the left and right edges. When G_x is equal to zero, the edge is not detected because the edge is perfectly horizontal and hence it is neither a left edge nor a right edge. However, perfectly horizontal edges are not of interest because since the aircraft follows the centreline, the edges of the centreline will normally have a non-zero horizontal gradient. The edges that result from the block effect caused by the CLLHE algorithm are suppressed by ignoring edges at the boundaries of the tiles.

The Sobel edge detector can produce thick edges. Hence the binary images outputted by the Sobel edge detector are thinned using the morphological thinning technique. This removes duplicate edge information in order to reduce the processing time of the Hough Transform and minimise redundant lines that can be identified by the Hough Transform.

3.3 Detection of Line Segments

The Hough Transform is used to detect the most dominant line segments in the binary images produced by the Sobel edge detector. Considering the Hough Transform equation in Equation 9, θ is incremented from -90° to 89° in steps of 1°.

$$\rho = x\cos\theta + y\sin\theta \tag{9}$$

where (x,y) are the Cartesian coordinates of the binary images and (ρ,θ) are the coordinates in the Hough space. Once the Hough Transform is used to construct the Hough accumulator, the peaks in the Hough space are identified to generate two lists of lines per binary image. One list consists of lines having the 5 most votes and that exceed 30% of the highest vote count. The other list consists of lines having the 25 most votes and that exceed 10% of the highest vote count. These lines are reconstructed back in the Cartesian space by inputting the ρ and θ values into the Hough Transform equation and finding the points in the binary images at which the equation holds true. Whenever the equation holds true, it means that a line segment is present at that point. Line segments that are separated by gaps smaller by 5 pixels are merged in order to make up for any discontinuities produced by the Sobel edge detector and line segments smaller than 15 pixels are discarded because these are considered to be a result of noise. Therefore, ultimately, this stage of the algorithm will produce two lists of line segments per binary image i.e. two lists for left edges and two lists for right edges.

3.4 Centreline Detection

After detecting the most dominant line segments, a pair of line segments which best represents the centreline edges is selected from the list of line segments derived from the lines with the 5 highest vote counts. By working on a small set of dominant line segments, the chance of having line segments resulting from noise is smaller than that of a larger set. Erroneous line segments can be detrimental

to this component and this is the reason why the algorithm works on the smaller list of line segments. Line segments that are horizontal or almost horizontal (i.e. having θ between -90° to -81° or between 81° to 89°) are ignored and not used in this component because these normally belong to the holding position markings.

First, the pair of line segments that are closest to the bottom centre point in the image are found by calculating the distances between the line segments and the bottom centre point (320,480). Considering line segment AB and point C, the equation that is used depends upon the position of the perpendicular projection of C onto AB, denoted by r. The value of r is calculated by Equation 10.

$$r = \frac{AC \bullet AB}{\|AB\|^2} = \frac{(A_y - C_y)(A_y - B_y) - (A_x - C_x)(B_x - A_x)}{(B_x - A_x)^2 + (B_y - A_y)^2} \tag{10}$$

Next, the distance between point C and line segment AB is found using Equation 11.

$$d_{\min} = \begin{cases} \sqrt{(D_x - C_x)^2 + (D_y - C_y)^2}, & 0 \le r \le 1 \\ \sqrt{(A_x - C_x)^2 + (A_y - C_y)^2}, & r < 0 \\ \sqrt{(B_x - C_x)^2 + (B_y - C_y)^2}, & r > 1 \end{cases} \tag{11}$$

where $D = (A_x + r(B_x - A_x), A_y + r(B_y - A_y))$.

There may be situations in which the line segment representing the left edge of the centreline lies on the right of the one representing the right edge of the centreline. To identify such situations, the points at which the extended line segments intersect with the bottom of the image are calculated and compared. When these situations are identified, the line segment in the pair that is closest to the bottom centre point (320,480) is retained while the other one is discarded. The discarded line segment is then replaced by the next line segment lying closest to bottom centre point and on the correct side of the retained line segment. This ensures that the line segment representing the left edge of the centreline lies on the left of the one representing the right edge.

3.5 *Inverse Perspective*

The line segments that are derived from the lines having the 25 most votes are mapped from the image plane to the ground plane by using the Homographic Transform. Unlike the Inverse Perspective Mapping [21], the camera parameters do not have to be known, but the transformation matrix is found through a calibration procedure. The most common calibration method makes use of a checkboard of known dimensions. The coordinates of the squares in the image and on the board are inputted into Equation 12 in order to derive the Homography matrix.

$$\begin{bmatrix} x_1 & y_2 & 1 & 0 & 0 & 0 & -X_1x_1 & -X_1y_1 \\ 0 & 0 & 0 & x_1 & y_1 & 1 & -Y_1x_1 & -Y_1y_1 \\ x_2 & y_2 & 1 & 0 & 0 & 0 & -X_2x_2 & -X_2y_2 \\ 0 & 0 & 0 & x_2 & y_2 & 1 & -Y_2x_2 & -Y_2y_2 \\ \vdots & \vdots & \vdots & \vdots & \vdots & \vdots & \vdots & \vdots \\ x_n & y_n & 1 & 0 & 0 & 0 & -X_nx_n & -X_ny_n \\ 0 & 0 & 0 & x_n & y_n & 1 & -Y_nx_n & -Y_ny_n \end{bmatrix} \begin{bmatrix} a \\ b \\ c \\ d \\ e \\ f \\ g \\ h \end{bmatrix} = \begin{bmatrix} X_1 \\ Y_1 \\ X_2 \\ Y_2 \\ \vdots \\ X_n \\ Y_n \end{bmatrix} \tag{12}$$

where (x,y) are the image plane coordinates, (X,Y) are the ground plane coordinates and a, b, c, d, e, f, g and h are the elements of the Homography matrix.

The Homography matrix remains constant as long as the camera parameters are left unchanged. During the execution of the algorithm, the endpoints of the line segments are mapped from the image plane to the ground plane by inputting them into Equation 13:

$$\begin{bmatrix} XW \\ YW \\ W \end{bmatrix} = \begin{bmatrix} a & b & c \\ d & e & f \\ g & h & 1 \end{bmatrix} \begin{bmatrix} x \\ y \\ 1 \end{bmatrix} \tag{13}$$

where the matrix with elements a, b, c, d, e, f, g and h is the Homography matrix.

3.6 Line Clustering

Up to the centreline detection stage, the algorithm describes the centreline edges using a pair of line segments. This is only suitable for straight centrelines. In order to be able to described curved centrelines, the algorithm groups other line segments that appear to belong to the centreline edges. The pair of line segments found earlier are used as a starting point for finding the other line segments that lie on the centreline edges. First, the search is repeated iteratively from one segment to another in the upward direction. Then, it is repeated in the downward direction (once again starting from the pair of line segments).

Considering the example illustrated in Figure 5, line segment *a* is considered to belong to the line segment pair representing the centreline edges and therefore it is used as a starting point. The algorithm then searches upwards from line segment *a* in order to find the next line segment that appears to belong to the same edge and finds line segment *b*. Once again, the algorithm searches upwards, this time from line segment *b* and finds line segment *c*. However, when the search is repeated upwards from line segment *c*, it does not find any other line segments that appear to belong to the same edge so the upward search stops there. This procedure is then repeated in the downward direction. The algorithm searches downward of line segment *a* and finds line segment *d*. However, it fails to find any line segment below line segment *d*, so the downward search stops there. All of the line segments that appear to belong to the same edge are added to the cluster.

The algorithm selects the line segments by comparing various characteristics and ensuring that they meet certain criteria. The first property that is checked is the difference in their orientation. The angles of successive line segments must not differ by more than 45°. Next, four other conditions are checked and if any of these is satisfied, the line segment under consideration is selected and added to the cluster. In case that there is more than one line segment that match these criteria, only the one closest to the last selected line segment is added to the cluster.

One of the four conditions is whether the line segments intersect. The algorithm determines if two line segments intersect by comparing the relative positions of the line segment endpoints. Considering two intersecting line segments AB and CD, one endpoint of line segment CD should lie on the left of AB while the other endpoint should lie on the right of AB. If this is not the case, then the line segments do not intersect. The positions of the end points relative to each other are found by using the cross product. For example, to find the position of endpoint C relative to line segment AB, the cross product is calculated as in Equation 14.

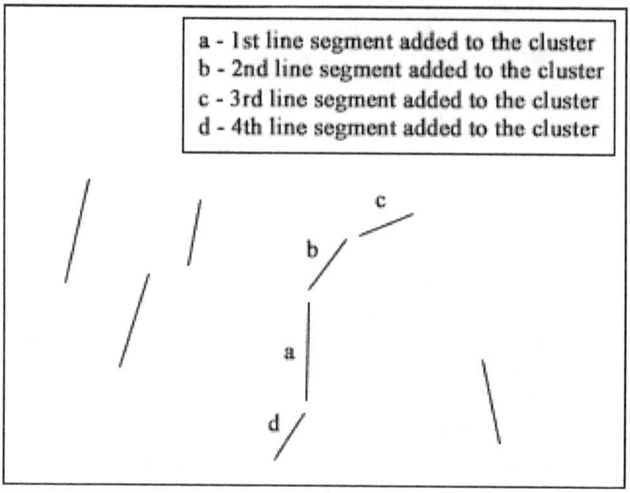

Fig. 5 An illustration depicting the procecedure that clusters line segments belonging to the same edge

To find the position of D relative to line segment AB the same equation is used, but C is replaced with D. Opposite signs indicate that points C and D lie on opposite sides while identical signs indicate that points C and D lie on the same side [22].

$$p = (B - A) \times (C - A) \tag{14}$$

The second condition checked is whether the distance between the endpoints of the line segments is smaller than 5 pixels. If the search is upwards, the distance is measured from the upper endpoint of the last selected line segment to the lower endpoint of the line segment under consideration. If the search is downwards, the distance is measured from the lower endpoint of the last selected line segment to the upper endpoint of the line segment under consideration. The distance between the points is calculated by using Pythagoras' Theorem.

The third conditions checked is whether the line segments are collinear or almost collinear. This is done by using the equation that calculates the distance between a point and a line. The line segment is treated as an extended line. If the search for the next line segment is upwards, then the distance calculated is that between the last selected line segment and the lower endpoint of the line under consideration. Otherwise if the search is downwards, the distance calculated is that between the last selected line segment and the upper end point of the line under consideration. The maximum accepted distance is 5 pixels.

The last of the four conditions is similar to the third but the acceptable distance is increased as the line segments get farther from each other and is set equal to the distance measured whilst checking for the second condition. If the fourth condition is met, the line segments must satisfy another condition that checks whether the line segments form a turn. This is the case when a line passing through the closest endpoints of the two line segments has both line segments on the same side.

3.7 Curve Fitting

Curves are fitted through the clustered line segments by using the Weighted Least Squares Fit (WSLF). This fitting technique provides smooth polynomial curves for representing the centreline edges and allows priority to be given to the lower part of the curve by assigning higher weights to points in that part. The WLSF takes points as inputs so points are sampled from the clustered line segments and inputted into the WLSF. The points are sampled by repeatedly splitting the clustered line segments into two until they are smaller than 1 pixel. When this occurs, the endpoints of the line segments are used as the points for the WLSF. The advantage of sampling using this technique is that there are no problems associated with infinite gradients and divisions by 2 can be translated into logical shifts to the right. Therefore this technique is preferred to having to derive the line equations. The x and y-coordinates of the sampled points are swapped when inputted into the WLSF so that the resulting polynomial is a function of the y-coordinate. Fifth order weights are assigned to the points depending on their vertical position. The lower they are, the higher the weight that is assigned to them. Consequently, the weight of a point in row r is r^5. This ensures that the lower part of the curve is characterised well because this part of extreme importance since the cross-track and heading errors are derived from it. The curves are fitted as third order polynomials which are sufficiently flexible for characterising bends and does no not result in sub-optimal fits (that usually result when fitting curves of high order polynomials). When the polynomials describing the left and right edges of the centreline are fitted, the one describing the actual centreline is found by adding corresponding polynomial coefficients and dividing them by two. This gives the polynomial equation of the curve in the middle of those describing the centreline edges.

3.8 Measurement of Cross-track and Heading Errors

After obtaining the polynomial equation describing the taxiway centreline, the cross-track and heading errors are measured from it. The cross-track error is the horizontal distance between the bottom of the curve and the middle of the image. It is calculated by substituting variable y with the height of the image (i.e. 480) and subtracting half the image width (i.e. 320) from the result. The resulting equation is Equation 15:

$$x_t = c_3 h^3 + c_2 h^2 + c_1 h + c_0 - \frac{w}{2} \tag{15}$$

where c_0, c_1, c_2 and c_3 are the polynomial coefficients, h is the height of the image, w is the width of the image and x_t is the cross-track error. The heading error is indicated by the tangent to the bottom of the curve. The slope of the tangent is found by differentiating the curve equation and then substituting variable y with the height of the image. The arctangent function is used on the resulting slope in order to find the heading error in terms of an angle. The calculation of the heading error is summarised by Equation 16:

$$\theta_e = \tan^{-1}(3c_3h^2 + 2c_2h + c_1) \tag{16}$$

where c_1, c_2 and c_3 are the polynomial coefficients, h is the height of the image and θ_e is the heading error.

3.9 *Tracking Filter*

The Kalman Filter is used to track the cross-track and heading errors in order to filter noise. The Kalman filter reduces noise by comparing two measurements with each other. In this work, the cross-track and heading errors measured by the algorithm are compared to a mathematical model. The mathematical model of the cross-track error is derived from the SUVAT equations. The mathematical model of the heading error is derived from equations describing angular motion and results in the same coefficients. The model of the Kalman filter for both cross-track and heading errors is presented in Equation 17:

$$x_k = \begin{bmatrix} 1 & t & 0.5t^2 \\ 0 & 1 & t \\ 0 & 0 & 1 \end{bmatrix} x_{k-1} + w_{k-1} \tag{17}$$

$$z_k = \begin{bmatrix} 1 & 0 & 0 \end{bmatrix} x_k + v_k \tag{18}$$

where x_k is the prediction value, x_{k-1} is the previous prediction value, w_{k-1} is the process noise, z_k is the measurement value and v_k is the measurement noise. The process noise and measurement noise covariance matrices (Q and R respectively) were chosen by experimentation. These were set as in Equation 19 and Equation 20.

$$Q = \begin{bmatrix} 0.001 & 0 & 0 \\ 0 & 1 & 0 \\ 0 & 0 & 1 \end{bmatrix} \tag{19}$$

$$R = [1] \tag{20}$$

3.10 *Adapting the Algorithm for Infrared Vision*

The components that have been described in this section above are intended for processing images captured by visible light cameras. In order to take advantage of infrared technology (considering that infrared cameras penetrate through poor visibility and are immune to shadows and variable illumination) the algorithm was adapted for these type of images.

When testing the algorithm on images captured by infrared cameras, in some scenarios artefacts were appearing in the images. These artefacts are obfuscated by blurring the images so that they are not detected as part of the centreline.

Also, since the desired information remained in the same range of grey levels (between 64 and 90) during testing, the contrast enhancement technique was replaced by another one that simply stretches the histogram. Therefore, the contrast enhancement is not affected by pixels that are not of interest. Equation 21 is used for implementing this technique.

$$g(x, y) = \frac{f(x, y) - f_{min}}{f_{max} - f_{min}} \times 255 \tag{21}$$

where f_{max} denotes the upper boundary of the wanted range (i.e. 90), f_{min} denotes the lower boundary (i.e. 64), f(x,y) is the original pixel intensity and g(x,y) is the mapped pixel intensity.

Light emitting objects generally have intensities that are considerably higher than passive objects making up the rest of the image. This gap in intensities was exploited to suppress edges caused by lights. The lights are identified by binarising the image with a threshold of 128. The binarised image is dilated with a structuring element having the shape of a disc whose radius is 10 pixels. The edges produced by the Sobel edge detector that superimpose the dilated region are removed.

4 Results

As an initial evaluation, the algorithm was tested on videos captured from simulations in X-Plane. A Boeing 737 was taxied around the taxiways in an environment simulating Malta International Airport and the session was stored. This session was then replayed under different simulated weather conditions (broken, cirrus, clear, foggy, low visibility, overcast, scattered and stormy) and at different times of the day (noon, midnight, 6am and 6pm). The recorded videos were each 225 seconds long and had a native resolution of 1280×960 but were saved at a resolution of 640×480. Since the horizon was visible in the captured videos, the area above the horizon was ignored by the algorithm so that it did not interfere with the results.

The cross-track and heading errors measured by the algorithm were compared to ones measured manually and the differences were tabulated. These values were measured manually by opening the video frames in an image editing software and using an inbuilt ruler to measure the cross-track error and an onscreen protractor to measure the heading error. Statistical data of the differences observed in 200 successive frames in a simulation involving clear visibility is presented in Table 1. The results indicate that the algorithm managed to attain centimetric accuracy and hence has the desired performance.

Table 1 Statistical data of the discrepancies in cross-track and heading errors measured by the algorithm against ones measured manually when processing the synthetic video simulated in clear weather

	Discrepancy in cross-track error (mm)	Discrepancy in heading error (°)
Maximum	50.54	4.45
Minimum	1.23	0.07
Average	32.80	1.13
Standard deviation	13.58	0.89
95th percentile	49.89	2.58

The algorithm was also tested on video captured from simulations generated in the lowest visibility in X-Plane i.e. 0.10 statute miles. Statistical data of the differences between the cross-track and heading errors measured by the algorithm against ones measured manually in 200 frames is presented in Table 2. The results show that the algorithm works well in poor visibility conditions. The algorithm actually performs better than in clear weather conditions. This is due to the fact that the coefficients of the algorithm were mostly based on tests performed in poor visibility and at night.

Table 2 Statistical data of the discrepancies in cross-track and heading errors measured by the algorithm against ones measured manually when processing the synthetic video simulated in the lowest visibility setting

	Discrepancy in cross-track error (mm)	Discrepancy in heading error (°)
Maximum	45.97	3.85
Minimum	1.14	0.01
Average	23.19	1.06
Standard deviation	13.26	0.82
95th percentile	44.81	2.56

The algorithm was also tested on real videos captured at Malta International Airport in field trials using a cameras mounted on a van driving around the airfield. These videos were obtained from trials conducted during the EC FP7 ALICIA project. The footage was taken on 13th August 2013 and on 14th August 2013 at night, and was recorded by two types of cameras: visible light and infrared. Two visible light videos were captured by a Go Pro Hero 3 camera at a resolution of 1280×720 and at a rate of 25 frames per second. These had a duration of 1418 seconds and 1439 seconds. Another two infrared videos were recorded using a Flare SC7000 Thermal

IR camera with a resolution of 320×256 recording at a rate of 25 frames per second. These video streams had a duration of 893 seconds and 914 seconds. All videos were resized to a resolution of 640×480 before they were input to the algorithm and the area above the horizon was ignored so that it did not affect the results.

When testing the algorithm on the visible light videos, the differences between the cross-track and heading errors measured by the algorithm against ones measured manually were tabulated. Statistical data of the differences in 200 frames is presented in Table 3. The results show that the algorithm still performs with centimetric accuracy, and therefore it performs adequately.

Table 3 Statistical data of the discrepancies in cross-track and heading errors measured by the algorithm against ones measured manually when processing the visible light videos captured at Malta International Airport

	Discrepancy in cross-track error (mm)	Discrepancy in heading error (°)
Maximum	16.06	1.63
Minimum	0.09	0.02
Average	4.77	0.70
Standard deviation	4.17	0.44
95th percentile	14.08	1.60

Statistical data of the differences belonging to video sequences in which the centreline is curved is presented in Table 4. The algorithm still performs well, but understandably the accuracy decreases.

Table 4 Statistical data of the discrepancies in cross-track and heading errors measured by the algorithm against ones measured manually when processing video sequences from the visible light video captured at Malta International Airport in which the centreline markings are curved

	Discrepancy in cross-track error (mm)	Discrepancy in heading error (°)
Maximum	13.48	2.50
Minimum	0.41	0.04
Average	5.96	0.56
Standard deviation	3.68	0.48
95th percentile	12.51	1.47

Results not presented herein also show that the algorithm only works well when the centreline markings are lighter than the ground. During testing, in some of the frames, the ground was lighter than the centreline and the curves representing the centreline edges were being fitted on edges of other markings. Another problem that was noticed was that at night, whenever the centreline lights were present, the contrast between the lights of the ground increased while that between the centreline and the ground dropped. As a result, the Sobel edge detector was only detecting the edges of the lights and curves were being fitted on the edges of the lights.

To test the algorithm adapted to handle infrared videos, statistical data of the differences in 200 frames was extracted, comparing the cross-track and heading errors estimated by the algorithm against ones measured manually. Results are presented in Table 5 and indicate that the adapted algorithm also performs well and its accuracy is well within a few centimetres.

Table 5 Statistical data of the discrepancies in cross-track and heading errors measured by the algorithm against ones measured manually when processing the infrared videos captured at Malta International Airport

	Discrepancy in cross-track error (mm)	Discrepancy in heading error (°)
Maximum	18.05	9.65
Minimum	0.17	0.31
Average	6.21	3.41
Standard deviation	4.65	2.28
95th percentile	16.56	7.36

5 Conclusion

In this work, an algorithm that can determine the position of the aircraft with respect to the taxiway centreline has been presented. The algorithm identifies the centreline markings through image processing techniques, models them using curve fitting techniques and then uses this information to measure the cross-track and heading errors. The algorithm was originally developed for processing visible light images, and was later adapted to process infrared imagery.

Results show that the original algorithm and the adapted one work adequately in straight lines, bends, in low visibility and at night, indicating that the objective of centimetric accuracy may be achieved in real operations.

Further work will include improvements to the algorithm to process visible light imagery so that it can automatically identify situations in which the ground is lighter than the centreline markings and, in these cases, it will automatically swap the binary images indicating left and right edges. Also, the centreline lights could be suppressed so that they do not affect the algorithm designed for visible light imagery.

References

[1] PlaneCrashInfo.com, Accident statistics (August 28, 2014), `http://www.planecrashinfo.com/cause.htm`

[2] Kulesa, G.: Weather and aviation: How does weather affect the safety and operations of airports and aviation, and how does FAA work to manage weather-related effects? In: The Potential Impacts of Climate Change on Transportation (2003)

[3] Clynch, J.R.: GPS Accuracy Levels. (2001), http://www.oc.nps.edu/oc2902w/gps/gpsacc.html (October 15, 2014)

[4] Belabbas, B., Hornbostel, A., Sadeque, M.Z.: Error analysis of single frequency gps measurements and impact on timing and positioning accuracy (2005)

[5] Fox, K.C.: Impacts of Strong Solar Flares. (2013), http://www.nasa.gov/mission_pages/sunearth/news/flare-impacts.html (October 15, 2014)

[6] Authority, A.M.S.: Differential Global Positioning System. (October 15, 2014), http://www.amsa.gov.au/navigation/services/dgps/

[7] F. A. Administration. WAAS - How It Works. (2010), http://www.faa.gov/about/office_org/headquarters_offices/ato/service_units/techops/navservices/gnss/waas/howitworks/ (October 15, 2014)

[8] Zammit, C., Zammit-Mangion, D.: An enhanced automatic taxi control algorithm for fixed wing aircraft (2014)

[9] Bertozzi, M., Broggi, A.: GOLD: a parallel real-time stereo vision system for generic obstacle and lane detection. IEEE Trans. Image Process. 7, 62–81 (1998)

[10] Wang, Y., Teoh, E.K., Shen, D.: Lane detection and tracking using B-Snake. Image and Vision Computing 22, 269–280 (2004)

[11] Yu, B., Jain, A.K.: Lane boundary detection using a multiresolution hough transform. In: Proceedings of the International Conference on Image Processing, pp. 748–751 (1997)

[12] Hota, R.N., Syed, S., Bandyopadhyay, S., Krishna, P.R.: A Simple and Efficient Lane Detection using Clustering and Weighted Regression. In: COMAD (2009)

[13] Gonzalez, R.C., Woods, R.E.: Digital image processing, pp. 671–672. Prentice Hall, Upper Saddle River (2007)

[14] Gonzalez, R.C., Woods, R.E.: Digital image processing, pp. 755–760. Prentice Hall, Upper Saddle River (2007)

[15] Criminisi, A., Reid, I., Zisserman, A.: A plane measuring device. Image and Vision Computing 17, 625–634 (1999)

[16] W. MathWorld. Least Squares Fitting, http://mathworld.wolfram.com/LeastSquaresFitting.html (October 15, 2014)

[17] Esme, B.: Kalman Filter For Dummies (2009), http://bilgin.esme.org/BitsBytes/KalmanFilterforDummies.aspx (October 15, 2014)

[18] Zuiderveld, K.: Contrast limited adaptive histogram equalization. In: Paul, S.H. (ed.) Graphics Gems IV, pp. 474–485. Academic Press Professional, Inc. (1994)

[19] Gonzalez, R.C., Woods, R.E.: Digital image processing, p. 189. Prentice Hall, Upper Saddle River (2007)

[20] Gonzalez, R.C., Woods, R.E.: Digital image processing, pp. 741–747. Prentice Hall, Upper Saddle River (2007)

[21] Muad, A.M., Hussain, A., Samad, S.A., Mustaffa, M.M., Majlis, B.Y.: Implementation of inverse perspective mapping algorithm for the development of an automatic lane tracking system. In: 2004 IEEE Region 10 Conference, TENCON 2004, pp. 207–210 (2004)

[22] Cormen, T.H., Leiserson, C.E., Rivest, R.L., Stein, C.: Introduction to algorithms, pp. 1015–1019. MIT Press, Cambridge (2009)

A New Observer for Range Identification in Perspective Vision Systems

Victor Gibert, Laurent Burlion, Abdelhamid Chriette,
Josep Boada-Bauxell, and Franck Plestan

1 Introduction

Automatic guidance of flying vehicle usually needs external information. *GPS* (Global Positionning System) is a worldwide technology which provides to the guided system its deviations with respect to its guidance objective. Nevertheless, this technology is not available everywhere (indoor evironment), everytime (in case of failure) and not precise enough to ensure critical operations as landing a civil aircraft. Current civil aircraft are able to land autonomously on a runway thanks to *ILS* (Instrument Landing System) or differential GPS. However, these equipment are expensive and can fail. In frame of the future aircraft, manufacturers like *AIRBUS* company study the possibility to make aircraft landing everywhere (unequipped or unknown runway) without using informations from external systems.

In order to overcome the use of external needs, an embedded solution that provides deviations is needed. In parallel, image processing and camera technology have made a technological leap in the last decade. Hence, the use of a camera to perform visual servoing becomes an interesting solution to cope with precision and availability requirements.

Visual servoing consists in using vision as a sensor in order to control the motion of the system. Tutorial in [3] explains the different ways to use visual servoing. Two main classes of visual servoing have been studied: the first and oldest one

Victor Gibert · Josep Boada-Bauxell
Airbus Operations S.A.S., Toulouse, France
e-mail: {victor.gibert,josep.boada-bauxell}@airbus.com

Laurent Burlion
ONERA, DCSD, Toulouse, France
e-mail: laurent.burlion@onera.fr

Abdelhamid Chriette · Franck Plestan
LUNAM Université, Ecole Centrale de Nantes - IRCCyN, Nantes, France
e-mail: {Abdelhamid.Chriette,Franck.Plestan}@irccyn.ec-nantes.fr

J. Bordeneuve-Guibé et al. (eds.), *Advances in Aerospace Guidance, Navigation and Control*, DOI: 10.1007/978-3-319-17518-8_23

is Pose Based Visual Servoing (*PBVS*) [1] whereas a more recent one is Image Based Visual Servoing (*IBVS*) [22]. In *PBVS* scheme, vision is used to estimate the pose (position and attitude) of the camera. This estimation can be obtained by using more than one camera [23], by knowing dimensions of seen objects [14] or by taking into account the camera motions [4][6][7][12][8][15][16]. Once the estimate is complete, standard guidance laws can be set up by using estimated deviations.

On the other side, *IBVS* scheme expresses the objective and the current state in the image frame. This scheme can use, for example, the full image for an homography comparison [13][21] or is directly using the measurement of visual features coordinates [2][5][11][18] and make the current features match with their corresponding desired ones on the image plane.

IBVS takes advantage on PBVS because of the overcoming of estimation process and is less impacted by calibration errors. However, PBVS takes advantage on IBVS because it allows to use existing and certified guidance laws; in IBVS, new guidance laws need to be designed. In PBVS, stereo-vision appears difficult in this case of study because the distance from runway is important, calibration must be very precise and not impacted by vibrations and two camera bring weight and space problem. This paper considers a generic runway whose size and markers are not known: geometric reconstruction solutions using these informations can not then be applied. If the motion of the aircraft is known, the use of dynamics of visual features between several images is sufficient to estimate deviations *w.r.t.* the runway.

Among these visual servoing strategies, *PBVS* using known motions appears to be a potential candidate for aircraft landing; thus, estimated deviations are used as measurement in order to guide the aircraft in final approach. Actually, aircraft dynamic can be considered always known thanks to IRS sensors availability.

The aim of the article is to estimate the deviations of the camera *w.r.t.* to the runway. In this purpose, the single available informations are the knowledge on rational and translational velocities provided by inertial sensors and visual measurement. The visual informations, provided by image processing algorithms, correspond to the perspective projection of a 3D point in the image plane.

The three main contributions of this paper can be presented as follows : a new nonlinear observer solution is proposed based on state coordinates transformation; a comparison with previously published solutions is made on a common example of literature [4]; and the proposed estimator has been applied on a realistic scenario corresponding to a civil aircraft landing. The paper shows the advantages of the new range identification estimator compared with previous solutions. First, the proposed solution provides a general observer formulation for non-linear problem with a generic correction term. This correction term admits several type of non-linear estimators. Second, using an expression in a transformed state coordinates renders the observer design simple compared with [16]. Compared with previous solutions [4][8][15], the dynamics of estimates is easy to control. Finally, the proposed solution presents good robustness properties with respect to noisy measurement.

The paper is organized as follows. Section 2 presents a standard range identification formulation. Then, Section 3 proposes a new pose estimation method with detailed observability analysis and design method of the observer. Next, section 4 presents results obtained on a standard example which are presented and compared with other existing solutions. Results obtained with the proposed observer on a realistic aircraft motion during landing phases are also presented. Finally, section 5 concludes this paper.

2 Problem Statement

The problem under interest consists in using an embedded monocular camera in order to estimate three-dimensional deviations *w.r.t.* a point of interest from two-dimensional image measurement.

Denoting $\boldsymbol{x}$ the three-dimensional coordinates of a point attached to the ground express in the camera frame $\mathscr{C}$, its dynamics reads as

$$\dot{\boldsymbol{x}} = \underbrace{\begin{bmatrix} a_{11}(t) & a_{12}(t) & a_{13}(t) \\ a_{21}(t) & a_{22}(t) & a_{23}(t) \\ a_{31}(t) & a_{32}(t) & a_{33}(t) \end{bmatrix}}_{\mathbf{A}} \boldsymbol{x} + \underbrace{\begin{bmatrix} b_1(t) \\ b_2(t) \\ b_3(t) \end{bmatrix}}_{\mathbf{B}} \tag{1}$$

with **A** and **B** respectively the rotational and translational motion matrices which can be time-varying and supposed known.

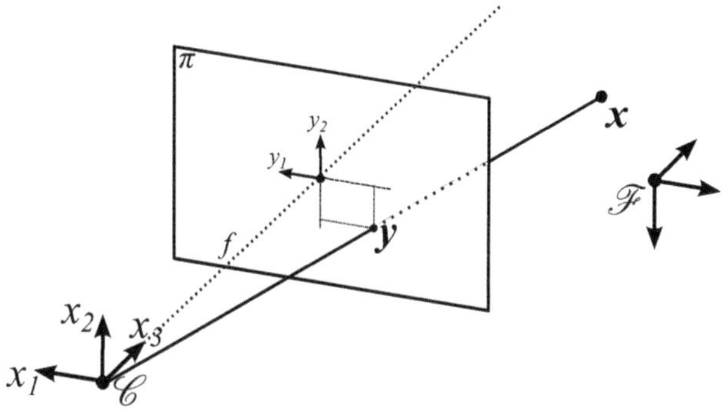

Fig. 1. Perspective projection of the runway in the camera plane.

Assuming a calibrated pinhole camera model (see Figure 1), thanks to image processing algorithms, a perspective projection $\boldsymbol{y} = [y_1 \; y_2]^T$ in the image plane (π) is obtained from the unmeasurable state $x = [x_1 \; x_2 \; x_3]^T$. The relationship between y and x reads as

$$\boldsymbol{y} = C(\boldsymbol{x}) = f \begin{bmatrix} \frac{x_1}{x_3} & \frac{x_2}{x_3} \end{bmatrix}^T . \tag{2}$$

with f the focal length of the camera[1].

[1] Without loss of generality, one can consider that the focal length $f = 1$.

The objective of the following observer is to estimate the state coordinate $[x_1\ x_2\ x_3]^T$ from the measurement of the image space coordinate $[y_1\ y_2]^T$ and the motions of the camera.

3 A New Range Identification Technique

Several approaches as [16][6][4] propose a range identification method in perspective system. This section proposes a new approach compared with the existing range identification methods in perspective system. The proposed solution allows an easy design of the correction term compared with previous observer and, as shown in a sequel, a better robustness versus noisy measurement. The new observer consists in using the canonical observability formulation with a state coordinate transformation that provides the capacity to easily design the observer. With this scheme, high gain or high order sliding-mode approaches could be applied.

3.1 Observability Analysis

Let us define the following function:

$$\Psi(x) = \begin{bmatrix} y_1 \\ \dot{y}_1 \\ y_2 \end{bmatrix} = \begin{bmatrix} \frac{x_1}{x_3} & \dot{\left(\frac{x_1}{x_3}\right)} & \frac{x_2}{x_3} \end{bmatrix}^T \tag{3}$$

Consider $M_x \subset I\!R^n$ the operating physical domain in which $\boldsymbol{x}$ is evolving.

Definition 1. [17] The system (1)-(2) is locally observable (*i.e.*, observable $\forall x \in M_x$) if $\Psi(x)$ is a state coordinates transformation, *i.e.*, $\Psi(x)$ is invertible $\forall x \in M_x$. ■

Given the complexity of Ψ, it is difficult (even with formal computation software) to analytically establish its invertibility. Thus, this latter will be numerically evaluated through the evaluation of its Jacobian $\frac{\partial \Psi}{\partial x}$.

Corollary 1. *System (1)-(2) is locally observable in the sense of Definition 1 if*

$$\left(\frac{\partial \Psi}{\partial x}\right)_{x \in M_x} \text{invertible} \Leftrightarrow \left|\frac{\partial \Psi}{\partial x}\right|_{x \in M_x} \neq 0. \tag{4}$$

■

With the transformation Ψ defined in (3), the observability condition is fulfilled if

$$b_1 - b_3 y_1 > 0. \tag{5}$$

withb_i from (1).

Remark 1. *The choice for the function Ψ is not unique. For example, it would be also possible to choose $\Psi = [y_1\ \ y_2\ \ \dot{y}_2]^T$*

3.2 Observer Design

Given that $\Psi(x)$ is invertible under the proposed operating conditions (5), it defines a state coordinates transformation $\zeta = \Psi(x)$. Then, it is trivial to show that the nonlinear system (1)-(2) is locally equivalent to

$$\dot{\zeta} = \underbrace{\begin{bmatrix} 0 & 1 & 0 & 0 \\ 0 & 0 & 0 & 0 \\ 0 & 0 & 0 & 0 \end{bmatrix}}_{F} \zeta + \underbrace{\begin{bmatrix} 0 \\ \Phi_1(\zeta) \\ \Phi_2(\zeta) \end{bmatrix}}_{\Phi(\zeta)} \tag{6}$$

Proposition 1. *An observer for system (6) reads as*

$$\dot{\hat{\zeta}} = F\hat{\zeta} + \Phi(\hat{\zeta}) + \kappa(y, \hat{\zeta}) \tag{7}$$

with $\hat{\zeta}$ the estimated state of ζ and the function $\kappa(y, \hat{\zeta})$ called "correction term" and forcing $\hat{\zeta} \to \zeta$. ■

It is obvious that the correction term $\kappa(y, \hat{\zeta})$ is not unique and can be obtained by several different methods depending on the desired features (robustness, finite time convergence, etc.,). Note that the term κ depends only on "known" variables, *i.e.* measurements y and estimated state $\hat{\zeta}$. Given that estimation error dynamics reads as (with $e = \hat{\zeta} - \zeta$)

$$\dot{e} = Fe + \Phi(\hat{\zeta}) - \Phi(\zeta) + \kappa(y, \hat{\zeta}), \tag{8}$$

$\kappa(y, \hat{\zeta})$ has to force the observer to converge (exponentially or in a finite time) to the real system in spite of the initial error $e(0)$. From $\hat{\zeta} = \Psi(\hat{x})$, one gets

$$\dot{\hat{\zeta}} = \frac{\partial \Psi}{\partial \hat{x}} \dot{\hat{x}} \rightarrow \dot{\hat{x}} = \left[\frac{\partial \Psi}{\partial \hat{x}}\right]^{-1} \dot{\hat{\zeta}}. \tag{9}$$

Then, by a similar way than [19], an observer for system (1)-(2) reads as

$$\dot{\hat{x}} = A\hat{x} + B + \left[\frac{\partial \Psi}{\partial x}\right]^{-1} \kappa(y, \hat{x}) \tag{10}$$

The structure of the observer having been defined, the correction term $\kappa(y, \hat{x})$ has to be designed. The proposed observer in (10) allows to choose among different correction term as high gain [10] or sliding mode observer [20] (more details in [9]). To provide a simple design of the correction term, the high gain observer has been chosen for this paper.

High-Gain Observer [10]: The observer (10) for the system (1)-(2) admits a correction term $\kappa(y, \hat{x})$ defined as

$$\kappa(y, \hat{x}) = \Lambda^{-1} K(y - C(\hat{x})) \tag{11}$$

with

$$\Lambda = \begin{bmatrix} \lambda_1 & 0 & 0 \\ 0 & \lambda_1^2 & 0 \\ 0 & 0 & \lambda_2 \end{bmatrix} \tag{12}$$

$$K = \begin{bmatrix} K_1 & 0 \\ K_2 & 0 \\ 0 & K_3 \end{bmatrix}. \tag{13}$$

and with λ_1, λ_2, K_1, K_2 and K_3 strictly positive constant so that $F - KC$ is Hurwitz.

The poles were specified in order to correspond to a characteristic polynomial with three parameters, α, ξ and ω, as $(s+\alpha)(s^2+2\xi\omega s+\omega^2) = 0$. It remains to $K_1 = 2\xi\omega$, $K_2 = \omega^2$ and $K_3 = \alpha$.

4 Simulation Results

4.1 Academic Example

Numerical simulations are presented using the proposed observer (10). A comparison with existing approaches for range identification [4][6][8][16] allows assessing the performance of this new estimator solution. Consider the example given in example [4] for the affine system (1)-(2), with matrices **A** and **B** defined as

$$\mathbf{A} = \begin{bmatrix} -0.2 & 0.4 & -0.6 \\ 0.1 & -0.2 & 0.3 \\ 0.3 & -0.4 & 0.4 \end{bmatrix} \text{ and } \mathbf{B} = \begin{bmatrix} 0.5 \\ 0.25 \\ 0.3 \end{bmatrix} \tag{14}$$

with the initial conditions

$$\begin{bmatrix} x_1(0) & x_2(0) & x_3(0) \end{bmatrix}^T = \begin{bmatrix} 1 & 1.5 & 2.5 \end{bmatrix}^T \tag{15}$$

and

$$\begin{bmatrix} \hat{x}_1(0) & \hat{x}_2(0) & \hat{x}_3(0) \end{bmatrix}^T = \begin{bmatrix} 0.4 & 0.6 & 1 \end{bmatrix}^T \tag{16}$$

The desired rate of convergence of the estimate error is obtained with $\xi = 0.99$ and $\omega = 7$ corresponding to a 5% response time of 1 seconds with strong damping and $\alpha = 1$. These parameters roughly correspond to $\lambda = 30$ in [16], to the same parameters than in [4] except $\delta_i = 0.003$ and to the same parameters in [8] and in [6].

Simulation results for estimation of $\hat{x}_3$ with observers [4][6][8][16] in case of no measurement noise are shown in Figure 2.

Most observers have similar behavior except the range identification proposed by [8]. Note that, estimation error dynamics present a non-null tangent at the origint only with [16]. Taking into account the influence of observer design parameters on the error estimation dynamics, the proposed solution appears to be a good candidate to range identification problem in perspective systems.

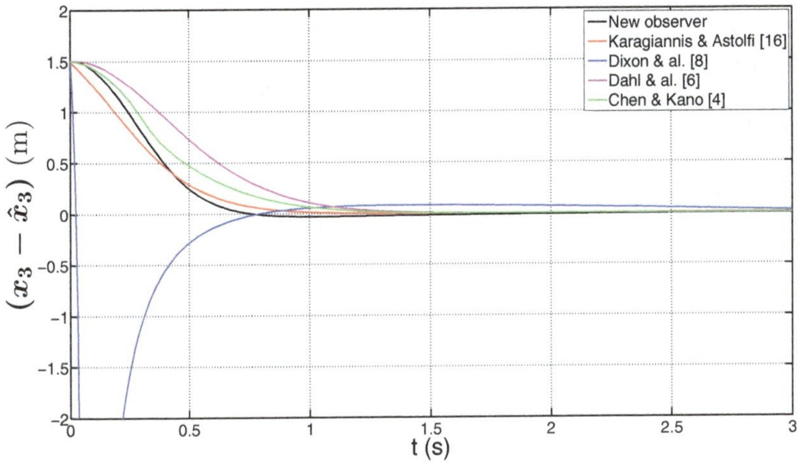

Fig. 2. Comparison between range identification methods [4][6][8][16] and the proposed solution without noisy measurement.

To investigate the effects of noisy measurements on the estimation results, a 1% random noise of the signals is added on the measured informations y. Figure 3 shows the results for estimation of $\hat{y}_3$ under noisy conditions.

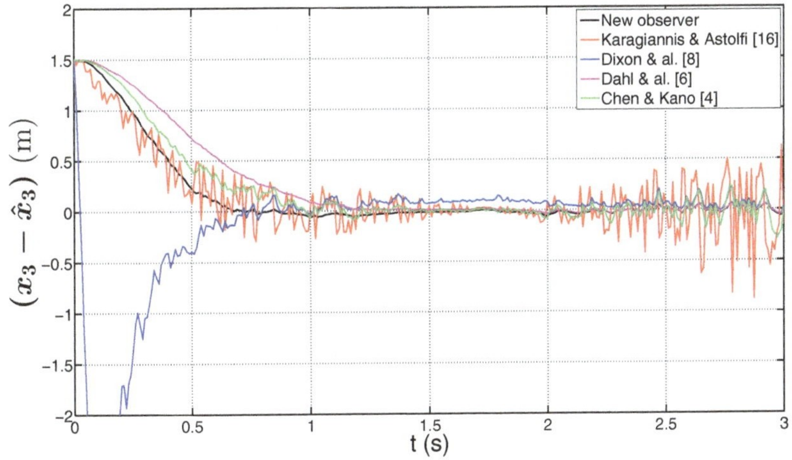

Fig. 3. Comparison between range identification methods [4][6][8][16] and the proposed solution in presence of noisy measurement.

In presence of noise, the effects on estimations are not the same for all observers. Although the dynamic estimation is approximately the same for each estimator, the proposed solution is showing a very good robustness to noisy measurement compared with other observer. A similar robustness to noisy measurement is obtained with [6]. Indeed, this observer is also based on a coordinate transformation.

To conclude, the proposed observer is providing a new scheme for range identification in perspective vision system. Dynamics of the estimation rate could be easily tuned and robustness to noisy measurement appears to be one of the features of this estimation solution.

4.2 Example Based on a Fixed Wing Civil Aircraft

Previous section results have been obtained on a very simple example with constant motion parameters. Hereafter, the proposed observer is applied on a realistic time-varying motion of a fixed wing civil aircraft. The coordinates of a tracking point $\boldsymbol{x}$ in the camera frame are expressed by

$$\dot{\boldsymbol{x}} = \underbrace{\begin{bmatrix} 0 & -\omega_z & \omega_y \\ \omega_z & 0 & -\omega_x \\ -\omega_y & \omega_x & 0 \end{bmatrix}}_{\mathbf{A}} \boldsymbol{x} + \underbrace{\begin{bmatrix} v_x(t) \\ v_y(t) \\ v_z(t) \end{bmatrix}}_{\mathbf{B}} \tag{17}$$

with ω_i and V_i respectively the rotational and translational velocities expressed in the camera frame[2].

In final approach, the desired trajectory, named glide path, is ending on the runway at a 3D point x (see Figure 4). The missing informations which need to be estimated are the deviations $x = [x_1\ x_2\ x_3]^T$ *w.r.t.* the aircraft.

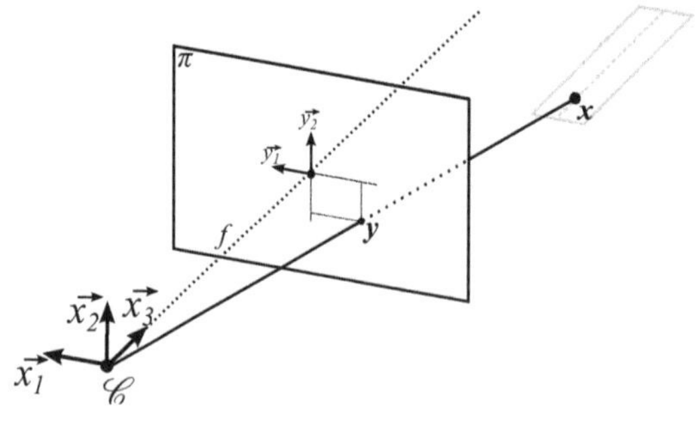

Fig. 4. Perspective projection of the glide path ending point on the runway x

[2] An embedded camera attached to the aircraft is used; here, the aircraft frame can be supposed merging with it.

Consider an aircraft, reaching the airport with a rough known position. At $5km$ from the runway, one can consider the image processing able to deliver the perspective projection coordinates of x in the image plane. The aircraft is neither aligned with the runway and nor already on the glide path. In the simulations, the aircraft will reach the desired trajectory and will track it during the end of the approach.

The observer will be initialized with wrong deviations $[\hat{x}_1(0)\ \hat{x}_2(0)\ \hat{x}_3(0)]$. Two initialised states will be simulated corresponding to a $\pm 50\%$ error between the estimated and current state vectors. Figure 5 shows the trajectory followed by x in the camera frame and the estimation of the 3D coordinates of x during the landing with the two initialized states.

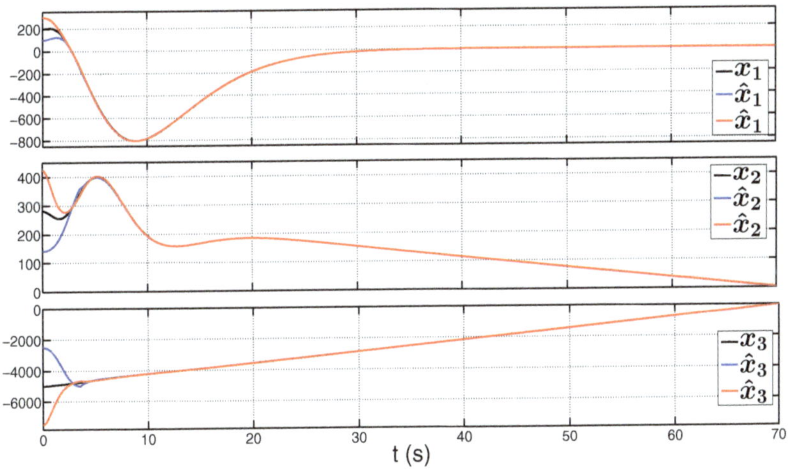

Fig. 5. Estimation results during a civil aircraft landing. Aircraft trajecory (in black), estimation with -50% initial error (in blue) and estimation with $+50\%$ initial error (in red).

Simulation results confirm that the proposed observer provides a good estimation with a time varying motion of the camera. Estimation convergence is fast enough compared with the time to landing and the estimation error converges to zero.

Remark 2. *Previous results have been obtained along a trajectory independent on the estimation results. In visual servoing, the estimated deviations feed the guidance law. Nevertheless, the estimated deviations of x, expressed in the camera frame, should be expressed in the inertial frame (attached to the runway). In this purpose, a change coordinate must be done using the rotational matrix $R = R_\phi R_\theta R_\psi$ which uses the orientation of the aircraft frame (i.e. camera frame) with the inertial frame. Note that the rotational matricies depends on the roll angle ϕ, the pitch angle θ and the heading difference between the aircraft and the runway Δ_ψ. The Inertial Measurement Unit (IMU) is providing ϕ and θ whereas Δ_ψ need to be measured. Particular visual features could be used to provide Δ_ψ as shown in appendix.* ∎

5 Conclusion and Future Perspectives

This paper proposes a new pose estimation solution using a nonlinear observer. The presented solution provides good performances and appears to be more robust versus noisy measurement compared with previous solution. The proposed observer has been also applied on a realistic landing scenario.

Future works will focus on observability limitations during landing approach and calibrations errors effects. Furthermore, delays caused by image processing computation or numerical discretization caused by embedded computers will be taken into account.

Appendix

The measurement of Δ_ψ could be obtained from the image by using the visual feature d_F (see Figure 6). Indeed, this visual feature depends only on f (the focal length), ϕ, θ and Δ_ψ. It gives

$$d_F = f\left(\frac{\tan\Delta_\psi}{\cos\theta} + \tan\phi\tan\theta\right). \tag{18}$$

From (18), one can compute Δ_ψ with

$$\Delta_\psi = \tan^{-1}\left(\cos\theta\left(\frac{d_F}{f} - \tan\phi\tan\theta\right)\right). \tag{19}$$

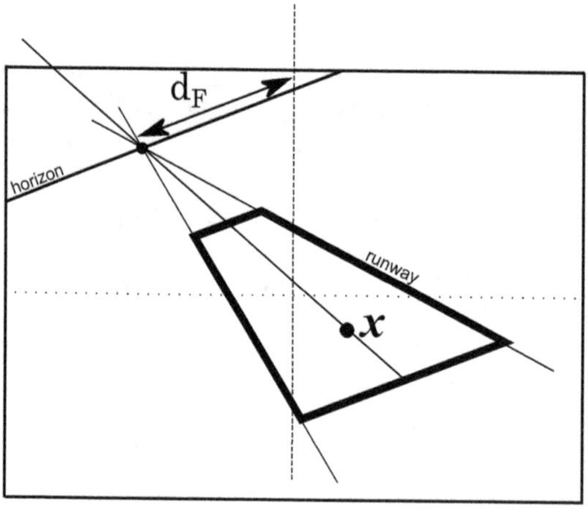

Fig. 6. Visual features d_F corresponding to the distance between the vanishing point of runway side lines and the middle of the image along the horizon line.

References

1. Agin, G.: Computer vision systems for industrial inspection and assembly. Computer 13(5), 11–20 (1980)
2. Bourquardez, O., Chaumette, F.: Visual servoing of an airplane for alignment with respect to a runway. In: IEEE International Conference on Robotics and Automation, Roma, Italia (2007)
3. Chaumette, F., Hutchinson, S.: Visual servo control. i. basic approaches. IEEE Robotics Automation Magazine 13(4), 82–90 (2006)
4. Chen, X., Kano, H.: A new state observer for perspective systems. IEEE Transactions on Automatic Control 47(4), 658–663 (2002)
5. Coutard, L., Chaumette, F., Pflimlin, J.M.: Automatic landing on aircraft carrier by visual servoing. In: 2011 IEEE/RSJ International Conference on Intelligent Robots and Systems (IROS), pp. 2843–2848 (2011)
6. Dahl, O., Nyberg, F., Holst, J., Heyden, A.: Linear design of a nonlinear observer for perspective systems. In: IEEE International Conference on Robotics and Automation (ICRA), Barcelona, Spain (2005)
7. De Luca, A., Oriolo, G., Giordano, P.: On-line estimation of feature depth for image-based visual servoing schemes. In: 2007 IEEE International Conference on Robotics and Automation, pp. 2823–2828 (2007)
8. Dixon, W.E., Fang, D., Dawson, D.M., Flynn, T.J.: Range identification for perspective vision systems. IEEE Transactions on Automatic Control 48(12), 2232–2238 (2003)
9. El Tannoury, C., Moussaoui, S., Plestan, F., Romani, N., Pita-Gil, G.: Synthesis and application of nonlinear observers for the estimation of tire effective radius and rolling resistance of an automotive vehicle. IEEE Transactions on Control Systems Technology 21(6), 2408–2416 (2013)
10. Gauthier, J.P., Hammouri, H., Othman, S.: A simple observer for nonlinear systems, application to bioreactors. IEEE Transactions on Automatic Control 37(6), 875–880 (1992)
11. Gibert, V., Puyou, G.: Landing of a transport aircraft using image based visual servoing. In: 9th IFAC Symposium on Nonlinear Control Systems (NOLCOS), Toulouse, France (2013)
12. Giordano, P., De Luca, A., Oriolo, G.: 3d structure identification from image moments. In: IEEE International Conference on Robotics and Automation, Pasadena, California, USA, pp. 93–100 (2008)
13. Gonçalves, T., Azinheira, J., Rives, P.: Homography-based visual servoing of an aircraft for automatic approach and landing. In: IEEE International Conference on Robotics and Automation, Anchorage, Alaska, USA (2010)
14. Gui, Y., Guo, P., Zhang, H., Lei, Z., Zhou, X., Du, J., Yu, Q.: Airborne vision-based navigation method for uav accuracy landing using infrared lamps. Journal of Intelligent & Robotic Systems 72(2), 197–218 (2013)
15. Jankovic, M., Ghosh, B.K.: Visually guided ranging from observations of points, lines and curves via an identifier based nonlinear observer. Systems and Control Letters 25(1), 63–73 (1995)
16. Karagiannis, D., Astolfi, A.: A new solution to the problem of range identification in perspective vision systems. IEEE Transactions on Automatic Control 50(12), 2074–2077 (2005)
17. Krener, A.J., Respondek, W.: Nonlinear observers with linearizable error dynamics. SIAM Journal on Control and Optimization 23(2), 197–216 (1985)
18. Le Bras, F., Hamel, T., Barat, C., Mahony, R.: Nonlinear image-based visual servo controller for automatic landing guidance of a fixed-wing aircraft. In: European Control Conference, Budapest, Hungary (2009)

19. Lebastard, V., Aoustin, Y., Plestan, F.: Estimation of absolute orientation for a bipedal robot: Experimental results. IEEE Transactions on Robotics 27(1), 170–174 (2011)
20. Levant, A.: Higher-order sliding modes, differentiation and output-feedback control. International Journal of Control 76(9-10), 924–941 (2003)
21. de Plinval, H., Morin, P., Mouyon, P., Hamel, T.: Visual servoing for underactuated vtol uavs: a linear homography-based framework. International Journal of Robust and Nonlinear Control (2013)
22. Sanderson, A., Weiss, L.: Adaptive visual servo control of robots. In: Pugh, A. (ed.) Robot Vision, International Trends in Manufacturing Technology, pp. 107–116 (1983)
23. Trisiripisal, P., Parks, M.R., Abbott, A.L., Liu, T., Fleming, G.A.: Stereo analysis for vision-based guidance and control of aircraft landing. In: 44th AIAA Aerospace Science Meeting and Exhibit., Reno, Nevada, USA (2006)

Part III

Atmospheric Applications

Flocking Algorithm for Fixed-Wing Unmanned Aerial Vehicles

Cezary Kownacki and Daniel Ołdziej

Abstract. The problem of swarms autonomous flying has been extensively studied for many years, giving a variety of great applications and contributing a lot of knowledge to the theory of swarms. In many cases, the researchers try to imitate animals which are perfectly adopted to moving collectively. Therefore, it would be a great idea to create a flock of UAVs flying like a herd of pigeons. Hence, the paper presents the algorithm of aerial flocking, which is a step towards this idea. The algorithm assumes a hierarchical and decentralized structure of the flock based on two flocking rules: of cohesion and repulsion. These rules of aerial flocking combined with the leadership in the flock, similarly as it is in a herd of pigeons, allow achieving a coherent swarm of fixed-wing UAVs. To prove this conclusion, both numerical and experimental results are presented.

1 Introduction

One of amazing instincts which are possessed by a wide range of animals is the ability to create the coherent flock, moving collectively in one direction [3, 16 and 17]. Those instincts are commonly called as the flocking behaviors or rules, and they can be observed in normal life of a great number of species of fish, birds and even mammals (e.g. shoal of herrings, sprats, or flocks of pigeons). The first detailed definition of the flocking principles for birds was formulated by C.W. Reynolds [3]. Basing on the observation of birds behaviors, he found out four fundamental rules which are essential for a collective motion: cohesion, repulsion, migration and alignment rule. The cohesion rule is responsible for grouping a number of individuals in the same space around the flock's center of mass. Hence, this rule is crucial for achieving the coherence of the flock, and keeping all

Cezary Kownacki · Daniel Ołdziej
Bialystok University of Technology, ul. Wiejska 45C, 15-351, Białystok, Poland
e-mail: {c.kownacki,d.oldziej}@pb.edu.pl

J. Bordeneuve-Guibé et al. (eds.), *Advances in Aerospace Guidance, Navigation and Control*, DOI: 10.1007/978-3-319-17518-8_24

individuals together. The repulsion rule secures the flock against inner collisions between neighbors that can happen while all individuals are moving close to one another. Both the cohesion and the repulsion rules act directly on the mutual interactions inside the flock and determine spacing between all individuals. Two other rules decide on the manner in which the flock moves. The alignment rule causes parallel movement of nearby individuals as the effect of the alignment of local velocity vectors. The rule of migration determines the global direction for the entire flock. The alignment and migration rules act together but they use different operating ranges in the flock: they can be local or global. The purpose of these two rules is to move all individuals collectively in the same direction which is determined by an overriding objective of the flock.

The model of birds' flock presents attractive control principles that could be adopted to the swarms of autonomous robots [8, 12 and 13], especially it concerns unmanned aerial vehicles [2, 4, 5, 7, 9, 14 and 15]. Looking at a large number of UAVs flying like pigeons would be a very amazing experience. This challenge was realized by Tomas Vicsek's team [14] who formulated the flocking algorithm for autonomous robots [15]. The algorithm was implemented in real UAVs. Unfortunately, presented results are valid only for objects that are capable of moving in an arbitrary direction, independently of the actual orientation, which was also admitted in the same work [14]. Additionally, in order to contribute to the integrity of the flock, the global positional constraint was applied, instead of the cohesion rule. This means that the algorithm cannot be applied in fixed-wing vehicles which suffer from the limited turn radius or the limited climb rate. Another consequence of the limited maneuverability of the fixed-wings is impossibility to concentrate all vehicles in a limited ring-shaped arena. However, the aerial flocking can be also applicable for this kind of vehicles, which was proved by Hauert et al. [2]. The experiment with 10 autonomous fixed-wing aircraft was presented to evaluate the relation between the limited turn radius and the limited range of WiFi communication. The application shows mostly the migration, the alignment and the cohesion rules without real repulsion, since all vehicles were flying at different altitudes with 10- meter spacing.

The main goal of our research is to formulate a model of aerial flocking dedicated for fixed-wing UAVs which would be based on only two rules: the cohesion and the repulsion combined with the leadership. Such combination simplifies the implementation of the overall flocking algorithm by reducing the number of control principles to two rules. However, the alignment and the migration rules could be still realized indirectly by the leader-follower interactions which tend to create the equilibrium point between the cohesion and the repulsion. Moreover, the leadership inside the flock is a normal phenomenon for animals and such similarity is a good reason to assume the hierarchical structure of our model [14]. For example, a herd of pigeons has the hierarchical structure, which was investigated by Tomas Vicsek et al. [16 and 17]. In the model, we assume that a communication between the UAVs is global, therefore, calculating a global center of mass is possible. This results in the global range of the cohesion rule that contributes to the integrity of the flock. Obviously, an appropriate long-range radio equipment is necessary. Therefore, 868-MHz radio modems with the bandwidth reduced to 2400 bps, operating in the broadcast mode were used in the experiment. This type of equipment secured the research against exceeding 10% of duty cycle, otherwise unnecessary gap in the transmission can appear. There is a practical aspect of the long-range communication inside the flock that allows to maintain a wider spacing between robots. As a result, the flock will cover a wider area of land to be monitored during reconnaissance missions. However, there is also a disadvantage of this solution - the low flexibility in resizing the flock and transmission delays growing with the number of vehicles in the flock. Those issues were solved partially by designing an algorithm of broadcast queuing, which is robust against the change in the number of robots.

The numeric simulations deliver promising results that present the coherent flight of three fixed-wing UAVs. To test the flocking algorithm in reality, we implemented it into commercial autopilots as the high-level control and we prepared test flights to observe the elementary flocking behaviors. The results present flight trajectories while the UAVs are repulsed from one another applying the rule of the leadership, or one UAV is attracted to the global center of mass. The results prove that our flocking algorithm is able to be successfully implemented into real fixed-wing UAVs, to create a flock of such vehicles.

2 Flocking Algorithm

The model of the flock of fixed-wing UAVs has a hierarchical structure achieved by the introduction of the leadership which is selected among all UAVs in the flock [10, 11]. Other vehicles, creating the following part of the flock, are controlled by the rules of cohesion in the global term and repulsion in the local term. The leader is the only UAV controllable by the ground control station and it does not apply any flocking rule. The leader broadcasts only its speed and position to the rest of the flock. If it is assumed, that the flock's inner communication is global and long-range, each UAV in the flock is able to calculate the global center of mass of the flock. The center is used to aggregate all UAVs in the flock and contribute to the integrity of the flock. Hence, the cohesion is the rule valid globally in a distance greater than the specified threshold distance D_C. Considering the fact that the position of the leader is taken into account in calculations of the global center of mass (GCoM) and the ground control station decides about the flight direction of the leader, the ground control station will control the position of GCoM indirectly. And if the cohesion rule is active for all UAVs, the flock will start to track GCoM, which in turn will follow the leader. Hence, the conclusion is that the global cohesion combined with the leadership, which is controlled via GCS, complements the missing rule of the migration. In the contrary, the repulsion rule has only the local range, and it means that the nearest vehicles are repulsed the most strongly to avoid collisions and maintain safe spacing inside the flock. Figure 1 presents the interactions inside the flock according to the formulated flocking algorithm.

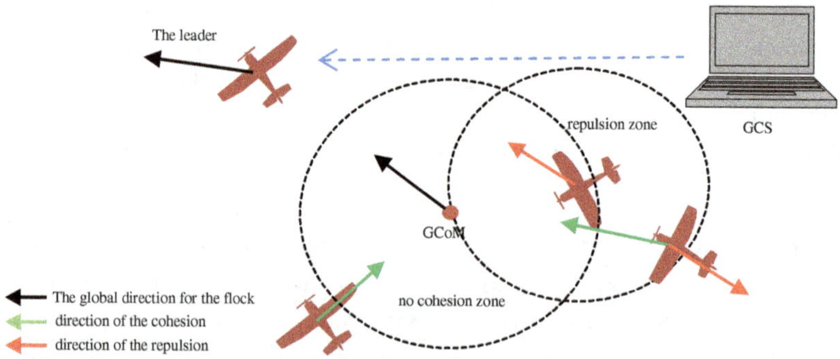

Fig. 1 The illustration of the flocking interactions based on the proposed flocking algorithm [10, 11].

As it was mentioned, the model of the flock of fixed-wing UAVs has the hierarchal structure but simultaneously, it is also decentralized. The control laws are realized locally and separately by the flight computer of each UAV. All local control laws are divided into two levels. A low-level control is based on PID loops, and it handles the vehicle's dynamics and supervises realization of the desired flight direction, requested by a high-level control. The high-level control switches the control between the local navigation and the flocking algorithm. It depends on conditions related to the cohesion rule and the repulsion rule, which are respectively: the distance to the GCoM and the distance to the nearest neighbor. In the case of the leader, the high-level control is always switched to the local navigation. The overall diagram of the UAV's local control is presented in Figure 2.

The flock's state is related unambiguously to the location of the GCoM and the value of the average speed of the flock. The location of the GCoM is a weighted average sum of coordinates of all UAVs in the flock. It must be weighted because the leader's impact on the location of the GCoM should be at least equal to the total impact of all other UAVs in the flock. Such approach prevents from the situation when the leader simply flies away from the flock. Definitions of the GCoM and the average speed of the flock are given by equations [10, 11]:

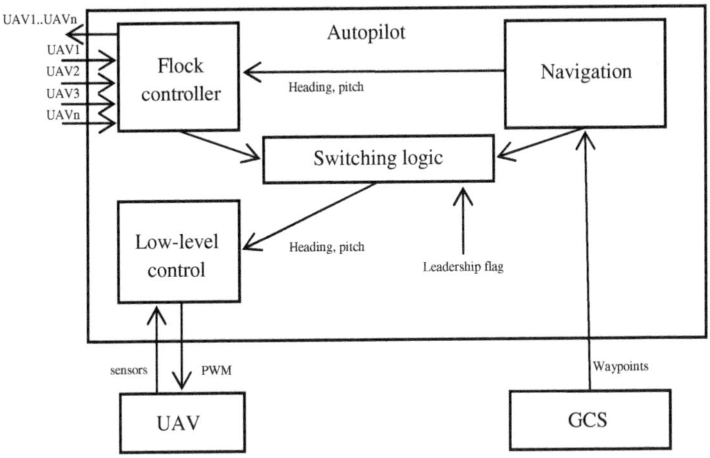

Fig. 2 The diagram of the local control realized by the flight computer i.e. autopilot [10,11].

$$\overrightarrow{GCoM} = \left(\frac{(n-1)\cdot \overrightarrow{P_{UAV1-L}} + \overrightarrow{P_{UAV2}} + \ldots + \overrightarrow{P_{UAVn}}}{2\cdot(n-1)} \right), \tag{1}$$

$$V_{Avg} = \frac{V_{UAV1} + V_{UAV2} + \ldots + V_{UAVn}}{n},$$

$$\overrightarrow{P_{UAVn}} = \begin{bmatrix} X_E \\ Y_N \\ h \end{bmatrix}.$$

Where: $\overrightarrow{P_{UAVn}}$– coordinates of the n-th UAV (X_E – longitude, Y_N – latitude, h – altitude), n – number of UAVs creating the flock, $\overrightarrow{GCoM}$ – coordinates of the flock's GCoM, V_{UAVn} – airspeed of the n-th UAV, V_{Avg} – the flock's average speed, *UAV1-L* – a label for the leader.

The average speed of the flock is used in the alignment of speeds of all vehicles in the flock. Each vehicle takes the average speed as its own desired speed. Thus, the average speed together with the cohesion plays the important role contributing to the integrity of the flock.

As alluded above, the cohesion rule is global for the entire flock. Thus, each vehicle must know positions of all other UAVs in the flock to calculate the GCoM. Obviously, as a consequence all UAVs have to create a communication network to be able to broadcast their positions and speeds. It is possible and it was done in the research.

Now, let's explain the issue of the cohesion rule. The cohesion rule should direct an UAV towards the GCoM. According to this, the desired direction of flight overlies with the line connecting the center of mass of the UAV with the GCoM. To define this line and the direction of flight, a versor whose point of anchor is located in the UAV's center of mass and which targets toward the GCoM can be formulated. The versor can be called the cohesion vector. The idea of the cohesion rule is explained in Figure 3.

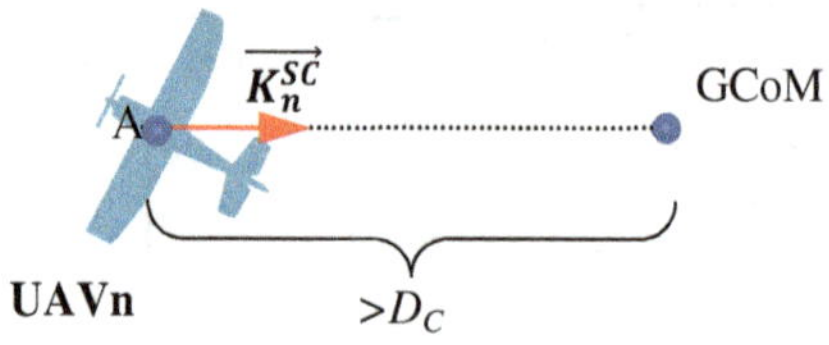

Fig. 3 The horizontal projection of the idea of the cohesion rule. *A* – the UAV's center of mass, GCoM – the global center of mass, D_C – the distance above which the cohesion rule is active [10, 11].

The definition of the cohesion vector is given by the following equations [10, 11]:

$$\overline{K_n^C} = \begin{cases} \frac{1}{|AGCoM|} \cdot \overline{AGCoM} & |AGCoM| \geq D_C \\ 0 & |AGCoM| < D_C \end{cases}, \tag{2}$$
$$\overline{AGCoM} = \overline{GCoM} - \overline{P_{UAVn}},$$

Where: $\overrightarrow{AGCoM}$ – the cohesion vector defined by subtraction of coordinates of both the n-th UAV (P_{UAVn}) and GCoM, |AGCoM| - is a distance between GCoM and the *n*-th UAV, D_C – the maximum permitted distance from GCoM.

In contrast to the cohesion rule, the repulsion rule has a local range, since its main objective is to maintain a safe spacing between the UAVs. Therefore, this rule is used to repulse two closest neighbors in the flock. But if we are considering a global communication in the flock, and each UAV is able to calculate the GCoM, they will be also able to determine repulsion directions from all other UAVs. Thus, the individual strength of the repulsion should be related to the value of distance between two different vehicles in the flock. The strongest repulsion is related to two nearest neighbors, while the smallest repulsion is related to two the farthest neighbors. Whereas, the direction of the repulsion lies on the line connecting two centers of mass of a pair of UAVs and respectively for each UAV in the considered pair, it oriented reversely to the second UAV. The directions are determined separately for each pair of UAVs in the flock but with different strengths of the repulsion, which is a result of various spacing of UAVs inside the pairs. Next, each UAV in the flock computes the resultant direction of the repulsion by taking into account only those pairs to which it belongs, and assigned strengths of the repulsion to them. The concept of the repulsion rule for one pair of UAVs is shown in Figure 4.

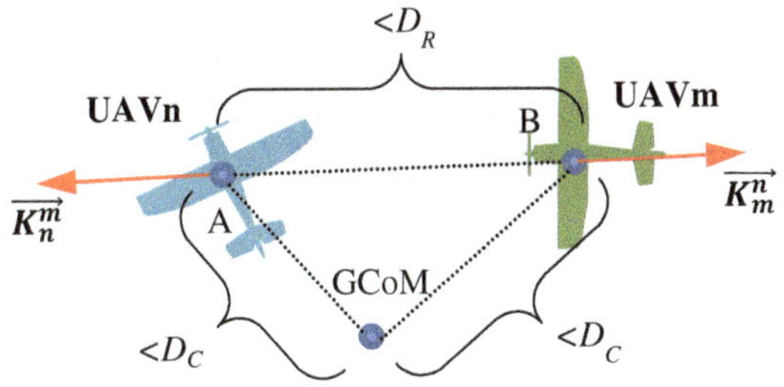

Fig. 4 The horizontal projection of the idea of the repulsion rule. *A, B* – the center of mass of two nearest UAVs, GCoM – the global center of mass, D_C – the distance above which the cohesion rule is active, D_R – the distance below which the repulsion rule is active [10, 11].

The directions of the repulsion for each pair of UAVs are determined by following versors $\overrightarrow{K_n^m}$ and $\overrightarrow{K_m^n}$. These versors can be also called the repulsion vectors which is defined by equations [10,11]:

$$\begin{aligned} &\overline{K_n^m} = \begin{cases} \frac{1}{|BA|}\cdot \overline{BA} & |BA| \le D_R \\ 0 & |BA| > D_R \end{cases}, \\ &\overline{BA} = P_{UAVn} - P_{UAVm}, \\ &\overline{K_m^n} = \begin{cases} \frac{1}{|AB|}\cdot \overline{AB} & |AB| \le D_R \\ 0 & |AB| > D_R \end{cases}, \\ &\overline{AB} = P_{UAVm} - P_{UAVn}. \end{aligned} \tag{3}$$

Where: $\overrightarrow{BA}$ – the vector defined as the difference between P_{UAVm} coordinates (*UAVm*) and P_{UAVn} coordinates (*UAVn*) , $|AB|$ - the distance between *UAVn* and *UAVm*, D_R – the minimum permitted distance between two UAVs.

When the cohesion and the repulsion rules are active simultaneously, the resultant direction of flight is computed as a sum of the direction of the cohesion and all directions of the repulsion. The resultant direction of flight, which takes into account also the determined strengths of the repulsion for each pair of UAVs, is formulated by the following equations [10, 11]:

$$\overrightarrow{K_n} = \overrightarrow{K_n^{SC}} + \sum_{m=1,m\neq n}^{N} \alpha_n^m \cdot \overrightarrow{K_n^m}. \tag{4}$$

$$\alpha_n^m = \left(1 - \frac{|A_n B_m|}{D_R}\right) \quad \alpha_n^m \in (0,1). \tag{5}$$

Where: α_n^m – the m-th strength of the repulsion assigned to the vector $\overrightarrow{K_n^m}$ for the n-th UAV, $\overrightarrow{K_n^{SC}}$ – the cohesion vector for the n-th UAV, $\overrightarrow{K_n^m}$ – the repulsion vector for the n-th UAV, N – the quantity of the flock.

The α_n is a function of the distance between two UAVs whose values are within the range of 0 (for the furthest neighbors) to 1 (for the closest neighbors), and it corresponds to the ratio of the distance to the threshold D_R. The scheme of simultaneousness of the repulsion and the cohesion for a pair of UAVs is presented in Figure 5.

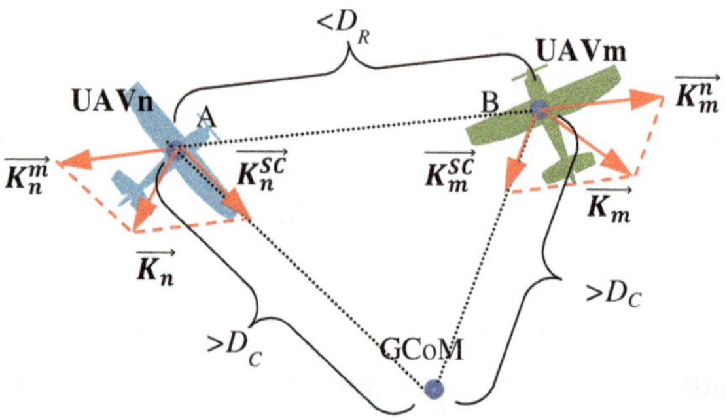

Fig. 5 The horizontal projection of the idea of simultaneousness of the repulsion and the cohesion. *A, B* – the centers of mass of two nearest UAVs, GCoM – the global center of mass, D_C – the distance above which the cohesion rule is active, D_R – the distance below which the repulsion rule is active [10, 11]

The versor $\overrightarrow{K_n}$ is the final, local result of the overall flocking algorithm designed for fixed-wing UAVs because it combines the cohesion and the repulsion for a single vehicle. The last thing to do is to nominate command signals which will be processed by the low-level controls. This depends on the structure of low-level controls, specific for each autopilot system. In our case, the autopilot requires the information about the following flight parameters: desired heading angle, desired pitch angle and desired airspeed. These parameters, forming the control vector u, can be easily extracted from the versor $\overrightarrow{K_n}$, basing on the following equations [10, 11]:

$$\overrightarrow{K_n} = \begin{bmatrix} x \\ y \\ z \end{bmatrix} \neq 0, \tag{6}$$

$$\Psi_d = arctan\left(\frac{x}{y}\right), \tag{7}$$

$$\Theta_d = arctan\left(\frac{z}{\sqrt{x^2+y^2}}\right), \tag{8}$$

$$V_d = \frac{1}{N} \cdot \sum_{i=0}^{N-1} V_i, \tag{9}$$

$$u = \begin{bmatrix} \Psi_d \\ \theta_d \\ V_d \end{bmatrix}. \tag{10}$$

Where: $\overrightarrow{K_n}$– the resultant directional versor, the result of the flocking algorithm, Ψ_d– the desired heading angle, θ_d – the desired pitch angle, V_d – the desired airspeed, N – the quantity of the flock, V_i – the airspeed of i-th vehicle in the flock, u – the control vector.

Because the switching logic decides whether the low-level control is supervised by the local waypoint navigation or by the flocking algorithm, the desired angles of roll and pitch related to the versor $\overrightarrow{K_n}$ are realized by the low-level control only in the case when any flocking rule is activated. Otherwise, these angles are determined by the local navigation to realize locally predefined flight program. It could be a making circles over the area of reconnaissance missions.

To discover the possibilities of the flocking algorithm, we prepared numerical simulations basing on the exemplary flock of three UAVs, including the leader [10]. The results of the simulations present coherence flights of three UAVs and some of them are presented in Figure 6.

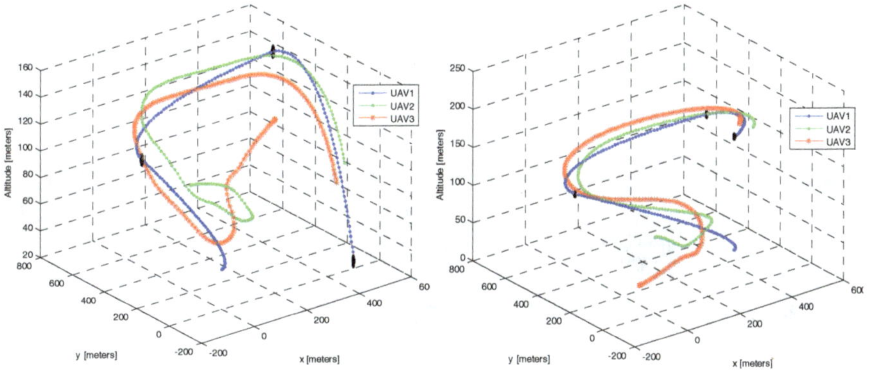

Fig. 6 The results of numerical simulations of flights of three UAVs through three waypoints (black points), where UAV1 (blue) is the leader, and UAV2, UAV3 (red and green) are the followers. The UAVs start their collective flight in different locations and at different altitudes to show the principle of the cohesion [10].

Figure 6 presents the trajectories of the three UAVs which are flying collectively, passing through three successive waypoints defining the leader's desired flight path. The trajectories do not intersect mutually and it means that the repulsion rule secures the UAVs against collisions. Moreover, we are able to observe how the combination of the cohesion and the repulsion stabilizes spacing between the UAVs and it results in the fact that the trajectories become almost parallel to one another. Obviously, it can be also related to the fact that in the simulations we did not include external disturbances like wind, or we used only kinematics of a fixed-wing vehicle. Such simplification can be accepted if we recall that the flocking algorithm, similarly to the navigation, is a supervisor control in relation to the low-level control which is responsible for handling the vehicle's dynamics. From this point of view, the received results are promising and make it possible to conclude that the flocking algorithm is applicable to fixed-wing UAVs. Thus, the next step is an experiment that will apply the flocking rules to real vehicles.

3 Experiment

To verify the flocking algorithm in reality, we prepared an experiment that uses two fixed-wing aircraft which are based on two Twinstars with installed autopilots and required radio equipment. The autopilots are commercial products but we modified their firmware by extending it with the flocking algorithm and also implemented an additional serial port as the support for a global communication inside the flock. Thus, the autopilot's primary serial port is used to communicate with the ground control station (GCS) using 2.4 GHz radio modems and the additional port enables the global broadcasting inside the flock by long-range 868MHz radio modems. Hence, each autopilot is connected with a pair of radio modems (2.4 GHz and 868 MHz). Additionally, each vehicle supports standard RC control on 2.4 GHz. Any other equipment, especially a separate onboard computer, is not necessary to realize our algorithm. The UAVs used in the experiment are presented in Figure 7.

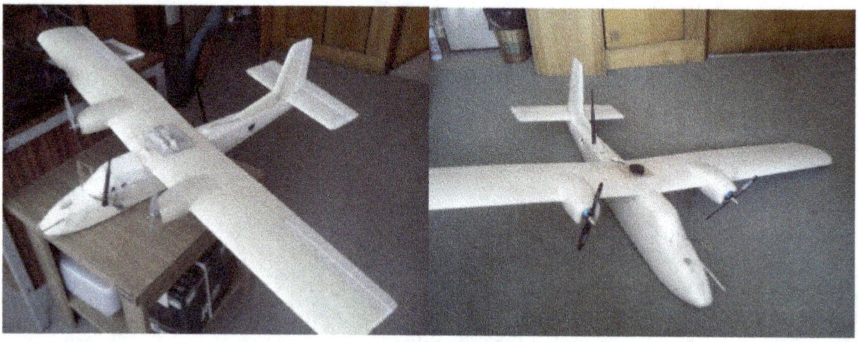

Fig. 7 Two fixed-wing UAVs which were used in the experiment. Antennas of 868 MHz radio modems are visible at the nose and the tail. The rudder of the UAV on the left picture has a built-in antenna for 2.4 GHz modem. In the second UAV 2.4GHz antenna is under the left wing

The UAV in the right picture is selected as the leader and the second one will represent the part of the flock which tracks the leader. To receive more emphatic evidence that the flocking rules are applied and operate correctly, we decided to keep the leader on the ground. The second UAV flies over and around the leader according to preprogrammed flight path, which should be modified by the flocking algorithm. In the experiment, flight tests were done separately for the cohesion and repulsion rules. Their schemes are presented in Figure 8.

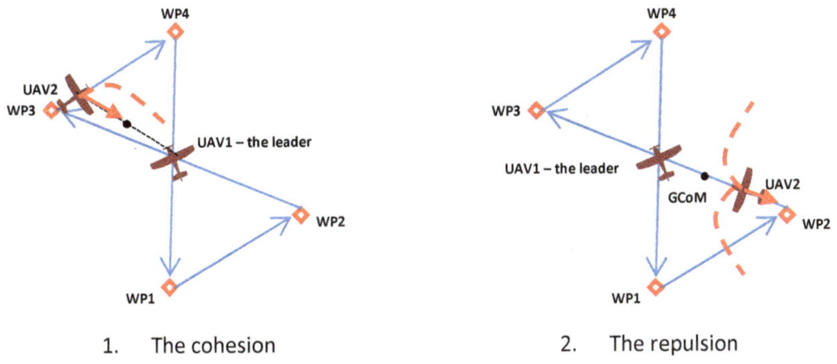

Fig. 8 Schemes of the cohesion test and the repulsion test which were done during the experiment. The black dots are momentary positions of the global center of mass, the diamonds represent four successive waypoints that create the programmed flight path for the UAV2. The red arrows are calculated flight directions from the flocking algorithm and the dashed red lines are expected flight paths related to these directions.

In both tests, the UAV2 was flying over the same route created by four waypoints, WP1, WP2, WP3 and WP4. The directions of flight between waypoints are indicated by arrows (Fig. 8). The position of the leader location is exactly in the middle of the section between WP2 and WP3, and simultaneously, this is a takeoff point for the UAV2. To improve presentation of the results, which are based on the horizontal projections of trajectories, the leader simulates its altitude to be about 80 meters. This makes turns of the UAV2, inducted by the algorithm, more visible, since both UAVs and the GCoM are located on similar altitudes.

During and just after the takeoff, the flocking algorithm was deactivated by setting both values of the thresholds D_R and D_C to 0.3 meter. Next, these values are changed respectively to induce one of the rules: the cohesion or the repulsion rule. In the cohesion test, the thresholds D_R and D_C were set respectively to 0.3 meter and 380 meters at the moment when the UAV was in the WP3 point. As a consequence, the UAVs should fly towards the GCoM and the leader instead of flying to the WP4 point. Because of the limited turn radius, the UAV should start to circle around the leader. In the repulsion test both values of the thresholds D_R and D_C are set to 380 meters, at the moment when UAV passes the WP2 or WP4 point. Thus, the UAV2 should make a turn instead of flying towards the next waypoint, i.e. the WP3 or WP1 point. Only such scenarios of the experiment will give emphatic evidence of the realization of the flocking algorithm while using real UAVs.

4 Results

Figure 9 presents the flight path of the UAV2, saved in the cohesion test. In yellow color, we highlight the section of the flight path when the UAV2 is attracted to the GCoM lying in the middle between two UAVs. We are able to observe that the UAV2, instead of flying towards the waypoint (the red diamond), directs to the leader's location (the UAV1), which lies on the same line as the GCoM. The UAV2 circled around the leader's location, and moreover, it managed to fly over it.

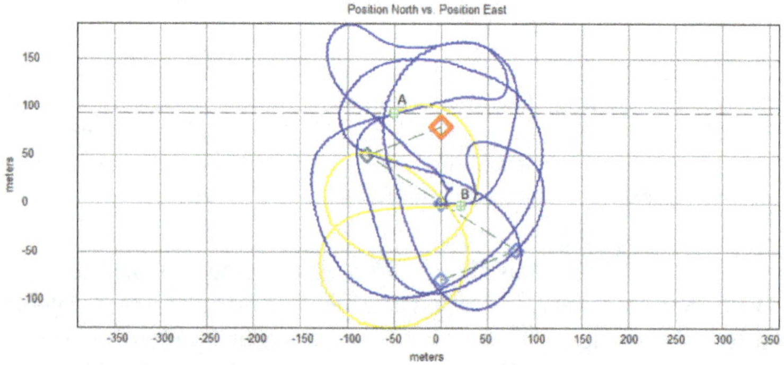

Fig. 9 The flight path of the UAV2 in the cohesion test. The section of the flight path in yellow color, between the points A and B, is related to the moments when the cohesion rule is active. The red diamond is a waypoint to be reached by the UAV2 after passing the A point

Figure 10 presents desired angles of heading, roll and pitch from the flock control (FC) versus desired heading angle from the waypoint navigation and desired angles of roll and pitch applied by the low-level control. We can notice that between two dashed red lines related to points A and B in Figure 9, the desired heading from FC is different from zero and from the desired heading given by the navigation. These differences explain clearly the behavior of the UAV2 which realizes the cohesion rule instead of the navigation. Moreover, the plots of desired angles of roll and pitch from FC overlap with the plots of desired angles of roll and pitch used by the low-level control, mostly between those dashed red lines. This means that the low-level control realizes the flocking rules. Otherwise, the low-level control realizes the waypoint navigation.

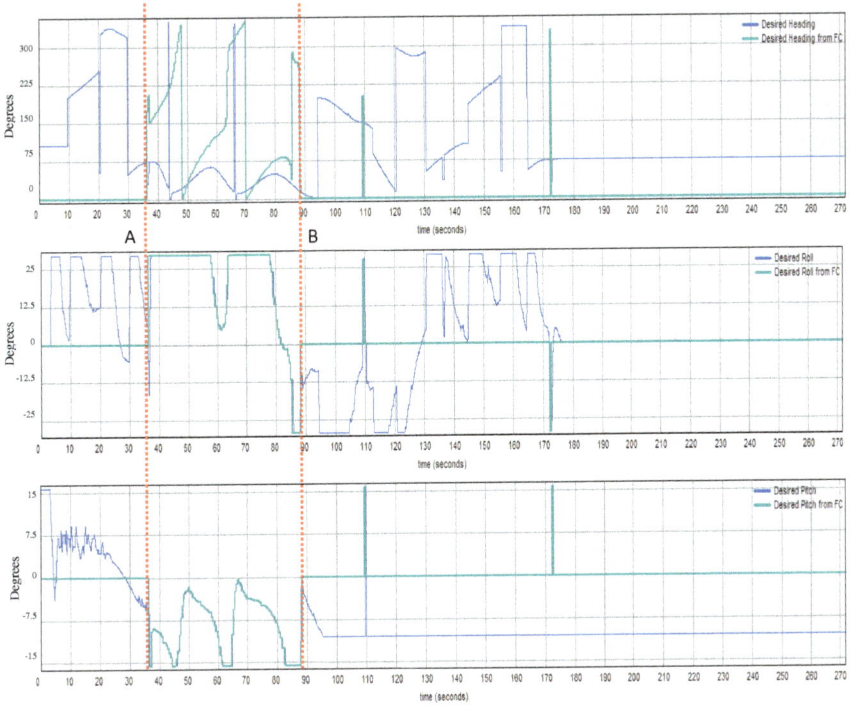

Fig. 10 The plots of desired angles of heading, roll and pitch from the flocking algorithm (FC) vs. the plots of desired angle of heading from the waypoint navigation and desired angles of roll and pitch applied by the low-level control. Dashed red lines relate to the points A and B in figure 9

Figure 11 presents the results of the repulsion test. The sections of the flight path in yellow color are related to the moments when the UAV2 is repulsed from the leader. The sections of the flight path in green color are related to the phases of takeoff and landing which were performed manually so the repulsion rule was disabled. The section in orange color presents the situation when the UAV2 was able to keep distance to the leader higher than D_R without a significant change of heading angle, as the result of higher flight altitude (the red rectangle on figure 12). Hence, the repulsion rule did not work evidently in that part of the flight. Whereas, the section in red color seems to be quite problematic because the UAV2 was flying towards and over the leader even if the repulsion rule was active.

The problem can be observed clearly in Figure 13 as rapid changes of the sign of the desired roll angle. It looks like the UAV2 could not decide in which direction it should make a turn to achieve the reverse direction of flight. Because the error of heading angle (an angle between the actual heading and the heading related to the direction of the repulsion) is changing from -180^0 to 180^0 circularly, in the situation when the UAV2 must take a turn to fly in a completely reverse direction, the error might be changed directly from about 180^0 to about -180^0, and reversely.

Probably the stationarity of the leader contributes to this situation which is the result of a singularity in angles of the vehicle's attitude, and in the mentioned heading error. The singularity causes unattended switching of the sign of the desired roll angle, at its maximum, when the heading angle error is once about 180^0 and next about -180^0. Exactly the same situation can happen only if two UAVs are flying to make head-on collision. Hence, the simplest solution is to redefine the relation between the heading error and the desired roll angle by moving the moment of sign change of the desired roll angle to other region of the range of the heading error. It is necessary because the singularity cannot be eliminated completely. Let us assume that the positive sign of the roll angle is related to the range of the heading angle error from 0^0 to $180^0 - \alpha$, and the negative one is related to the range from 0^0 to -180^0 and from α to 180^0. The problem of head-on collisions will be solved because both UAVs will be able to pass by each other in closer distance, equal to $D_R \cdot \sin(\alpha)$, but without collision.

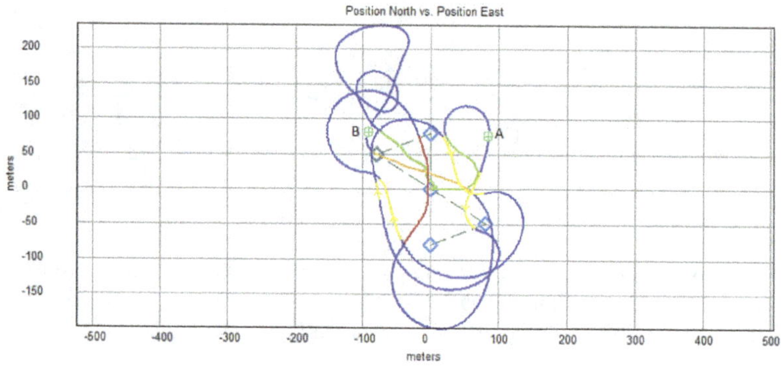

Fig. 11 The flight path of the UAV2 in the repulsion test. The sections of the flight path in yellow color are related to the moments when the repulsion rule is active, the sections in green color are the phases of takeoff and landing – the repulsion rule is disabled, the section in orange color is related to higher flight altitude so the UAV2 could keep a safe distance to the leader without a significant change of heading angle. Points A and B are the moments respectively of activation and of deactivation of the repulsion rule

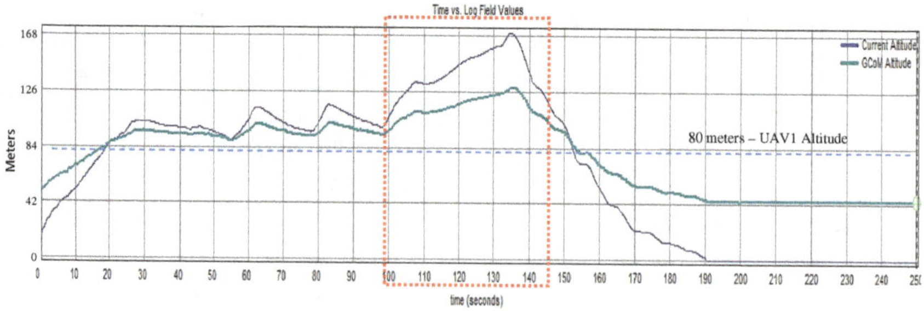

Fig. 12 The flight altitude of the UAV2 vs. the altitude of GCoM. The red rectangle corresponds to the section of the flight path in orange color, in Figure 11.

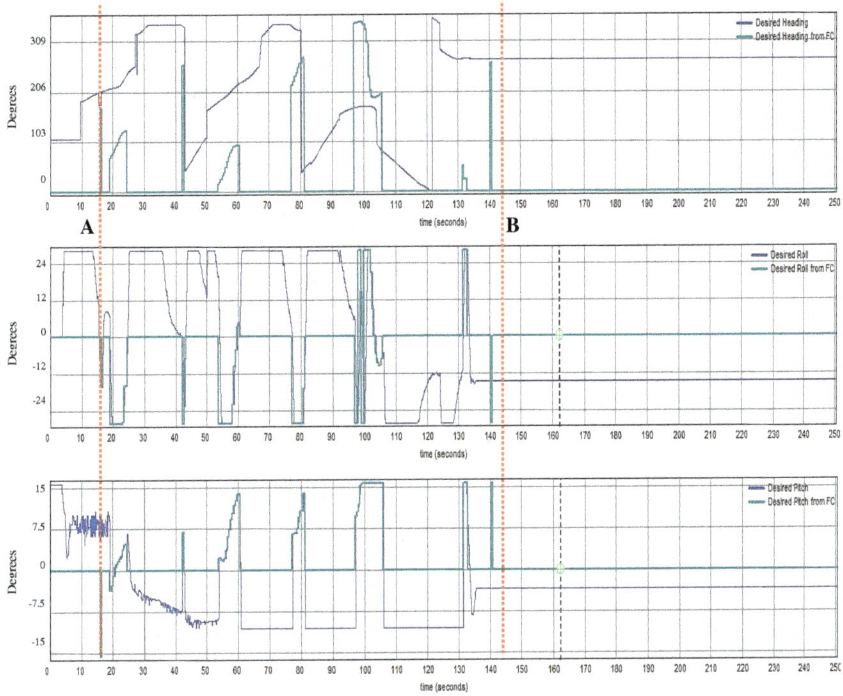

Fig. 13 The plots of desired angles of heading, roll and pitch from the flocking algorithm (FC) versus the plots of desired angle of heading from the waypoint navigation and desired angles of roll and pitch used by the low-level control. Dashed red lines relate to points A and B in Figure 11

In Figure 13 we can observe that between two dashed red lines, related to the moments of activation and deactivation of the repulsion rule, the desired heading angle from FC is different from zero and from the desired heading angle given by the waypoint navigation. Similarly to the cohesion test, the plots of roll and pitch angles are overlapping in the region restricted by the lines. This means that the repulsion rule works correctly and determines the control signals for the low-level control.

In two separate tests we managed to observe real flocking behaviors applied in the fixed-wing UAVs. There is a high probability that the flocking algorithm with activated rules of cohesion and repulsion, will allow us to create a flock, formed by more vehicles than just two, which will fly collectively like a herd of pigeons. Moreover, objectives of the flock's flight will be supervised by an operator of a GCS. Hence, the approach combining flocking rules with the leadership offers an interesting possibility of a practical use of swarms of UAVs.

5 Conclusions

The behaviors of flocking that can be easily identified in many species of animals are a very attractive approach in order to achieve coherent flocks of autonomous robots, especially unmanned aerial vehicles. The most quaint problem is how to apply those rules to vehicles that are not able to change its orientation in a place, as it is in the case of fixed-wing UAVs. Thus, we propose the approach that uses only two behaviors: of the cohesion and the repulsion, and combines them with leader-followers interactions. The formulated algorithm has the hierarchical, decentralized structure, similarly as it is in flocks of animals. The numerical simulations demonstrate a potential possibility of collective flight of the fixed-wing UAVs and they predict the applicability of the flocking algorithm in a real UAVs. Hence, to reveal the potential applicability of the algorithm, we prepare adequate flight experiments which should prove that fixed- wing UAVs are able to demonstrate flocking rules.

The results that we received in the flights show a satisfying response of the flocking algorithm to relative locations of two UAVs and their GCoM. The cohesion test illustrates the attraction of the UAV2 to the GCoM, since the distance between the UAV2 and the GCoM was higher than the threshold D_C constantly. In the contrary, the repulsion test shows the UAV2 being repulsed from the leader only when the distance between both of them is lower than the threshold D_R. Thus, both thresholds are substantial parameters which impacts the integrity and inner spacing between vehicles inside the flock. Therefore, they also have a practical importance, as they can be used to resize the flock for a specific mission purpose.

During the repulsion test we discovered a singularity in the heading control of the flocking algorithm which may contribute to collisions in the flock. It happens when the heading error is about 180^0 and it is the result of a circular range of the heading error. The issue will be eliminated by a modification of the relation between the heading error and the desired roll angle.

Summing up, we managed to formulate the flocking algorithm which adopts the flocking rules to use them in fixed-wing UAVs. Moreover, in the experiments on the algorithm we applied equipment consisting only of commercial autopilots together with a pair of radio modems that do not affect much the total weights of the aircraft. Hence, it will not be exaggerated if we state that our research ended up with a significant success.

Acknowledgments. The research was funded by funds of the Polish Ministry of Science and Higher Education for science in 2010-2013 as a developmental project No O R00 0029 11.

References

1. Karimoddini, A., Lin, H., Chen, B.M., Lee, T.H.: Hybrid formation control of the Unmanned Aerial Vehicles. Mechatronics 21(5), 886–898 (2011)
2. Hauert, S., Leven, S., Varga, M., Ruini, F., Cangelosi, A., Zufferey, J.-C., Floreano, D.: Reynolds flocking in reality with fixed-wing robots: communication range vs. maximum turning rate. In: IEEE/RSJ International Conference on Intelligent Robots and Systems (IROS), pp. 5015–5020 (2011)

3. Reynolds, C.W.: Flocks, herds and schools: a distributed behavioral model. SIGGRAPH Computer Graphics 21, 25–34 (1987)
4. Crowther, B.: Rule-based guidance for flight vehicle flocking. Proceedings of the Institute of Mechanical Engineers, Part G: Journal of Aerospace Engineering 218(2), 111–124 (2004)
5. Basu, P., Redi, J., Shurbanov, V.: Coordinated flocking of UAVs for improved connectivity of mobile ground nodes. In: Proceedings of the IEEE Military Communications Conference, vol. 3, pp. 1628–1634. IEEE Press (2004)
6. Chang, D., Shadden, S., Marsden, J., Olfati-Saber, R.: Collision avoidance for multiple agent systems. In: Proceedings of the 42nd IEEE Conference on Decision and Control, pp. 539–543. IEEE Press (2003)
7. De Nardi, R.: Flocking of UAVs Software model and limited vision simulations. Ph.D. dissertation, University of Essex (2004)
8. Welsby, J., Melhuish, C.: Autonomous minimalist following in three dimensions: A study with small-scale dirigibles, Proceedings of Towards Intelligent Mobile Robots. Technical Report Series, Manchester University, Department of Computer Science (2001)
9. Kownacki, C.: Koncepcja sterowania rojem bezzałogowych obiektów latających. Pomiary Automatyka Kontrola 58(8), 704–711 (2012) (in Polish)
10. Kownacki, C.: Modelowanie i symulacja lotu roju bezzałogowych obiektów latających. In: Gosiewski, Z. (ed.) Loty grupowe i wyrzutnie bezzałogowych aparatów, pp. 249–270. Oficyna Wydawnicza Politechniki Białostockiej (2013) (in Polish)
11. Kownacki, C.: Badania symulacyjne HWIL lotu roju dwóch bezzałogowych obiektów z wykorzystaniem zaawansowanych autopilotów. In: Gosiewski, Z. (ed.) Loty grupowe i wyrzutnie bezzałogowych aparatów, pp. 271–301. Oficyna Wydawnicza Politechniki Białostockiej (2013) (in Polish)
12. Vaughan, R., Sumpter, N., Handerson, J., Frost, A., Cameron, S.: Experiments in automatic flock control. Robotics and Autonomous Systems 31(1), 109–117 (2000)
13. Fine, B.T., Shell, D.A.: Unifying microscopic flocking motion models for virtual, robotic, and biological flock members. Autonomous Robots 35(2-3), 195–219
14. Vásárhelyi, G., Virágh, C., Tarcai, N., Szörényi, T., Somorjai, G.G., Nepusz, T., Vicsek, T.: Outdoor flocking and formation flight with autonomous aerial robots. In: IROS 2014 (2014) (accepted)
15. Virágh, C., Vásárhelyi, G., Tarcai, N., Szörényi, T., Somorjai, G., Nepusz, T., Vicsek, T.: Flocking algorithm for autonomous flying robots. Bioinspiration & Biomimetics 9(2), 025012 (2014)
16. Flack, A., Ákos, Z., Nagy, M., Vicsek, T., Biro, D.: Robustness of flight leadership relations in pigeons. Animal Behaviour 86(4), 723–732
17. Nagy, M., Ákos, Z., Biro, D., Vicsek, T.: Hierarchical group dynamics in pigeon flocks. Nature 464(7290), 890–893

Airborne Doppler LiDAR Sensor Parameter Analysis for Wake Vortex Impact Alleviation Purposes

Jana Ehlers and Nicolas Fezans

Abstract. This paper presents a sensitivity study of a wake vortex impact alleviation system based on an airborne forward-looking Doppler LiDAR sensor. The basic principle of the system is to use this sensor to measure the wind remotely ahead of the aircraft. On the basis of these measurements the system estimates whether a wake vortex is located in front of the aircraft. If this is the case, the wake vortex characteristics are identified and the control deflections countervailing the wake-induced aircraft response are computed and applied. An integrated simulation environment comprising a full nonlinear 6-DoF A320 model (with control laws), wake vortex models, and the wake impact alleviation algorithms was developed. The LiDAR sensor subsystem has many design parameters that influence the overall performance in a complex way, which makes it difficult to derive adequate requirements. The presented parameter study provides first insights into the role of each parameter as well as some adequate parameter combinations.

1 Introduction

Wake vortices are an inevitable phenomenon in air traffic. They evolve from the pressure difference between the lower and upper side of the wing during the lift generation and roll up to form a pair of two strong rotating flow fields. For an aircraft flying into these rotating flow fields, the wake vortices can pose a serious safety threat and can lead to dangerous aircraft reactions usually involving a sudden rolling motion or a loss of altitude. As wake vortices are under most atmospheric conditions not visible the aircraft reaction occurs unexpectedly for the pilot

Jana Ehlers · Nicolas Fezans
DLR (German Aerospace Center), Institute of Flight Systems,
Lilienthalplatz 7, 38108 Braunschweig, Germany
e-mail: `{Jana.Ehlers,Nicolas.Fezans}@dlr.de`

J. Bordeneuve-Guibé et al. (eds.), *Advances in Aerospace Guidance, Navigation and Control*, DOI: 10.1007/978-3-319-17518-8_25

and increases his workload. In extreme cases a wake vortex encounter can cause structural damages to the aircraft, lead to incidents with injuries of the passengers and crew [1-3] or even a crash of the aircraft [4,5]. In order to prevent the encounter of dangerous wake vortices, ICAO introduced separation minima for approach and departure. However, the drawback of this safety measure is the resulting air traffic capacity limitation. Finding ways to reduce these separation requirements while at least maintaining the same safety standards will certainly bring significant benefits for congested airports.

A possible approach to achieve this is to reduce the impact of a wake vortex on the encountering aircraft by equipping this aircraft with a specific control system for wake vortex encounters. Looye et al. [6] and Rafi and Steck [7], included wake vortices in their control design. They considered them as one of the disturbances the flight controller should be able to handle. Another approach is a control system which is based on a remote wind sensor [8-10].

The generic term "remote wind sensor" is chosen because different sensor types could be used. It could for instance be a so-called Doppler LiDAR sensor. Doppler LiDAR sensors permit to measure one component of the wind at a remote location. A LASER source is used to illuminate the location at which the measurement shall be made. At this location, particles (e.g. aerosols or even molecules) are scattering this light back to the sensor which also contains a detector. The backscattered light properties provide information on the corresponding particles. The present work only considers the relative wind velocity measurements that can be deduced from the Doppler frequency shift between the original LASER source and the received backscattered light. The Doppler frequency shift is proportional to the rate of change of the distance between the sensor and the particles, which means that the relative velocity component that is measured is the component that is in the direction of the line joining the sensor to the particles location (assuming that the light source and the detector are colocated). This direction is commonly called "Line-of-Sight" (LoS). The amount of light that is backscattered decreases with the square of the measurement distance and consequently the measurement uncertainty increases correspondingly.

The location where the measurement is made depends on the time elapsed between the moment the light was emitted and the moment at which the received backscattered light was observed by the detector. Indeed, both the pulse and the observation typically last for a few tenths of nanoseconds, which implies that the measured location is not a point but rather has a very long frustoconical shape. As the beam divergence is usually very low, an approximation of this domain with a cylinder or even with a line can usually be made. The resulting averaging effect that results from not measuring in only one point is often referred by the term blur, by analogy with photo imaging.

The investigations of [8-10] were based on the assumption that sensors capable of measuring the wind in 3D would be available. It seems that LoS-only Doppler LiDAR might be available for airborne application in the near future, whereas 3D remote wind sensors would probably remain too experimental and costly in the foreseeable future. Even when only considering LoS velocity measurements, there are plenty of choices to be made for the technical characteristics of the Doppler LiDAR sensors: wavelength, type of pulses (shape, duration), LoS directions (scanning geometry, field-of-view), number of domains being measured along the LASER beam line, detector technologies, etc. These parameters influence the measurement quality (e.g. noise and precision) and/or the collected information itself (spatial resolution, blurring effect, measurement location). The present study aims at helping to find a good trade-off between the possible Doppler LiDAR configurations by considering the aircraft dynamics as well as the flight control system and the whole sensor measurement processing chain.

When measuring the wind only in LoS direction, two of the three components are necessarily missing and from a flight control point of view, the remaining LoS direction usually contains only marginal part of the relevant information. In [11] the first author of the present paper and two colleagues proposed a solution to this problem that was called "Wake Impact Alleviation Control" (WIAC). This system consists of two main steps. First an Online Wake Identification (OWI) algorithm is used to reconstruct the wake vortex disturbance from several LoS measurements. Then the identified wake vortex model is used to predict the control-deflection-free wake-induced aircraft response that the control system countervails by commanding the adequate control surface deflections. This wake impact alleviation concept was found very promising [11] based on a simplified simulation setup. The wake impact alleviation control system was applied in a complete 6-DoF simulation environment wherein the aircraft encounters a wake vortex. However, in the simulations of [11] the online wake identification algorithm was not coupled with the 6-DoF simulation, but had been performed beforehand in a separate simplified 3-DoF simulation of the same encounter whereupon the wake vortex disturbance did not cause any aircraft reaction. Consequently, the influence of the aircraft reaction on the measurement of the Doppler LiDAR sensor was neglected. Additionally, once a wake vortex pair matching well the measurements was found, this solution was kept and not updated.

The current paper presents the fully integrated wake identification and impact alleviation control system OWIDIA, which consists of OWI and WIAC. An extensive parameter study was performed during summer 2014 with a total of 18432 full 6-DoF simulations with online wake identification performed every 200 ms (simulation time) using a small-scale grid computer (up to 16 simulations in parallel). Some of the first insights gained from these data on the performance are presented in this paper and will serve to improve the current design of the whole system.

2 Wake Impact Alleviation Approach

The concept of the wake impact alleviation approach considered in this paper is shown in Fig. 1. The overall principle is identical to the approach presented in [11]. The main difference consists in the implementation of the online wake identification, which is now fully integrated into the control system. The purpose of the control system is to compensate the disturbance wind velocities of the wake vortex by specific control command deflections which prevent a wake-induced aircraft reaction. The disturbance wind velocities and the commanded control surface deflections of the WIAC consequently represent the input and output of the complete process as illustrated in Fig. 1.

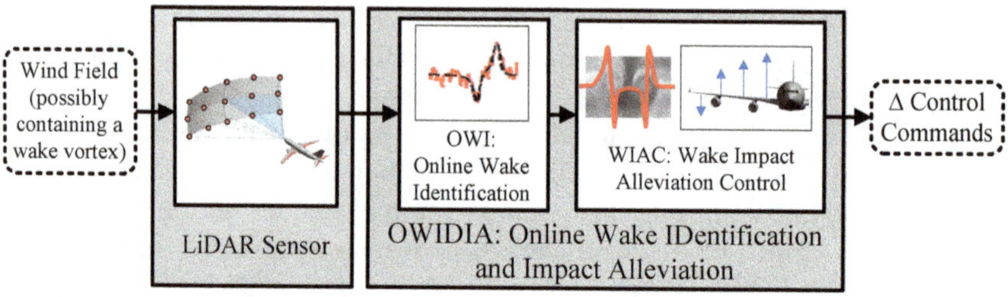

Fig. 1 Concept for wake impact alleviation

The LiDAR sensor measures the wind velocities at several locations within a relatively short range ahead of the aircraft. As already mentioned, each measurement is a line-of-sight velocity, which is then passed to the online wake identification algorithm. The OWI identifies a wake vortex model which provides the best match with the provided line-of-sight measurements and hands the parameters of this model to the wake impact alleviation control module. This module derives the wake induced aircraft disturbance on the basis of the identified wake model and calculates the required control surface deflection to countervail the wake-induced aircraft response. Further details on the different elements of the OWIDIA system are given in the following.

2.1 *Online Wake Identification*

The LiDAR sensor only provides LoS measurements, in which most of the information about the wake vortex wind velocities is lost. The online wake identification needs to reconstruct both the two lost components of the wind measurements and the wind at the locations where no measurement was made. The principle of the online wake identification is adapted from [12,13] and was also described in [11]. For the integrated implementation in the OWIDIA system within the full 6-DoF simulation framework, some extensions and modifications of the OWI were realized. Fig. 2 gives an overview over the new process of the OWI and its integration into the flight control.

The OWI is usually called with a lower update rate than the LiDAR sensor and the rest of the flight control system. The measurements of the sensor over a time frame of 2 seconds are stored in a measurement buffer. Depending on the sensor characteristics this can be one or several measurements at a time. The location of each measurement is stored in an airframe-carried north-east-down (NED) coordinate system. As the aircraft moves the coordinates of each measurement relative to the aircraft are updated in the measurement buffer. The software modules that are managing the measurement buffer and computing the control commands both run at every major step of the simulation (i.e. with 100 Hz in the applied simulation environment). Both modules should also provide a very similar performance if they were executed with larger sample time. The OWI is executed with a lower frequency because it is assumed that it requires a larger computation time. For the studies presented here we chose an update rate of 5 Hz for the OWI because this would correspond to a realistic trade-off between responsiveness to new measurements and available computation time. Reaching hard real-time requirements for an update rate of 5 Hz seems achievable, but is not a priority at the current stage of this work. The OWI update rate can be varied independently from the update rate of the LiDAR measurements as these are buffered beforehand.

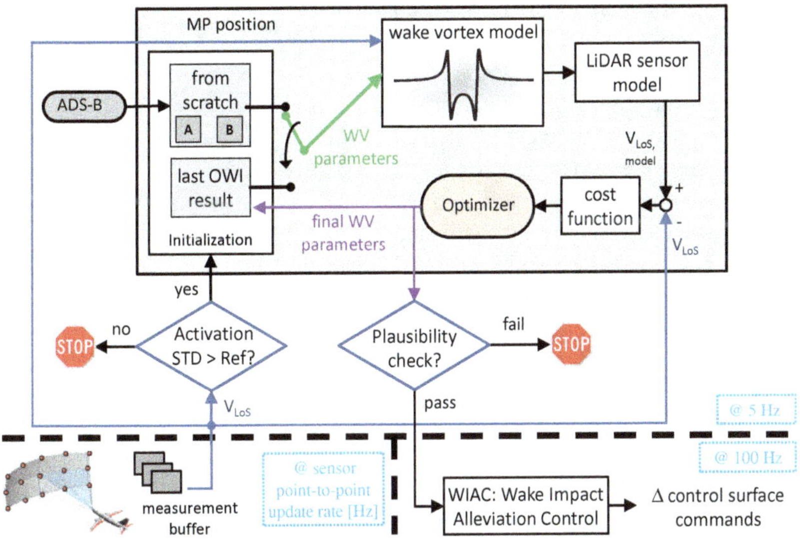

Fig. 2 Online wake identification workflow

Every time the OWI is called, first a very simple criterion tries to determine if there are strong indications of a possible wake vortex in the current content of the measurement buffer. This permits to avoid an execution of the OWI if no wake vortex is present, which is the case during most of the flight time. It, thus, represents an additional safety net for most irrelevant situations (no wake vortex) and saved computing time during the parameter study. This very simple criterion is based on the comparison of the standard deviation of all line-of-sight measurements stored in the data buffer with the reference measurement noise. The identification algorithm in the OWI is only started if the current standard deviation of the LoS measurements in the buffer is larger than 120% of the reference value. If the standard deviation of the sensor measurement noise is reliably known this value can be used directly as a reference value. There are also possibilities to estimate it online. In the current implementation the reference standard deviation of the LoS measurement is defined as the standard deviation at the beginning of the simulation right after the measurement buffer has been filled for the first time. The simulation is set up such that there are no significant wake-induced velocity measurements contained in the buffer at that time. The standard deviation of the LoS measurements thus corresponds approximately to the measurement noise.

Fig. 3 exemplarily shows a time history of the standard deviation of the line-of-sight wind velocities stored in the measurement buffer for a 30° lateral encounter. When the data buffer is filled for the first time the standard deviation of the stored LoS velocities is 0.87 m/s. When no wake vortex is present the standard deviation keeps similar values. Between 16 s and 20 s the standard deviation rises significantly due to the wind velocities caused by the wake vortex and contained in the measurement buffer at that time. The LoS velocity criterion is activated between 16.5 s and 19.7 s (black solid line above the dotted-dashed blue line in Fig. 3). When this happens the OWI algorithm starts.

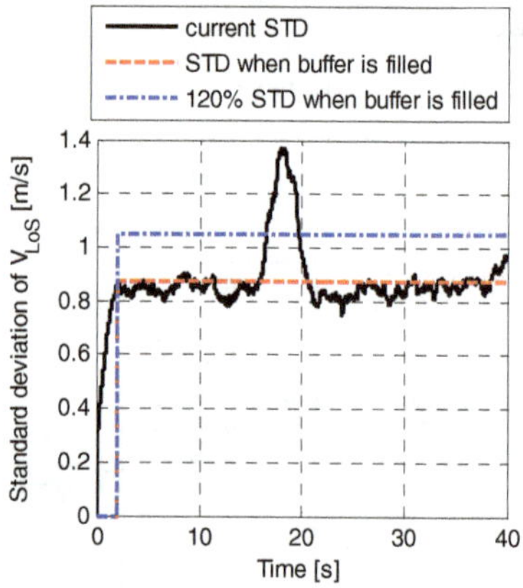

Fig. 3 LoS velocity trigger for a 30° lateral encounter

This very simple criterion is such that the actual signal (wake vortex signature in the measurements) should cause an increase of at least 20% compared to the pure noise level. Such a criterion is certainly too restrictive for cases with relatively low signal-to-noise-ratios (SNR). “Low amplitude” (i.e. low circulation) wake vortices are not relevant for the designed system. However it might also be interesting to consider sensors with high spatial resolution but high noise levels. It is clear that the criterion used during this study is too simple to handle this type of cases and that it will have to be improved to investigate these cases, too.

The OWI problem is a maximum-likelihood estimation problem, which consists in finding parameter values of a wake vortex model such that the LoS velocities derived from the identified wake vortex model and the measured LoS velocities show the best possible match. This maximum likelihood process is solved using an optimization algorithm as shown in Fig. 2. The wake vortex model used in the current work is the well-known Burnham-Hallock model [15]. A strong dependence of the results on the model used (among the classical ones) is not expected. The resulting model-based LoS velocities of all measurement point positions in the data buffer are then compared to the actual LoS velocities of the real measurement.

2.1.1 Optimization Algorithm

The maximum likelihood problem can be easily solved with many classical optimization algorithms. For more explanation of the way to formulate such maximum likelihood problems and to solve them, refer to [14]. In [11], for practical reasons, the measurements from different times were grouped for each LoS direction. In the present implementation, measurements were not grouped in any way and it can be shown that the maximum-likelihood cost function is equivalent to a weighted least square fitting problem in that case.

When dealing with data buffers not containing any vortex, the maximum likelihood problem is usually badly conditioned and the algorithms will need many small steps to finally end up with a solution where the identified vortex either has a very small circulation or is located well outside of the domain containing the measurements. Even if none of these solutions is an issue, it is considered safer to detect this situation before even starting the algorithm. This was the main motivation for introducing the aforementioned activation criterion for the estimation process.

The algorithm used here is a Broyden-Fletcher-Goldfarb-Shanno and the line search step is performed by a nonstandard algorithm that combines several ideas from the classical line search algorithms depending on the situation. As already mentioned most standard algorithms will converge to the right solution if enough information is contained in the buffer: the differences are mainly a question of execution speed and the currently used solution was not particularly optimized.

2.1.2 Identified Parameters

The parameters of the wake vortex model which are optimized by the OWI are:

- the vortex circulation Γ(which defines the strength of the vortex),
- the distance between the two vortex cores b'
- and four geometry parameters specifying location and orientation of the wake vortex with respect to the aircraft.

The geometry parameters of the wake vortex are displayed in Fig. 4 and 5. The wake vortex location is defined by means of two points on the vortex centerline. These points are described in a particular coordinate system which is called identification (ID) coordinate system here. The ID-coordinate system is located in the horizontal x-y-plane behind the aircraft and its x-directions points into the direction of the body-fixed x-axis projected into the horizontal plane. The y-axis points to the right of the aircraft and the z-axis points downwards. The befenit of using a coordinate system that is not attached to the wake is that it significantly reduces the number of coordinate transformations that must be repeatedly made during the wake estimation process. The coordinates of the LoS measurements in this coordinate system are fixed during the entire estimation process.

In this coordinate system two points P_1 and P_2 are specified which define the location of the wake vortex. Point P_1 is located on the vortex centerline at the position where the projections of the body-fixed x-axis of the aircraft and the vortex centerline into the horizontal x-y-plane cross each other. The y-position of P_1 is zero by definition. Point P_2 is defined as the position of the vortex center line for $x_{ID} = 0$. The coordinates x_1, z_1, y_2 and z_2 of the points P_1 and P_2 uniquely describe the location and orientation of the wake vortex and are used as identification parameters for the OWI in the current implementation.

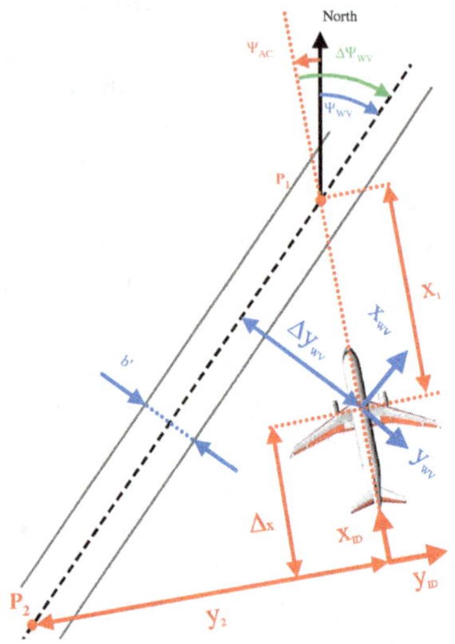

Fig. 4 Parameterization (top view)

In Fig. 4 and 5 it can be noticed that the origin of the ID-coordinate system is located a distance Δx behind the center of gravity of the aircraft. This shift of the origin with respect to the center of gravity was introduced because the parameterization chosen has a singularity when the vortex centerline goes through the point $(x_{ID},y_{ID})=(0,0)$. By shifting the point (0,0) far enough behind the aircraft, the corresponding vortex becomes practically uninteresting and irrelevant. In the current implementation the origin of the ID-coordinate system is located 200 m behind the center of gravity of the aircraft. Measurements made using a forward-looking sensor are basically around P_1 (assuming there is a vortex to be detected), such that by varying x_1 and z_1 the position of the vortex within the measurement zone is varied, whereas a variation of y_2 and z_2 then lead to a change in orientation. The properties of the optimization problems with this choice of coordinate system and with the coordinates that were previously used in [11] were not compared.

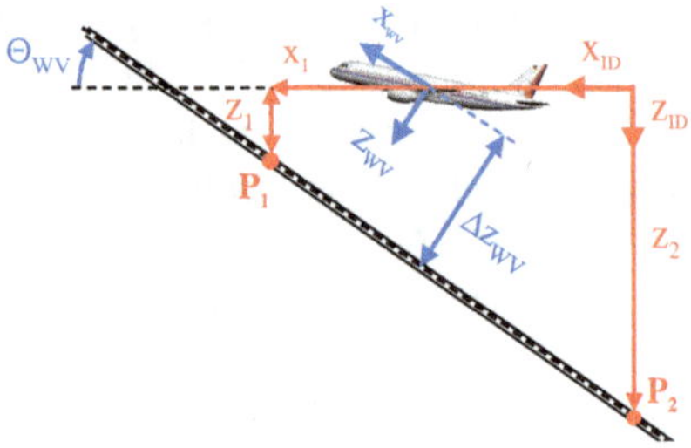

Fig. 5 Parameterization (side view)

2.1.3 Initialization of the OWI Algorithm

The OWI algorithm works iteratively and needs starting values for the identification parameters. Depending on whether the OWI has recently found a valid identification result or not these initial values are determined in two different ways.

If the OWI has not identified a wake vortex shortly before, the initial parameters for the optimization routine are defined from scratch on the basis of the current aircraft dynamics and by means of input data from the generator aircraft which are available with standard Automatic Dependent Surveillance – Broadcast (ADS-B).

The initial values of the vortex circulation Γ_0 and the lateral vortex separation b'_0 are always set to 250 m²/s and 40 m for all wake encounter scenarios as suggested in [12]. An adequate initial value for parameter x_1 corresponds to a distance that is slightly smaller than the distance to the farthest measurement point which is present in the buffer. To further increase the robustness for cases with no or only small influence of the wake vortex, the OWI is run with two different initial values for x_1. For the second initial value of x_1 a large distance of 500 m is added to the first initial value of x_1. This helps to detect the absence of a wake vortex because the second initial value of x_1 corresponds to a remote location of the wake vortex, which has practically no influence on the measurements. The corresponding OWI result has a lower cost function value and is thus selected as OWI output.

The initial value of z_1 is derived under the assumption that the point P_1 is located on the body-fixed x-axis of the aircraft in the distance of the average measurement range of the LiDAR $x_{mean,LiDAR}$. As for the simple activation criterion, the choices made for x_1 and z_1 should rather be seen as a first-shot but yet practical and functioning solution, which could relatively easily be improved and become more elegant. The basic idea behind these choices is to consider that if a vortex enters the measurement domain due to the aircraft translational motion, then this vortex is expected to be crossing the flight path of the aircraft approximately at the location defined by these initial x_1 and z_1.

The initial guess of the position of point P_2 is determined by means of external input data via ADS-B. It is assumed that the generator aircraft is equipped with standard ADS-B and transmits its current azimuth and flight path angle. The initial values for the lateral and vertical position of point P_2 can be derived easily from the position of P_1 and these two angles.

These parameters are used as initial values for the optimization process if the OWI has not identified a wake vortex for a certain timeframe. If the OWI converged to a plausible result (cf. Section 2.1.4), the identified parameters are selected as initial values for the next OWI call. The only modification of the identified values of the last OWI call is the adaptation of the wake vortex position. As the OWI is called with a relatively low update rate, the aircraft has moved significantly when the optimization is started again. For an aircraft flying with 150 kt during approach and an OWI call rate of 5 Hz, as it is implemented for the studies presented here, the aircraft moves approximately 15 m between two OWI calls. The identified wake vortex position of the previous OWI call is thus corrected by the aircraft motion since this OWI call. The translation made between two successive OWI executions plays the exact same role as the "propagation steps" of Kalman or particle filters.

In case of using the last OWI result as an initial guess for the next OWI call it is not necessary to start the OWI with two different initial values. Only one optimization procedure is executed on the basis of the previously identified wake vortex. While the aircraft is encountering the same wake vortex, the initial parameters based on the last OWI result usually represent a better initial guess than estimated values on the basis of ADS-B data. After the aircraft has passed the wake vortex, however, it might encounter the wake generated by another aircraft. To get appropriate initial values for a potential new encounter the identified parameters are not used for the initialization anymore, if the OWI result has not been updated for 4 seconds. In this case it is assumed that the aircraft has passed the wake vortex and the initial OWI parameters are initialized from scratch with the help of ADS-B data again (cf. Fig. 2).

2.1.4 Plausibility Check of Identified Wake Vortex

After having identified a wake vortex, the physical plausibility of the identified wake vortex is checked. Six criteria are applied to evaluate if the output of the OWI is physically plausible. This plausibility check adds a safety net between the identification process and the use of the identified vortex by the flight control system. The validity criteria concern the vortex circulation and separation as well as the position and orientation of the wake vortex. The wake elevation Θ_{WV} (i.e. inclination of the vortex centerline with the respect to the horizontal plane) and the wake azimuth Ψ_{WV} (i.e. orientation of the vortex centerline with respect to north) are derived from the coordinates x_1, z_1, y_2 and z_2. In detail the six validity criteria are the following:

1. **Limits for circulation: 100 m²/s < Γ < 500 m²/s**
   The limits for the identified vortex circulation are specified as suggested in [12] in order to filter out results caused by atmospheric turbulence without the presence of wake vortices as well as unrealistically high results.
2. **Limits for lateral vortex separation : 10 m < b′< 100 m**
   According to [12], the lower bound of 10 m is applied in order to filter out wakes of small aircraft or wakes with advanced decay. The upper bound of 100 m corresponds to the wingspan of aircraft at the upper limits of the “heavy” category with some extra margin for moderate Crow instability.
3. **Limits for wake elevation: -10° < Θ_{WV} < 20°**
   The lower bound of -10° for valid wake elevation is selected as the sum of the estimated maximum descent angles of the generator aircraft and the wake vortex as suggested in [12]. The upper bound is selected as 20° according to the assumed maximum climb rate of the generator aircraft [12].
4. **Limits for wake azimuth: $|\Psi_{WV} - \chi_{ADS\text{-}B}| < 15°$**
   It is assumed that the deviation of the actual wake azimuth from the course angle of the generator aircraft transmitted via ADS-B shall be smaller than 15°. Identified wake vortices with larger deviation from the generator flight path azimuth are thus defined as invalid.
5. **Plausibility check for vertical position of wake vortex centerline**
   The vertical position of the wake vortex is verified if the vortex centerline plus a certain tolerance lies within the vertical range of the stored measurement locations. The reason for the addition of a tolerance is that the wake vortex still

has a significant effect at a certain distance above and below the vortex centerline. The tolerance is chosen as 75% of an assumed lateral vortex core distance of 50 m, i.e. 37.5 m.

6. **Plausibility check for lateral position of wake vortex centerline**
   Concerning the lateral position of the identified wake vortex it is checked whether the projection of the centerline onto the horizontal plane intersects (with a tolerance of 37.5 m) the domain that contains the measurements (projected as well). This domain was approximated by a rectangle that contains all the measurements.

If all of these criteria are fulfilled the identified wake vortex of the OWI is considered as valid and is provided to the wake impact alleviation. If no new OWI result is available the control command generation for the wake impact alleviation control is based on the last available valid OWI result.

2.2 *Wake Impact Alleviation Control Command Generation*

The control command generation of the wake impact alleviation control system works identically to the process described in [11]. An overview of the workflow is illustrated in Fig. 6.

As already mentioned the OWI output contains the information on the identified wake vortex which the wake impact alleviation must countervail. Each time the control commands are computed, the relative position of the wake (with respect to the aircraft) must be updated beforehand. To compute the wake-induced forces and moments an aerodynamic interaction model (AIM) is used. The aircraft is divided into strips for which the disturbance wind velocities are determined. However, in order to account for the time delay due to processing time and actuator delays, the positions at which the wind velocities are calculated, are not the current positions of the strips, but their estimated positions at the time "now + total time delay". The wind velocities at each strip are directly provided by a wake vortex model using the identified parameter values. Based on this information the AIM computes the additional forces and moments induced by the wake. A more detailed description about the AIM model and its validation is given in [16].

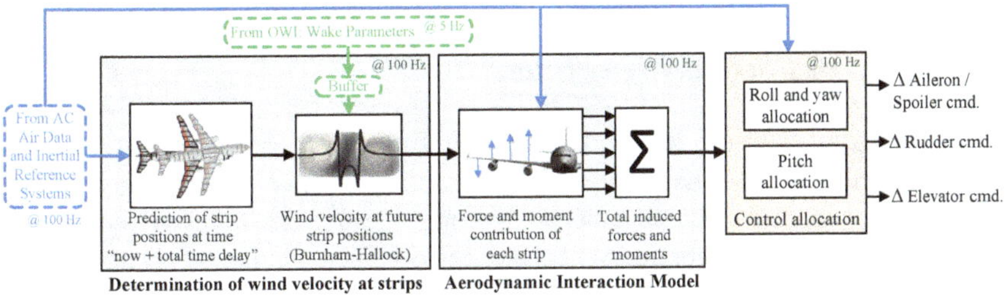

Fig. 6 Workflow of wake impact alleviation control command generation

The goal of the wake impact alleviation control system is to compensate for these wake-induced forces and moments. Common aircraft configurations only use ailerons, elevators and rudder(s) as control surfaces with which only three degrees of freedom (roll, pitch and yaw) can be controlled. Consequently, a complete compensation of all six disturbance forces and moments is not achievable. Instead only the wake-induced moments are compensated by the WIAC to alleviate rotational aircraft movements resulting from the wake vortex. The required control surface deflections to counteract the wake-induced moments are determined by analytically inverting the aerodynamic control surface efficiency matrix of the aircraft. The determined control surface commands are added to the control surface deflections commanded by the regular flight control system.

3 Assessment of Influence of LiDAR Sensor Characteristics on OWI and WIAC

The presented approach for the alleviation of the wake-induced aircraft response obviously only works if the identified wake is sufficiently representative of the wind field ahead of the aircraft. From the best of the authors' knowledge, there is no commonly accepted criterion on which the definition of a required OWI performance could be based. In order to overcome this issue, a simplified sensitivity study was performed in [11], in which the WIAC performance for various wake parameter estimation errors was investigated. The identification of the parameters was not performed online, but "the estimation loop was opened" and the identified parameters were directly fed into the WIAC as constant input values. The simulations carried out in the present study were all performed with the wake identification in the closed-loop. The parameters whose sensitivities are being analyzed are the Doppler LiDAR parameters instead of the wake parameter estimation errors as in [11]. The goal of this study is to assess the performance of the whole OWIDIA system (OWI + WIAC) and thereby to represent a first major step towards identifying the best sensor parameter combinations (ultimately permitting to derive a complete set of requirements for the sensor specialists / manufacturers). This study shall also help detecting the limits of the system or potential problems that still need to be addressed.

3.1 Setup of the Sensitivity Study

The challenge of defining required LiDAR sensor characteristics results from the fact that a LiDAR sensor has a large number of different parameters whose effects are not independent from each other and strongly nonlinear. Many parameters of the Doppler LiDAR sensor influence the performance of the complete system.

In order to limit the number of simulations to a reasonable number some parameters were not varied and for those that were varied only a limited number of values was considered.

In order to identify which of the various parameter combinations represent feasible and desirable LiDAR settings an automatized and distributed simulation framework has been set up, in which the LiDAR sensor parameters are systematically varied and combinations of ranges of different sensor parameters are studied for different encounter scenarios. The goal of the sensitivity study is to find the possible interesting sensor characteristics and not yet to make a thorough assessment of a given system configuration. Therefore, the WIAC is not considered for many different encounter scenarios with a given set of sensor characteristics, but rather for many sensor characteristics and a few encounter configurations. In all cases considered hereafter an Airbus A320 aircraft encounters the wake of an Airbus A340 aircraft during approach. The wake vortex is always located 2 m above the center of gravity of the encountering aircraft and the vertical encounter angle is 0°. The lateral encounter angle $\Delta\Psi_{WV}$ is varied between 5°, 10°, 15° and 30°. Larger encounter angles are not considered because the encounter time is too short to induce a critical aircraft reaction. The ranges of the LiDAR parameters considered in the sensitivity analysis are shown in Table 1. Hereafter the location of each measurement is called measurement point (MP) and is defined as the middle point of the corresponding measurement volume.

Table 1 LiDAR parameters for sensitivity study

Parameter	Range of values
minimum measurement range $range_{min}$ [m]	60; 90; 120; (150)
lateral Scan angle range Ψ_{scan} [°]	+/-16; +/-30; +/-40
vertical Scan angle range Θ_{scan} [°]	+/-10
# MP along measurement axis, N_a (2 in Fig. 8)	1; 3; 5
# horizontal MP axes, N_h (3 in Fig. 8)	3; 5; 7; 9
blur depth [m]	15; 30
full screen update rate [Hz]	5;10

The parameters defining the geometry of the LiDAR measurement are shown in Fig. 7. The length of each measurement volume in axial direction is defined by the parameter blur depth. Along one measurement axis there can be more than one measurement point. The position of the measurement point located closest to the LiDAR sensor is described by the parameter $range_{min}$. Further measurement

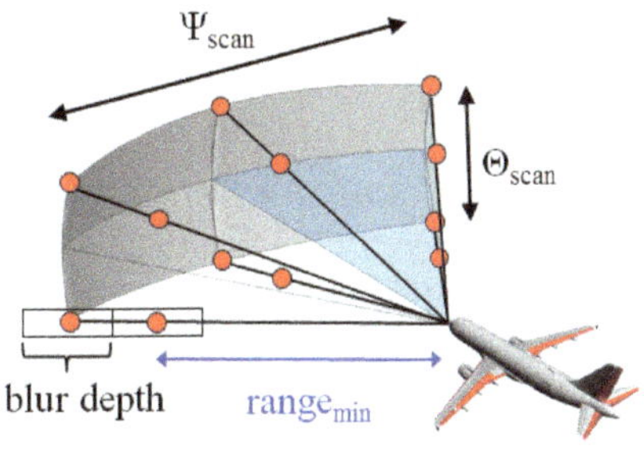

Fig. 7 Geometric LiDAR parameters

points are always located subsequently with a distance corresponding to the blur depth after the prior measurement point. The total number of measurement points in a full scan cycle results from the combination of the number of measurement points along each axis N_a and the number of measurement axes. The number of measurement axes in horizontal direction N_h is varied between the values shown in Table 1. In vertical direction, the number of 3 vertical axes is kept constant for all simulations. The field-of-view of the LiDAR sensor is defined by the maximum scan angles in lateral (Ψ_{scan}) and vertical (Θ_{scan}) direction. The update rate of the measurements is defined by the "full screen update rate". This parameter specifies the frequency with which all measurement points in the scanning sphere are updated. Nevertheless, each measurement axis is updated subsequently. That means the point-to-point update rate varies if the number of measurement axes is varied. All measurement points along one axis are updated simultaneously.

The minimum range parameter in parentheses in Table 1 is not included in the results that are displayed in Section 3.2. For a minimum range of 150 m it turned out that the measurement noise, which increases with increasing measurement distance, is too high and the optimization algorithm has rarely been executed for almost all sensor characteristics.

The dependency of the measurement errors and noise levels on the Doppler LiDAR parameters is very difficult to model. The LiDAR model that is used in the simulation environment includes the main dependencies in a generic fashion. In addition to the parameters that are varied here further parameters such as the detector properties (e.g. resolution and sensitivity) and the LASER source properties (e.g. quality, output power) would also influence the measurement errors. Moreover the measurement errors, as considered in the current paper, are the errors on the determination of the line-of-sight speed measurements, which is the result of both the physical effects and of the algorithmic postprocessing (e.g. for interpreting the interference patterns observed with the detector). This introduces many factors related to very specific technological choices in the sensor design. These factors shall be considered when modelling a very specific system, but would introduce too many parameters for the present study. As a consequence, the sensor model that is used here represents the relative variations due to the considered design parameters in addition to an absolute measurement error level that was estimated based on available data and results of previous works. The present study provides an insight into the kind of trade-off that has to be made between the design parameters rather than a particular "best" sensor configuration.

3.2 Results of the Sensitivity Study

The major effect of the wake vortex disturbance on the aircraft response during a wake vortex encounter with small lateral encounter angles between 5° and 30°, as they are considered here, is a rolling motion. The reduction of the maximum absolute bank angle reached during the wake vortex encounter can thus provide an indication on the overall effectiveness of the wake impact alleviation control system. It has to be kept in mind that the WIAC only generates control commands if the OWI has identified a wake vortex. As shown in Fig. 2, there are two possibilities why the OWI might provide no output. The first reason is that the OWI is not executed because the measurement noise is very high, causing the wind velocities resulting from the wake vortex to disappear in the background noise. The second reason is that the identified wake vortex parameters are rejected by the plausibility check (cf. Section 2.1.4).

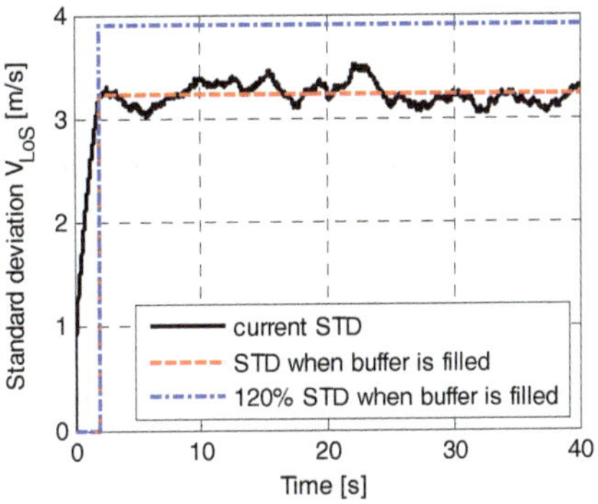

Fig. 8 Wake vortex encounter with very high measurement noise: no activation.

Fig. 8 exemplarily shows a wake vortex encounter during which the standard deviation of the LoS velocity of all measurement points stored in the data buffer is very high. In this case the LiDAR sensor has such a measurement noise that the standard deviation does not significantly increase when the vortex is contained in the data buffer. The threshold of 120% of the initial standard deviation of the LoS velocities is never reached and therefore the OWI is not executed during the whole simulation.

Fig. 9 illustrates the correlation between the measurement noise level, the number of plausible OWI results, and the measurement point densities. The measurement point density is defined as the number of measurement points stored in the data buffer (i.e. considered by the OWI) divided by the volume of the convex hull of all these measurement locations. The measurement noise level is assumed to be equal to the initial standard deviation of the LoS velocities (i.e. equal to reference measurement noise as used in the activation criterion). For visibility purposes Fig. 10 only displays LiDAR settings with a blur depth of 15 m. Red circles indicate that the optimization was not executed because the peaks of the standard deviation due to the wake vortex never reached the threshold of 120% of

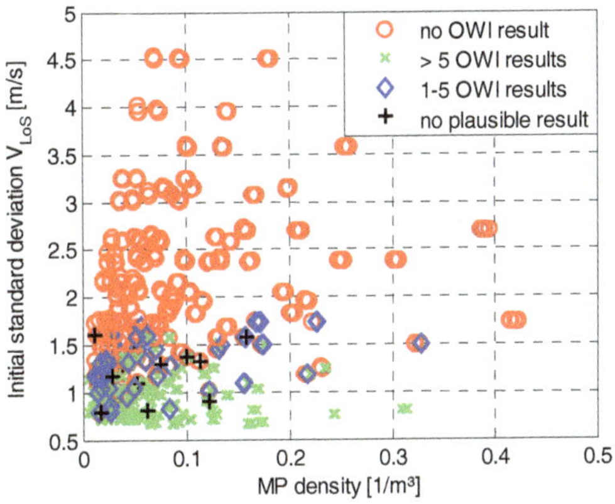

Fig. 9 Influence of initial standard deviation on execution of OWI algorithm

the initial standard deviation. Black + symbols mark cases where the optimization was executed (eventually several times) but none of the results ever passed the plausibility check (i.e. wake impact alleviation was never active). Blue diamonds symbolize wake vortex encounters with at least one and at most five valid OWI results and green crosses mark wake vortex encounters where more than 5 OWI results were valid and used for the wake impact alleviation during the encounter. It can be noticed that if the initial standard deviation of the considered LoS velocities is low enough (approximately 1 m/s and below) the wake vortex can be detected and the OWI usually finds valid results. In most of the cases more than five OWI results are available. But in some cases, with less favorable LiDAR characteristics, the OWI only provides 1 to 4 results during the encounter. Furthermore, there are cases in which the OWI result does not pass the plausibility check. When the measurement noise (i.e. the initial standard deviation) increases the OWI is less often executed and even when executed the plausibility check is failed more often. On average the red circles are located above the blue diamonds and these in turn above the green crosses. The boundaries are not clearly separated because more parameters than just the noise are varied in the different cases causing different SNR for the same noise level. If the measurement noise gets very large and the initial standard deviation of the stored LoS velocities is larger than 2 m/s the wake identification algorithm is never executed. In the majority of the investigated cases the SNR levels were too low and the activation threshold was never reached. As a consequence, Fig. 10 might give the impression that the presented wake impact alleviation system is not useful for a wake vortex encounter because it is rarely active. But it has to be kept in mind that the main cause for not executing the WIAC of the wake is the currently used activation criterion. An improved criterion should be developed and used in future studies. For LiDAR settings with a blur depth of 30 m, cases not shown in Fig. 10, the general observations are the same. To evaluate the behavior of the OWI and the WIAC for different sensor characteristics, in the following, only those cases will be considered in which the sensor settings allowed at least one valid OWI.

As the control commands are directly derived from the identified wake vortex disturbance, it is essential to get an adequately identified wake vortex model from which the wake-induced moments can be correctly derived. Fig. 10 shows the average deviation between the rolling moment induced by the actual wake in the flight dynamic simulation and the detected rolling moment on the basis of the wake identification for the different sensor settings. This average deviation can be used as

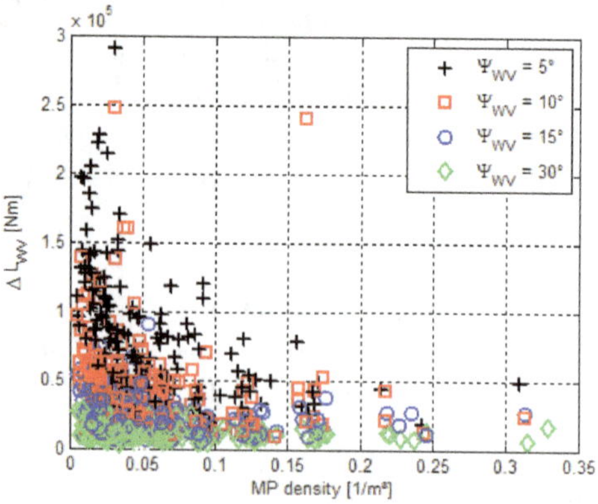

Fig. 10 Mean absolute error of detected compared to actually occurring wake-induced rolling moment

an identification quality index that focuses on the most important degree of freedom (roll) and considers at each point in time only the wind field errors that are currently relevant for the aircraft roll motion. It can be noticed that, apart from one outlier for a 10° lateral encounter, an increase of measurement point density leads to a reduction of the error of the detected moment. The aforementioned outlier represents a case in which the OWI has found a result of bad quality but that just passed the validity check. The WIAC version applied here has no forgetting factor (i.e. the WIAC continues to use the last valid OWI result until the wake vortex was left behind the aircraft) and thus used this bad OWI result over a long time. Even though this was an unlikely case, which occurred only once in 18432 simulations, both the OWI validity check and the WIAC implementation can and should certainly be improved. The outlier case is excluded from the subsequent analysis.

An increase in measurement point density leads to a higher spatial resolution such that its correlation with reduced moment errors was expected. With the parameters that are varied there are two main possibilities to increase the density: reducing the size of the domain (e.g. with a smaller field-of-view and by reducing the minimum range) and increasing either the number of LoS directions or the number of points along each LoS. All these ways of increasing the density except for the reduction of the minimum measurement range have a negative influence on the measurement noise.

Increasing the noise level degrades the OWIDIA performance; however the positive effects of an increased spatial resolution and/or a reduction of the minimum range seem to dominate the negative effects due to the other parameters (e.g. number of MP per axis, reduction of measurement depth or increase of full scan update rate). It would be interesting to pursue the parameter variation study in that particular direction. The absolute errors between the detected and the actual wake-induced rolling moment are larger for the smaller encounter angles. This is due to the fact that the wake-induced moments are smaller and that they are acting during a shorter time if the encounter angle increases. The relative errors (not shown here) are similar between the various encounter angles.

Fig. 12 shows the relative reduction of the maximum bank angle due to the OWIDIA system with respect to the maximum bank angle occurring without wake impact alleviation for different encounter angles and LiDAR settings. During all encounters there are no pilot inputs and the autopilot is not engaged. The only active control is an Airbus A320-like Normal Law. This law is not the original Airbus Normal Law but should behave very similarly. The upper three subplots show the variations of the encounter angle Ψ_{WV} and of some of the LiDAR settings for the set of simulation included in Fig. 12 (cf. Table 1). Two more parameters (blur depth and full scan update rate) are coded by different colors in the lower two plots. These two plots show the relative bank angle reduction for 60 and 90 m minimum range respectively. Each + sign corresponds to a single simulation. A relative value of 0.6 means that a reduction of 40% was achieved compared to the same encounter without OWIDIA. The results are split into categories labeled by letters and are visually separated by dashed vertical lines. Within each category all LiDAR parameters but the color-coded ones (blur depth and full scan update rate) stay the same and the wake encounter angle is varied. For readability reasons the $\Psi_{WV} = 30°$ cases are not shown. In these cases the relative bank angle reduction is lower on average but very robust. However, the absolute values of the bank angle are so small that the 30° encounter is practically much less relevant.

Simulations with 120 m minimum range are not displayed in Fig. 12 because the study revealed that these LiDAR settings are not useful for the current OWIDIA implementation. In most 120 m range cases the noise level was too high to allow a meaningful interpretation of the results obtained in these cases. Moreover, it also turned out that a timeframe of 2 s to store measurements in the data buffer is too short, especially for the smallest encounter angle of 5° and large LiDAR scan angles of 30° or 40°. In this case, the measurement points are located beyond the wake vortex when the aircraft comes close to the encounter. Hence an analysis of the 120 m minimum range LiDAR would only make sense if the simulation was conducted with an improved OWIDIA, e.g. with a larger buffer size and an improved activation criterion for high noise levels.

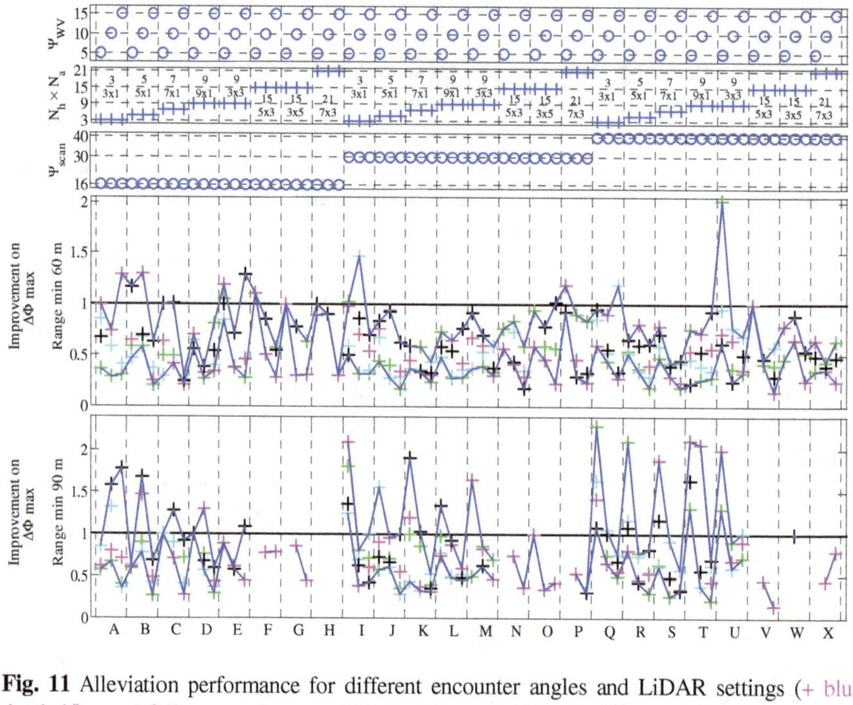

Fig. 11 Alleviation performance for different encounter angles and LiDAR settings (+ blur depth 15 m and full scan update rate 5 Hz | + 15 m and 10 Hz | + 30 m and 5 Hz | + 30 m and 10 Hz)

Fig. 11 illustrates that some simulations show an increase of the maximum bank angle: $\Delta\Phi$ max > 1, i.e. above the horizontal black line. The majority, however, showed significant improvements thanks to the OWIDIA system.

On average, LiDAR settings with a minimum range of 60 m provide better results than those with 90 m minimum range. For 60 m minimum range the best results are obtained for products N_h x N_a of 7 (7x1) as well as the neighboring categories (5x1 and 9x1). This applies to all three scan angles Ψ_{scan}. It results from the fact that the short minimum range and a reasonable number of measurement points provide a good trade-off between measurement noise and spatial resolution. For both 60 and 90 m cases and the same encounter angle Ψ_{WV}, the vertical spread between different blur depths and full scan update rates (i.e. the different colored + signs) is sometimes very large. This seems to occur for parameter combinations that are close to or beyond the boundaries of the domain in the parameter space in which the OWIDIA system performs well. The wide spreads tend to occur more often in cases with 90 m minimum range and at small encounter angles. Further investigation of these cases shall help improving the OWIDIA system.

Overall, the best wake impact alleviation and most robust (against blur depth and full screen update rate) OWIDIA performance can be found in category K with 60 m range, i.e. $\Psi_{scan} = 30°$, $N_h = 7$, $N_a = 1$. Categories D, L, N, R and S with 60 m range also exhibit very good results for all blur depth and full screen update rate values. Further categories provide good alleviation results for some LiDAR settings and bad results in a few cases. For the blur depth and full scan update rate, no clear trend was found regarding their influence on the alleviation performance.

All in all, the OWIDIA system can reduce the maximum bank angle during a typical wake vortex encounter by 50-60% on average and up to 80% in some cases. The application of the system with appropriate LiDAR settings would allow a huge gain in terms of safety. This level of wake impact alleviation performance could support the reduction of wake vortex based separation minima for following aircraft equipped with the OWIDIA system.

4 Summary

A first evaluation of the complete wake impact alleviation system was presented. It uses an explicit wake identification module based on Doppler LiDAR measurements. The cross-dependencies between the modules (sensor, OWI, WIAC) are very complex and the derivation of an adequate requirement set for each module is very challenging. The overall system can only be evaluated using an environment that takes all the cross-dependencies as well as a full flight dynamics model (including flight control laws) into account. Such an environment was developed and used in this work. The main performance driver for the complete wake impact alleviation system is the LiDAR sensor. The purpose of the presented sensitivity study was to provide first insights into the domains of adequate parameter combinations. Future studies shall further refine these first results. The present study has already provided very valuable information on the system behavior and performance. It also indicates which parts of the system should be enhanced.

Parameter combinations leading to a good compromise between measurement noise and spatial resolution seem to provide the most robust performances.

However, it must be kept in mind that the applied simplified activation criteria filtered out a significant number of cases with high spatial resolution, even though these cases might have led to good impact alleviation performance. Consequently, the observed trend might have been biased through the statistical differences between the compared populations. This part of the parameter space should also be investigated in future studies, which requires an improved activation criterion. Furthermore, the behavior of the system in presence of old and deformed vortices as well as regular atmospheric turbulence should also be investigated.

Overall, a wake alleviation performance of about 60% (in terms of maximum bank angle reduction) seems achievable with the proposed system and various sets of a priori realistic sensor characteristics.

References

1. TSB, Encounter with Wake Turbulence Air Canada Airbus A319-114 C-GBHZ Washington State, USA 10 January 2008, Aviation Investigation Report A08W0007 (April 8, 2010)
2. NTSB, Brief of Accident LAX99LA291, Washington, D.C. (May 17, 2001)
3. ATSB, Wake turbulence event Sydney Airport, NSW 3 November 2008, Aviation Occurrence Investigation AO-2008-077 (November 2009) ISBN 978-1-74251-009-5.
4. NTSB, In-Flight Separation of Vertical Stabilizer American Airlines Flight 587 Airbus Industrie A300-605R, N14053 Belle Harbor, New York November 12, 2001, NTSB/AAR-04/04, Washington, D.C. (2004)
5. NTSB, Brief of Accident LAX94FA073, Washington, D.C. (October 06, 1994)
6. Looye, G., Lombaerts, T., Kier, T.: Design and Flight Testing of Feedback Control Laws. Research Report DLR-FB-2012-02, The DLR Project Wetter & Fliegen, p. 162-170 (2012)
7. Rafi, M., Steck, J.: Response and Recovery of an MRAC Advanced Flight Control System to Wake Vortex Encounters. In: AIAA Infotech@ Aerospace Conference, Boston (August 2013)
8. Schwarz, C., Hahn, K.-U.: Automated Pilot Assistance for Wake Vortex Encounters. In: CEAS, Berlin, Germany, September 10-13 (2006)
9. Hahn, K.-U., Fischenberg, D., Niedermeier, D., Horn, C.: Wake Encounter Flight Control Assistance Based on Forward-Looking Measurement Processing. In: AIAA Guidance, Navigation, and Control Conference, Toronto, Ontario Canada, August 2-5 (2010)
10. Kloidt, S.: Beiträge zum Entwurf eines Flugregelungssystems zur Reduktion des Wirbelschleppeneinflusses (English: Contributions to the Design of a Flight Control System for the Reduction of the Wake Vortex Impact). Ph.D Thesis, TU Berlin, Germany (2007)
11. Ehlers, J., Fischenberg, D., Niedermeier, D.: Wake Identification Based Wake Impact Alleviation Control. In: AIAA AVIATION, Atlanta, GA, US, June 16-20 (2014)
12. Fischenberg, D.: Online Wake Identification Algorithms Using Forward Looking LIDAR Sensor Measurements. DLR Report IB111-2013/11, DLR German Aerospace Center, Braunschweig, Germany (February 2013)

13. Fischenberg, D.: Strömungsermittlungsverfahren/Flow Determination Method/Procédé de détermination d'écoulement. Patent No. EP 2 340 438 B1, European Patent Office (2013)
14. Jategaonkar, R.V.: Flight Vehicle System Identification: A Time Domain Methodology. Progress in Astronautics and Aeronautics. AIAA, Reston (2006)
15. Burnham, D.C., Hallock, J.N.: Chicago Monoacoustic Vortex Sensing System. Wale Vortex Decay, vol. 4. National Information Service, Springfield (1982)
16. Fischenberg, D.: A method to validate wake vortex encounter models from flight test data. In: 27th International Congress of the Aeronautical Sciences, ICAS 2010, Nice, France (2010)

Traffic Management along Air Streams through Space Metering

Mastura Ab Wahid, Hakim Bouadi, Antoine Drouin, Benjamas Panomruttanarug, and Felix Mora-Camino

Abstract. Via the current performances of aeronautical communication, navigation and surveillance systems, free flight and traffic management through trajectory negotiation have become a reality. However, the adoption of free flight in congested airspace leads to an increase of the number of potential traffic conflicts which are solved by diverting aircraft from their original flight plan, limiting the benefits of free flight. For high density traffic, air corridor concept and time-based flow management have recently been proposed. In the present paper, it is proposed to organize main traffic flows in congested airspace along air streams which are characterized by a three-dimensional (3D) common reference track and lateral lanes with a dynamic slot structure. There aircraft position is processed in a local space indexed axial coordinates system which should ease the management of traffic separation and surveillance. This change results in the need to develop new 3D space indexed guidance modes to perform position tracking, as well as to design and assign standard trajectories to enter into, evolve inside and exit from the air stream while insuring time and space separation between aircraft.

Mastura Ab Wahid · Antoine Drouin · Felix Mora-Camino
MAIAA Lab., French Civil Aviation Institute (ENAC), University of Toulouse, France
e-mail: {mastura.ab-wahid, felix.mora}@enac.fr,
drouin@recherche.enac.fr

Hakim Bouadi
Robotics and Production Laboratory, Polytechnic Military School, Bordj el Bahri, Algeria
e-mail: hakimbouadi@yahoo.fr

Benjamas Panomruttanarug
Department of Control, King Mongkuts University of Technology, Bangkok, Thailand
e-mail: benjamas.pan@kmutt.ac.th

J. Bordeneuve-Guibé et al. (eds.), *Advances in Aerospace Guidance, Navigation and Control*, DOI: 10.1007/978-3-319-17518-8_26

1 Introduction

The air transportation traffic has realized a sustained increase over the last decades, leading to airspace near saturation in some large areas of developed and emerging countries. According to [1] the global air traffic has doubled in size once every 15 years since 1977, and it will continue to do so. The design of new navigation and guidance systems with improved accuracy for spatial and temporal trajectory tracking has made possible free flight creating an appealing opportunity for airlines. However, in the case of high traffic density regions, the adoption of free flight should result, even through a three-dimensional plus time (3D+T) trajectory negotiation process with air traffic management (ATM), in an increasing number of conflicts [2] which are solved by modifying these aircraft trajectories. Currently, new concepts such as air corridors and time-based flow management (TBFM) are under development and early implementation in United States [3, 4]. TBFM should regulate aircraft through speed adjustments with discrete metering points through all flight phases in order to monitor the level of traffic demand in airspace sectors and deliver traffic down to the runway smoothly.

Here we propose that in high traffic density regions, air traffic can be organized along main airstreams using the full navigation and guidance capability of modern on-board systems. By organizing air traffic in such a manner, a common air stream reference trajectory (ASRT) can be used by aircraft flying through a high density air traffic space to maintain an accurate position within their assigned dynamic slot on a given lane. The adoption of such spatial reference will enforce the efficient use of the available time-space capacity along the air streams, and ease the on-board traffic separation task for an organized traffic along such fixed 3D reference. The flight plan of an aircraft will be composed of air stream and free flight segments depending upon if the crossed space is considered to be either high or low traffic density.

The main contributions of this communication are conceptual. The air stream concept is discussed as well as its essential differences from a classical airway or an airspace flow corridor. Additionally, a local space indexed frame is defined to allow the local axial coordinates positioning of an aircraft flying along an air stream lane. Aircraft flight guidance equations are revisited to better account for the surrounding traffic. Also, a space indexed version of these equations is developed. This leads to the formulation of space indexed guidance problems along the air stream lanes. An illustration of the proposed approach in the case of a straight air stream with parallel lanes is displayed where standard lane change trajectories are defined, synchronization conditions for merging are established and different heuristics to assign conflict free trajectories to lane changing aircraft.

An air stream is defined as a set of air lanes organized around and along a common 3D reference trajectory (ASRT). There aircraft are supposed to adopt the reference speed and flight level of their assigned lane and are guided in a common space index basis to maintain their position in their assigned dynamic slot. Like classical airways, air streams propose a common space for aircraft adopting similar navigation and guidance objectives for a portion of their flights but they can join or quit the air stream at any stage. Contrarily to airways and airspace flow corridors [5, 6, 7], air streams have no nominal shape (width, height or radius). In many situations, the central line of an airway or airspace flow corridor could be adopted as an ASRT, although air stream reference trajectories may present turns and may be changed according to different factors such as traffic demand and next day forecasted weather conditions. Aircraft with different performances or adopted cost indexes and speeds can be present in the same air stream. To be allowed in an air stream, aircraft equipment requirements are similar to that of airspace flow corridors: transportation aircraft must be equipped with required navigation performance (RNP), self-separation capability and on-board automated separation assurance. Self-separation on a lane is performed by dynamic position adjustments where the automatic dependent surveillance-broadcast (ADS-B) technology can provide position and speed information. Lane change manoeuvres within the air stream are performed without intervention by a central controller when an aircraft adopts a new reference airspeed. The on-board automated separation assurance system incorporates different levels of protection against a collision, including conflict detection and resolution, where the last protection against a collision is the Traffic Collision Avoidance System (TCAS). In air streams, the pilots will remain responsible for insuring the safe separation with nearby aircraft by maintaining situational awareness, performing standard manoeuvers and reacting to conflict resolution advices.

Here it is considered that the common reference track of the air stream, the ASRT, is a 3D curve given by a smooth parametric mapping which produces the geocentric coordinates of its points:

$$s \in [s_1, s_2] \subset \mathbb{R}^+ \rightarrow (\lambda(s), \mu(s), r(s)) \in [0, 2\pi] \times [-\frac{\pi}{2}, \frac{\pi}{2}] \times \mathbb{R}^+ \tag{1}$$

where for the point associated to a s-value, $\lambda(s)$, $\mu(s)$ and $r(s)$ are respectively the geocentric longitude and latitude and the distance to the center of the Earth, taken as its center of gravity. Here s is defined as the curvilinear abscissa along the ASRT by:

$$s - s_1 = \int_{\text{track}} \sqrt{(dX^2 + dY^2 + dZ^2)} = \int_{\text{track}} \sqrt{dr^2 + r^2.(d\mu^2 + \cos\mu^2.d\lambda^2)} \tag{2}$$

where $s_1 = 0$ at the initial point of the air stream reference track. It is supposed that functions λ, μ and r are smooth, injective and correspond to a flyable trajectory for a transportation aircraft. Typical examples of such curves are the orthodromia curves and the loxodromia curves which locally can often be assimilated with straight horizontal lines. Here it is assumed that the air stream traffic will follow trajectories positioned radially along this reference track.

2 Aircraft Guidance Dynamic Equations

Since in this study the concern is with aircraft trajectory, the flight dynamics equations are considered from the point of view of guidance. The following assumptions are done: The Earth centered Earth fixed (ECEF) is inertial, the modulus of the gravity vector is constant in the close atmosphere, the atmosphere is at rest relative to the Earth, the mass of the aircraft is considered to be a constant and the aircraft is assumed to be a rigid body.

2.1 *Time Indexed Guidance Equations*

The guidance dynamic equations are given first in the centered local Earth frame (CLEF) which can be considered an inertial frame since it is rotating at a very slow rate with respect to ECEF. Let $\dot{X}$, $\dot{Y}$, $\dot{Z}$ be the components of the inertial speed V of the aircraft at position $\underline{P} = \left(X\ Y\ Z\right)^T$ and at time t in the ECEF frame and let :

$$\underline{W}(\underline{P},t) = \left(w_X(\underline{P},t)\ w_Y(\underline{P},t)\ w_Z(\underline{P},t)\right)^T \tag{3}$$

be the wind speed at position $\underline{P} = (X,Y,Z)$ and at time t in the ECEF frame. The modulus of the airspeed at position (X,Y,Z) and at time t, V_a , is such as:

$$V_a(\underline{P},t) = \sqrt{\left(\dot{X} - w_X\right)^2 + \left(\dot{Y} - w_Y\right)^2 + \left(\dot{Z} - w_Z\right)^2} \tag{4}$$

Then the angle of attack α and the sideslip angle β are respectively given by complex expressions of $\underline{P}$, $\underline{\dot{P}}$, $\underline{W}$ and $\underline{a}(t)$, with $\underline{a}(t) = \left(\phi(t)\ \theta(t)\ \psi(t)\right)^T$ [12]:

$$\begin{aligned} \alpha(t) &= \alpha\left(\underline{P}(t),\underline{\dot{P}}(t),\underline{W}(\underline{P}(t),t),\underline{a}(t)\right) \quad \text{and} \\ \beta(t) &= \beta\left(\underline{P}(t),\underline{\dot{P}}(t),\underline{W}(\underline{P}(t),t),\underline{a}(t)\right) \end{aligned} \tag{5}$$

The attitude angles θ and ϕ as well as the heading angle ψ are related with the angular rates p, q and r , with $\underline{\omega} = \left(p\ q\ r\right)^T$, by the Euler equations:

$$\underline{\dot{a}} = \begin{pmatrix} 1 & \sin\phi.\tan\theta & \cos\phi.\tan\theta \\ 0 & \cos\phi & -\sin\phi \\ 0 & \sin\phi/\cos\theta & \cos\phi/\cos\theta \end{pmatrix} \underline{\omega}(t) \tag{6}$$

Let the rotation matrix between the Body frame and the centered local Earth frame, $R_B^L(\phi,\theta,\psi)$, be given by:

$$R_B^L = \begin{pmatrix} c(\theta)c(\psi) & s(\phi)s(\theta)c(\psi) - c(\phi)s(\psi) & c(\phi)s(\theta)c(\psi) + s(\phi)s(\psi) \\ c(\theta)s(\psi) & & \\ -s(\theta) & s(\phi)c(\theta) & c(\phi)c(\theta) \end{pmatrix} \tag{7}$$

where here $c(.)$ stands for $\cos(.)$ and $s(.)$ stands for $\sin(.)$. The components of the aerodynamic forces in the body frame are given by:

$$F_X = \frac{1}{2}\rho_{\text{air}}(X,Y,Z)V_a^2 S_{\text{ref}}(-C_D \cos\alpha\cos\beta - C_Y\cos\alpha\sin\beta + C_L\sin\alpha) \quad (8a)$$

$$F_Y = \frac{1}{2}\rho_{\text{air}}(X,Y,Z)V_a^2 S_{\text{ref}}(-C_D \sin\beta + C_Y\cos\beta) \quad (8b)$$

$$F_Z = \frac{1}{2}\rho_{\text{air}}(X,Y,Z)V_a^2 S_{\text{ref}}(-C_D \sin\alpha\cos\beta - C_Y\sin\alpha\sin\beta - C_L\cos\alpha) \quad (8c)$$

where C_D, C_Y and C_L are respectively the dimensionless aerodynamic coefficients of the drag, the side force and the lift which depend mainly on the angle of attack α and the side-slip angle β, and through the Mach number, on the airspeed and the flight level. S_{ref} is the reference surface of the aircraft and ρ_{air} is the density of air.

Then the external forces applied to the supposed rigid aircraft are expressed in the centered local Earth frame (CLEF) as:

$$\begin{pmatrix} F_N \\ F_E \\ F_D \end{pmatrix} = R_B^L(\phi,\theta,\psi)\begin{pmatrix} F_X + T_h.\cos\varepsilon \\ F_Y \\ F_Z + T_h.\sin\varepsilon \end{pmatrix} + m\begin{pmatrix} g_N(\underline{P}(t)) \\ g_E(\underline{P}(t)) \\ g_D(\underline{P}(t)) \end{pmatrix} \quad (9)$$

where T_h is the total engine thrust,ε is the engine alignment angle and g_N, g_E and g_D are the components in the CLEF of the acceleration of gravity at point $\underline{P} = (X,Y,Z)$.

The force equations can be written in the CLEF as:

$$\dot{u} = \frac{1}{m}F_N(\underline{P},\underline{\dot{P}},\underline{a},\underline{W}) \quad (10a)$$

$$\dot{v} = \frac{1}{m}F_E(\underline{P},\underline{\dot{P}},\underline{a},\underline{W}) \quad (10b)$$

$$\dot{w} = \frac{1}{m}F_D(\underline{P},\underline{\dot{P}},\underline{a},\underline{W}) \quad (10c)$$

where m is the mass of the aircraft. The thrust dynamics are supposed given by a first order linear model:

$$\dot{T}_h = \frac{1}{\tau_E}(T_h - T_{hc}) \quad (11)$$

where τ_E is a time constant and T_{hc} is the commanded thrust.

The rotation matrix from the local Earth frame to the ECEF frame, R_L^E is given by:

$$R_L^E = \begin{pmatrix} -\sin\mu\cos\lambda & -\sin\lambda & -\cos\mu\cos\lambda \\ -\sin\mu\sin\lambda & \cos\mu & -\cos\mu\sin\lambda \\ \cos\mu & 0 & -\sin\mu \end{pmatrix} \quad (12)$$

where λ and μ are respectively the current longitude, and geocentric latitude of the aircraft which are related to the coordinates X, Y and Z of the aircraft in the ECEF frame by relations:

$$\lambda = \mathrm{atan}(Y/X) \quad \text{and} \quad \mu = \mathrm{atan}(Z/\sqrt{X^2+Y^2}) \tag{13}$$

and the rotation matrix from the local Earth frame to the ECEF frame is now written $R_L^E(X,Y,Z)$. Then the speed of the aircraft with respect to the ECEF frame at time t is given by:

$$\dot{\underline{P}} = R_L^E(X,Y,Z)\begin{pmatrix} u \\ v \\ w \end{pmatrix} \tag{14}$$

Here, where mainly transportation aircraft are of interest, the yaw rate will be considered associated with bank angle ϕ and the ground speed to insure equilibrated turns and will not be taken as an independent input. Then an input-output representation is displayed in Fig. 1. According to nonlinear inverse control theory [13], and according to relations (6),(10) and (14), the relative degrees of outputs X, Y and Z with respect to independent inputs p, q and T_{hc} are all equal to 3 and the corresponding input-output representation will present no internal dynamics [14]. Considering the homeomorphism between (X,Y,Z) and (s,ρ,σ), the relative degree of outputs s, ρ and σ with respect to the same inputs will be also equal to 3.

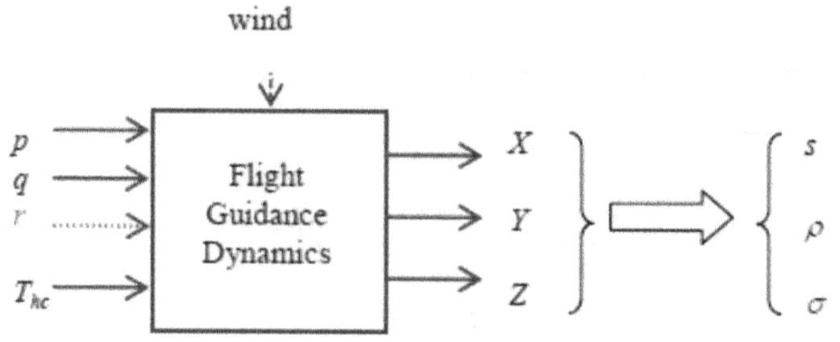

Fig. 1 Guidance along an ASRT

2.2 *Space Indexed Variables*

According with the air stream relative positioning , the curvilinear abscissa along the reference track can be adopted as independent variable. This provides important benefits such as a a common special reference for the different aircraft, overfly times become explicit control objectives while time and space separation constraints can be implemented. The rate of change of any flight variable var with respect to the curvilinear abscissa of the aircraft track is such that:

$$\mathrm{var}^{[1]} = \frac{d}{ds}(\mathrm{var}) = \frac{d}{dt}(\mathrm{var})\frac{dt}{ds} = \frac{1}{V_S}\frac{d}{dt}(\mathrm{var}) \tag{15}$$

where V_S is the current speed of the track of the aircraft on the ASRT. Then, second and third derivatives are such as:

$$\text{var}^{[2]} = \frac{1}{V_S^2}\left(\frac{d^2}{dt^2}(\text{var}) - \frac{d}{dt}(\text{var})\frac{\dot{V}_S}{V_S}\right) \tag{16a}$$

$$\text{var}^{[3]} = \frac{1}{V_S^3}\left(\frac{d^3}{dt^3}(\text{var}) - 3\frac{d^2}{dt^2}(\text{var})\frac{\dot{V}_S}{V_S} + \frac{d}{dt}(\text{var})\left(3\frac{\dot{V}_S^2}{V_S^2} - \frac{\ddot{V}_S}{V_S}\right)\right) \tag{16b}$$

Then the time equation is written:

$$t^{[1]} = \frac{1}{V_S(s)} \tag{17}$$

To insure a bijective mapping between space and time, $t(s)$, V_S must remain of constant sign: positive for an upward flight along the ASRT and negative otherwise. Then it is possible to consider functions $t(s)$ and $s(t)$. Also, the second and third derivatives of time with respect to space are given by:

$$t^{[2]} = -\frac{\dot{V}_S}{V_S^3} \tag{18a}$$

and

$$t^{[3]} = \frac{1}{V_S^3}\left(3\frac{\dot{V}_S^2}{V_S^2} - \frac{\ddot{V}_S}{V_S}\right) \tag{18b}$$

3 3D+T Guidance along a Lane of an Air Stream

In this section, space indexed objectives for the guidance function of an aircraft attached to a given lane of an air stream are first introduced, then these objectives are traduced into the time frame and a solution approach based on dynamic inversion of nonlinear flight guidance equations is proposed to produce guidance control laws meeting the space indexed objectives.

3.1 Formulation of the Space Indexed Guidance Problem

Suppose that the ASRT of an airstream is given by equation (1) and that an aircraft has to follow the 3D+T reference trajectory of the center of a slot along a lane embedded in this air stream and given in the local axial coordinates by equations such as:

$$s \in [s_1, s_2] \rightarrow (t_c(s), \rho_c(s), \sigma_c(s)) \in (\mathbb{R}^+)^3 \tag{19}$$

where $t_c(s)$, $\rho_c(s)$ and $\sigma_c(s)$ are real positive valued smooth functions over $[s1,s2]$, and where s is the curvilinear abscissa of the ASRT. Here it is assumed that $t_c(s)$ is an increasing function with inverse $S_c(t)$ and that:

$$\rho_c(s) \in [0, \rho_{\max}], \quad \sigma_c(s) \in [0, 2\pi] \tag{20}$$

Here it is supposed that space and time separation constraints have been satisfied when defining the nominal trajectories assigned to the different aircraft using the air stream. This will minimize the possibility of traffic conflict and avoid the need for relative guidance capability with its pitfalls (in general aircraft string instability [15]).

The space-time guidance error when the aircraft is above track abscissa s is defined as:

$$\varepsilon_t(s) = t(s) - t_c(s), \quad \varepsilon_\rho(s) = \rho(s) - \rho_c(s) \quad \text{and} \quad \varepsilon_\sigma(s) = \sigma(s) - \sigma_c(s) \tag{21}$$

and a major objective of guidance here is to make these errors stable towards 0 with a given space interval for convergence. Observing that these variables are related with the guidance inputs through dynamics of degrees equal to 3, to insure this control objective, it is assumed that the above guidance errors follow linear differential equations of the third order:

$$\varepsilon_t^{[3]}(s) + k_{1t}\varepsilon_t^{[2]}(s) + k_{2t}\varepsilon_t^{[1]}(s) + k_{3t}\varepsilon_t(s) = 0 \tag{22a}$$
$$\varepsilon_\rho^{[3]}(s) + k_{1\rho}\varepsilon_\rho^{[2]}(s) + k_{2\rho}\varepsilon_\rho^{[1]}(s) + k_{3\rho}\varepsilon_\rho(s) = 0 \tag{22b}$$
$$\varepsilon_\sigma^{[3]}(s) + k_{1\sigma}\varepsilon_\sigma^{[2]}(s) + k_{2\sigma}\varepsilon_\sigma^{[1]}(s) + k_{3\sigma}\varepsilon_\sigma(s) = 0 \tag{22c}$$

where k_{1i}, k_{2i}, k_{3i}, for $i = t, \rho$ and σ are real parameters such as the roots of the associated polynomial $s^3 + k_{1i}s^2 + k_{2i}s + k_{3i}$ are conveniently positioned in the complex plane.

3.2 Solution of the Space Indexed Guidance Problem

From (17) and (18), it appears that equation (22a) induces for the track speed V_S of the aircraft a time indexed second order nonlinear differential equation:

$$\ddot{V} = 3\frac{\dot{V}_S^2}{V_s} - k_{1t}V_s^2\dot{V}_s + k_{2t}V_s^3 + k_{3t}V_s^4(t - t_c) - h(V_{Sc}, \dot{V}_{Sc}, \ddot{V}_{Sc})V_s^4 \tag{23}$$

with

$$h(V_{Sc}, \dot{V}_{Sc}, \ddot{V}_{Sc}) = -\frac{\ddot{V}_{Sc}}{V_{Sc}^4} + 3\frac{\dot{V}_{Sc}}{V_{Sc}^5} - k_{1t}\frac{\dot{V}_s}{V_{Sc}^3} + k_{2t}\frac{1}{V_{Sc}} \tag{24}$$

which is driven by the desired track speed given by:

$$V_{Sc}(t) = 1/\left(\frac{dt_c}{ds}\right)\Bigg|_{s,\ t_c(s)=t} \tag{25}$$

and then through derivation, a third order nonlinear equation for the track abscissa:

$$d_s(s,\dot{s},\ddot{s},\dddot{s},s_c,\dot{s}_c,\ddot{s}_c,\dddot{s}_c,\underline{k}_t)=0 \tag{26}$$

From (15) and (16), equations (22b) and (22c) induce two time indexed nonlinear differential equations which can be written in a generic way as:

$$\begin{cases} d_\rho\left(\rho,\dot{\rho},\ddot{\rho},\dddot{\rho},\rho_c,\dot{\rho}_c,\ddot{\rho}_c,\dddot{\rho}_c,s,\dot{s},\ddot{s},\ddot{s},\dddot{s},\underline{k}_\rho\right)=0 \\ d_\sigma\left(\rho,\dot{\rho},\ddot{\rho},\dddot{\rho},\rho_c,\dot{\rho}_c,\ddot{\rho}_c,\dddot{\rho}_c,s,\dot{s},\ddot{s},\ddot{s},\dddot{s},\underline{k}_\sigma\right)=0 \end{cases} \tag{27}$$

Then, writing:

$$\underline{P}(t)=\left(X(t)\ Y(t)\ Z(t)\right)^T \tag{28}$$

and taking into account relation between the Cartesian representation in the ECEF frame and the ASRT axial representation ,we get:

$$\begin{cases} d_\rho\left(\underline{P},\underline{\dot{P}},\underline{\ddot{P}},\underline{\dddot{P}},\underline{P}_c,\underline{\dot{P}}_c,\underline{\ddot{P}}_c,\underline{\dddot{P}}_c,s,\dot{s},\ddot{s},\ddot{s},\dddot{s},\underline{k}_\rho\right)=0 \\ d_\sigma\left(\underline{P},\underline{\dot{P}},\underline{\ddot{P}},\underline{\dddot{P}},\underline{P}_c,\underline{\dot{P}}_c,\underline{\ddot{P}}_c,\underline{\dddot{P}}_c,s,\dot{s},\ddot{s},\ddot{s},\dddot{s},\underline{k}_\sigma\right)=0 \end{cases} \tag{29}$$

where

$$\underline{P}_c(t)=T_S\left(s_c(t),\rho_c(s_c(t)),\sigma_c(s_c(t))\right) \tag{30}$$

The guidance problem consists in finding the adequate piloting variables (p, q and T_{hc}) so that the aircraft accurately follows its nominal space indexed trajectory within the air stream. The adoption of a space indexed inverse control technique to solve this guidance problem implies space indexed linear dynamics for the guidance errors, including overfly time errors [16]. Then, writing:

$$\underline{U}(t)=\left(u(t)\ v(t)\ w(t)\right)^T \tag{31}$$

and derivating twice equation (14), $\underline{\dddot{P}}$ is written as:

$$\underline{\dddot{P}}=\ddot{R}_L^E(\underline{P},\underline{\dot{P}},\underline{\ddot{P}})\underline{U}+2\dot{R}_L^E(\underline{P},\underline{\dot{P}})\underline{\dot{U}}+R_L^E(\underline{P})\underline{\ddot{U}} \tag{32}$$

where the piloting variables p, q, r as well as the thrust control value, appear. Then, writing:

$$\underline{a}(t)=\left(\phi(t)\ \theta(t)\ \psi(t)\right)^T \quad \text{and} \quad \underline{\omega}(t)=\left(p(t)\ q(t)\ r(t)\right)^T \tag{33}$$

we get successively:

$$\underline{U}={R_L^E}^{-1}\underline{\dot{P}},\quad \underline{\dot{U}}=\phi_1(\underline{P},\underline{\dot{P}},\underline{a},T_h,\underline{w}) \quad \text{and} \quad \underline{\ddot{U}}=\phi_2(\underline{P},\underline{\dot{P}},\underline{\ddot{P}},\underline{a},\underline{\omega},T_h,T_{hc},\underline{w},\underline{\dot{w}}) \tag{34}$$

where phi_1 and phi_2 are smooth non linear functions. Then equation (25) can be rewritten as:

$$\dddot{\underline{P}} = f_P\left(\underline{P}, \underline{\dot{P}}, \underline{\ddot{P}}, \underline{a}, \underline{\omega}, T_h, T_{hc}, \underline{w}, \underline{\dot{w}}, \underline{\ddot{w}}\right) \quad (35)$$

The track speed V_S is given by:

$$\dot{s} = V_S(t) = \frac{\partial f}{\partial X}\dot{X} + \frac{\partial f}{\partial Y}\dot{Y} + \frac{\partial f}{\partial Z}\dot{Z} = < \mathrm{grad}(f(\underline{P})).\underline{\dot{P}} > \quad (36)$$

where the partial derivatives of function f are computed at the current position $P(t) = \left(X(t)\ Y(t)\ Z(t)\right)^T$. Then the third derivative of s can be written as:

$$\dddot{s} = f_s(\underline{P}, \underline{\dot{P}}, \underline{\ddot{P}}, \underline{a}, \underline{\omega}, T_h, T_{hc}, \underline{w}, \underline{\dot{w}}, \underline{\ddot{w}}) \quad (37)$$

Then equations (26) and (29) can be rewritten:

$$\begin{cases} D_s(\underline{P}, \underline{\dot{P}}, \underline{\ddot{P}}, \underline{P}_c, \underline{\dot{P}}_c, \underline{\ddot{P}}_c, \underline{a}, \underline{\omega}, T_h, T_{hc}, \underline{w}, \underline{\dot{w}}, \underline{\ddot{w}}, \underline{k}_t) = 0 \\ D_\rho(\underline{P}, \underline{\dot{P}}, \underline{\ddot{P}}, \underline{P}_c, \underline{\dot{P}}_c, \underline{\ddot{P}}_c, \underline{a}, \underline{\omega}, T_h, T_{hc}, \underline{w}, \underline{\dot{w}}, \underline{\ddot{w}}, \underline{k}_\rho) = 0 \\ D_\sigma(\underline{P}, \underline{\dot{P}}, \underline{\ddot{P}}, \underline{P}_c, \underline{\dot{P}}_c, \underline{\ddot{P}}_c, \underline{a}, \underline{\omega}, T_h, T_{hc}, \underline{w}, \underline{\dot{w}}, \underline{\ddot{w}}, \underline{k}_\sigma) = 0 \end{cases} \quad (38)$$

where D_i are smooth linear functions.

Once relation (38) is found invertible with respect to p, q and T_{hc}, it will be possible, using estimates of the wind and its first and second derivatives, to compute guidance inputs in accordance with the space indexed control objectives displayed by relations (22). In order to numerically achieve this inversion, neural techniques such as the ones displayed in [17] can be of interest to face the resulting complex computation. Then at a given time, the resulting control signals will be such as:

$$\begin{cases} p_c = C_p(\underline{P}, \underline{\dot{P}}, \underline{\ddot{P}}, \underline{P}_c, \underline{\dot{P}}_c, \underline{\ddot{P}}_c, \underline{a}, T_h, \underline{\tilde{w}}, \underline{\tilde{\dot{w}}}, \underline{\tilde{\ddot{w}}}, \underline{k}_t) \\ q_c = C_q(\underline{P}, \underline{\dot{P}}, \underline{\ddot{P}}, \underline{P}_c, \underline{\dot{P}}_c, \underline{\ddot{P}}_c, \underline{a}, T_h, \underline{\tilde{w}}, \underline{\tilde{\dot{w}}}, \underline{\tilde{\ddot{w}}}, \underline{k}_\rho) \\ T_{hc} = C_{Th}(\underline{P}, \underline{\dot{P}}, \underline{\ddot{P}}, \underline{P}_c, \underline{\dot{P}}_c, \underline{\ddot{P}}_c, \underline{a}, T_h, \underline{\tilde{w}}, \underline{\tilde{\dot{w}}}, \underline{\tilde{\ddot{w}}}, \underline{k}_\sigma) \end{cases} \quad (39)$$

where the C_i functions result from the inversion of (38) and where the wind speed and its time derivatives are replaced by their estimates $\underline{\tilde{w}}$, $\underline{\tilde{\dot{w}}}$ and $\underline{\tilde{\ddot{w}}}$.

Also, to check if the 3D+T trajectory given by relation (19) is flyable, the temporal evolution of the flight parameters generated by the inversion of (38) when P is chosen equal to P_c can be analyzed for a given wind scenario with respect to the flight domain and the performance of the autopilot.

4 Illustration of the Proposed Approach

Here a straight and level air stream reference trajectory is considered with several parallel lanes at the same altitude (see figure 2). The kth lane of the ASRT is composed of space slots of width Δ moving at a constant speed V_k ; Here it is assumed that two lanes k and h are separated by a constant distance D_{kh}.

Here is considered the case in which a transport aircraft initially flying in lane h with a ground speed V_h , is to merge in a free slot of the kth lane. Let $\delta V_{kh} = V_k - V_h$.

Let the initial position of the center of slot i of lane h at time t_0 be x_{h0}^k , its current position at time t is given by:

$$x_h^i(t) = x_{h0}^i + V_h(t - t_0) \tag{40}$$

while the position of the following slot at time t on the same lane is given by:

$$x_h^{i+1}(t) = x_h^i(t) - \Delta \tag{41}$$

4.1 Reference Trajectory between Lanes

The reference trajectory of the merging aircraft is supposed to be composed of a succession of segments where only one maneuver is performed at a time. First the aircraft waits on its original lane for the right time and position to turn towards the target lane, then on the straight segment between turns speed is changed to V_i, then another turn is performed to reach the center of a free slot on lane i. Here, to limit the number of parameters characterizing the maneuver, the turns are supposed to be symmetrical (same radius and angle). This is described in figure 2.

To perform the maneuver, the following assumptions are taken: aircraft a_1 on lane h decides to shift to lane k at position corresponding to s_0, it knows the separation D_{hk} between lanes h and k as well as their reference speeds. The decisions parameters are the common turn radius, R_m, the common turning angle δ_m, and the

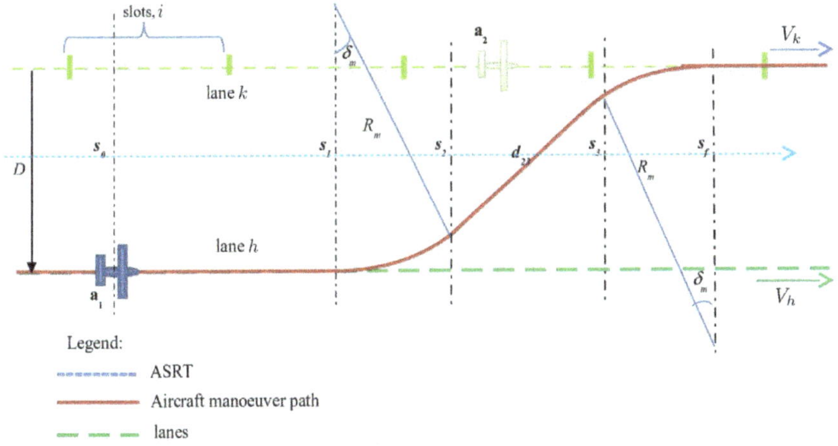

Fig. 2 Merge maneuver in an ASRT

position s_1 at which the aircraft leaves its original lane h. It is supposed that the straight segment between turns (from s_2 to s_3) is long enough to allow the change of speed from V_h to V_k so that the final turn to merge into the destination lane is already performed at speed V_k.

4.2 *Characterization of the Reference Trajectory*

From s_0 to s_1, the aircraft a_t flies a straight segment at constant speed V_h. The maneuver starts at s_1, the aircraft performs at constant speed V_h a left equilibrated turn of angle $\delta_m(< \frac{\pi}{2})$ and radius R_m such as:

$$R_m = \frac{V_h^2}{g.\sin\phi_m} \tag{42}$$

here ϕ_m is a standard turn bank angle such as $\phi_m \leq \phi_{max}$, where ϕ_{max} is a maximum bank angle value and:

$$R_m \geq R_{min} \quad \text{with} \quad R_{min} > \frac{V_h^2}{g.\sin\phi_{max}} \tag{43}$$

Then:

$$s_2 = s_1 + R_m \sin\delta_m \tag{44}$$

In order to merge safely into lane k , the aircraft performs a nominal change of speed from V_h to $V_k = V_h + \delta V_{hk}$ from s_2 to s_3 . Here it is supposed that this nominal change of speed is characterized by a constant space rate *acc* (m/s/m), such as:

$$acc = \frac{\delta V_{hk}\sin\delta_m}{D - 2R_m(1-\cos\delta_m)} \quad \text{with} \quad acc_{min} \leq acc \leq acc_{max} \tag{45}$$

where acc_{min} and acc_{max} are the minimum and maximum speed space rate of change. The abscissa s_3 is then given by:

$$s_3 = s_2 + \frac{D - 2R_m(1-\cos\delta_m)}{\tan\delta_m} \tag{46}$$

From s_2 to s_f, the aircraft performs at constant speed V_k a right turn of angle δ_m and radius R_m to adopt the ASRT track at the center of a free space slot:

$$s_f = s_3 + R_m \sin\delta_m \tag{47}$$

Here also:

$$R_m \geq R_{min} \quad \text{with} \quad R_{min} > \frac{V_h^2}{g.\sin\phi_{max}} \tag{48}$$

Then s_f is parameterized by s_1, D, V_{hk} , V_j, R_m and δ_m where δ_m , R_m and s_1 are design parameters to be chosen. Therefore s_f is given by:

$$s_f = s_1 + \frac{D - 2R_m(1-\cos\delta_m)}{\tan\delta_m} + 2R_m \sin\delta_m \tag{49}$$

The ground speed of the aircraft will vary along its reference merging trajectory:

$$V(s) = \begin{cases} V_h & \text{if} \quad s_0 \leq s \leq s_2 \\ V_h + \delta V_{hk} \sin\delta_m \frac{s-s_2}{D-2R_m(1-\cos\delta_m)} & \text{if} \quad s_2 \leq s \leq s_3 \\ V_k & \text{if} \quad s_3 \leq s \leq s_f \end{cases} \tag{50}$$

The corresponding reference trajectory $(t_c(s), \rho_c(s), \sigma_c(s))$ with $s \in [s_0, s_f]$, is given by the following expressions:

$$t_c(s) = \begin{cases} t_1 + \frac{R_m}{V_h} \arcsin\frac{s-s_1}{R_m} & \text{if} \quad s_1 \leq s \leq s_2 \\ t_2 + \frac{\delta V_{hk} \sin\delta_m}{D-2R_m(1-\cos\delta_m)} \ln\left(1 - \frac{\delta V_{hk} \sin\delta_m (s-s_2)}{V_h(D-2R_m(1-\cos\delta_m))}\right) & \text{if} \quad s_2 \leq s \leq s_3 \\ t_3 + \frac{R_m}{V_k} \arcsin\frac{s-s_3}{R_m} & \text{if} \quad s_3 \leq s \leq s_f \end{cases} \tag{51}$$

with

$$t_1 = t_0 + \frac{(s_1 - s_0)}{V_h} \tag{52}$$

$$t_2 = t_1 + \frac{R_m \delta_m}{V_h} \tag{53}$$

$$t_3 = t_2 + \frac{\delta V_{hk} \sin\delta_m}{D - 2R_m(1-\cos\delta_m)} \ln\left(1 - \frac{\delta V_{hk} \cos\delta_m}{V_h}\right) \tag{54}$$

Therefore, the time to reach the final merging position is given by:

$$t_f = t_0 + \frac{s_1 - s_0}{V_h} + \frac{R_m \delta_m}{V_h} + \frac{\delta V_{hk} \sin\delta_m}{D - 2R_m(1-\cos\delta_m)} \ln\left(1 - \frac{\delta V_{hk} \cos\delta_m}{V_h}\right) + \frac{R_m \delta_m}{V_k} \tag{55}$$

The distance of the aircraft to the ASRT, $\rho(s)$, is given by:

$$\rho_c(s) = \begin{cases} D & \text{if} \quad s_0 \leq s \leq s_1 \\ D - R_m\left(1 - \cos\left(\arcsin\frac{s-s_1}{R_m}\right)\right) & \text{if} \quad s_1 \leq s \leq s_2 \\ D - R_m(1-\cos\delta_m) - (s - s_2)\tan\delta_m & \text{if} \quad s_2 \leq s \leq s_3 \\ R_m\left(1 - \cos\left(\arcsin\frac{s-s_f}{R_m}\right)\right) & \text{if} \quad s_3 \leq s \leq s_f \end{cases} \tag{56}$$

while the reference azimuth angle $\sigma(s)$ remains constant and equal to $\frac{\pi}{2}$.

Let K_k be the set of free slots on lane k, an efficient management of the air stream will make the aircraft to merge to the center of earliest free slot i_m on lane k such as:

$$i_m = \min\{i \in K_k\} \quad \text{where} \quad \exists R_m \leq R_{min} \quad \text{and} \quad \exists \delta_m \in \left]0, \frac{\pi}{2}\right[$$
$$\text{such as} \quad s_f(s_1, R_m, \delta_m) = x_k^{i_m} \quad (57)$$

5 On-Line Traffic Management along an Air Stream

Here is considered an air stream composed of different parallel lanes such as those of Fig. 3 which are characterized by their position with respect to the ASRT and their reference ground speeds which take into account the wind speed so that they are attached in fact to reference Mach numbers. To each flight is attached an aircraft with specific performances. It is supposed that the ongoing traffic is composed of two kinds of flights, those which are already assigned their preferred lane, set J_a , and those which are not, said transient flights. Let J_t be the set of these transient flights. The first kind of flights is occupying time-space slots along lanes parallel to the ASRT. Let $L = \{L_1, L_2, \ldots, L_{|L|}\}$ be the set composed of these lanes, including the ASRT and let Δ_k be the set of free slots along lane k where:

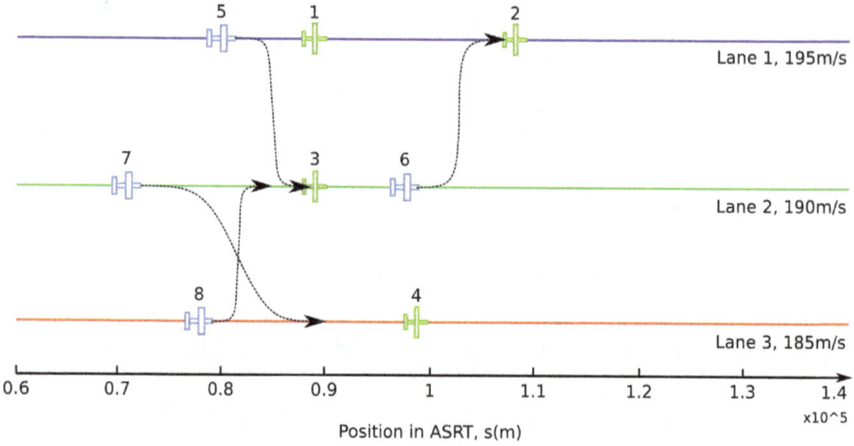

Fig. 3 Example of transient flights along an ASRT

$$\left(t_k^i(s)\ \rho_k^i(s)\ \sigma_k^i(s)\right) \quad i \in \Delta_k \quad k \in L \quad (58)$$

is the reference trajectory of the ith free slot of lane k. Here with a constant speed assigned to each lane:

$$t_k^i(s) = t_k^i(s_0^{ki} + (s - s_0^k)/V_k \quad (59)$$

where V_k is the reference ground speed of lane k and $t_k^i(s_0^k)$ is the time at which slot i enters lane k at position s_0^{ki}.

The problem considered here is to propose to each transient flight a conflict free trajectory allowing it to join its preferred lane as soon as possible. Here it is supposed that the merge trajectories are of the class considered in the previous paragraph.

Let o_j and d_j be the origin and the destination lanes of flight $j, j \in J_t$. Then let M_j be the set of feasible (conflict free trajectories with respect to the already assigned ones) merging trajectories from lane o_j to lane d_j for flight j and starting after position s_0^j. For a given flight, each of these trajectories, indexed by m, is attached to a time-space moving slot on the destination lane d_j which is reached at time $t_{d_j}^{jm}$ and at abscissa $s_{d_j}^{jm}$.

5.1 Assignment Heuristic

The considered problem is an assignment problem between flights and free slots on desired lanes where the total waiting times on the original lanes for transient flights could be a measure of the effectiveness of the management of the flights within the airstream. This assignment problem is a complex combinatorial one and its exact on-line solution may be unfeasible even for rather small instances of the problem [14]. Then heuristic approaches seem appropriate to generate on-line assignment solutions. Different heuristics can be considered, however among them, the greedy ones look to be the simpler to be put into operation. Either a time strategy or a space strategy can be adopted. Here two examples of greedy heuristic assignment methods according to a time strategy are considered: The min-time heuristic ranks the aircraft in set J_t increasingly with respect to m_j^f given by:

$$m_j^f = \underset{m \in M_j}{\operatorname{argmin}}\, t_{d_j}^m \tag{60}$$

The index of the first flight to be assigned, j^* , is given by:

$$j^* = \underset{j \in J_t}{\operatorname{argmin}}\, m_j^f \tag{61}$$

where flight j^* is assigned the merge trajectory $m_{j^*}^f$, J_t is then updated by deleting j^*. Note that J_t must be incremented any time a new flight enters the airstream. The set of conflict free trajectories M_j are updated for $j \in J_t$. The risk with this heuristic is that the trajectory assignment of some flights may be postponed repeatedly, making these flights support additional operations costs. Differently, the max-wait heuristic ranks the aircraft in set J_t decreasingly according to their waiting time within this set and assigns to the first of them, j^*, its earliest conflict free merge trajectory $m_{j^*}^f$. Now the risk is to assign trajectory merging trajectories to flights only when they have been waiting for a long time. An hybrid heuristic could be to adopt the min-time heuristic, but whenever the waiting time of an aircraft becomes higher than some given upper bound, it has to be treated in priority. Similar heuristics could be proposed adopting the spatial index $s_{d_j}^{jm}$ within a spatial strategy. Observe that the performances resulting from the temporal and the spatial strategies should not be equivalent since the speed of merging aircraft does not remain constant during the maneuver When considering aircraft j of set J_t, the ith free slot of d_j will generate a candidate merging trajectory if there is a solution to the equations:

$$t_f = t_{d_j}^j \text{ and } s_f = s_{d_j}^j \tag{62}$$

where s_f is given by (49), t_f is given by (50) and:

$$t_{d_j}^i(s_{d_j}^j) = t_{d_j}^i(s_0^{d_j i}) + (s_{d_j}^j - s_0^{d_j})/V_A^{d_j} \tag{63}$$

This candidate trajectory will be conflict free if it remains far from any other planned trajectory in the air stream. A possible way to express this condition is such as:

$$\forall g \in J_a \text{ with } s_0^{gj} s_f^{gj} : \forall s \in [s_0^{gj} s_f^{gj}] d_{gj}(s) > d_{\min} \tag{64}$$

where

$$s_0^{gj} = \max\{s_0^g, s_0^j\} \text{ and } s_f^{gj} = \min\{s_f^g, s_f^j\} \tag{65}$$

and

$$d_{gj}(s) = \sqrt{\begin{array}{l}(\rho_g \cos\sigma_g - \rho_j \cos\sigma_j)^2 + (\rho_g \sin\sigma_g - \rho_j \sin\sigma_j)^2 \\ + \frac{1}{4}(V_A^g + V_A^{d_j})^2 (t_g - t_j)^2\end{array}} \tag{66}$$

and where $d_{\min}$ is a minimum safe distance.

5.2 *Illustration of Traffic Assignment*

Here is considered a scenario which considers three lanes of an air stream with different reference speeds and same altitude:

- In the considered section of the first lane there are three aircraft, one of them intending to shift to the second lane.
- In the second lane there are also three aircraft, one of them intending to shift to the first lane and another intending to shift to the third lane.
- In the third lane, one of the two present aircraft intend to shift to the second lane.

Figure 3 displays the relative position of these aircraft at initial time, as well as their intends within the air stream. The slots on the lanes are numbered backwards starting from the more engaged flight into each lane.

In order to perform the assignment of the free slots to the standard shift maneuvers, a greedy heuristic based on the min-time approach is developed. The main step of the resulting assignment algorithm are displayed below:

1. Rank increasingly the transient flights according to their minimum final maneuver time, m_j^f. Let j^* be the first of the list.
2. Assign to flight j^* the maneuver associated to $m_{j^*}^f$ and update the sets J_a, J_t :

$$J_a = J_a \cup j^* \text{ and } J_t = J_t / j^* \tag{67}$$

3. If $J_t = \varnothing$ then Exit
4. Update the sets M_j with $j \in J_t$, if $\forall j \in J_t : M_j = \varnothing$ then Exit otherwise go back to step 1.

Observe that when $M_j = \varnothing$, flight j has no opportunity on its target lane and must remain on its original lane. Note also that this algorithm can be run on line by adding flight entry and exit events.

Table 1 below summarizes the initial position of each flight and their intends. Table 2 shows the first calculation of the assignment without delay and their ranking.

Table 1 Initial situation in ASRT

Flight	Initial Lane	Slot Position	Target Lane	Initial Lane Speed (m/s)	Target Lane Speed (m/s)	Distance between Lanes (m)	Initial Position s0 (m)
1	1	3	1	195	195	0	12626
2	1	1	1	195	195	0	22727
3	2	2	2	190	190	0	12626
4	3	1	3	185	185	0	22607
5	1	5	2	195	190	10000	7390
6	2	1	1	190	195	10000	17431
7	2	4	3	190	185	10000	2516
8	3	4	2	185	190	10000	7548

It can be seen that the assignment can be done for flight 6, 7 and 8. However for flight 5, since the target slot position has already been occupied by flight 3, some delay time will be proposed. To avoid any conflict during the maneuver, flight 7 will be ranked third after flight 8 and there will be a delay since the assignment will have taken place after flight 8. Table 3 shows the final proposed assignment. From Table 3, it can be seen that flight 5 cannot be assigned. Even after a delay of 230.35 seconds has been proposed, flight 5 targets slot 3 in lane 2 which has been occupied by flight 8. Adding to the delay might solve the problem or proposing an extra lane as a queue lane or waiting lane. This proposed lane can have a variable speed that would be adjusted to be higher or lower to assist the transient aircraft to switch lane in between the maneuver.

Table 2 First ranking between transient flights

Flight	Initial Lane	Initial Slot Position	Target Lane	Target Slot Position	Maneuver Duration (s)	Earliest Completion Time (s)	Ranking of Transient Flights
1	1	3	1	3	0	-	-
2	1	1	1	1	0	-	-
3	2	2	2	2	0	-	-
4	3	1	3	1	0	-	-
5	1	4	2	3	104.28	104.28	4
6	2	1	1	2	104.28	104.28	1
7	2	4	3	4	107.07	107.07	3
8	3	3	2	3	107.07	107.07	2

Table 3 Final ranking between transient flights

Flight	Initial Lane	Initial Slot Position	Final Lane	Final Slot Position	Start of Maneuver (m)	End of Maneuver (m)	Delay (s)
1	1	3	1	3	-	-	-
2	1	1	1	1	-	-	-
3	2	2	2	2	-	-	-
4	3	1	3	1	-	-	-
5	1	4	2	-	-	-	230.35
6	2	1	1	2	98165	118360	0
7	2	4	3	4	89307	109505	111.1
8	3	3	2	3	79838	100030	2

6 Conclusion

In this study a new approach to traffic organization in dense areas compatible with the current performances of aeronautical communication, navigation and surveillance systems, has been proposed. The objective is to complement the free flight and traffic management through trajectory negotiation approaches, by introducing new high density traffic links organized in a way limiting traffic conflicts. We have also proposed to organize traffic in congested airspace along main air streams which are characterized by a central 3D track which acts as a common space indexed reference for the involved aircraft. Preliminary concepts have been introduced in this paper to support this idea and many complementary studies should be developed to prove the effectiveness of the proposed approach. Guidance objectives have been expressed with respect to the common spatial reference offered by an ASRT and the synthesis of guidance control laws based on nonlinear inversion has been considered. Finally, the case of merging traffic along a common track has been considered, leading to the complete parametrization of the resulting trajectory. This [TODO??idea, proposal??] allows to easily manage traffic inside the air stream by using the common spatial reference to assign conflict free trajectories to flights inside the air stream.

Acronyms

3D three-dimensional. 1–3, 18
3D+T three-dimensional plus time. 2,7,10

ADS-B automatic dependent surveillance-broadcast. 3
ASRT air stream reference trajectory. 2, 3, 6–14, 17, 18
ATM air traffic management. 2

CLEF centered local Earth frame. 4, 5

ECEF Earth centered Earth fixed. 4–6, 9

RNP required navigation performance. 3

TBFM time-based flow management. 2
TCAS Traffic Collision Avoidance System. 3

References

1. Paielli, R.A., Erzberger, H.: Conflict probability estimation for free flight, NASA Technical Memorandum 110411 (October 1996)
2. Ramamoorthy, K., Crassidis, J.L., Singh, T.: Potential functions for en-route air traffic management and flight planning. In: AIAA Guidance, Navigation, Control. Providence, Rhodes Island (2004)
3. Joint Planning and Development Office, Concepts of operations for the next generation air transportation system, Washington, D.C. (2010)
4. Ky, P., Miaillier, B.: Sesar: towards the new generation of air traffic management systems in Europe. Journal of Air Traffic Control 48(1), 11–14 (2006)
5. Hoffman, R., Prete, J.: Principles of airspace tube design for dynamic airspace configuration. In: Proceedings of the 8th AIAA-ATIO Conference, AIAA-2008-8939 (2008)
6. Yousefi, A., Zadeh, A.N., Tafazzoli, A.: A dynamic allocation and benefit assessment of NextGen flow corridors. Transportation Research Part C 33, 297–310 (2013)
7. Ye, B., Hu, M., Shortle, J.F.: Collision risk-capacity tradeoff analysis of an route corridor model. Chinese Journal of Aeronautics 27(1), 124–135 (2014)
8. Fedchin, S.S.: Aircraft Navigation. National Aeronautics and Space Administration, Washington (1969)
9. Nourel Din, A., Karamat, T.B., Georgy, J.: Fundamentals of Inertial Navigation. Satellite-based Position and their integration. Springer (2013)
10. Baleri, G.: Datum Transformations of NAV420 Reference Frames; Crossbow Technology, Inc., `http://www.xbow.com`
11. Grewal, M.S., Weill, L.R., Andrews, A.P.: Global Positioning Systems, Inertial Navigation and Integration. John Wiley & Sons, New York (2001)
12. Mora-Camino, F.: Flight Control Systems, Lecture Notes, ENAC (2014)
13. Bouadi, H., Mora-Camino, F.: Aircraft trajectory tacking by nonlinear spatial inversion, AIAA-GNC, Minneapolis, Minnesota, USA, 10.2514/6. 2012-4613 (2012)
14. Shen, T.J.: Les Réseaux Neurones Affines et Leur Application à la Commande Automatique du Vol, PhD thesis, INPT, Toulouse, France (1995)

15. Weitz, L.A.: Investigating string stability of a time-history control law for Interval Management. Transportation Research Part C: Emerging Technologies 33, 257–271 (2013)
16. Bouadi, H., Choukroun, D., Mora-Camino, F.: Aircraft longitudinal guidance based on a spatial reference. In: Chu, Q., Mulder, B., Choukroun, D., van Kampen, E.-J., de Visser, C., Looye, G. (eds.) Advances in Aerospace Guidance, Navigation and Control, vol. 135, pp. 427–442. Springer, Heidelberg (2013)
17. Lu, W.C., Duan, L., Hsiao, F.B., Mora-Camino, F.: Neural Guidance Control for Aircraft Based on Differential Flatness. Journal of Guidance, Control, and Dynamics 31(4), 892–898 (2008)

Integrated Design and Control of a Flying Wing Using Nonsmooth Optimization Techniques

Yann Denieul, Joël Bordeneuve, Daniel Alazard,
Clément Toussaint, and Gilles Taquin

Abstract. In this paper we consider the problem of simultaneously stabilizing a civil flying wing aircraft and optimizing the control surfaces physical parameters, such as control surfaces sizes and actuators bandwidth. This flying wing configuration is characterized by unstable longitudinal modes, badly damped lateral modes, and a lack of control efficiency despite large movables. The question is then to determine the energy penalty associated to the control of these unstable modes, and more precisely to optimize the control surfaces architecture in order to minimize the control-associated energy. Our approach uses latest nonsmooth optimization techniques, which allows more possibilities on requirements specifications and controller structure compared to other approaches such as LMI-based optimizations. Results show a consistent behaviour for tuned parameters of the control surfaces.

1 Introduction and Motivation

Among other disruptive concepts for the future of civil aviation, the flying wing configuration has been studied for years [9]. This aircraft architecture combines several theoretical advantages compared to a "conventional" design. The main benefit provided by the flying wing concept is an enhanced aerodynamic efficiency, by eliminating all the devices (fuselage and tail planes) that do not create lift. As a result a better lift distribution along the span is achieved. The empty weight of

---

Yann Denieul · Joel Bordeneuve · Daniel Alazard
University of Toulouse-ISAE, 10, Av. Edouard Belin, 31055 Toulouse, France
e-mail: {yann.denieul,joel.bordeneuve,daniel.alazard}@isae.fr

Clément Toussaint
ONERA, 2, Av. Edouard Belin, 31055 Toulouse, France
e-mail: clement.toussaint@onera.fr

Gilles Taquin
Airbus Operations SAS, 316 route de Bayonne, 31060 Toulouse, France
e-mail: gilles.taquin@airbus.com

J. Bordeneuve-Guibé et al. (eds.), *Advances in Aerospace Guidance, Navigation and Control*, DOI: 10.1007/978-3-319-17518-8_27

the aircraft is also reduced, as is the overall wetted area. However this configuration faces some huge challenges which are yet to overcome. One of the main remaining challenge is the Handling Qualities resolution [13]. Indeed the flying wing has very special features that make it difficult to properly control and stabilize:

- No tailplane to trim the aircraft and damp the pitch oscillation.
- Forward aerodynamic center leading to a strong longitudinal instability.
- Control surfaces with poor pitch authority due to a weak longitudinal lever arm w.r.t. the center of gravity.
- Coupled control surfaces: trailing edge elevons are capable of providing pitching and rolling moments.
- Lack of lever arm of the vertical surfaces (if any) leading to a lack of yawing authority and a badly damped Dutch Roll mode.

Traditionnally, Aircraft Conceptual Design is concerned only with disciplines such as Aerodynamics, Performance, Weight and Handling Qualities. Control laws are designed afterwards, when the geometrical design is frozen. However for the flying wing case, such a sequential approach may be too restrictive: because of the previously mentioned specificities, active stabilization control laws are needed. A strong coupling exists between the plant — the flying wing— and the stabilizing controller; and it has been shown in [6] that such a coupling may lead to suboptimal design when the plant and controller are designed separately. The goal of our work is then (a) to identify potential physical limitations induced by the need for active stabilization and (b) to optimize the control architecture in order to minimize a physical criterion, e.g. energy needed to stabilize and control the aircraft.

This problem, known as *Plant-Controller Optimization*, *Co-design* or *Integrated Design and Control*, has been adressed in a variety of domains, such as Astronautics [2], Aeronautics [11], Chemistry [5] or Autonomous Underwater vehicles [15]. But whereas most of these studies have been using the Linear Matrix Inequalities (LMI) framework to solve the combined plant-controller optimization problem, the novelty of our approach is to use nonsmooth optimization techniques presented in [3]. Such a formulation allows for defining an arbitrary fixed-order controller, specifying physical parameters as controller parameters to be optimized and using a wider range of design specifications: constraints can be handled on the H_∞ form, but also as pole placement constraints. Moreover H_2 norm objectives can be taken into account.

This paper is organised as follows: in Section 2 the general model of the aircraft of interest is presented. In Section 3 the co-design optimization problem is set up. Then in Section 4 first results illustrating our approach are presented.

2 Problem Setup

In this section, the Flight Mechanics model of the Airbus Flying Wing model is presented. This paper only deals with longitudinal dynamics; however eventually both longitudinal and lateral dynamics will be considered, for sizing cases concerning the elevons are multi-axes maneuvers.

2.1 Flight Dynamics Equations

The longitudinal Flight Mechanics equations can be written in the aerodynamic reference system $R_a(x_a, y_a, z_a)$ (see Figure 1):

$$m\dot{V} = -\frac{1}{2}\rho V^2 S C_D - mg\sin(\gamma) + F \tag{1}$$

$$-mV\dot{\gamma} = -\frac{1}{2}\rho V^2 S C_L + mg\cos(\gamma) \tag{2}$$

$$B\dot{q} = \frac{1}{2}\rho V^2 S l C_m \tag{3}$$

$$\theta = \alpha + \gamma \tag{4}$$

where:

- concerning aircraft parameters, m and B denote the aircraft mass and inertia around the y_a axis respectively, S and l are reference surface and length, corresponding to the wing aera and mean aerodynamic chord respectively.
- angles are classically defined as follows: γ, α and θ denote the flight path angle, angle of attack and aircraft pitch attitude respectively.
- concerning aerodynamic parameters, V denotes the aerodynamic speed, ρ denotes the air density, and C_D, C_L and C_m denote the drag, lift and pitching moment coefficients respectively.

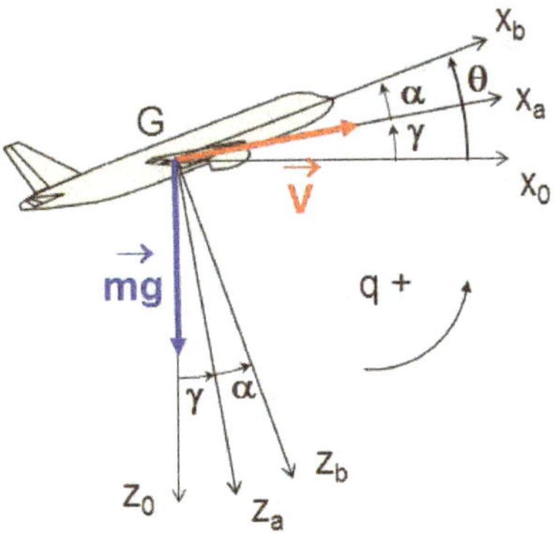

Fig. 1 Aerodynamic frame scheme.

Following classical assumptions on small angles approximations, this system of equations can be linearized and converted into the following state-space representation:

$$\begin{bmatrix} \delta\dot{V} \\ \delta\dot{\alpha} \\ \delta\dot{q} \\ \delta\dot{\theta} \end{bmatrix} = \begin{bmatrix} x_V & x_\alpha & x_q & x_\theta \\ -z_V & -z_\alpha & 1-z_q & 0 \\ 0 & m_\alpha & m_q & 0 \\ 0 & 0 & 1 & 0 \end{bmatrix} \begin{bmatrix} \delta V \\ \delta\alpha \\ \delta q \\ \delta\theta \end{bmatrix} + \begin{bmatrix} x_{\delta x} & x_{\delta m_i} \\ 0 & -z_{\delta m_i} \\ 0 & m_{\delta m_i} \\ 0 & 0 \end{bmatrix} \begin{bmatrix} \Delta\delta x \\ \Delta\delta m_i \end{bmatrix} \tag{5}$$

where the states are $\delta V, \delta\alpha, \delta q, \delta\theta$, which are the variations around an equilibrium of the airspeed, angle of attack, pitch rate, and attitude respectively. The different terms of the matrices are developped in the Appendix. Concerning the controls, $\Delta\delta x$ denotes the thrust command, and $\Delta\delta m_i$ denotes the i-th control surface command, the control surfaces architecture being developped in Section 2.2. In order to get the flight path angle γ as an output, and knowing the classical relation for longitudinal flight $\gamma = \theta - \alpha$, the output vector is chosen as follows:

$$\begin{bmatrix} \delta V \\ \delta\gamma \\ \delta\alpha \\ \delta q \end{bmatrix} = \begin{bmatrix} 1 & 0 & 0 & 0 \\ 0 & -1 & 0 & 1 \\ 0 & 1 & 0 & 0 \\ 0 & 0 & 1 & 0 \end{bmatrix} \begin{bmatrix} \delta V \\ \delta\alpha \\ \delta q \\ \delta\theta \end{bmatrix} \tag{6}$$

2.2 Control Surfaces Architecture

The control surfaces architecture deserves a special attention. The initial configuration of this flying wing includes five control surfaces on each side of the trailing edge. It has been demonstrated in previous studies [13] that in order to comply with maneuvrability specifications these surfaces should be multicontrol, i.e. elevons. Concerning yaw effectors, it has been shown [14] that control with crocodile flaps only is not satisfactory, and vertical surfaces are needed. The control surfaces general layout is visible on Figure 2. Using Figure 2 nomenclature, the general control vector is therefore: $u = [\Delta\delta x, LDQ1, \ldots LDQ5, RDQ1, \ldots RDQ5, LDR, RDR]'$, where $\Delta\delta x$ is the throttle command.

Now in the study presented here the two rudders are grouped as a single equivalent effector *DR*, for we are only concerned with longitudinal motion and control. Moreover in this paper we only deal with the control of the short period mode, as it will be explained in Section 2.3. Therefore we will not consider the throttle as a command. Including this control, for instance in order to design an autopilot for this aircraft, is a matter for future work. The final control vector considered in this study is finally: $u = [LDQ1, \ldots LDQ5, RDQ1, \ldots RDQ5, DR]'$. The control vector is eventually an 11th-dimension vector.

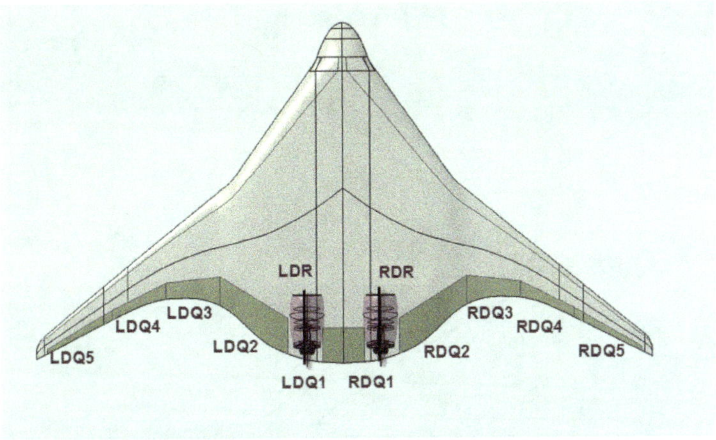

Fig. 2 Top view of the Airbus flying wing and its associated control surfaces architecture.

2.3 *Longitudinal Dynamics Analysis*

In this paper we focus on one particular challenge of this flying wing configuration: its longitudinal instability. Other issues, such as lateral instability, will be treated in a future work.

On an aircraft the x-wise relative position of the aerodynamic center w.r.t. the CG (center of gravity) is strongly linked to the aircraft longitudinal stability [18]. Now for aerodynamic planform optimization reasons, and due to the lack of any horizontal tail, on our flying wing configuration the aerodynamic center is located quite forward the CG. This leads to a consequent longitudinal instability, that needs to be controlled with active stabilization control laws.

Actually the short period mode leading to the longitudinal instability can be analyzed by extracting the $[\Delta\alpha, q]$ model from the state-space representation described in 2.1. The limit of stability associated to this subsystem occurs for a CG located at the so-called manoeuvre point, whose position is strongly linked to the aerodynamic center. A relative location of the manoeuvre point and CG can be found on Figure 3. The forward position of the manoeuvre point, especially in low speed, denotes a dynamic instability.

The poles of the complete longitudinal model are visible on Figure 4. The varying parameters are the aircraft mass, speed and altitude. The maximum unstable pole is obtained at low speed, light mass, and corresponds to a frequency of approximately $1.2\ rad/s$.

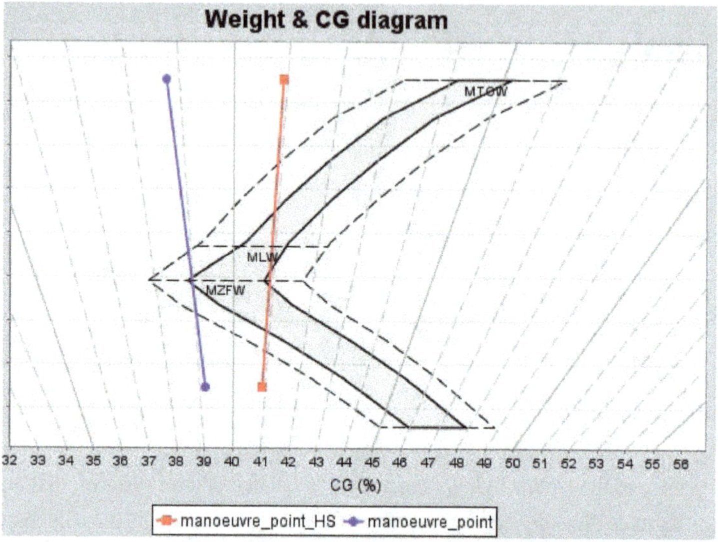

Fig. 3 Weight *&CG* diagram with manoeuvre point location for high and low speed.

3 Integrated Design and Control

In this section the general problem of integrated design and control is formulated. First the general equations are set in subsection 3.1, then this problem is adapted to our case in subsection 3.2.

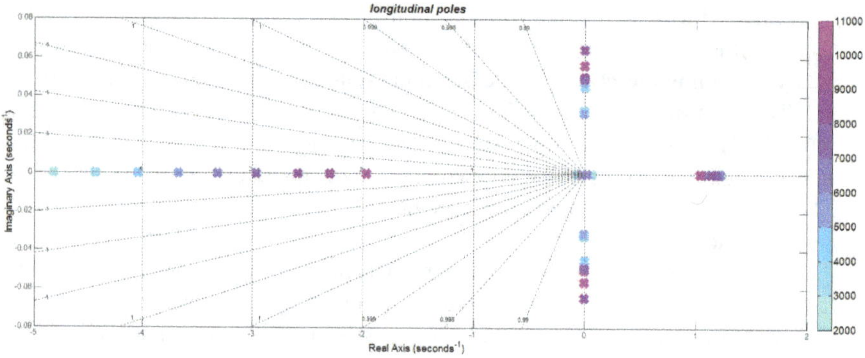

Fig. 4 Poles of the longitudinal model for M=0.55 and different altitudes. Altitude is given in meters. The complex conjugate poles correspond to the badly damped phugoid mode, which are easily controllable by an autopilot using the throttle command. The real modes correspond to the short period oscillation.

3.1 General Formulation

Following [6], the general combined plant / controller optimization problem is a multiobjective problem which can be stated on the form:

$$\begin{aligned} \min_{\xi,u(t),x(t),t_0,T} \quad & \max\{e(\xi),\Phi(x(t),T)+\int_{t_0}^{T} L(x(t),u(t),t)dt\} \\ \text{subject to}: & \; h(\xi)=0,\; g(\xi)\leq 0,\; \dot{x}=f(x(t),u(t),t,\xi), \\ & \eta(u(t),t,\xi)\leq 0,\; \psi(x(t),t)=0,\; x(t_0)=x_0 \end{aligned} \tag{7}$$

where ξ represents the physical parameters to be optimized, x are the states of the system to be controlled, u is the command vector.t_0 and T are the initial and final times, respectively.L is a controller cost functional. $h(\xi)$ and $g(\xi)$ represent respectively the equality and inequality constraints on physical parameters, Φ and ψ are final state objective and constraint respectively, and $e(\xi)$ is the objective function associated to the physical parameters. Finally η represents constraints on command vector. For an exhaustive description of the general optimal control problem formulation, please refer to [7].

3.2 Problem Specificities

In this section the previously described plant/controller optimization problem is adapted to our specific use-case.

- First of all for a fixed flight point and fixed physical parameters the state-space representation of the aircraft is supposed linear time-invariant with state feedback —indeed at this pre-sizing conceptual stage all the states are supposed known:

$$\begin{aligned} \dot{x}(t) &= A(\xi)x(t)+B(\xi)u(t) \\ y &= x \end{aligned} \tag{8}$$

- Moreover the structure of the compensator is restricted to a static state-feedback:

$$u=-Kx \tag{9}$$

- The cost function associated to the control objective is a linear quadratic, infinite-horizon criterion:

$$\min_{u(t)} J(u)=\frac{1}{2}\int_{0}^{+\infty}[x^{T}Qx+u^{T}Ru]dt \tag{10}$$

  Such a criterion aims at minimizing the energy of the states x and of the control input u, weighted by two matrices Q and R respectively.
- Constraints on physical variables are of the form:

$$\forall i \in [1; n_{param}],\ \xi_{i_{min}} \leq \xi_i \leq \xi_{i_{max}} \tag{11}$$

$$\sum_{i=1}^{n_{param}} c_i \xi_i = C \tag{12}$$

For instance, if physical parameters include control surfaces sizes, constraints on these parameters would be that the total span of the control surfaces should not exceed the aircraft total span. In the example described in this paper such constraints will include bounds on the actuators bandwidth (see Equation 17). Indeed we are focusing in this work on an actuator bandwidth optimization, however future work will also include control surfaces sizing.

- Following previous work on integrated aircraft/controller design [11], constraints on control law performance are specified as follows:
  - Closed-loop stability.
  - Sufficient stability margins.
  - Adequate Handling Qualities performance. This may include sufficient stability, and appropriate maneuverability in order to comply with certification maneuvers.

To our knowledge, most plant/ controller optimizations applied to aerospace problems in the past years have been using the LMI framework [12] [19][8] [16]. In this approach all handling qualities constraints are cast as H_∞-norms of a transfer functions, e.g. $\|W^{-1}(s)T_{w\to z}\| \leq 1$ means that the closed-loop transfer $T_{w\to z}$ fits the frequency domain template $W^{-1}(s)$. The main interest of this approach lies in the fact that when correctly translated into an LMI formulation, the problem can be convexified and solved through dedicated LMI solvers.

However such a formulation quite differs from usual specifications used by engineers for defining acceptable handling qualities. Most of the time these specifications are expressed in terms of modes characteristics, such as minimal damping or frequency. Moreover controllers found by LMI solvers are unstructured and full-order. This means that these controllers have a n-th order internal dynamics, n being the order of the plant to control. This is not acceptable for our problem, for aircraft closed-loop characteristics should be obtained at the conceptual design stage with rather simple controllers, such as pitch and yaw dampers.

Therefore we propose a different approach: solve the plant/controller problem through nonsmooth optimization techniques for control synthesis developped in [4] and applied in [3]. This has two main advantages: on one side the possibility to tune fixed low-order controllers compliant with industrial applications and on the other side greater possibilities concerning closed-loop requirements. Two kinds of specifications available with these tools are of particular interest for our study: (a) the possibility to specify characteristics on closed-loop poles such as minimum damping, frequency or decay, and (b) the possibility to specify a multiobjective problem of the form H_2/H_∞. The H_∞ channel accounts for performance, and the H_2 channel minimizes the energy to control the aircraft. In the next section we will develop this approach and the first results it enables for co-design.

4 First Results

In this part the process we have set up is explained and first results are shown.

4.1 Control Problem Setup

According to the longitudinal model presented in 2.3, we focus on the flight point where the instability is maximal, corresponding to an unstable pole of frequency approximately 1.2 rad/s.

Following results presented in [1], we choose to set the H_∞ problem as a weighting on the acceleration sensitivity function. A disturbance w acts on the pitch acceleration $\dot{q}$. The desired behaviour for this acceleration is specified through a weighting function of the form $W_1 = \frac{s^2+2\xi\omega s+\omega^2}{s^2}$ with appropriate values of ξ and ω. The expected behaviour for $\dot{q}$ is therefore of the second-order form, and the shape of desired disturbance rejection profile on the acceleration W_1^{-1} is plotted on Figure 5.

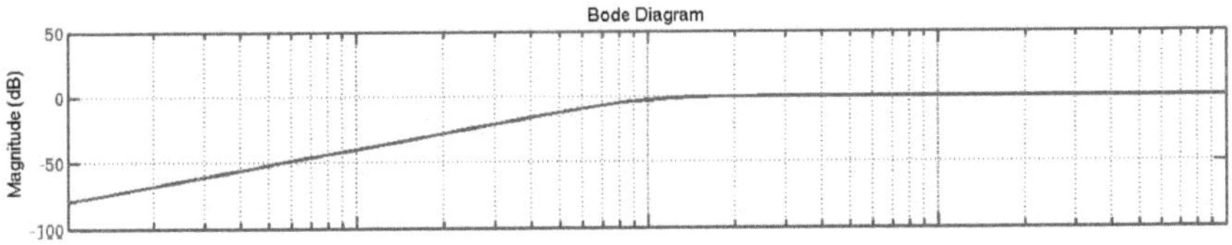

Fig. 5 Desired disturbance rejection profile on pitch acceleration W_1^{-1}.

The non-detectable double integrators $\frac{1}{s^2}$ of the weight W_1 can be removed by computing a minimal realization of the whole standard form.

The H_2/H_∞ problem is then set as follows: the H_∞ constraint is $||T_{w\to z_{inf}}||_\infty \le \gamma_\infty$, with γ_∞ being a value slightly above 1; in practice we will assume $\gamma_\infty = 1.5$. z_{inf} is the output channel defined on Figure 6. This physically means that a sinusoidal perturbation of any frequency on the pitch acceleration shall not perturbate the closed-loop acceleration of a factor more than γ_∞. This channel shapes the pitch acceleration desired closed-loop behaviour. Then a minimization of the energy used to control the aircraft is performed through minimizing the following objective function: $||T_{w\to u}||_2$. This problem setup is depicted on Figure 6. Finally the controller is assumed to be a state-feedback static compensator such as: $u = -Kx$ with $x = [\alpha\ q\ \theta]^T$. K is therefore a 11×3 matrix, for there are 11 controls (see Section 2.2).

The initial control problem may then be written of the form:

$$\min_K \quad ||T_{w\to u}||_2 \tag{13}$$

$$\text{subject to:} \quad ||T_{w\to z_{inf}}||_\infty \le \gamma \tag{14}$$

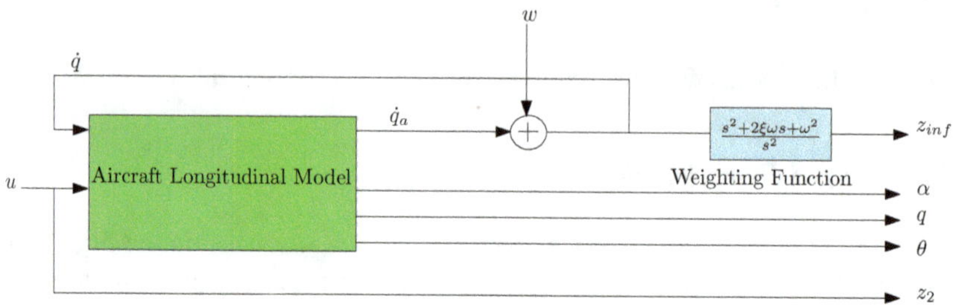

Fig. 6 H_2/H_∞ problem standard form with aircraft model (green) and weighting function (blue).

4.2 Implementation of the Co-design Problem

Now that the initial control problem is well defined, let us introduce some meaningful physical variables which should be optimized conjointly with the controller in order to solve the control problem. The underlying question is: how far is it possible to control our aircraft, and at which expense on the control system? A good rule of thumb, which may be found in [17], states that the actuators bandwidth to control the system should be at least ten times above the fastest mode. We aim at a more precise condition, and finally we should be able to evaluate the gains or penalties of considering an unstable aircraft configuration at a conceptual design phase.

In this paper we introduce actuators dynamics as first-order low-pass filters, which bandwidths are design parameters. More precisely, the 11 actuators blocks are modelled as follows:

$$\frac{y_{act}}{u_{act}}(s) = \frac{\omega_i}{\omega_i + s}, \; i = 1...11 \tag{15}$$

where y_{act} and u_{act} are the actuators outputs and inputs respectively, as defined on Figure 7. $\Omega = [\omega_1, \dots \omega_{11}]^T$ is the vector of design parameters. In order to minimize these bandwidths (one needs the actuators to be as slow as possible: fast actuators mean high required energy, heavier and bigge actuators), a cost function is also added to the previous problem. An H_2-norm of the derivative of the actuators outputs is chosen: $||T_{u_{act} \to \dot{y}_{act}}||_2$. In the future a more physically meaningful function, such as the effectors kinematic energy, could be chosen.

The previously described control problem then becomes:

$$\min_{K,\Omega} \max \; \{W_2 ||T_{w \to u}||_2, W_3 ||T_{u_{act} \to \dot{y}_{act}}||_2\} \tag{16}$$

$$\text{subject to: } \; ||T_{w \to z_{inf}}||_\infty \leq \gamma, \; 0 \leq \Omega \leq \Omega_{max} \tag{17}$$

where W_2 and W_3 are weightings associated to each objective function, and $0 \leq \Omega \leq \Omega_{max}$, which should be understood element-wise, specifies bounds on the actuators bandwidths. This problem has $3 \times 11 + 11 = 44$ variables.

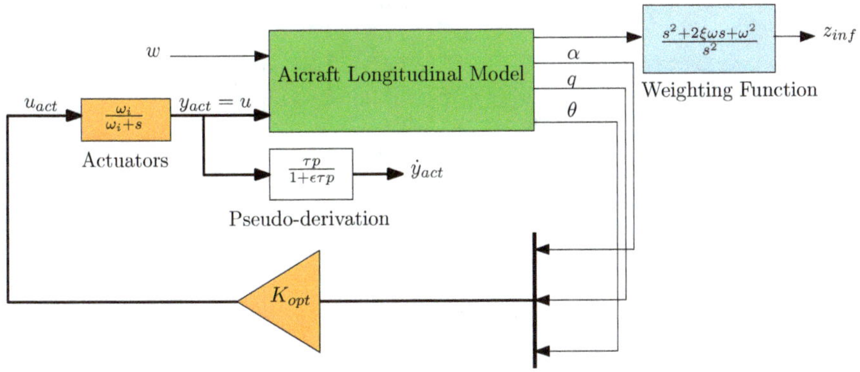

Fig. 7 Closed-loop problem for co-design approach. Tunable blocks are coloured in orange.

From an implementation point of view, it has been chosen to work with the `systune` and `slTunable` [10] routines for several reasons:

- It allows for mixed H_2/H_∞ synthesis and multiobjective optimization.
- It allows for structured parameters, such as fixed-order controllers and first-orders filters, which is appreciable for our problem. Moreover bounds on the variables are easily applicable.
- The `slTunable` syntax allows for directly specifying the closed-loop structure, as well as the tunable blocks and their structure. Moreover it allows dealing with a single `Simulink` model for synthesis and simulation, as it is performed on Figure 7.
- The constraints specifications are not limited to specifications on frequencies, but may also handle pole placement constraints. This may be more suitable to Handling Qualities purpose and will be adressed in a future work.

4.3 First Co-design Results

In this section the first results of the co-design approach are presented. A first synthesis is performed with variable bandwidths initialized to 20 rad/s, as well as with random initializations. The upper bound on the bandwidths is chosen to be also at 50 rad/s. The following controller is obtained (corresponding physical channels are shown as a reminder):

$K_{opt} =$				
	0.5816	1.1211	6.0981	*LDQ1*
	1.3919	2.2578	14.5759	*LDQ2*
	1.0279	1.7584	10.5422	*LDQ3*
	0.9229	1.6367	10.0334	*LDQ4*
	0.2329	0.3630	1.5500	*LDQ5*
	0.5824	1.1203	6.0976	*RDQ1*
	1.3904	2.2612	14.5797	*RDQ2*
	1.0262	1.7594	10.5529	*RDQ3*
	0.9235	1.6349	10.0309	*RDQ4*
	0.2287	0.3664	1.5525	*RDQ5*
	-0.0017	0.0028	0.0021	*DR*
	α	q	θ	

At first sight, this controller looks rather consistent for:

- All pairs of elevons are commanded symmetrically (1 with 6, 2 with 7, etc), which seems obvious for a longitudinal kinematics, but which had not been specified as a particular structure for the controller.
- Gains have the right signs: for instance, for a positive $\delta\alpha$, one needs to deflect the elevons downwards, therefore positively.
- Magnitude of the gains follows elevons respective efficiencies. For instance, the second elevon has the largest surface, and it can be shown (see Figure 9) that it has the largest longitudinal efficiency. Therefore the controller chooses to use it accordingly.
- The rudder — last row — is set to almost zero.

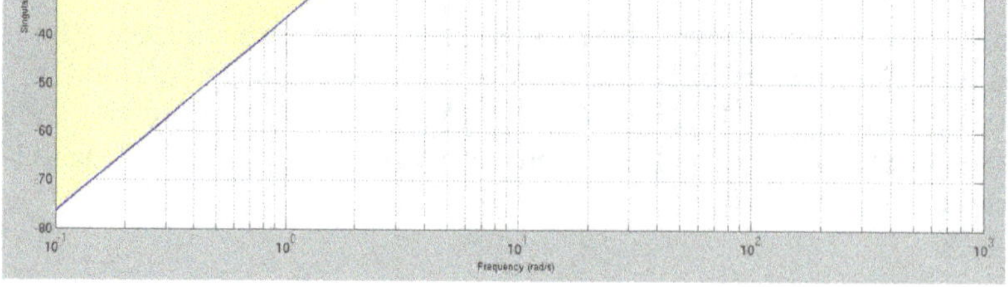

Fig. 8 Frequency-domain response of $T_{w\to z_{inf}}(K_{opt})$ (blue) and W_1^{-1} (coloured)

The optimal H_2 control objective is $||T_{w\to u}||_2 = 1.38$. As a comparison the optimal value given by an LQ synthesis for minimum energy control and infinite actuator bandwidth is 0.6, and the mixed H_2/H_∞ problem with also infinite actuator bandwidth gives 1.26. Therefore adding actuators dynamics only slightly increase the objective. The closed-loop sensitivity function frequency response w.r.t. the desired weighting function is visible on Figure 8.

Then the actuators bandwidth, which are also tunable parameters, are also tuned in an interesting way. In order to meet the requirements, most bandwidths tend to be increased, however different bandwidths are allocated to the different control surfaces. The results can be found on Figure 9.

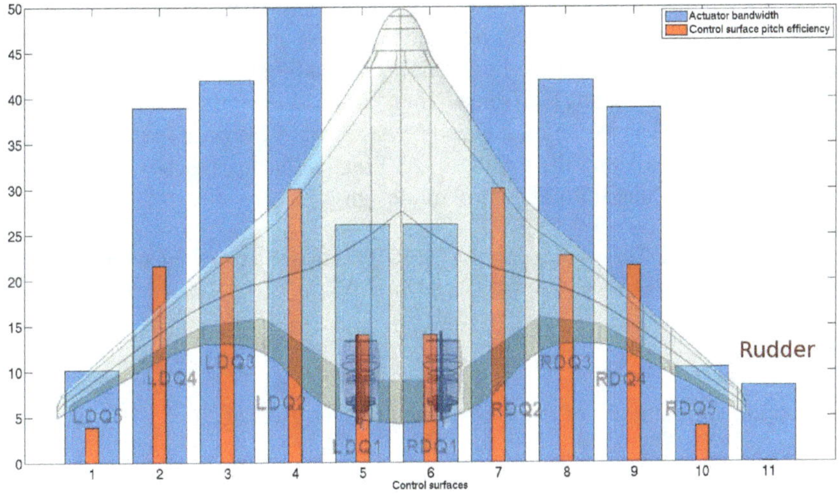

Fig. 9 Actuators tuned bandwidths (blue) and elevons pitch efficiencies (red).

Similarly to what was found for controller gains, bandwidths are tuned symetrically for symetrical elevons, which is physically consistent for longitudinal motion. Then the optimizer tends to use in a more powerful way — thus allocating more bandwidth— control surfaces that are the most efficient in creating pitching rate. For comparison purpose we plotted on Figure 9 control surfaces pitch efficiencies, corresponding to the the pitch rate row in the control matrix B. Finally for the most effective elevon —the second one, which is also the largest one— the upper bound of 50 rad/s is achieved; that tends to indicate that meeting the requirements is challenging on this configuration.

To conclude this part, it should be mentioned that here all actuators bandwidths were set as free variables and their symetrical tuning was checked afterwards only for method validation purpose. Symetrical results tend to indicate a consistent behaviour of the optimization. However, as in reality identical actuators are expected for symetrical control surfaces, further work should impose this constraint instead of expecting it as an output. As a result the complexity of the problem would be reduced.

5 Conclusion and Perspectives

In this work, a new method to integrated design and control is proposed. This method relies on nonsmooth optimizations techniques, and is applied to an unstable civil flying wing for longitudinal control. A mixed H_2/H_∞ synthesis is performed and actuators bandwidths are tuned simultaneously. This approach appears to be promising for conceptual aircraft design because of the diversity of criteria it may handle, and the fast calculations it allows. Future work may include different kinds of requirements, such as pole placement constraints, which is more suitable to handling qualities requirements. Then more "physical" objective functions, such as energy or mass minimization, should be handled. Also commonality of actuators for identical control surfaces will be imposed, hence leading to a simplification of the problem without loosing physical sense. Then future *co-design* will include not only actuators bandwidth as variables, but also physical control surfaces parameters such as their relative span along the trailing edge. Finally, this approach will be applied to longitudinal and lateral flight control co-design.

Acknowledgements. This work is part of a CIFRE PhD thesis in cooperation between ISAE and the Future Projects Office of Airbus Opérations SAS.

Appendix

Developping the elements of the state-space matrices gives:

$$
\begin{aligned}
x_V &= \tfrac{-\rho VSC_x}{m} + \tfrac{\partial F}{\partial V}, & x_\alpha &= -2gkC_{L\alpha},\\
x_q &= \tfrac{-2gLk}{V} C_{L_q}, & x_\theta &= -g,\\
z_V &= \tfrac{-2g}{V^2}, & z_\alpha &= \frac{\rho VS}{2m} C_{L\alpha},\\
z_q &= \tfrac{\rho SL}{2m} C_{L_q}, & m_\alpha &= \frac{\rho V^2 S}{B} C_{m\alpha},\\
m_q &= \tfrac{\rho VSL^2}{2B} C_{m_q}, & x_{\delta x} &= \frac{1}{m}\frac{\partial F}{\partial \delta x},\\
x_{\delta m_i} &= -2gkC_{L_{\delta m_i}}, & z_{\delta m_i} &= \frac{\rho VS}{2m} C_{L_{\delta m_i}},\\
m_{\delta m_i} &= \tfrac{\rho V^2 SL}{2B} C_{m_{\delta m_i}}
\end{aligned}
$$

References

1. Alazard, D.: Reverse Engineering in Control Design. John Wiley & Sons (2013)
2. Alazard, D., Loquen, T., de Plinval, H., Cumer, C.: Avionics/control co-design for large flexible space structures. In: AIAA Guidance, Navigation, and Control (GNC) Conference, Guidance, Navigation, and Control and Co-located Conferences. American Institute of Aeronautics and Astronautics (2013), `http://dx.doi.org/10.2514/6.2013-4638`

3. Apkarian, P.: Tuning controllers against multiple design requirements. In: 2012 16th International Conference on System Theory, Control and Computing (ICSTCC), pp. 1–6. IEEE (2012)
4. Apkarian, P., Noll, D., Rondepierre, A., et al.: Nonsmooth optimization algorithm for mixed h2/h∞ synthesis. In: Proc. of the 46th IEEE Conference on Decision and Control, pp. 4110–4115 (2007)
5. Bansal, V., Ross, R., Perkins, J., Pistikopoulos, E.: The interactions of design and control: double-effect distillation. Journal of Process Control 10(2-3), 219–227 (2000)
6. Fathy, H., Reyer, J., Papalambros, P., Ulsov, A.: On the coupling between the plant and controller optimization problems, vol. 3, pp. 1864–1869 (2001)
7. Leitmann, G.: The calculus of variations and optimal control, vol. 24. Springer (1981)
8. Liao, F., Lum, K.Y., Wang, J.L.: An LMI-based optimization approach for integrated plant/output-feedback controller design. In: Proceedings of the 2005 American Control Conference, pp. 4880–4885. IEEE (2005)
9. Liebeck, R.: Design of the blended-wing body subsonic transport. 2005-06. von Karman Institute for Fluid Dynamics (2005)
10. MATLAB: version 2013a. Robust Control Toolbox. The MathWorks Inc., Natick, Massachusetts, USA (2013)
11. Niewhoener, R.J., Kaminer, I.: Linear matrix inequalities in integrated aircraft/controller design. In: Proceedings of the 1995 American Control Conference, vol. 1, pp. 177–181. IEEE (1995)
12. Niewoehner, R., Kaminer, I.: Integrated aircraft-controller design using linear matrix inequalities. Journal of Guidance, Control, and Dynamics 19(2), 445–452 (1996)
13. Saucez, M.: Handling qualities of the airbus flying wing resolution. PhD thesis, ISAE-Airbus, Toulouse (2013)
14. Saucez, M., Boiffier, J.L.: Optimization of engine failure on a flying wing configuration. AIAA, Minneapolis (2012)
15. Silvestre, C., Pascoal, A., Kaminer, I., Healey, A.: Plant/controller optimization with applications to integrated surface sizing and feedback controller design for autonomous underwater vehicles (AUVs) (1998)
16. Sridharan, S., Echols, J.A., Rodriguez, A.A., Mondal, K.: Integrated design and control of hypersonic vehicles. In: American Control Conference (ACC), pp. 1371–1376. IEEE (2014)
17. Stein, G.: The practical, physical (and sometimes dangerous) consequences of control must be respected, and the underlying principles must be clearly and well taught. IEEE Control Systems Magazine 272(1708/03) (2003)
18. Taquin, G.: Flight mechanics- master of science supaero (2009)
19. Yang, G.H., Lum, K.Y.: An optimization approach to integrated aircraft-controller design. In: Proceedings of the 2003 American Control Conference, vol. 2, pp. 1649–1654. IEEE (2003)

Structured Control Law Design and Robustness Assessment for the Automatic Launch of Small UAVs

Jan Bolting, Jean-Marc Biannic, and François Defaÿ

Abstract. Automatic launch is an important capability towards the truly autonomous flight of Unmanned Aerial Vehicles (UAVs) that does not require the presence of an expert pilot, as it is often the case today. In this work a complete approach to the design and robustness assessment of a set of control laws for the automatic launch of fixed-wing UAVs is presented.

The proposed control system consists of an airspeed tracking loop and a nested lateral guidance loop. Important nonlinearities such as actuator saturations and signal delays are taken into account for controller synthesis and robustness evaluation. Due to the high risk inherent to flight testing the launch phase, extensive Monte Carlo simulations over the space of model uncertainties and initial launch conditions have been performed on the nonlinear model of a flying-wing type UAV, including atmospheric turbulence. Time consuming Monte Carlo simulations are complemented by testing for robust stability and identifying worst-case performance configurations using Structured Singular Value (μ) analysis methods.

1 Introduction

Automatic take off is an important capability towards the fully automatic flight of Unmanned Aerial Vehicles (UAVs). It reduces the need for the presence of a skilled

Jan Bolting · François Defaÿ
Institut Supérieur de l'Aéronautique et de l'Espace (ISAE)
10 Avenue Edouard Belin 31055 Toulouse, France
e-mail: {jan.bolting,francois.defay}@isae.fr

Jean-Marc Biannic
Office National d'Études et de Recherches Aérospatiales (ONERA)
2 Avenue Edouard Belin, 31055 Toulouse Cedex 4, France
e-mail: Jean-Marc.Biannic@onera.fr

J. Bordeneuve-Guibé et al. (eds.), *Advances in Aerospace Guidance, Navigation and Control*, DOI: 10.1007/978-3-319-17518-8_28

pilot and thus the overall operating costs and enhances usability. At the same time, it is one of the most critical mission phases, since - being executed at low altitudes - insufficient control performance easily leads to the loss of the aircraft and total mission failure.

The launch maneuver is generally characterized by rapidly changing dynamics due to variations in dynamic pressure during the initial acceleration and the transition to the desired launch airspeed. Furthermore the low-cost approach and shorter development cycles of small UAVs tend to lead to dynamic models of lower quality, since generally less time and effort is invested in system identification or wind tunnel campaigns, in addition to low-quality on board sensors. This poses an additional challenge for any model-based control design technique, which should take into account model uncertainties.

While control laws for flight maneuvers in higher altitudes can be tested in relative safety without major danger to the vehicle or operating personnel, unsuccessful tests of automatic launches are prone to be associated with higher costs due to damage or loss of the vehicle. That being said, for the launch problem control techniques are desirable that enable the control designer to estimate strong performance bounds and stability guarantees beforehand, thus minimizing the risk encountered when performing the first flight tests.

Linear control techniques provide the desirable property that with the framework of μ analysis a computationally cheap means of estimating worst-case bounds on performance and stability is available.

Among modern linear control techniques, H_∞ methods provide some features that make them the method of choice for this application. Performance requirements as well as the suppression of exogenous disturbances - such as turbulence - and important constraints such as actuator rate and deflection saturations can be incorporated in a systematic way into the controller synthesis. H_∞ control techniques have been successfully applied before to the problem of controller synthesis for fixed-wing UAVs.

The authors of [1] present attitude controllers for a ship-launched UAV based on H_∞ and μ synthesis methods. In [2], a single-loop altitude tracking controller is designed with both H2 and H_∞ methods. The authors of [3] use H_∞ methods to design baseline launch trajectory controllers for a ship-launched UAV.

In this article, a complete approach to the design and robustness verification by simulations of a set of control laws for the automatic launch of a small fixed-wing UAV is presented.

Section 2 gives a short overview of H_∞ synthesis and Structured Singular Value analysis techniques. Section 3 introduces the dynamic models used in this work. In section 4, the reasoning behind the chosen control structure is presented, followed by details of the controller synthesis in section 5. The robustness of the closed loop system is evaluated by μ analysis techniques, presented in section 6, as well as Monte Carlo simulations, presented in section 7. Section 8 provides an overview of the results and future work.

2 Review of H_∞ Synthesis and μ Analysis

This section gives a short overview of H_∞ controller synthesis and the Structured Singular Value μ used for robustness analysis.

2.1 H_∞ Controller Synthesis

The application of H_∞ control techniques to the multi-objective tuning of MIMO systems is rooted in the modern control paradigm. As depicted in fig. 1, the control problem is rearranged into an augmented plant model $\mathbf{P}(s)$ and the controller $\mathbf{K}(s)$, driven by the plant outputs $\mathbf{y}$ and driving the control inputs $\mathbf{u}$. The remaining inputs to the closed loop system, such as disturbances, reference inputs and measurement noise, are combined into the generalized vector of exogenous inputs $\mathbf{w}$. Signals whose smallness can be interpreted as performance indicator, such as control activity or deviations from a desired reference model, form the generalized output vector $\mathbf{z}$. Doing so, the task of any controller design algorithm reduces to the task of ensuring the smallness of $\mathbf{z}$ by adjusting $\mathbf{K}(s)$ in the presence of disturbance inputs $\mathbf{w}$. Different measures of smallness are available, the most popular ones being the H_2 (equivalent to LQG techniques in the time domain) and H_∞ norm. The H_∞ norm of the closed loop transfer function matrix $\mathbf{T} = \mathscr{F}_l(\mathbf{P}(s), \mathbf{K}(s))$ is defined as

$$||\mathbf{T}(s)||_\infty = \sup_\omega \bar{\sigma}(\mathbf{T}(j\omega)) \tag{1}$$

Thus the H_∞ norm of a transfer matrix is the largest value of the maximum singular value over all frequencies, i.e. the highest amplification of a signal in $\mathbf{w}$ induced by all possible input directions over all frequencies.

By using appropriate, possibly frequency-dependent weighting functions applied to the exogenous inputs $\mathbf{w}$ and the outputs $\mathbf{z}$, the minimization of $||\mathbf{T}(s)||_\infty$ leads to a controller $\mathbf{K}(s)$ that provides a trade-off between conflicting performance requirements, e.g. reference tracking for low frequencies and measurement noise suppression for higher frequencies. Usually scaling is applied to normalize inputs and outputs, leading to a performance criterion of

$$||\mathbf{T}(s)||_\infty \leq 1 \tag{2}$$

2.1.1 Fixed-Structure H_∞ Controller Synthesis

Recent advances in non-smooth optimization (see e.g. [4]) have enabled the computation of fixed-structure H_∞ controllers, i.e. the controller structure can be defined beforehand e.g. as a function of the available feedbacks, which might be defined by the UAS' onboard sensor suite.

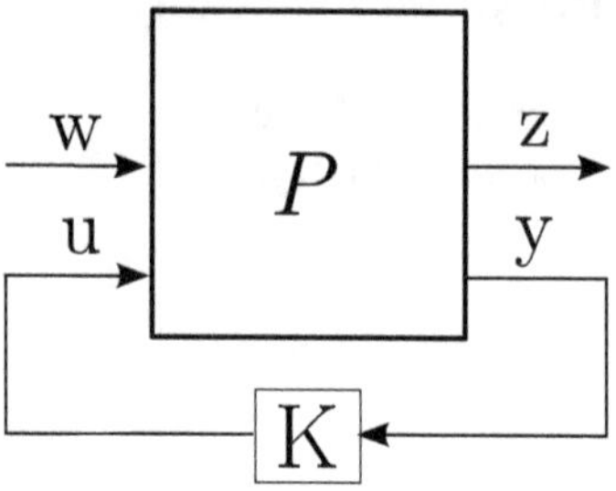

Fig. 1 General synthesis structure

2.2 The Structured Singular Value

The Structured Singular Value μ provides an analytical tool to evaluate the stability and performance of linear systems subject to time-invariant uncertainties. Consider the uncertain closed loop system in fig. 2a with the uncertainties pulled out of the system by an upper Linear Fractional Transformation (LFT) into the block-diagonal matrix Δ. The closed loop transfer function matrix $\mathbf{T}(s)$ can then be partitioned as in eq. (3)

$$\mathbf{T}(s) = \begin{bmatrix} \mathbf{T}_{11}(s) & \mathbf{T}_{12}(s) \\ \mathbf{T}_{21}(s) & \mathbf{T}_{22}(s) \end{bmatrix} \tag{3}$$

leading to a transfer function from **w** to **z** of

$$\frac{\mathbf{z}}{\mathbf{w}} = [\mathbf{T}_{22}(s) + \mathbf{T}_{21}(s)\Delta(I - \mathbf{T}_{11}(s)\Delta)^{-1}\mathbf{T}_{12}(s)] \tag{4}$$

Since the the submatrix $\mathbf{T}_{22}(s)$ represents the nominal system dynamics, nominal stability and performance is guaranteed by

$$||\mathbf{T}_{22}(s)||_\infty \leq 1 \tag{5}$$

i.e. the standard H_∞ synthesis problem.

The contribution of Δ is covered by the second term of the right side of eq. (4). For some Δ, $(I - F_{11}\Delta)^{-1}$ might become singular, leading to instability.

The Structured Singular Value μ provides a measure for robust stability, taking into account the structure of Δ, by finding the largest Δ ($||\Delta||_\infty$ is used as measure of size) that leads to instability. It is defined as

$$\mu(\mathbf{M}(s)) = \frac{1}{min\{||(\Delta)||_\infty \in \boldsymbol{\Delta}, \det(\mathbf{I} - \mathbf{M}(s)\Delta) = 0\}} \tag{6}$$

where $\mathbf{M}(s)$ is a matrix of transfer functions and $\boldsymbol{\Delta}$ is the set of admissible block-diagonal perturbations Δ.

Thus μ is the reciprocal of the largest singular value of the Δ that leads to instability. Since Δ itself is norm-bounded to $\|\Delta\|_\infty \leq 1$, a μ inferior to 1 implies that there is no $\Delta \in \boldsymbol{\Delta}$ that causes instability. Thus the condition for robust stability becomes

$$\mu(\mathbf{T}_{11}(s)) \leq 1 \tag{7}$$

By adding a virtual additional norm-bounded structure Δ_2 between the exogenous inputs $\mathbf{w}$ and the outputs $\mathbf{z}$ (see fig. 2b), the robust performance problem an be cast into a robust stability problem, using the same measure of μ, leading to the condition

$$\mu(\mathbf{T}(s)) \leq 1 \tag{8}$$

for robust performance and stability (see [5, 6] for a detailed derivation). Since no closed-form solution for the computation of μ is available so far, it is approximated by its upper and lower bounds by numerical methods. In this work, the Matlab Toolbox SMART developed at ONERA ([7]) has been used.

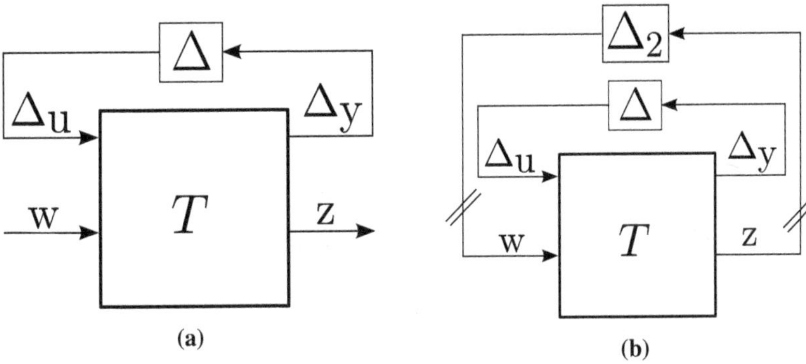

Fig. 2 General analysis structure for a) robust stability and b) robust performance

3 Vehicle Model

For the purposes of this work, the model of a small ($b = 1.42m$) flying-wing type UAV published in [8] has been used. Its lack of separate elevator and rudder control surfaces makes it largely under-actuated and as such more challenging to control. This configuration is fairly popular especially in the research community due to its simplicity and structural robustness.

For controller synthesis and linear analysis, the nonlinear model is linearized about the steady state launch state. Only relevant states are considered, leading to reduced longitudinal (eq. (9)) and lateral (eq. (10)) state space models.

Fig. 3 Zagi UAV, from [8]

$$\begin{aligned} \left(\dot{V}_a\ \dot{\alpha}\ \dot{\theta}\ \dot{q}\right)^T &= \mathbf{A}_{\text{lon}}\left(V_a\ \alpha\ \theta\ q\right)^T + \mathbf{B}_{\text{lon}}\left(\delta_e\ \delta_{en}\ \mathbf{v}_w\ \omega_w\right)^T \\ \left(V_a\ \alpha\ \theta\ q\right)^T &= \mathbf{C}_{\text{lon}}\left(V_a\ \alpha\ \theta\ q\right)^T + \mathbf{D}_{\text{lon}}\left(\delta_e\ \delta_{en}\ \mathbf{v}_w\ \omega_w\right)^T \end{aligned} \tag{9}$$

$$\begin{aligned} \left(\dot{y}\ \dot{v}_y\ \dot{\beta}\ \dot{\phi}\ \dot{p}\ \dot{r}\right)^T &= \mathbf{A}_{\text{lat}}\left(y\ v_y\ \beta\ \phi\ p\ r\right)^T + \mathbf{B}_{\text{lat}}\left(\delta_a\ v_w\ \omega_w\ \right)^T \\ \left(y\ v_y\ \phi\ p\ r\right)^T &= \mathbf{C}_{\text{lat}}\left(y\ v_y\ \beta\ \phi\ p\ r\right)^T + \mathbf{D}_{\text{lat}}\left(\delta_a\ v_w\ \omega_w\ \right)^T \end{aligned} \tag{10}$$

3.1 Actuator Model

An actuator of type Hitec HS-55 has been identified in the frame of previous work and is approximated as second order system

$$\frac{\delta_i}{\delta_{i,c}} = \frac{1}{2.53 \cdot 10^{-4} s^2 + 0.025s + 1} \tag{11}$$

and is assumed to be constrained by deflection and rate limits of $\pm\frac{\pi}{4}$ and $\pm 5.8\frac{rad}{s}$ respectively. Actuator saturations are a fundamental problem in linear control, since the break linearity and can lead to instability if the control input to the plant is no longer the same as that demanded by the controller. See section 5.3 for how saturations are taken into account for controller synthesis.

3.2 Delay Model

The combined computing and signal transport delay has been approximated by a 3rd order Padé approximation. The $R_{n-1,n}(s)$ form proposed in [9] is preferred over the common equal nominator/denominator form (for instance implemented in Matlab's *pade.m*), for its smaller approximation error and better behavior close to $t = 0$ (see fig. 4).

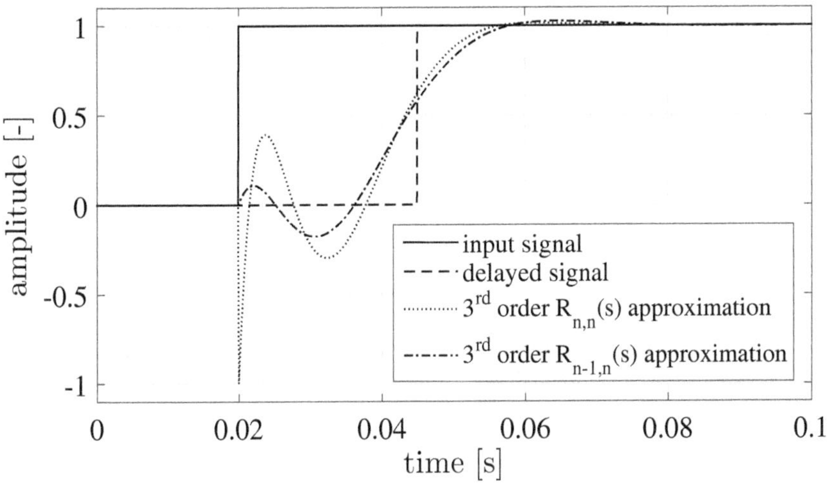

Fig. 4 Third order $R_{n,n}(s)$ and $R_{n,n}(s)$ Padé approximations of a time delay of 0.025 s. Note the spike at $t = 0$ in the more common $R_{n,n}(s)$ approximation and the smoother behavior of the $R_{n-1,n}(s)$ approximation. The author of [9] shows that the RMS error between the actual delayed signal and the $R_{n-1,n}(s)$ form is smaller as well.

4 Control Structure Design

The lateral and longitudinal dynamics are considered as uncoupled and dealt with separately, as is common practice. From an automatic launch controller it is generally expected to maintain a positive climb rate without violating the minimal and maximum airspeed, while keeping the vehicle inside some defined launch corridor e.g. to avoid ground obstacles such as trees. The longitudinal constraints are most simply satisfied by controlling the airspeed using the elevator while keeping the throttle at some fixed launch preset.

Regarding lateral guidance, proportional navigation is known to work well for small deviations from the operating point (see e.g. [10]). Since the UAV is launched on the desired launch vector, the initial lateral deviation is zero, thus proportional navigation is applied to maintain the lateral position on the launch vector.

H_∞ controllers have been used both in nested control loop setups (e.g. [11])) and as single-loop controllers ([2]). Nested control loops provide well known practical advantages, such as simple testing by successive loop closure and re-usability of inner loop controllers. On the other hand, our experiences suggest that development complexity is significantly increased. Nested control loops rely on the principle of timescale separation, i.e. the separation of subsystems with slower (e.g. position) and significantly faster states (e.g. attitude). Taking a closer look at the timescales at work in the linearized longitudinal dynamics (see section 3), one sees that the fast states that have a major impact on the controlled state V_a are the AoA α and the pitch angle θ. The elevator acts on angular accelerations $\dot{q}$, which are integrated to q and subsequently to Θ. For short periods of time, the flight path angle γ can be considered as constant, thus

$$\dot{\alpha} \propto \dot{\theta} \tag{12}$$

since

$$\alpha = \theta - \gamma \tag{13}$$

Thus δ_e acts on V_a via the chain of 3 integrators $\dot{q}(\delta_e) \xrightarrow{\int} q \xrightarrow{\int} (\theta, \alpha) \rightarrow \dot{V}_a \xrightarrow{\int} V_a$.

In the lateral case, the lateral acceleration $\ddot{y} = \dot{v}_y$ is mainly driven by ϕ, i.e. by inclining the lift vector. Thus δ_a acts on y via the chain of 4 integrators $\dot{p}(\delta_a) \xrightarrow{\int} p \xrightarrow{\int} \phi \rightarrow \dot{v}_y \xrightarrow{\int} v_y \xrightarrow{\int} y$.

That being said, for airspeed control a single controller acting on the elevator deflection δ_e as control input has been designed, since timescale separation is less strong than in the lateral case.

Early simulations showed a large overshoot in the airspeed loop when the engine is enabled. To improve the reaction time of the airspeed loop, the throttle signal is added as additional input to the controller. To reduce steady state tracking errors, an integral term is included in the control law, leading to a control gain $\mathbf{K} \in \mathbb{R}^{1\times 6}$ and the following control law

$$\delta_e = \mathbf{K}_{V_a} \begin{pmatrix} \int (V_a - V_{a_c}) dt \\ V_a - V_{a_c} \\ \Theta \\ \alpha \\ q \\ \delta_{en} \end{pmatrix} \tag{14}$$

For lateral position control, two nested control loops are selected, since timescale separation is more distinct due to the additional integrator. What is more, previous flight experiments have shown that low-cost GPS receivers can be prone to loss of signal for typically one GPS epoch due to the accelerations of the launch. The nested architecture of the lateral controller allows to switch to a degraded law by disabling the outer loop and keeping

$$\phi_c = 0 \tag{15}$$

thus maintaining some kind of lateral stabilization without position feedback. The inner controller acts on the roll subsystem[1] Since the outer loop acts as integrator on ϕ, no integral action is used, leading to the control law

[1] The UAV considered here has no rudder.

$$\delta_a = \mathbf{K}_\phi \begin{pmatrix} \phi - \phi_c \\ q \end{pmatrix} \tag{16}$$

The outer loop controller acts on ϕ_c as virtual control input. Integral action is added to reduce steady state errors:

$$\phi_c = \mathbf{K}_y \begin{pmatrix} \int(y - y_c)dt \\ y - y_c \\ v_y \\ r \end{pmatrix} \tag{17}$$

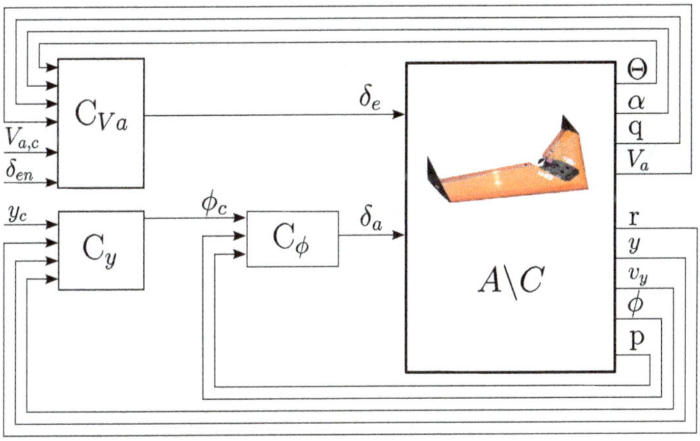

Fig. 5 Control law architecture for automatic launch

5 Controller Synthesis

This section provides details of the controller synthesis models. Where applicable, the airspeed control problem as depicted in fig. 6 is used as an example. The controller synthesis is driven by three groups of requirements: the closed loop system should a) track the reference signals while b) rejecting the effects of atmospheric turbulence, without c) driving the actuators into saturation. Controllers have been computed with the Matlab function *hinfstruct* ([12]).

Since the H_∞ performance and stability criterion (eq. (2)) is expressed in terms of unity input and output signals, all input and output signals are normalized by diagonal scaling matrices $(\mathbf{S}_{in}, \mathbf{S}_{out})$. Note that for controller tuning, only relevant transfer functions from elements of $\mathbf{w}$ to elements of $\mathbf{z}$ are included, forming a tuning matrix $\mathbf{T}_\infty$.

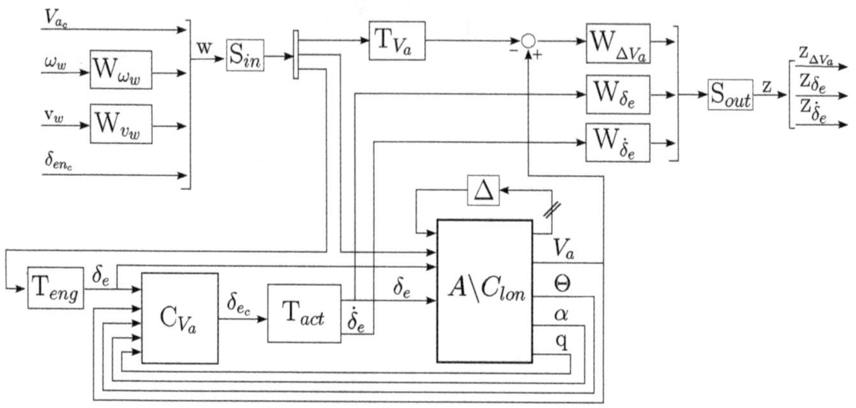

Fig. 6 Interconnection structure for H_∞ synthesis of V_a law

5.1 Reference Tracking

Time domain tracking requirements are incorporated into the controller synthesis by introducing reference models ($\mathbf{T}_{V_a}$) and including the model mismatch ($\mathbf{z}_{\Delta V_a}$) into the output vector $\mathbf{z}$. The reference models for the longitudinal closed loop, the inner lateral closed loop and the lateral outer closed loop have been selected as second order systems, parameters are given in table 1. The signals to be tracked are either constant or generated by an outer loop controller, thus below the bandwidth of the inner loop. Their frequency content is expressed by second-order weights on the model mismatch ($\mathbf{W}_{\Delta V_a}$) that exhibit two times the cutoff frequency of the reference model to enforce its magnitude roll-off.

The throttle command δ_{en_c} is added to the vector of exogenous inputs $\mathbf{w}$. The transfer function $\delta_{en_c} \to V_a$ is added to the tuning matrix $\mathbf{T}_\infty$ to enforce decoupling between the throttle input and the airspeed, leading to a feed-forward gain acting on the elevator, decreasing the initial airspeed overshoot. To take into account the reaction time of the engine, a first-order engine model is added to the synthesis model.

5.2 Turbulence Attenuation

Perturbations on airspeed and angular rates induced by atmospheric turbulence are incorporated as additional system inputs. They enter the system through shaping filters that reproduce the Dryden turbulence model and, at the same time, have the function of weights.

5.3 Actuator Constraints

The actuators' deflection and rate limits of $\pm\frac{\pi}{4}$ and $\pm 5.8\frac{rad}{s}$ are taken into account by appropriate scaling factors and additional scalar weights ($\mathbf{W}_{\delta_e}, \mathbf{W}_{\dot{\delta}_e}$). As a result, from a time domain point of view, the contribution of the actuator activity to the overall size of z exceeds 1 once one of these limits is violated in steady state for any combination of unity sine signals of any frequency entering the system. The actuator rate is an output of the actuator model and as such readily available.

A feedback delay of nominally 0.025 s is included in the synthesis models as a 3rd order Padé approximation. It lumps together sensor dynamics, computation time and control signal protocol delay

Table 1 Reference models for controller synthesis

System	Reference model
Longitudinal $V_{a,c} \rightarrow V_a$	$\frac{1}{s^2+1.4s+1}$
Lateral $\phi_c \rightarrow \phi$	$\frac{39.4}{s^2+12.4s+39.4}$
Lateral $y_c \rightarrow y$	$\frac{0.25}{s^2+s+0.25}$

6 Structured Singular Value Analysis

The Structured Singular Value is used to verify stability over the set of uncertainties $\boldsymbol{\Delta}$ by computing the worst case H_∞ norm. A finite worst case norm guarantees robust performance. What is more, once the worst case Δ is identified, the worst-case time-domain performance can be evaluated by simulating the nonlinear model subject to the worst-case parameter set. This procedure provides a computationally cheap means to evaluate the performance bounds before conducting extensive and therefore computationally demanding Monte Carlo simulations.

A total of 19 parameters of the nonlinear model are assumed to be subject to uncertainties: the dynamic pressure, representing the airspeed envelope, the aerodynamic derivatives (e.g. $C_{m\alpha}$) and constants (e.g. C_{m0}), the damping and natural frequency of the actuators, the time constant of the electric engine and the feedback delay. Since parameter bounds are not part of the model used in this work, relative bounds are taken from [13], where system identification results for a similar small UAV based on a similar aerodynamic model are presented.

Upper and lower parameter bounds are derived from stochastic parameter uncertainties (result of the identification process) by

$$\theta_{i,n} - 3\sigma_{\theta_{i,n}} \leq \theta_i \leq \theta_{i,n} + 3\sigma_{\theta_{i,n}} \tag{18}$$

where $\sigma_{\theta_{i,n}}$ is the standard deviation of the parameter θ_i and $\theta_{i,n}$ its nominal value. For parameters for which no uncertainty bounds are available, a rather pessimistic range of $\pm 25\%$ is assumed. The feedback delay is assumed as $0.02s \leq t_f \leq 0.03s$ based on existing avionics hardware.

6.1 Model Decomposition

To enable analysis of robust performance and stability with Structured Singular Value methods, the longitudinal and lateral system models are reformulated as Linear Fractional Transformations (LFT), integrating parameter uncertainties as real diagonal perturbation Δ. The LFT representations are generated numerically[2] by sampling the parameter space (using Matlab toolboxes developed at ONERA, see [14, 15]) and linearizing the nonlinear model for each parameter set. To keep the computation needed to do so to a reasonable level, the longitudinal and lateral systems are each decomposed into their respective slow and fast subsystems. This allows to sample only the parameters exhibiting a significant impact on the respective subsystem. To identify these parameters, they are ranked by means of a sensitivity analysis. To do so, each one of the parameters θ is varied by a small amount and the "closeness" of the resulting linear model to the unperturbed model is used to rank its impact. As a measure of closeness, both the H_2 norm

$$||\Delta \mathbf{T}(s)||_2 = ||\mathbf{T}(\theta)(s) - \mathbf{T}(\Delta\theta)(s)||_2 \tag{19}$$

and the H_∞ norm

$$||\Delta \mathbf{T}(s)||_\infty = ||\mathbf{T}(\theta)(s) - \mathbf{T}(\Delta\theta)(s)||_\infty \tag{20}$$

have been found to lead to the same ranking. $\mathbf{T}(s)$ is the closed loop system, θ is the nominal parameter set and $\Delta\theta$ the perturbed one. The closed loop system has to be considered to include the impact of elements of θ on the input matrix $\mathbf{B}$.

In general terms, the longitudinal and lateral state space systems are decomposed into two subsystems each, connected in a feedback structure, i.e. the states of one subsystem system act as inputs to the other system. To do so, the matrices of the dynamic equation $\dot{\mathbf{x}} = \mathbf{A}\,\mathbf{x} + \mathbf{B}\,\mathbf{u}$ are partitioned as in eq. (21), and the inner and outer system are extracted as given in eqs. (22) and (23)

$$\begin{pmatrix} \dot{\mathbf{x}}_\mathrm{i} \\ \dot{\mathbf{x}}_\mathrm{o} \end{pmatrix} = \begin{bmatrix} \mathbf{A}_\mathrm{ii} & \mathbf{A_{io}} \\ \mathbf{A}_\mathrm{oi} & \mathbf{A_{oo}} \end{bmatrix} \begin{pmatrix} \mathbf{x}_\mathrm{i} \\ \mathbf{x}_\mathrm{o} \end{pmatrix} + \begin{bmatrix} \mathbf{B}_\mathrm{i} \\ \mathbf{B}_\mathrm{o} \end{bmatrix} \delta \tag{21}$$

$$\dot{\mathbf{x}}_\mathrm{i} = \mathbf{A}_\mathrm{ii}\mathbf{x}_\mathrm{i} + \begin{bmatrix} \mathbf{B}_\mathrm{i} & \mathbf{A}_\mathrm{io} \end{bmatrix} \begin{pmatrix} \delta \\ \mathbf{x}_\mathrm{o} \end{pmatrix} \tag{22}$$

[2] For models largely based on constant aerodynamic derivatives, LFTs can alternatively at least partly be calculated analytically. Numerical tools offer however independence from details of the model representation (aerodynamic coefficients might for instance be modeled by lookup tables or neural networks) and are thus preferred.

$$\dot{\mathbf{x}}_{\mathrm{o}} = \mathbf{A}_{\mathrm{oo}}\mathbf{x}_{\mathrm{o}} + \begin{bmatrix}\mathbf{B}_{\mathrm{o}} & \mathbf{A}_{\mathrm{oi}}\end{bmatrix} \begin{pmatrix} \delta \\ \mathbf{x}_{\mathrm{i}} \end{pmatrix} \tag{23}$$

Since linear combinations of LFTs are LFTs themselves, the resulting LFTs can easily cast into one single LFT. To give an example, for the longitudinal system this decomposition leads to a fast inner system (eq. (24)) that represents the short period mode and a slow outer system (eq. (25)) that comprises the airspeed V_a as only state.

$$\begin{pmatrix} \dot{\alpha} \\ \dot{\theta} \\ \dot{q} \end{pmatrix} = \mathbf{A}_{\mathrm{i}} \begin{pmatrix} \alpha \\ \theta \\ q \end{pmatrix} \begin{bmatrix}\mathbf{B}_{\mathrm{i}} & \mathbf{A}_{\mathrm{io}}\end{bmatrix} \begin{pmatrix} \delta_e \\ \delta_{en} \\ \mathbf{v}_w \\ \omega_w \\ V_a \end{pmatrix} \tag{24}$$

$$\dot{V}_a = \mathbf{A}_{\mathrm{o}} V_a + \begin{bmatrix}\mathbf{B}_{\mathrm{o}} & \mathbf{A}_{\mathrm{oi}}\end{bmatrix} \begin{pmatrix} \delta_e \\ \delta_{en} \\ \mathbf{v}_w \\ \omega_w \\ \alpha \\ \theta \\ q \end{pmatrix} \tag{25}$$

7 Monte Carlo Simulations on 6DOF Model

Once the μ analysis shows robust stability and acceptable robust performance, Monte-Carlo simulations of the nonlinear model provide additional confidence. They reveal the effect of important nonlinearities that are neglected by linear representations, such as actuator saturations, discrete sampling of control laws, aerodynamic stall and performance of the linear controllers for larger deviations from their operating point.

Monte-Carlo simulations exploit the fact that for a large number of simulations of a system, the statistical properties of the outcome approach the theoretical ones, which are often not tractable for sufficiently complex systems such as a UAV. For the automatic launch problem, the most interesting information Monte-Carlo simulations can provide is an estimate of the probability of launch failure, defined by the relative number of simulations for which $h_{min} \leq 0$.

A uniform distribution of all uncertain simulation parameters is sampled. The samples for two example parameters are visualized in fig. 7. Two simulation scenarios have been run: Firstly, one that comprises only variations of the aircraft model parameters, thus clearly revealing the robustness of the controllers towards model uncertainty. In a second scenario, turbulent headwind representing 20 % of the nominal launch speed has been added, as well as varying initial airspeed to take into account imperfections of the launching mechanism or the human thrower.

A step of $5\,\frac{m}{s}$ is added to the commanded airspeed and a step of $5\,m$ to the commanded lateral position at $t = 12\,s$.

To cope with the resulting large dataset, for each sample, the maximum, minimum and nominal values of the controlled variable and the actuator activity are extracted, leading to worst-case envelopes. Note that the actuator deflections $\delta_{a/e}$ and rates $\dot{\delta}_{a/e}$ are normalized with their respective limits and that for both saturated values are given. As is to be expected, turbulence leads to decreased tracking performance and generally larger error bounds on V_a, as depicted in fig. 8 and fig. 9.

While in calm air, the lateral deviation stays very small, due to decoupled lateral and longitudinal dynamics, in turbulent air the control law keeps the UAV inside a corridor of about $\pm 10\,m$. Note that turbulence attenuation is improved for higher airspeeds.

In both scenarios, while actuator deflection saturation is avoided, commanded actuator rates inevitably saturate. This is partly due to the response of the proportional controller gains to step inputs, i.e. infinite error rates, partly due to turbulence, see fig. 9 around $t = 15\,s$. While these short-term rate saturations do not affect stability, the control system would benefit from some kind of anti-windup scheme.

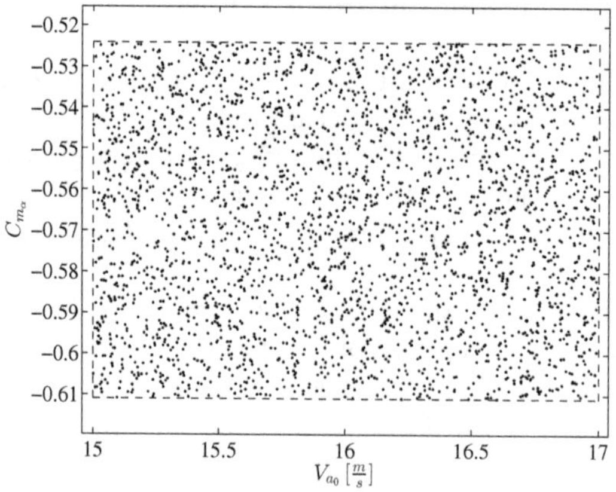

Fig. 7 Monte Carlo samples for two parameters, $n_s = 3 \cdot 10^3$

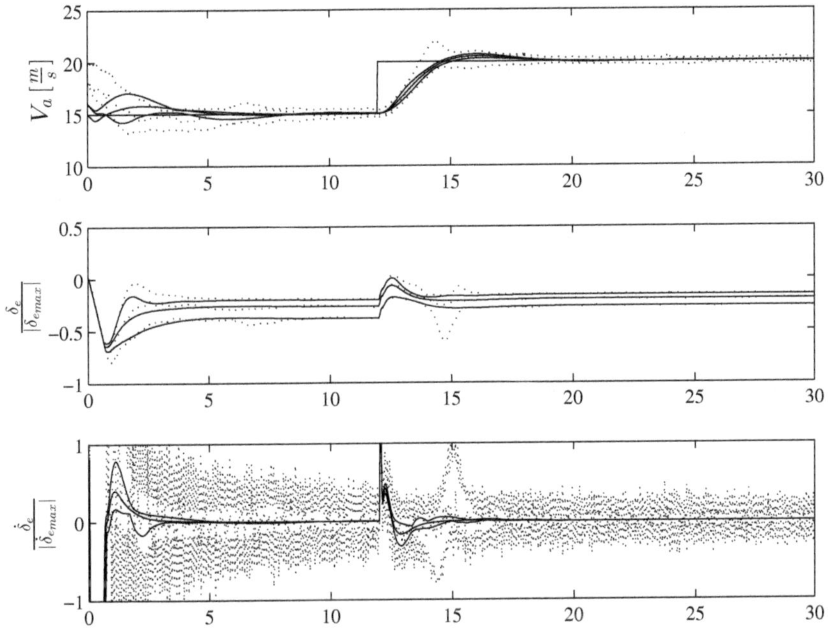

Fig. 8 Monte Carlo envelopes airspeed tracking law, $n_s = 3 \cdot 10^3$ ($\cdots$ turbulent headwind, — calm air)

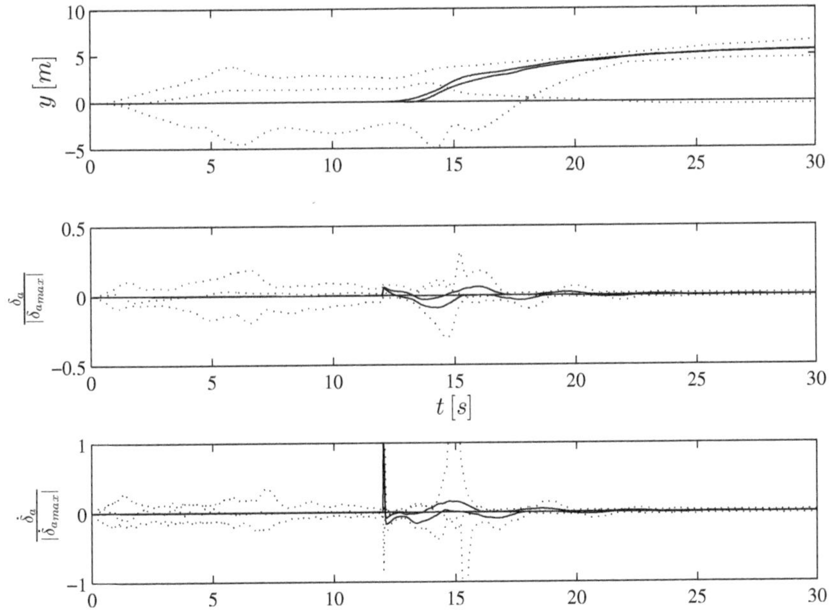

Fig. 9 Monte Carlo envelopes lateral position tracking law, $n_s = 3 \cdot 10^3$ ($\cdots$ turbulent headwind, — calm air)

8 Conclusion

A set of control laws for the automatic launch of small UAVs is presented. The combination of structured H_∞ controller synthesis methods, μ analysis to identify worst-case model variations and Monte-Carlo Simulations to validate stability and acceptable performance proved to be a viable tool chain. Satisfactory airspeed tracking and lateral guidance could be achieved in the presence of realistic aircraft model variations and turbulent headwind. Good robustness towards dynamic pressure variations is achieved, as an envelope of $14\,\frac{m}{s}$ to $22\,\frac{m}{s}$ is covered with static controller gains.

Further work will include closing the controller tuning loop by including μ into the controller synthesis as well as further exploring the operating envelope towards cross- and tailwind conditions. Flight tests will be conducted with a particular focus on the validity of performance bounds predicted by simulations.

References

[1] Meng, L., Li, L., Veres, S.: Empirical aerodynamic modeling for robust control design of an oceanographic uninhabited aerial vehicle. In: 2010 International Conference on Electronics and Information Engineering (ICEIE), vol. 2, pp. V2–409–V2–415 (August 2010)

[2] Santoso, F., Liu, M., Egan, G.: H2 and H-infinity robust autopilot synthesis for longitudinal flight of a special unmanned aerial vehicle. IET Control Theory and Applications 2(7), 583–594 (2008)

[3] Crump, M., Bil, C.: An autonomous control technique for launching ship based unmanned aerial vehicles (UAVs) in extreme conditions. In: Proceedings of the 23rd International Congress of Aeronautical Sciences (2002)

[4] Apkarian, P., Noll, D.: Nonsmooth H-infinity Synthesis. IEEE Transactions on Automatic Control 51(1), 71–86 (2006)

[5] Stein, G., Doyle, J.: Beyond singular values and loop shapes. Journal of Guidance, Control and Dynamics (1991)

[6] Skogestad, S., Postlethwaite, I.: Multivariable Feedback Control: Analysis and Design. John Wiley & Sons (2005)

[7] Roos, C.: Systems modeling, analysis and control (SMAC) toolbox: An insight into the robustness analysis library. In: 2013 IEEE Conference on Computer Aided Control System Design (CACSD), pp. 176–181 (August 2013)

[8] Beard, R.W., McLain, T.W.: Small Unmanned Aircraft: Theory and Practice. Princeton University Press, Princeton (2012)

[9] Vajta, M.: Some Remarks on Pade - Approximations. In: 3rd TEMPUS-INTCOM Symposium (September 2000)

[10] Park, S.: Avionics and Control System Development for Mid-Air Rendezvous of Two Unmanned Aerial Vehicles. PhD thesis, MIT (2004)

[11] Civita, M.L., Papageorgiou, G., Messner, W., Kanade, T.: Design and Flight Testing of a High-Bandwidth H-infinity Loop Shaping Controller for a Robotic Helicopter. In: Guidance, Navigation, and Control and Co-located Conferences. American Institute of Aeronautics and Astronautics (August 2002)

[12] Balas, G., Chiang, R., Packard, A., Safonov, M.: Matlab 2012b Robust Control Toolbox Reference. The Mathworks, Inc. (September 2012)
[13] Jung, D., Tsiotras, P.: Modelling and hardware-in-the-loop simulation for a small unmanned aerial vehicle. In: AIAA Infotech at Aerospace, Rohnert Park, CA, AIAA Paper 07-2768 (May 2007)
[14] Magni, J.-F.: Presentation of the linear fractional representation toolbox (LFRT). In: Proceedings of the 2002 IEEE International Symposium on Computer Aided Control System Design, pp. 248–253 (2002)
[15] Roos, C., Hardier, G., Biannic, J.-M.: Polynomial and rational approximation with the APRICOT Library of the SMAC Toolbox. In: Proceedings of the IEEE Multiconference on Systems and Control (2014)

Behavior Trees with Stateful Tasks

Andreas Klöckner

Abstract. The behavior tree formalism as introduced recently to the application of mission management of unmanned aerial vehicles does provide for internal memory of mission plans. This is an important drawback for even simple plans such as waypoint sequences, because the information about visited waypoints must be stored outside of the plan execution engine. In this paper, two approaches are presented in order to provide tasks with states inside behavior trees: The first allows to embed regular state machines in a specialized behavior tree task. The second provides new memory and reset tasks in order to store information directly in the tree. Both approaches are shown to solve the waypoint following plan and promise to be applicable to a much broader range of mission management problems.

1 Introduction

Current research aims at developing a number of different capabilities for unmanned aerial systems (UASs). Research groups engage in fields such as collision avoidance, formation flying or physical interaction with the environment. These capabilities do not only grow more and more diverse, but also integrate a number of low-level skills of the systems.

Additionally, practitioners seek to use UASs for an increasing range of different missions. Solar platforms e. g. are supposed to fly non-stop for several days and receive different missions as specified by the user during the flight. The operating cost of such a system is mainly determined by the personnel needed to operate the aircraft. Current solar aircraft constantly require multiple crew members to monitor the aircraft, weather, and traffic conditions.

Andreas Klöckner
DLR German Aerospace Center, Institute of System Dynamics and Control,
Münchener Str. 20, D-82234 Oberpfaffenhofen
e-mail: andreas.kloeckner@dlr.de

J. Bordeneuve-Guibé et al. (eds.), *Advances in Aerospace Guidance, Navigation and Control*, DOI: 10.1007/978-3-319-17518-8_29

In order to more efficiently deliver industrially relevant missions, the number and workload of the personnel must be decreased. This can be achieved by increasing the autonomous functions of the systems, which handle all the capabilities of the aircraft. Nonspecialist crew members must additionally be able to specify all conceivable missions targeted with the UAS. A scalable, intuitive, and flexible technique is thus needed for UAS autonomy.

Behavior trees were introduced to solve this challenge [5, 10]. They organize capabilities of the UAS in a tree of increasingly complex behaviors by using a standardized and simple interface. The formalism was first introduced for steering non-player characters in computer games [4]. It is argued that behavior trees combine a number of advantageous properties of state machines, scripting, and planning techniques [1].

However, behavior trees have a fundamental disadvantage: Behavior trees continuously adapt to changing input signals and do not have internal memory. They do thus not inherently provide means to implement behaviors requiring such internal state. This is especially important for UAS missions. These typically include behaviors such as following waypoint sequences, during which the fact of having reached a waypoint is only asserted by the sensor signals for limited duration. Mission plans implementing waypoint sequences thus require states to remember the visited waypoints. Unfortunately, this cannot be implemented easily in the standard behavior tree framework.

Researchers have invented ad-hoc solutions to this problem by describing specialized versions of the canonical behavior tree building blocks. While these are very practical solutions, they usually confound the logics of e. g. sequence with the memory introduced into the system. This paper thus describes two more general approaches in order to increase the system's modularity. The contributions of the paper are as follows:

- The formalism of conventional behavior trees is introduced in Sec. 2 and the shortcomings of state-free behavior tree implementations is shown at hand of a simple waypoint-following example in Sec. 3.
- In order to remedy this shortcoming two generic solutions are presented: Sec. 4 integrates state machines into regular behavior trees. In Sec 5, new task types handling memory within the behavior tree formalism are introduced.
- Both approaches are evaluated in the concluding Sec. 6.

2 Behavior Trees

Similar to hierarchical finite state machines, behavior trees use a hierarchy of operational modes to structure a complex mission. In a behavior tree, tasks are used instead of states and mode switches are triggered by internal statuses instead of external events.

A task is self-contained and goal-directed: it can be executed without a further framework in order to achieve a goal. A UAS mission e. g. would be composed of basic tasks, such as flying towards waypoints and probing the state of the aircraft.

These basic tasks are composed in a tree structure to arbitrary complexity using generic composite tasks.

In order for the tasks to be modularly interchangeable, all tasks have the same interface to their parent tasks: All tasks report a status to their parent node, which can be either Running, Success, or Failure in the most basic behavior tree implementation. The parent nodes can activate and deactivate their children depending on their internal logics. A complete mission plan is executed by activating the root node of the tree.

The basic task types as mentioned above can be classified as actions and conditions. They actually interact with the aircraft's systems and determine their status based on custom implementation. Composite tasks are more generic and determine their status based on their internal logic and the statuses of their children. The most common composite tasks are selectors and sequences. A very basic behavior tree system thus provides the following four types of tasks:

Actions provide interfaces to the aircraft system in order to change the its environment. They typically send low-level commands to the autopilot or payloads.

Conditions are used to test properties of the environment with boolean-valued functions. Examples are probing for minimal altitudes or a sufficient energy status. Conditions are specialized actions, because they cannot have Running status.

Selectors try to execute their children according to their priority and return Success, if one of their children is successful. Selectors are typically used to provide several alternatives to achieve a common goal. When comparing to logic, selectors can be regarded as an OR-operator.

Sequences activate all of their children one after another and return Success, only if all of their children are successful. They describe a series of tasks in order to achieve higher-level goals. Sequences correspond to the logical AND-operator.

Figure 1 demonstrates using behavior trees for the simple example of harvesting energy with a solar aircraft. The sequence in Fig. 1a activates its second child, if a surplus of energy is available. The second child consists of a selector to provide two strategies for maximizing the potential energy (see Fig. 1b). The first ensures that a given ceiling altitude is not exceeded and the second commands the aircraft to climb to that altitude.

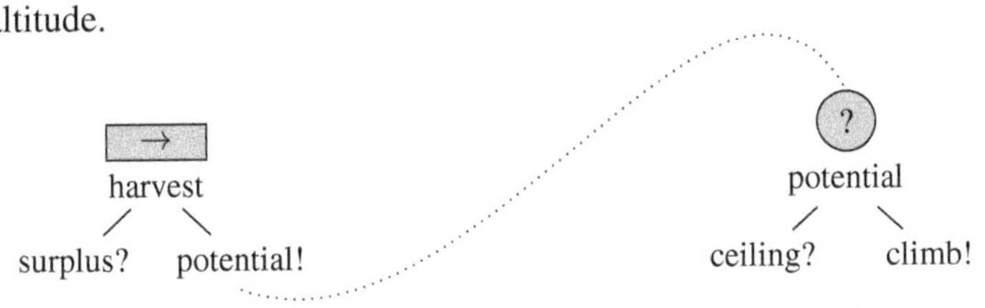

(a) A sequence is used to describe the harvesting strategy by gathering potential energy, if a surplus of energy is asserted.

(b) The selector maximizes the potential energy either by holding a ceiling altitude or by climbing.

Fig. 1 Simple tasks can be connected hierarchically in a behavior tree in order to describe the energy harvesting strategy of a solar-powered aircraft. See [7].

Note that the statuses within a behavior tree are continuously evaluated. Each composite task reacts immediately to changes in its sub-tasks. This is opposed to the semantics of e. g. state machines, which retain once active states unless specific changes in the inputs trigger a change of states. Behavior trees do not include such internal memory by default in order to provide the modular logics as outlined above. This makes them very reactive, but prevents implementing behavior requiring internal memory.

Extensions to the basic notion of a behavior tree laid out here are proposed in the literature. Additional composite task types are e. g. parallel tasks used to execute multiple children simultaneously. Semaphores can be used to guard shared resources and loops repeat a task multiple times. Millington's textbook [9] provides a more detailed overview of behavior trees and common additional task types.

3 Behavior Trees in Mission Management

Taking the technology from the computer game industry, behavior trees were introduced to the UAS community to modularize control and mission management systems [10, 5]. Behavior trees provide a number of advantages for these application compared to the state-of-the-art technology of finite state machines.

The growing number of capabilities provided by a UAS makes it difficult to maintain mission plans built as state machines for versatile aircraft, since every change requires rewiring wide parts of the mission plan. Behavior trees provide superior scalability in this case because of the standard interface of all tasks and because of the implicit switch logics. This makes it possible to modularly add, remove, and exchange arbitrary tasks at arbitrary locations in the tree without the need for global changes to the plan.

Additionally, the goal-directed semantics of behavior trees provide a very intuitive way of reading and building mission plans. Since each task can be used to achieve a sub-goal, higher-level goals can easily be composed by combining these tasks. Reading a behavior tree is intuitive on this very detailed level, but also on a high abstraction, where the actual leaf nodes are hidden from the user. This feature makes it also easy to provide an intuitive library of re-usable building blocks.

The close proximity of the basic composite tasks and logical operators additionally provides for means of validating mission plans [6, 2]. There are also approaches to behavior tree analysis building on translations to other formalisms such as hybrid dynamical systems [8].

Despite the conceptional advantages of behavior trees, there are certain things that cannot be done with standard behavior tree implementations. In particular, behavior trees in their canonical formulation do not have internal states and do not allow for proper initialization and termination of tasks. Behavior tree sequences e. g. restart a higher-priority sub-task, whenever it does not return Success anymore. This makes it hard to implement actual sequences such as simple waypoints in behavior trees.

Figure 2a shows a simple, rectangular waypoint plan to be flown by a UAS. This plan cannot easily implemented as a behavior tree because of the missing internal state. A naïve behavior tree implementation is shown in Fig. 2b. It contains the four waypoints as instances of a waypoint task taken from a task library in a sequence. A loop decorator is intended to repeat the waypoint sequence, once it has completed.

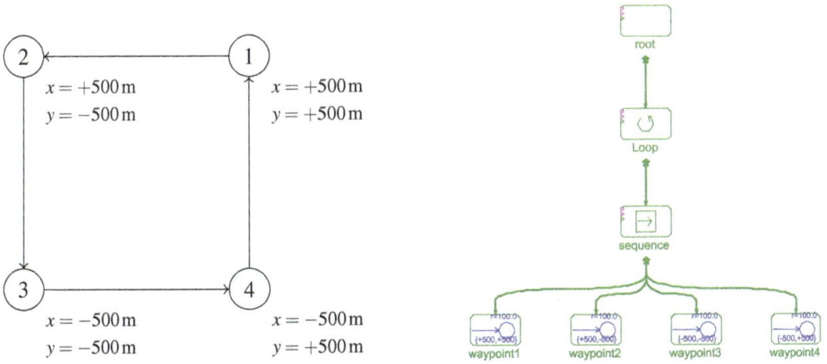

(a) The example mission consists of four waypoints arranged in a rectangular flight plan. The plan is to be flown by a small research UAS.

(b) An implementation using state-free behavior trees will prove unable to describe a waypoint plan, such as shown in Fig. 2a.

Fig. 2 Common segments of UAS missions consists of a set of waypoints to be visited by the UAS. These missions must be representable with a behavior tree for practical applications. This example shows the four waypoints used as a demonstration of this required capability.

The behavior tree is built using the Modelica BehaviorTrees library [7]. The library provides a generic framework to graphically compose behavior trees (see Fig. 3a). It also contains extensions to the behavior trees formalism such as the additional Accept status for tasks, which can be activated (see [5]). Application specific libraries can be derived easily in an object-oriented fashion. The waypoint task in Fig. 3b is e. g. composed of a condition to assert a proximity of 100 m to the waypoint and a steering task commanding a heading angle to the UAS. The signals are exchanged with the UAS model through embedded blackboard components.

As mentioned before, the implementation with standard behavior tree tasks fails to complete the mission as intended. Figure 4 shows the results of a simulation with a simplified UAS model. The UAS correctly approaches the first waypoint and then activates the second waypoint task. However, as soon as the proximity radius of the first waypoint is left by the UAS, the plan re-engages the first waypoint task. This leads to the UAS flying in circles at the boundary of the first waypoint's proximity radius.

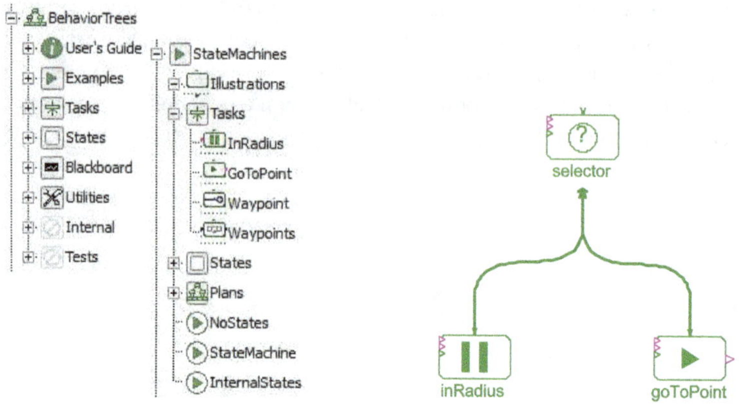

(a) The Modelica BehaviorTrees library provides a general behavior trees framework. Task libraries as e. g. for this paper can be derived easily.

(b) The waypoint following task is implemented with a selector. It checks, if the UAS is inside a proximity radius of 100 m to the waypoint, and approaches the waypoint otherwise.

Fig. 3 All plans in this paper are built with the Modelica BehaviorTrees library [7].

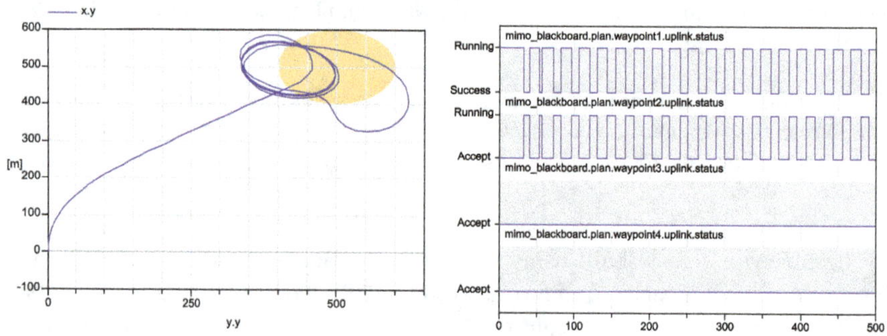

(a) The UAS flies to the first waypoint and then starts to move around this waypoint in an infinite circle.

(b) The behavior is caused by re-activating the first waypoint, when the second leads the UAS out of the proximity circle of the first waypoint.

Fig. 4 The state-free conventional behavior tree is unable to implement the intended mission correctly, because the state-free sequence falls back to higher priority sub-tasks, when their success conditions are no longer fulfilled.

4 Embedding State Machines Inside Behavior Trees

For state transitions such as between waypoints, the typical engineering approach is to use state machines. However, using state machines to model an entire mission plan means to abandon the superior modularity of behavior trees. It is therefore desirable to allow for a systematic integration of state machines inside behavior trees.

In order to allow for proper initialization and termination of tasks, a more comprehensive status cycle was already introduced in prior work [5, 7]. This modification effectively embeds a state machine in every task allowing to detect activation and deactivation of the task. Since regular behavior tree actions may contain arbitrary custom code, it is an obvious extension to also allow arbitrary state machines inside a behavior tree task. A similar approach integrates both formalisms in a third execution environment [11].

Figure 5 presents such an embedded state machine for the application described above. It simply consists of the four waypoint states. Each of the waypoint states generates a steering signal for the UAS simulation. The state machine is modeled by the synchronous elements of the Modelica language [3]. It could thus be exchanged by an arbitrary state machine described in the Modelica framework.

The waypoint states in Fig. 5b are hierarchically embedded in a *running* state. This hierarchy layer is used to generate correct behavior tree statuses for passing them up the tree. Another state, *accept* is used to allow the state machine to be stopped by the tree logic. States for generating any other allowed status are also provided in the BehaviorTrees library. The top-level states are switched using the active flag generated by the parent tasks in the tree. The tree is shown in Fig. 5a together with the blackboard interface to the UAS.

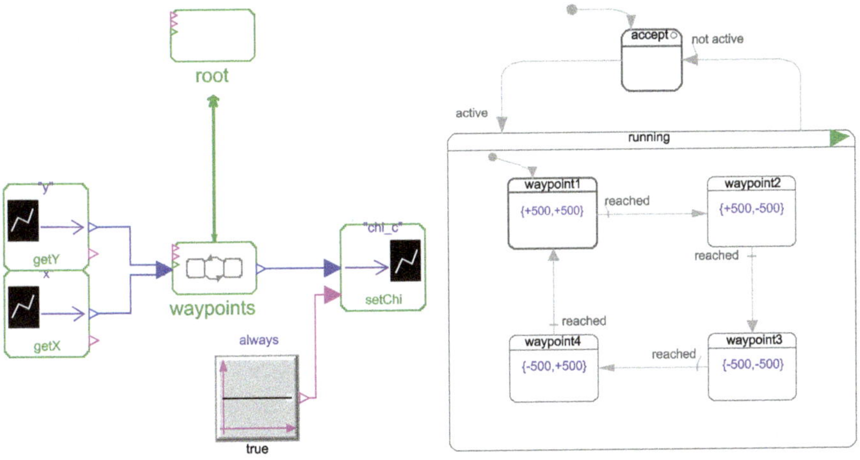

(a) A specialized behavior tree task is introduced in order to embed a regular state machine inside the behavior tree. The mission plan is additionally equipped with blocks used to communicate the state machine's commands with the aircraft model.

(b) The embedded state machine consists of the four waypoint following states and top-level states related to the behavior tree interface. The top-level states generate a valid status to be used in the behavior tree. They are switched by the behavior tree's active flag.

Fig. 5 In order to allow states in a behavior tree, it is combined with the common technique of state machines.

The resulting trajectory, when simulating the mission plan from Fig. 5 with the UAS model, is shown in Fig. 6. The UAS now correctly follows the prescribed waypoints in a square of each 1000 m height and width. Figure 6b also shows the active states during the simulation. The states iterate the four waypoints as intended. It should be noted that the Modelica state machine formalism only allows for one transition per clock cycle. Therefore, the waypoint plan remains in its *accept* state for one cycle at the beginning of the simulation. In the example, the state machine is clocked with an interval of 1 s.

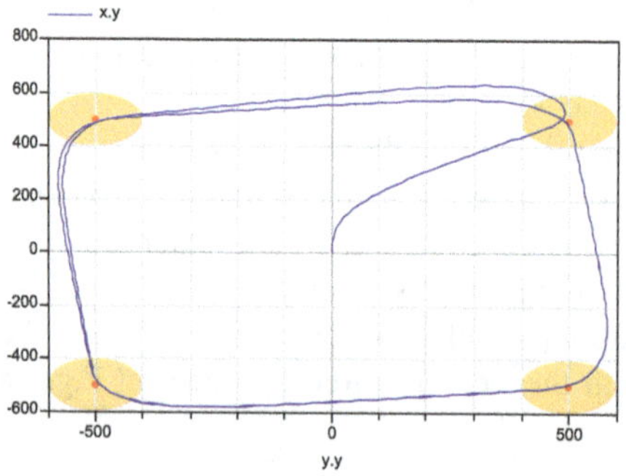

(a) The UAS repeatedly follows the four waypoints as intended on a square with each 1000 m height and width.

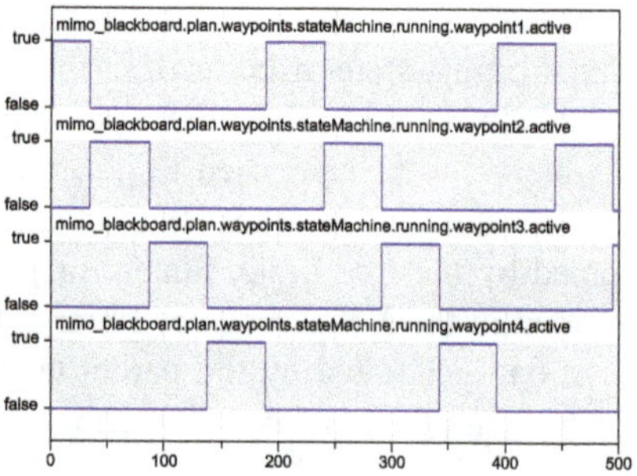

(b) The state machine switches between the four waypoint states as expected. The state machine takes one clock cycle to start.

Fig. 6 The mission plan of integrated behavior trees and state machines is able to correctly follow the intended mission plan. The state is stored exclusively in the embedded state machine.

5 Stateful Tasks for Behavior Trees

Although the solution described above provides for the desired functionality, it breaks the modularity of continuously using behavior trees. For planning purposes it is more convenient to provide the necessary states inside the behavior tree formalism. Several authors thus describe special variants of the common composite tasks (selectors and sequences) with stateful behavior. These variants are e. g. called "sequence*" [8] and exist in addition to the regular state-free variants.

Since these extensions mix standard nodes such as a sequence with a memory behavior, their modularity can be improved by separating the two behaviors. To this end, this paper introduces a memory and a reset decorator:

Memory decorators remember their status, after their child has entered a successful or failed status. They thus modify the return status of their child task such that it is prevented from being reactivated by the behavior tree.

Reset tasks send a new reset signal to their child task, when they are activated. This signal is distributed through all composite nodes to the underlying memory tasks. The reset task can thus be used to reset the memory tasks located in their child branch.

These new tasks can be used together with a regular sequence to model the behavior of previously introduced sequence* tasks (see Fig. 7b). However, the memory and reset tasks can be used in a more arbitrary way, thus allowing more flexible behaviors. In order to continuously evaluate the first child in Fig. 7b, the first memory task could e. g. be removed. The composed sequence* task can now be used in order to allow actual sequences in a behavior tree (see Fig. 7a).

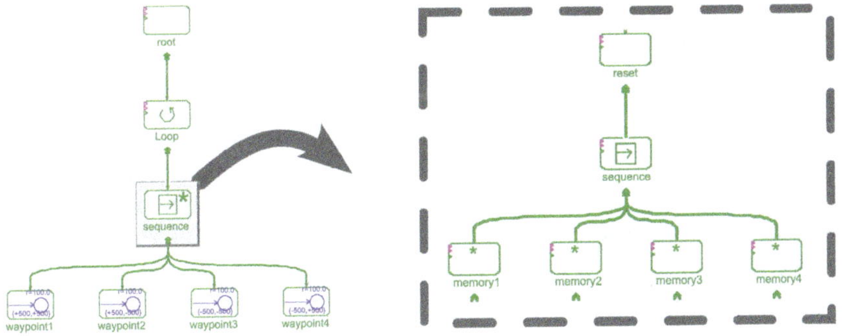

(a) The conventional sequence task is replaced in the behavior tree mission plan by a stateful sequence* task.

(b) The sequence* task consists of a regular sequence task with subordinate memory tasks and a superior reset task.

Fig. 7 Stateful behavior tree tasks are introduced in the waypoint plan as the sequence*, memory and reset tasks.

Figure 8 shows the results of a simulation with the mission plan defined in Fig. 7. The sequence of waypoints is correctly followed repeatedly as intended. Figure 8b additionally shows the statuses of the four waypoint tasks. It can be seen, that the tasks return Accept after the proximity radius of the waypoint is left. These events are marked with red circles in Fig. 8b. However, the superior memory task remembers the Success status returned previously and prevents re-activating the waypoint tasks. Only, when the loop task restarts the complete sequence at about $t = 180\,\mathrm{s}$, all memory tasks are reset and allow to steer towards the first waypoint again.

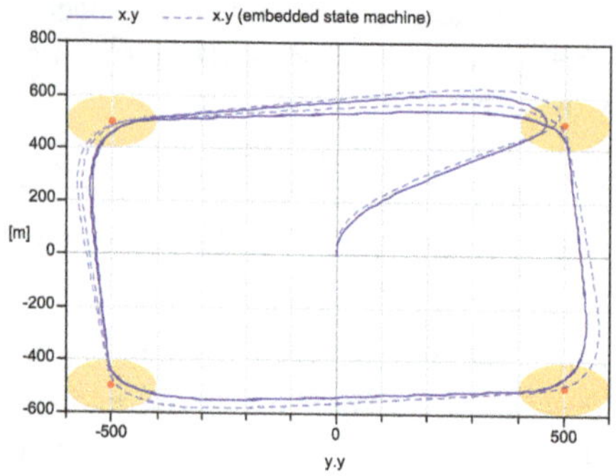

(a) The UAS follows the four waypoints as expected. The dashed line reproduces results using an embedded state machine for comparison.

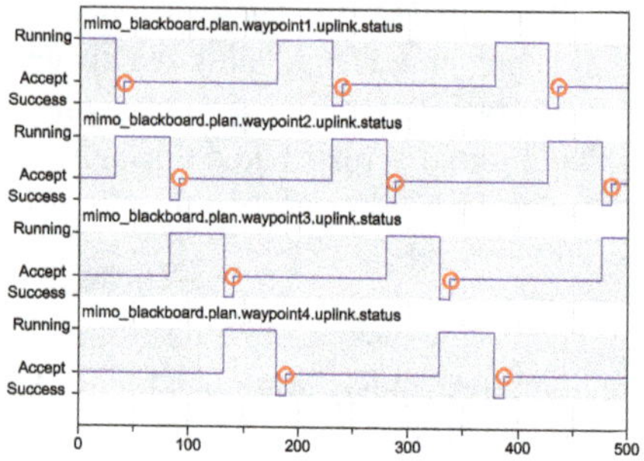

(b) The waypoint statuses change between Accept, Running, and Success. Reactivation is prevented by memory tasks in the sequence* task.

Fig. 8 Using the stateful tasks effectively solves the waypoint following mission plan, while still providing for a modular behavior tree framework.

6 Conclusions

In the present paper, the technology of behavior trees was introduced and the need for stateful tasks within behavior trees was motivated. Two solutions were proposed in order to introduce such stateful tasks into the behavior tree formalism. The first integrates state machines within behavior trees and the second introduces the new memory and reset tasks.

Both solutions provide for accurate means to implement stateful behaviors such as waypoint plans with behavior trees. An example waypoint plan is successfully followed with both plans. The solutions are completely compatible to the existing BehaviorTrees library [7] and Modelica state machines [3].

The first solution allows to integrate arbitrary state machines into arbitrary behavior trees. This makes available the full power of well-understood state machines to the mission designer. However, this technique forces the state machine to be clocked. It thus introduces delays in the control flow. Additionally, the introduced time events might prove problematic for complex UAS models and long-term mission simulation.

The new task types of the second solution use the same interface as all other behavior tree tasks. They can thus literally introduce states in any place of a regular behavior tree. This approach even increases the modularity of stateful tasks in a behavior tree over the first approach. However, the theoretical properties of the technique such as termination or dead-locks have not yet been investigated as is the case for well-studied state machines.

Future research will target using the new approaches for industrially relevant applications. More detailed extensions of the behavior tree formalism will need to be developed when facing new challenges. In parallel, the theoretical properties of the approaches will need to be investigated in order to prove their applicability to the high standards of the aircraft industry.

References

1. Champandard, A.J.: Behavior Trees for Next-Gen Game AI. In: Game Developers Conference, Lyon, France, Decembre 3-4 (2007), http://aigamedev.com/insider/presentation/behavior-trees/
2. Colledanchise, M., Marzinotto, A., Ögren, P.: Performance analysis of stochastic behavior trees. In: 2014 IEEE International Conference on Robotics and Automation (ICRA), pp. 3265–3272 (May 2014), doi:10.1109/ICRA.2014.6907328
3. Elmqvist, H., Gaucher, F., Mattsson, S.E., Dupont, F.: State machines in modelica. In: Otter, M., Zimmer, D. (eds.) Proceedings of 9th International Modelica Conference, Munich, Germany, September 3-5. Linköping Electronic Conference Proceedings, vol. 76, pp. 37–46. Modelica Association and Linköping University Electronic Press (2012), doi:10.3384/ecp1207637, ISBN: 978-91-7519-826-2, ISSN: 1650-3686, eISSN: 1650-3740
4. Isla, D.: Handling complexity in the Halo 2 AI. In: Game Developers Conference (2005), http://www.naimadgames.com/publications/gdc05/gdc05.doc
5. Klöckner, A.: Behavior Trees for UAV Mission Management. In: Horbach, M. (ed.) INFORMATIK 2013: Informatik Angepasst an Mensch, Organisation und Umwelt, Koblenz, Germany, September 16-20. GI-Edition-Lecture Notes in Informatics (LNI) - Proceedings, vol. P-220, pp. 57–68. Gesellschaft für Informatik e.V (GI), Köllen Druck + Verlag GmbH, Bonn (2013), http://www.gi.de/service/publikationen/lni/ ISBN 978-3-88579-614-5
6. Klöckner, A.: Interfacing Behavior Trees with the World Using Description Logic. In: AIAA Guidance, Navigation and Control Conference, Boston, MA, August 19-22. AIAA. AIAA 2013-4636 (2013), doi:10.2514/6.2013-4636
7. Klöckner, A.: The Modelica BehaviorTrees Library: Mission Planning in Continuous-Time for Unmanned Aircraft. In: Tummescheit, H., Årzén, K.-E. (eds.) Proceedings of the 10th International Modelica Conference, Lund, Sweden, March 10-12. Linköping Electronic Conference Proceedings, vol. 96, pp. 727–736. Modelica Association and Linköping University Electronic Press (2014), doi:10.3384/ECP14096727, ISBN: 978-91-7519-380-9, ISSN: 1650-3686, eISSN: 1650-3740
8. Marzinotto, A., Colledanchise, M., Smith, C., Ögren, P.: Towards a unified behavior trees framework for robot control. In: 2014 IEEE International Conference on Robotics and Automation (ICRA), pp. 5420–5427 (May 2014), doi:10.1109/ICRA.2014.6907656
9. Millington, I., Funge, J.: Artificial intelligence for games, 2nd edn. Morgan Kaufmann, Burlington (2009) ISBN 978-0-12-374731-0
10. Ögren, P.: Increasing Modularity of UAV Control Systems using Computer Game Behavior Trees. In: AIAA Guidance, Navigation and Control Conference, Minneapolis, Minnesota, August 13-16. AIAA. AIAA 2012-4458 (2012), doi:10.2514/6.2012-4458
11. Sagredo-Olivenza, I., Gómez-Martín, M.A., González-Calero, P.A.: Un modelo integrador de máquinas de estados y árboles de comportamiento para videojuegos. In: Camacho, D., Gómez-Martın, M.A., González-Calero, P.A. (eds.) Proceedings 1st Congreso de la Sociedad Española para las Ciencias del Videojuego (CoSECivi), Barcelona, Spain, June 24. Workshop Proceedings, vol. 1196, pp. 1613–1673. Sun SITE Central Europe, CEUR (2014), http://ceur-ws.org/Vol-1196/cosecivi14_submission_27.pdf

Functional Interior Point Programming Applied to the Aircraft Path Planning Problem

Stephane Puechmorel and Daniel Delahaye

Abstract. Multiple aircraft trajectory planning is a central problem in future air traffic management concepts where some part of the separation task, currently assumed by human controllers, will be delegated to on-board automated systems. Several approaches have been taken to address it and fall within two categories: meta-heuristic algorithms or deterministic methods. The framework proposed here models the planning problem as a optimization program in a space of functions with constraints obtained by semi-infinite programming. A specially designed innovative interior point algorithm is used to solve it.

1 Introduction

The path planning problem for multiple robots evolving in a possibly dynamic environment with obstacles is a very active area of research among the automatics and robotics communities, and is addressed in many classical references [5, 6]. For application to aircraft trajectories, nonholonomic constraints have to be considered since bounds on velocity and curvature are dictated by flight dynamics and aircraft operations. In such a context, even for two dimensional motion, it is known that an exact shortest path computation for a single mobile avoiding polyhedral obstacles is NP-Hard [2]. In order to make the problem tractable, only approximate solutions are to be sought after. One possible way of dealing with the intrinsic complexity is to resort to solvers based on metaheuristics. Since the state space exploration is made on a random fashion, there is no insurance on the quality of the solution obtained after a finite number of iterations and it is not even possible for some algorithms to prove convergence in a probabilistic sense. However, due to the nature

Daniel Delahaye · Stephane Puechmorel
ENAC, 7 avenue Edouard Belin, 31055 Toulouse
e-mail: delahaye@recherche.enac.fr,
stephane.puechmorel@enac.fr

J. Bordeneuve-Guibé et al. (eds.), *Advances in Aerospace Guidance, Navigation and Control*, DOI: 10.1007/978-3-319-17518-8_30

of the metaheuristic approach, a wide class of constraints and optimality criteria can be taken into account. In the field of air traffic management, one can impose that the solution is built using only maneuvers used by human controllers. Successful applications of stochastic optimization algorithms can be found in [3, 4]. The second way of addressing the issue of computational complexity is to allow sub-optimality of the solution. Turning the original problem into a continuous optimization program under constraints allows traditional algorithms to be used and provides a locally optimal solution. As the NP-hardness of the original formulation cannot be avoided, global optimality cannot be reached generally, unless some kind of exhaustive search among all local optima is performed. Hybridization between the two approaches can be done in order to randomly sample the set of local optima, yielding in turn an improved admissible solution. The algorithm that will be presented here falls in the second category, and is built using a special geometrical property of complex configuration spaces. The path planning program is designed in a functional way, with criterion based on a cost associated to individual trajectories and constraints expressed using semi-infinite programming. The paper is organized as follows: in a first section, general results on complex configuration spaces will be briefly recapped, as they will provide the theoretical framework on which the interior point formulation will be built. The second section will introduce the optimization program associated to the path planning problem, and a relation with a recently introduced complex harmonic navigation function will be pointed out. Finally, a possible algorithmic implementation of the solver yielding an approximate solution to the original problem will be detailed.

2 Complex Configuration Spaces

Given a set of N mobiles with planar motion, the complex configuration space $\mathscr{C}^N$ is defined to be the Cartesian product of n copies of the complex plane $\mathbb{C}$ with the set $\Delta_N = \{(z_1, \dots, z_N), \exists i \neq j, z_i = z_j\}$ of simultaneous positions removed. Any path connecting two points in $\mathscr{C}^N$ is an admissible collision-free planning of N trajectories. It is easy to show that $\mathscr{C}^N$ is a path connected space using a sequential planning argument: let $(z_1^0, \dots, z_N^0)$ be any initial configuration and let $(z_1^1, \dots, z_N^1)$ be the desired final situation. Let $z_1, \dots z_N$ be fixed. Then it exists a path joining z_1^0 and z_1^1 in $\mathbb{C} - \{z_1, \dots z_N\}$ since this space is path-connected. The same procedure can then be applied iteratively to the remaining points. The resulting path is a sequence of moves along coordinates axis in $\mathscr{C}^n$ and assumes that only one mobile moves at a time.

Geometrical insights about $\mathscr{C}^N$ can be gained from the knowledge of its cohomology group, obtained in [1]. It turns out that the generators of this group will be used to define the constraints in the optimization program constructed later. They consist of degree one forms:

$$\omega_{k,l} = \frac{dz_k - dz_l}{z_k - z_l}, 1 \leq k \leq N, 1 \leq l \leq N, k \neq l$$

Please note that in the original paper [1] a factor $(i2\pi)^{-1}$ appears in the expression: it is removed here as the real part of $\omega_{k,l}$ is of primary interest. Let a C^1 path $\Gamma\colon [0,1] \to \mathscr{C}^N$ be given. Then:

$$\int_\Gamma d\omega_{k,l} = \int_{[0,1]} \frac{\Gamma_k'(t) - \Gamma_l'(t)}{\Gamma_k(t) - \Gamma_l(t)} dt$$

with $\Gamma_j, j = 1 \ldots N$ denoting the j-th component of Γ. Rewriting the right hand side, it comes:

$$\int_\Gamma d\omega_{k,l} = \int_{[0,1]} \frac{\left(\Gamma_k'(t) - \Gamma_l'(t)\right)\left(\overline{\Gamma_k(t)} - \overline{\Gamma_l(t)}\right)}{|\Gamma_k(t) - \Gamma_l(t)|^2} dt$$

Given any two complex numbers $z_1 = x_1 + iy_1, z_2 = x_2 + iy_2$, the real part of the product $\overline{z_1}z_2$ is the inner product of the vectors (x_1, y_1) and (x_2, y_2) and imaginary part their determinant. The expression:

$$\frac{\left(\Gamma_k'(t) - \Gamma_l'(t)\right)\left(\overline{\Gamma_k(t)} - \overline{\Gamma_l(t)}\right)}{|\Gamma_k(t) - \Gamma_l(t)|^2}$$

thus admits an interpretation as an expansion rate (resp. rotation rate) for the vector $d_{kl}(t) = \Gamma_k(t) - \Gamma_l(t)$ and in turn the integral:

$$\int_\Gamma d\omega_{k,l}$$

will give $\log(\|d_{kl}(1)\|) - \log(\|d_{kl}(0)\|)$ as its real part and $2\pi\theta$ for its imaginary part, with θ the winding number of the path d_{kl}. Considering the plane $z_k = z_l$ that is a subset of Δ_N, one can interpret $\|d_{kl}(t)\|$, for $t \in [0,1]$, as twice the distance of the path $t \mapsto \Gamma_k(t)$ (resp. $t \mapsto \Gamma_l(t)$) to the constraint $z_k = z_l$. In the spirit of interior point algorithms, it is natural to consider $-\log(\|d_{kl}(t)\|)$ as a barrier function for the constraint $z_k \neq z_l$, with the major difference that it gives rise to a mapping instead of a single value.

2.1 Basic Assumptions for the Path Planning Problem

As the path planning problem is targeted towards air traffic applications, some restrictions on the manœuvers that an aircraft can do are coming from flight dynamics and passengers comfort. First of all, velocity has to be bounded below and from above, with a quite narrow interval of efficiency dictated by engines performance. Second, curvature cannot be made arbitrary high and it is advisable to limit its total integrated value so as to minimize its detrimental effect on passengers comfort.

It is not intended in this work to consider aircraft in terminal manœuvering areas, nor the climb and descent phases: as a consequence, the path planning algorithm will not perform any change in altitude, yielding a problem that conforms to the complex configuration space modelling. Furthermore, only collision avoidance is considered, the compliance with separation norms will be addressed in a future work.

Finally, it is assumed that the planner is used in a free flight context where the aircraft are not bound to predefined routes and at a tactical level with a time horizon not exceeding 20 minutes. In such a case, one can assume that the level of uncertainty is low enough to allow a deterministic approach to be taken. The effect of wind will not be included in the model. This may be unrealistic at first glance since it is one of the most influential factor on aircraft trajectories, but given the ability of future FMS systems to infer and broadcast the wind experienced along the flight path, it seems reasonable that the wind field will be known with a sufficient degree of accuracy to adjust the initial planning.

Perfect knowledge of the aircraft positions within the airspace of interest is assumed, no communication issues are considered.

3 Path Planning as a Penalized Optimization Program

To turn the path planning problem into an optimization program amenable to interior point algorithms, it is needed to define first a usable criterion and second a mean of getting a tractable set of barrier functions that will be used to penalize the criterion. A major concern for the last point is the functional nature of the state space: a mean of turning the continuous time constraints into a vector of real value must be sought after.

3.1 Applicable Criterion on a Set of Trajectories

For en-route traffic, overall flight cost is the dominant factor that airlines want to optimize. Since most of the time aircraft are flying near their efficient altitude and at constant velocity, it can be related to trajectory length or flight time. A second consideration that can be taken into account is the search for a flight path minimizing total curvature, as every turns induces an increase in fuel consumption and has a detrimental effect on passengers comfort.

Based on the previous remarks, a natural choice for the the optimization criterion is the cumulative length. Let a smooth path $\Gamma\colon [0,1] \to \mathscr{C}^N$ in the configuration space for N planes be given. Its cumulative length is defined as:

$$L(\Gamma) = \sum_{j=1}^{N} \int_0^1 \left\| \Gamma_j'(t) \right\| dt$$

In order to take into account possible priorities between aircraft, a weight may added:

$$L(\Gamma) = \sum_{j=1}^{N} w_j \int_0^1 \|\Gamma_j'(t)\| \, dt$$

with $w_j \in [0,1]$ and $\sum_{j=1}^{N} w_j = 1$. Let ε be a positive value and $H\colon]-\varepsilon,\varepsilon[\times[0,1] \to \mathscr{C}^N$ a smooth mapping such that for all $t \in [0,1]$, $H(0,t) = \Gamma(t)$ and for all $s \in]-\varepsilon,\varepsilon[$, $H(s,0) = \Gamma(0)$, $H(s,1) = \Gamma(1)$. The derivative:

$$\frac{\partial H}{\partial s}(0,\cdot)\colon [0,1] \to \mathbb{C}^n$$

may be interpreted as a tangent vector at base point Γ in the manifold of smooth paths of $\mathscr{C}^N$ [7]. The derivative of L applied to it can be computed from the definition:

$$\frac{\partial L(H(s,\cdot))}{\partial s}_{|s=0} = \sum_{j=1}^{N} w_j \int_0^1 \frac{\left\langle \frac{\partial^2 H_j}{\partial s \partial t}(0,t), \frac{\partial H_j}{\partial t}(0,t) \right\rangle}{\left\| \frac{\partial H_j}{\partial t}(0,t) \right\|} dt$$

using an integration by parts and, since the endpoints of H are fixed:

$$\frac{\partial H}{\partial s}(0,0) = \frac{\partial H}{\partial s}(0,1) = 0$$

it comes:

$$\frac{\partial L(H(s,\cdot))}{\partial s}_{|s=0} = -\sum_{j=1}^{N} w_j \int_0^1 \left\langle \frac{\partial H_j}{\partial s}(0,t), \kappa_i(t) N_i(t) \right\rangle dt$$

where κ_i is the curvature of the path Γ_i and N_i its unit normal vector. The expression obtained is the classical first order variation for the length of a smooth path [7] , summed over all components of the path Γ. In the language of manifolds,

$$\frac{\partial L(H(s,\cdot))}{\partial s}_{|s=0}$$

gives the value obtained by applying the form dL to the tangent vector

$$\left(\Gamma, \frac{\partial H}{\partial s}(0,\cdot)\right)$$

3.2 *Turning Collision Avoidance Constraints into Semi-infinite Barrier Functions*

The results on complex configuration spaces presented in the first section make natural the choice of an integral form of the barrier function [9]. For the constraint $z_i \neq z_j$ it will be defined as:

$$-\int_0^1 \log \left\| \Gamma_i(t) - \Gamma_j(t) \right\| dt$$

Please note that such an integral does not necessary take an infinite value if $\Gamma_i(t_0) = \Gamma_j(t_0)$ and thus differs from the finite dimensional setting. However, it turns out that there is a close relationship between curvatures of the component paths Γ_i and the norms $\left\| \Gamma_i(t) - \Gamma_j(t) \right\|$ that can be used to make the barrier functions effective.

Let $H\colon]-\varepsilon, \varepsilon[\times [0,1] \to \mathscr{C}^N$ be a smooth mapping satisfying the same properties than in 3.1. The penalized criterion is:

$$\sum_{j=1}^{N} w_j \int_0^1 \left\| \Gamma_j'(t) \right\| dt - \sum_{j \neq k} \lambda_{jk} \int_0^1 \log \left\| \Gamma_j(t) - \Gamma_k(t) \right\| dt$$

where the λ_{jk} are strictly positive penalty weights. Let:

$$\Theta(\Gamma)_{jk} = \int_0^1 \log \left\| \Gamma_j(t) - \Gamma_k(t) \right\| dt$$

be the j,k barrier function. The first order variation formula for $\Theta(\Gamma)_{jk}$ gives:

$$\frac{\partial \Theta_{jk}(H(s,\cdot))}{\partial s}\Big|_{s=0} = \int_0^1 \frac{\left\langle \frac{\partial H_j}{\partial s}(0,t) - \frac{\partial H_k}{\partial s}(0,t), \Gamma_j(t) - \Gamma_k(t) \right\rangle}{\left\| \Gamma_j(t) - \Gamma_k(t) \right\|^2} dt$$

Reordering terms, the variation formula for the complete criterion is:

$$-\sum_{j=1}^{N} w_j \int_0^1 \left\langle \frac{\partial H_j}{\partial s}(0,t), \kappa_i(t) N_i(t) \right\rangle dt -$$

$$\sum_{j=1}^{N} \sum_{k \neq j} \lambda_{jk} \int_0^1 \frac{\left\langle \frac{\partial H_j}{\partial s}(0,t) - \frac{\partial H_k}{\partial s}(0,t), \Gamma_j(t) - \Gamma_k(t) \right\rangle}{\left\| \Gamma_j(t) - \Gamma_k(t) \right\|^2} dt$$

Assuming that the path Γ is minimal with respect to penalized criterion, the first order necessary optimality conditions yields, for all $j = 1 \dots N$:

$$w_j \kappa_j(t) N_j(t) = \sum_{k \neq j} \lambda_{jk} \frac{\Gamma_k(t) - \Gamma_j(t))}{\left\| \Gamma_j(t) - \Gamma_k(t) \right\|^2}$$

Some important facts may be derived from the expression above. First of all, since $N_i(t)$ is of unit norm, it comes:

$$w_j |\kappa_j(t)| \leq \sum_{k \neq j} \frac{\lambda_{jk}}{\left\| \Gamma_j(t) - \Gamma_k(t) \right\|}$$

So that there is a control on the curvature given by the penalty weights. In the special case of a path planning problem with only two mobiles, that turns out to be associated in the frame of air traffic applications with two planes encounters, the relation becomes an equality:

$$w_j|\kappa_j(t)| = \frac{\lambda_{jk}}{\|\Gamma_j(t) - \Gamma_k(t)\|}, j = 1\ldots 2, k \neq j$$

In such a case, bounding the curvature of the paths will in turn bound the minimal distance separating the two mobiles: it improves over the simple collision avoidance guarantee by allowing separation norms to be enforced.

The general situation with an arbitrary number of mobiles is not so simple to deal with. However, from the expression linking $\kappa_j(t)N_j(t)$ and the sum of terms coming from the barrier functions, it is clear that a collision is not possible with two mobiles if the curvature is bounded (otherwise, the corresponding term will go to infinity while the remainder will be bounded). Only cases involving encounters with 3 or more mobiles and a special symmetry may violate this fact.

Addressing this issue can be done by noticing that the barrier functions were derived from the complex mapping $\log(z_i - z_j)$. It is in fact the complex potential generated by a simple sheet uniform distribution on the plane $z_j = z_j$. When encounters involving $p > 2$ mobiles are considered, the forbidden area in $\mathscr{C}^N$ is an intersection of $p-1$ hyperplanes and has an expression:

$$\Delta_{i_1,\ldots,i_k} = \{z : z_{i_1} = z_{i_2} = \cdots = z_{i_k}\}$$

where the sequence $i_1, \neq i_2 \ldots, \neq i_p$ is extracted from the complete set of indices $1\ldots N$ and denotes the mobiles in interaction. The complex potential generated by a simple sheet distribution on $\Delta_{i_1,\ldots,i_k}$ can be obtained using the procedure described below. First of all, the projection of a point z in the configuration space onto $\Delta_{i_1,\ldots,i_k}$ is given by:

$$P_{i_1,\ldots,i_k}(z) = \left(z_1,..,z_{i_1-1},h,z_{i_1+1},..,z_{i_k-1},h,z_{i_k+1},..,z_n\right)$$

where $h = k^{-1}\sum_{j=1}^{k} z_{i_j}$ is the mean value of the components belonging to the forbidden set. The complex potential generated by a simple sheet distribution is then:

$$\left(\sum_{j=1}^{k} z_{i_j} - h\right)^{2-k} = \|z - Pz\|^{2(2-k)}$$

This potential remains harmonic, but not pluriharmonic as the one based on the log.

Taking the integral of $\|z - Pz\|^{2(2-k)}$ along a path Γ in $\mathscr{C}^N$ yields the additional barrier function needed to deal with the encounter situation described by $\Delta_{i_1,\ldots,i_k}$. It is clear from the expression of this set that 2^{N-1} terms will have to be taken into account in the final penalty term. It can be viewed as a consequence of the NP-Hardness of the original problem and shows how the intrinsic combinatorics of the optimization program will appear indirectly. For practical applications, is it extremely uncommon to consider encounters involving more than 4 mobiles: on a the French airspace, it happens only a few times a year. Exceeding 5 occurs only on simulated traffic. The choice made was thus not to include the extra barrier terms and to postpone the complete formulation to a future work, where a conflict detection algorithm will be launched in a pre-processing phase in order to keep only the really needed high order barrier functions.

4 Interior Point Solver

The solver is implemented using a very simple procedure that consists in discretizing the trajectories at points regularly located in the interval $[0,1]$. The flight path of aircraft $i, i = 1 \dots N$ is then described as a sequence $x_{i,j} = \Gamma_i(j/m)$ where m is the number of samples on each trajectory. Following the derivations made in [8], a maximum curvature $\kappa_{\max}$ can be imposed using the relation:

$$\frac{\left\|x_{i,j+1} - x_{i,j-1}\right\|}{\left\|x_{i,j} - x_{i,j-1}\right\|} \geq \sqrt{4 - \kappa_{\max}^2 \left\|x_{i,j} - x_{i,j-1}\right\|}$$

Under the near constant velocity assumption:

$$\left\|x_{i,j} - x_{i,j-1}\right\| \approx \frac{L_i}{m}$$

with L_i the total length of Γ_i, and the expression reduces to a lower bound condition on $\left\|x_{i,j+1} - x_{i,j-1}\right\|$. The integrals involved in the expression of the functional criterion turn to finite sums, in a way amenable to standard non-linear optimization algorithms. The expression of the gradient of the penalized criterion at a single vertex $x_{i,j}$ may be obtained using the following approximation of the curvature and normal vector, again assuming constant velocity and putting $l_i = L_i/m$:

- $|\kappa_i(j/m)| = \left\|x_{i,j+1} - 2x_{i,j} + x_{i,j-1}\right\| / l_i$
- $N_i(j/m) = \frac{x_{i,j+1} - 2x_{i,j} + x_{i,j-1}}{\left\|x_{i,j+1} - 2x_{i,j} + x_{i,j-1}\right\|}$

The gradient is then, with respect to the i-th trajectory:

$$w_i \kappa_i(j/m) N_i(j/m) + \sum_{k \neq i} \lambda_{i,k} \frac{x_{i,j} - x_{k,j}}{\left\|x_{i,j} - x_{k,j}\right\|^2}$$

and summing up all the contribution gives rises to the complete gradient of the penalized criterion, that is used within a classical finite dimensional optimizer.

5 Conclusion and Future Work

A framework based on a functional description of the path planning problem has been described. Specially designed barrier functions inspired both by complex potential theory and semi-infinite programming allows to turn the original problem with an infinite number of constraints into a more tractable one, involving only a finite number of integrals. Bounding the curvature below allows to ensure collision avoidance. A possible implementation of the algorithms based on sampling trajectories in an even fashion was described. The complete procedure is in an early stage of development and a large amount of work still has to be performed:

- Pre-process the input traffic so as to identify encounters geometries in order to be able to use higher order barrier functions without incurring too much computational complexity.
- Investigate alternative representations of trajectories, especially those based on expansions on functional basis.
- Perform an extensive benchmark test on both synthetic and real traffic so as to assess the performance of the algorithm. A special attention must be paid to sensitivity analysis and robustness assessment against random perturbations.
- Integrate the algorithm within a realistic trafic simulator so as to quantify the effects of flight path tracking errors.

References

1. Arnol'd, V.: The cohomology ring of the colored braid group. Mathematical notes of the Academy of Sciences of the USSR 5(2), 138–140 (1969), http://dx.doi.org/10.1007/BF01098313, doi:10.1007/BF01098313
2. Canny, J., Reif, J.: New lower bound techniques for robot motion planning problems. In: Proceedings of the 28th Annual Symposium on Foundations of Computer Science, SFCS 1987, pp. 49–60. IEEE Computer Society, Washington, DC (1987), http://dx.doi.org/10.1109/SFCS.1987.42, doi:10.1109/SFCS.1987.42
3. Delahaye, D., Durand, N., Alliot, J.M., Schoenauer, M.: Genetic algorithms for air traffic control system. In: 14th IFORS Triennal Conference (1996)
4. Durand, N., Alliot, J.M.: Ant colony optimization for air traffic conflict resolution. In: 8th USA/Europe Air Traffic Management Research and Developpment Seminar (2009)
5. Latombe, J.C.: Robot Motion Planning: Edition en anglais. The International Series in Engineering and Computer Science. Springer (1991), http://books.google.com.au/books?id=Mbo_p4-46-cC
6. LaValle, S.: Planning Algorithms. Cambridge University Press (2006), http://books.google.com.au/books?id=-PwLBAAAQBAJ
7. Michor, P., Mumford, D.B.: Riemannian geometries on spaces of plane curves. Journal of the European Mathematical Society 8(1), 1–48 (2006), http://dx.doi.org/10.4171/JEMS/37, doi:10.4171/JEMS/37
8. Motee, N., Jabdabaie, A., Pappas, G.: Path planing for multiple robots: an alternative duality approach. In: 2010 American Control Conference (2010)
9. Shapiro, A.: Semi-infinite progamming, duality, discretization and optimality conditions. Journal of Optimization 58(2), 133–161 (2009), doi:10.1080/02331930902730070

Nonlinear Visual Servoing Control for VTOL UAVs with Field of View Constraint

Henry de Plinval and Laurent Burlion

Abstract. This paper proposes a new nonlinear visual servoing control law for Vertical Take-Off and Landing (VTOL) Unmanned Aerial Vehicles (UAVs) which forces the observed target to be maintained inside the videocamera field of view. The considered vehicles are underactuated -there are less control inputs than degrees of freedom- and the control task is performed in a minimal sensors configuration, where only images from a videocamera, inertial and linear velocity measurements are available. An input saturation technique is applied to a recently defined nonlinear control design for vision based UAV stabilization, which forces the target to remain in view of the UAV during the flight, instead of assuming that this property will be true, as most existing approaches do. A stability analysis of the obtained closed-loop is presented. Simulation results show the relevance of the proposed approach.

1 Introduction

This paper proposes a control laws for underactuated thrust-propelled vehicles, that is rigid bodies with one body-fixed thrust control and full torque actuation [1]. A particular example of such vehicles is VTOL UAVs (i.e. Vertical Take-Off and Landing Unmanned Aerial Vehicles). These UAV can be used for surveillance purposes, or for infrastructures inspection (bridges, power lines, etc). In this area, estimating the vehicle's position and orientation is key for the design of feedback laws to make the system autonomous. Specifically, accurate relative position with respect to the environment is an important challenge in this domain. Since the payload for such vehicles is limited, the choice of sensors is a key for such positioning, oftentimes leading to low quality measurements. This relates to the more generic field of

Henry de Plinval · Laurent Burlion
Onera - The French Aerospace Lab., 2, avenue Edouard Belin, 31055 Toulouse
e-mail: {henry.de_plinval, laurent.burlion}@onera.fr

J. Bordeneuve-Guibé et al. (eds.), *Advances in Aerospace Guidance, Navigation and Control*, DOI: 10.1007/978-3-319-17518-8_31

sensor-based control [2], where the relation between the measured signal used in the feedback law and the Cartesian coordinates is often poorly known. A large literature exists for sensor-based control for fully actuated mechanical systems, including aircraft as in [3], and some results have also been proposed for nonholonomic wheeled vehicles [4]. As for underactuated systems, the associate litterature is less aboundant. A particular case of sensor based control, which is addressed as an application case in this paper, consists in visual servoing of a UAV in front of a textured planar structure, based on measurements provided by a videocamera. In this case, various papers have proposed control design methods guaranteeing semi-global stability[1] (e.g., [5, 6, 1]). The difference between these works and the one presented here is that they do assume that the vehicle position is accurately measured, while we investigate the case where significant uncertainties on the position vector are present. This result is related to recent work [1] and involves control in presence of input saturation [7]. It is also related to [8], although the result used here is stronger since it adresses a larger class of measurement uncertainties. The application to visual servoing presented here has links with [9, 8], where (semi)-global controllers are also derived based on homography measurements. In particular, uncertainties on the position measurements are also considered in [8]. Due to assumptions on the environment, however, the class of these uncertainties is much smaller than the one here considered.

The second control theoretic field tackled in this paper is that of output constraint. Namely, the capability to force a particular variable (called "output", but which does not need to be one of the measured variables) to remain inside a given range of interest, while stabilizing the system. Such a question arises in a number of applications, and poses a severe theoretical challenge which is, to the best of the authors knowledge, unsolved yet in its more general setting. Note that input constraint problem is somewhat less difficult and has received large attention in the past few decades (see e.g. [10]).

Roughly speaking, two routes for output constraint have been investigated: either checking whether the constraints will be violated by prediction over a time horizon, or avoiding constraint violation at each step in a direct way. Regarding linear systems, prediction is easier, and special LMI methods do exist[11, 12]. There also exists a direct link between state and input constraints problems, see [13, 14].

One of the authors of this paper has recently [15] proposed a generalization of this idea to nonlinear systems, by transforming an output constraint into an input one. When comparing with existing results [16, 17, 18, 19] this method does not rely on prediction and/or does not require the nonlinear system to be in any special form. The case of a landing aircraft maintaining the ground inside its field of view has been investigated by the authors in [20].

As for the more restrictive field of UAVs visual servoing, the output constraint considered here -namely that the object of interest be maintained inside the videocamera field of view- is most often considered an assumption. Some studies however

[1] From now on, we use this term to denote asymptotic stability with convergence domain containing all position/velocity initial errors and a neighborhood of the identity matrix in $\mathbb{SO}(3)$ for orientation errors.

have considered implementing control laws so as to ensure this property. Thus, in [21], where this is performed through optimal paths computation and homography matrix switches. In [22], authors propose a controller for a rigid body while enforcing that the target object be maintained inside the field of view through backstepping motion. These works propose smart strategies to avoid the object of interest getting out of the camera field of view. However, as far as underactuated VTOL UAVs control is concerned, these strategies cannot be directly implemented, since they would lead to infeasible trajectories: such a vehicle cannot rotate at a given spot in space, and its linear / rotational velocities are not independent degrees of freedom.

In this paper, the visual servoing application is dealt with through exploitation of the so-called homography matrix. This matrix, which represents the image transformation corresponding to a change in the camera point of view, is used both for control law design and in order to maintain the target object inside the videocamera field of view. Previous studies have used this matrix in similar contexts. In [23], this matrix is used to reconstruct the 3D structure based on two views. [24] uses a similar approach in the context of vision based car platooning. In [25], a complete visual map-less navigation is built upon the use of homographies. Studies have also investigated the analytical decomposition of the homography matrix in terms of position / orientation, as [26], a task which is fulfilled in [27]. Extensions have been proposed to the notion of homography, as the super-homography in [28]. In [29], a framework is proposed to control a robot based on information extracted from the homography matrix.

This paper is organized as follows: Section 2 presents useful background, including notations. In Section 3, the control design is presented, including a stability analysis for the obtained closed-loop system. Section 4 proposes simulation results. Conclusions and future direction end the paper.

2 Preliminary Background

2.1 Notations

The transpose of a matrix M is denoted as M^T. The $n \times n$ identity matrix is denoted as I_n. Given a smooth function f defined on an open set of $\mathbb{R}$, its derivative is denoted as f'. For any square matrix M, $M_s := \frac{M+M^T}{2}$ and $M_a := \frac{M-M^T}{2}$ respectively denote the symmetric and antisymmetric part of M.

We now introduce two definitions of functions which have essentially the same role but are defined exactly the same way for technical reasons (see [31] for details). Given $0 < \delta < \Delta$, a function $sat_{\delta,\Delta} : \mathbb{R}^+ \longrightarrow \mathbb{R}^+$ of class C^1 is called a *saturation function* if:

i) $sat_{\delta,\Delta}(\tau^2) = 1$ for $0 \le \tau \le \delta$
ii) $\tau sat_{\delta,\Delta}(\tau^2) \le \Delta$ for any $\tau \ge 0$
iii) the function $\tau \longmapsto \tau sat_{\delta,\Delta}(\tau^2)$ is non decreasing on $\mathbb{R}^+$

A function $h : \mathbb{R}^+ \longmapsto \mathbb{R}^+$ of class C^1 is called a *saturating function* if:

i) h is strictly positive and bounded on $\mathbb{R}^+$
ii) $\tau h(\tau^2) \leq 1$
iii) $\tau h(\tau) \longrightarrow \infty$ when $\tau \longrightarrow \infty$
iv) $h'(\tau) \leq 0$ for any $\tau \geq 0$
v) the function $\tau \longmapsto \tau h'(\tau)$ is bounded on $\mathbb{R}^+$

Examples of saturation and saturating function are given by

$$\begin{aligned} sat_{\delta,\Delta}(\tau) &= \begin{cases} 1 \; if \; \tau \leq \delta^2 \\ \frac{\Delta}{\sqrt{\tau}} - \frac{(\Delta-\delta)^2}{\sqrt{\tau}(\sqrt{\tau}+\Delta-2\delta)} \; if \; \tau > \delta^2 \end{cases} \\ h(\tau) &= \frac{1}{\sqrt{1+\tau}} \end{aligned} \tag{1}$$

Let $\mathscr{R}$ (resp. $\mathscr{N}$) denote the set of real numbers (resp. natural integers). In this paper, we are interested in nonlinear systems of dimension $n \in \mathscr{N}$. For $i \in [1,n]$, we note e_i the vector of the Euclidean basis of $\mathscr{R}^n$. Given $\sigma \in \mathscr{C}^1(\mathscr{R}^n,\mathscr{R})$, $L_f\sigma := \sum_{i=1}^n f_i\partial_i\sigma$ denotes its Lie-derivative with respect to f.

Throughout this paper, we will use the following useful notations:

Given $r \in \mathscr{N}$ real numbers $K_1,\ldots,K_r$, we note

$$K_{j,r} := \prod_{i=j}^{r} K_i$$

and we also pose (for convenience):

$$\forall j \in \mathscr{N},\; K_{j+1,j} := 1$$

Given two real numbers $z_{\min} < z_{\max}$, we note:

$$z \longmapsto Sat_{z_{\min}}^{z_{\max}}(z) = \max(z_{\min}, \min(z_{\max}, z))$$

the "classical" saturation function of a variable between $z_{\min}$ and $z_{\max}$.

2.2 Dynamics of Thrust-Propelled Underactuated Vehicles

This paper is concerned with a special class of underactuated UAVs, namely thrust-propelled underactuated vehicles. This class is modelled as rigid bodies moving in $3D$-space under the action of one body-fixed force control and full torque actuation [1]. Almost all VTOL UAVs can be put inside this category (quadrotors, ducted fans, helicopters, etc). The model used for their dynamics is the following:

$$\begin{cases} \dot{p} &= \xi \\ \dot{\xi} &= -uRe_3 + ge_3 \\ \dot{R} &= RS(\omega) \\ J\dot{\omega} &= J\omega \times \omega + \Gamma \end{cases} \tag{2}$$

with p the position vector of the vehicle's center of mass, expressed in an reference (inertial) frame, R the rotation matrix from the body frame to the reference frame, ω the angular velocity vector expressed in the body frame, $S(.)$ the matrix-valued function associated with the cross product, i.e. $S(x)y = x \times y,\ \forall x,y \in \mathbb{R}^3$, u the normalized thrust input, i.e. $u = \frac{T}{m}$ where m is the mass and T the thrust input, $e_3 = (0,0,1)^T$, J the inertia matrix, Γ the torque vector, and g the gravity constant.

In the following, we make the classical assumption that the angular velocity can be used as a control variable, through a standard time separation and high-gain argument. Indeed, once a desired angular velocity ω^d has been defined, the torque control input Γ can be computed through a high gain controller:

$$\Gamma = -J\omega \times \omega + kJ\left(\omega^d - \omega\right)$$

with k chosen large enough. Therefore, we shall consider instead the following subsystem

$$\begin{cases} \dot{p} = \xi \\ \dot{\xi} = -uRe_3 + ge_3 \\ \dot{R} = RS(\omega) \end{cases} \tag{3}$$

2.3 Image Model

A UAV whose dynamic is described by Eq. (3) with a videocamera facing a planar target is considered. Suppose that a "reference" picture of this target taken from a reference pose (e.g. position and orientation) is known. Fig. 1 depicts this reference pose through reference frame $\mathcal{R}^*$. This pose is assumed to be a possible equilibrium for the dynamics of this underactuated vehicle, meaning that the z axis of $\mathcal{R}^*$ is vertical. Otherwise, asymptotic stablization of this pose would not be possible. Finally, the optical center of the camera is assumed to correspond to the vehicle's center of mass and the optical axis corresponds to the x-axis of $\mathcal{R}^*$. The "current" picture of the target taken from the current pose represented by the frame $\mathcal{R}$ is available during the flight, and compared to the reference one. Based on the comparison between these two images (the reference one, and the current one), the so-called *homography matrix* is computed (more details on homography matrices is available in e.g. [30]). This matrix, which allows to transform the target's points coordinates from the reference pose to the current pose, encompasses the relative translation and orientation relating the two frames $\mathcal{R}$ and $\mathcal{R}^*$. Its expression is given by

$$H = R^T - \frac{1}{d^*} R^T p n^{*T} \tag{4}$$

where d^* is the distance from the UAV reference position to the target plane and n^* is the normal to the target plane expressed in the reference frame. In the considered scenario, both variables are unknown since they cannot be extracted from the visual data. Thus, they are not available for the control design.

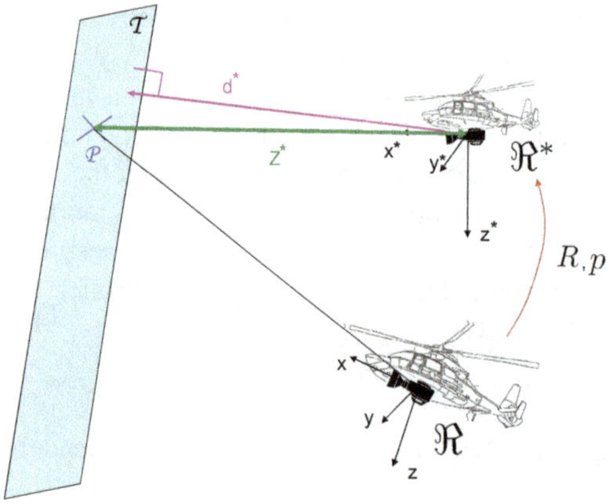

Fig. 1 Problem scheme

2.4 *A Nonlinear Visual Servoing Control Design*

[31] proposes a nonlinear control design for vehicles of the class described by Eq.(3), when the following set of measurements is available:

$$\begin{cases} \pi &= R^T M p \\ \gamma &= g R^T e_3 \\ v &= R^T \dot{p} \\ \omega \end{cases} \tag{5}$$

with M some 3×3 constant matrix. π is an uncertain body frame position measurement, the uncertainty being associated to the M matrix; γ is the gravity vector projection in the body frame, v is the linear velocity vector and ω is the angular velocity vector, both expressed in body frame. The presence of the rotation matrix R^T between the reference and body frames in the expressions of π and γ emphasizes the fact that these variables are measured in the body frame. This is a typical situation with embedded sensors. The visual servoing task considered in this paper is a typical case where these measurements are available. Most studies on feedback control for underactuated UAVs assume that M is the identity matrix, so that the relation between the measurement function and the cartesian coordinates is perfectly known.

Let us assume for now on that the observed target is vertical, which is of interest in many inspection applications. This assumption is equivalent to $n_3^* = 0$. It is now possible to define measurements of the above form so as to be able to apply the control scheme proposed in [31] in order to add a feature which will force the target to remain inside the videocamera field of view. Let:

$$\begin{cases} \pi = He_2 \times He_3 - He_1 \\ \gamma = gHe_3 \end{cases} \tag{6}$$

From the assumption $n_3^* = 0$, one can verify that:

$$\begin{cases} \pi = R^T M(\frac{n^*}{d^*})p \\ \gamma = gR^T e_3 \end{cases} \tag{7}$$

with $M(\tau) = \tau_1 I + S(\tau_2 e_3)$.

This equation shows that the desired measurements are available in the considered visual servoing task. The following theorem ([31]) presents a control design able to stabilize the vehicle in such a situation.

Theorem 1. *Consider the system described by Eq.(3), and let $sat_{\delta,\Delta}$ and h denote respectively a saturation and saturating function. Assume that M is positive definite and consider any gain values $k_1, k_2 > 0$ such that*

$$\begin{cases} k_2^2 \lambda_{min}(M_s) > k_1 ||M_a|| \, ||M|| sup_\tau (h(\tau) + 2\tau |h'(\tau)|) \\ k_2 \delta \qquad\qquad > k_1 \\ k_1 + k_2 \Delta \quad\; < g \end{cases} \tag{8}$$

Define a dynamic augmentation:

$$\dot{\eta} = \eta \times \omega - k_3(\eta - \pi), \quad k_3 > 0 \tag{9}$$

together with the control:

$$\begin{cases} \omega_1 = -\frac{k_4 |\bar{\mu}| \bar{\mu}_2}{(|\bar{\mu}| + \bar{\mu}_3)^2} - \frac{1}{|\bar{\mu}|^2} \bar{\mu}^T S(e_1) R^T \dot{\mu} \\ \omega_2 = \frac{k_4 |\bar{\mu}| \bar{\mu}_1}{(|\bar{\mu}| + \bar{\mu}_3)^2} - \frac{1}{|\bar{\mu}|^2} \bar{\mu}^T S(e_2) R^T \dot{\mu} \\ u \;\; = \bar{\mu}_3 \end{cases} \tag{10}$$

where $\bar{\mu}, \mu$, and the feedforward term $R^T \dot{\mu}$ are given by

$$\begin{aligned} &\bar{\mu} := \gamma + k_1 h\left(|\eta|^2\right) \eta + k_2 sat_{\delta,\Delta}\left(|v|^2\right) v \\ &\mu := R\bar{\mu} \\ &R^T \dot{\mu} = -k_1 k_3 \left[h(|\eta|^2) I_3 + 2h'(|\eta|^2) \eta \eta^T\right] (\eta - \pi) \\ &\quad + k_2 \left[sat_{\delta,\Delta}(|v|^2) I_3 + 2sat'_{\delta,\Delta}(|v|^2) v v^T\right] (\gamma - u e_3) \end{aligned}$$

Then,

i) there exists $k_{3,m} > 0$ such that, for any $k_3 > k_{3,m}$, the closed-loop system (3)-(10) *together with* (9) *is asymptotically stable and locally exponentially stable with convergence domain given by $\{\bar{\mu}(0) \neq -|\bar{\mu}(0)|e_3\}$.*

ii) if M_s and M_a commute, the same conclusion holds with the first inequality in (8) *replaced by:*

$$k_2^2\lambda_{min}(M_s) > k_1\|M_a\|\left(\|M_a\|\sup_\tau h(\tau) + \|M_s\|\sup_\tau 2\tau|h'(\tau)|\right) \tag{11}$$

Proof: The proof can be found in [31].

Note that the need for linear velocity measurement has been alleviated through the use of optical flow measurement, but this has not been included in the present study for the sake of clarity, that result leading to more a complex derivation for the control design. Theorem 1 applies with $M = M\left(\frac{n^*}{d^*}\right)$. Thus, the control law defined in Theorem 1 stabilizes the vehicle provided that:

$$\begin{cases} n_1^* > 0 \\ k_1, k_2 > 0 \\ k_2\delta > k_1 \\ k_1 + k_2\Delta < g \\ n_1^* d^* k_2^2 > k_1|n_2^*|\left(|n_2^*| + \frac{2n_1^*}{3\sqrt{3}}\right) \end{cases} \tag{12}$$

Note that the first condition, which ensures that M is positive definite, essentialy means that the camera is not pointing its back towards the target at the reference pose- which is a very natural assumption from an application viewpoint. When (loose) bounds are known for d^*: $d_{\min} \leq d^* \leq d_{\max}$ and $n_1^* \geq n_{1\min}$, and recalling that $|n^*| = 1$, the last condition of equation (12) can be replaced by:

$$n_{1\min} d_{\min} k_2^2 > k_1\left(1 + \frac{2}{3\sqrt{3}}\right) \tag{13}$$

3 Control Design

We now present the modifications proposed to the above described control law so as to force the target to remain sinde the videocamera field of view.

3.1 A Method to Constrain a Variable Inside a Given Range

Stabilizing a nonlinear system while maintaing an output inside a given range is a very challenging task and, in general, an open theoretical field. This section recalls a recent method [15] to convert this output constraint problem into an input saturation problem. Although it is well-known that saturating an input can lead to instability, it is not always the case, not to mention the various solutions proposed in recent years to enlarge the stability domain.

Let us consider the following class of nonlinear systems

$$\begin{cases} \dot{x} = f(x) + g(x)u \\ z = \sigma_1(x) + \sigma_2(x)u \end{cases} \tag{14}$$

where the "output" variable z is to be maintained inside a given interval $[z_{\min}, z_{\max}]$ through the input u. The functions f, g, σ_1, σ_2 are $\mathscr{C}^\infty$, $x \in \mathscr{R}^n$ is fully measured, the control u and the constrained output z belong to $\mathscr{R}$.

The method is based on an iteration on the relative degree r of the variable z, $r = d_u^{rel} z$.

When $d_u^{rel} z = 0$ (i.e when $\sigma_2 \neq 0$) the output directly depends on the input so that limiting the input u and limiting the output z is equivalent:

$$z \in [z_{\min}, z_{\max}] \Leftrightarrow u \in \left[\frac{z_{\min} - \sigma_1}{\sigma_2}, \frac{z_{\max} - \sigma_1}{\sigma_2}\right]$$

When $d_u^{rel} z = 1$, we have :

$$\dot{z} = L_f \sigma_1 + (L_g \sigma_1)\, u \quad \textit{where } L_g \sigma_1 \neq 0$$

Let us consider, for any given constant $K > 0$, applying the following constraint :

$$\dot{z} \in [-K(z - z_{\min}), -K(z - z_{\max})] \tag{15}$$

z is forced to remain inside $[z_{\min}, z_{\max}]$ as long as its initial value is inside this interval. Let us note that this is equivalent to apply the following input saturation :

$$(L_g \sigma_1)\, u \in [-L_f \sigma_1 - K(z - z_{\min}), -L_f \sigma_1 - K(z - z_{\max})] \tag{16}$$

Now, this expression does convert into an input constraint, and can also be applied iteratively with the output relative degree.

Let us consider the case $d_u^{rel} z = 2$. First, given $K_1 > 0$, we would like to apply

$$\dot{z} \in [-K_1(z - z_{\min}), -K_1(z - z_{\max})] \tag{17}$$

However, this constraint cannot directly be applied, since the input u does not appear in the expression of $\dot{z}$, since $d_u^{rel} z = 2$. We thus denote:

$$\begin{cases} \dot{z}_{\min} := -K_1(z - z_{\min}) \\ \dot{z}_{\max} := -K_1(z - z_{\max}) \end{cases}$$

which are the desired bounds for $\dot{z}$. If $\dot{z}$ is forced inside these bounds, then z will be forced inside its desired range. Now, $\dot{z}$ is a new output variable whose relative degree is 1. The previous input constraint can thus be applied, and force $\dot{z}$ in its desired range, so that z, in turn, will be force inside its target range. More precisely, given $K_2 > 0$, let us force:

$$\ddot{z} \in [-K_2(\dot{z} - \dot{z}_{\min}), -K_2(\dot{z} - \dot{z}_{\max})] \tag{18}$$

This constraint is equivalent to a constraint on the control u since $d_u^{rel} z = 2$. This technique amounts to keeping the trajectory $t \longmapsto (z(t), \dot{z}(t), \ddot{z}(t))$ between two hyperplanes of the space $(z, \dot{z}, \ddot{z})$ (as depicted on Fig. 2).

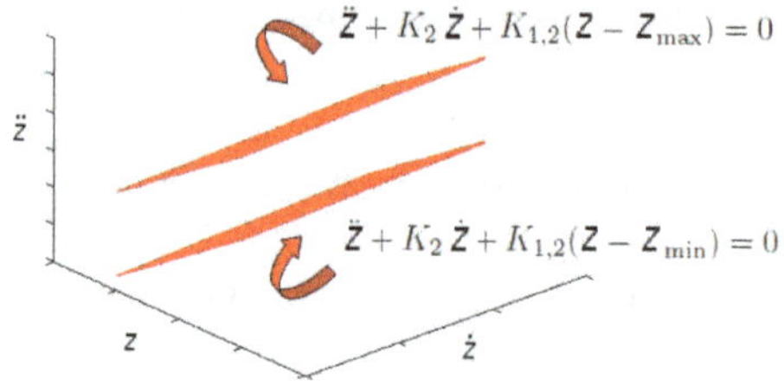

Fig. 2 Relative degree 2 case

We now state the theorem for the case of multiple inputs, multiple outputs.

Theorem 2. *Let us consider the class of nonlinear systems (14) for which the full state is measured (i.e $y = x$) and z and u both belong to $\mathscr{R}^2$. Let us consider four real numbers which verify $z^1_{\min} < z^1_{\max}$ and $z^2_{\min} < z^2_{\max}$. We note $z = [z_1, z_2]^T$, $u = [u_1, u_2]^T$, $g = [g_1\ g_2]$ and $\sigma(x) = [\sigma_1(x), \sigma_2(x)]$.*

Suppose z_1 (resp. z_2) is of relative degree $r_1 \in \mathscr{N}_{>0}$ (resp. r_2) and that the following 2×2 matrix $M_z(x)$ is invertible for all $x \in \mathscr{R}^n$

$$M_z(x) := \begin{bmatrix} L_{g_1} L_f^{r_1-1}\sigma_1(x) & L_{g_2} L_f^{r_1-1}\sigma_1(x) \\ L_{g_1} L_f^{r_2-1}\sigma_2(x) & L_{g_2} L_f^{r_2-1}\sigma_2(x) \end{bmatrix}$$

Suppose also that there exist $K^1_1, \ldots, K^1_{r_1} > 0$ and $K^2_1, \ldots, K^2_{r_2} > 0$ such that the initial state $x(0)$ satisfies the following conditions :
$\forall k \in [1,2], \forall j \in [0, r_k - 1]$,

$$K^k_{1,j} z_{k,\min} \leq \sum_{i=0}^{j} K^k_{i+1,j} L^i_f \sigma_k(x(0)) \leq K^k_{1,j} z_{k,\max}$$

then defining the auxiliary functions:

$$\begin{cases} h^k_1(x) = K^k_{1,r_k} z^k_{\min} - \Sigma^{r_k}_{i=0} K^k_{i+1,r_k} L^i_f \sigma_k(x) \\ h^k_2(x) = K^k_{1,r_k} z^k_{\max} - \Sigma^{r_k}_{i=0} K^k_{i+1,r_k} L^i_f \sigma_k(x) \end{cases}$$

the folllowing state-dependent input saturation does enforce z_1 (resp. z_2) to remain in the set $[z_{1,\min}, z_{1,\max}]$ (resp. $[z_{2,\min}, z_{2,\max}]$):

$$\begin{bmatrix} u_1^{sat} \\ u_2^{sat} \end{bmatrix} = [M_z(x)]^{-1} \begin{bmatrix} Sat_{h_1^1(x)}^{h_2^1(x)} \left(e_1^T M_z(x) \begin{matrix} u_1 \\ u_2 \end{matrix} \right) \\ Sat_{h_1^2(x)}^{h_2^2(x)} \left(e_2^T M_z(x) \begin{matrix} u_1 \\ u_2 \end{matrix} \right) \end{bmatrix}$$

Proof: The proof can be found in [15].

3.2 Application to VTOL UAVs Visual Servoing

In this section, we want to make use of the previously recalled result to the case of VTOL UAVs visual servoing.

We first need to cast the field-of-view constraint in terms of the UAV state variables. First note that, from the definition of the homography matrix, He_1 is the direction in the body frame of the point which was straight in front of the camera at the reference pose. More precisely, the coordinates of the point of the observed object which is right in the center of the videocamera field of view at the reference pose are: $\left[\frac{(He_1)_2}{(He_1)_1}; \frac{(He_1)_3}{(He_1)_1}\right]$ As a result, maintaining this point inside the videocamera field of view is equivalent to forcing the following constraints, with d_y, d_z begin the videocamera field of view in both directions:

$$\begin{aligned} |(He_1)_2| &\leq d_y (He_1)_1 \\ |(He_1)_3| &\leq d_z (He_1)_1 \end{aligned}$$

Let us note $h = He_1$ for simplicity. The inequalities to be fulfilled are:

$$\begin{aligned} -d_y h_1 \leq h_2 \leq d_y h_1 \\ -d_z h_1 \leq h_3 \leq d_z h_1 \end{aligned}$$

Note that the object being in front of the camera leads to ensuring $h_1 > 0$. Now, these inequalities can be forced through the output constraint method.

Direct application of Theorem 2 leads to the following inequalities, to be force on the outputs derivatives:

$$\begin{aligned} k[-d_y h_1 - h_2] \leq \dot{h}_2 \leq k[d_y h_1 - h_2] \\ k[-d_z h_1 - h_3] \leq \dot{h}_3 \leq k[d_z h_1 - h_3] \end{aligned}$$

Wih $\dot{h}_2 = \omega_1 h_3 - \omega_3 h_1 - n_2 \frac{v}{d^*}$ and $\dot{h}_3 = \omega_2 h_1 - \omega_1 h_2 - n_3 \frac{v}{d^*}$. Now, ensuring the first constraint through the yaw dynamics and the second one through the pitch dynamics, one gets:

$$\begin{aligned} \omega_{2_{min}} \leq \omega_2 \leq \omega_{2_{max}} \\ \omega_{3_{min}} \leq \omega_3 \leq \omega_{3_{max}} \end{aligned}$$

with:

$$\begin{cases} \omega_{2_{min}} = \frac{\left(k[-d_z h_1 - h_3] + \omega_1 h_2 + n_3 \frac{v}{d^*}\right)}{h_1} \\ \omega_{2_{max}} = \frac{\left(k[d_z h_1 - h_3] + \omega_1 h_2 + n_3 \frac{v}{d^*}\right)}{h_1} \\ \omega_{3_{min}} = -\frac{\left(k\left[d_y h_1 - h_2\right] - \omega_1 h_3 + n_2 \frac{v}{d^*}\right)}{h_1} \\ \omega_{3_{max}} = -\frac{\left(k\left[-d_y h_1 - h_2\right] - \omega_1 h_3 + n_2 \frac{v}{d^*}\right)}{h_1} \end{cases}$$

Some of the terms appearing in the saturation to be applied to ω are unknown, namely the terms $n_i \frac{v}{d^*}$. Given bounds on these quantities $|n_i \frac{v}{d^*}| \leq \varpi$, we can modify the bounds to be forced on ω according to the following conservative values:

$$\begin{cases} \omega_{2_{min}} = \frac{(k[-d_z h_1 - h_3] + \omega_1 h_2 + \varpi)}{h_1} \\ \omega_{2_{max}} = \frac{(k[d_z h_1 - h_3] + \omega_1 h_2 - \varpi)}{h_1} \\ \omega_{3_{min}} = -\frac{\left(k\left[d_y h_1 - h_2\right] - \omega_1 h_3 - \varpi\right)}{h_1} \\ \omega_{3_{max}} = -\frac{\left(k\left[-d_y h_1 - h_2\right] - \omega_1 h_3 + \varpi\right)}{h_1} \end{cases}$$

Note that with the unknown term bounded by ϖ, these time-varying intervals may be empty, so that the saturation may not be applicable. This depends on the level of the bound ϖ. Note also that, from an applicative viewpoint, a fine approximation of these terms $|n_i \frac{v}{d^*}|$ may be used. Indeed:

- n_i being a coordinate of a unit vector, this quantity is known to be lower than 1. Moreover, since it is directly related to the orientation of the observed target, for some applicative scenario, one can imagine that this orientation is known with some degree of approximation. In this case, refined bounds may be used in the above scheme, thus improving the control.
- $\frac{v}{d^*}$ is a quantity called the optical flow, and can be extracted from image processing on a series of images. As a result, this value may actually be known exactly, if this image processing is available, thus reducing the conservatism of the above method.

Finally, recall that the actual control input is the torque control Γ, which is defined by a high gain controller based on the desired value for ω. Now, with the proposed method used to constrain the videocamera field of view, it is possible to define the saturation to be applied to the control input Γ based on that applied to ω, which is one of the main interest of the method.

3.3 Stability Analysis

The output constraint method presented above guarantees that abiding by the proposed input saturations will force the output to be maintained inside its defined bounds, that is: saturating the input guarantees the observed target will be forced inside the videocamera field of view. On the other hand, the closed-loop system stability under input saturation is not guaranteed. This is what we want to investigate in this section.

To perform such an analysis, we consider the effect the input saturation has on the UAV orientation: the goal of the saturation is to force the orientation, through the inputs ω_2 and ω_3, to maintain the target inside the videocamera field of view. Let us consider the orientation to which this saturation leads. The action on ω_2 and ω_3 will force the pitch and yaw stay inside some limits which, in turn, depend on the current pose of the UAV. These limits are given by the homography matrix, and more precisely, the goal to constrain He_1. At the desired pose, the homography matrix equals the identity, $H = I$, so that, for small displacements, the term He_{11} which appears in the field of view constraint, is close to 1. As a result, the field of view constraint is fulfilled by maintainaing He_{12}, He_{13} in their ranges, that is $[-d_y; d_y]$ and $[-d_z; d_z]$, respectively.

The expression for $H = R^T - \frac{1}{d^*} R^T pn^{*T}$ shows that this imposes links between $R^T e_1$ and $\frac{n_1^*}{d^*} R^T p$. That is: depending on the value of $R^T p$ -which encompasses position and orientation errors from the desired pose-, the saturation will force the pitch and yaw angles, and thus the term $R^T e_1$, to stay inside some limits. These limits, in turn, depend on the term $\frac{n_1^*}{d^*} R^T p$.

Regarding the yaw dynamics, and the related ω_3 constraint, as emphasized above in the background section, it is assumed to be independent from the other dynamics. As a result, the saturation will force the yaw angle to stay inside some -varying- limits which, in turns, has no effect on the stability of the closed-loop. This saturation cannot provide any instability to the system. If the other one -related to the pitch actuation- does not either, then, upon convergence of the rest of the dynamics, this saturation will not prevent reaching the equilibrium point.

As for the pitch actuation, the above expression for He_1 shows that the saturation will for ce the pitch angle such that the term $\left(R^T e_1\right)_3$ stays inside some limits depending on $\frac{n_1^*}{d^*} \left(R^T p\right)_3$. Now, the variable $\left(R^T p\right)_3$ is the vertical position error as expressed in body frame, which is the direction of actuation of the thrust input. And this input is not concerned with the field of view constraint saturation. Thus, the thrust input is set to its nominal value, that is $u = \bar{\mu}_3$. As a result, the equation along this direction are not affected by the field of view saturation. Thus, this dynamics is stable. Upon convergence of the error in this direction, the field of view constraint boils down to constraining the pitch angle into bounds around zero, which do not prevent convergence of the closed-loop system. It should be emphasized, however, that by the time the body-frame vertical dynamics has converged, the UAV trajectory may have been very far from what it would have been without saturation.

As a result, under input saturation, the closed-loop system remains stable.

Let us notice that this result is very much linked to the configuration of the visual servoing task: in the considered case, the target is vertical. As a result, in the presence of a body frame vertical offset from the desired position, a pitch order is given in order to maintain the target inside the videocamera field of view, which, in turn, does not have an effect on the body frame vertical dynamics which caused the field of view problem. This is depicted on Fig. 3 thereafter. Similarly for the yaw: in the presence of a horizontal offset parallel to the wall, a yaw order is given to maintain the target in the field of view, which, in turn, does not affect the yaw dynamics.

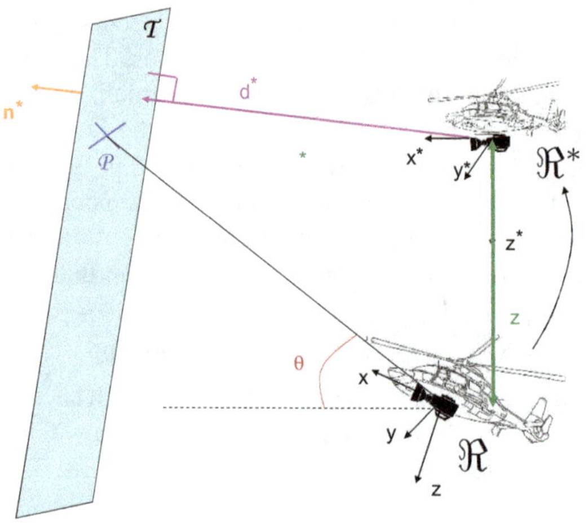

Fig. 3 Vertical target configuration: a vertical offset is treated through a pitch order, which does not affect the vertical dynamics

Now, would the target be on the ground, the situation would be by nature unstable due to the nonminimum phase of the visual outputs in that case. This can be understood from a picture: when the UAV is offset in one direction from the vertical to the target, in order to force the target inside the videocamera field of view, the UAV attitude needs to be shifted in a direction which actually moves it even further from the vertical to the target. This shows the unstable nature of that configuration, in both directions (roll and pitch, x and y), which is depicted on Fig. 4. Note however, that this analysis does not mean that the closed-loop under saturation will always be unstable. For some initial conditions, indeed, some time of this unstable closed-loop dynamics might lead to a new configuration where the unsaturated controller does maintain de target inside the videocamera field of view: thus, there might exist trajectories following the unstable saturated closed-loop for a finite time before catching back with the original, unsaturated, controller. The stability analysis, in this second case, seems much more complicated.

The simulation results presented in the following section show this efficiency.

4 Simulation Results

In this section, we present simulation results obtained with the proposed method. The initial conditions are as follows:

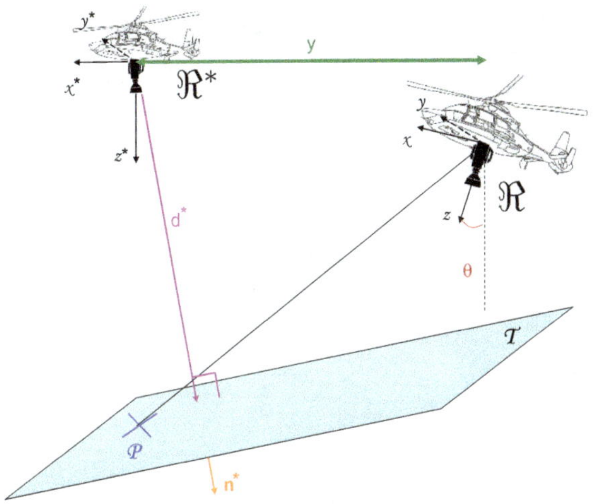

Fig. 4 Horizontal target configuration: a horizontal offset is treated by a pitch (or roll) order, which does affect the horizontal dynamics

$$\begin{cases} \theta_0 & = 29,7\,^\circ \\ \psi_0 & = 16,5\,^\circ \\ \phi_0 & = 27,8\,^\circ \\ p_0 & = \left[-1.8604\ -0.8511\ 1.7114\right]^T \\ v_0 & = \left[1.3790\ -1.0582\ -0.4686\right]^T \\ d^* & = 1,\ n^* = \left[0.9861\ 0.1659\ 0\right]^T \end{cases}$$

Fig. 5 and 6 present this application of the method. Fig. 5 shows the compared trajectories obtained with and without the field of view constraint. The position, orientation and angular velocities evolutions are plotted, in plain lines without the constraint, and in dashed line with this constraint. One can see that the system is still stable with the constraint active. Fig. 6 shows the trajectory of the target object inside the visual field of view, without (blue) and with (red) the constraint, the videocamera field of view being depicted in black. The figure shows the efficiency of the method: the blue curve escapes outside the videocamera field of view, while the red one is contained in it.

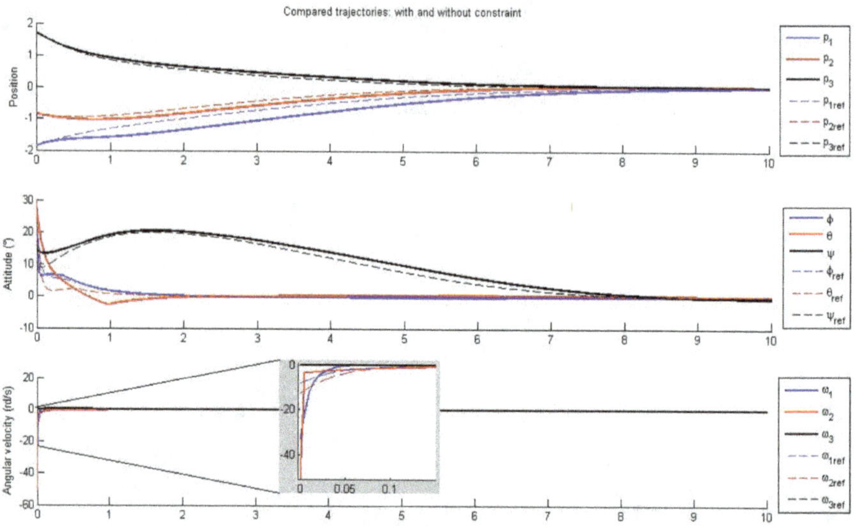

Fig. 5 Trajectories obtained without (dashed lines) and with (plain lines) output constraint

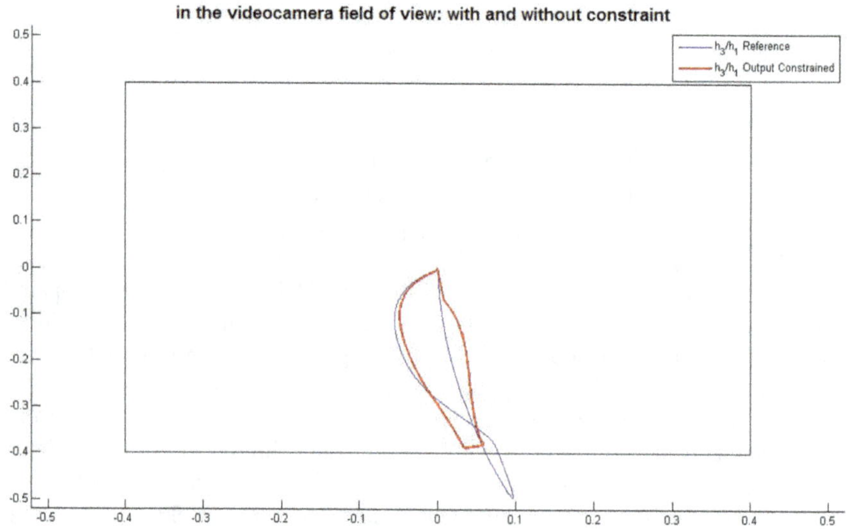

Fig. 6 Target object trajectory inside field of view without (blue) and with (red) constraint

5 Conclusion

In this paper, a nonlinear visual servoing control design for VTOL UAVs has been proposed, which includes a field of view constraint. The case of a UAV facing a vertical target with minimal measurements -namely, images from a videocamera, inertial measurements and velocity measurement- has been considered. Simulations have shown the relevance of the proposed method. Future work will propose anti-windup schemes to increase the stability domain.

References

1. Hua, M.D., Hamel, T., Morin, P., Samson, C.: A control approach for thrust-propelled underactuated vehicles and its application to VTOL drones. IEEE Trans. on Automatic Control 54, 1837–1853 (2009)
2. Weiss, L.E., Sanderson, A.C., Neuman, C.P.: Dynamic sensor-based control of robots with visual feedback. IEEE Journal of Robotics and Automation (1987)
3. Burlion, L., de Plinval, H., Mouyon, P.: Backstepping based visual servoing for transport aircraft automatic landing. In: Proc. of the IEEE Conference on Control Applications, Antibes, pp. 1461–1466 (2014)
4. Fang, Y., Dawson, D.M., Dixon, W.E., de Queiroz, M.S.: Homography-based visual servoing of wheeled mobile robots. In: CDC (2002)
5. Isidori, A., Marconi, L., Serrani, A.: Robust autonomous guidance: an internal model approach. Advances in industrial control. Springer (2003)
6. Hauser, J., Sastry, S., Meyer, G.: Nonlinear control design for slightly non-minimum phase systems: Application to V/STOL. Automatica 28, 651–670 (1992)
7. Teel, A.R.: Global stabilization and restricted tracking for multiple integrators with bounded controls. SCL 18, 165–171 (1992)
8. Metni, N., Hamel, T., Derkx, F.: A UAV for Bridges Inspection: Visual Servoing Control Law with Orientation Limits. In: 5th Symposium on Intelligent Autonomous Vehicles, IAV 2004 (2004)
9. Shakernia, O., Ma, Y., Koo, T., Sastry, S.: Landing an Unmanned Air Vehicle: Vision based motion estimation and nonlinear control. Asian Journal of Control 1, 128–145 (1999)
10. Bernstein, D.S., Michel, A.N.: A chronological bibliography on saturating actuators. International Journal of Robust and Nonlinear Control 5, 375–380 (1995)
11. Hu, T.: Nonlinear feedback laws for practical stabilization of systems with input and state constraints. In: Proc. of the 47th Conference on Decision and Control, Cancun, Mexico, pp. 3481–3486 (2008)
12. Turner, M.C., Postlethwaite, I.: Output violation compensation for systems with output constraints. IEEE Transactions on Automatic Control 47(9), 1540–1546 (2002)
13. Gilbert, E.G., Tan, K.T.: Linear systems with state and control constraints: the theory and application of maximal output admissible sets. IEEE Transactions on Automatic Control 36(9), 1008–1020 (1991)
14. Rojas, O.J., Goodwin, G.C.: A simple anti-windup strategy for state constrained linear control. In: Proc. of the 15th IFAC World Congress, Barcelona, Spain (2002)
15. Burlion, L.: A new Saturation function to convert an output constraint into an input constraint. In: 20th Mediterranean Conference on Control and Automation, pp. 1217–1222 (2012)

16. Allgöwer, F., Zheng, A.Z.: Nonlinear model predictive control: assessment and future directions for research. Progress in Systems and control Series. Birkhäuser Verlag, Basel (2000)
17. Ngo, K.B., Mahony, R., Zhong-Ping, J.: Integrator backstepping using barrier functions for systems with multiple state constraints. In: Proc. of the 44th Conference on Decision and Control, Seville, Spain, pp. 8306–8312 (2005)
18. Tee, K.P., Ge, S.S., Tay, E.H.: Barrier Lyapunov functions for the control of output-constrained nonlinear systems. Automatica 45(4), 918–927 (2009)
19. Bürger, M., Guay, M.: Robust Constraint Satisfaction for Continuous-time Nonlinear Systems in Strict Feedback Form. IEEE Transactions on Automatic Control 55(11), 2597–2601 (2010)
20. Burlion, L., de Plinval, H.: Keeping a Ground Point in the Camera Field of View of a Landing UAV. In: Proceedings of the 2013 IEEE International Conference on Robotics and Automation (ICRA), pp. 5763–5768 (2013)
21. Lopez-Nicolas, G., Bhattacharya, S., Guerrero, J.J., Sagues, C., Hutchinson, S.: Switched Homography-Based Visual Control of Differential Drive Vehicles with Field-of-View Constraints. In: IEEE International Conference on Robotics and Automation (2007)
22. Cunha, R., Silvestre, C., Hespanha, J.P., Aguiar, A.P.: Vision-based control for rigid body stabilization. Automatica, 1020–1027 (2011)
23. Zhang, Z., Hanson, A.R.: 3D Reconstruction Based on Homography Mapping. In: ARPA Image Understanding Workshop, pp. 0249–6399 (1996)
24. Benhimane, S., Malis, E., Rives, P., Azinheira, J.R.: Vision-based Control for Car Platooning using Homography Decomposition. In: IEEE Conference on Robotics and Automation, pp. 2161–2166 (2005)
25. Guerrero, J.J., Martinez-Cantin, R., Sagués, C.: Visual map-less navigation based on homographies, Journal of Robotic Systems, 569–581 (2005)
26. Vargas, M., Malis, E.: Visual servoing based on an analytical homography decomposition. In: IEEE Conference on Decision and Control and European Control Conference, pp. 5379–5384 (2005)
27. Malis, E., Vargas, M.: Deeper understanding of the homography decomposition for vision-based control, INRIA technical report (2007)
28. Simond, N., Laurgeau, C.: Vehicle Trajectory from an Uncalibrated Stereo-Rig with Super-Homography. In: IEEE International Conference on Robotics and Systems, pp. 4768–4773 (2006)
29. Benhimane, S., Malis, E.: Homography-based 2D Visual Tracking and Servoing. International Journal of Robotic Research, 661–676 (2007)
30. Benhimane, S., Malis, E.: Homography-based 2D visual tracking and servoing. The International Journal of Robotics Research, 661–676 (2007)
31. de Plinval, H., Morin, P., Mouyon, P.: Nonlinear control of underactuated vehicles with uncertain position measurements and application to visual servoing. In: Proc. of the American Control Conference, Montreal, Canada (2012)
32. de Plinval, H., Morin, P., Mouyon, P., Hamel, T.: Visual servoing for underactuated VTOL UAVs: A linear, Homography-Based Approach. In: IEEE International Conference on Robotics and Automation (2011)

Automatic Landing of a High-Aspect-Ratio Aircraft without Using the Thrust

Maxim Lamp and Robert Luckner

Abstract. This paper describes a landing procedure, the algorithms and the automatic flight control functions for the longitudinal motion of a high-aspect-ratio aircraft. Such aircraft have a high aerodynamic performance and are typically equipped with airbrakes. The automatic landing is performed without using the thrust, taking the prevailing wind conditions into account. This landing procedure is intended for emergency landing after engine failure or can be used to reduce noise in the vicinity of airports. After explaining the problem and limitations of landing an aircraft without thrust under varying wind conditions, the landing procedure is explained. The main idea is to adjust the glide path continuously according to the current wind conditions instead of using a fixed glide path. The algorithms and functions to calculate the glide slope angle, to command the glide path and the airspeed, and to control the aircraft are described. The control structure for the longitudinal motion is based on the principles of total energy control. As thrust control is not available, the airbrakes in combination with the elevator are used to control the airspeed and the glide path at the same time. This control algorithm is also used for the flare. The algorithms and functions for longitudinal control and for flare control are explained. They were designed for a CS-23 class 1 aircraft and implemented and tested in flight in an automatic flight control system, which was developed in the project LAPAZ and integrated into the utility aircraft STEMME S15. Finally flight test results will be discussed, that indicate the validity of the algorithms.

Maxim Lamp · Robert Luckner
Berlin Institute of Technology, Department of Flight Mechanics, Flight Control and Aeroelasticity
e-mail: {Maxim.Lamp,Robert.Luckner}@ilr.tu-berlin.de

J. Bordeneuve-Guibé et al. (eds.), *Advances in Aerospace Guidance, Navigation and Control*, DOI: 10.1007/978-3-319-17518-8_32

Nomenclature and Abbreviations

E	[Nm]	energy
F	[N]	force
F_G	[N]	weight force
H	[m]	Height, Altitude
H_{GS}	[m]	glide slope height
H_{msl}	[m]	Height above mean sea level
$\hat{H}_{msl}$	[m]	complementary filtered height above mean sea level
H_{rwy}	[m]	Height of runway threshold above mean sea level
K	varies	gain
P	[Nm/s]	power
T	varies	transformation matrix
$\hat{V}$	[m/s]	complementary filt. velocity
V_K	[m/s]	flight path velocity
V_A	[m/s]	true airspeed
V_W	[m/s]	wind speed
V_{S1}	[m/s]	stall speed in landing conf.
V_{cas}	[m/s]	calibrated airspeed
V_{eas}	[m/s]	equivalent airspeed
V_{gnd}	[m/s]	ground speed
g	[m/s^2]	acceleration due to gravity
m	[kg]	mass
t	[s]	time
u	[m/s]	horiz. velocity component
w	[m/s]	vert. velocity component
x_{rwy}	[m]	distance from aircraft to runway threshold
$z_{GRP,RCP}$	[m]	distance betw. gear reference point and measurement pos. of GPS height
Θ	[rad]	pitch angle
α_W	[rad]	wind angle of attack
γ	[rad]	flight path angle
γ_a	[rad]	air path inclination angle
γ_{GL}	[rad]	glide angle
γ_{GS}	[rad]	glide slope angle
η_S	[%]	airbrakes deflection

<u>Indizes</u>

FL	flare
TD	touchdown
c	command
g	earth refer. coord. syst.
k	flight path refer. coord. syst.

1 Introduction

Electronic flight control systems are becoming interesting for general aviation and utility aircraft. Currently a number of projects run with the aim to develop and certify fly-by-wire or full automatic flight control systems for such aircraft. Unlike aircraft of the CS-25 class they are often equipped with only one engine. Even if the flight control system can be designed redundant, the availability of thrust in this case is only simplex. After an engine failure, landing cannot be guaranteed. The reaching of the runway is then a function of the flight altitude, the glide ratio and the prevailing wind condition. Even if the glide path can be intercepted, the capability to follow the glide path depends on the glide slope angle and on the aerodynamic performance of the aircraft. For those aircraft a method is presented, in which the glide path is adapted to the prevailing wind conditions. By this the usable range of aircraft performance can be extended. After glide path interception, the aircraft can reach the runway and land safe. The presented landing method can be used as an emergency landing method in the case of an engine failure. It is similar to a manual glide landing and reduces the noise in the vicinity of airports.

If after an engine failure the elevator is the only control element, an independent control of the flight path and the airspeed at the same time is not possible. However, a significant number of single engine aircraft exist with high-aspect-ratio wings (as motor gliders), that are typically equipped with airbrakes. The airbrakes can be used in combination with the elevator to control the flight path and airspeed independently.

2 Limits of Landing Approaches with Fixed Glide Slope Angles

The glide angle γ_{GL} of an aircraft is defined by its aerodynamic performance. It is a function of the lift-to-drag ratio and thus of the airspeed and the aircraft configuration. Fig. 1 shows the stationary glide angles for a high-aspect-ratio utility aircraft, with flaps in landing position and with extended landing gear.

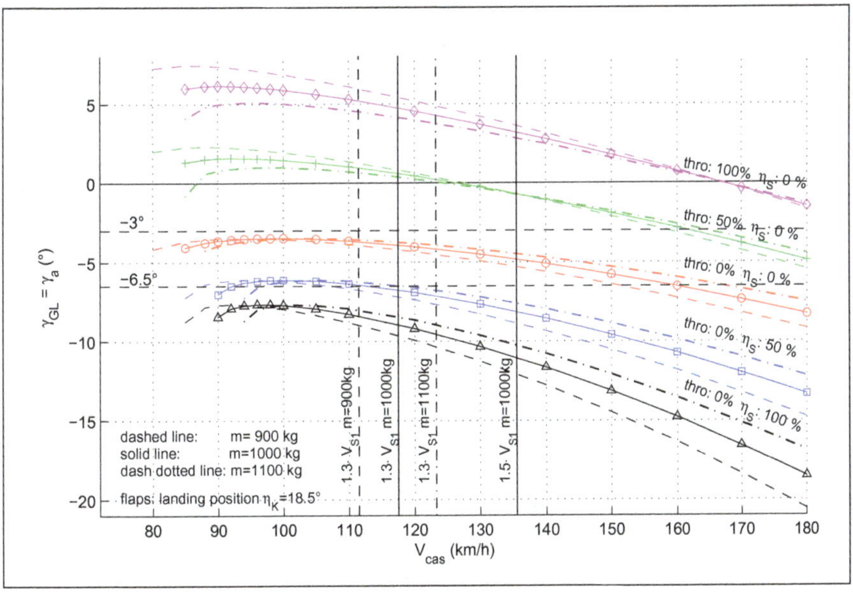

Fig. 1 Glide angle γ_{GL} versus airspeed V_{cas} as a function of aircraft mass m and power setting (throttle- and airbrakes setting (η_S)); aircraft equipped with pods, landing gear extracted

In [1] the minimum landing approach speed is defined by $V_{cas} = 1.3 \cdot V_{S1}$. At this speed and without thrust and with retracted airbrakes, the maximum glide angle for the reference aircraft is $\gamma_{GL,max} = -3.5°$. The glide angle γ_{GL} corresponds to the air path inclination angle γ_a in straight steady flight and is equal to the flight path angle $\gamma_{GL} = \gamma_a = \gamma$ if no wind occurs. To follow a glide path with a glide slope angle of $\gamma_{GS} = -3°$, which is the mean inclination for an instrument landing system (ILS) approach, in this example thrust is needed.

Steeper flight paths can be flown without thrust. For example, to follow a glide slope an angle of $\gamma_{GS} = -6.5°$ in windless condition, no thrust is needed. Instead of this, the airbrakes have to be extended to the 50% position (m=1000 kg). Using this glide slope angle, the airbrakes can be used from the 50% position in both directions, to compensate disturbances, like wind. By this, together with the elevator as control element, airspeed and glide path deviations can be controlled at the same time.

From fig. 2 the dependencies of the flight path velocity V_K and of the wind angle of attack α_W can be derived as functions of the wind components u_{Wg} and w_{Wg} and of the airspeed V_A and the flight path angle γ:

$$V_K = V_A \cdot \cos\alpha_W + u_{Wg} \cdot \cos\gamma - w_{Wg} \cdot \sin\gamma \tag{1}$$

$$\sin\alpha_W = -\frac{u_{Wg}}{V_A} \cdot \sin\gamma - \frac{w_{Wg}}{V_A} \cdot \cos\gamma \quad . \tag{2}$$

Fig. 2 Kinematic relations of velocities and angles for the longitudinal motion in the vertical plane

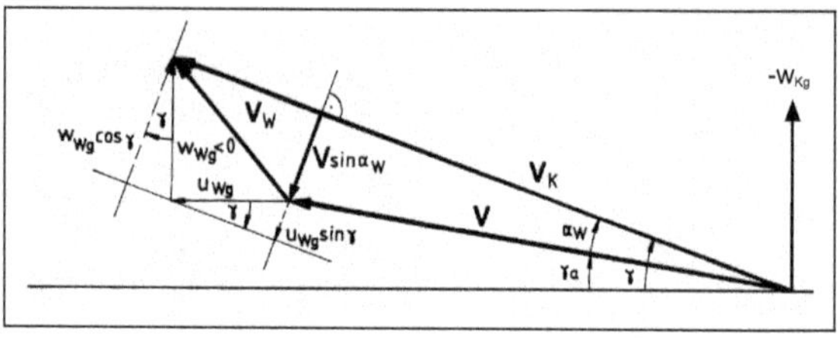

For descent where the flight path angle is negative ($\gamma < 0$) and with a positive airspeed V_A eq. (2) shows, that α_W is proportional to the horizontal and the vertical wind components in the way:

$$\alpha_W \sim u_{Wg} \quad \text{and} \quad \alpha_W \sim -w_{Wg} \quad . \tag{3}$$

Using the angle equation $\gamma_a = \gamma - \alpha_W$ and with $\gamma < 0$ the correlation between the air path angle γ_a and the wind can be shown:

$$\gamma_a \sim -u_{Wg} \quad \text{and} \quad \gamma_a \sim w_{Wg} \, . \tag{4}$$

If the glide slope angle γ_{GS} and the airspeed V_A shall be maintained, tailwind ($u_{Wg} > 0$) as well as upwind ($w_{Wg} < 0$) lead to a reduction of γ_a in steady flight, which means it gets steeper. Vice versa increasing headwind as well as downwind result in a flattening of γ_a. The lift-to-drag ratio have to be increased by retracting the airbrakes (see fig. 1). Vice versa the airbrakes have to be extended for increasing tailwind or upwind. Combinations of horizontal and vertical wind components can add up in terms of their effect on α_W and γ_a or cancel each other, see eq. (2).

Fig. 3 shows this relations graphically in the speed polar. The gray backed sections enclose the compensatable wind components. It results, starting from the configuration with 50% airbrake extension, by parallel shifting of the lines with constant flight path angles γ until the configurations with full retracted or extended airbrakes are reached for the same airspeed. In the depicted speed polar a headwind component of approximately $u_{Wg} = -14\,\mathrm{m/s}$ can be compensated by retracting the airbrakes. Larger headwinds or in combination with downwind will drive the aircraft to its performance limit. Without thrust the flight path angle cannot be further increased. As a result the flight glide slope angle during approach cannot be maintained and the runway will not be reached.

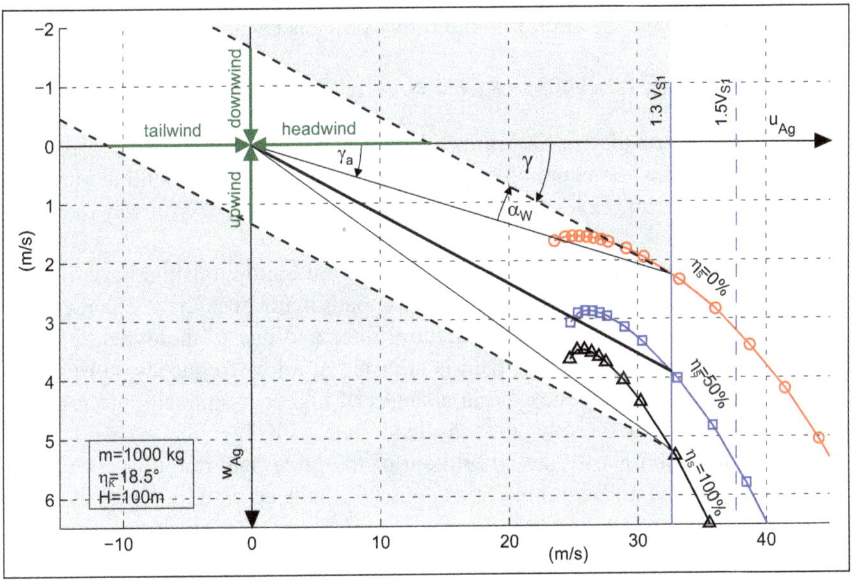

Fig. 3 Speed polars; compensatable wind components by using the airbrakes and for constant path angle γ and airspeed $V_A \approx u_{Ag}$

One option to increase the range of possible wind compensation by airbrakes, is to increase the approach speed. A second option is to use other control elements to adjust the lift-to-drag ratio, like the flaps. Here the flaps are fixed in the landing position, so this option will not be used.

The vertical speed component $w_{Kg} = -\sin\gamma \cdot V_K$ (see fig. 2) becomes a function of the wind if γ and V_A are kept constant because the flight path velocity $V_K = f(u_{Wg}, w_{Wg})$ then depends on the wind, according to eq. (1). This effect on vertical speed leads to different initial conditions for the flare.

3 Glide Path with Adjusted Glide Slope Angle

The main idea is, to adapt the glide slope angle γ_{GS} to the estimated prevailing wind conditions, while keeping the air path angle γ_a constant during the approach. That means, the aerodynamic trim condition mainly stays constant. The glide path will then be adjusted as a function of the current wind speed.

With $u_{Ag} = \cos\gamma_a \cdot V_A$ and $w_{Ag} = -\sin\gamma_a \cdot V_A$ the flight path angle γ results as:

$$\tan\gamma = -\frac{w_{Kg}}{u_{Kg}} = \frac{\sin\gamma_a \cdot V_A - w_{Wg}}{\cos\gamma_a \cdot V_A + u_{Wg}}. \tag{5}$$

Headwind and downwind lead to steeper, tailwind and upwind to shallower flight path angles. Using this resulting flight path angle[1] as glide slope angle $\gamma = \gamma_{GS}$, the glide slope will be adjusted to the wind conditions. The glide path height H_{GS} for the current distance of the aircraft to the runway x_{rwy} is calculated by:

$$H_{GS} = -\tan\gamma_{GS} \cdot x_{rwy} + H_{rwy} + 15\,\text{m} + z_{GRP,RCP}. \tag{6}$$

H_{rwy} is the height of the runway threshold above mean sea level, $z_{GRP,RCP}$ is the distance between the main landing gear and the position of the sensor that measures the height above mean sea level[2]. The glide path height above the runway threshold is 15 m.

The glide slope angle and thus the glide path are continuously adjusted to the current wind condition. By this the total glide path is not straight, it changes with the wind. As the aircraft motion is a natural filter and due to the dynamic of the flight control functions, the glide path is adjusted at a low frequency, normally in the range of phugoid frequency. Wind changes of higher frequencies are regarded as disturbances and are directly compensated by the airbrakes.

Fig. 4 shows exemplarily the adjustment of the glide path due to a discrete increase of the headwind. Beginning from point ①, the headwind increases by a step

Fig. 4 Transition between glide paths after adjusting the glide path due to a discrete increase of headwind

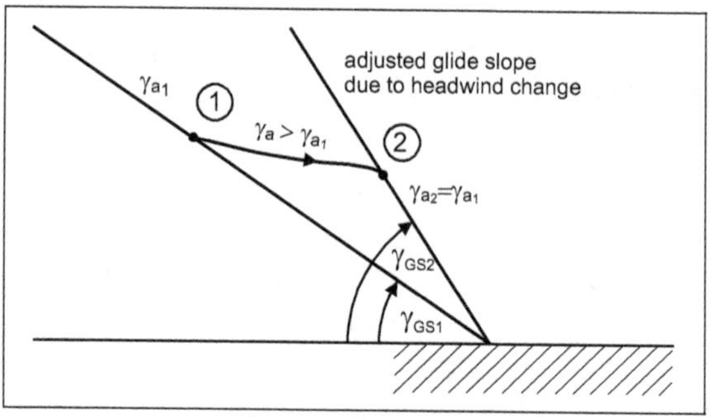

and the calculated glide path gets steeper. The flight control laws (FCL) command the resulting higher glide path height at the current distance to the runway threshold. As a result a temporarily shallower flight path angle γ is commanded to reach

---

[1] The wind components are calculated on-board by using the sensor values of the inertial and the aerodynamic measurement systems.

[2] The height above mean sea level is measured by the GPS (EGNOS based).

the new commanded height. The new, steeper glide path will be reached at point ②. During the transition between point ① to ② the air path angle γ_a also gets shallower. This is achieved at constant air speed by retracting the airbrakes. After tracking the new glide path the airbrakes are moved back to the 50% reference position and the flight resumes with the previous aerodynamic condition of point ①.

4 Description of the Landing Procedure

Fig. 5 schematically shows the landing procedure in the vertical plane for the longitudinal motion of the aircraft. The aircraft flies to the glide path with the maximum glide path angle in order to reach the glide path at the maximum possible height[3]. The glide slope angle γ_{GS} and the glide path are continuously adjusted as functions of the current wind condition by using eq. (5) and (6). With the assumption, that vertical wind variations occur only for short time periods and that those effects can be compensated by using the airbrakes, the vertical wind component is neglected in the calculation of the glide slope angle γ_{GS}.

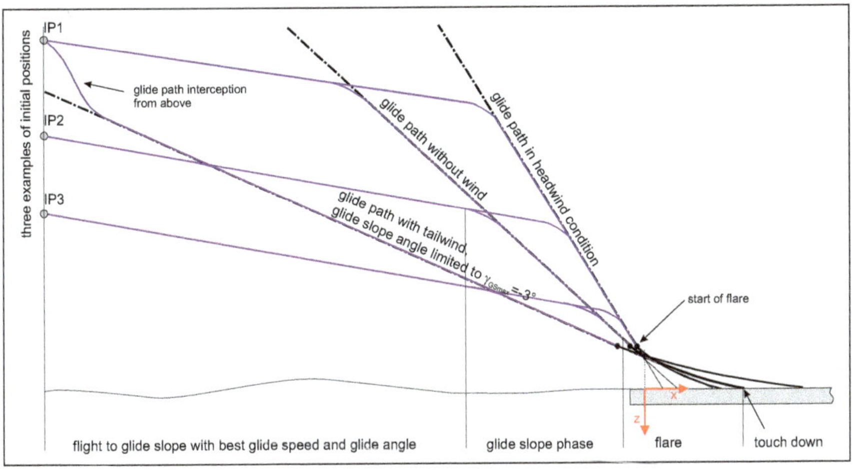

Fig. 5 Sketch of the landing procedure for three different initial positions IP1, IP2, IP3; Remark: The flight paths to the interception points are simplified depicted as parallel. As the best glide path angle is a function of the wind, correctly the flight path for headwind is steeper than for tailwind.

[3] The airspeed of best glide path angle depends on the wind condition. For headwind it is higher with a steeper glide path angle. A more detailed description of the function for calculating and commanding this airspeed is given in [7]. It is assumed, that the initial altitude is high enough to reach the runway under the prevailing wind condition. The minimum required altitude is a function of the prevailing wind condition, the aircraft distance to the runway and the maximum achievable glide path angle.

After reaching the glide path, the flight control law for glide path tracking is activated. The aircraft follows the varying glide path to the runway. At the switching height for flare, the flare control law is activated and the aircraft flares until touchdown.

The described algorithms and functions for the automatic landing can be used for landing approaches with as well as without available engine thrust. Only the approach until the glide path differs dependent on the availability of thrust. With thrust, the aircraft flies from a defined initial approach point to the glide path at a commanded height within a range[4] of 300 till 500m. The glide path is defined in that way, that no thrust is required on it and the airbrakes has to be used. So, after the glide path interception the approach along the glide path with and without available thrust is the same.

5 Flare Control

In this section the flare control algorithms are briefly described. A more detailed description is given in [7].

Vertical Flare Profile

The vertical flare profile is defined by an exponential curve. The time dependent functions for the height and vertical speed are:

$$H(t) = (H_{\mathrm{FL}} - H_{\mathrm{c}}) \cdot e^{-t/\tau_{\mathrm{flare}}} + H_{\mathrm{c}} \quad (7)$$

$$\dot{H}(t) = -\frac{1}{\tau_{\mathrm{flare}}} (H_{\mathrm{FL}} - H_{\mathrm{c}}) \cdot e^{-t/\tau_{\mathrm{flare}}} \,. \quad (8)$$

H_{c} is the constant final height, τ_{flare} the time constant and H_{FL} the initial height for the flare with t=0. This profile can be implemented by a first order control function with a proportional gain:

$$\dot{H}_{\mathrm{flare,c}} = (H_{\mathrm{c}} - H) \cdot K_{\mathrm{H,flare}} \quad , K_{\mathrm{H,flare}} = \frac{1}{\tau_{\mathrm{flare}}} \,. \quad (9)$$

The proportional gain $K_{\mathrm{H,flare}}$ allows fitting the exponential flight path curve, the final height command H_{c} can be used to design the flare duration and by this the vertical touchdown speed. H_{c} is set below the runway surface to prevent an exponential flare and to guarantee a touchdown within a defined runway range. Both parameters were optimized in simulation runs for a nominal vertical touchdown speed of $\dot{H}_{\mathrm{TD}} = -0.4\,\mathrm{m/s}$.

For constant air path angle ($\gamma_a = $ const.) and airspeed ($V_A = $ const.) the vertical speed w_{Kg} on the glide path only depends the vertical and not on the horizontal wind speed:

---

[4] The current altitude of the aircraft is limited to the height range between 300 and 500m and set as constant command.

$$w_{Kg} = -\sin\gamma_a \cdot V_A + w_{Wg} \,. \tag{10}$$

If no vertical wind is present, the aircraft approaches the runway before flare with a defined and constant vertical speed. By comparing w_{Kg} on the glide slope with the commanded vertical speed for the flare $\dot{H}_{\text{flare,c}}$ (eq. (9)), which is only time-dependent, the switching condition to flare law can be realized. This leads on the one hand to a transient free transition. On the other hand the switching height is always the same due to the exponential time dependent relation of the height regarding to the vertical speed. The switching condition is:

$$\dot{H}_{\text{flare,c}} \geq \dot{H}_{\text{GS,c}} \qquad \text{with} \qquad \dot{H}_{\text{GS,c}} \approx \gamma_c \cdot V_{\text{gnd}} \,. \tag{11}$$

For the switching condition the vertical speed command on the glide path $\dot{H}_{\text{GS,c}}$ is used instead of the vertical speed w_{Kg}. This stabilizes the switching height in turbulent atmosphere, where the vertical speed changes often. Fig. 6 shows the switching heights and positions depending on the horizontal wind. While the switching height stays constant, the switching distance to the runway threshold varies. Due to the steeper glide path and the slower ground speed at headwind conditions, the touchdown point is closer to the runway threshold. The flare duration and the calculated vertical touchdown speed are the same in all wind conditions.

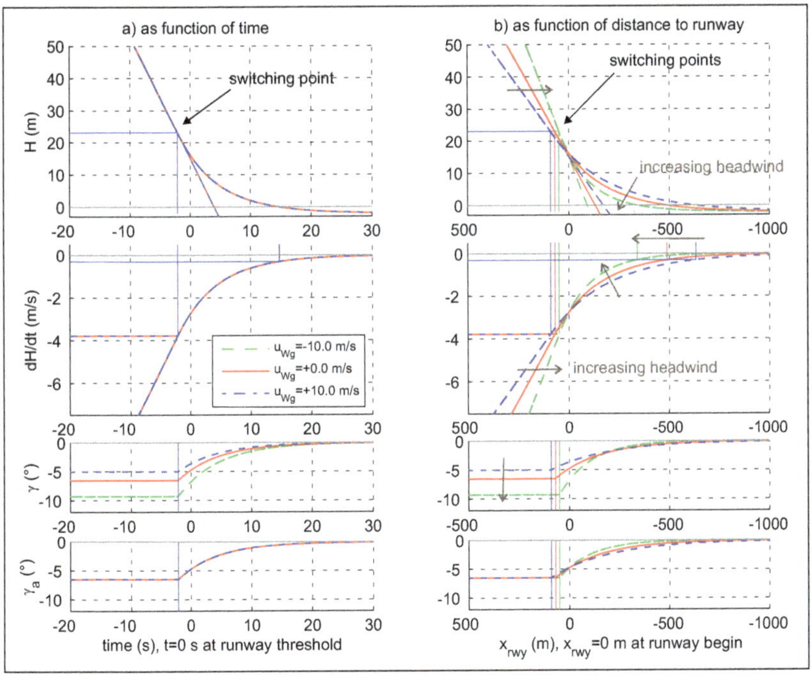

Fig. 6 Calculated switching heights and positions, depending on horizontal wind, V_A constant during flare

Speed Profile during Flare

During flare the airspeed is reduced. It is controlled as well. By this changing wind conditions in the boundary layer and the ground effect on the aerodynamic forces can be taken into account. This is important especially for light aircraft with high glide ratio. Those aircraft on the one hand are sensitive to wind changes and a sudden reduction of headwind can lead to a stall. On the other hand, the airspeed has to be reduced enough to prevent lift off after touch down. The control law for the airspeed is:

$$\dot{V}_{\mathrm{flare,c}} = (V_{\mathrm{flare,c}} - V_{\mathrm{cas}}) \cdot K_{\mathrm{V,flare}} \, . \tag{12}$$

$K_{\mathrm{V,flare}}$ is a proportional gain, V_{cas} the calibrated airspeed. The speed command $V_{\mathrm{flare,c}}$ is calculated as:

$$V_{\mathrm{flare,c}}(t) = y_{\mathrm{norm}}(t) \cdot (V_{\mathrm{casFL}} - V_{\mathrm{casTD}}) + V_{\mathrm{casFL}} \, . \tag{13}$$

V_{casFL} is the initial airspeed at the beginning of the flare, V_{casTD} is the target touch-down speed, y_{norm} is a normalized function. It is calculated by a second order polynomial, time dependent function:

$$y_{\mathrm{norm}}(t) = c_2 \cdot t^2 + c_1 \cdot t + c_0 \, . \tag{14}$$

The parameters c_j of the normalized function are specified such that the airspeed reduction is similar to the uncontrolled speed reduction with fixed airbrakes and that the airspeed is reduced to the touchdown speed of $V_{\mathrm{casTD}} = 1.05 \cdot V_{\mathrm{S1}}$ just before touchdown. By this it is achieved that the airbrakes are extended shortly before touchdown. This leaves sufficient energy reserve in the system in the near of the ground, where large wind shear gradients may occur. Fig. 7 shows the normalized time-dependent function for reducing the airspeed.

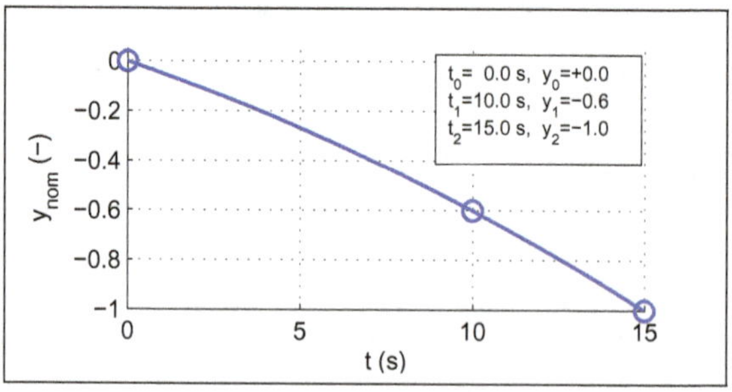

Fig. 7 Normalized function for airspeed speed command reduction during flare

6 Flight Control Laws for Longitudinal Control

For longitudinal control the Total Energy Control System (TECS), which is described in [4], is used in a modified form. The TECS consists of two control cascades. The functions of the inner cascade, the so called *TECS core*, are based on the energy principles. It is split into the control of the total energy rate[5] and the control of the energy distribution to kinetic and potential energy. It provides a decoupled, controlled system for the outer control cascades. Furthermore it provides a speed protection function and an algorithm that provides a transient free switching between the different outer loop modes. The outer control cascades contain the control functions for the vertical path and speed modes. The outputs of the outer cascades, the acceleration command $\dot{V}_{Ac}$ and the air path angle[6] $\tilde{\gamma}_{a,c}$, are the input commands to the TECS core. A detailed description of the TECS can be found in [4] and [5]. Fig. 8 shows the structure of the longitudinal path control.

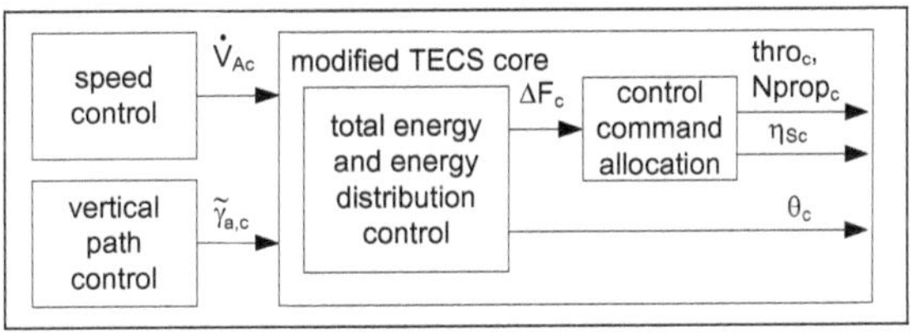

Fig. 8 Schematic overview of longitudinal path control

Modified TECS Core

The TECS core of [4] has been modified as described in [7] and [8] as follows:

- adding of a function to split the thrust command into a throttle and an airbrake command,
- adding of a function to compensate lift changes due to airbrakes deflections,
- adding of a function to compensate lift changes due to speed changes,
- modification of priority switching between speed and flight path priority,
- modification of anti-windup control strategies,
- modification of the total energy control strategy from *PI*-control to *P*-control.

Fig. 9 shows the modified TECS core. The inputs to the TECS core are acceleration command $\dot{V}_{Ac}$ and the glide angle command $\tilde{\gamma}_{a,c}$. Outputs are the pitch angle command Θ_c to the next inner loop, the airbrakes deflection command $\eta_{S,c}$ and the commands to the throttle and propeller revolutions[7].

[5] The total energy is the sum of kinetic and potential energy of the aircraft.

[6] The air path angle command is calculated by the vertical speed command divided by the airspeed: $\tilde{\gamma}_{a,c} = \dot{H}_c / V_A$. Vertical wind components are neglected.

[7] The aircraft is equipped with a piston engine with a propeller with adjustable blade pitch. The blade pitch is controlled by an external controller to achieve the commanded propeller revolutions.

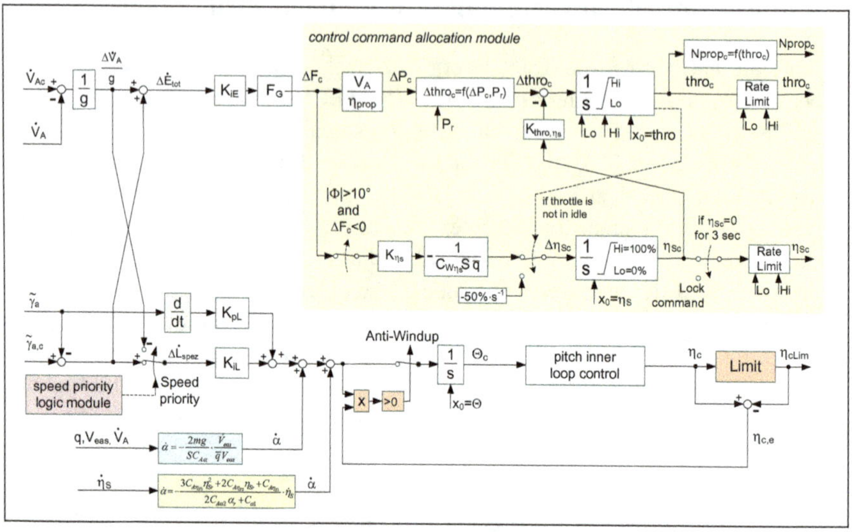

Fig. 9 Modified TECS core

Vertical Flight Control Laws

The basic height control for the approach to the glide path is realized by:

$$\gamma_c = (H_c - H_{msl}) \cdot K_H \cdot V_{gnd}^{-1} . \tag{15}$$

The path angle command γ_c is calculated by the deviation between the commanded height H_c and the height above mean sea level H_{msl}. During the approach to the glide path and without available thrust, H_c is equal to the glide path height at the current aircraft position, according to eq. (6). Before the glide path interception, H_c is normally higher than the current aircraft height. If it is lower, for example, if the landing mode is activated near the runway, then the switching condition (eq. (16)) for glide path interception gets true, the glide path control law is activated immediately and the glide path is intercepted from above, see fig. 5 case IP1 in tailwind.

The glide path will be intercepted and the glide path control law gets active, if the following condition is true:

$$(H_{GS} - H_{msl}) \cdot K_H \cdot V_{gnd}^{-1} - \gamma \quad \leq -\gamma_{GS} \tag{16}$$

By using this switching condition and the glide path law (eq. (18)), the interception can be performed from above and from below the glide path and with the given dynamic of the height control and without overshoot[8].

The vertical distance, at which the switching condition in eq. (16) becomes true, varies dependent on the flight path angle γ, the glide slope angle γ_{GS}, the ground

[8] It is assumed that the dynamic of height control is asymptotically stable and that the height deviation ΔH can be reduced within the aircraft performance limits.

speed V_{gnd} and the controller dynamics which is given by the gain K_H. It can be calculated by:

$$\Delta H = (-\gamma_{\text{GS}} + \gamma) \cdot \frac{V_{\text{gnd}}}{K_H} \quad . \tag{17}$$

For steeper glide slopes (in headwind) the vertical distance ΔH becomes larger, for shallower glide slopes it becomes smaller. Fig. 10 shows the switching conditions schematically.

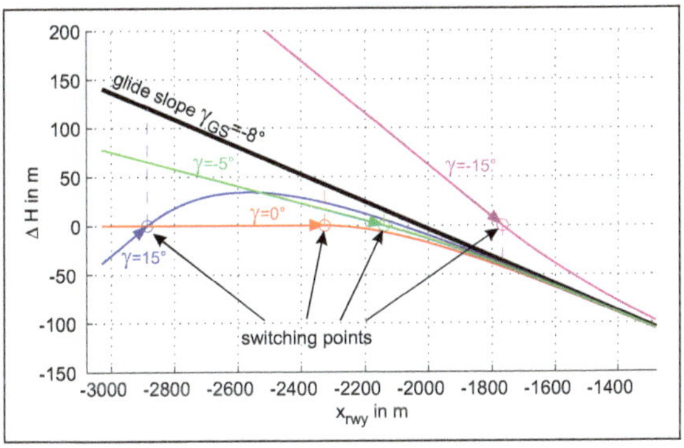

Fig. 10 Switching condition for interception of glide path, dependent on γ; ΔH=0 m for γ=0°

During approach to the glide path the speed priority in the TECS core is active. That means, the elevator controls the airspeed, while the total energy is controlled by the airbrakes. The sink rate depends on the airspeed and the corresponding lift-to-drag ratio. With commanded airspeed of best gliding, the aircraft sinks with the best glide angle.

The glide path control law is similar to the basic height control law for glide slope approach (eq. (15)). The difference is the feed-forward addition of the current glide slope angle γ_{GS}. This addition reduces of the deviation from the glide path, which would otherwise occur. The command for the glide path control is calculated by:

$$\gamma_{\text{c}} = (H_{\text{GS}} - H_{\text{msl}}) \cdot K_{\text{H}} \cdot {V_{\text{gnd}}}^{-1} + \gamma_{\text{GS}} \, . \tag{18}$$

Both path angle commands γ_{c} for height control (eq. (15)) and for glide path control (eq. 18) are afterwards transformed to an air path angle command $\widetilde{\gamma}_{a,\text{c}}$ by multiplication with the ground speed to airspeed ratio to correspond to the specified input to the TECS core:

$$\widetilde{\gamma}_{a,\text{c}} = \gamma_{\text{c}} \cdot \frac{V_{\text{gnd}}}{V_A} \, . \tag{19}$$

The flare control law for the vertical profile is specified in eq. (9). The calculated vertical speed command is transformed to an air path angle command by division with the airspeed V_A. With the runway height above ground H_{rwy} and $z_{\text{GRP,RCP}}$, the flare control law is:

$$\tilde{\gamma}_{a,c,\text{flare}} = \Delta H_{\text{msl}} \cdot K_{\text{H,flare}} \cdot V_A^{-1} \qquad \text{with} \tag{20}$$

$$\Delta H_{\text{msl}} = \left[H_{\text{c,flare}} - H_{\text{msl}} + H_{\text{rwy}} + z_{\text{GRP,RCP}}\right] . \tag{21}$$

During the flare the height above mean sea level is used as control variable. To consider measurement errors in H_{msl}, the height deviation ΔH_{msl} is compared to the corresponding height deviation ΔH_{gnd}, which is the deviation between the height above ground[9] and the corresponding height command. Furthermore, in the flare law a function is implemented that lowers the final height command if the aircraft is close to the ground. As the desired vertical sink rate at touchdown is very low and the glide ratio is high, this ensures the touchdown and prevents from lift off again. More details of this function and of the height correction function are given in [7]. Fig. 11 shows the block diagram for the vertical flight control modes.

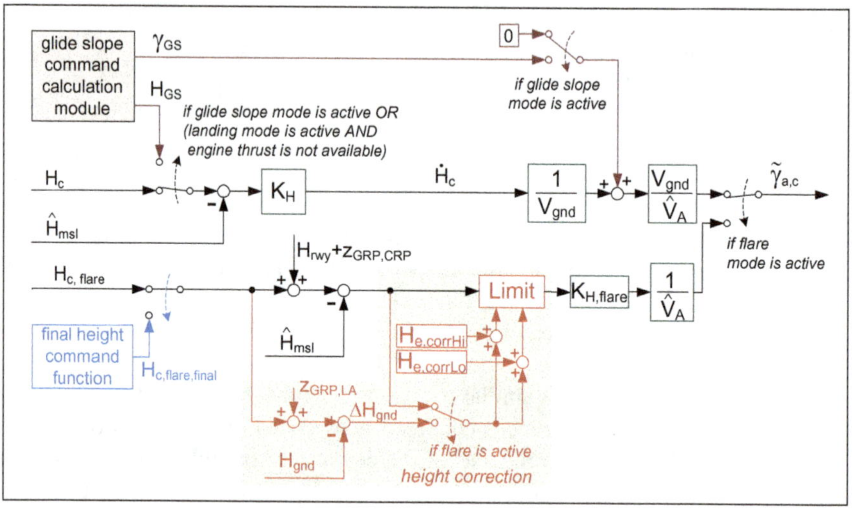

Fig. 11 Block diagram of vertical flight path control

Speed Control Laws

As introduced for flare in eq. (12), the speed control function is:

$$\dot{V}_{\text{c}} = (V_{\text{c}} - V_{\text{cas}}) \cdot K_{\text{V}} . \tag{22}$$

The speed control laws for flare and for the other speed modes only differ in the speed command V_{c} and the gain K_{V}. During the flight to the glide path and without available thrust, the speed of best glide ratio is commanded. On the glide path the approach speed is commanded. During flare, the flare speed command profile of eq. (13) and (14) is used. Fig. 12 shows the block diagram for the speed control laws.

[9] The height above ground is measured in the demonstrator aircraft by redundant laser altimeters.

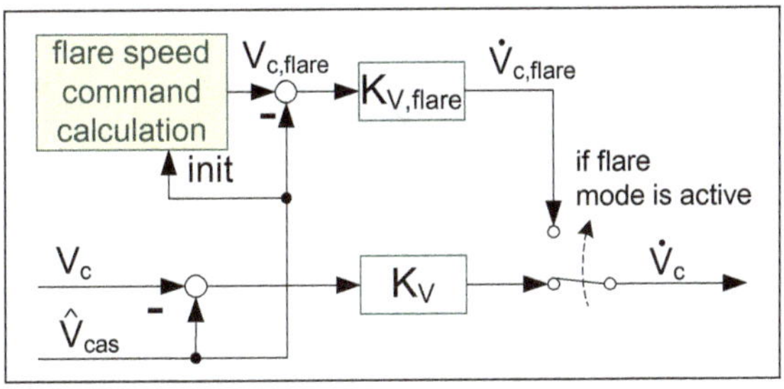

Fig. 12 Block diagram of speed control

7 Simulation and Flight Test Results

The described landing procedure and the algorithms and functions were developed and tested in simulations on PC and then tested in flight in 38 automatic landings since 2014. Compared to the above description, the flight control functions were slightly modified. This modifications increase the safety. They are:

- The commanded approach airspeed on the glide path was increased to $1.5 \cdot V_{S1}$ in order to have a larger energy reserve and safety margin to the stall speed.
- Furthermore an additional safety value is added to the approach speed as a function of the headwind.
- To keep the airbrakes in the 50% position on the glide path, the value of the nominal air path angle without wind is increased to $\gamma_a = -8.0°$. In headwind conditions, where the approach speed is added by an additional value, the air path angle is adjusted, to maintain the flight condition with the airbrakes in the 50% position.
- The glide slope angle γ_{GS} was fixed at heights beneath 100 m above the runway threshold, in order to have a stabilized approach. The constant γ_{GS} makes it easier for the pilot to monitor the approach.

Flight Test Results

Fig. 13 shows the results of three flight tests. The depicted flights took place at almost optimal wind conditions with a light headwind. At these tests the thrust generally was available, so the glide slope was reached at nearly the same height, which improves the analysis of the aircraft behaviour on the glide slope. However further flight test were done without thrust at all. Those are analysed in [7].

After glide path interception the airbrakes are commanded to the 50% position. Due to the small headwind, the glide slope angle γ_{GS} is slightly steeper than the nominal value of $-8°$ for windless condition. The approach speed is approximately $V_{cas} = 138\,\text{km/h}$.

Below 100 m a constant glide slope angle is commanded. The aircraft sinks with a vertical speed of round about $\dot{H} = -5.5\,\text{m/s}$. The flare begins at a height of 35 m. During flare the sink rate is reduced to $\dot{H}_{TD} = -0.6\,\text{m/s}$. The airspeed reduces within 15 s to $V_{cas} = 102\,\text{km/h}$, which is $1.09 \cdot V_{S1}$. The pitch angle increases from $\Theta = -8°$ on the glide path to $\Theta_{TD} = 4.5°$ at touchdown. The flight condition at the touch down satisfies the requirements, which are marked with green and yellow colours (green for satisfactory and yellow for adequate).

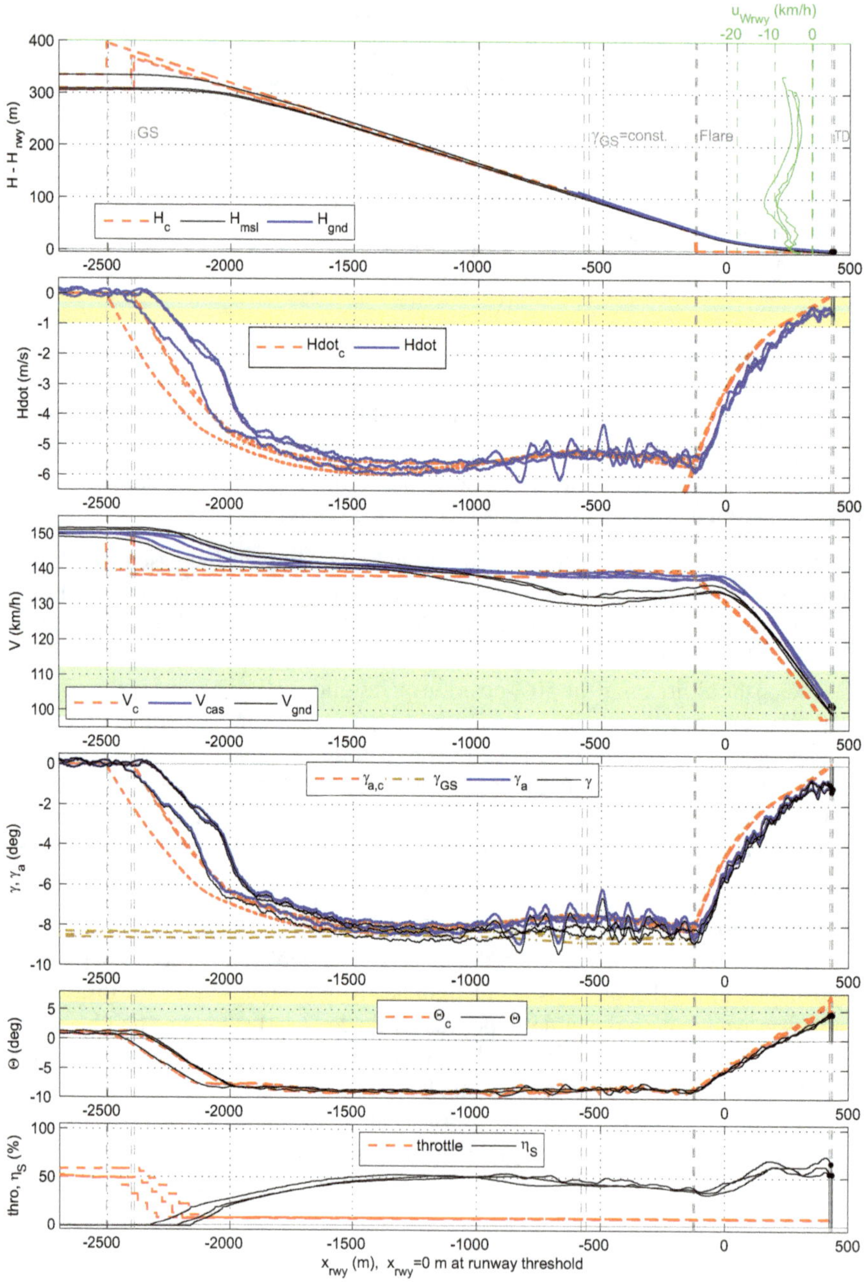

Fig. 13 Flight test results of three automatic landings with the demonstrator

Simulation Results

In simulations automatic landings were tested under extreme wind conditions. Fig. 14 shows examples of simulation results for approaches with a headwind and a tailwind compared to a landing without wind. For the wind the logarithmic wind profile is used, which is defined in CS-AWO as *Normal Wind Shear*[10]. The maximum wind velocities, which the automatic landing function has to cope with, are a headwind of 50 km/h and a tailwind of 10 km/h at the wind reference height of 10 m above the ground. At the glide path interception height of 300 m, the maximum headwind is $u_{Wg} = 82\,\text{km/h}$. The maximum tailwind has a value of 15 km/h.

In the simulations shown in fig. 14 the approach to the glide slope took place with available thrust, so the aircraft flies to the glide slope at the same heights which makes the comparison of the behaviour on the glide slope easier.

Without wind and with tailwind the approach speed is 138 km/h. For maximum headwind the approach airspeed command is increased by 14 km/h which leads to an approach speed of 152 km/h. The air path angle γ_a is adjusted to the approach speed, so that the glide slope can be maintained with the airbrakes in the 50% position. The glide path is adjusted accordingly. For the headwind condition and in the height of 300 m the glide slope angle is $\gamma_{GS} = -20°$. It flattens a bit due to the decrease of wind speed when approaching the ground.

Because of the high γ_a and V_A in the headwind condition, the sink rate on the glide path is high. During flare its value is reduced within 15 s from $\dot{H} = -7.7\,\text{m/s}$ to the vertical touchdown speed[11] of $\dot{H}_{TD} = -0.6\text{m/s}$. The airspeed is reduced during flare from $V_{cas} = 152\,\text{km/h}$ to $V_{cas} = 101\,\text{km/h}$, which is $1.08 \cdot V_{S1}$.

In the maximum headwind condition on the glide path the airbrakes are not in the 50% position, the deflection is 60% at the beginning. This is because γ_{GS} is held constant below 100 m. Due to the decreasing headwind at lower altitudes, the air path angle γ_a increases a bit and the airbrakes have to be extended a bit more. During the flare they are first retracted a bit to compensate the decreasing headwind. Close to the ground they are first extended and then as the airspeed comes closer to the desired touchdown speed, they are retracted again. The pitch angle rises during the flare from $\Theta = -11.6°$ to $\Theta_{TD} = +5.2°$.

In summary, all flight parameters stayed during those extreme wind conditions within the required limits.

---

[10] The wind shear model was taken from CS-AWO [2], book 2, paragraph 3.1.2.1 .

[11] The maximum vertical touchdown speed for the STEMME S15 is $\dot{H}_{TD} = -1.0\,\text{m/s}$.

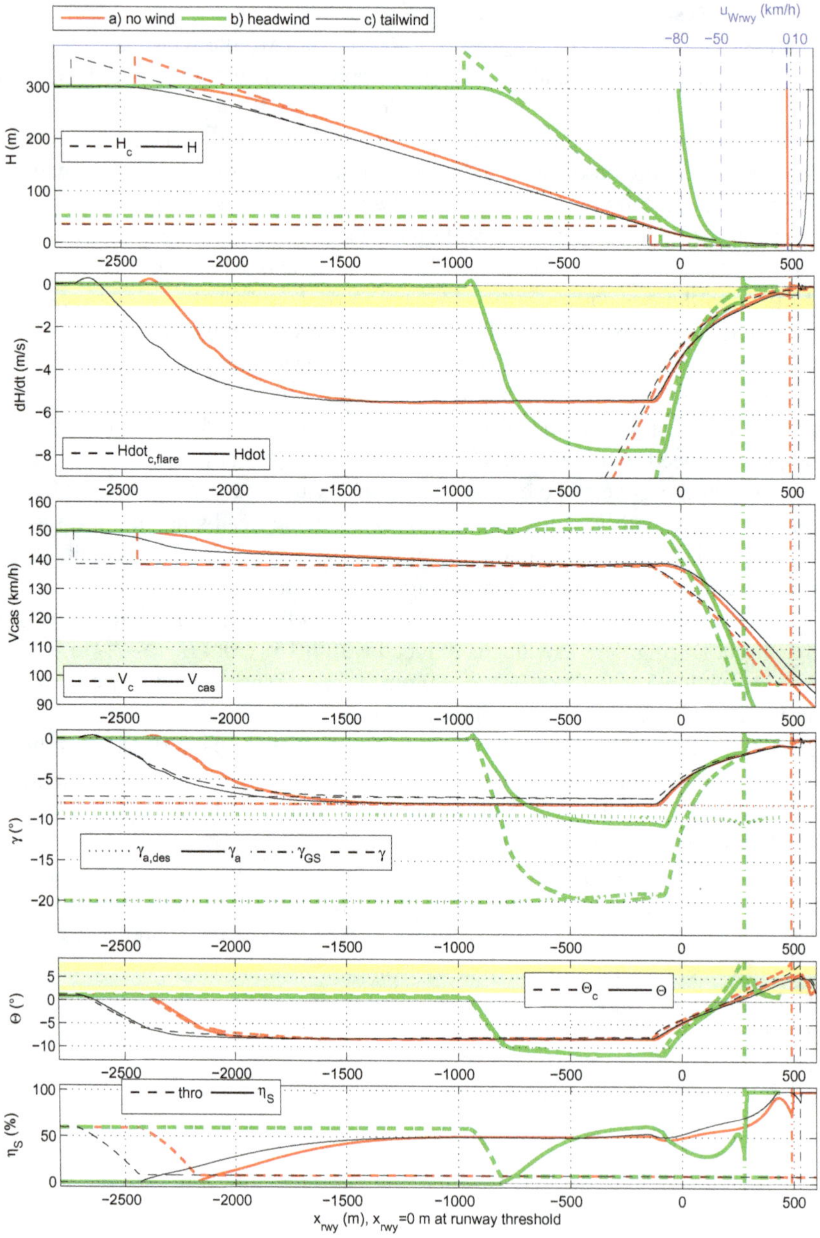

Fig. 14 Simulation results for landing approaches under different wind conditions

8 Conclusion

A method for an automatic landing of a high-aspect-ratio aircraft has been described. It uses the elevator for speed and flight path control as well the airbrakes for total energy control - according to the TECS principle. The thrust is fixed in idle or it is not available, e.g. in case of an engine failure. Such landing methods are applied by glider pilots. They are also used by powered gliders and have to be used in case of complete loss of all engines.

By adaptation of the commanded glide path to the prevailing wind conditions, a vertically bended approach trajectory develops that can be followed by the aircraft even under extreme head- and tailwind conditions.

The autoland control functions have been successfully validated in flight simulations and in flight test with a high-aspect-ratio utility aircraft STEMME S15. Further improvements are possible if a reliable wind prediction is added.

References

1. CS-23, Certification Specifications for Normal, Utility, Aerobatic, and Commuter Category Aeroplanes, European Aviation Safety Agency (EASA), ED Decision 2003/14/RM
2. CS-AWO, Certification Specifications for All Weather Operations, European Aviation Safety Agency (EASA), Decision No. 2003/6/RM
3. Kaminer, I., O'Shaughnessy, P.R.: 4D-TECS Integration for NASA TSRV Airplane, NASA Contractor Report 4231 (1989)
4. Lambregts, A.A.: Vertical Flight Path and Speed Control Autopilot Design Using Total Energy Principles, AIAA (1983) No 83-2239
5. Lambregts, A.A.: TECS Generalized Airplane Control System Design - An Update. In: 2nd CEAS Specialist Conference on Guidance, Navigation & Control, EuroGNC (2013)
6. Lamp, M., Luckner, R.: The Total Energy Control Concept for a Motor Glider. In: 2nd CEAS Specialist Conference on Guidance, Navigation & Control, EuroGNC (2013)
7. Lamp, M.: Automatische Segelfluglandung eines Flugzeuges großer Streckung unter Verwendung der Bremsklappen, dissertation, TU Berlin (to be published)
8. Lamp, M., Luckner, R.: Flight Control Law Development for the Automatic Flight Control System LAPAZ. In: 1st CEAS Specialist Conference on Guidance, Navigation & Control, EuroGNC (2011)
9. Reichmann, H.: Streckensegelflug. Motorbuchverlag, Stuttgart (1975) ISBN 3-87943-371-2

Part IV

Space Applications

Bearings-Only Rendezvous with Enhanced Performance

Jonathan Grzymisch, Walter Fichter, Damiana Losa, and Massimo Casasco

Abstract. Employing only bearing/angular measurements for navigation during the far to medium range rendezvous with a non-cooperative target has several advantages with respect to directly measuring the range using active sensors such as RADAR or LIDAR. Angular measurements can be acquired using simple sensors such as a single optical camera, significantly reducing the mass and power requirements. Nevertheless, several challenges arise form the lack of a direct range measurement, which renders the problem instantaneously unobservable. The execution of known maneuvers is thus necessary to introduce observability in the estimation problem, which results in the navigation performance being directly dependent on the trajectory followed. A few single-maneuver schemes have been proposed to enhance bearings-only navigation performance. Nonetheless, little research has been published on the use of on-line trajectory optimization methods accounting for observability on the complete rendezvous trajectory. This paper presents the non-linear simulation results of a Model Predictive Control architecture for rendezvous that simultaneously enhances bearings-only observability in order to improve navigation performance. A detailed simulation environment provided by Thales Alenia Space France is used to show that the proposed scheme based on linearized equations displays satisfactory performance in a higher fidelity non-linear environment, when observability is considered in the trajectory optimization.

Jonathan Grzymisch · Walter Fichter
Institute of Flight Mechanics and Control, University of Stuttgart, Germany
e-mail: jonathan.grzymisch@ifr.uni-stuttgart.de

Damiana Losa
Research and Technology Department, Thales Alenia Space, France

Massimo Casasco
European Space Technology Centre, European Space Agency, The Netherlands

J. Bordeneuve-Guibé et al. (eds.), *Advances in Aerospace Guidance, Navigation and Control*, DOI: 10.1007/978-3-319-17518-8_33

1 Introduction

In-orbit rendezvous is a well established technology with significant heritage from missions such as in-orbit assembly of space stations, planetary/lunar exploration and return as well as several servicing missions, notably many performed during the space shuttle era. Nevertheless, when the target is non-cooperative, skin tracking sensors such as RADAR or LIDAR are required due to the lack of repeaters or reflectors on the target. These sensors consume high power and are thus massive when they are designed to operate in the far to medium range, since the power reflected back from the skin of a far away target diminishes as function of the fourth power of the distance [6]. In addition, for missions servicing defunct satellites or Active Debris Removal (ADR) in Earth orbit, high accuracy ground tracking of the targeted debris is an alternative navigation method for the far to medium range rendezvous phases. However, costly dedicated tracking campaigns just before the rendezvous are needed in order to achieve the required accuracy [1]. For these reasons, employing only bearing or angular measurements for navigation with non-cooperative targets has several advantages with respect to directly measuring the range using active sensors, especially in the far to medium range. These measurements can be obtained using simple and compact sensors such as a single optical camera, which would significantly reduce the mass and power requirements, enabling rendezvous missions for smaller and less expensive platforms.

Several challenges arise when using only bearing measurements, since the lack of a direct range measurement renders the problem instantaneously unobservable. Therefore, in order to introduce the additional information needed fully estimate the relative position, the execution of known maneuvers is required, which renders the problem observable [20, 11]. However, this results in the achievable navigation performance being directly dependent on the trajectory followed, creating an intrinsic coupling between guidance and navigation, which is not usual when employing ranging sensors. For this reason, a few authors have proposed maneuvering schemes to enhance bearings-only navigation performance during a rendezvous [19, 10]. However, the proposed single maneuvers introduce optimal observability, but are detached from the overall rendezvous trajectory design. In addition, several observability metrics have been proposed [22, 13, 8], which could be employed inside a trajectory optimization scheme in order to yield complete rendezvous trajectories that enhance bearings-only observability. This would guarantee observable trajectories with reduced risk of navigation filter divergence, as well as enhance the overall navigation performance by aiding filter convergence. However, little research has been published on the use of on-line trajectory optimization methods incorporating bearings-only observability. In fact, the authors of [3] describing the results of the recent angles-only rendezvous experiment (ARGON), performed at the end of the PRISMA mission, claimed the need for such an optimization of the guidance profile to enhance bearings-only observability. Model Predictive Control (MPC) methods, such as those proposed by [15] for minimum fuel rendezvous, can be used to implement such a trajectory optimization on-line. Our previous work in [12] introduces these concepts, presenting a comprehensive analysis of the trade-off between observability and fuel consumption, and its impact on bearings-only navigation performance. Nonetheless, the analysis presented there is based on linearized equations of motion, and proposes an optimization transcription which is limited in terms of its applicability to more realistic non-linear constraints.

This paper presents the non-linear simulation results of a Model Predictive Control (MPC) formulation to generate fuel-optimal rendezvous trajectories that simultaneously enhance bearings-only observability in order to guarantee observable trajectories and improve navigation performance. Monte-Carlo simulations using a non-linear spacecraft simulator provided by Thales Alenia Space France [14] demonstrate that an efficient trajectory optimization scheme based on linearized equations of motion can display satisfactory performance in a higher fidelity non-linear environment, when an observability metric is weighed into the optimization. In addition, the proposed implementation uses curvilinear coordinates and a superior optimization transcription as compared to our previous work. This results in half the optimization decision variables, making the problem not only faster to solve, but also compatible with advanced Non-Linear Programming (NLP) solvers such as ESA's WORHP [18], and amenable to the inclusion of other non-linear constraints, for example to ensure passive trajectory safety.

2 Background

2.1 In-Orbit Relative Motion

The homogeneous solutions to the in-orbit relative motion equations for circular orbits in Cartesian coordinates are known as the Clohessy-Wiltshire equations [6]. These equations can also be written in closed form in other coordinate systems, such as relative orbital elements [7] or curvilinear coordinates [5]. There is also a similar homogeneous closed form solution for relative motion in elliptical orbits [21]. Moreover, there are particular solutions for a variety of maneuvers such as impulsive or constant thrust [6], which can be used to model chemical and electrical propulsion systems respectively.

Combined, these solutions describe the relative trajectory of two spacecraft in close proximity. The relative position at any time t after a maneuver occurring when the state is $\mathbf{x}_m$ can be written as follows:

$$\mathbf{x}(t) = \Phi(t)\,\mathbf{x}_m + G(t)\,\mathbf{u} \tag{1}$$

where the state vector ($\mathbf{x}$), state transition (Φ) and input (G) matrices are defined in the Appendix for the case of curvilinear coordinates and impulsive maneuvers. Note that the time dependencies (t) will be omitted for conciseness.

2.2 *Optimal Bearings-Only Observability*

The several authors have shown that to provide observability to the in-orbit bearings-only navigation problem, a maneuver must change the resulting angular measurement profile from its natural evolution without maneuvers [20, 11]. That is, the sequence of angular measurements after a maneuver must be different to the sequence that would have resulted if no maneuver had occurred. As a result, these authors later proposed observability metrics that can be used to compute single maneuvers that "maximize" this difference, in order to optimize observability [19, 10].

The observability metric proposed in [10] "maximizes" the difference in the angular measurement profiles by *minimizing* the dot product between the relative position vector after a maneuver ($\mathbf{x}_p = \Phi_s \mathbf{x}_m + G_r\mathbf{u}$) and the natural evolution of this position without maneuvers ($\bar{\mathbf{x}}_p = \Phi_s \mathbf{x}_m$). This metric is recalled below [10]:

$$J_m(\mathbf{u}, \mathbf{x}_m, t) = (\Phi_s \mathbf{x}_m)^T (\Phi_s \mathbf{x}_m + G_r\mathbf{u}) \tag{2}$$

Note that the applicability of this metric was shown to be constrained to the positive results only, since opposite position vectors $\mathbf{x}_p$ and $\bar{\mathbf{x}}_p$ provide mirrored angular measurements, which are different and therefore observable [10]. Thus, Eq. (2) is constrained, as follows [10]:

$$J_m(\mathbf{u}, \mathbf{x}_m, t) > 0 \tag{3}$$

Equation (2) is a single-maneuver observability metric which indicates grater observability with decreasing values of $J_m(\mathbf{u}, \mathbf{x}_m, t)$. Maximum observability is attained when this value tends towards zero, due to the constraint in Eq. (3). In the following sections, this metric will be used to optimize observability for each maneuver along the entire rendezvous trajectory.

3 Rendezvous Trajectory Optimization

In order to generate rendezvous approach trajectories that are also optimal for bearings-only observability, the observability metric shown in Equation (2) is used as an optimization objective inside a fuel-optimal rendezvous trajectory optimization scheme. The problem is then posed as the weighed minimization of a fuel cost J_f and an observability cost J_o, as was previously proposed in [12]. This section proposes a new direct shooting transcription method [2] for the bearings-only rendezvous problem, which is a transcription resulting in less than half independent variables. This makes the problem not only much simpler for the optimizer, but also compatible with advanced NLP solvers such as ESA's WORHP [17], that are needed if non-linear constraints are to be included.

Without non-linear constraints on the trajectory geometry, discritizing the time-varying dynamics over n time-steps results in a convex quadratic optimization problem for fuel/observability fixed-time rendezvous. The resulting quadratic problem is then solved using the Quadratic Programming (QP) solver included in the MATLAB® Optimization Toolbox.

3.1 Problem Discretization

The trajectory can be discretized into n steps separated by a fixed time step ΔT, where chaser maneuvers are allowed at each time step $i = 1, 2, \ldots, n$. Thus, the position of the chaser spacecraft at each discrete point i can be written as follows:

$$\mathbf{x}_i = \Phi(i\Delta T)\mathbf{x}_0 + \sum_{k=0}^{i-1} G((i-k)\Delta T)\mathbf{u}_k \tag{4}$$

The maneuvers $\mathbf{u}_i$ performed at each time step are the free input quantities that define the trajectory, the optimization decision variables. Thus, they can be collected in a large input vector $\mathbf{U}$ and the summation in Eq. (4) can be written in matrix form:

$$\sum_{k=0}^{i-1} G((i-k)\Delta T)\mathbf{u}_k = \mathbb{G}_i\mathbf{U}_i \tag{5}$$

where the large input matrix $\mathbb{G}_i$ is defined at time step i as follows:

$$\mathbb{G}_i = \left[G(i\Delta T)\ G((i-1)\Delta T)\ G((i-2)\Delta T)\ \ldots\ G(\Delta T) \right] \tag{6}$$

and the maneuver vector is:

$$\mathbf{U}_i = \begin{bmatrix} \mathbf{u}_0 \\ \mathbf{u}_1 \\ \vdots \\ \mathbf{u}_{i-1} \end{bmatrix} \tag{7}$$

For the entire trajectory, the optimization decision vector contains all n maneuvers and is defined as $\mathbf{U} \equiv \mathbf{U}_n$.

3.2 Fuel Cost

The fuel cost function is the the sum of all thrusts during the trajectory:

$$J_f = \sum_{i=0}^{n-1} |\mathbf{u}_i| \tag{8}$$

Equation (8) is a piece-wise linear function that can not be implemented into a discrete programming optimization framework. Yet, it can be converted into a linear function using slack variables [16]. Defining a new set of slack variables $\mathbf{S} = [\mathbf{s}_0\ \mathbf{s}_1\ \ldots\ \mathbf{s}_n]$, the optimization decision vector can be augmented to include them $\mathbf{U}_s = [\mathbf{U}\ \mathbf{S}]^T$. Hence, the cost function becomes the sum of all $\mathbf{s}_i$, which can be written in linear form as follows:

$$J_f = F_f^T \mathbf{U}_s \tag{9}$$

where $F_f = [\mathbf{0}\ \mathbf{1}]^T$. Each slack variable $\mathbf{s}_i$ is then subject to the following constraints, which render them equal to the absolute value of $\mathbf{u}_i$ when they are minimized:

$$\mathbf{u}_i - \mathbf{s}_i \leq \mathbf{0} \tag{10}$$

$$-\mathbf{u}_i - \mathbf{s}_i \leq \mathbf{0} \tag{11}$$

The collection of these constraints for all times steps i can be written as follows:

$$A_f \mathbf{U}_s \leq \mathbf{b}_f \tag{12}$$

where

$$A_f = \begin{bmatrix} I & -I \\ -I & -I \end{bmatrix} \tag{13}$$

$$\mathbf{b}_f = \begin{bmatrix} \mathbf{0}\ \mathbf{0} \end{bmatrix}^T \tag{14}$$

and I is a $3n \times 3n$ identity matrix.

3.3 Observability Cost

The observability objective function in Eq. (2) can be used to demand observability at each time step along the trajectory. This is equivalent to demanding each maneuver to provide maximum possible observability at the next discretized time step, i.e., demanding optimal observability throughout the complete trajectory.

Therefore, the trajectory observability cost can be defined as the sum of the cost for each time step i from Eq. (4). Expanding this equation, the summation becomes:

$$J_o = \sum_{i=0}^{n} \left(\mathbf{x}_i^T \Phi_s^T \Phi_s \mathbf{x}_i + \mathbf{x}_i^T \Phi_s^T G_r \mathbf{u}_i \right) \tag{15}$$

In order to use this equation in the optimization, the cost J_o must only be a function of the maneuvers $\mathbf{u}_i$. Thus, the relative state at each point of the trajectory $\mathbf{x}_i$ must be substituted from Eq. (4) into Eq. (15). After some algebra and expanding of the summations, the cost J_o can be written as follows:

$$J_o = \sum_{i=0}^{n} (Q_i + L_i \mathbf{u}_i + C_i) \tag{16}$$

where

$$Q_i = \sum_{j=0}^{i-1} \left(\mathbf{u}_j^T G((i-j)\Delta T)^T \Phi_s^T G_r \mathbf{u}_i \ldots \right.$$

$$\left. + \sum_{k=0}^{i-1} \mathbf{u}_j^T G((i-j)\Delta T)^T \Phi_s^T \Phi_s G((i-k)\Delta T) \mathbf{u}_k \right) \quad (17)$$

$$L_i = \mathbf{x}_0^T \Phi(i\Delta T)^T \Phi_s^T G_r + 2\mathbf{x}_0^T \sum_{j=i+1}^{n-1} \Phi(j\Delta T)^T \Phi_s^T \Phi_s G((j-i)\Delta T) \quad (18)$$

$$C_i = \mathbf{x}_0^T \Phi(i\Delta T)^T \Phi_s^T \Phi_s \Phi(i\Delta T) \mathbf{x}_0 \quad (19)$$

Equation 16 is a quadratic function of the maneuvers $\mathbf{u}_i$ and the initial conditions $\mathbf{x}_0$. The constant term with respect to the optimization decision variables C_i can be omitted from the optimization and a reduced objective function $J_o^{'}$ can be defined. In matrix form, this can be written as a quadratic function of the decision vector $\mathbf{U}$, which allows direct inclusion into a quadratic programming (QP) optimizer:

$$J_o^{'} = \frac{1}{2} \mathbf{U}^T H_o \mathbf{U} + F_o^T \mathbf{U} \quad (20)$$

where

$$F_o = \begin{bmatrix} F_0 \; F_1 \; F_2 \; \ldots \; F_{n-1} \end{bmatrix} \quad (21)$$

$$H_o = \begin{bmatrix} H_{0,0} & H_{0,1} & \ldots & H_{0,n-1} \\ H_{1,0} & H_{1,1} & \ldots & H_{1,n-1} \\ \vdots & \vdots & \ddots & \vdots \\ H_{n-1,0} & H_{n-1,1} & \ldots & H_{n-1,n-1} \end{bmatrix} \quad (22)$$

and

$$F_i = \mathbf{x}_0^T \Phi(i\Delta T)^T \Phi_s^T G_r + 2\mathbf{x}_0 \sum_{j=i+1}^{n-1} \Phi(j\Delta T)^T \Phi_s^T \Phi_s G((j-i)\Delta T) \quad (23)$$

$$H_{i,j} = G((j-i)\Delta T)^T \Phi_s^T G_r + \sum_{k=0}^{i-1} G((i-j)\Delta T)^T \Phi_s^T \Phi_s G((i-k)\Delta T) \quad (24)$$

The matrix G is defined to be zero ($G(t) \equiv 0$) when its argument is negative ($t \leq 0$).

3.4 Composite Optimization Cost

The observability and fuel costs can be combined in a weighted sum, where the weight w describes the trade-off between the objectives, as follows:

$$J_w = (w-1)J_f + wJ_o \tag{25}$$

Since the value of the J_f and J_o objectives can differ by orders of magnitude at the optimal solution, the objectives must be normalized in order to aid optimization convergence [4]. Thus, the objectives J_f and J_o are divided by their value at the fuel optimal trajectory solution $\mathbf{Y}_f$, pre-computed using an LP optimization routine.

Finally, for any given mission, the trade-off w which takes a value from 0 to 1, must be selected depending on the mission performance objectives and available fuel. A comprehensive analysis for its selection is presented in [12].

3.5 Relative Motion Constraints

The discrete relative motion dynamics from Eq. (4) are enforced by "direct shooting" integration. Writing the equation for the desired final relative position/velocity $\mathbf{x}_f$ at time step n using Eq. (4) and isolating the maneuver term yields the following set of linear equality constraints:

$$\mathbb{G}_n\mathbf{U} = \mathbf{x}_f - \Phi(n\Delta T)\mathbf{x}_0 \tag{26}$$

3.6 Geometric Trajectory Constraints

In a similar manner, other constraints can be included in the optimization. For example, geometrical constraints on the trajectory can be included by computing the discrete trajectory $\mathbf{X} = [\mathbf{x}_0\ \mathbf{x}_1\ \ldots\ \mathbf{x}_n]^T$:

$$\mathbf{X} = \mathbb{P}\mathbf{x}_0 + \mathbb{G}\mathbf{U} \tag{27}$$

where

$$\mathbb{P} = \left[I\ \Phi(\Delta T)\ \Phi(2\Delta T)\ \ldots\ \Phi((n-1)\Delta T)\right]^T \tag{28}$$

$$\mathbb{G} = \begin{bmatrix} 0 & 0 & 0 & \ldots & 0 \\ G(\Delta T) & 0 & 0 & \ldots & 0 \\ G(2\Delta T) & G(\Delta T) & 0 & \ldots & 0 \\ \vdots & \vdots & \ddots & \ddots & \vdots \\ G((n-1)\Delta T) & G((n-2)\Delta T) & \ldots & G(\Delta T) & 0 \end{bmatrix} \tag{29}$$

and I is the identity matrix of size equal to that of the matrix Φ.

Subsequently, the geometrical constraints can be defined as follows:

$$\mathbf{X}_l \leq \mathbb{G}\mathbf{U} \leq \mathbf{X}_u \tag{30}$$

where

$$\mathbf{X}_{u,l} = \begin{bmatrix} \mathbf{x}_0^{u,l} \\ \mathbf{x}_1^{u,l} \\ \vdots \\ \mathbf{x}_n^{u,l} \end{bmatrix} - \mathbb{P}\mathbf{x}_0 \tag{31}$$

and the vectors $\mathbf{x}_i^u$ and $\mathbf{x}_i^l$ correspond to the upper and lower limits for each point i in the discrete trajectory. For example, in order to set a boundary limit so that the chaser spacecraft does not cross the position of the target ($\mathbf{x}_i^l = \mathbf{0}$ for all i).

3.7 *Thrust Constraints*

Thrust magnitude constraints can be included for the physical limitations of thrusters. These can be written in the following inequality:

$$A_u\mathbf{U} \leq \mathbf{b}_u \tag{32}$$

where

$$A_u = \begin{bmatrix} I \\ -I \end{bmatrix} \tag{33}$$

$$\mathbf{b}_u = \begin{bmatrix} \mathbf{u}_{max} \; \mathbf{u}_{min} \end{bmatrix}^T \tag{34}$$

4 Model Predictive Control Architecture

Using the above optimization transcription, a Model Predictive Control (MPC) architecture is proposed, where the optimization is solved on-line using current navigation filter estimates as initial conditions for the optimization. Similar MPC architectures have been employed for rendezvous with dynamic avoidance constraints [16]. Nevertheless, in our proposed architecture the optimization is always solved to the end of the rendezvous, resulting in a shrinking-horizon MPC. The proposed MPC algorithm solves the optimization at different scheduled times $t_i = t_0, t_1, \ldots, t_k$ along the trajectory in order to update the time-stamped maneuver plan $\mathbf{U}_{opt}$ the chaser must perform in order to follow the optimal trajectory. The algorithm can be summarized as shown in Algorithm 1.

Finally, based on the analysis presented in [12], the trade-off $w = 0.75$ was selected, which yields the best compromise between arrival navigation performance and fuel consumption. Note that in order to prevent excessive fuel consumption, observability is only enhanced when navigation errors are lager, to aid navigation filter convergence. Since all impulsive maneuvers provide observability during an approach trajectory [11], demanding additional observability will no longer improve navigation performance once the filter has converged. Thus, the weight is set to zero when the navigation filter covariance (P) indicates that navigation errors are below a given threshold:

for $t_i = t_0, t_1, \ldots, t_k$ **do**

Initialize optimization with current navigation estimates $\mathbf{b}_{BCs} = \left[\hat{\mathbf{x}}_i^T \; \mathbf{x}_f^T\right]^T$

Adjust remaining rendezvous time ($t_{rem} = t_f - t_i$) for arrival at (t_f)

Update the number of discretization steps for the optimization $n = \frac{t_{rem}}{\Delta T}$

Solve the following optimization problem to update $\mathbf{U}_{opt}$:

$$\begin{aligned} \min_{\mathbf{U}_s} \quad & J_w \\ \text{subject to} \quad & \mathbb{G}_n \mathbf{U} = \mathbf{x}_f - \Phi(n\Delta T)\mathbf{x}_0 \\ & \mathbb{G}\mathbf{U} \geq \mathbf{X}_l \\ & A_f \mathbf{U}_s \leq \mathbf{b}_f \\ & A_u \mathbf{U} \leq \mathbf{b}_u \end{aligned} \tag{35}$$

Execute the next maneuver according to the maneuver plan $\mathbf{U}_{opt}$

end

Algorithm 1. Rendezvous Model Predictive Control

$$w = \begin{cases} 0.75 & \text{if} \;\; \|diag(P)\|_2 \geq P_{thr} \\ 0 & \text{if} \;\; \|diag(P)\|_2 < P_{thr} \end{cases} \tag{36}$$

Note that during an impulsive rendezvous approach maneuvers are performed only a few times per orbit, leaving enough computational time for the optimization. Hence, depending on the discretization step used, this approach is extremely feasible for a real-time implementation.

5 Non-linear Simulations

Monte Carlo simulations on a detailed non-linear simulation environment provided by Thales Alenia Space France (TAS-F) were used to demonstrate that the proposed MPC architecture can achieve bearings-only rendezvous with enhanced navigation performance, when observability is taken into account in the optimization. For this purpose, a bearings-only navigation filter was implemented as a classical Cartesian EKF using the Hill Equations, as described in [9]. The proposed MPC was used to close the control loop, as shown in Figure 1. The detailed simulation environment provided by TAS-F includes absolute motion equations for both the chaser and target, a J2 gravity model as well as atmospheric, solar and magnetic disturbances corresponding to a typical (~2 Ton) spacecraft and similarly sized debris. Bearings-sensor errors as well as thruster pointing and thrust magnitude errors were modelled as stochastic white noise. Very conservative errors were used for the simulations in order to demonstrate feasibility, but finer performance can be achieved if better sensors and actuators are available. The parameters used for the simulations

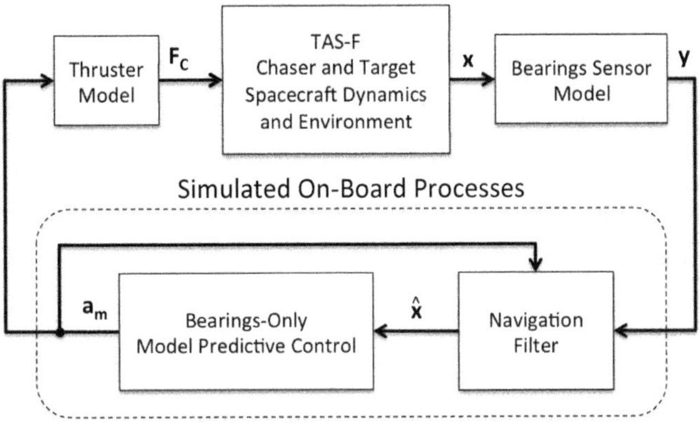

Fig. 1 Detailed Simulation Model

are summarized in Table 1. The reference mission was selected as a rendezvous between hold points at 30 km and 1 km in front of the target. The former hold point was identified as the distance below which relative navigation is required when no high accuracy ground tracking is available. The use of bearings-only navigation is then most promising for non-cooperative targets due to the mass savings as compared to skin-tracking sensors. Similarly, when closer than the arrival hold point, the target shape can be identified and other navigation techniques with similar benefits can provide better navigation accuracy, such as vision-based navigation.

5.1 Simulation Results

In order to statistically compare the performance of the proposed observability-enhanced MPC ($w = 0.75$) with a traditional fuel-optimal ($w = 0$) rendezvous guidance, 500 Monte-Carlo simulations were run for each scenario. For each simulation, the following were scattered: chaser initial position, navigation filter initialization errors, sensor and thruster modelling errors including chaser mass estimate.

The in-track projection of the trajectories resulting from the use of the proposed Enhanced Observability MPC are shown on the left hand side of Figure 2, and those for the Fuel-Optimal MPC on the right hand side of the same figure. Cross track projections are not shown as the scatter of the trajectories is symmetric about the y axis. As these figures show, the addition of observability in the optimization results in trajectories with larger excursions, which enhance observability in order to improve navigation performance. As it can seen by the distribution of the arrival positions on the right hand side of Figure 2, without the addition of observability the navigation filter is not able to always converge. For this reason, following the maneuvers planned by the optimizer results in the chaser not being able to successfully reach the target for about 10% of the fuel-optimal MPC simulations.

Table 1 Simulation Parameters

Parameter	Value	Units
Simulation Environment		
Target Orbital Elements $[a\, e\, i\, \Omega\, \omega\, \nu]$	$[7.1764\times10^6$ 0 98.4 0 0 0]	-
Target Orbital Period	$\sim$100.1	min
Simulation Time Step (T_s)	10	sec
Bearings Sensor		
Measurement Noise	200	arcsec/axis (3σ)
Maneuver Knowledge Errors		
Thruster Pointing Errors	0.5	deg/axis (3σ)
Thrust Magnitude Error	1	% (3σ)
Thrust Rise/Fall Time	400	ms
Max Thrust	22	N
Min Thrust	0.22	N
Rendezvous Mission		
Nominal Initial State $\mathbf{x}_0 = [x\, y\, z\, \dot{x}\, \dot{y}\, \dot{z}]$	[30000 0 0 0 0 0]	m & m/s
Spread of Initial States $\boldsymbol{\sigma}_{\mathbf{x}_0} = [x\, y\, z\, \dot{x}\, \dot{y}\, \dot{z}]$	[3000 300 300 0.3 0.3 0.3]	m & m/s
Target State $\mathbf{x}_f = [x\, y\, z\, \dot{x}\, \dot{y}\, \dot{z}]$	[1000 0 0 0 0 0]	m & m/s
Rendezvous Duration	4	Orbits
MPC Optimization Parameters		
Discretization Time Step (ΔT)	440	sec
Re-Optimization Rate (T_r)	440	sec
Maneuver Duration (T_m)	440	sec
Max Maneuver Amplitude ($\mathbf{a}_{max}$)	2	m/s
Along-Track Position Constraint (x^l)	500	m
Observability Weight (w)	0.75	-
Observability Enable Threshold ($\mathbf{P}_{tr}$)	0.5	% of Range
Navigation Filter Initialization		
Spread of Initial Errors ($\boldsymbol{\sigma}_0$)	[3000 300 300 0.3 0.3 0.3]	m & m/s

In order to observe the effects augmenting observability on the overall GNC performance, navigation errors and arrival dispersions were computed for both sets of Monte-Carlo simulations. Note that in order to obtain a more objective measure of the overall performance for comparison, simulations with filter divergence were removed from the results of the fuel-optimal case.

The navigation performance (1-σ) at arrival to the desired hold point is shown in Figure 3 for both cases. As it can be seen from this figure, augmenting observability in the optimization prevents filter divergence and significantly improves the achievable navigation performance with respect to following fuel-optimal trajectories.

The 1-σ navigation errors along for the complete trajectory can be seen in Figure 4. This figure shows that when observability is accounted for in the optimization, the navigation filter achieves final convergence faster, resulting in lower errors throughout the trajectory. Note that the initial divergence in navigation errors

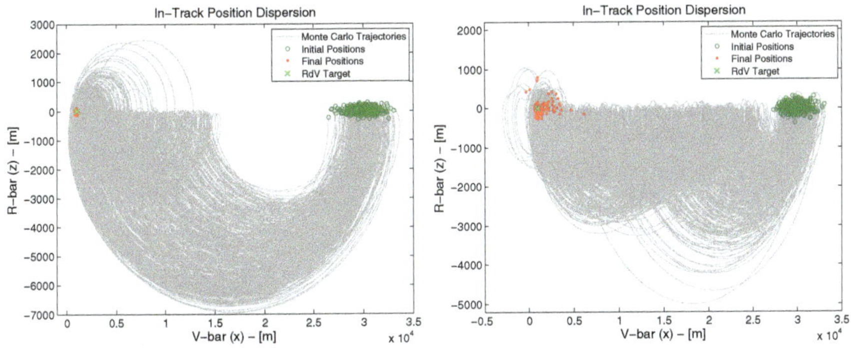

Fig. 2 Monte Carlo Trajectories: Enhanced Observability (left) and Fuel-Optimal (right) Guidance

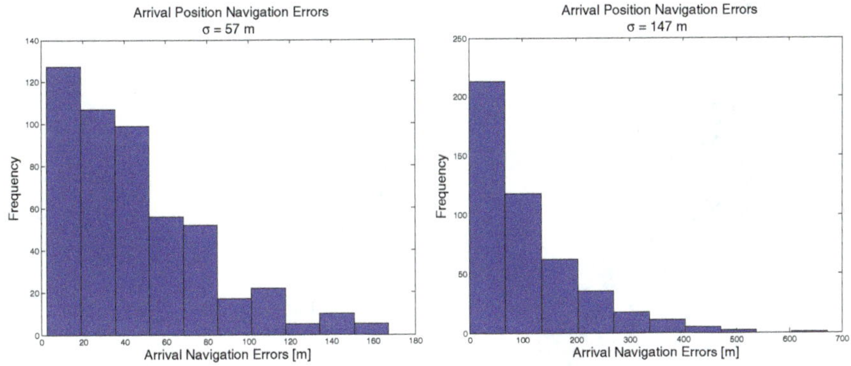

Fig. 3 Arrival Navigation Errors: Enhanced Observability (left) and Fuel-Optimal (right) Guidance

for the fuel-optimal case is due to the fact that planning fuel-optimal trajectories results in an initial drift period, without many maneuvers. Since maneuvers are needed to render the range estimation observable, the lack of maneuvers implies a lack in range information, which results in an initial worsening of the navigation errors.

Improved navigation performance has a direct impact on the spread of the arrival dispersions. That is, the position error at arrival with respect to the desired final position at the end of the rendezvous phase. Figure 5 shows these errors, which are significantly improved when observability is accounted for in the trajectory optimization. In addition, this figure also shows that the use of linearized relative motion equations inside the GNC algorithms provides satisfactory results in a non-linear environment, when observability is taken into account in the trajectory optimization.

Augmenting observability implies additional fuel consumption for this purpose, which can be seen in Figure 6. For this mission scenario, the fuel consumption is almost doubled when observability is enhanced. Nevertheless, this not only results

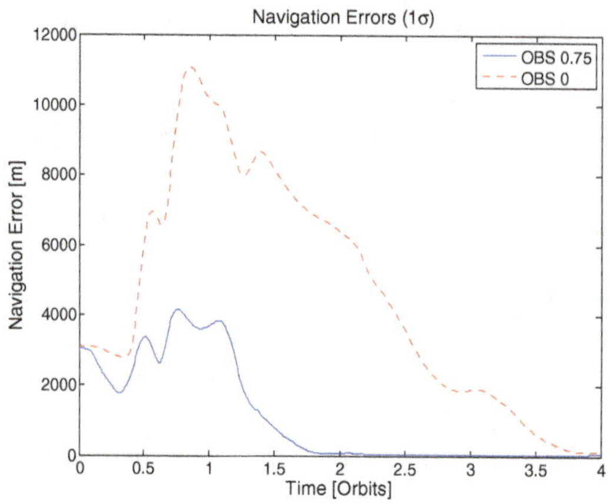

Fig. 4 Navigation Errors Along the Trajectory (1-σ)

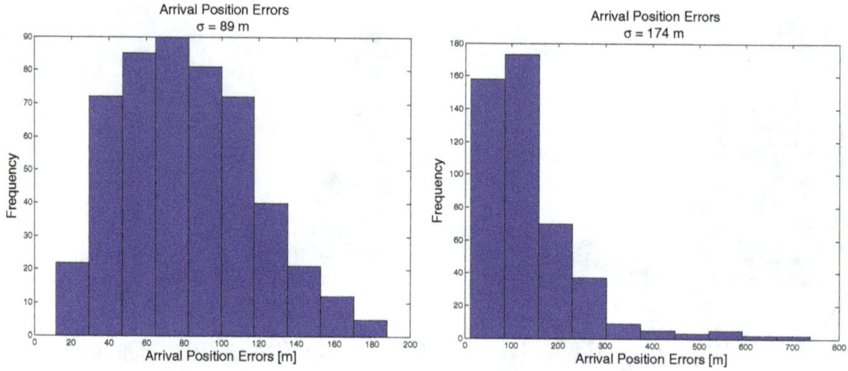

Fig. 5 Position Targeting Errors: Enhanced Observability (left) and Fuel-Optimal (right) Guidance

in superior navigation performance, but since bearings-only observability is guaranteed throughout the resulting trajectories, this prevents navigation filter divergence with respect to following fuel-optimal trajectories. This factor alone implies that taking observability into account in the trajectory optimization might be an enabling factor for rendezvous using bearings-only navigation.

Table 2 summarizes the results of the non-linear simulations, showing that the proposed Observability-Enhanced MPC algorithm results in superior overall performance and prevents navigation filter divergence during bearings-only rendezvous.

Finally, it is worth noting the effects of maneuver knowledge errors on the GNC performance. Since maneuvers are used to introduce the observability required to estimate the relative range, these essentially provide the range information for the estimation problem. Thus, larger errors in the knowledge of the resulting accelerations will deteriorate the achievable navigation performance. Further simulations

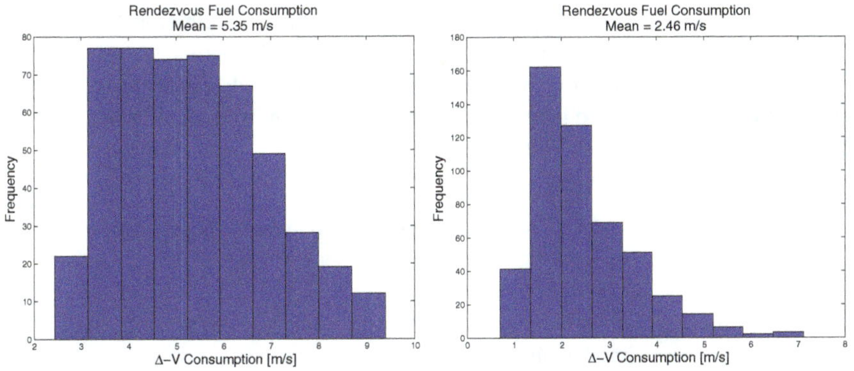

Fig. 6 Fuel Consumption: Enhanced Observability (left) and Fuel-Optimal (right) Guidance

Table 2 Simulation Results Summary

Results* (1σ)	**Enhanced Observability**	**Fuel Optimal**
Navigation Filter Divergence (out of 500)	0	36
Mean Fuel Consumption [m/s]	5.35	2.46
Arrival Position Errors [m]	89	174
Arrival Navigation Errors [m]	57	147

*Diverged simulations removed to compute performances

not shown in this paper indicate that if the maneuver knowledge error is above about $1-3\%$, navigation filter divergence can not be prevented even with enhanced observability. Since this uncertainty could result from the knowledge of the chaser's mass, other methods such as higher accuracy orbit determination of the chaser might be needed in order to improve the maneuver knowledge accuracy.

6 Conclusions

This paper presents the non-linear simulation results of a bearings-only rendezvous guidance and control scheme based on an on-line trajectory optimization that enhances observability in order to improve navigation performance. The proposed implementation solves a direct shooting type optimization based on linearized relative motion equations inside the control loop, resulting in a shrinking-horizon model predictive control architecture. Monte-Carlo simulations in a non-linear simulation environment provided by Thales Alenia Space France show that the addition of observability in the optimization is essential to ensure navigation filter convergence. Moreover, the performance achievable by such a GNC architecture for rendezvous is essentially doubled when observability is taken into account in the trajectory optimization, as compared to a fuel optimal guidance.

Even though satisfactory performance was achieved, these results could be further improved by the use of higher fidelity navigation filters and other methods to obtain more accurate maneuver knowledge, such as orbit determination of the chaser spacecraft either by ground tracking or on-board GNSS. Nonetheless, these results show the feasibility of performing autonomous bearings-only rendezvous using an on-board GNC architecture based on linearized equations and a calibrated thruster model, by augmenting observability in order to guarantee navigation filter convergence and improve navigation performance.

Acknowledgement. The authors would like to thank ESA (Guidance, Navigation and Control Section at the European Space Research and Technology Centre) and Thales Alenia Space France (Research and Technology/Science and Observation within the Research and Development Department) for providing the funding that made this research possible, through the ESA Networking/Partnering Initiative (NPI) Program.

Appendix

Local Orbital Frame

The Local Orbital Frame ($\mathfrak{F}^{\mathbf{lo}}$), is fixed to the target position $\mathbf{r}_t$ and rotates with respect to the inertial frame at the instantaneous orbital rate of the target $\boldsymbol{\omega}$, as show in Figure 7. In this frame the **z**-axis ("relative altitude" or "R-Bar") always points in the radial direction; the **y**-axis ("cross-track" or "H-Bar") is opposite to the orbital angular momentum; and the **x**-axis ("downrange" or "V-Bar") completes the triad.

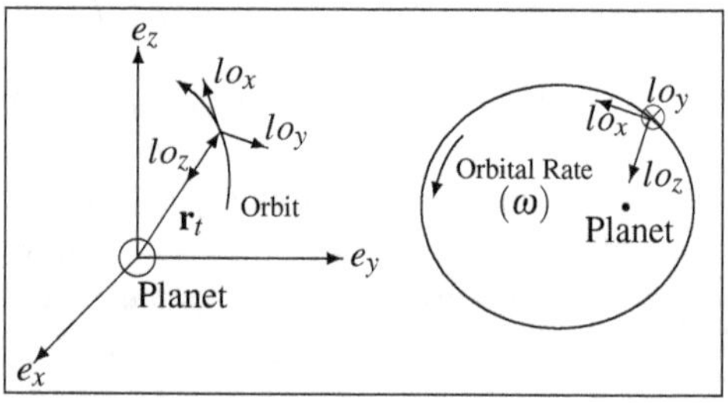

Fig. 7 Local Orbital Frame ($\mathfrak{F}^{\mathbf{lo}}$) Definition

Curvilinear Coordinates

Similarly to its definition in [5], the curvilinear coordinate frame follows the motion and orientation of the Local orbital frame ($\mathfrak{F}^{\mathbf{lo}}$). Nonetheless, the **x** and **y** axis are defined to curve along a circular orbit with radius equal to that of the target's orbit. Thus, the relative position coordinates along the curved path are defined by the angles ϑ and φ for the x and y relative distances, respectively. The relative altitude $\delta r = r_t - r_c$ is defined in the opposite **z** direction, as shown in Figure 8.

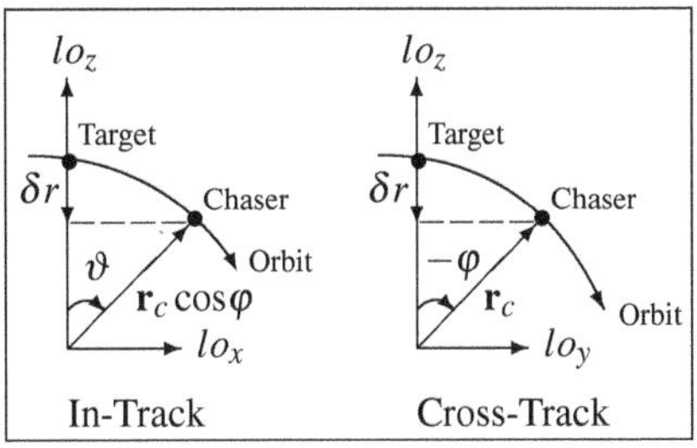

Fig. 8 Curvilinear Coordinates

The relation between the relative curvilinear coordinates and their cartesian counterparts can then be extracted from the geometry in Figure 8, yielding the following:

$$\varphi = \sin^{-1}\left(-\frac{y}{r_c}\right) \tag{37}$$

$$\vartheta = \sin^{-1}\left(\frac{x}{r_c \cos\varphi}\right) \tag{38}$$

$$\delta r = z \tag{39}$$

Finally, the transformations from Curvilinear to Cartesian coordinates and vice-versa ($[\vartheta, \varphi, \delta r] \Leftrightarrow [x, y, z]$) is found by solving Equations (37) to (39) for the corresponding quantities, and taking their time derivatives for the velocities.

Relative Motion Equations

In order to define the state transition and maneuver solution matrices for the linear equations of relative motion in Eq. (1), a relative state vector is defined in the $\left(\mathfrak{F}^{\mathbf{lo}}\right)$:

$$\mathbf{x} = [x\; y\; z\; \dot{x}\; \dot{y}\; \dot{z}]^T \tag{40}$$

The state transition matrix for linearized relative motion in circular orbits is then:

$$\Phi(t) = \begin{bmatrix} 1 & 0 & 6t\omega - 6s_{\omega t} & \frac{4s_{\omega t}}{\omega} - 3t & 0 & \frac{2(1-c_{\omega t})}{\omega} \\ 0 & c_{\omega t} & 0 & 0 & \frac{s_{\omega t}}{\omega} & 0 \\ 0 & 0 & 4-3c_{\omega t} & \frac{2(c_{\omega t}-1)}{\omega} & 0 & \frac{s_{\omega t}}{\omega} \\ 0 & 0 & 6\omega(1-c_{\omega t}) & 4c_{\omega t}-3 & 0 & 2s_{\omega t} \\ 0 & -\omega s_{\omega t} & 0 & 0 & c_{\omega t} & 0 \\ 0 & 0 & 3\omega s_{\omega t} & -2s_{\omega t} & 0 & c_{\omega t} \end{bmatrix} \tag{41}$$

where ω is the absolute orbital rate of the target, t is the propagation time for the solution, and $s_{\omega t} = \sin(\omega t)$ and $c_{\omega t} = \cos(\omega t)$.

The general discrete input matrix solution must be written as a convolution integral of the time-varying maneuver $\mathbf{u}(t)$ and the position-velocity portions of the state transition matrix Φ, selected by the matrix B:

$$G(t)\,\mathbf{u} = \int_0^t \Phi(\tau)\,B\mathbf{u}(\tau)\,d\tau \tag{42}$$

where

$$B = \begin{bmatrix} 0\,0\,0\,1\,0\,0 \\ 0\,0\,0\,0\,1\,0 \\ 0\,0\,0\,0\,0\,1 \end{bmatrix}^T \tag{43}$$

Explicitly specifying the maneuver allows the solution of the integral. For example, the discrete solution for impulsive maneuvers at time zero can be written as the convolution $G(t)\,\mathbf{u}$, where $\mathbf{u}$ defines the magnitude/direction of the impulse, and the input transition matrix G is:

$$G(t) = \begin{bmatrix} \frac{4s_{\omega t}}{\omega} - 3t & 0 & \frac{2(1-c_{\omega t})}{\omega} \\ 0 & \frac{s_{\omega t}}{\omega} & 0 \\ \frac{2(c_{\omega t}-1)}{\omega} & 0 & \frac{s_{\omega t}}{\omega} \\ 4c_{\omega t} - 3 & 0 & 2s_{\omega t} \\ 0 & c_{\omega t} & 0 \\ -2s_{\omega t} & 0 & c_{\omega t} \end{bmatrix} \tag{44}$$

Closed form solutions in the form $G(t)\,\mathbf{u}$ also exist for other maneuver types, such as constant thrust, making the formulations in this paper also applicable.

As noted in [5], if the state vector is defined in Curvilinear coordinates as in Equation (45), the state transition and maneuver solution matrices for linearized relative motion in circular orbits are identical to the corresponding Cartesian Clohessy-Wiltshire equations [5]. Thus, only the initial conditions are different, which must be transformed as explained in the previous section.

$$\mathbf{x} = \left[r_t\vartheta,\ -r_t\varphi,\ \delta r,\ r_t\dot{\vartheta},\ -r_t\dot{\varphi},\ \delta\dot{r} \right]^T \tag{45}$$

Bearings-Only Measurement Angles

The measurement angles are defined in the $\left(\mathfrak{F}^{\mathbf{lo}}\right)$ frame, as depicted in Figure 9. Note that θ is defined from 0° to 180° and ϕ from −90° to +90°.

The bearings-only measurement equations provide the measurement angles to the target as a function of the relative position coordinates x, y and z, as follows:

$$\mathbf{y} = \mathbf{h}(\mathbf{x}) \equiv \begin{bmatrix} \theta \\ \phi \end{bmatrix} = \begin{bmatrix} \arctan(y/x) \\ \arcsin\left(z/\sqrt{x^2+y^2+z^2}\right) \end{bmatrix} \tag{46}$$

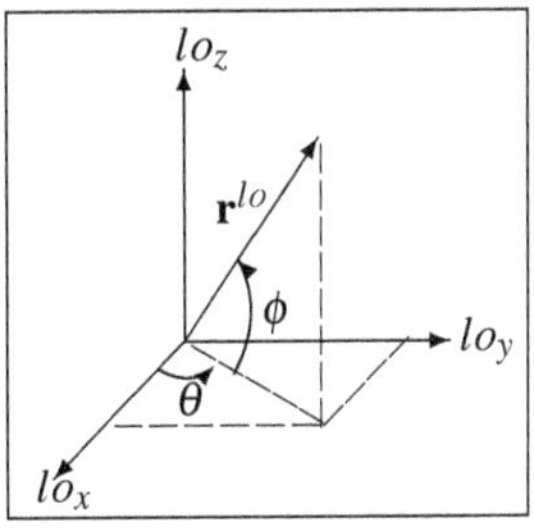

Fig. 9 Measurement angles in the $\mathfrak{F}^{\mathbf{lo}}$ frame

References

1. Aida, S., Patzelt, T., Leushacke, L., Kirschner, M., Kiehling, R.: Monitoring and Mitigation of Close Proximities in Low Earth Orbit. In: 21st International Symposium on Space Flight Dynamics, vol. 49 (2009)
2. Betts, J.T.: Survey of Numerical Methods for Trajectory Optimization. Journal of Guidance, Control, and Dynamics 21(2), 193–207 (1998)
3. D'Amico, S., Ardaens, J.-S., Gaias, G., Benninghoff, H., Schlepp, B., Jørgensen, J.L.: Noncooperative Rendezvous Using Angles-Only Optical Navigation: System Design and Flight Results. Journal of Guidance, Control, and Dynamics 36(6), 1576–1595 (2013)
4. Das, I., Dennis, J.E.: A closer look at drawbacks of minimizing weighted sums of objectives for Pareto set generation in multicriteria optimization problems. Structural Optimization 14(1), 63–69 (1997)
5. de Bruijn, F., Gill, E., How, J.: Comparative Analysis of Cartesian and Curvilinear Clohessy-Wiltshire Equations. Journal of Aerospace Engineering, Sciences and Applications 3(2), 1–15 (2011)
6. Fehse, W.: Automated Rendezvous and Docking of Spacecraft. Cambridge Aerospace Series. Cambridge University Press, Cambridge (2003)
7. Gaias, G., D'Amico, S., Ardaens, J.-S.: Angles-only Navigation to a Non-Cooperative Satellite using Relative Orbital Elements. In: AIAA/AAS Astrodynamics Specialist Conference, Reston, Virigina. American Institute of Aeronautics and Astronautics (August 2012)
8. Gaias, G., D'Amico, S., Ardaens, J.S.: Angles-Only Navigation to a Noncooperative Satellite Using Relative Orbital Elements. Journal of Guidance, Control, and Dynamics 37(2), 439–451 (2014)
9. Grzymisch, J., Fichter, W., Casasco, M., Losa, D.: A Spherical Coordinate Parametrization for an In-Orbit Bearings-Only Navigation Filter. In: Advances in Aerospace Guidance, Navigation and Control, pp. 215–231 (2013)
10. Grzymisch, J., Fichter, W.: Analytic Optimal Observability Maneuvers for In-Orbit Bearings-Only Rendezvous. Journal of Guidance, Control, and Dynamics 37(5), 1658–1664 (2014)
11. Grzymisch, J., Fichter, W.: Observability Criteria and Unobservable Maneuvers for In-Orbit Bearings-Only Navigation. Journal of Guidance, Control, and Dynamics, 1–10 (January 2014)
12. Grzymisch, J., Fichter, W.: Optimal Rendezvous Guidance with Enhanced Bearings-Only Observability. Journal of Guidance, Control, and Dynamics (2014) (under review)
13. Li, J., Li, H., Tang, G., Luo, Y.: Research on the Strategy of Angles-Only Relative Navigation for Autonomous Rendezvous. Science China Technological Sciences 54(7), 1865–1872 (2011)

14. Losa, D., Le-Peuvedic, C.: Simulator of a Chaser and Target Satellite Dynamics in Earth Environment. Thales Alenia Space TR 01 (2014)
15. Richards, A., How, J.: Performance Evaluation of Rendezvous Using Model Predictive Control. In: AIAA Guidance, Navigation, and Control Conference and Exhibit, Austin, Texas, pp. 1–9 (August 2003)
16. Richards, A., Schouwenaars, T., How, J.P., Feron, E.: Spacecraft Trajectory Planning with Avoidance Constraints Using Mixed-Integer Linear Programming. Journal of Guidance, Control, and Dynamics 25(4), 755–764 (2002)
17. Wassel, D., Buskens, C.: Modeling and Optimization in Space Engineering. Springer Optimization and Its Applications, vol. 73 (2013)
18. Wassel, D., Wolff, F., Vogelsang, J., Büskens, C.: The ESA NLP-Solver WORHP – Recent Developments and Applications. In: International Conference on Astrodynamic Tools and Techniques (2012)
19. Woffinden, D.C., Geller, D.K.: Optimal Orbital Rendezvous Maneuvering for Angles-Only Navigation. Journal of Guidance, Control, and Dynamics 32(4), 1382–1387 (2009)
20. Woffinden, D.C., Geller, D.K.: Observability Criteria for Angles-Only Navigation. IEEE Transactions on Aerospace and Electronic Systems 45(3), 1194–1208 (2009)
21. Yamanaka, K., Ankersen, F.: New State Transition Matrix for Relative Motion on an Arbitrary Elliptical Orbit. Journal of Guidance, Control, and Dynamics 25(1), 60–66 (2002)
22. Yim, J.R., Crassidis, J., Junkins, J.L.: Autonomous orbit navigation of two spacecraft system using relative line of sight vector measurements. Advances in the Astronautical Sciences AAS 04-257, 1–14 (2004)

Structured Accelero-Stellar Estimator for Microscope Drag-Free Mission

Christelle Pittet and Pascal Prieur

Abstract. MICROSCOPE is CNES fourth microsatellite based on MYRIADE platform. Its mission is dedicated to the test of the Equivalence Principle (EP) with an improved accuracy of 10^{-15}. Therefore the satellite attitude requirement is very stringent and the payload accelerometric measurements have to be used in the control and estimation loop. This paper deals with one of the most critical issue in MICROSCOPE attitude and acceleration control: the accelero-stellar hybridization in rotating mission mode. Non-smooth H infinity optimization algorithms have been used to design structured and fixed order estimators, leading to the same performances as the full-order reference estimators, with a huge gain in computation efficiency. These structured estimators have then been selected to replace the reference ones and will be operating on board the satellite in 2016. This paper presents the reference estimators, gives the design procedure used for the structured ones, and shows their compared performances and robustness.

1 Introduction

MICROSCOPE is CNES fourth scientific microsatellite based on the MYRIADE product line, but the AOCS differs a lot from the previous microsatellites DEMETER, PARASOL or PICARD, which already ended their missions. Indeed, Microscope mission is dedicated to the test of the Equivalence Principle (EP) with an improved accuracy of 10^{-15} m/s², and therefore requires the 270-kg satellite accelerations to be finely controlled [2, 7]. The project is in final development phase, and is expected to be launched on a 700km dusk dawn orbit in 2016 for a two year mission.

To reach the required accuracy, a specific AOCS mode has been designed, allowing for air drag compensation, in such a way that the satellite follows the

Christelle Pittet · Pascal Prieur
CNES, 18 av Edouard Belin, 31401 Toulouse, France
e-mail: {christelle.pittet,pascal.prieur}@cnes.fr

J. Bordeneuve-Guibé et al. (eds.), *Advances in Aerospace Guidance, Navigation and Control*, DOI: 10.1007/978-3-319-17518-8_34

payload Test Masses in a pure gravitational motion. In this mode the attitude measurement is provided by an innovative hybridization between a star tracker and the scientific accemerometers [1]. The estimation filter for the spinning sub-mode has been designed using H infinity optimization under LMI constraints with pole placement. Because of a strong coupling between the axes due to the specific spinning guidance profile, the resulting dynamic Kalman filter could not be reduced and is of high order. The mission was planned to be launched in 2009, but some problems occurred with the electric microthrusters development, leading to a change of propulsion system. Because of the heavy impact of the propulsion on the platform and AOCS design, a new phase B study was then decided, delaying the launch date of several years.

In the meantime, new algorithms for solving H infinity problems appeared in the control community, based on non-smooth optimization [5, 6, 8] with available and efficient numerical tools. The main advantage of the new optimization technique is the possibility to design structured controllers of fixed order, which is a great improvement in the design process because the reduction of high order controllers frequently lead to a loss in the performances or stability margins. In 2012, CNES funded two industrial research and technology projects, one with Thales Alenia Space, one with Airbus Defence and Space, to evaluate the benefits of the non-smooth optimization algorithms with respect to existing industrial synthesis process [4]. The results of the studies were so good that the new design procedure is now used on projects in development. CNES was also interested in testing the new algorithm on its own projects and we chose to apply it to Microscope estimator, trying to reduce the order without decreasing the performances.

The paper aims at presenting the result of this new synthesis and comparing it with the reference estimator. First the AACS (Attitude and Acceleration Control System) reference estimator for spinning sub-mode is presented. In the third section, the structure of the reference estimator is analyzed to orient the reduced order estimator structure, then the estimator synthesis is explained. The new controller is finally compared to the reference one. Concluding remarks and future work perspectives end the paper.

2 Hybridization Reference

In this section, Microscope's spinning sub-mode requirements for drag-free control are presented and the full-order reference estimator synthesis is recalled.

2.1 Attitude Control Requirements

The payload designed by ONERA is composed of two differential electrostatic accelerometers, one for reference, one for the Equivalence Principle violation testing, each one including two proof-masses [7]. A violation of the EP will appear as a difference in the electrostatic forces necessary to maintain the masses on the same orbit. To be able to measure the differential forces with the required accuracy, the satellite must compensate the non-gravitational forces so that the residual accelerations in mission mode must be less than 10^{-12} m/s² in the bandwidth of scientific interest. To limit angular to linear coupling, additional pointing requirements at F_{EP} are stated.

Because they are the more stringent ones, we will focus on the spinning submode requirements: in this mode, the satellite spins slowly opposite to orbital movement. Then $F_{EP} = F_{orb}$ (orbital frequency: 0.001 rad/s) + F_{spin}. In addition, the spinning frequency is a high level mission parameter which can be chose between (π+1/2) Forb and (π+3/2) F_{orb}. The observation duration is 20 orbits (about 120000s). During the observation window, the attitude requirements are threefold. The first is a Mean Pointing Error (MPE) requirement analytically computed through the low-pass filtering of the pointing error. The second is a Tone Error (TE) requirement corresponding to the amplitude of a sinusoidal residual pointing error at $F_{EP} \pm F_{orb}/20$. It is the result of the summation of sinusoidal residuals over a range of frequencies around F_{EP} (from F_{spin} to $4F_{EP}$) attenuated by a transmission shape. The last is a Random Error (RE) requirement over a centered range of $\pm 10^{-4}$ Hz around F_{EP}. It is computed as an equivalent uniform noise over the frequency range. Because it is the most stringent one, the TE requirement at $F_{EP} \pm F_{orb}/20$ has been taken for design driver.

Table 1 Spinning mode attitude requirements

Attitude requirements	X	Y	Z
MPE (μrad)	200	100	100
TE @ F_{EP} (μrad)	0.166	0.166	0.166
RE @ F_{EP} (μrad.Hz$^{-1/2}$)	159	159	159

These control requirements are be allocated to one control part and one estimation part, but because of the complexity of the estimation and the good accuracy of the control part, no allocation is made a priori. The objective is to get the best achievable estimation budget with sufficient margins for the control part.

It is important to notice that these requirements have to be met for each spinning frequency.

2.2 Reference Full-Order Estimator Design

The standard MYRIADE estimation function in mission mode is based only on star tracker measurements [3, 9] For MICROSCOPE, the thermoelastic disturbance between the star tracker and the accelerometer occurring at F_{EP} could reach as much as 45 μrad in the worst case conditions. Then it is obvious that the star tracker only cannot achieve less than 0.166 μrad at F_{EP} and that another sensor has to be used for attitude measurement at F_{EP}. The solution is to use the payload accelerometer itself, which is expected to have only a 3.10^{-12} rad/s² sine error at F_{EP}, that is 0.083 µrad at F_{EP}=0.006rad/s. Then, to meet TE requirement, a hybridization between star tracker and accelerometer must be done and the accelerometer must provide the measurement at F_{EP}. Taking this into account, the star tracker error residue has been allocated to 0.02 µrad, which means an attenuation of -67 dB at F_{EP} frequency.

Several hybridization strategies have been studied and compared to get the best trade-off concerning the time response of the estimator and the TE satisfaction. For the spinning sub-mode, a sine-stop estimator was selected using a control synthesis scheme. The details of the synthesis can be found in [1]. Only the main steps and results are recalled here.

The basic idea is to use a control problem formulation to design a Kalman filter with a dynamic updating filter K(s). The estimator architecture can be equivalently rewritten as a closed loop between a plant G(s) and a controller K(s), where G(s) is the model of the system to estimate (in our case, the satellite dynamics with acceleration input and attitude output), and K(s) is the updating gain of the estimator. Using this property, the updating gain can be found as the solution of the synthesis problem of finding a stabilizing K(s) for the given plant G(s). The estimator scheme is illustrated on Fig. 1.

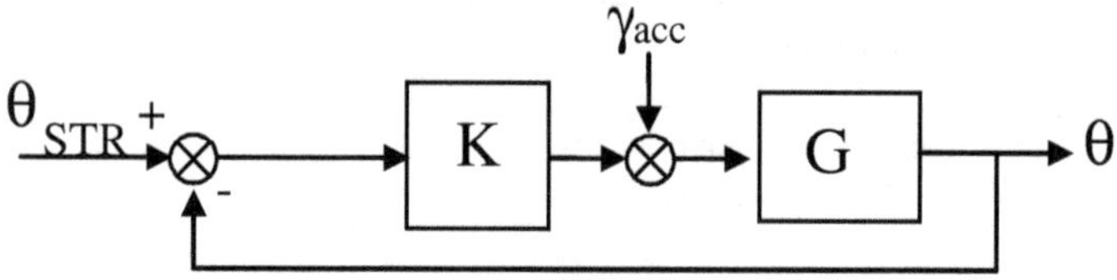

Fig. 1 Control equivalent scheme for estimator

The dynamics equations of system G are given by Eq. 1 in continuous time, with θ the attitude Euler angles, ω the satellite angular velocity vector, γ the satellite angular acceleration vector, γ_{ACC} the measured acceleration vector, θ_{STR} the measured attitude vector, w the accelerometer noise and v the star tracker noise, Ω_{spin} is the spin frequency around X axis, and I_3 the identity matrix of order 3.

$$\begin{pmatrix} \dot{\vec{\theta}} \\ \dot{\vec{\omega}} \end{pmatrix} = \begin{bmatrix} -\Omega_c^* & I_3 \\ 0 & 0 \end{bmatrix} \begin{pmatrix} \vec{\theta} \\ \vec{\omega} \end{pmatrix} + \begin{bmatrix} 0 \\ I_3 \end{bmatrix} \vec{\gamma} + \begin{bmatrix} -\vec{\Omega}_c \\ 0 \end{bmatrix}$$
$$\vec{\gamma}_{ACC} = \vec{\gamma} + \vec{w} \qquad (1)$$
$$\vec{\theta}_{STR} = \vec{\theta} + \vec{v}$$
$$\Omega_c^* = \begin{pmatrix} 0 & 0 & 0 \\ 0 & 0 & -\Omega_{spin} \\ 0 & \Omega_{spin} & 0 \end{pmatrix}$$

The dynamic filter K(s) is composed of two filters in series:

- a pre-computed notching filter W(s) of required attenuation level at F_{EP} : with a star tracker sine error of 45 μrad at F_{EP} and an allocation of 0.02 μrad, the attenuation level must be of -67 dB at least.

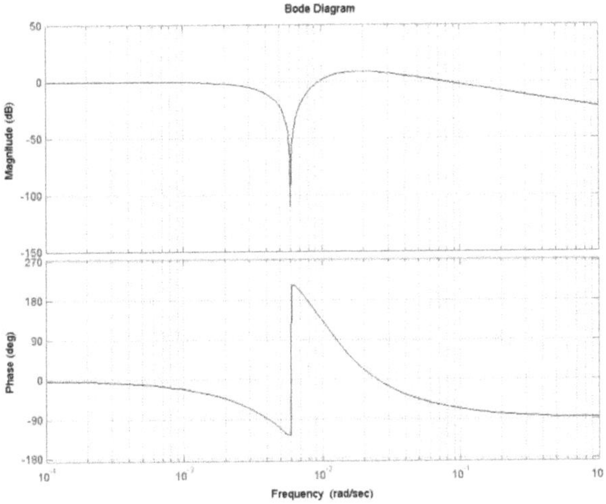

Fig. 2 Notching filter W(s) Bode diagram

- a stabilizing filter $K_w(s)$; this filter is designed for the chosen W(s) through minimization of $\|z/\theta_{STR}\|_\infty$ under pole placement in a conic region constraint. A static gain weighting function P(s) is added on $K_w(s)$ output to avoid the notch mode compensation. Indeed, in case of full-order synthesis, the optimization scheme allows the controller to use additional degrees of freedom to compensate some flexible modes of the model to control. As the notching filter is inside this model to control, some controllers could compensate the zero with a pole, destroying then the accuracy. Filter $K_w(s)$ design scheme is illustrated on Fig. 3.

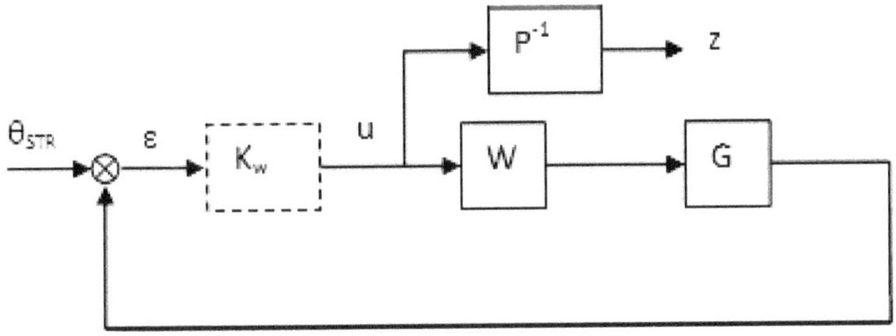

Fig. 3 Reference estimator design scheme

The Bode diagram of the reference estimator for X-axis is shown on Fig. 2. The notch at F_{EP} ensures the robustness to frequency variation of the thermoelastic disturbances within the $F_{EP}\pm F_{orb}/20$ band, which is the bandwidth of scientific interest. The maximum gain in the TE band is -69 dB, meeting the rejection requirement.

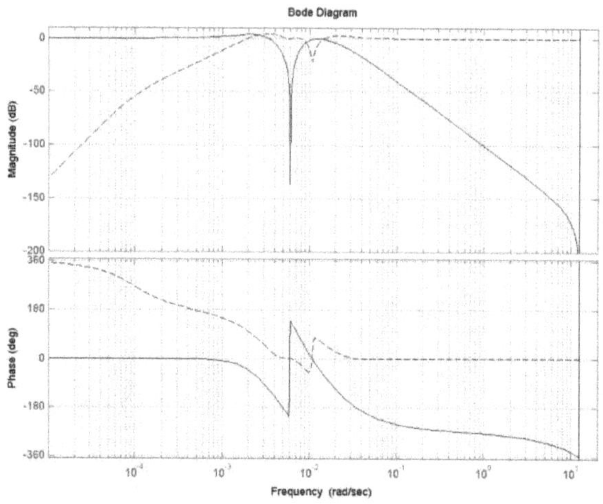

Fig. 4 H∞ reference estimator for X-axis (θ/θ_{STR} plain, θ/γ_{ACC} dotted)

The design of K_w has been made with standard convex optimization tools for H infinity optimization. Thus $K_w(s)$ order is equal to the sum of G's order, W's order and P's order. This results in a high order final estimator which is composed of system G(s) in closed loop with the updating gain K(s) = W(s) K_W(s).. The spin axis X is decoupled, so one SISO estimator has been designed for this axis independently from the two others. Because of the high coupling between the cross axes Y and Z, we performed a MIMO synthesis. The resulting order is given in Table 2.

Table 2 Reference full-order estimator order (before reduction)

Order	X	Y/Z
G	2	4
W	5	10
K_w (G+W)	5+2 = 7	4+10=14
K (K_w+W)	7+5 = 12	14+10 = 24
Estimator (K+G)	12+2 = 14	24 + 4 = 28

The full order reference estimator was only designed for one nominal F_{spin} frequency inside the spinning frequencies interval. Taking into account the other values in the interval, we showed that one estimator cannot cover the whole interval with the required accuracy; then several estimators have to be designed for a gridding of spinning frequencies and the closest to spinning frequency set point should be selected through remote control during the satellite mission. All estimators must be pre-computed and stored in the satellite data bank. Because of the high order of the reference estimator, reducing the order is becoming a real issue for the mission.

3 Structured Estimator Design

In this section, the reference full-order estimator structure is analyzed and the fixed order and structured estimator design is presented.

3.1 Structure Choice

To orient the design of the structured and fixed order estimator, we first analyzed the structure of the reference estimator:

- The spin axis estimator is decoupled and SISO, allowing for order reduction. The final resulting order was 6, with W(s) of order 5 and $K_{wx}(s)$ of first order. The measurement equation on X axis is a double integrator. It cannot be stabilized by a static gain. Thus the minimal order for $K_{wx}(s)$ is 1.
- Regarding the cross axes, the estimator gain $K_{wyz}(s)$ mimics the structure of the measurement equations. $G_{yz}(s)$ and $K_{wyz}(s)$ can be rewritten as :

$$G_{yz}(s) = \begin{bmatrix} G_{yy}(s) & G_{yz}(s) \\ G_{zy}(s) & G_{zz}(s) \end{bmatrix}, \quad K_{wyz}(s) = \begin{bmatrix} K_{yy}(s) & K_{yz}(s) \\ K_{zy}(s) & K_{zz}(s) \end{bmatrix}$$

with $G_{yy} = G_{zz}$, $G_{yz} = -G_{zy}$ and $K_{yy} = K_{zz}$, $K_{yz} = -K_{zz}$, as illustrated on Fig. 5 and Fig. 6.

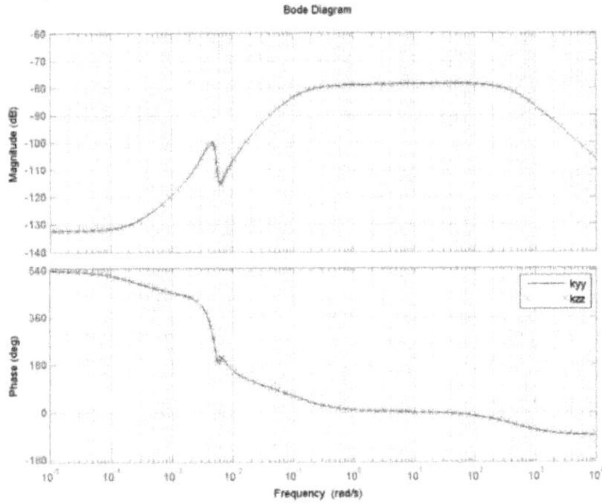

Fig. 5 Diagonal transfers Bode diagrams

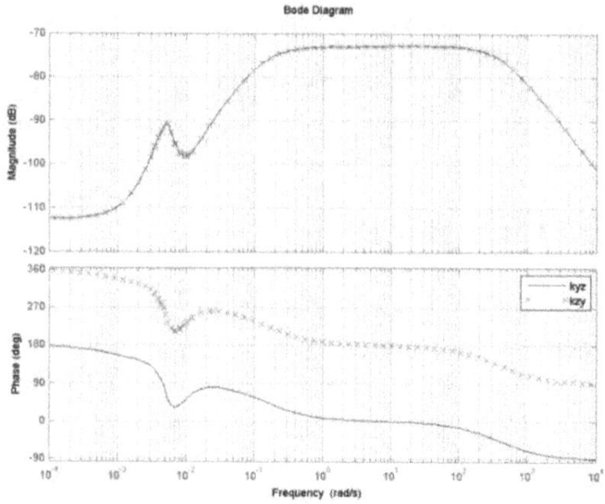

Fig. 6 Cross transfers Bode diagrams

The components of K_w are of phase lead type as expected, but we can see an additional low frequency stable pole. Thus the reduced filters will probably be of order 2 at least.

The final structure and order of the filters has been chosen through trials/errors design iterations, based on the previous analysis and the design scheme presented hereafter.

3.2 H Infinity Design Scheme

We did not use the same H infinity design scheme for the structured one. Indeed, the weighting function added on the reference estimator design scheme to prevent the compensation of the notching filter is no longer useful. Because the structured estimator is of reduced order, it does not have enough degrees of freedom to do so. A new weighting function was introduced on the sensitivity function to ensure a large enough bandwidth for the filter: indeed the convergence time must be kept lower than 3 orbits, which is not easy because of the low frequency of the notching filter. The choice of the suitable weighting functions is the result of a trials/errors iterative procedure. The advantage of structured design via non-smooth optimization is that the estimator order does not depend on the system (G(s), W(s)) or weighting functions orders. Thus we are not restricted in the complexity of the weighting functions. Nevertheless, we chose to keep it as simple as possible. The final choice is illustrated on Fig. 7.

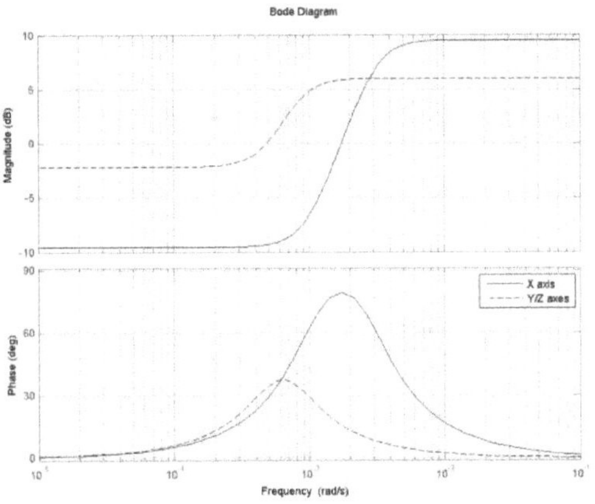

Fig. 7 Weighting function Bode Diagram

The design scheme is given Fig. 8.

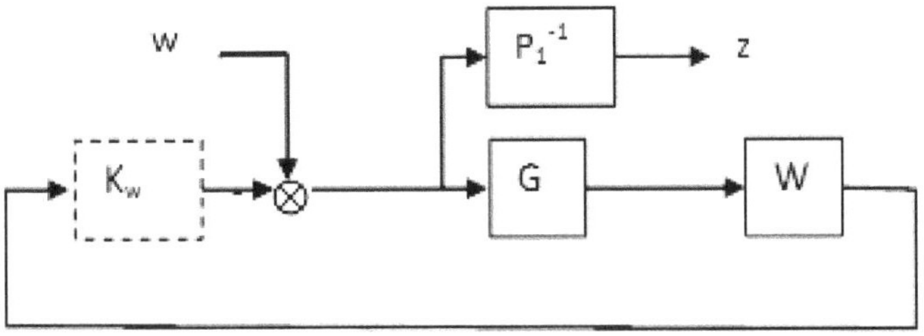

Fig. 8 Structured estimator design scheme

The non-smooth H infinity optimization tool Hinfstruct included in Matlab 2012a Robust Control toolbox™ does not allow for pole placement in a conic region. As a replacement we put spectral radius and stability offset constraints to avoid very low dynamics of the estimator. Finally the minimized $\|z/w\|_\infty$ value was constrained to be larger than 0.7.

The result of the iteration between structure choice and design is the following:

- For X axis, K_{wx} is a first order phase lead controller. The result is equivalent to the full-order estimator after reduction, with the difference that the performance cannot be affected by the reduction.

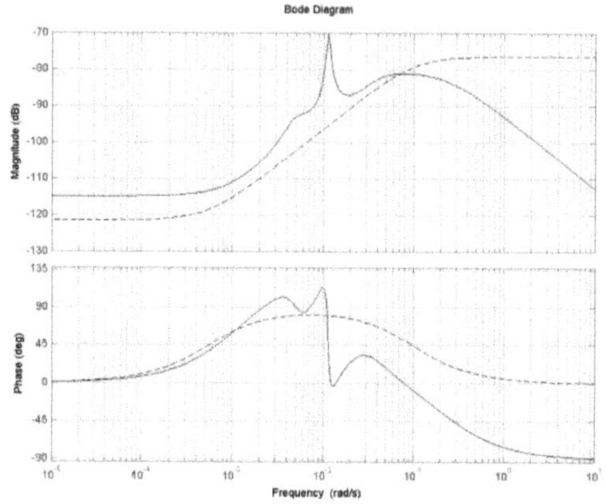

Fig. 9 Reference (plain) and structured (dotted) updating filter for X axis

- For Y/Z axes, $K_{yy} = K_{zz}$ is a second order phase lead controller and $K_{yz} = -K_{zy}$ is a very small static gain (around -110 dB). The full-order estimator order after reduction was 12, the structured estimator one is only 4.

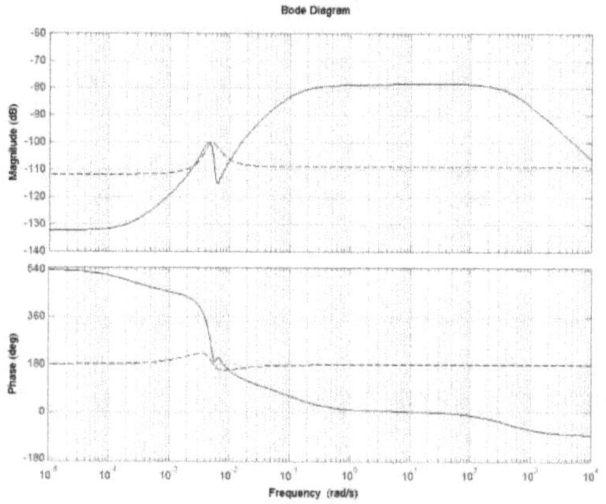

Fig. 10 Reference (plain) and structured (dotted) updating filter K_{yy} for Y/Z axes

3.3 Robustness to F_{spin} Variation

Because the adequate rotating frequency will only be stated in flight, the adequate estimator is not necessary the one designed with the nominal rotating frequency. To avoid retuning the filters during in flight operations, we choosed to pre-design a series of filters that will be loaded on-board ready to be selected.

As for the reference estimator, the structured estimator cannot cover the whole F_{spin} interval. Thus one has to design several structured filters. The constraint is then to make sure to get the same structure and the same order for each of these filters.

This could be achieved by setting the structured estimator designed for the lower bound of the interval as initial controller for the optimization procedure. This frequency corresponds to the most difficult case with respect to the large bandwidth for the star tracker transfer θ/θ_{STR} that we want to achieve. Indeed, the optimization algorithm tends to find a controller such that the cut-off frequency of θ/θ_{STR} is set close after the notching mode frequency. The lower the notching mode is, the lower the cut-off frequency is also.

Using the "worst case" structured estimator as initialization, the optimization algorithm will only perform small adaptations of the coefficients, keeping the same structure and order as the initial controller.

Twenty five filters have been designed to cover the spinning frequency interval. The number of estimators has been chosen by analysing the width of the notch at -67 dB. The minimum number is chosen by dividing the rotating frequency interval by this value. The final number allows to get some margin with the required accuracy. The results on Y/Z axes for the updating filter are illustrated on Fig. 11.

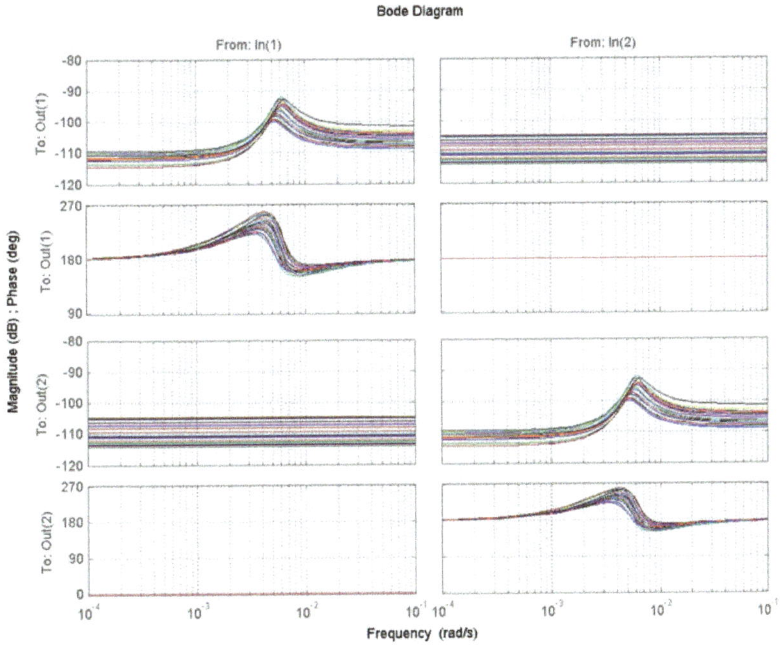

Fig. 11 Structured updating filters for 25 spinning frequencies (In(1), Out(1) are for Y axis, In(2), Out(2) are for Z axis).

Remark : Only one filter will be used for one measurement session, no switching is allowed inside a measurement session. Thus the stability of the switching from one structured estimator to another is not an issue here (between two sessions, another submode will be used with other controllers and estimators. The transitions are checked through Monte-Carlo simulations, not presented here).

4 Simulation Results

The structured estimator notch with more than -67 dB attenuation ensures that the star tracker error residue meets the 0.02μrad allocation. We now have to verify the behavior of the structured estimator in the global AOCS control loop and compare it to the reference estimator. The objective is to get the same performance level with a reduced order estimator. The main challenge is in the switching between a sub-mode where the notching estimators are not used and the sub-mode where they are: indeed, one has to ensure that the accelerometer will not saturate during the transient, and the estimator reaches the steady state level as quick as possible (3 orbit periods is an objective) .Because the dynamics of the structured estimator is not tuned as easily as in the full-order case (no pole placement routine available by the time of the design), the transient of the new estimator has to be carefully analysed.

The simulation showed hereafter illustrates the estimation performance but also its time convergence. The simulation scheme includes complex and representative models of every element (satellite's 6-axis nonlinear and coupled dynamics and kinematics, transport model from centre of gravity to center of measurement, cold gas thrusters, star tracker and accelemeter models). It is thus different from the one used for the estimators design. This type of models are used for Monte Carlo validation campaigns which have been successfully run with the new estimators, and whose results are not given here.

Fig. 12 illustrates one of the sub-mode switching configuration: the simulation starts in a calibration mode where the satellite oscillates around X axis. At t = 2960s, the satellite enters a spinning session. The attitude estimates are regulated

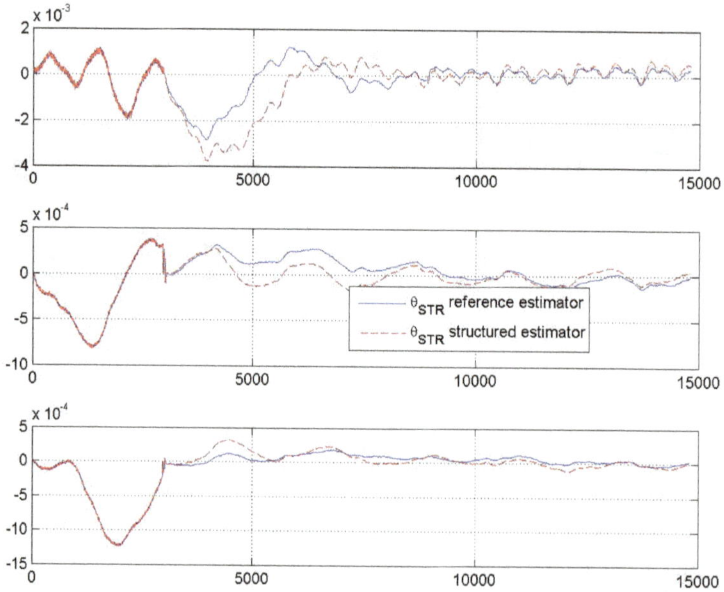

Fig. 12 Calibration to spinning sub-mode switching: filtered STR measurement (rad)

to zero by the cold gas thrusters control, so the estimation performance is observable on the star tracker measurement. We can see at the transition that the structured controller transient is larger on axes X and Z, and better on axis Y. At the end of the simulation (before the 3 orbits convergence time), both estimators give similar results.

5 Conclusion

In this paper we have presented a new synthesis for a structured and reduced order accelero-stellar estimator for Microscope spinning sessions. By analyzing the reference estimator structure, we orient the search for the suitable order and structure of the new controller. The design scheme and procedure has been adapted to the new non-smooth optimization tool, and we showed that the resulting structure could be robust to varying spinning frequency. Finally the new estimator has been compared to the reference one, and we showed that its performances were meeting the requirements.

Because of the great reduction of the estimator order without degradation of the performances, checked through phase C exhaustive simulation campaigns, the structured estimators have been selected to replace the reference estimators in the embedded Microscope AOCS software. The structure allows also improvement in the implementation scheme (zero blocks of matrices can be eliminated from the databank, a simple scalar flag allows to adapt the filters for 6h local time of ascending node or 18h orbit in case of launcher change). The structured estimator parameters number in the databank is then only 132, whereas it was equal to1024 with the reference estimators.

Finally, this example of application shows the interest of the new non-smooth optimization algorithms for H infinity synthesis, even in the estimation case. The provided estimation design scheme is generic and could be used for other systems.

References

1. Pittet, C.: Accelero-stellar hybridization for MICROSCOPE drag-free mission. In: Proc. ACA 2007, Aerospace Control Application, Toulouse (France), June 25-29 (2007)
2. Pittet, C., Prieur, P., Torres, A., Peus, A., Fallet, C.: MICROSCOPE: Myriade AOCS adaptation for a drag-free mission. In: 6th GNC ESA, Loutraki, Greece (2005)
3. Pittet, C., Fallet, C.: Gyroless attitude control of a flexible microsatellite. In: Proc. DCSSS 2002, Cambridge, UK (2002)
4. Falcoz, A., Pittet, C., Dinh, M., Bennani, S.: Synthesis of structures controllers for performance and numerical implementation mastering. In: 2014 GNC ESA Conference, Porto, Portugal
5. Apkarian, P.: Nonsmooth μ-synthesis. International Journal of Robust and Nonlinear Control 21(13), 1493–1508 (2011)
6. Apkarian, P., Noll, D.: Nonsmooth H(synthesis. Transactions on Automatic Control 51(1), 71–86 (2006)

7. Dubois, J.-B., André, Y., Prieur, P., Pouilloux, B.: MICROSCOPE, a femto-g accelerometry mission: technologies and mission overview. In: Proc. 4S (Small Satellite Systems and Services), Chia Laguna, Sardinia, Italy (2006)
8. Gahinet, P., Apkarian, P.: Structured H (synthesis in matlab). In: Proc. IFAC, Milan, Italy (2011)
9. Le Du, M., Maureau, J., Prieur, P.: Myriade: an adaptative concept. In: Proc. 4th GNC ESA, Frascati (2002)

Lidar-Aided Camera Feature Tracking and Visual SLAM for Spacecraft Low-Orbit Navigation and Planetary Landing

Franz Andert, Nikolaus Ammann, and Bolko Maass

Abstract. This paper explores the state estimation problem for an autonomous precision landing approach on celestial bodies. This is generally based on sensor fusion from inertial and optical sensor data. Independent of the state estimation filter, a remaining problem is the provision of position updates without the use of known absolute support information as it appears when the vehicle navigates within unknown terrain. Visual odometry or simultaneous localization and mapping (SLAM) approaches typically provide relative position. This is quite suitable, but it can be adverse due to error accumulation. The presented method combines monocular camera images with laser distance measurements to allow visual SLAM without errors from increasing scale uncertainty. It is shown that this reduces the accumulated error in comparison to sole monocular visual SLAM. Further, the presented method integrates the matching to known landmarks if they are available in the beginning of a landing approach so that the relative optical navigation can be initialized without systematic errors. Finally, tests with a simulated moon landing are performed and it is shown that the method is capable of navigating down to the ground impact.

1 Introduction

Future space exploration and landing missions strive for an investigation of local surface phenomena on moons, planets and other celestial bodies, requiring safe and precise navigation close to them. In this paper, the aspired mission (see sec. 2) is the

Franz Andert · Nikolaus Ammann
German Aerospace Center (DLR), Institute of Flight Systems,
38108 Braunschweig, Germany
e-mail: {franz.andert,nikolaus.ammann}@dlr.de

Bolko Maass
German Aerospace Center (DLR), Institute of Space Systems, 28359 Bremen, Germany
e-mail: bolko.maass@dlr.de

J. Bordeneuve-Guibé et al. (eds.), *Advances in Aerospace Guidance, Navigation and Control*, DOI: 10.1007/978-3-319-17518-8_35

landing on the moon as an example to investigate into the technology for a variety of future spacecraft applications. One of the required key technologies is a complex navigation system which provides an accurate state estimation, independent of external support like telemetry towards a ground control station. This basically means that the navigation system can only use information generated automatically and onboard the vehicle itself, meaning sensor data fusion, state estimation, and automatic vehicle guidance and control. As a part of this complex system, this paper is here focused on the integration of camera and laser distance measurements to provide positioning information. Optical-based position information is fused with data from other (i.e. mainly inertial) sensors. The resulting output will be a trajectory estimate for a lunar lander vehicle, compared to a reference path from the used simulation environment.

2 DLR's ATON Project: Mission Overview

Goal of the project "Autonomous Terrain-Based Optical Navigation for Landers" (ATON) is the development of technologies for an autonomous and safe landing within space scenarios. The aspired mission is the landing on the Moon within a pre-defined area of interest. The sensor configuration includes inertial measurements (acceleration and rotation rates), camera for star tracking (attitude estimation, see [16]), and optical ground-faced sensors (camera, laser ranging) for optical-aided navigation, landing site detection and obstacle avoidance. Fig. 1 gives a schematic overview about the type of lander that is simulated. The project comprises software-based tests with pre-calculated reference data (e.g. ray-traced images from the trajectory definition) and closed-loop tests with robotic hardware and real sensors [9].

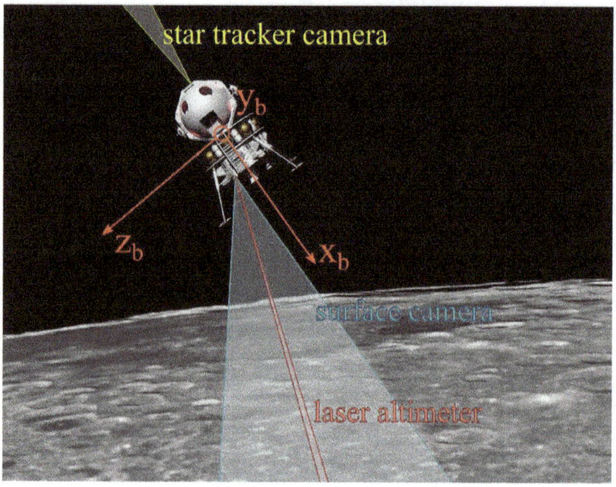

Fig. 1 Schematic optical sensor configuration of the lander system.

The navigation down to the landing site is divided into three phases as depicted in Fig. 2. In all phases, an Unscented Kalman Filter provides state estimation, based on inertial and star tracking camera data. Additionally, position updates and path corrections are given, depending on the particular phase.

Beginning with the rather slow descent from parking orbit, phase I includes absolute optical navigation with map correlation. The principle is to detect ground objects (i.e. craters) in the camera image and to match them with a map that stores their Moon-fixed absolute 3D positions. The algorithm consists basically of crater-like object detection, ellipsoid fitting, and camera resectioning, which it is described in [13].

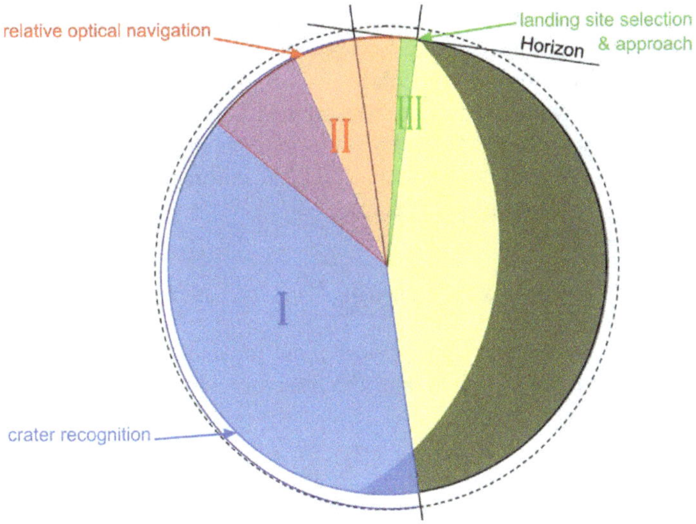

Fig. 2 Side view of the Moon, with three navigation phases from orbit descent initiation to landing.

While moving downwards, the amount of recognizable and known craters in the camera images will decrease. Hence, absolute crater navigation becomes more and more difficult, and the navigation phase II with relative optical navigation based on arbitrary tracked image features is executed. This is basically done with visual odometry methods. One of the algorithms that has been developed for this project is based on angular movement information derived from optical flow measurements [21]. Here, the integration into the state estimator is completed. However, the algorithm has some drawbacks especially due to the error accumulation from drifting image features.

As visible in Fig. 2, there will be an overlap between the absolute and relative navigation phases, meaning that the state estimation filter may use both crater matching and relative navigation if possible, but it is known that absolute navigation cannot be used after some time. Phase II will then last until the final landing. Parallel to that, the final landing phase is initiated. Within this phase, camera and laser sensors will be additionally used for ground appearance detection and to generate the desired landing coordinate within a flat area without obstacles and shadows.

3 Visual SLAM in the Context of Lunar Landing

3.1 Problem Statement

This paper tackles the further optimization of optical navigation (i.e. within the landing phases I and II) since this turned out to be a significant issue for the whole landing approach. To solve this problem, a visual navigation algorithm which has been successfully implemented on unmanned helicopters [2] is adapted to the preconditions of lunar landing. This means to replace satellite navigation with crater matching to get absolute positions and to replace absolute distance and scale measurement from stereo imagery with laser range measurements. The solution how this can be integrated into a visual SLAM algorithm is described in the next sections. The paper repeats some basics of common mapping and localization principles and focuses on the special terms for the presented scenario.

3.2 General Notations and Image Projection Basics

First of all, table 1 gives a brief overview about the used symbols for coordinate systems, image and object points, and the transformations used in the equations throughout this paper.

Table 1 Nomenclature used within the visual SLAM equations.

Symbol	Description
Index: c	Cartesian camera-fixed coordinate frame
Index: l	Cartesian lidar-fixed coordinate frame (z is the direction of laser beam)
Index: b	Cartesian Vehicle (body)-fixed coordinate frame
Index: g	Cartesian Geodetic coordinate frame, Moon-fixed, Moon-centered
$\mathbf{p}$	2D image point vector $(u,v)^\top$, undistorted and normalized pixel coordinates
z_l	Laser range measurement
$\mathbf{q}_a$	3D point vector $(x_a,y_a,z_a)^\top$ in coordinate system a (without index: geodetic)
$\mathbf{c}_g$	Geodetic 3D point vector $\mathbf{q}_g$ describing crater position from catalogue map
$\mathbf{t}_{ba}$	Translation vector from coordinate system a to b (notated in frame a)
$\mathbf{R}_{ba}, \mathbf{r}_{ba}$	Rotation matrix resp. vector (i.e. quaternion without real part) from a to b
$\Sigma_\mathbf{x}, \sigma_x$	Covariance matrix of vector $\mathbf{x}$, standard deviation of scalar x

For image projections, a calibrated camera and the (planar) pinhole camera model is used. The image points refer to a normalized image frame with the principal point

$\mathbf{p}_0 = (0,0)^\top$ and focal length $f = 1$. The projection of objects $\mathbf{q}_g$ to image features $\mathbf{p}$ is denoted

$$\lambda \begin{pmatrix} u \\ v \\ 1 \end{pmatrix}_c = \mathbf{R}_{cg} \begin{pmatrix} x - x_0 \\ y - y_0 \\ z - z_0 \end{pmatrix}_g \tag{1}$$

with the geodetic camera position $\mathbf{t}_{cg} = (x_0, y_0, z_0)^\top$ and rotation matrix $\mathbf{R}_{cg}$. A camera pose $[\mathbf{R}_{cg}, \mathbf{t}_{cg}]$ is composed by the camera alignment on the vehicle $[\mathbf{R}_{cb}, \mathbf{t}_{cb}]$ and the vehicle pose $[\mathbf{R}_{bg}, \mathbf{t}_{bg}]$ which is a part of the vehicle state. Eventually, the goal is to estimate this vehicle pose. It is

$$\begin{aligned} \mathbf{R}_{cg} &= \mathbf{R}_{cb} \cdot \mathbf{R}_{bg}, \\ \mathbf{t}_{cg} &= \mathbf{t}_{bg} + (\mathbf{R}_{bg}^\top \cdot \mathbf{t}_{cb}). \end{aligned} \tag{2}$$

If now the camera alignment on the vehicle is assumed to be constant and known from hand-to-eye calibration (usually done in the laboratory; see additionally e.g. [2] for a principle how to perform this step during flight if absolute positioning is available), the pixel coordinate $\mathbf{p}$ is eventually returned by a function dependent on the object point $\mathbf{q}_g$ and the vehicle pose $[\mathbf{R}_{bg}, \mathbf{t}_{bg}]$. This relationship is later used for optimization and to estimate the vehicle pose from image projections.

3.3 Data Input from Visual Features and Laser Distances

This section is about the used input data from optical sensors and available a-priori maps. The input data comprises crater features (image regions which look like a crater) and their relationship to a crater map, commonly tracked features (arbitrary textured image regions where the optical movement is measured) from the same camera, and laser distance measurements.

3.3.1 Crater Features for Absolute and Relative Localization

Craters are an ubiquitous physical feature of the surfaces of most celestial bodies without atmosphere (moons, larger asteroids) and they share a common trait: On all but the largest and smallest scales, their rims are almost always mostly circular. This means that imaging crater rims yields (for undistorted monocular camera images) a set of image space ellipses that, when detected, can be used as a feature set to support the visual SLAM method. Detected crater features are matched with a catalogue of known craters. For the moon landing scenario, a height map is available from the Japanese Kaguya Moon exploration mission [3] where the properties of craters and their relationship can be extracted. If matches between the image crater features and the catalogue are available, the corresponding points $\mathbf{p}$ and crater map objects $\mathbf{c}_g$ (here as landmarks with absolutely known position and covariance, but without systematic integration errors) are taken for absolute localization as described.

To this end, ATON employs the crater detector described in [13]. To summarize, the method comprises a detection stage that searches the camera image for regions of interest (ROI) characterized by containing two neighboring connected areas of each homogenous, but mutually opposite, intensity. If these neighboring areas of light and shadow are then also arranged in the image in a way consistent with the locally dominant illumination direction, they are assumed to be caused by illumination of the typical bowl shape of a crater (cf. fig. 3a).

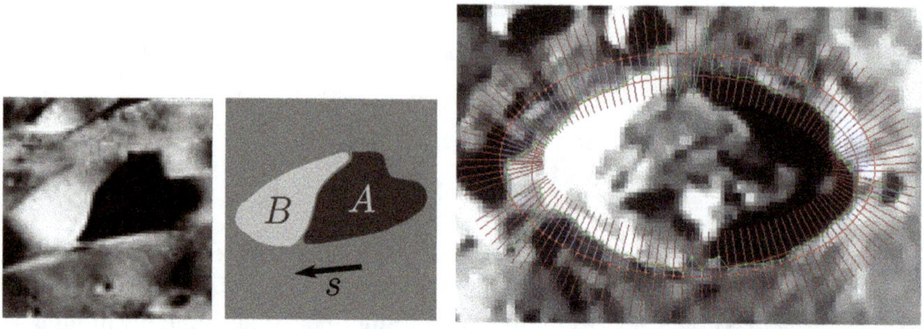

a) Contrasting areas in correct arrangement. b) Locking ellipse rim to local image gradient.

Fig. 3 Crater detection: region of interest (ROI) search method and ellipse rim locking.

After deriving an initial ellipse from this region of interest by principal component analysis (PCA) and applying some plausibility filters, a locking stage will exploit some of the present image gradient information in an attempt to better approximate the crater rim, using a cheap radial line search based on the local normals of the initial ellipse guess (cf. fig. 3b).

Applying the above summarized method to a camera image, a set of image ellipses $\mathbf{E} = \{E_1, \ldots, E_n\}$ is obtained. Each ellipse is parametrized by $E_i = (u, v, a, b, \varphi)_i$, where u and v are the image coordinates of the ellipse center, a and b are the semimajor and semiminor axes, and φ is the ellipse's rotation. Of these parameters, the points $\mathbf{p} := (u, v)$ can be used as image-space feature locations for the visual SLAM method, while the remaining parameters (a, b, φ) can be used to extend the features' descriptors to reliably track them over multiple images.

While the detection process itself may introduce errors of the order of a pixel into $\mathbf{p}$, features of this kind are tied to actual physical structures of the observed celestial body. This leads to the beneficial property that they are not subject to drift like features extracted and tracked by, for example, Lukas-Kanade methods are, and thus are reliably trackable over long time spans.

Beyond providing image-space feature points $\mathbf{p}$, in conjunction with a catalog of known crater positions $\mathbf{c}_g$, radii r and surface normals $\mathbf{n}_g$ on the surface of the observed celestial body, the detected crater ellipses can also be used to provide the features' absolute positions in the camera reference frame by matching them against the catalog. For this, a coarse initial guess for the camera pose needs to be known, where imaging errors from camera pose uncertainty on the order of a few pixels are easily tolerable. Matching image points to the catalog points requires the following intermediate steps:

1. Computation of the image point of projection $\mathbf{p}'$ of the actual center $\mathbf{c}_g$ of the crater rim circle C whose projection the ellipse parameterized by $E = (\mathbf{p}, a, b, \varphi)$ is (cf. fig. 4). This is performed by solving for the projective conic that intersects the image plane in the ellipse E, and finding the plane(s) that in turn intersect this conic in a perfect circle C. The projection of that circle's center point onto the image plane is $\mathbf{p}'$. There is an ambiguity of reflection about the conics center line here since exactly two planes fulfill the above requirement in the non-degenerative case. This ambiguity can be resolved trivially by reasoning from the estimated camera pose, however.

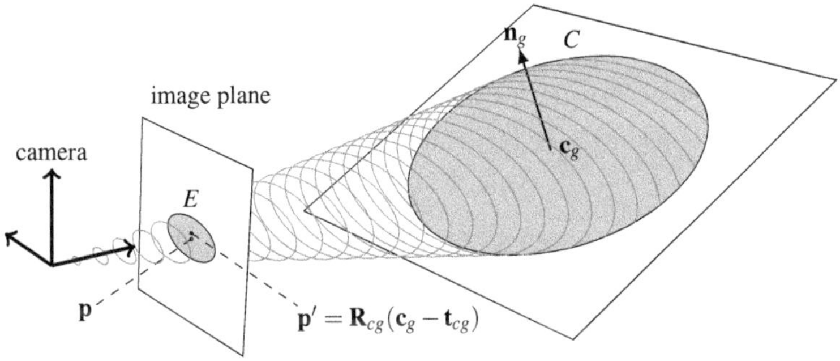

Fig. 4 Projection of circle's center differs form Ellipse center.

2. Establishing correspondences between the catalog craters and the extracted image ellipses. This is accomplished by using the estimated camera pose $[\mathbf{R}_{cg}, \mathbf{t}_{cg}]$ and projecting the catalog craters into the image plane, yielding a second set of ellipses. Provided that the estimated camera pose was sufficiently accurate, matches between these projected ellipses and the ellipse set resulting from the crater detection can be found by looking for closest matches between the two in the metric space of the five ellipse parameters.

The above derived image point $\mathbf{p}'$ and the known cartesian geodetic crater rim center point $\mathbf{c}_g$ complete the 3D feature pair to be used to support the visual SLAM algorithm.

3.3.2 Feature Tracking

Methods how to match or to track homologous image points are widely known [6]. The output are generally sets of points $\mathbf{p}$ for each camera image, with identifiers showing the correspondence to the points of the other images from the sequence. The implementation used here is rather simple, it uses the common Lucas-Kanade tracker [12] to get correspondences between succeeding images. The initialization of features comes from a Shi-Tomasi corner detector [19]. While tracking, tracked features are removed if they come too close to the edge, too close to other features, or if the texture quality has decreased. New corner features are automatically added if required (i.e. if the current amount undershoots a threshold).

In contrast to descriptive matchers (such as SIFT, SURF or the described crater ellipsoids), the Lucas-Kanade tracker is very fast. However, tracking is a similarity match of image regions from one image to the other within the sequence, and thus the features itself may drift. This property is considered by the output, which means that the tracking algorithm provides a set of image features **p** whose covariance is modeled by

$$\Sigma_{\mathbf{p}} = \begin{pmatrix} \sigma_u^2 & 0 \\ 0 & \sigma_v^2 \end{pmatrix}, \quad \text{with} \quad \sigma_{u/v} = (0.25 + 0.05 \cdot n)/f_{u/v}, \tag{3}$$

where f_u and f_v are the focal lengths in u- and v-direction and n is the number of tracking steps after the first identification of this feature. The chosen error model assumes a feature error of 0.25 pixels in the beginning which is increased by 0.05 pixels per tracking step. This takes into account that, in contrast to crater features, a tracking drift is existent here. The result is a common set of tracked features ready for visual SLAM (Fig. 5).

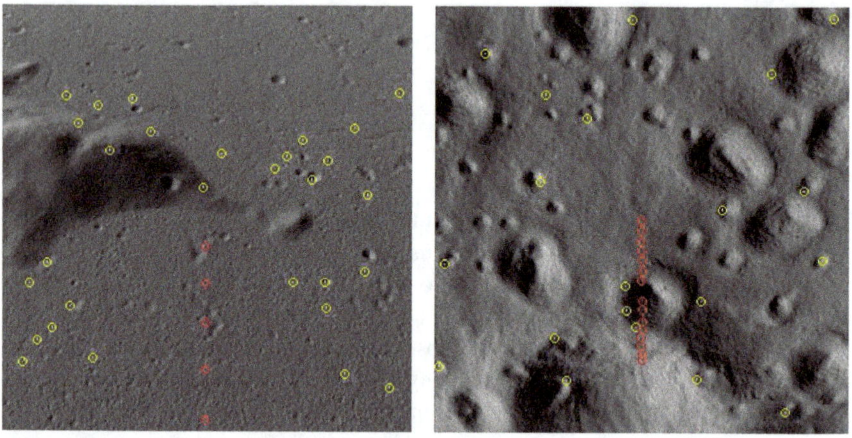

Fig. 5 Tracked image features of a simulated moon landing camera image sequence. Example images taken from 11 km height (left) and later from 300 m height (right). Features inserted with initial laser distance are marked red (cf. sec. 3.4.3), the others yellow.

3.3.3 Laser Distance Measurements

The simulated moon lander includes a laser range finder. There are experiments with scanning and flash lidar systems (with horizontal and vertical field of view) throughout the whole project, but here only one single measurement is taken for ease of use and to be open for many possibilities of current and future hardware. The sensor is considered as an altimeter which points about parallel to the optical axis of the camera and measures the diagonal distance z_l between sensor and moon surface. The measuring scope is from zero to about 30 kilometers. The measuring uncertainty is assumed to be constantly $\sigma_{z_l} = 5.0\,\mathrm{m}$.

3.4 *Camera Localization and Environment Mapping*

3.4.1 Overview

Simultaneous Localization and Mapping generally consists of a step where a sensor pose (and, with that, a vehicle pose) is estimated by the registration between current sensor data and an environment map, and a step where this map is generated and updated out of the sensor data. Since localization and mapping do highly depend on each other, it marks a typical chicken-and-egg problem and thus requires an adequate initialization of the pose or the map.

In the presented scenario, it is possible to either do an initialization with known camera poses from the absolute crater navigation, and the relative visual navigation can be initialized during the overlap of the landing phases I and II (cf. Fig. 2). It would be also possible to directly use identified matches between crater map and image feature positions as an initial input for the localization initialization. However, in both cases, the map will be the eventual a-priori knowledge and the difference between both variants is whether the visual SLAM algorithm takes either the camera pose input from the absolute crater navigation algorithm or calculates the initial poses by itself. The latter case is a tighter integration of absolute and relative navigation within one visual navigation algorithm (Fig. 6).

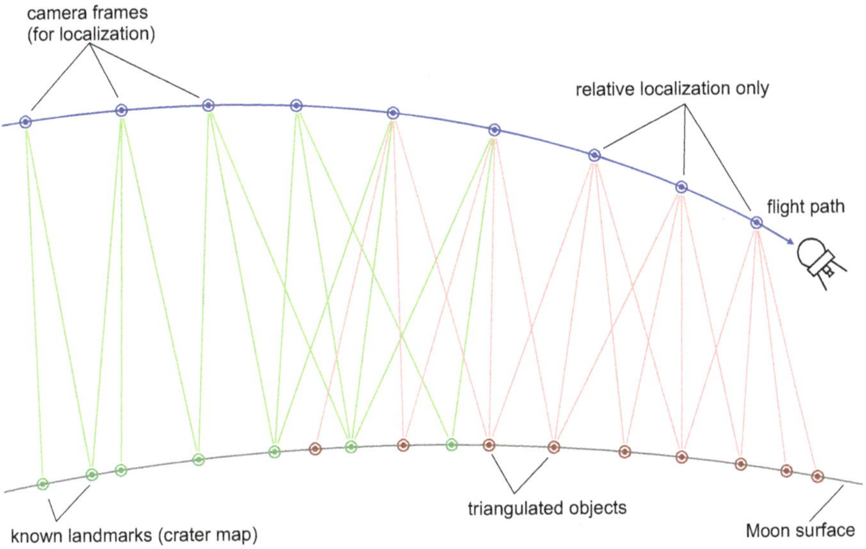

Fig. 6 Camera localization from known landmarks and a-priori unknown objects triangulated during the flight. A sub-set of the camera images (key frames) is used to map them.

3.4.2 Absolute and Relative Localization

If a map is available (and here, this is applicable from the beginning due to the present crater map), localization generally means camera resectioning from 2D-3D point correspondences [17]. From the algorithm perspective, there is no difference between absolute and relative localization: the difference is the source of 3D objects (i.e. truly known landmarks vs. objects mapped during the flight) and their possible uncertainty and systematic error. Resectioning algorithms are widely known, and here, the implementation described in [2] is used. The steps are:

1. Input is a set of 2D-3D point correspondences $\{\mathbf{p}, \mathbf{q}_g\}$ including covariance matrices. The amount of points should be at least three, but it is advantageous to use a set of 30 points or more. The 3D points form the map and refer to visible ground objects. The 2D points are the pixel coordinates of these objects projected onto the image plane. Internal camera calibration and undistortion of raw pixel coordinates is required.
2. If no initial camera pose is given (which is the case in the beginning), linear algorithms can provide an initial pose estimate from the given map. This implementation uses a Direct Linear Transform [10] together with RANSAC [5] for outlier removal. Later during the flight, the previous pose estimate or the current state prediction from inertial data can be used as an initialization for step 3.
3. Based on given initial values, the projections $\mathbf{q}_c$ and their reprojection $\mathbf{p}'$ onto the image plane are calculated. From eq. 1, a reprojection $\mathbf{p}'$ is

$$\lambda \begin{pmatrix} \mathbf{p}' \\ 1 \end{pmatrix} = \mathbf{R}_{cg} \left(\mathbf{q}_g - \mathbf{t}_{cg} \right), \tag{4}$$

the covariance $\Sigma_{\mathbf{p}'}$ is calculated by

$$\lambda^2 \begin{pmatrix} \Sigma_{\mathbf{p}'} & \\ & 1 \end{pmatrix} = \mathbf{R}_{cg} \Sigma_{\mathbf{q}_g} \mathbf{R}_{cg}^{\top}. \tag{5}$$

The reprojection error d between $\mathbf{p}'$ and the measurement $\mathbf{p}$ is now a kind of Mahalanobis distance with

$$d^2 = (\mathbf{p} - \mathbf{p}')^{\top} \left(\Sigma_{\mathbf{p}} + \Sigma_{\mathbf{p}'} \right)^{-1} (\mathbf{p} - \mathbf{p}'). \tag{6}$$

This kind of reprojection error considers the object and image feature uncertainty to prefer objects and features with low uncertainty values.

Now let a projection $\mathbf{q}_g$ be a function value dependent on the vehicle pose $(\mathbf{t}_{bg}, \mathbf{r}_{bg})$. With that, the reprojection error d_i for the i-th feature is also parametrized by that. A non-linear optimization algorithm (here: Levenberg-Marquardt, LM) can now minimize d_i over all available point correspondences i to estimate the six pose parameters $(\mathbf{t}_{bg}, \mathbf{r}_{bg})$. If needed, a second LM step is run where items with large residuals have been removed before.
4. Based on the triangulation functions and the objects' uncertainties (inliers only), the algorithm also returns an uncertainty estimation for the pose estimate. For ease of use, only a position covariance matrix is provided here. It is

$$\Sigma_{\mathbf{t}_{bg}} = m \cdot \left(\sum_{\mathbf{q}_g} \Sigma_{\mathbf{q}_g}^{-1} \right)^{-1}, \tag{7}$$

meaning covariance intersection (i.e. averaging) of an amount of m object uncertainties.

For each new camera image, the output is a pose estimate in vehicle body-fixed coordinates (and its position uncertainty), being used as a correction input for the state estimation filter (sec. 4). The calculation time mainly depends on the number of point correspondences, but it is fast enough to be real-time capable on standard hardware with 30 camera frames per second and at least 100 image features.

3.4.3 Mapping with Monocular Images and Laser Distance Measurements

For image features where no corresponding 3D landmarks are available, 3D objects are mapped from at least two known points of view. The used algorithm is basically an iterative L_2 triangulation [7, 11], followed by a non-linear optimization with the Levenberg-Marquardt method. See [2] for details to get the $\mathbf{q}_g$ from each at least two homologous point correspondences $\mathbf{p}$. The cited paper also describes the calculation of the uncertainties $\Sigma_{\mathbf{q}_g}$ based on the Jacobian of the triangulation function and a covariance intersection of the vehicle pose uncertainties. A detailed description is skipped since monocular multi-view triangulation is an established principle for many applications.

In this case of monocular mapping, the absolute scale of the map is obtained by the camera viewpoints whose distance must be known. In the beginning of relative navigation, the distance between different viewpoints is available from landmark localization and thus not influenced by scale errors. However, if now mapping and localization is performed without any landmark hints, errors in localization have a direct influence on errors in further mapping, which will again increase the uncertainty of the localization steps. This accumulating error is typical for all SLAM methods where no landmarks are used. In the case of monocular SLAM, the correctness of object triangulation and camera resectioning is highly dependent on a correct scale initialization since scale is not obtained anymore. It is obvious that scale uncertainty is increasing over time so that monocular SLAM is going to accumulate errors faster than methods where absolute scale measurements are available.

Actually, it was observed from test image sequences that the scale factor tends to decrease, which means that distances between succeeding images are estimated too small and that the map objects draw near the vehicle path. This can terminate wrongly with zero scaling which corrupts the whole visual navigation process. With that, it is necessary to estimate the scale during the relative navigation phase. Learning from unmanned aircraft navigation at low altitudes, this can be solved with stereo imagery [18] or pressure sensors [1] providing absolute ground distance or elevation measurements. In the presented scenario with an orbit descent from 11 km flight altitude, a long-range laser distances can be used for this task. The sensor of choice can be a one-point altimeter (as used here), but also scanning and flash lidar systems may be used in the same way as described below.

Laser-aided monocular mapping assumes synchronous distance measurements and images, achieved by triggering or adequate interpolation, and known alignment parameters between vehicle, camera and laser sensor from laboratory calibration (see e.g. [8]). From a single point range measurement with distance z_l, the object point is

$$\mathbf{q}_g = \mathbf{R}_{lg}^{\top} \begin{pmatrix} 0 \\ 0 \\ zl \end{pmatrix} + \mathbf{t}_{lg} \tag{8}$$

Following eq. 1, this object is projected onto the image plane and thus the distance of this pixel is known (see Fig. 7). In practice, the laser spot hits always the same pixel if the alignment between the sensors is constant and if the distance is large enough (approx. >100 m for the setup here). The main benefit is a corresponding pair of $\mathbf{p}$ and $\mathbf{q}_g$ from a single point of view for further mapping and localization. For features that are tracked beginning at this pixel position, an initial object correspondence is known without scale ambiguity. Later triangulations from mapping steps can include this information by minimizing the object distance error in addition to the (usual) object reprojection errors on the image plane. The uncertainty of the point $\mathbf{q}_g$ is modeled with

$$\Sigma_{\mathbf{q}_g} = \mathbf{R}_{cg}^{\top} \begin{pmatrix} \left(z_l^2 \cdot \Sigma_{\mathbf{p}}\right) & \begin{matrix} 0 \\ 0 \end{matrix} \\ \begin{matrix} 0 & 0 \end{matrix} & \sigma_{zl}^2 \end{pmatrix} \mathbf{R}_{cg} + \Sigma_{\mathbf{t}_{bg}} + z_l^2 \cdot \Sigma_{\mathbf{r}_{bg}} \tag{9}$$

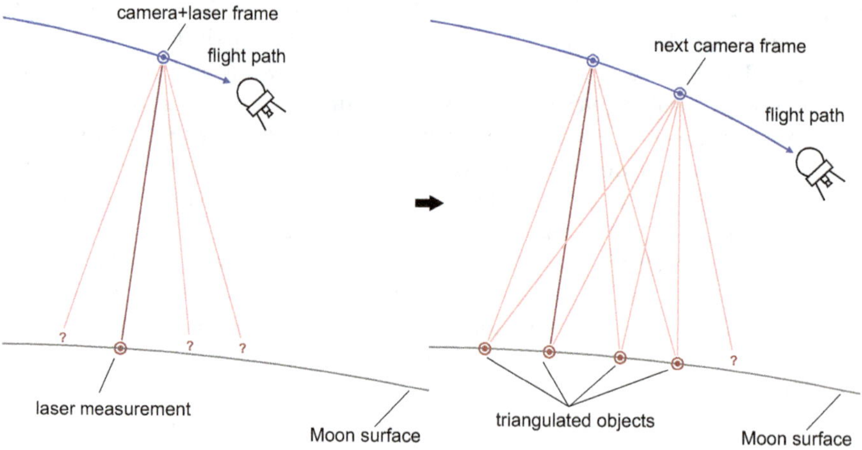

Fig. 7 Laser-aided mapping. For the image feature with available distance measurement, the object is obtained from one camera pose (left). Later, multi-view triangulation is used for refinement and the triangulation of all other image features without distance measuring (right).

where σ_{z_l} is the uncertainty of the measured distance and $\Sigma_{\mathbf{p}}$ is the image feature uncertainty. The addition term $z_l^2 \cdot \Sigma_{\mathbf{r}_{bg}}$ models an error caused by the uncertainty of the vehicle rotation.

Since the map contains at least some objects without scale ambiguities, localization is improved automatically. This should be the common case, at least if the pixel $\mathbf{p}_c$ generated from z_l is trackable (i.e. this region contains significant image texture) and if the point is not later removed as an outlier (e.g. from false tracking).

4 Integration into the Navigation System

The feature tracking and visual SLAM algorithms are integrated into the spacecraft navigation system as depicted in fig. 8. The concept is designed with different modules which can be run separately using the shown connections for data sharing. The main filter output is computed by the *state estimator* module where an Unscented Kalman Filter (error state filter) is run. It predicts the state by inertial data and uses corrections from the star tracking camera (absolute attitude) and the *visual SLAM* module (position data). The *image processing* module comprises the attitude calculation from the upwards-looking camera images, and the ground feature and crater detection from the downwards-looking camera and laser range finder. Visual SLAM and crater matching work as described in the previous chapter. The *filter states* module summarizes the on-board data generation, mainly the flight state itself as output and to be used for further predictions, and the continuously updating 3D objects map used within visual SLAM.

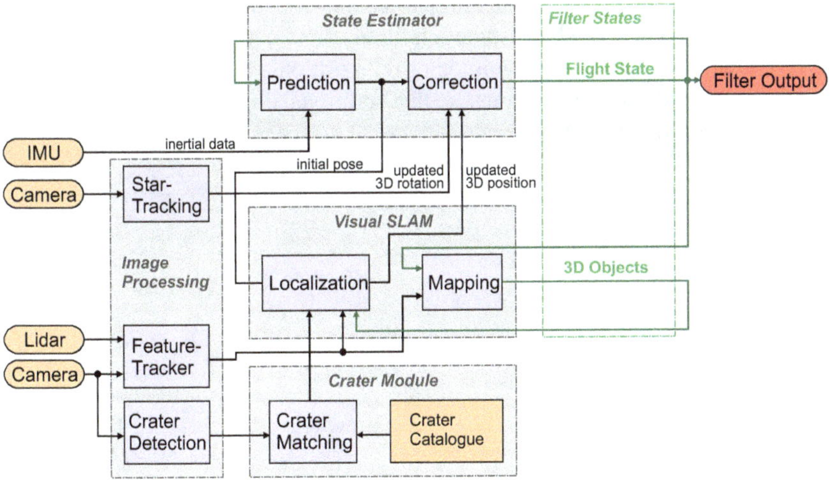

Fig. 8 Data processing modules used for optical-aided spacecraft navigation.

5 Test and Evaluation

5.1 Simulation Setup

The navigation system is tested in an open-loop simulation environment based on a reference trajectory from the powered descent initiation to landing. The trajectory begins in approximately 11 km above ground with a flight speed of 1600 m/s. After a flight time of around 52 minutes, the scheduled landing site at the moon is at -5° latitude, 8° longitude (mean earth / polar axis coordinates, see [4]) and the lander will be flying with a heading of 180° towards it.

From this trajectory, sensor data is generated, e.g. inertial data with a noisy and biased IMU model [21], star tracker attitude data, and laser distance measurements. Based on the lunar height map from the Kaguya mission, camera images are rendered by the ray tracing algorithm from [14]. For the test here, about 90 000 images are generated (1024×1024 pixels, 40°×40° field of view, 30 frames per second, see again Fig. 5 for example images) for a flight at a specific time, considering sunlight conditions and also radiometric effects (e.g. background radiation noise). Total flight time of this trajectory is 3143 s. The open-loop simulation can now be run in a post-processing step: Images and sensor data are taken to estimate the flight path which is compared to the ground truth of the simulation.

5.2 Test Run

The figures 9– 11 show the estimated positions with regard to the ground truth available from the simulation. They also compare raw visual SLAM estimations (i.e. object mapping directly with image-based localization) with full filter integration (i.e. localization feeds the state estimation filter, and mapping with filter update positions). Due to the order of magnitude of the positions, there are visible overlaps between estimated and true positions, so that additional error plots may give a better view on the resulting performance.

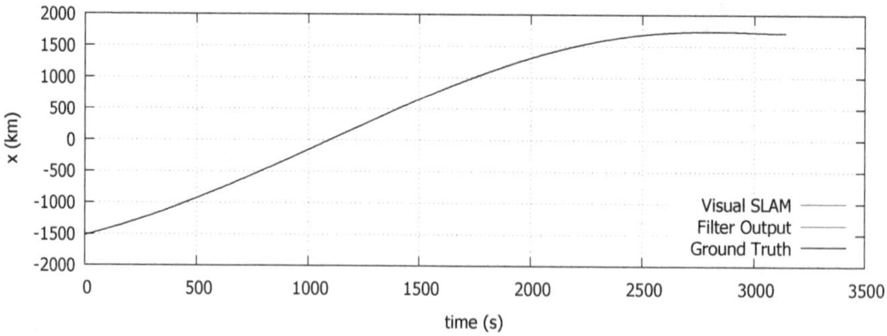

Fig. 9 True and estimated position (x), Moon-fixed, Moon-centered coordinates.

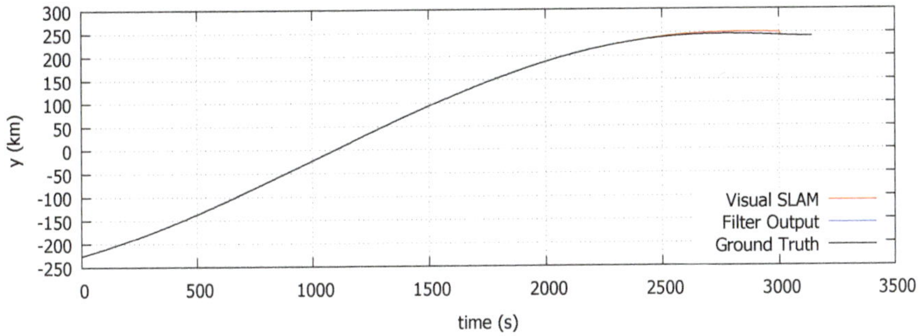

Fig. 10 True and estimated position (y), Moon-fixed, Moon-centered coordinates.

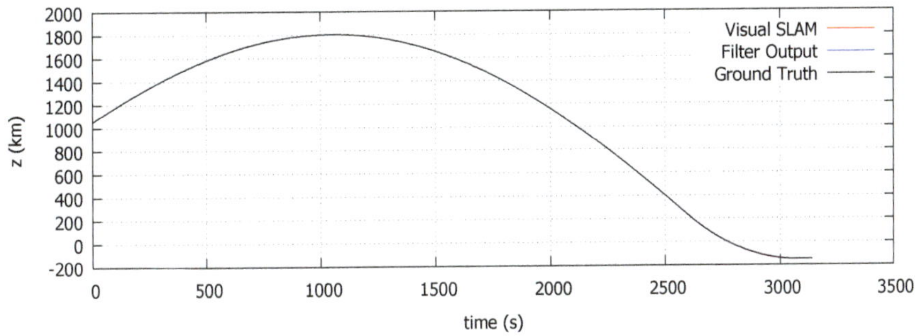

Fig. 11 True and estimated position (z), Moon-fixed, Moon-centered coordinates.

These error plots are shown in the figures 12–14. They also compare raw visual SLAM estimations with the filter integration (error to ground truth), and they show the estimated standard deviations, derived from the filter covariance output. Plots with rotational data are skipped since star tracking provides attitude with an uncertainty below 0.01° without drift and error accumulation.

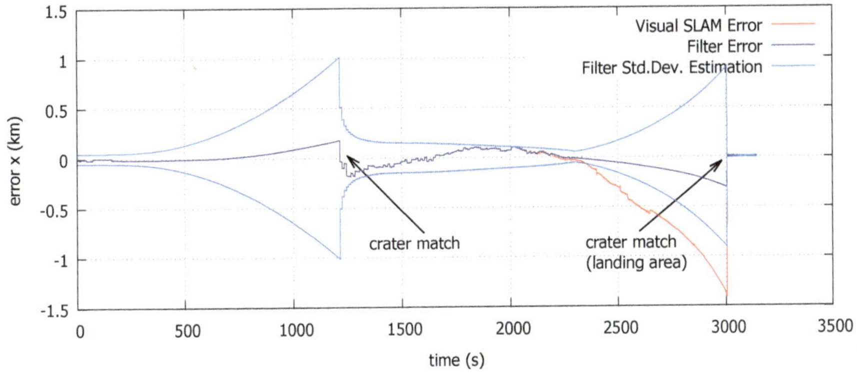

Fig. 12 Error (x) of the visual SLAM and filter estimate and filter output of standard deviation.

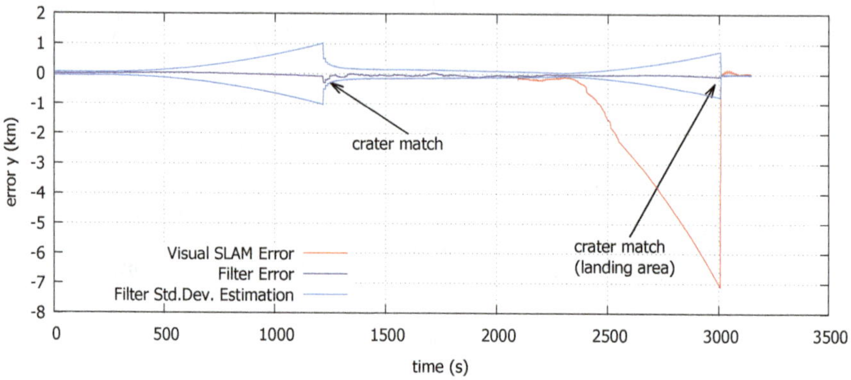

Fig. 13 Error (y) of the visual SLAM and filter estimate and filter output of standard deviation.

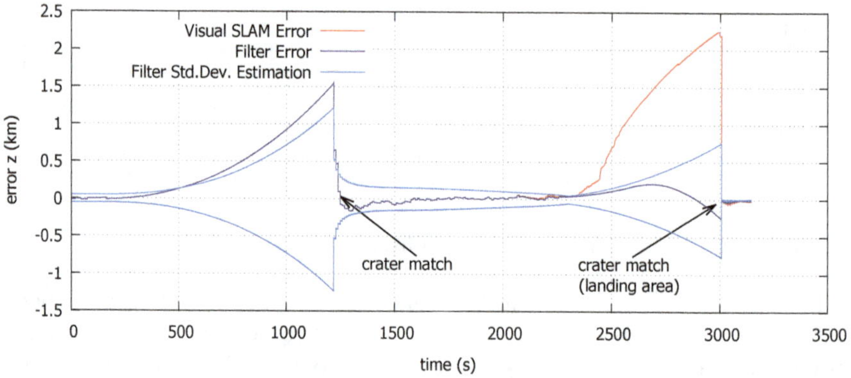

Fig. 14 Error (z) of the visual SLAM and filter estimate and filter output of standard deviation.

5.3 Evaluation of the Results

As already mentioned, the position plots (figs. 9– 11) reveal strong overlaps between estimation and true position, which is an indicator that visual SLAM and the overall filter provide suitable results. Large deviations exist on the y-axis, visible in fig. 10 between 2500 s and 3000 s. More details on the error is shown by the plots in the following figures 12–14.

The test run reveals two successful position corrections based on the crater matching algorithm. The first absolute position correction is more in the beginning at approximately 1200 s from a flight altitude of about 8 kilometers and would be available on a real lunar landing. The second correction results from a match with the landing spot with craters assumed to be known here. For a landing on such a known site, the vehicle positioning uncertainty is in the order of magnitude of the landing site resolution, here few meters on each axis. It is clear that precise absolute vehicle positioning depends on such an availability of a map of the landing area (e.g. from the LRO's narrow angle camera, see [20]). Note that that this may not applicable in scenarios on arbitrary landing sites or if the lander misses the desired area where a high-resolution map is available.

However, the presented filter reaches an accuracy of 300 m (x), 60 m (y), and 250 m (z) before landing site matching. Within the filter, standard deviation estimation also corresponds with the real errors from their time response. In most cases, the real error is highly below its estimate, except for the z-coordinate before the first crater match. The curves also show the typical drift behavior of inertial systems where no corrections are available. It is also shown that sole visual SLAM may return poor results. A possible reason for high visual errors (here between 2400 s and 3000 s) could be the attitude of the camera towards the ground: While the camera looks diagonally towards the moon surface in the beginning of the flight where features can be tracked and mapped over a long flight distance, the vehicle turns later (pitch rotation) so that the camera looks vertically towards the moon surface and with less image overlap. In such cases, 3D object triangulation is restricted to viewpoints with shorter distances, which highly reduces the triangulation quality. In the presented scenario, sole visual SLAM causes accumulation errors of up to 7 km (y-axis) which is significantly higher than the resulting error of the coupled filter. Hence, good inertial measurements are still mandatory for suitable flight state prediction.

6 Conclusion

The project ATON is on the way to establish a navigation procedure for spacecraft landers, tested in a simulation environment. As a part of this project, this paper presents an optical method for absolute and relative positioning and its application to lunar landing. The algorithms comprise visual localization from 2D-3D point correspondences and laser distances. These corresponding items can be known objects like craters and their image projections (absolute navigation), or arbitrary but trackable image features of the moon surface whose the 3D object coordinates are triangulated during the tracking phase (relative navigation). Both principles are combined within one visual simultaneous localization and mapping module which generates position updates (i.e. inertial drift corrections) for a state estimation filter.

For a test, the filter including the vision algorithms is fed with a simulated moon landing data sequence, consisting of surface camera images, diagonal surface distance from laser altimeter, attitude from a star tracker, and high-frequent inertial data. The test also simulates real-time capabilities, that is why the computationally complex crater identification and matching algorithm only returns two absolute position corrections within one hour flight time. Contrary to that, relative navigation can be run with the camera frame rate to produce continuous updates, but this is influenced by accumulation errors as expected so that the overall system cannot perform without crater-based updates. Depending on the availability of a high-quality map of the landing spot, the system is capable of a precise hit, or to perform a landing with an uncertainty of few hundred meters. This means that precise landings can be performed in mapped areas, e.g. on the lunar north pole.

Based on the presented data processing, other investigations of this project are dealing with closed-loop integrations on different test facilities. First of all, this means an integration with flight control within a simulation environment, but also the use of hardware test beds. More precisely, the software is run on a robotic test site where a camera can be automatically moved inside a room with a moon-like facade and different lighting conditions. In future experiments, this will be supplemented with tests with an unmanned helicopter where a landing with the sensors is reproduced.

Acknowledgements. This work is part of the collaborative project "Autonomous Terrain-Based Optical Navigation" for Landers (ATON), managed by the DLR-Institute of Space Systems, Bremen, Germany, in cooperation with the DLR-Institutes of Flight Systems, Optical Sensor Systems, Robotics and Mechatronics, and Simulation and Software Technology.

References

1. Achtelik, M., Achtelik, M., Weiss, S., Siegwart, R.: Onboard imu and monocular vision based control for MAVs in unknown in- and outdoor environments. In: IEEE International Conference on Robotics and Automation, pp. 3056–3063 (2012)
2. Andert, F., Ammann, N., Püschel, J., Dittrich, J.: On the safe navigation problem for unmanned aircraft: Visual odometry and alignment optimizations for UAV positioning. In: International Conference on Unmanned Aircraft Systems (ICUAS), pp. 734–743 (2014)
3. Araki, H., Tazawa, S., Noda, H., Ishihara, Y., Goossens, S., Sasaki, S., Kawano, N., Kamiya, I., Otake, H., Oberst, J., Shum, C.: Lunar global shape and polar topography derived from Kaguya-LALT laser altimetry. Science 323(5916), 897–900 (2009)
4. Davies, M.E., Colvin, T.R.: Lunar coordinates in the regions of the apollo landers. Journal of Geophysical Research 105(E8), 20:277–20:280 (2000)
5. Fischler, M.A., Bolles, R.C.: Random sample consensus: A paradigm for model fitting with applications to image analysis and automated cartography. Graphics and Image Processing 24(6), 381–395 (1981)
6. Fraundorfer, F., Scaramuzza, D.: Visual odometry – part II: Matching, robustness, optimization, and applications. IEEE Robotics & Automation Magazine 19(2), 78–90 (2012)
7. Hartley, R., Sturm, P.: Triangulation. Computer Vision and Image Understanding 68, 146–157 (1997)
8. Krause, S., Evert, R.: Remission based improvement of extrinsic parameter calibration of camera and laser scanner. In: 12th International Conference on Control, Automation, Robotics & Vision, pp. 829–834 (2012)
9. Krüger, H., Theil, S.: TRON – hardware-in-the-loop test facility for lunar descent and landing optical navigation. In: 18th IFAC Symposium on Automatic Control in Aerospace (2010)
10. Kwon, Y.H.: Direct linear transform method (1998), url: `http://www.kwon3d.com/theory/dlt/dlt.html`
11. Lu, F., Hartley, R.: A fast optimal algorithm for l_2 triangulation. In: Asian Conf. on Computer Vision (2007)

12. Lucas, B.D., Kanade, T.: An iterative image registration technique with an application to stereo vision. In: International Joint Conference on Artificial Intelligence, pp. 674–679 (1981)
13. Maass, B., Krüger, H., Theil, S.: An edge-free, scale-, pose- and illumination-invariant approach to crater detection for spacecraft navigation. In: 7th International Symposium on Image and Signal Processing and Analysis (ISPA), pp. 603–608 (2011)
14. Paproth, C., Schlüßler, E., Scherbaum, P., Börner, A.: Sensor++: Simulation of remote sensing systems from visible to thermal infrared. In: International Archives of the Photogrammetry, Remote Sensing and Spatial Information Sciences, XXII ISPRS Congress, pp. 257–260 (2012)
15. Sabiron, G., Chavent, P., Burlion, L., Kervendal, E., Bornschlegl, E., Fabiani, P., Raharijaona, T., Ruffier, F.: Toward an autonomous lunar landing based on low-speed optic flow sensors. In: Chu, Q., et al. (eds.) Advances in Aerospace Guidance, Navigation and Control: Selected Papers from the CEAS EuroGNC Conf., pp. 681–699. Springer (2013)
16. Samaan, M., Theil, S.: Development of a low cost star tracker for the SHEFEX mission. Aerospace Science and Technology 23(1), 469–478 (2012)
17. Scaramuzza, D., Fraundorfer, F.: Visual odometry – part I: The first 30 years and fundamentals. IEEE Robotics & Automation Magazine 18(4), 80–92 (2011)
18. Shen, S., Mulgaonkar, Y., Michael, N., Kumar, V.: Vision-based state estimation and trajectory control towards high-speed flight with a quadrotor. In: Robotics: Science and Systems (2013)
19. Shi, J., Tomasi, C.: Good features to track. In: IEEE Conference on Computer Vision and Pattern Recognition, pp. 593–600 (1994)
20. Tran, T., Rosiek, M.R., Beyer, R.A., Mattson, S., Howington-Kraus, E., Robinson, M., Archinal, B.A., Edmundson, K., Harbour, D., Anderson, E., the LROC Science Team: Generating digital terrain models using LROC NAC images. In: Proc. of Special joint symposium of ISPRS Technical Commission IV & AutoCarto / ASPRS/CaGIS 2010 Fall Specialty Conference (2010)
21. Verveld, M.J.: Relative optical navigation for a lunar lander mission. In: Chu, Q., et al. (eds.) Advances in Aerospace Guidance, Navigation and Control: Selected Papers from the CEAS EuroGNC Conf., pp. 661–679. Springer (2013)

Camera-Based Tracking for Rendezvous and Proximity Operation of a Satellite

Nassir W. Oumer and Giorgio Panin

Abstract. This paper focuses on vision-based detection and tracking of a nozzle of a satellite for rendezvous and proximity operation at very close range. For this purpose, on-board cameras can provide an effective solution in accuracy and robustness during the approach. However, the illumination conditions in space are especially challenging, due to the direct sunlight exposure, and to the glossy surface of a satellite. We propose an efficient tracking method that can be realized on standard processor, robustly dealing with the above issues exploiting model and image edges. The algorithm has been validated at the facility of the European Proximity Operations Simulator of DLR, using a ground simulation system that is able to reproduce sunlight conditions through a high power floodlight source, satellite surface properties using reflective foils, as well as complex motion trajectories with ground truth data.

1 Introduction

Rendezvous and docking of a satellite provides tremendous benefits. Among the common applications are transfer of supplies to the International Space Station, formation flying of cluster of satellites. More recently, major space agencies such as DLR, ESA, NASA and JAXA have paid close attention to the capability for extending the life of a satellite, as ageing satellites in orbit have been steadily increasing during time. In the frame work of on-orbit servicing, the life of the satellite can be either extended by refuelling and taking over control, or de-orbiting to reuse congested precious orbits such as GEO. For this purpose of rendezvous, the position and orientation (pose) of the *client* (malfunctioned satellite) in space has to be estimated, for a successful approach and docking. This can be achieved through various

Nassir W. Oumer · Giorgio Panin
Robotics and Mechatornics Center, German Aerospace Center
e-mail: {nassir.oumer,giorgio.panin}@dlr.de

J. Bordeneuve-Guibé et al. (eds.), *Advances in Aerospace Guidance, Navigation and Control*, DOI: 10.1007/978-3-319-17518-8_36

rendezvous sensors. The most popular rendezvous sensors during past missions are videometer, advanced video guidance sensor, rendezvous and docking sensor (RVS) and Laser Mapper, which rely on LIDAR and as well as radio frequency.

In spite of the fact that active sensors based on LIDAR such as laser camera system [3] are effective for rendezvous even under harsh lighting conditions, their high weight and power consumption inhibits their intensive use for future missions, as passive, low weight and low power cameras are now widely available. Cameras have been used, along with secondary optical instruments such as laser range finder, in some space missions such as *Rosetta* and *Gemini*. Moreover, advances in computer vision and computing power motivate camera-based rendezvous. However, specular reflections due to the directional sunlight and the glossy surface of the satellite, is still quite challenging for camera-based motion estimation and tracking.

Several researchers have attempted to address this problem through various approaches [4, 2, 1, 6, 15, 16]. Miravet et. al [2] demonstrated the use of monocular, camera-based relative navigation sensor for rendezvous and docking of a non-cooperative client, by the Orbital Life Extension Vehicle (OLEV). The OLEV image processing module operates on-ground, on downloaded images acquired from the satellite at 1s interval. A wire-frame 2d model-based tracking method, using edge detection and the simplex downhill algorithm for pose estimation, is employed.

Camera-based pose estimation using a nozzle of the apogee engine, by simulating a geostationary satellite, is presented in [1], which is based on a closed-form analytical solution from 3D location of circular features. In this approach, the normal vector to the 3D circle surface and its position are algebraically determined from detected image ellipses. However, this approach did not address the reflective nature of the surface due to multilayer insulation (MLI), consequently analytic solutions can easily fail. In [4], model-based tracking in the context of robotic visual servoing for space applications, is thoroughly presented. The most related work [6], which exploits a line model of the client to estimate orientation and position from monocular sequences. Nevertheless, none of the previous works so far addressed the challenges of detection and tracking of space object with MLI surface at critical close range. In fact, key-point tracking [15] and hybrid key-point and edge tracking[16] at very close range has been performed, without addressing detection and full 6 DOF pose of the target.

In this paper, we present camera-based detection and tracking of a nozzle of a satellite to assist rendezvous and capture at very close range, under realistic space environment simulations. Although the visual tracking can be achieved with monocular camera, we exploit the data from the two views of the stereo cameras. The system is based on edge detection and tracking related to [7, 5] and Euclidean pose estimation with an efficient geometric model representation. A recognition and tracking method that utilizes key points and edges of an image for robot manipulation is provided in [10]. The proposed method is successfully demonstrated with shapes such as a tea box, book, cup and car door under ambient lighting condition. The disadvantage of the method is that, it requires off-line keyframes for pose initialization, which are not available for satellite nozzle ahead of the tracking. Furthermore, the assumption of a plane to plane transformation for the correspondences of key points is restrictive since the object may not be non-planar, and the key points can be on different plane.

Our method combines two robust modules for tracking, using image data and object geometry. Moreover, we validate and verify the proposed method, based on ground truth sequences that simulate a predominantly specular and irregular surface (corresponding to the insulation layer), under strong sunlight conditions.

While monocular camera tracking is valid in case of failure of other camera (redundancy of hardware), our method takes advantage of a better accuracy in depth provided by calibrated stereo, with a suitable baseline and field of view for the operational range. The paper is organized as follows: in Section 2 we describe the algorithmic setup, while Section 3 and 4 focus more in detail on the detection and tracking modules, respectively. Finally, we present results in Section 5 and conclude in Section 6.

2 Overview of the Localization Approach

The localization system (Fig. 1) consists of: a calibrated stereo camera setup, a geometric model of client nozzle, and modules for detection, tracking, loss detection and recovery. The main functionalities consist of:

- *Stereo camera setup*: both for detection and tracking, a pin-hole camera model is assumed to be available from off-line camera calibration, and used for re-projection of model features.
- *Global detection and loss recovery*: this function is used at the onset of close-range rendezvous, in order to detect the nozzle of the satellite and localize it in 3D space, by using stereo images and the known geometry. Global search is

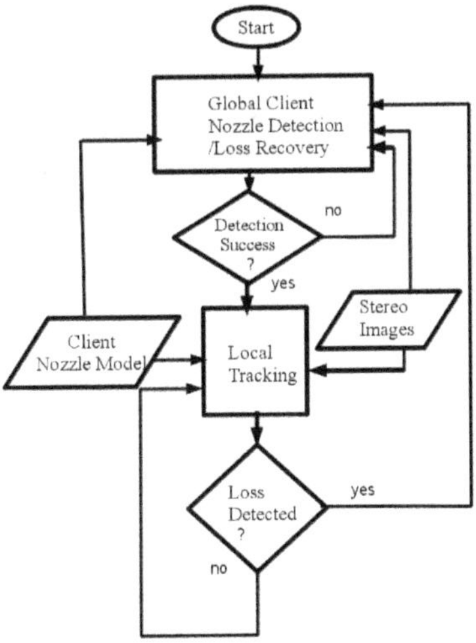

Fig. 1 Client nozzle localization and tracking system.

generally more complex and less accurate procedure, whose result is only used to initialize the local tracker, as well as to recover from a temporary target loss.

- *Model-based tracking*: after initialization, the location of the client nozzle is updated during time by the tracking module, which also uses stereo images and the same geometry, however performing faster and more accurate local searches.
- *Loss detection*: after tracking, automatic detection of a target loss is provided by residual statistics at estimated pose, by monitoring the percentage of outlier matchings on a single-frame basis. An alternative, in presence of a Kalman filter, could be the covariance matrix of the output state, that grows when large residuals are observed during consecutive frames.
- *Model*: a simple circular model of the client nozzle, with known radius, is used both for detection and tracking. For tracking efficiency, contour points are sampled at equispaced intervals, and re-projected on both images to perform local matching and pose estimation.

(a) (b) (c)

Fig. 2 Nozzle detection process on stereo images: edge thinning and labeling on left image(a), detected hypothetical circles on left image(b), detected nozzle after stereo matching(c).

3 Detection

The detection module is responsible for global recognition of the client nozzle in the incoming images. Therefore, localization robustness against clutter and noise, as well as sufficient accuracy, are vital in order to successfully initialize the tracker.

In our case, the client produces strong background clutter, due to reflection of the glossy and irregular surface under direct sunlight. The cluttered edge map (see Fig.2a) poses challenges to both localization modules.

For detection purposes, we assume a frontal position and rectified stereo images, while both assumptions are relaxed during tracking, that work with arbitrary cameras and nozzle attitudes. Then, we perform the following *bottom-up* processing: edge detection, linking, circle detection, stereo matching, triangulation of the center and least square error (LSE) refinement.

Binary edge maps are first obtained by the Canny detector [9]. Subsequently, edge linking [13] is perfomed to recover connected, 1-pixel wide lines, which are broken at detected junction points (where two or more lines cross), obtaining a set of labeled contours. This procedure may be computationally intensive, especially in a cluttered scenario like ours, however it provides a very good and reliable information for shape detection.

Each connected line is then tested for circularity, that is, how well it fits an arc of a circle. In a first implementation, we considered thresholding the residual error after a standard LSE fitting of a cricle (L_2 norm).

However, standard LSE is not robust against outliers, that in our case are given by false edges, erroneously linked to the correct ones.

A better strategy considers instead the L_1 norm of the geometric re-projection error

$$AGE : \arg\min_{x_c, y_c, r} \sum_i \left| \sqrt{(x_i - x_c)^2 + (y_i - y_c)^2} - r \right| \tag{1}$$

where (x_c, y_c) and r are the center and radius of the circle respectively, and (x_i, y_i) are the measurements.

Despite the generally higher difficulty of L_1 optimization, in this case the result can be obtained efficiently through a recently developed algorithm [11] (by the authors referred to as absolute geometric error, or AGE).

We verified that this method performs at best when inliers have a low measurement noise, which is the case of the Canny detector that, at least in relatively low-noise images, shows a pixel-scale accuracy.

A further improvement is given by first applying a RANSAC strategy (3 point circle fitting) for outlier detection (by thresholding against a reasonable percentage, e.g. 30%), before performing the L_1 (AGE) optimization. Although this may increase the computational cost, especially in case of high clutter or long contour lines, we actually observed that on the average RANSAC takes very few iterations to converge, again because of the low inlier noise, thus resulting in a negligible additional cost. A clear advantage of RANSAC is the fact that it explicitly selects inliers, resulting not only in an improved robustness, but also allowing to rule out non-circles, by thresholding the percentage of outliers.

After fitting, we threshold candidate arcs against a minimum spanned angle of 45 deg, and against an allowed range of radii, that are computed from the range of observable nozzle depths. That means, the radius of the 3D circle of the nozzle is projected on to the image plane, using known maximum and minimum distance of the tracking range. Thus, the allowed range of the radii r_{min} and r_{max} are respectively

$$r_{min} = f \frac{R}{Z_{max}} \tag{2}$$

$$r_{max} = f \frac{R}{Z_{min}} \tag{3}$$

where R, radius of the 3D circle, f focal length, Z_{min} and Z_{max} are closest and furthest distance of the nozzle from the camera respectively. The remaining circles are matched pair-wise between stereo images, by further setting a few thresholds to exclude impossible matchings: in particular, for a given nozzle radius and camera parameters, we test against the epipolar constraint between the two centers (same y, up to a few pixels), the allowed disparity range (corresponding to observable depths), the similarity of estimated radii, and the known radius of the nozzle (which becomes a simple constraint, between the image radii and center disparity).

After all of the pruning, either none (in case missing detection) or very few candidate circles will remain, each one leading to a hypothetical pose, first computed by triangulation of the image centers. Those hypotheses most of the times cluster together around the correct pose, due to the fact that the detected nozzle edge splits into multiple arcs. Therefore, in order to obtain a more robust and accurate pose estimation, we merge together all candidate arcs (on both images), and perform a final nonlinear LSE, by minimizing the overall re-projection error, initialized by the average of triangulated centers.

We finally remark that while selecting the above criteria, we keep quite conservative, because false positives are not acceptable by the system (they would badly initialize the tracker), while occasional missing detections may occur, at least when the target is for some time under a strong illumination or shadow. In Fig. 2, intermediate results of the detection process are shown. Notice that size of images have been rescaled for better visualization.

4 Tracking

Once the nozzle position has been globally initialized with sufficient precision, the tracking module refines and updates the estimate in real-time, still relying on the Canny edge map, but this time minimizing geometric re-projection errors, after sampling the model into a set of points, to be projected onto both images.

This procedure, other than being faster for real-time purposes (because it is a local optimization, as opposed to a global, bottom-up search), it is also more accurate and robust. Moreover, it does not rely anymore on the assumption of parallel cameras, thus avoiding the need for image rectification which has been used above to simplify epipolar matching and triangulation.

Projected model points are associated to candidate edge pixels along the respective normals, also checking against the matching edge directions (as measured by a Sobel filter), up to a reasonable threshold of about ± 15 deg.

By assuming Gaussian noise statistics for detected edges, this is a classical nonlinear least-squares problem, that can be solved by Gauss-Newton optimization [12, 7], improved for robustness by an M-estimator.

Hereafter, we provide more details about this module. We also notice that the procedure may be improved by using a Kalman filter with a simple dynamical prior (for example a constant velocity model, since we are dealing with controlled and smooth trajectories). However, in the present implementation we decided to rely on the maximum-likelihood solution provided by the above edge fitting, which has relatively a large convergence region, at the same time avoiding bias from mismatched dynamic or parameter identification issues.

In a more general setting, let us first consider the Euclidean group $SE(3)$ of rigid body motion, given by

$$T = \begin{bmatrix} R & \mathbf{t} \\ \mathbf{0} & 1 \end{bmatrix} \tag{4}$$

where R is a (3×3) rotation matrix, and $\mathbf{t}$ a translation vector. A standard singularity-free parametrization (at least around the previous estimate of T) is achieved by using the tangent space to the manifold, given by

$$T_t = T_{t-1} \delta T(\delta \mathbf{p}_t) \tag{5}$$

where the incremental transform is singularity-free around $\delta T(\delta \mathbf{p} = 0)$. The tangent space to $SE(3)$ at T_{t-1} is given by the Lie algebra $se(3)$, that provides δT through the exponential mapping

$$\delta T(\delta \mathbf{p}) = \exp\left(\sum_{d=1}^{6} G_d \delta p_d\right) \tag{6}$$

where G_d are the canonical (4×4) generators of $se(3)$ [8].

Each 3D model point $\mathbf{x}$ (sampled on circular rim of the nozzle), expressed in homogeneous coordinates, is projected onto a given camera $\mathbf{y}$ by the respective (3×4) projection matrix K

$$\mathbf{y} = Proj(KT\delta T(\delta \mathbf{p})\mathbf{x}) \tag{7}$$

where $Proj()$ is a perspective projection which transforms a point from homogeneous to Euclidean 2D coordinates. We minimize the objective function

$$\underset{\delta \mathbf{p}}{\text{minimize}} \quad \|\mathbf{s} - \mathbf{y}(\delta \mathbf{T}(\delta \mathbf{p}))\|^2 \tag{8}$$

along the edge normals, where $\mathbf{s}$ is a vector of matching edge points.

Pose parameters $\delta \mathbf{p}$ are, then updated by minimizing linearised least square problem

$$\|J\delta \mathbf{p} - \mathbf{r}\|^2 \tag{9}$$

where, $\mathbf{r}$ is residual and J is Jacobian matrix of the residual. The pose is updated iteratively with the new T matrix, computed by (5), until the increment is sufficiently small.

The general formulation can be applied also to cases with reduced degrees of freedom, because of symmetry: for example, since the axial rotation of the nozzle cannot be estimated, the Jacobian column related to one of the generators G_d can be suppressed. Moreover, for our scenario rotation is controlled quite precisely through attitude measurements such as star tracker, thus we only show results concerning translation, by constraining the least square problem as

$$\begin{aligned} \underset{\delta \mathbf{p}}{\text{minimize}} \quad & \|J\delta \mathbf{p} - \mathbf{r}\|^2 \\ \text{subject to} \quad & C\delta \mathbf{p} = 0 \end{aligned} \tag{10}$$

where C is a matrix of Lie group generators associated to the rotation components. However, we demonstrate results of the full 6 DOF pose tracking to show robustness of the method to roto-translation motion. We note that Camera based methods have inherent difficulty in recovering the pose of symmetrical objects about the axis of symmetry(for example Nozzle).

As we have seen, in our scenario several background clutters exist because of specular reflections, resulting in false positives on the edge map that may be very close to the nozzle edge, and very strong as well. Robustness to outliers is critical both for detection and tracking. However, the latter case is even worse, because local matching is performed independently point-wise, without the ambiguity reduction provided by edge linking (Sec. 3).

Also in this context, we could apply a RANSAC strategy (under the stereo re-projection error). However, for real-time efficiency we adapt the M-estimator that adaptively thresholds outliers based on robust statistics, while reweighting inliers according to the Tukey bi-weight function.

5 Results and Discussion

The proposed method has been validated and verified on a terresterial experimental setup, consisting of simulated space object and environment. The mock-up system reproduces at least the following three conditions: space illumination, both from the Sun and from the servicer satellite (with specified ratios), optical surface properties of the client relative motion between servicer and client, during approach and docking. The spectrum of the sunlight is simulated with a high power floodlight system, which is highly directional and illuminates the client surface from a desired direction. The surface of the client mock-up is covered with a highly reflective multilayer insulation (MLI) as shown in Fig.3. Due to the bumpy surface and the directional light source, there exist strong specular reflections, mostly located on irregular corners and edges, with an unstable shape and size. In order to realize relative motion, the two satellites are mounted on large 6 DOF robots. Several trajectories, consisting of translation and rotation at different velocitites, are generated.

The ideal scenario consists of pure translation, with linear trajectory including acceleration/deceleration phases, possibly superimposed to camera rotational disturbances, arising from the external attitude controller. Moreover, each trajectory is further tested at different directions of the sun (90, 66 and 15 deg, with respect to the camera optical axis), and under various level of illumination (provided as ratio of sunlight illumination to target illumination system). Thus, the set of motion sequences is composed of all possible conditions, from underexposed to overexposed lighting, that can be expected in space.

Fig. 3 Mock-up of client satellite and servicer mounted on the two 6 DOF robots for rendezvous and capture.

Images of size 1024×1024 were captured at a rate of 1Hz by a calibrated stereo setup with 50 cm baseline and 41.6 deg field of view, suitable for the range of distances. For the verification of our method, motion of the robot and image capturing have been synchronized, and ground truth data were collected from the sensor readings of the robot, after hand-eye calibration (using DLR Callab toolbox, [14]) to recover the constant tool center point (TCP) to camera transformations. The current implementation is single-threaded C++ on Intel(R) Xeon(R) Desktop, with 2.80 GHz processor. The computation time of the detection and tracking are 500ms and 100ms respectively, which are well below the acceptable real time constraint (1Hz) for the application of on-orbit servicing.

5.1 Nozzle Detection

Various methods of circle fitting and their combinations are implemented, to evaluate their performance for nozzle localization under strong background clutter. The evaluation is based on errors generated in 3d localization, as well as percentage of correct detections under conservative parameter settings. We fix the same thresholding parameters for all fitting methods and test trajectories. Thresholding parameters can be easily selectetd; however, the arc angle and threshold of reprojection error require a careful tuning to avoid false detection. In our experiment, a minimum arc angle of 45 deg and an overall reprojection error of 6 pixels is used for all the experiments.

The combined RANSAC circle fitting (with three non-collinear points) and AGE (RANSACL1, in Fig.4) shows best performance in localizing the nozzle of the satellite over all test sequences because of their complementary robustness property. Here we demonstrated those detection errors only in Z-axis (the optical axis) due to limited space. This combination is better in that it provides less position errors and a few false detections (false positive) consistently. It is worth noting in Fig. 4, that RANSACL1 and RANSACL2 localized the nozzle position accurately (relatively a few errors), however errors of RANSACL2 in the first five frames, are too high to initialize the local tracker (absolute errors more than 100mm in Z-axis).

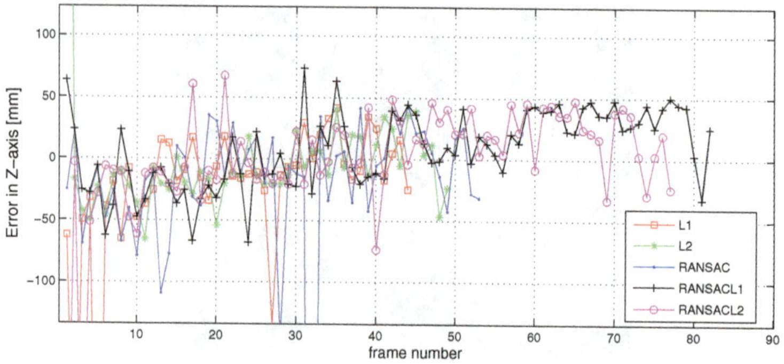

Fig. 4 Localization error; Comparison of circle fitting methods for detection of satellite nozzle under specular reflection: Least square error(L2), absolute error(L1), 3 non-collinear point circle fitting(RANSAC), and combination of RANSAC with L1 and L2. RANSACL1 and RANSACL2 are demonstrated to perform the best compared to the rest of the methods shown in this graph, particularly RANSACL1 is more robust, and able to detect the nozzle over relatively long sequence.

5.2 Local Tracking

We present some of the results, selected among the large set of trajectories, to show robustness and accuracy of local tracking of the nozzle center during very close range rendezvous. The reprojection of the nozzle in frontal view at estimated pose is well aligned with the desired image edge in spite of worse illumination condition in Fig.5.

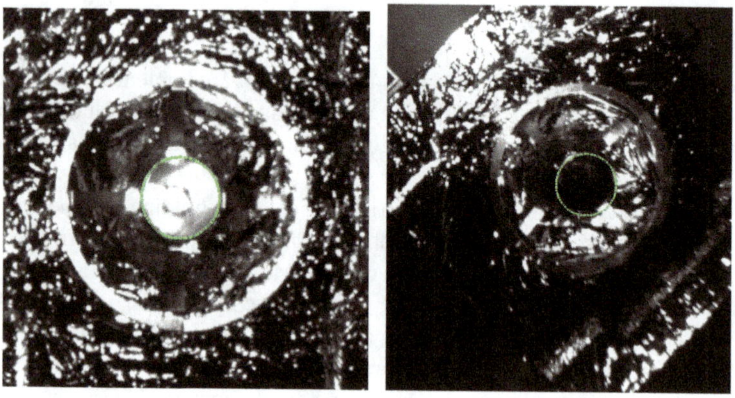

(a) (b)

Fig. 5 Reprojected nozzle at estimated pose is well aligned with the image nozzle under; a)strong illumination and b) weak illumination condition.

Tracking errors under two different illumination conditions and motion trajectories are shown in Fig. 6a and Fig. 6b, the latter illuminated with only target illumination system of the servicer (without the Sun simulator). Fig. 7a and b demonstrate

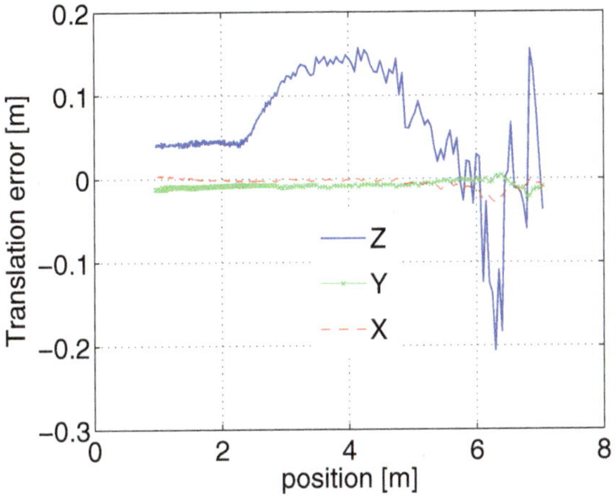

(a)

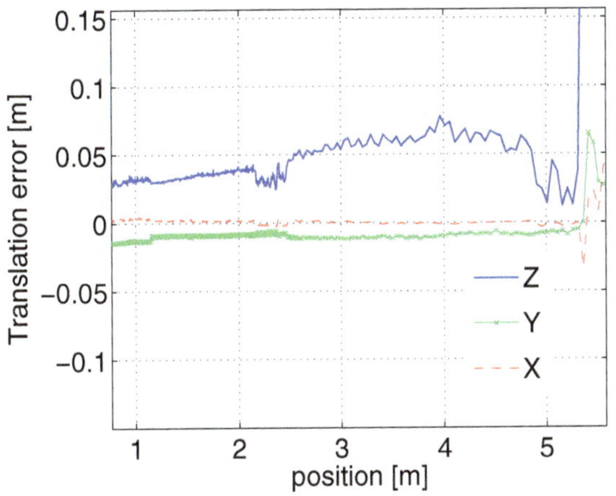

(b)

Fig. 6 Tracking errors for different trajectories and illumination conditions (provided as the ratio of sunlight illumination to target illumination system). (a) Pure translation; the Sun is perpendicular to the line of sight of camera, 80% illumination. (b) Same configuration as in (a), except only target illumination system (no sun light).

tracking accuracy in presence of attitude disturbances (the camera rotates about the y-axis, with a sinusoidal pattern of about 1 deg amplitude): Fig. 7a is due to the result of translation in x- and y-directions at constant velocity, while Fig. 7b shows errors during station keeping at 5 m from the client.

As we expect, the errors shown in all plots decrease as the servicer approaches the client (also depending on the illumination conditions), because of less clutter and larger disparities. In fact, at close distance we can distinctly observe the visible

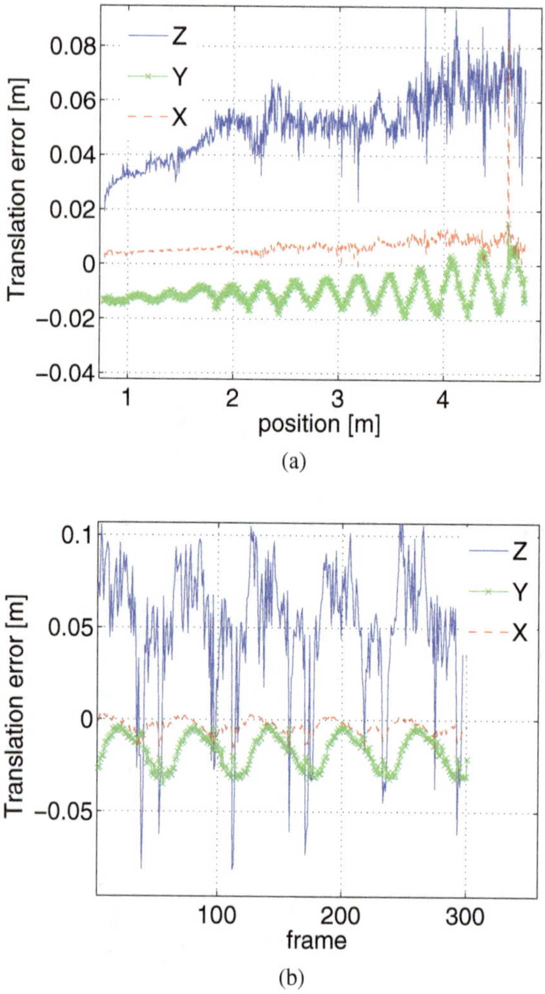

Fig. 7 Tracking errors for different trajectories and illumination conditions (provided as the ratio of sunlight illumination to target illumination system). (a) Translation along (x,y) and rotation about x; the Sun position is 66 deg to the line of sight of camera, and 90% illumination. (b) Roto-translation about y alternating with sinusoidal oscillation; fixed at 5m in Z-axis, with the Sun position 15 deg to the line of sight of camera and 90% illumination.

circular feature, under better lighting conditions and less background clutter, so that accuracy of localization is greatly improved along all axes.

On the other hand, we observe swinging errors in some test scenarios, though it is predominant in oscillatory motion consisting of small rotation (Fig. 7a and b). The swinging appears when the tracker encounters dominant edges due to reflection, and the actual edges of the nozzle in the image are too few. The adverse effect of virtual edges because of reflection during pose refinement can be observed in Canny edge map in Fig. 2, in which some of the false edges are comparable to desired edges of the satellite nozzle, even after several prunnings. Using stereo images for local tracking provides several benefits; firstly, we can acquire more information about the scene in two views and increase an accuracy and a robustness particularly at very close range. Secondly, one of the stereo cameras can be used in case either of them fails to function[1], at least for local tracking. We remark that that in Fig.6a, the tracked sequence is relatively longer (up to 7) than the rest of the sequences (for example, Fig.6b). This is because, the trajectory is pure translation without oscillation and good illumination, which favours distinctive nozzle at relatively far distance. In general, the nozzle is not any more distinctive feature for tracking at distances above 5m when the client motion is oscillatory. Thus, we should use the whole satellite model instead of the nozzle for higher ranges unlike the close range in which the full shape of the satellite may not be visible in the camera field of view.

5.3 Limitation

Despite good accuracy and robustness of the proposed method, there exist limitations in both localization methods. The nozzle detector fails to detect some frames (a maximum of two frames after last detection) of the image sequence, which are typically low contrast images, with shadows and several background clutters. However, detection is required only when there exists loss of tracking, which occurs rarely. Hence, it is safe to initialize the loss of tracking for relatively slow motion of the target satellite.

The local tracking, on the other hand, is more robust and accurate, although the accuracy of the tracker degrades with low contrast image and shaking motion (see Fig.7b). Moreover, the last pose is used to predict the current pose, assuming again slow client motion dynamics.

6 Conclusion

Assuming that attitudes of two satellites are aligned, we developed a model-based detection and tracking method based on stereo images, for fully automatic localization of a satellite nozzle during very close-range rendezvous and capture operations. Our method has shown to be robust and accurate under various space lighting conditions and motion trajectories. Several tests have been performed on a full scale satellite mock-up with ground-truth and realistic environment conditions. For future work, an extensive evaluation of the localization method at wider range and complex motion will be performed.

Acknowledgement. The authors would like to thank Dr. Toralf Boge of DLR, for his support in realizing test trajectories at EPOS, and Dr. Quirin Mühlbauer of Kayser-Threde GmbH, for providing camera system.

[1] Redundancy of critical-components because of radiation.

References

1. Xu, W., Liang, B., Gao, D., Xu, Y.: A Universal On-orbit Servicing System Used in the Geostationary Orbit. Advances in Space Research 48, 95–119 (2011)
2. Miravet, C., Pascual, L., Krouch, E., Delcura, J.M.: An Image-based Sensor System for Autonmous Rendezvous with Uncooperative Satellites. In: 7th International ESA Conference on Guidance, Navigation and Control Systems (June 2008)
3. Samson, C., English, C., Deslauriers, A., Christie, I., Blais, F., Ferrie, F.: Neptec 3D laser camera system : From space mission STS-105 to terresterial applications, Canadian Aeronotics and Space Journal (2004)
4. Dionnet, F., Marchand, E.: Robust Stereo Tracking for Space applications. In: Proceedings of the IEEE/RSJ International Conference on Intelligent Robots and Systems, San Diego, CA, USA, October 29-November 2 (2007)
5. Comport, A.I., Marchand, E., Pressigout, M., Chaumette, F.: Real-Time Markerless Tracking for Augmented Reality: The Virtual Visual Servoing Framework. IEEE Transaction on Visualization and Computer Graphics 12(4), 615–628 (2006)
6. Petit, A., March, E., Kanani, K.: Vision-based Space Autonomous Rendezvous: A Case Study. In: EEE/RSJ Int. Conf. on Intelligent Robots and Systems, pp. 619–624 (2011)
7. Drummond, T., Cipolla, R.: Real-Time Visual Tracking of Complex Structures. IEEE Trans on Pattern Analysis and Machine Intelligence, 932–946 (2002)
8. Drummond, T., Cipolla, R.: Visual Tracking and Control using Lie Algebras. IEEE Computer Society Computer Vision and Pattern Recognition 2, 2652–2659 (1999)
9. John Canny, F.: A computational Approach to Edge Detection. IEEE-PAMI 8(6), 679–698 (1986)
10. Choi, C., Christensen, H.I.: Real-time 3D Model-based Tracking Using Edge and Keypoint Features for Robotic Manipulation. In: IEEE International Conference on Robotics and Automation, May 3-8, pp. 4048–4055 (2010)
11. Ladrón De Guevara, I., Muñoz, J., Cózar, O.D., Blázquez, E.B.: Robust Fitting of Circle Arcs. Journal of Mathematical Imaging and Vision 40(2), 147–161 (2011)
12. Harris, C., Stennet, C.: RAPiD a video-rate object tracker. Proceedings of the British Machine Vision Conference, 73–77 (September 1990)
13. Lowe, D.G.: Fitting parametrized Three-Dimensional Models to Images. IEEE Transaction Pattern Analysis and Machine Intelligence 13(5), 441–450 (1991)
14. Strobl, K.H., Hirzinger, G.: Optimal Hand-Eye Calibration. In: IEEE/RSJ International Conference on Intelligent Robots and Systems, Beijing, China, pp. 4647–4653 (2006)
15. Oumer, N.W., Panin, G.: 3D point tracking and pose estimation of a space object using stereo images. In: 21st International Conference on Pattern Recognition, Tsukuba, Japan, pp. 796–800 (2012)
16. Oumer, N.W., Panin, G.: Monocular 3D Pose Tracking of a Specular Object. In: 9th International Conference on Computer Vision Theory and Applications, Lisbon, Portugal, January 5-8, vol. 3, pp. 458–465 (2014)

Linear Dynamic Modeling of Spacecraft with Open-Chain Assembly of Flexible Bodies for ACS/Structure Co-design

Jose Alvaro Perez, Daniel Alazard, Thomas Loquen, Christelle Cumer, and Christelle Pittet

Abstract. This work presents a method to build a linear dynamic model of open-chain assembly of spacecraft flexible appendages for future Attitude Control System (ACS)/Structure co-design. This kind of modeling takes into account the flexible interactions between all the spacecraft substructures, called bodies or appendages, to finally provide the loads (forces and torques) induced to the main body. More generally, this method can be applied to any open mechanical chain, such as segments of robotic arms, segments of antenna mast or masts linking solar panels to the main hub. Therefore, the dynamics model of the entire spacecraft can be derived easily in order to design the spacecraft ACS. The method is based on the Craig-Bampton modal synthesis, from which a state-space representation is obtained.

Introduction

The significance of structural flexibility in the attitude control and stability design has been recognised since the first launches of artificial satellites. Typically, a satellite is composed of a main body or hub, B, and several bodies, called appendages,

Jose Alvaro Perez · Thomas Loquen · Christelle Cumer
ONERA, 2 Avenue Edouard Belin, 31055 Toulouse, France
e-mail: {Jose-Alvaro.perez_gonzalez, thomas.loquen, christelle.cumer}@onera.fr

Daniel Alazard
ISAE,10 Avenue Edouard Belin, 31055 Toulouse, France
e-mail: daniel.alazard@isae.fr

Christelle Pittet
CNES, 18 Avenue Edouard Belin, 31400 Toulouse, France
e-mail: christelle.pittet@cnes.fr

J. Bordeneuve-Guibé et al. (eds.), *Advances in Aerospace Guidance, Navigation and Control*, DOI: 10.1007/978-3-319-17518-8_37

attached to this main body: solar arrays, antennae, deployable booms and other payloads which the satellite uses to accomplish its mission.

Despite of the importance of appendage flexible modes in satellite dynamics, the rigid model has been the most used for the majority of investigations. However, this assumption starts being questionable as spacecraft dimensions increase with large chains of rigid and flexible bodies. At this point, perturbations caused by structure flexible modes become significant for spacecraft dynamics, making necessary their inclusion not only in the validation process but also in the design process.

A great number of studies have been carried out to model appendage flexible modes effects when attached to spacecraft main body, as for example [1], [2] and [3]. A general detailed review of dynamics of multibody systems is presented in [4], mainly focused in general applications and not to the spacecraft domain. Nevertheless, these approaches always stay in the theoretical domain, with non-linearities which cause difficulties to create a *Linear Time Invariant* (*LTI*) representation. Moreover, these methods do not use parameters provided by Finite Element Model (FEM) analysis for such purpose. In [5] and [6] a *LTI* representation is found, including some parameters obtained by FEM analysis, but they do not consider the possibility of linking several flexible appendages between them.

At the same time, there is an increasing relevance as well concerning new co-design (also called integrated design in the literature) methodologies which allow a simultaneous design of the Attitude Control System (ACS) and spacecraft structure properties. Specifically, co-design methods have a relevant importance when leading with flexible multibody systems because of the intense Control-Structure Interaction (CSI). An innovative technique developed at the ONERA Toulouse Research Center allows performing integrated control/structure design using spacecraft *LTI* representations, as it is explained in [7] and [8].

So that such a technique can be applied in the flexible multibody spacecraft, an accurate *LTI* representation of the spacecraft is needed. This work aims at finding *LTI* form of a flexible body which can be linked to other bodies, rigid or flexible, without losing generality, in order to apply it to future ACS design of satellites with long masts or solar array chains. Even more, this model aims at being able of introducing parametric uncertainties or variations for future ACS/Structure co-design application. Such representation must meet the following requirements:

- Respect of the dynamical behavior of the appendage; i.e., a representation which conserves the required number of structure modes.
- Independence of other external parameters; i.e., the possibility to independently analyze each flexible body regardless its connections to the other bodies.
- Use of the adequate inputs and outputs for ACS, such as forces, torques and acceleration between interfaces.
- Direct access to physical parameters to take into account uncertainties of the appendages for ACS/structure co-design.

To accomplish this task, this investigation proposes a method to model an **Interconnected Flexible Appendage** (IFA) model in open chain-like assembly. In the Section 1, the Craig-Bampton modal synthesis is presented in order to introduce the reader to the dynamic substructuring process. Next, the IFA model is built from the equations. In Section 3, a way to interconnect the different IFA models is presented. Finally, the possibility of introducing uncertainties into the model is analyzed and conclusions are presented.

1 Dynamic Substructuring Using the Craig-Bampton Method

The Interconnected Flexible Appendage (IFA) model parameters will be obtained from the modal synthesis method for substructures developed by Craig-Bampton in [9], which has been used for practical applications in several studies such as those of [10], [11] and [12]. However, such applications lack the state space representation. This study aims at providing such needed state-space representation, as accomplished in [5] and [6], but adding the possibility of appendage concatenation.

The uncoupling accomplished by the Craig-Bampton method for substructuring is the main reason for using it in this work. The transformation provided by Craig-Bampton allows to separate the interface motion from the interior displacements. Thus, substructure dynamics can be condensed at its interface and, with the correct connections, the flexible modes can be embedded inside the substructure automatically.

The Craig-Bampton transformation consists of two steps. The first one is a coordinate transformation in order to uncouple interior-interface displacements (see Section 1.1). The second step consists on truncating high frequency elastic modes expressed in modal coordinates in order to capture the fundamental low frequency response modes of the substructure (see Section 1.2).

1.1 Uncoupling of Interface-Interior Displacements

Let consider a typical FEM formulation for structural dynamics, which allows to approach the substructure dynamical behaviour with a limited number of degrees of freedom (*dof*):

$$\mathbf{M}\ddot{\mathbf{q}} + \mathbf{D}\dot{\mathbf{q}} + \mathbf{K}\mathbf{q} = \mathbf{F}(t) \tag{1}$$

where $\mathbf{M}$ is the mass matrix, $\mathbf{D}$ the damping matrix, $\mathbf{K}$ the stiffness matrix and $\mathbf{F}(t)$ the external loads vector. For analysis purposes, matrix $\mathbf{D}$ is not taken into account for the moment. However, it can be easily implemented as explained later in this section and in [12].

Following the dynamic substructuring explained in [9], the set of substructure *dof*, noted as $\mathbf{q}$ (spacecraft appendage), such as the one shown in Figure 1, can be partitioned in two sets: interfaced or supported boundary nodes (j index) and the interior elastic nodes (i index). Therefore the corresponding dynamic matrices can be rewritten as follows:

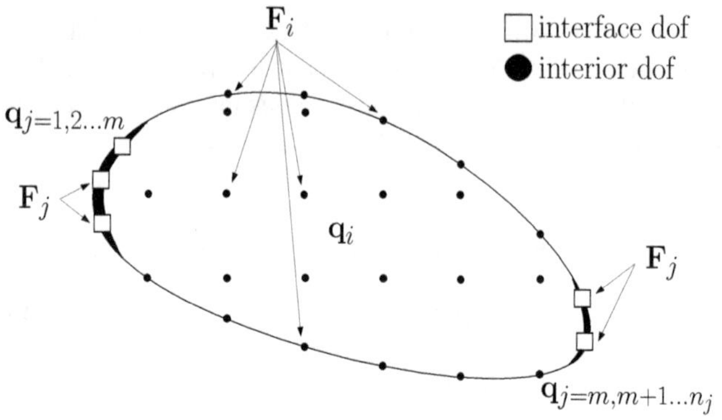

Fig. 1 Generic mechanical substructure.

$$\mathbf{q} = \begin{bmatrix} \mathbf{q}_i \\ \mathbf{q}_j \end{bmatrix} \quad \mathbf{F} = \begin{bmatrix} \mathbf{F}_i \\ \mathbf{F}_j \end{bmatrix} \quad \mathbf{K} = \begin{bmatrix} \mathbf{K}_{ii} & \mathbf{K}_{ij} \\ \mathbf{K}_{ij}^T & \mathbf{K}_{jj} \end{bmatrix} \quad \mathbf{M} = \begin{bmatrix} \mathbf{M}_{ii} & \mathbf{M}_{ij} \\ \mathbf{M}_{ij}^T & \mathbf{M}_{jj} \end{bmatrix} \tag{2}$$

The Craig-Bampton transformation transforms the set of elastic physical coordinates $\mathbf{q}_i$ to a set of modal coordinates η_m. Then, the set of physical coordinates $\mathbf{q}$ is transformed to a hybrid set of physical coordinates at the interface, $\mathbf{q}_j$, and modal coordinates at the interior, η_m, as stated in equation (3). Thus, the coordinate transformation can be written as:

$$\mathbf{q} = \begin{bmatrix} \mathbf{q}_i \\ \mathbf{q}_j \end{bmatrix} = \begin{bmatrix} \phi_{ij} & \varphi_{im} \\ \mathbf{I} & \mathbf{0} \end{bmatrix} \begin{bmatrix} \mathbf{q}_j \\ \eta_m \end{bmatrix} = \Psi \begin{bmatrix} \mathbf{q}_j \\ \eta_m \end{bmatrix} \tag{3}$$

with the Craig-Bampton Transformation Ψ which can be partitioned as:

$$\Psi = \begin{bmatrix} \Phi_j & \Phi_m \end{bmatrix}; \qquad \Phi_j = \begin{bmatrix} \phi_{ij} \\ \mathbf{I} \end{bmatrix} \quad \Phi_m = \begin{bmatrix} \varphi_{im} \\ \mathbf{0} \end{bmatrix} \tag{4}$$

where Φ_j is usually referred to as the Interface Node Functions and Φ_m to as the Fixed Base Mode Shapes [10]. Understanding the physical meaning of these matrices is important to learn how to use properly the Craig-Bampton method:

- The Interface Node Functions, Φ_j, where ϕ_{ij} is the Static Constraint Modes Matrix, relate physical displacements at the interface, $\mathbf{q}_j$, to physical displacements of the elastic degrees of freedom, $\mathbf{q}_i$. It is essential as it describes the static response of the substructure to excitations coming from neighbouring substructures through the interface degrees of freedom. It can be obtained by the following expression, obtained with the static problem of equation (1):

$$\phi_{ij} = -\mathbf{K}_{ii}^{-1}\mathbf{K}_{ij} \tag{5}$$

Rigid modes are embedded in these functions. Thus, for a substructure with an isostatic interface (exactly 6 degrees of freedom) the Static Constraint Modes Matrix will lapse into the rigid modes matrix at the interface point.

- The Fixed Base Mode Shapes, $\boldsymbol{\Phi}_m$, with φ_{im} relating the modal responses η_m to the physical displacements of the elastic degrees of freedom $\mathbf{q}_i$. It is obtained from the equations of motion with the interface degrees of freedom ($\mathbf{q}_j = 0$) constrained and with no force acting in the interior points ($\mathbf{F}_i = 0$):

$$\mathbf{K}_{ii}\varphi_{im} = \omega^2 \mathbf{M}_{ii}\varphi_{im} \tag{6}$$

Solving the substructure eigenvalues ω^2 and mode shapes φ_{im} the transformation of physical displacements to modal responses is accomplished by the relation $\mathbf{q}_i = \boldsymbol{\Phi}_{ij}\mathbf{q}_j + \varphi_{im}\eta_m$.

The Craig-Bampton method rewrites the substructure motion equation (1) from the set of physical coordinates to a set of coordinates consisting of physical coordinates of physical interface points and modal or generalized coordinates, so the equation of motion of a linear substructure is:

$$\begin{bmatrix} \mathbf{M}_{ii} & \mathbf{M}_{ij} \\ \mathbf{M}_{ij}^T & \mathbf{M}_{jj} \end{bmatrix} \begin{bmatrix} \ddot{\mathbf{q}}_i \\ \ddot{\mathbf{q}}_j \end{bmatrix} + \begin{bmatrix} \mathbf{K}_{ii} & \mathbf{K}_{ij} \\ \mathbf{K}_{ij}^T & \mathbf{K}_{jj} \end{bmatrix} \begin{bmatrix} \mathbf{q}_i \\ \mathbf{q}_j \end{bmatrix} = \begin{bmatrix} \mathbf{F}_i \\ \mathbf{F}_j \end{bmatrix} \tag{7}$$

Equation (7) can be rewritten using the Craig-Bampton Transformation Matrix Ψ and obtaining a new linear system in terms of the physical boundary displacements $\mathbf{q}_j$ and generalized coordinates η_m. Therefore, introducing (3) in equation (7), premultiplying by Ψ^T the equation becomes:

$$\begin{bmatrix} \bar{\mathbf{M}}_{jj} & \bar{\mathbf{M}}_{jm} \\ \bar{\mathbf{M}}_{jm}^T & \mathbf{m}_m \end{bmatrix} \begin{bmatrix} \ddot{\mathbf{q}}_j \\ \ddot{\eta}_m \end{bmatrix} + \begin{bmatrix} \bar{\mathbf{K}}_{jj} & \mathbf{0} \\ \mathbf{0} & \mathbf{k}_m \end{bmatrix} \begin{bmatrix} \mathbf{q}_j \\ \eta_m \end{bmatrix} = \begin{bmatrix} \phi_{ij}^T\mathbf{F}_i + \mathbf{F}_j \\ \varphi_{im}^T\mathbf{F}_i \end{bmatrix} \tag{8}$$

As stated before, now it can be easily appreciated that the interior modes become uncoupled from each other thanks to the transformation matrix properties. Equation (8) shows that the FEM software has to provide a set of particular sub-matrices in order to achieve the Craig-Bampton decomposition. These sub-matrices are:

- The Interface Mass Matrix, $\bar{\mathbf{M}}_{jj} = \boldsymbol{\Phi}_j^T \mathbf{M} \boldsymbol{\Phi}_j$, represents the mass properties translated to the interface points.
- The Interface Stiffness Matrix, $\bar{\mathbf{K}}_{jj} = \boldsymbol{\Phi}_j^T \mathbf{K} \boldsymbol{\Phi}_j$, expresses the stiffness associated with the displacement of one interface degree of freedom while the others are held fixed. Note that for an isostatic substructure (Static Modes Matrix lapsing into rigid modes matrix) this matrix is zero.
- The Coupling Matrix between static constraint modes and fixed base modes, $\bar{\mathbf{M}}_{jm} = \boldsymbol{\Phi}_j^T \mathbf{M} \boldsymbol{\Phi}_m$, that expresses the interaction between the intern flexible modes and the interface points.
- $\mathbf{m}_m$ is the generalized mass matrix and is almost always set equal to the identity matrix by the majority of FEM softwares ([10]).

- $\mathbf{k}_m$ is the generalized stiffness matrix and contains the natural frequencies of the fixed base modes in its diagonal:

$$\mathbf{k}_m = \begin{bmatrix} \omega_1^2 & 0 & \cdots & 0 \\ 0 & \omega_2^2 & \cdots & 0 \\ \vdots & \vdots & \ddots & \vdots \\ 0 & 0 & \cdots & \omega_m^2 \end{bmatrix} \tag{9}$$

If damping effects are considered when modeling appendage dynamics, the analysis can be performed as in [12] or [10], where the damping matrix is assumed to have the following structure after Craig-Bampton transformation:

$$\bar{\mathbf{D}} = \begin{bmatrix} \bar{\mathbf{D}}_{jj} & \bar{\mathbf{D}}_{jm} \\ \bar{\mathbf{D}}_{jm}^T & \mathbf{d}_m \end{bmatrix} \tag{10}$$

where $\bar{\mathbf{D}}_{jj}$ and $\bar{\mathbf{D}}_{jm}$ are nearly chosen equal to zero and the $\mathbf{d}_m$ matrix is written as follows:

$$\mathbf{d}_m = \begin{bmatrix} 2\omega_1\xi_1 & 0 & \cdots & 0 \\ 0 & 2\omega_2\xi_2 & \cdots & 0 \\ \vdots & \vdots & \ddots & \vdots \\ 0 & 0 & \cdots & 2\omega_m\xi_m \end{bmatrix} \tag{11}$$

Note that ξ stands for the equivalent viscous damping defined as the ratio of damping and ω the corresponding fixed base mode.

1.2 Model Reduction

Finally it is interesting to reduce substructure complexity by removing the less significant modes, which leads to the second and final step in the Craig-Bampton method. The final set of modal solutions, η_m, can be truncated to some smaller set, like η_k with $k < m$, by removing the higher frequency modes since their contribution to the total low frequency excitation is negligible (see [10] and [6]). This truncation leads to a reduced model of the substructure:

$$\begin{bmatrix} \bar{\mathbf{M}}_{jj} & \bar{\mathbf{M}}_{jk} \\ \bar{\mathbf{M}}_{jk}^T & \mathbf{m}_k \end{bmatrix} \begin{bmatrix} \ddot{\mathbf{q}}_j \\ \ddot{\eta}_k \end{bmatrix} + \begin{bmatrix} \bar{\mathbf{K}}_{jj} & \mathbf{0} \\ \mathbf{0} & \mathbf{k}_k \end{bmatrix} \begin{bmatrix} \mathbf{q}_j \\ \eta_k \end{bmatrix} = \begin{bmatrix} \phi_{ij}^T\mathbf{F}_i + \mathbf{F}_j \\ \varphi_{ik}^T\mathbf{F}_i \end{bmatrix} \tag{12}$$

This will lead to a less numerically complex model which will reduce computation time when introduced in robust control and co-design utilities. As an example, for a 3D beam analysis only the two first bending modes are taken for first order approximations. Rarely more than five modes are taken for AOCS control design. In the next section a new manipulation of these equations is presented in order to get the IFA model.

2 Interconnected Flexible Appendage LTI Model

When designing a linear model for an interconnected flexible appendage in a chain two cases have to be distinguished :

- The intermediate appendage, which can be between two appendages or one appendage and the main hub.
- The terminal appendage in the chain: the one placed at the end of the chain and connected to the main hub only by the chain of appendages that precedes it.

This is due to the particularities that the terminal appendage modeling presents. In this section the intermediate appendage modeling is presented followed by the case of terminal appendage modeling.

2.1 *Intermediate Appendage*

As introduced in Section 1.1, interface degrees of freedom $\mathbf{q}_j$ can be easily isolated from interior degrees of freedom $\mathbf{q}_i$ using the Craig-Bampton Transformation. The uncoupling provided by this transformation is a major advantage the model has to exploit.

For this work the flexible appendage is considered to have two connection points, named P and C, which represent the corresponding interfaces of the appendage; i.e, the appendage receives exterior loads and accelerations through these points, the rest of degrees of freedom are internal. The fact of considering that the appendage is modeled with two connection points leads to the following partition of the set of physical displacements $\mathbf{q}_j$ and its corresponding mass, stiffness and loads matrices:

$$\mathbf{q}_j = \begin{bmatrix} \mathbf{q}_p \\ \mathbf{q}_c \end{bmatrix} \qquad \bar{\mathbf{M}}_{jj} = \begin{bmatrix} \bar{\mathbf{M}}_{pp} & \bar{\mathbf{M}}_{pc} \\ \bar{\mathbf{M}}_{pc}^T & \bar{\mathbf{M}}_{cc} \end{bmatrix} \qquad \bar{\mathbf{K}}_{jj} = \begin{bmatrix} \bar{\mathbf{K}}_{pp} & \bar{\mathbf{K}}_{pc} \\ \bar{\mathbf{K}}_{pc}^T & \bar{\mathbf{K}}_{cc} \end{bmatrix} \qquad \mathbf{F}_j = \begin{bmatrix} \mathbf{F}_p \\ \mathbf{F}_c \end{bmatrix} \tag{13}$$

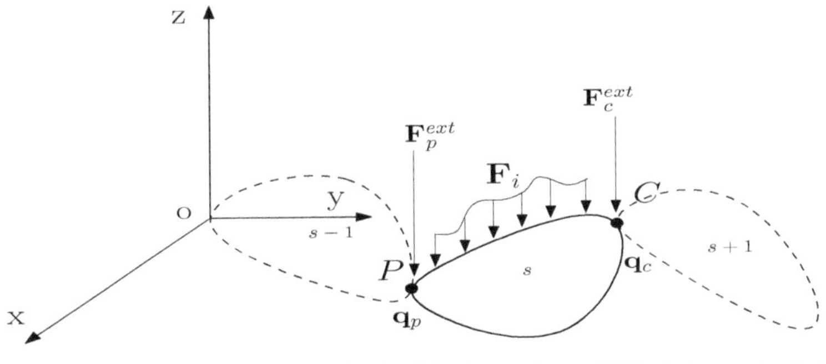

Fig. 2 Illustration of an Interconnected Flexible Appendage (IFA). Index s stands for the substructure being analyzed, while $s-1$ calls for the preceding appendage and $s+1$ for the next substructure in the open-chain assembly.

Figure 2 shows the studied appendage in the middle of a generic open chain of appendages. The appendage s is linked to a preceding appendage $s-1$ through the interface point P, and to a successive one $s+1$ through the interface point C. External applied loads may be included at the interface as well, such as $\mathbf{F}_p^{ext}$ and $\mathbf{F}_c^{ext}$.

The purpose is to establish a cause-effect interaction between the current flexible appendage and its neighborhoods through the interfaces P and C. This interaction is expressed developing a **two-port model** as the one shown in Figure 3 (see also [13] for further informations). This model is characterized as follows:

- **Inputs**: acceleration of point P, $\ddot{\mathbf{q}}_\mathbf{p}$, and the load transmitted by the next substructure $s+1$ at point C, noted as $\mathbf{F}_{s+1/s}$.
- **Outputs**: acceleration at point C, $\ddot{\mathbf{q}}_\mathbf{c}$, and the load transmitted to the preceding substructure through the point P, noted $\mathbf{F}_{s/s-1}$.

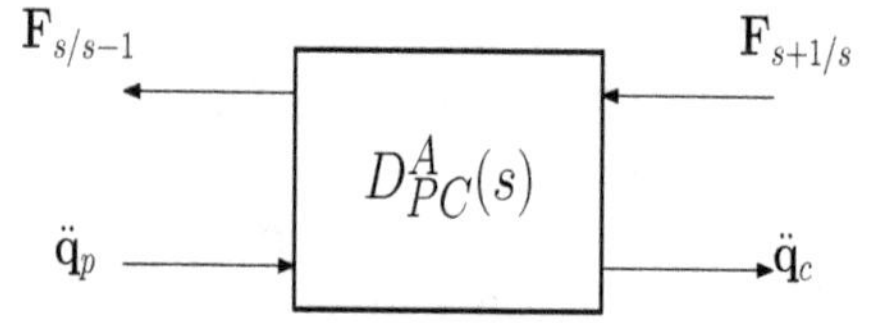

Fig. 3 Two-Port model sketch proposed for substructure modeling. Loads and accelerations are exchanged at the interface points. Index A stands for *appendage* and indexes P, C stand for the model between P and C interfaces.

The use of the two-port model allows to establish a "data flow" that permits to interconnect appendages systematically in their connection points through the exchange *Load* $\leftrightarrow$ *Acceleration*; i.e, an appendage named s perceives the acceleration at its "base" P and the loads at its opposite end C, providing in exchange the load transmitted through P and the resulting acceleration in C. In Section 3, the simple concatenation of these models between them is demonstrated.

Therefore, computation of opposite end acceleration $\ddot{\mathbf{q}}_c$ and transmitted load $\mathbf{F}_{s/s-1}$ are absolutely necessary for the interconnected flexible appendage model in the form of a double-port model. Since the nature of both physical manifestations is not the same, the computation of each one of them differs. In the next paragraphs a detailed method is explained for both physical parameters.

2.1.1 Acceleration Transfer Calculation

The acceleration of connection point C, $\ddot{\mathbf{q}}_c$ must be a function of the acceleration of connection point P, $\ddot{\mathbf{q}}_p$, and the load transmitted through C, $\mathbf{F}_{s+1/s}$, as imposed by the double-port modeling.

The dynamical behavior of the interfaces is given by the upper part of equation (12) and it is written as follows:

$$\bar{\mathbf{M}}_{jj}\ddot{\mathbf{q}}_j + \bar{\mathbf{M}}_{jk}\ddot{\eta}_k + \bar{\mathbf{K}}_{jj}\mathbf{q}_j = \mathbf{F}_j + \phi_{ij}^T\mathbf{F}_i \tag{14}$$

which in the case of two connection points P and C and no internal loads, $\mathbf{F}_i = 0$, becomes:

$$\begin{bmatrix} \bar{\mathbf{M}}_{pp} & \bar{\mathbf{M}}_{pc} \\ \bar{\mathbf{M}}_{pc}^T & \bar{\mathbf{M}}_{cc} \end{bmatrix} \begin{bmatrix} \ddot{\mathbf{q}}_p \\ \ddot{\mathbf{q}}_c \end{bmatrix} + \begin{bmatrix} \bar{\mathbf{K}}_{pp} & \bar{\mathbf{K}}_{pc} \\ \bar{\mathbf{K}}_{pc}^T & \bar{\mathbf{K}}_{cc} \end{bmatrix} \begin{bmatrix} \mathbf{q}_p \\ \mathbf{q}_c \end{bmatrix} = \begin{bmatrix} \mathbf{F}_p - \bar{\mathbf{M}}_{pk}\ddot{\eta}_k \\ \mathbf{F}_c - \bar{\mathbf{M}}_{ck}\ddot{\eta}_k \end{bmatrix} \tag{15}$$

Let the connection P be considered as the "base" of appendage movement. From equation (15) it can be inferred that the dynamical behaviour is influenced by the stiffness between the two connections points, since $\bar{\mathbf{K}}_{jj}$ is not diagonal, by the movement of connection point P and by the interior flexible modes $\ddot{\eta}_k$ acting as a load through the modal participation matrix, $\bar{\mathbf{M}}_{pk}$ and $\bar{\mathbf{M}}_{ck}$. For better computing of connection C acceleration (remove stiffness coupling between connection points and imposing P as a "base" of the movement) an Imbert transformation [6] is applied to the displacement in C, which is denoted by :

$$\mathbf{q}_c = \Phi_r \mathbf{q}_p + \varphi_l \zeta_l \tag{16}$$

and leads to the interface degrees of freedom transformation:

$$\begin{bmatrix} \mathbf{q}_p \\ \mathbf{q}_c \end{bmatrix} = \begin{bmatrix} \mathbf{I} & \mathbf{0} \\ \Phi_r & \varphi_l \end{bmatrix} \begin{bmatrix} \mathbf{q}_p \\ \zeta_l \end{bmatrix} = \Gamma \begin{bmatrix} \mathbf{q}_p \\ \zeta_l \end{bmatrix} \tag{17}$$

where $\Phi_r = -\bar{\mathbf{K}}_{cc}^{-1}\bar{\mathbf{K}}_{cp}$ is the rigid modes matrix of the interface, as now connection P acts as the only isostatic interface (as explained in Section 1.1). The fixed base mode shapes matrix of the interface, φ_l, is obtained by solving the eigenvalue problem $\bar{\mathbf{K}}_{cc}\varphi_l = \omega_l^2 \bar{\mathbf{M}}_{cc}\varphi_l$. These matrices form the transformation matrix Γ.

Pre-multiplying by Γ^T and applying coordinates transformation equation in (14) the following expression is found:

$$\begin{bmatrix} \bar{\mathbf{m}}_{rr} & \bar{\mathbf{m}}_{rl} \\ \bar{\mathbf{m}}_{rl}^T & \bar{\mathbf{m}}_{ll} \end{bmatrix} \begin{bmatrix} \ddot{\mathbf{q}}_p \\ \ddot{\zeta}_l \end{bmatrix} + \begin{bmatrix} \mathbf{0} & \mathbf{0} \\ \mathbf{0} & \bar{\mathbf{k}}_{ll} \end{bmatrix} \begin{bmatrix} \mathbf{q}_p \\ \zeta_l \end{bmatrix} = \begin{bmatrix} \mathbf{F}_p - \bar{\mathbf{M}}_{pk}\ddot{\eta}_k + \Phi_r^T \mathbf{F}_c - \Phi_r^T \bar{\mathbf{M}}_{ck}\ddot{\eta}_k \\ \varphi_l^T \mathbf{F}_c - \varphi_l^T \bar{\mathbf{M}}_{ck}\ddot{\eta}_k \end{bmatrix} \tag{18}$$

Equation (18) presents two main advantages. The first one is the achieved mathematical simplicity since there is no longer a coupling caused by the interface stiffness matrix. The second one and most important is the finding of a more intuitive parameter, the rigid body matrix $\bar{\mathbf{m}}_{rr}$, named D_P^A in [5], which can be expressed as follows when Φ_r is evaluated in a correct coordinate frame:

$$D_P^A = \mathbf{m}_{rr} = \begin{bmatrix} \mathbf{I} & \Phi_r^T \end{bmatrix} \bar{\mathbf{M}}_{jj} \begin{bmatrix} \mathbf{I} \\ \Phi_r \end{bmatrix} = \begin{bmatrix} m^s\mathbf{I} & m^s(*\mathbf{GP}) \\ -m^s(*\mathbf{GP}) & \mathbf{J}_G^s - m^s(*\mathbf{GP})^2 \end{bmatrix} \tag{19}$$

where m^s is the mass of the appendage, $\mathbf{J}_G^s$ is the appendage inertia matrix at its gravity center, and $*\mathbf{GP}$ is the anti-symmetric matrix associated to the gravity center position, G, from the connection point P, denoted by the vector $\mathbf{GP}$. Such a finding is exploited later in Section 4 since it is an important source for appendage parametrization and thus being able to use co-design tools.

Other parameters in equation (18) are the modal participation matrix $\bar{\mathbf{m}}_{rl}$, the generalised mass matrix, $\bar{\mathbf{m}}_{ll}$, and the generalised stiffness matrix, $\bar{\mathbf{k}}_{ll} = diag(\omega_l^2)$. It should be noticed that these matrices are associated to the flexible movement between the connection points P and C, not to the appendage interior displacements η_k.

Hereinafter, applied loads at connection point P, $\mathbf{F}_p$, are ignored. This decision lies in the assembly of the interconnected appendage, since once an appendage called s is connected to another appendage named $s+1$, it results that $C^s \equiv P^{s+1}$, what could lead to count twice the external loads at the interface points. To avoid this problem, it is necessary to ignore one load, and it has to be $\mathbf{F}_p$ since $\mathbf{F}_c$ is needed so that $\mathbf{F}_{s/s-1}$ can be introduced.

Therefore, imposing $\mathbf{F}_p = 0$ in the matrix equation (18) and knowing that the acceleration at connection C is obtained deriving equation (16), a state-space representation is found in order to compute $\ddot{\mathbf{q}}_c$:

$$\begin{cases} \begin{bmatrix} \dot{\zeta}_l \\ \ddot{\zeta}_l \end{bmatrix} = \begin{bmatrix} \mathbf{0} & \mathbf{I} \\ -\bar{\mathbf{m}}_{ll}^{-1}\bar{\mathbf{k}}_{ll} & \mathbf{0} \end{bmatrix} \begin{bmatrix} \zeta_l \\ \dot{\zeta}_l \end{bmatrix} + \begin{bmatrix} \mathbf{0} & \mathbf{0} & \mathbf{0} \\ -\bar{\mathbf{m}}_{ll}^{-1}\bar{\mathbf{m}}_{rl}^T & \bar{\mathbf{m}}_{ll}^{-1}\varphi_l^T & \bar{\mathbf{m}}_{ll}^{-1}\varphi_l^T\bar{\mathbf{M}}_{ck} \end{bmatrix} \begin{bmatrix} \ddot{\mathbf{q}}_p \\ \mathbf{F}_c \\ \ddot{\eta}_k \end{bmatrix} \\ \ddot{\mathbf{q}}_c = \begin{bmatrix} -\varphi_l\bar{\mathbf{m}}_{ll}^{-1}\bar{\mathbf{k}}_{ll} & \mathbf{0} \end{bmatrix} \begin{bmatrix} \zeta_l \\ \dot{\zeta}_l \end{bmatrix} + \begin{bmatrix} \Phi_r - \varphi_l\bar{\mathbf{m}}_{ll}^{-1}\bar{\mathbf{m}}_{rl}^T & \varphi_l\bar{\mathbf{m}}_{ll}^{-1}\varphi_l^T & \varphi_l\bar{\mathbf{m}}_{ll}^{-1}\varphi_l^T\bar{\mathbf{M}}_{ck} \end{bmatrix} \begin{bmatrix} \ddot{\mathbf{q}}_p \\ \mathbf{F}_c \\ \ddot{\eta}_k \end{bmatrix} \end{cases} \tag{20}$$

From equation (20) several remarks can be extracted. It can be seen that the direct transfer of acceleration between P and C depends mainly on the rigid transfer Φ_r, reduced by the effect of flexibility between them as the transmission term, renamed $\phi_l = \varphi_l\bar{\mathbf{m}}_{ll}^{-1}\bar{\mathbf{m}}_{rl}^T$, shows. It also depends on the modal coordinates of the appendage internal degrees of freedom, $\ddot{\eta}_k$, which are transmitted through a rescaled modal participation factor at C which can be named as $\tilde{\mathbf{M}}_{ck} = \varphi_l\bar{\mathbf{m}}_{ll}^{-1}\varphi_l^T\bar{\mathbf{M}}_{ck}$. Once the opposite end acceleration has been determined, the transmitted load $\mathbf{F}_{s/s-1}$ can be computed.

2.1.2 Load Transfer

The load transfer through the connection point has a different origin than the acceleration. In order to see which information is transmitted through the interconnected appendage interfaces, the analyze of the synthesized structure is needed; i.e., the coupling study of two or more appendages.

Let an open-chain assembly of substructures be supposed, with the superscripts $s-1$, s and $s+1$ distinguishing the substructures which are being coupled. Following the Craig- Bampton synthesis steps explained in [6] or [10] and evaluating appendage equation of motion at its interfaces, the following mathematical statement is inferred:

$$\bar{\mathbf{M}}_{jj}^s\ddot{\mathbf{q}}_j^s + \bar{\mathbf{K}}_{jj}^s\mathbf{q}_j^s + \bar{\mathbf{M}}_{jk}^s\ddot{\eta}_k^s = \begin{bmatrix} [\mathbf{0}\ -\mathbf{I}]\,[\bar{\mathbf{M}}_{jj}^{s-1}\ddot{\mathbf{q}}_j^{s-1} + \bar{\mathbf{K}}_{jj}^{s-1}\mathbf{q}_j^{s-1} + \bar{\mathbf{M}}_{jk}^{s-1}\ddot{\eta}_k^{s-1}] \\ \mathbf{F}_c^{ext} + [-\mathbf{I}\ \mathbf{0}]\,[\bar{\mathbf{M}}_{jj}^{s+1}\ddot{\mathbf{q}}_j^{s+1} + \bar{\mathbf{K}}_{jj}^{s+1}\mathbf{q}_j^{s+1} + \bar{\mathbf{M}}_{jk}^{s+1}\ddot{\eta}_k^{s+1}] \end{bmatrix} \tag{21}$$

where **I** and **0** are $dof \times dof$ size. From equation (21) is deduced that the input loads at connection point C for a given appendage s are the result of an external load $\mathbf{F}_C^{ext}$ and a load transmitted by the successive appendage $s+1$, $\mathbf{F}_{s+1/s}$, which is written as follows:

$$\mathbf{F}_{s+1/s} = -\begin{bmatrix}\mathbf{I} & \mathbf{0}\end{bmatrix}\left[\bar{\mathbf{M}}_{jj}^{s+1}\ddot{\mathbf{q}}_j^{s+1} + \bar{\mathbf{K}}_{jj}^{s+1}\mathbf{q}_j^{s+1} + \bar{\mathbf{M}}_{jk}^{s+1}\ddot{\eta}_k^{s+1}\right] \tag{22}$$

and consequently loads at connection point C are expressed as:

$$\mathbf{F}_c = \mathbf{F}_{s+1/s} + \mathbf{F}_c^{ext} \tag{23}$$

Taking this into account, the load that the appendage s has to transmit to the preceding appendage $s-1$ has the same form but adding -1 to the current indexes:

$$\mathbf{F}_{s/s-1} = -\begin{bmatrix}\mathbf{I} & \mathbf{0}\end{bmatrix}\left[\bar{\mathbf{M}}_{jj}^{s}\ddot{\mathbf{q}}_j^{s} + \bar{\mathbf{K}}_{jj}^{s}\mathbf{q}_j^{s} + \bar{\mathbf{M}}_{jk}^{s}\ddot{\eta}_k^{s}\right] \tag{24}$$

However, as the load computation has to be embedded in the state-space model (20), it is compulsory to apply the coordinate transformation given in equation (17) for using the same inputs. This transformation can be applied as follows:

$$\mathbf{F}_{s/s-1} = -\begin{bmatrix}\mathbf{I} & \mathbf{0}\end{bmatrix}(\Gamma^T)^{-1}\left\{\Gamma^T\bar{\mathbf{M}}_{jj}\Gamma\begin{bmatrix}\ddot{\mathbf{q}}_p\\ \ddot{\zeta}_l\end{bmatrix} + \Gamma^T\bar{\mathbf{K}}_{jj}^{s}\Gamma\begin{bmatrix}\mathbf{q}_p\\ \zeta_l\end{bmatrix} + \Gamma^T\bar{\mathbf{M}}_{jk}\ddot{\eta}_k\right\} \tag{25}$$

and in a more compact expression:

$$\mathbf{F}_{s/s-1} = -\begin{bmatrix}\mathbf{I} & \Phi_r^T(\varphi_l^T)^{-1}\end{bmatrix}\begin{bmatrix}\bar{\mathbf{m}}_{rr}\ddot{\mathbf{q}}_p + \bar{\mathbf{m}}_{rl}\ddot{\zeta}_l + (\bar{\mathbf{M}}_{pk} + \Phi_r^T\bar{\mathbf{M}}_{ck})\ddot{\eta}_k\\ \bar{\mathbf{m}}_{rl}^T\ddot{\mathbf{q}}_p + \bar{\mathbf{m}}_{ll}\ddot{\zeta}_l + \bar{\mathbf{k}}_{ll}\zeta_l + \varphi_l^T\bar{\mathbf{M}}_{ck})\ddot{\eta}_k\end{bmatrix} \tag{26}$$

After tedious calculations it can be demonstrated that the resulting induced load from appendage s to the precedent appendage $s-1$ is determined by the equation:

$$\begin{aligned}\mathbf{F}_{s/s-1} = {} & \left[(\mathbf{T}\bar{\mathbf{m}}_{ll}^{-1} + \Phi_r^T(\varphi_l^T)^{-1})\bar{\mathbf{k}}_{ll}\middle|\mathbf{0}\right]\begin{bmatrix}\zeta_l\\ \dot{\zeta}_l\end{bmatrix}\\ & + \left[\mathbf{T}\bar{\mathbf{m}}_{ll}^{-1}\bar{\mathbf{m}}_{rl}^T - \mathbf{G}\middle| -\mathbf{T}\bar{\mathbf{m}}_{ll}^{-1}\varphi_l^T\middle|\mathbf{T}\bar{\mathbf{m}}_{ll}^{-1}\varphi_l^T\bar{\mathbf{M}}_{ck} - \bar{\mathbf{M}}_{pk}\right]\begin{bmatrix}\ddot{\mathbf{q}}_p\\ \mathbf{F}_c\\ \ddot{\eta}_k\end{bmatrix}\end{aligned} \tag{27}$$

with **T** and **G** defined as:

$$\begin{aligned}\mathbf{T} &= \bar{\mathbf{m}}_{rl}^T - \phi_r^T(\varphi_l^T)^{-1}\bar{\mathbf{m}}_{ll}\\ \mathbf{G} &= \bar{\mathbf{m}}_{rr} - \phi_r^T(\varphi_l^T)^{-1}\bar{\mathbf{m}}_{rl}^T\end{aligned} \tag{28}$$

2.1.3 Interface Dynamic Model

Once the acceleration and loads transfers have been calculated, a complete state-space representation of interface displacements can be established. This state-space representation is denoted as **Interface Dynamic Model** (IDM) of the substructure.

The IDM of the appendage is obtained by combination of the state-space model (20) and the load output given in (28), obtaining the following representation:

$$\begin{cases} \begin{bmatrix} \dot{\zeta}_l \\ \ddot{\zeta}_l \end{bmatrix} = \begin{bmatrix} \mathbf{0} & \mathbf{I} \\ -\bar{\mathbf{m}}_{ll}^{-1}\bar{\mathbf{k}}_{ll} & \mathbf{0} \end{bmatrix} \begin{bmatrix} \zeta_l \\ \dot{\zeta}_l \end{bmatrix} + \begin{bmatrix} \mathbf{0} & \mathbf{0} & \mathbf{0} \\ -\bar{\mathbf{m}}_{ll}^{-1}\bar{\mathbf{m}}_{rl}^T & \bar{\mathbf{m}}_{ll}^{-1}\varphi_l^T & \bar{\mathbf{m}}_{ll}^{-1}\varphi_l^T\bar{\mathbf{M}}_{ck} \end{bmatrix} \begin{bmatrix} \ddot{\mathbf{q}}_p \\ \mathbf{F}_c \\ \ddot{\eta}_k \end{bmatrix} \\ \begin{bmatrix} \ddot{\mathbf{q}}_c \\ \mathbf{F}_{s/s-1} \end{bmatrix} = \begin{bmatrix} -\varphi_l\bar{\mathbf{m}}_{ll}^{-1}\bar{\mathbf{k}}_{ll} & \mathbf{0} \\ (\mathbf{T}\bar{\mathbf{m}}_{ll}^{-1} + \Phi_r^T(\varphi_l^T)^{-1})\bar{\mathbf{k}}_{ll} & \mathbf{0} \end{bmatrix} \begin{bmatrix} \zeta_l \\ \dot{\zeta}_l \end{bmatrix} + \mathbf{D}_{IDM} \begin{bmatrix} \ddot{\mathbf{q}}_p \\ \mathbf{F}_c \\ \ddot{\eta}_k \end{bmatrix} \end{cases} \quad (29)$$

with $\mathbf{D}_{IDM}$ being:

$$\mathbf{D}_{IDM} = \begin{bmatrix} \Phi_r - \varphi_l\bar{\mathbf{m}}_{ll}^{-1}\bar{\mathbf{m}}_{rl}^T & \varphi_l\bar{\mathbf{m}}_{ll}^{-1}\varphi_l^T & \varphi_l\bar{\mathbf{m}}_{ll}^{-1}\varphi_l^T\bar{\mathbf{M}}_{ck} \\ \mathbf{T}\bar{\mathbf{m}}_{ll}^{-1}\bar{\mathbf{m}}_{rl}^T - \mathbf{G} & -\mathbf{T}\bar{\mathbf{m}}_{ll}^{-1}\varphi_l^T & \mathbf{T}\bar{\mathbf{m}}_{ll}^{-1}\varphi_l^T\bar{\mathbf{M}}_{ck} - \bar{\mathbf{M}}_{pk} \end{bmatrix} \quad (30)$$

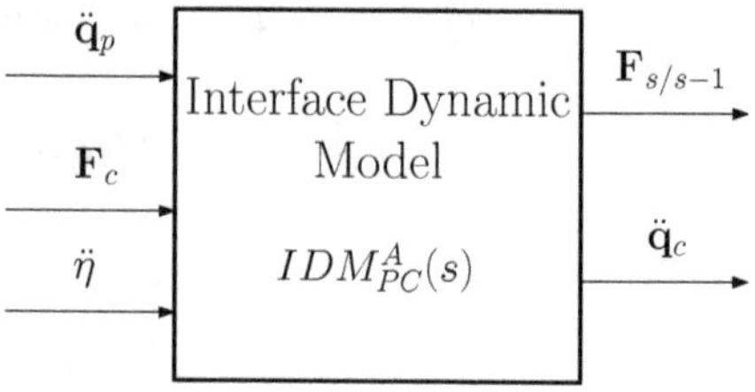

Fig. 4 Interface Dynamic Model block diagram which contains the state-space representation in (29)

The block diagram of the Interface Dynamic Model is depicted in Figure 4. As stated previously in this section, this model corresponds to a double-port system model (Figure 3). However, modal coordinates of interior displacements, η_k, are not computed yet. The effect of these displacements is included through the Interior Dynamics Model (iDM).

2.1.4 Interior Dynamic Model

The flexibility induced by the interior displacements in the IDM is not yet determined, since the modal coordinates η_k have no connexion. The lower part of equation (8) is used to model this effect, without no applied loads in the interior displacements ($\mathbf{F}_i$):

$$\bar{\mathbf{M}}_{jk}^T\ddot{\mathbf{q}}_j + \mathbf{m}_k\ddot{\eta}_k + \mathbf{k}_k\eta_k = \mathbf{0} \quad (31)$$

Hence it is possible to create another state-space representation which acts as a feedback within the IDM. This state-space representation is named the **Interior Dynamic Model** (iDM) of the substructure, and is described by the following equations (where $\mathbf{m}_k$ was supposed the identity matrix):

$$\begin{bmatrix}\dot{\eta}_k\\ \ddot{\eta}_k\end{bmatrix}=\begin{bmatrix}\mathbf{0} & \mathbf{I}\\ -\mathbf{k}_k & \mathbf{0}\end{bmatrix}\begin{bmatrix}\eta_k\\ \dot{\eta}_k\end{bmatrix}+\begin{bmatrix}\mathbf{0} & \mathbf{0}\\ -\bar{\mathbf{M}}_{pk}^T & -\bar{\mathbf{M}}_{ck}^T\end{bmatrix}\begin{bmatrix}\ddot{\mathbf{q}}_p\\ \ddot{\mathbf{q}}_c\end{bmatrix}$$
$$\ddot{\eta}_k=\begin{bmatrix}-\mathbf{k}_k & \mathbf{0}\end{bmatrix}\begin{bmatrix}\eta_k\\ \dot{\eta}_k\end{bmatrix}+\begin{bmatrix}-\bar{\mathbf{M}}_{pk}^T & -\bar{\mathbf{M}}_{ck}^T\end{bmatrix}\begin{bmatrix}\ddot{\mathbf{q}}_p\\ \ddot{\mathbf{q}}_c\end{bmatrix} \tag{32}$$

As it can be appreciated, the iDM contains all the information concerning substructure flexible modes, as it is composed of the generalized stiffness matrix. In addition, the matrices $\bar{\mathbf{M}}_{pk}^T$ and $\bar{\mathbf{M}}_{ck}^T$, describe the way these flexible modes are affected by the interface motion expressed by the acceleration at points P and C.

2.1.5 Assembly

The construction of the **Interconnected Flexible Appendage Model** (IFA) is achieved when the IDM is feedbacked with the iDM, as depicted in Figure 5. As a result, the desired double-port model is obtained with the established inputs (acceleration at point P and load caused by the next appendage) and outputs (acceleration at point C and load transmitted to the preceding appendage).

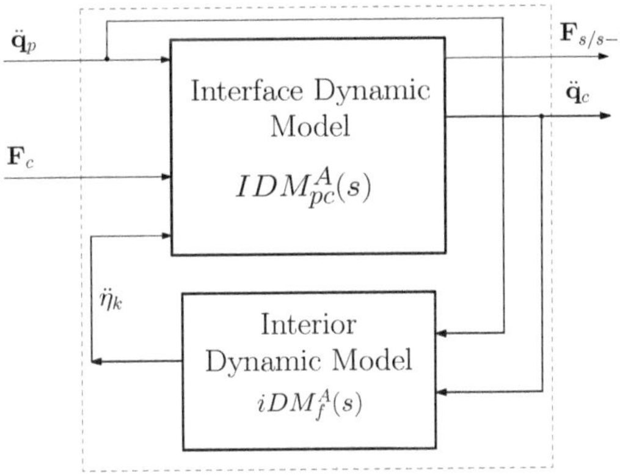

Fig. 5 Block Diagram representing the interior dynamics feedback to the interface dynamics.

The last step to complete the IFA model is to re-introduce the external loads at the interface point C, $\mathbf{F}_c$. As stated in Section 2.1.2, this load has an external origin, $\mathbf{F}_c^{ext}$ and an internal origin coming from the successive appendage, $\mathbf{F}_{s+1/s}$. The resultant IFA model is depicted in Figure 7. This model allows to use the necessary inputs/outputs to link the other substructures.

In addition, the introduction of the external loads at the end of each appendage has two advantages. Not only it serves as an input for perturbations along the chain, but also as an input for actuators that might be activated between appendages in order to rigidify the chain. It is important to highlight as well that typically in the analysis of spacecraft payloads all external loads are considered to be applied in the interface points and there are no applied loads to the non-interface points, which makes the simplification of $\mathbf{F}_i = \mathbf{0}$ even more suitable.

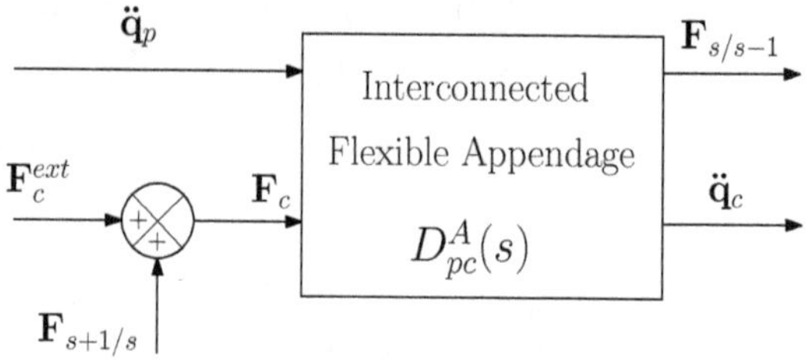

Fig. 6 Final Block Diagram of the Interconnected Flexible Appendage Model.

2.2 *Terminal Appendage*

A particular case is presented when the final appendage of the chain has to be modeled. This appendage has one connection point and thus parameters referring to connection *C* disappear. This fact corresponds to the problem of base excitation explained in [6]. Since this is an isostatic case, the Static Constraint Modes matrix lapses into the rigid modes matrix at the boundary point *P*, which leads to the same problem already treated in [5] and [6], where only rigid modes matrix are considered to express interface behavior.

In this case equation (8) becomes:

$$\begin{cases} \bar{\mathbf{M}}_{pp}\ddot{\mathbf{q}}_p + \bar{\mathbf{M}}_{pk}\ddot{\eta}_k = \mathbf{F}_p \\ \bar{\mathbf{M}}_{pk}^T\ddot{\mathbf{q}}_p + \mathbf{m}_k\ddot{\eta}_k + \mathbf{k}_k\eta_k = \mathbf{0} \end{cases} \tag{33}$$

Indeed, there are not sub-matrices referring to interface *C* and there is no interface stiffness matrix as the problem is isostatic. It can be inferred that $\bar{\mathbf{M}}_{pp}$ is the rigid mass matrix (as computed in equation (19)) and $\bar{\mathbf{M}}_{pk}$ the modal participation matrix at point *P*, both treated in references [5] and [6] as well.

The induced load to the preceding appendage is determined by the expression $\mathbf{F}_{s/s-1} = -\bar{\mathbf{M}}_{pp}\ddot{\mathbf{q}}_p - \bar{\mathbf{M}}_{pk}\ddot{\eta}_k$ as it can be deducted from the notions explained in Section 2.1.2. Taking this in consideration, the equations in (33) can be expressed in a state-space representation:

$$\begin{aligned} \begin{bmatrix} \dot{\eta}_k \\ \ddot{\eta}_k \end{bmatrix} &= \begin{bmatrix} \mathbf{0} & \mathbf{I} \\ -\mathbf{k}_k & \mathbf{0} \end{bmatrix} \begin{bmatrix} \eta_k \\ \dot{\eta}_k \end{bmatrix} + \begin{bmatrix} \mathbf{0} \\ -\bar{\mathbf{M}}_{pk}^T \end{bmatrix} \ddot{\mathbf{q}}_p \\ \mathbf{F}_{s/s-1} &= -\bar{\mathbf{M}}_{pk} \begin{bmatrix} -\mathbf{k}_k & \mathbf{0} \end{bmatrix} \begin{bmatrix} \eta_k \\ \dot{\eta}_k \end{bmatrix} + \begin{bmatrix} \bar{\mathbf{M}}_{pk}\bar{\mathbf{M}}_{pk}^T - \bar{\mathbf{M}}_{pp} \end{bmatrix} \ddot{\mathbf{q}}_p \end{aligned} \tag{34}$$

This model is equivalent as the one presented in [5], in which the only output is the load transmitted to the previous appendage and the only input is the acceleration at the interface point *P* of the appendage, as depicted in Figure 7. This last appendage acts as a feedback of the preceding appendage, as explained in Section 3.

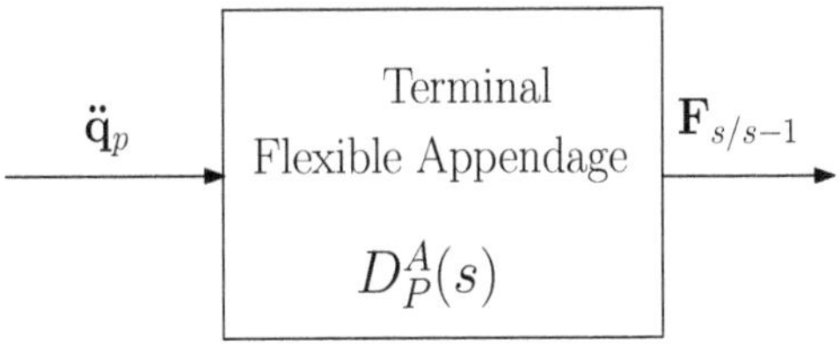

Fig. 7 The Terminal Appendage block diagram has one input, connection point acceleration, and one output, the load transmitted to the connection point.

3 Interconnection between Interconnected Flexible Appendage Models

IFA interconnections can be seen as *LFT* (Linear Fractional Transformation) multiplications, since the output of an appendage is the input for the next one while the next one has to provide its output to the preceding one. Particularly, output $\ddot{\mathbf{q}}_c$ of appendage s is the input for appendage $s+1$, while output $\mathbf{F}_{s/s-1}$ of appendage s is the input for appendage $s-1$. In the same way, input $\ddot{\mathbf{q}}_p$ of appendage s is the output of appendage $s-1$, and input $\mathbf{F}_{s+1/s}$ is the output of appendage $s+1$. Between the connection $\mathbf{F}_{s+1/s} \leftrightarrow \mathbf{F}_{s/s-1}$ the external load at the corresponding interface must be added.

A generic example of this process is given in Figure 8. Let consider a chain-like assembly of n appendages linked to the spacecraft main hub. The main hub inputs are the forces and torques at its gravity center G, and main hub outputs are the accelerations of point G and its angular acceleration. These outputs are taken by the first appendage *Appendage 1* as the accelerations of its P boundary, $\ddot{\mathbf{q}}_p$, after previous passage through the kinematic model τ_{PG} (the kinematic model which transports accelerations at spacecraft gravity center G to the anchorage point P where the appendage chain starts). The output $\ddot{\mathbf{q}}_c$ of *Appendage 1* must be linked to $\ddot{\mathbf{q}}_p$ input of *Appendage 2* since they are actually the same interface: the interface between 1 and 2. It occurs the same for the load that *Appendage 2* transmits, $\mathbf{F}_{s/s-1}$, which is linked to *Appendage 1* as its $\mathbf{F}_{s+1/s}$ input. Then, the output $\mathbf{F}_{s/s-1}$ of *Appendage 1* is summed to the spacecraft main hub loads, previous passage through τ_{PG}^T. The same process can be done for appendages $2,3\ldots n-1$. Finally, *Appendage n* closes the information loop by providing the corresponding load $\mathbf{F}_{s/s-1}$ to *Appendage n-1* caused by the acceleration $\ddot{\mathbf{q}}_P$ in its boundary P. It should be noticed that rotation matrix may be applied to the transfer of accelerations and loads between appendages if appendage frame do not coincide. To simplify the explanation of concatenation, these frame changes have not been considered in Figure 8.

As a result, an assembled appendage is found which contains all the sub-models $1,2\ldots n$ and their structural information. In addition, it takes into account the loads which might be applied at the boundaries between the substructures , and it is easy to have direct access to physical parameters as demonstrated in Section 4.

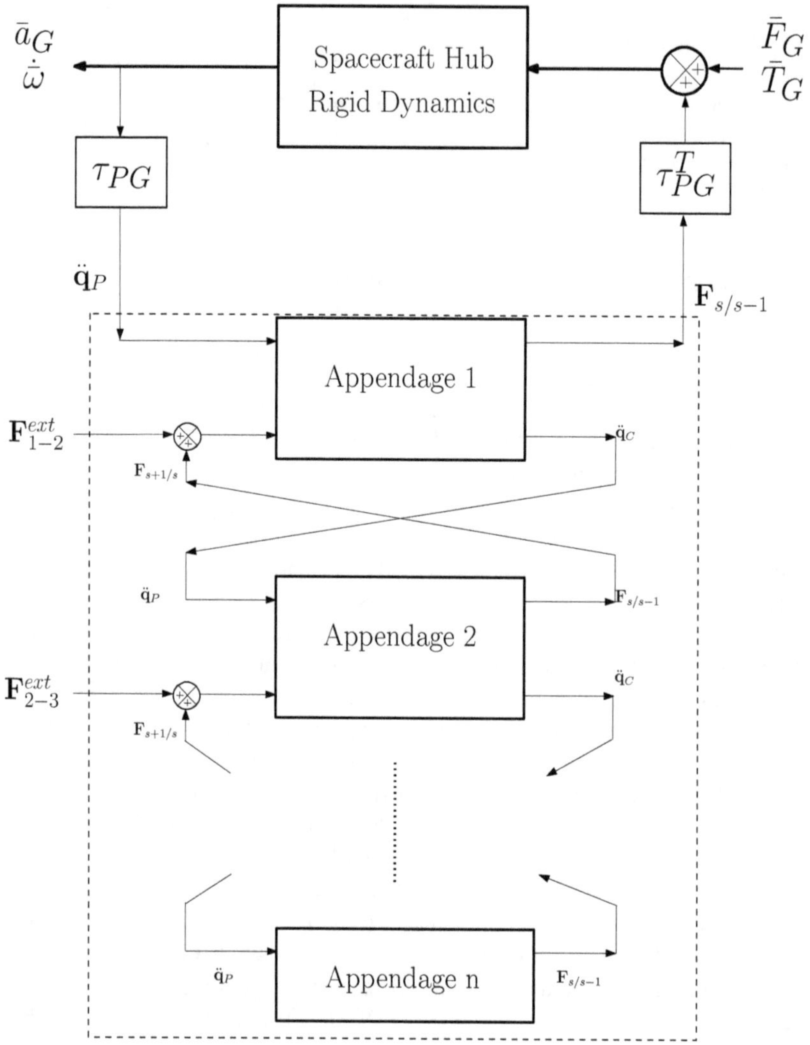

Fig. 8 Demonstration of how to concatenate *n* appendages modelled with the IFA method.

4 Model Parametrization for Co-design/Robust Control

The structure/control co-design method for which this modeling technique has been developed is based on structured $H_{\inf}$ synthesis ([14]). The sketch in Figure 9 shows the co-design approach. The method allows synthesizing a stabilizing controller $C(s)$ and optimizing tunable parameters Δ_i of the structure for a given system $P(s)$ ([7]). The block Δ_i is the uncertainty structured matrix commonly used in the ν-analysis robust control, but in the co-design method this matrix is used as a set of several parameters variations affecting $P(s)$ dynamics. The goal of structure/control co-design is to optimize such variations simultaneously with the controller.

Therefore, a modeling technique which allows extracting the system Δ_i block from the *LTI* representation is needed. In order to do this, it is necessary to have direct access to physical parameters inside the models. The direct access will allow the user to easily include physical parameter variations in the state space representation, simplifying the Δ_i block extraction. This direct access to parameters is ensured for the IFA model since it follows a clear parametric structure:

1. The Interface Dynamic Model (IDM) inside the IFA model ensures the access to physic parameters condensed at the connection points, such as appendage mass m^s, inertia $\mathbf{J}^s_G$ and gravity center position **GP**, found in equation (19) and then spread in the subsequent equations.
2. The Interior Dynamic Model (iDM) inside the IFA model ensures accessibility to flexibility parameters such as flexible mode frequencies in the form of ω^2, damping ratios ξ and modal participation factors at the boundary points, $\bar{\mathbf{M}}_{pk}$ and $\bar{\mathbf{M}}_{ck}$ (see Section 1.1).

Acceding to these physical parameters and imposing them the desired variation will allow extracting the subsequent Δ_i block of the entire system after appendage concatenation. These block Δ_i will be used as an input to be optimized by the H_{inf} synthesis method, given the optimal variations that ensure system performance and stability.

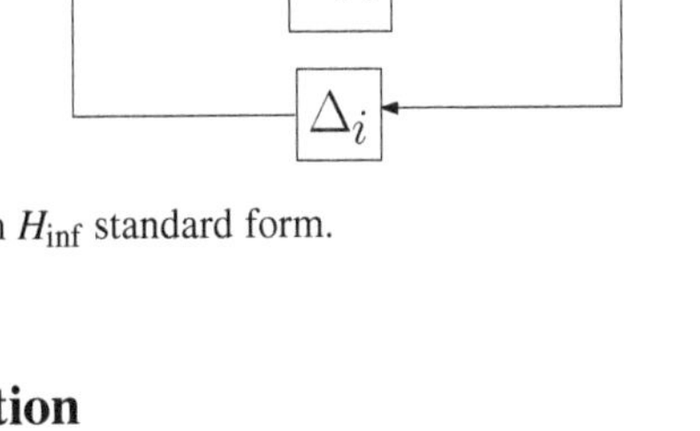

Fig. 9 Integrated Design H_{inf} standard form.

5 Model Validation

In order to verify the model truthfulness, a particular example of a 2D beam connected to a base is developed. Let a 2D beam of length L be substructured in three parts or child-beams A, B and C of lengths $L/2$, $L/4$ and $L/4$ respectively. Each substructure has identical cross-section properties (rectangular section of width b and height h) and identical material composition (density ρ and elastic module E constant). The IFA models, once concatenated, must present the same poles of the

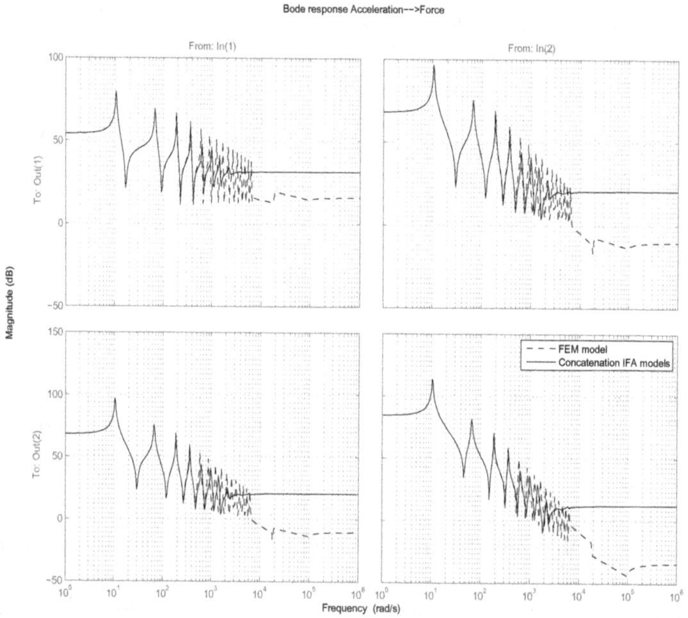

Fig. 10 Bode Diagram of the load transmitted to the connection point P. Study case: $b = 0.1$, $L = 10$, $h = 0.2$ $E = 69GPa$, $\rho = 2500$. No damping considered.

dynamic system *Acceleration* $\rightarrow$ *Load* of the entire beam of length L and must transmit an identical load to the connection point.

For each substructure the methods studied in Sections 1, 2 and 3 are applied in order to obtain the corresponding IFA model. Next, truncation as explained in Section 1.2 is performed, keeping the first $k = 2$ flexible modes for each appendage. Finally, models concatenation is executed as in Section 3.

The result of the final assembly is shown and validated in Figure 10, where the $\ddot{\mathbf{q}}_p \rightarrow \mathbf{F}_{chain/hub}$ of the open-chain of child-beams is compared with the entire beam model obtained by FEM analysis with more than 20 nodes. It can be appreciated that resonances and antiresonances occur at the same frequencies for the low-frequency band, denoting the identical dynamical behavior, and start to diverge after the fifth resonance for this numerical application, which was predictable since child-beams flexible modes have been truncated. In addition, static gains are equal, ensuring that the induced load to the hub is identical for both models.

As stated in previous sections, like in Section 1.2, for ACS design only a few low-frequency modes are needed, making this model largely suitable for the foreseen applications such as AOCS/Structure co-design and robust control.

6 Conclusions and Perspectives

In this work a new model for Interconnected Flexible Appendages (IFA) has been presented and tested. It allows to:

- take into account appendage flexible modes.
- make connections in chain-like assembly with other appendages.
- introduce loads applied at the interface connection points between appendages.
- interface the model with LFT representation to handle uncertain dynamic parameters in the modeling process.

All the operations for building the IFA models and connections between them are being implemented as a complement of the Matlab package *Spacecraft Dynamics Toolbox* mentioned in [5]. A full real-case application of this method is shown in [15]. As for the future, the use of IFA model in co-design and robust control utilities is foreseen, particularly for a preliminary study for a satellite with a long chain of flexible appendages.

References

1. Linkins, P.: Dynamics and control of flexible space vehicles. Tech. Rep. 32-1329, California Institute of Technology, Pasadena, California (1969)
2. Hughes, P.: Dynamics of flexible space vehicles with active attitude control. Celestial Mechanics 9, 21–39 (1974)
3. Pascal, M.: Dynamics analysis of a system of hinge-connected flexible bodies. Celestial mechanics 41(1-4), 253–274 (1988)
4. Shabana, A.: Flexible Multibody Dynamics: Review of Past and Recent Developments. Multibody System Dynamics 1, 189–222 (1997)
5. Alazard, D., Cumer, C., Tantawi, K.: Linear dynamic modeling of spacecraft with various flexible appendages and on-board angular momentums. In: 7th ESA Guidance, Navigation and Control Conference, Tralee, Ireland (June 2008)
6. Imbert, J.: Analyse des Structures par Elements Finis. ENSTA, CEPAD. Cepaudes ed. (1979)
7. Alazard, D., Loquen, T., de Plinval, H., Cumer, C.: Avionics/Control co-design for large flexible space structures. In: AIAA Guidance, Navigation, and Control (GNC) Conference, Boston, Massachusetts, USA (August 2013)
8. Alazard, D., Loquen, T., de Plinval, H., Cumer, C., Toglia, C., Pavia, P.: Optimal Co-Design for Earth Observation Satellites with Flexible Appendages. In: AIAA Guidance, Navigation, and Control (GNC) Conference,
9. Craig, R., Bampton, M.: Coupling of substructures for dynamic analysis. AIAA Journal 6(7) (1968)
10. Young, J.: Primer on Craig-Bampton CMS proceduremethod: an introduction to boundary node functions, base shake analyses, load transformation matrices, modal synthesis and much more. Tech. Rep., NASA (October 2000)
11. Howsman, T., Craig, R.: A substructure coupling procedure applicable to general linear time-invariant dynamic systems. Tech. Rep. NASA-CR-178510, NASA (May 1984)
12. Kraker, B.: Generalization of the craig-bampton CMS procedure for general damping. Tech. Rep., Technische Universiteit Eindhoven, Eindhoven (1993)

13. Alazard, D., Perez, J., Loquen, T., Cumer, C.: Two input/output port model approach for mechanical systems. In: AIAA Scitech Forum 2015, Kissimmee, Florida, USA (submitted 2015)
14. Gahinet, P., Apkarian, P.: Structured H-infinity synthesis using MATLAB. In: 18th IFAC World Congress, Milano, Italy (August 2011)
15. Perez, J., Pittet, C., Alazard, D., Loquen, T., Cumer, C.: A Flexible Appendage Model for Use in Integrated Control/Structure Spacecraft Design. In: IFAC Workshop on Advanced Control and Navigation for Autonomous Aerospace Vehicles, Seville, Spain (submitted, 2015)

Mechanical-Attitude Controller Co-design of Large Flexible Space Structures

Hari Hara Sudhan Murali, Daniel Alazard, Luca Massotti,
Finn Ankersen, and Chiara Toglia

Abstract. This paper provides an overview on the application and advantages of novel sub-structuring techniques in defining the multi-flexible body dynamics of an open mechanical chain. It employs single and double input-output port models that represent the six Degree Of Freedom (DOF) dynamics of terminal and intermediate appendages in the chain, with one and two connection points respectively. The interface that links the Finite Element Model (FEM) modal analysis output of the sub-structure to its parametric model in SIMULINK has also been explained. Also an analytical double port model of the six DOF dynamics of a uniform beam element with two connection points, called super-element, has been derived.

Different combinations of the sub-structure configurations are validated through the dynamics simulation of the Biomass satellite, selected as the 7-th Earth Explorer Core Mission of the European Space Agency. A robust reduced order H_∞ attitude controller that will meet the satellite performance requirement in terms of perturbation rejection, has been designed. Furthermore, the Linear Fractional Transformation (LFT) form of an uncertain parametric model is subjected to a non-smooth optimization that simultaneously tunes the attitude controller and the geometric parameters of an antenna section, so as to meet pointing requirements and save mass.

Hari Hara Sudhan Murali · Daniel Alazard
ISAE, 10 Av. Edouard Belin, 31055 Toulouse CEDEX 4, France

Luca Massotti, Finn Ankersen
European Space Agency -ESTEC, Keplerlaan 1, 2200 Noordwijk, The Netherlands

Chiara Toglia
Thales Alenia Space, Roma, Italia

J. Bordeneuve-Guibé et al. (eds.), *Advances in Aerospace Guidance, Navigation and Control*, DOI: 10.1007/978-3-319-17518-8_38

Nomenclature

$\mathscr{MB}$	: subscript referring to the main body.
$\mathscr{A}$	: subscript referring to the appendage.
$\mathscr{P}$	: parent appendage attached to $\mathscr{A}$ at point P.
$\mathscr{C}$	: child appendage attached to $\mathscr{A}$ at point C.
$\mathscr{R}_a = (P, \mathbf{x}_a, \mathbf{y}_a, \mathbf{z}_a)$	: body $\mathscr{A}$ reference frame at point P; $\mathbf{x}_a$, $\mathbf{y}_a$, $\mathbf{z}_a$ are orthogonal unit vectors.
A	: centre of mass of body $\mathscr{A}$.
$\overrightarrow{a}_P$	: absolute linear acceleration vector of point P.
$\overrightarrow{\omega}_P$	: absolute angular velocity vector of point P.
$\ddot{\mathrm{x}}_P$	: dual vector (6 components) of accelerations at point P.
$\overrightarrow{F}_{\mathscr{A}/\mathscr{P}}$	: interaction force by $\mathscr{A}$ on $\mathscr{P}$.
$\overrightarrow{T}_{\mathscr{A}/\mathscr{P},P}$	: interaction torque by $\mathscr{A}$ on $\mathscr{P}$ at point P.
$\mathrm{F}_{\mathscr{A}/\mathscr{P},P}$	: dual vector (6 components) of interactions at point P.
$\overrightarrow{a}_C$	: absolute linear acceleration vector of point C.
$\overrightarrow{\omega}_C$	: absolute angular velocity vector of point C.
$\ddot{\mathrm{x}}_C$	: dual vector (6 components) of accelerations at point C.
$\overrightarrow{F}_{\mathscr{C}/\mathscr{A}}$	: interaction force by $\mathscr{C}$ on $\mathscr{A}$.
$\overrightarrow{T}_{\mathscr{C}/\mathscr{A},C}$	: interaction torque by $\mathscr{C}$ on $\mathscr{A}$ at point C.
$\mathrm{F}_{\mathscr{C}/\mathscr{A},C}$	: dual vector (6 components) of interactions at point C.
$M_P^{\mathscr{A}}(\mathbf{s})$	: direct dynamic model of the body $\mathscr{A}$ at point P.
τ_{CP}	: kinematic model between the point C and the point P.
n_f	: number of flexible modes considered in the model.
$D_{P_0}^{\mathscr{A}}$	: residual mass matrix of appendage $\mathscr{A}$ at P.
ω_i	: frequency of the ith flexible mode.
ξ_i	: modal damping ratio if the ith flexible mode.
L_P	: $(n_f \times 6)$ matrix of modal participation factors at point P.
Φ_C	: $(6 \times n_f)$ projection of the n_f modal shapes on the 6 DOF $\delta\mathrm{x}_C$.
s	: LAPLACE variable.
$P(\mathrm{s})_{4:6,4:6}$	: subsystem of $P(\mathrm{s})$ between inputs 4 to 6 and outputs 4 to 6.
$\overrightarrow{u}.\overrightarrow{u}$	: scalar product of vectors $\overrightarrow{u}$ and $\overrightarrow{u}$.

1 Introduction

A spacecraft is a complex structure that encompasses multi-body mechanical chain of appendages, both flexible and rigid. All spacecraft, especially satellites must have pointing capabilites to a certain degree of accuracy, which can be degraded by vibrations induced by the flexibility of appendages during maneuvers, routine deployments or due to external disturbances like solar pressure and gravity gradient. However, contemporary satellite design uses more flexible structures for lighter launch and fuel mass and better agility. A proper trade-off must be achieved to guarantee a final design as light as possible and also with a good perfomance index, in terms of tracking, stabilization and robustness to external disturbances. The Biomass Earth observation satellite of the European Space Agency (ESA), has been designed to carry a P-band radar with circularly projected $12m$ diameter deployable flexible reflector, supported by a flexible deployment arm, and also a flexible solar panel. Such large flexible appendages can induce severe vibrations in the satellite. The study is intended to design an attitude controller for the satellite that will minimize the loss in pointing accuracy incurred during the tracking of reference attitude, and to investigate the possibility of reducing the mass of certain portions of the arm.

[1] details the derivation of the linear dynamic model at the center of gravity of the rigid bus of a spacecraft to which a stationary or rotating terminal flexible appendage has been rigidly fixed or pivoted. The substructure is defined by a direct dynamic model (that depends on the appendage specific mass and modal parameters), at the anchorage point. This approach has been extended to an intermediate appendage with two connection points [2] while the cantilever-free version of the appendage serves as the basis for modal analysis. The interactions with the neighbouring appendages are represented by the input and output vectors of reaction forces and accelerations (inter-model connections), in the appendage dynamic model. The first objective has been to develop an analytical model for an intermediate uniform beam that defines the sub-structure dynamics in terms of geometric and mechanical parameters, as detailed in Sect. 2. A further goal is to implement the aforementioned sub-structure models in MATLAB/SIMULINK, establish their interface with the appendage-specific FEM analysis data, apply the models to simulate the dynamic behaviour of the Biomass satellite and thereby validate them (see Sect. 3). The second problem is to design a decentralized attitude controller that will ensure a performance objective dictated by the perturbation rejection capability of the satellite. In Sect. 4, a detailed illustration of the simultaneous optimization of the H_∞ attitude controller and the geometric parameters of a uniform beam portion of the satellite antenna arm can be found. The aim of this exercise is to investigate the feasibility of reducing the mass of the arm portion by allowing an acceptable degradation in the pointing performance of the satellite. Prior research [3, 4] on control/structural co-design of spacecraft follow an iterative approach, which can lead to long runs of optimization as the number of tunable parameters increases especially in case of complex multi-body systems. In [5], a reduced order H_∞ controller is designed for a standard problem, and roll-off requirements and worst-case parametric configurations are implemented in a multi-model design approach. [6] details the co-design of the Biomass satellite where the controller is tuned towards better performance and robustness, while assuming a delay in the avionics of the satellite. Both the controller parameters and the delay index are treated as decision variables in order to select the lowest-cost avionics and to design the corresponding control law that meets the pointing requirements. The co-design procedure presented in this study differentiates between the constraints on the norms of the closed loop system and the weighted controller, and the objective to be satisfied, which is to minimize the mass of the arm linking the reflector to the bus. The main contribution of the study is the development of modelling tools that allow to build a model of complex

flexible space structure like Biomass, which is parameterized according to mechanical characteristics (like the section properties of the Biomass arm).

2 Sub-structure Configurations

In the design and dynamic analysis of complex mechanical structures, an important factor in sub-structuring is the choice of boundary conditions at the connection points, that can be fixed, free, loaded or connected to an external substructure. Several methods are in practice for modal synthesis with different boundary conditions, and the resultant modes vary in their accuracy and shape with respect to the modes of the exact model [7]. This section contains a brief review on the single and double input-output-port models for a sub-structure of arbitrary shape with one and two connection points respectively, defined only by its dynamic parameters, while its boundary conditions at the connection points have been externalized. Also a detailed account on the derivation of an analytical double input-output-port model for a uniform beam (super-element) can be found, where the dynamic model is parameterized by the geometrical and mechanical properties of the beam.

2.1 Hybrid-cantilever Model for Terminal and Intermediate Appendages

The appendage $\mathscr{A}$ shown in Fig. 1 is the basis for the derivations reviewed in this section. The hybrid-cantilever formulation [1] of a terminal (leaf) sub-structure with one connection point P, considers the rigid and flexible modes of its clamped-free form. This formulation is based on the effective mass approach and provides the transfer between interaction forces and accelerations (6 DOF) at the connection point P of the appendage $\mathscr{A}$ with the parent substructure $\mathscr{P}$:

$$\mathrm{F}_{\mathscr{A}/\mathscr{P},P} = -M_P^{\mathscr{A}}(\mathrm{s})\,\ddot{\mathrm{x}}_P\,. \tag{1}$$

This model, called one-port model, is enterely defined by the appendage own dynamic parameters $D_{P_0}^{\mathscr{A}}$, ω_i, ξ_i and L_P (see nomenclature). But the internal deformations of the appendage are lost in such a model and it cannot take into account that the substructure $\mathscr{A}$ can also be connected to a child substructure $\mathscr{C}$ at the point C.

For an intermediate appendage with two connection points P and C, [2] proposes a two-port model $M_{P,C}^{\mathscr{A}}(\mathrm{s})$ where accelerations and interactions forces at points P and C are both considered on inputs and outputs:

$$\begin{bmatrix} \ddot{\mathrm{x}}_C \\ \mathrm{F}_{\mathscr{A}/\mathscr{P},P} \end{bmatrix} = M_{P,C}^{\mathscr{A}}(\mathrm{s}) \begin{bmatrix} \mathrm{F}_{\mathscr{C}/\mathscr{A},C} \\ \ddot{\mathrm{x}}_P \end{bmatrix}\,. \tag{2}$$

In addition to the modal parameters $D_{P_0}^{\mathscr{A}}$, ω_i, ξ_i and L_P that appear in the one-port model $M_P^{\mathscr{A}}(s)$, the two-port model $M_{P,C}^{\mathscr{A}}(s)$ involves the kinematic model τ_{CP} between points P and C and the projection at the point C of the n_f clamped-free modal shapes Φ_C $(6 \times n_f)$.

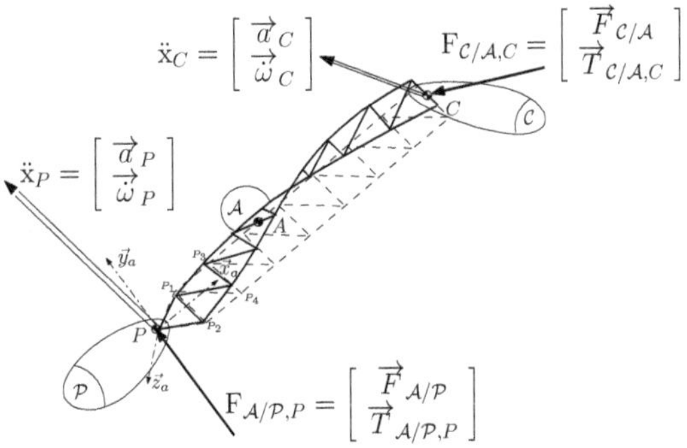

Fig. 1 Intermediate body $\mathscr{A}$ with parent $\mathscr{P}$ and $\mathscr{C}$ substructures on either sides

2.2 *Analytical Two-port Model of a Uniform Beam*

For a uniform beam $\mathscr{A}$, shown in Fig. 2, which is an intermediate appendage between points P and C and defined by parameters like density ρ, cross-sectional area s, length l, Young's modulus E, Poisson's ratio ν (where the shear modulus $G = \frac{E}{2(1+\nu)}$), second moment of area about the y and z axes : I_y and I_z respectively, polar second moment of area around the x-axis I_{px}, and arbitrary damping ratio for all flexible modes ξ, a two-port model $M_{P,C}^{\mathscr{A}}(s)$ similar to that of the intermediate appendage in Sect. 2.1 has been developed as a function of the parameter vector $\Theta = [\rho, s, l, E, G, I_y, I_z, I_{px}, \xi]^T$. The derivation assumes pure deflection in the planes $(P, \mathbf{x}_a, \mathbf{y}_a)$ and $(P, \mathbf{x}_a, \mathbf{z}_a)$ and torsion around the $\mathbf{x}_a$ axis, while longitudinal tension and compression along the $\mathbf{x}_a$ axis are neglected.

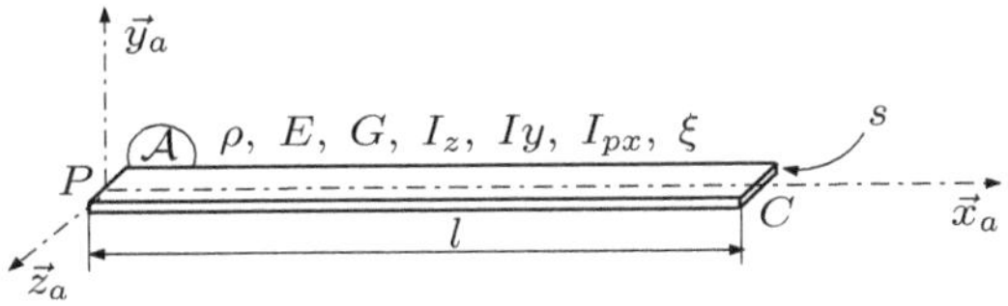

Fig. 2 Uniform beam as an intermediate body $\mathscr{A}$ with connection points P and C

2.2.1 Pure Bending in the $(P, \mathbf{x}_a, \mathbf{y}_a)$ Plane

A finite-element approach that exploits the uniformity of the beam (as commonly seen in the case of arms or links in a space robotic arm), uses a polynomial function of the abscissa x of the beam, of order 3 or above (here 5), to model its deflection $y(x,t)$ in the $(P,\mathbf{x}_a,\mathbf{y}_a)$ plane.

$$y(x,t) = a_0(t) + a_1(t)x + a_2(t)x^2 + a_3(t)x^3 + a_4(t)x^4 + a_5(t)x^5 \quad \forall t\,. \tag{3}$$

The beam element with a 5-th order deflection function is called a super-element, and in [8] it is seen that the first few modal frequencies and shapes for various boundary conditions are determined more accurately from a single super-element than a sequence of two or more elements that have a conventional 3-rd order deflection function for bending. The spatial derivative of equation (3) gives the slope of the deflection at point x which is equated to the angle of deflection at x in the plane of bending, and the double derivative of deflection gives the curvature of the deflected geometry at x.

$$\frac{\partial y}{\partial x}(x,t) = \phi(x,t) \;\; ; \;\; \frac{\partial^2 y}{\partial x^2}(x,t) = \frac{T_b(x,t)}{EI_z} \,,$$

where $T_b(x,t)$ is the bending moment at abscissa x, i.e. torque applied by the beam portion to the right of x, on the portion to its left. A matrix $\mathbf{F}$ maps the time-dependent coefficients $a(t)$ found in equation (3) to the time-dependent kinematic vector $q(t)$, such that $[q_i(t)]_{i=1,\cdots,6} = \mathbf{F}[a_j(t)]_{j=0,\cdots,5}$, where
$q(t) = [y_P(t),\ \phi_P(t),\ \frac{T_{bP}(t)}{EI_z},\ y_C(t),\ \phi_C(t),\ \frac{T_{bC}(t)}{EI_z}]^T$, see also Fig. 3.

$$\mathbf{F} = \begin{bmatrix} 1 & 0 & 0 & 0 & 0 & 0 \\ 0 & 1 & 0 & 0 & 0 & 0 \\ 0 & 0 & 2 & 0 & 0 & 0 \\ 1 & l & l^2 & l^3 & l^4 & l^5 \\ 0 & 1 & 2l & 3l^2 & 4l^3 & 5l^4 \\ 0 & 0 & 2 & 6l & 12l^2 & 20l^3 \end{bmatrix} \tag{4}$$

The deflection function $y(x,t)$ is then expressed as :

$$y(x,t) = \sum_{i=1}^{6} \varphi_i(x) q_i(t) \,, \tag{5}$$

where $\varphi_i(x) = [1\ x\ x^2\ x^3\ x^4\ x^5]\ \mathbf{F}^{-1}$. The kinetic energy of the beam element is given by

$$T = \frac{1}{2}\int_0^l \rho\, s \left(\frac{\partial y}{\partial t}(x,t)\right)^2 dx = \frac{1}{2}\dot{q}^T \mathbf{M}\dot{q} \,, \tag{6}$$

where the (i,j)-th element of the mass matrix $\mathbf{M}$ reads:

$$\mathbf{M}(i,j) = \int_0^l \rho\, s\, \varphi_i(x) \varphi_j(x) dx \,.$$

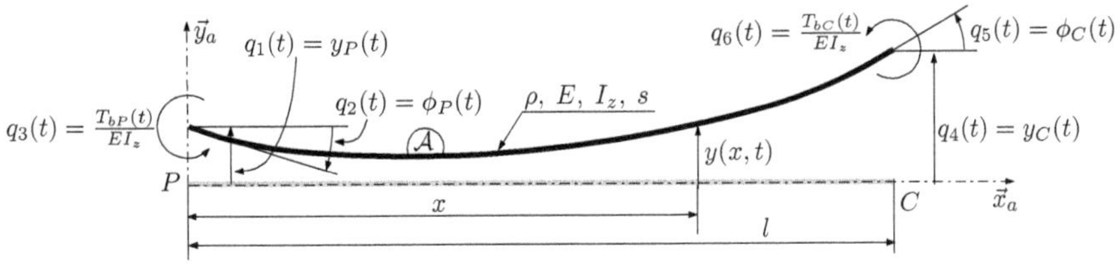

Fig. 3 Pure bending of the beam in the $(P, \mathbf{x}_a, \mathbf{y}_a)$ plane and definition of the six kinematic variables

The mass matrix derived from the super-element is:

$$\mathbf{M} = \frac{\rho s l}{55440}\begin{bmatrix} 21720 & 3732l & 281l^2 & 6000 & -1812l & 181l^2 \\ 3732l & 832l^2 & 69l^3 & 1812l & -532l^2 & 52l^3 \\ 281l^2 & 69l^3 & 6l^4 & 181l^2 & -52l^3 & 5l^4 \\ 6000 & 1812l & 181l^2 & 21720 & -3732l & 281l^2 \\ -1812l & -532l^2 & -52l^3 & -3732l & 832l^2 & -69l^3 \\ 181l^2 & 52l^3 & 5l^4 & 281l^2 & -69l^3 & 6l^4 \end{bmatrix} \tag{7}$$

The expression for elastic energy is:

$$V = \frac{1}{2}\int_0^l EI_z\left(\frac{\partial^2 y}{\partial x^2}(x,t)\right)^2 dx = \frac{1}{2}q^T\mathbf{K}q\,, \tag{8}$$

where the (i,j)-th element of the stifness matrix $\mathbf{K}$ reads:

$$\mathbf{K}(i,j) = \int_0^l EI_z\frac{d^2\varphi_i(x)}{dx^2}\frac{d^2\varphi_j(x)}{dx^2}dx$$

The stiffness matrix derived from the super-element is:

$$\mathbf{K} = \frac{EI_z}{70l^3}\begin{bmatrix} 1200 & 600l & 30l^2 & -1200 & 600l & -30l^2 \\ 600l & 384l^2 & 22l^3 & -600l & 216l^2 & -8l^3 \\ 30l^2 & 22l^3 & 6l^4 & -30l^2 & 8l^3 & l^4 \\ -1200 & -600l & -30l^2 & 1200 & -600l & 30l^2 \\ 600l & 216l^2 & 8l^3 & -600l & 384l^2 & -22l^3 \\ -30l^2 & -8l^3 & l^4 & 30l^2 & -22l^3 & 6l^4 \end{bmatrix} \tag{9}$$

From equation (9) the rank of matrix $\mathbf{K}$ is found to be 4, indicating that the Eigen value analysis of $\mathbf{M}^{-1}\mathbf{K}$ returns two rigid modes, corresponding to the translation of the beam along $\mathbf{y}_a$ axis and the rotation around the $\mathbf{z}_a$ axis, and four flexible modes whose frequencies are given by the square roots of the four non-zero Eigen values. The isolation of the kinematic variables corresponding to the rigid and flexible modes is achieved by change of variables, $q(t) = \mathbf{P}\tilde{q}(t)$, where

$$\tilde{q}(t) = [y_P(t),\ \phi_P(t),\ \frac{T_{bP}(t)}{EI_z},\ y_C(t) - y_P(t) - l\phi_P(t),\ \phi_C(t) - \phi_P(t),\ \frac{T_{bC}(t)}{EI_z}]^T$$

is the vector of relative kinematic parameters that includes the deformation at point P (deflection, slope and curvature) and the relative deformation at C (deflection, slope and curvature) with respect to P, see also Fig. 4. Applying the change of

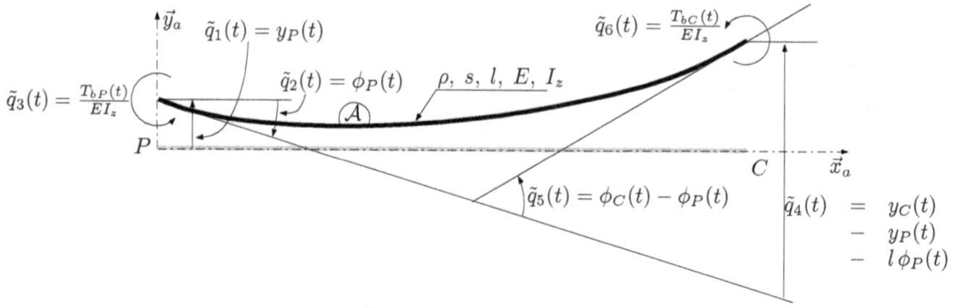

Fig. 4 Pure bending of the beam in the $(P, \mathbf{x}_a, \mathbf{y}_a)$ plane and definition of the six relative kinematic variables

variables in equations (6) and (8), the modified mass and stiffness matrices are:

$$\widetilde{\mathbf{M}} = \mathbf{P}^T \mathbf{M} \mathbf{P} = \frac{\rho s l}{55440} \begin{bmatrix} 55440 & 27720\,l & 462\,l^2 & 27720 & -5544\,l & 462\,l^2 \\ 27720\,l & 18480\,l^2 & 198\,l^3 & 19800\,l & -3432\,l^2 & 264\,l^3 \\ 462\,l^2 & 198\,l^3 & 6\,l^4 & 181\,l^2 & -52\,l^3 & 5\,l^4 \\ 27720 & 19800\,l & 181\,l^2 & 21720 & -3732\,l & 281\,l^2 \\ -5544\,l & -3432\,l^2 & -52\,l^3 & -3732\,l & 832\,l^2 & -69\,l^3 \\ 462\,l^2 & 264\,l^3 & 5\,l^4 & 281\,l^2 & -69\,l^3 & 6\,l^4 \end{bmatrix} \tag{10}$$

$$\widetilde{\mathbf{K}} = \mathbf{P}^T \mathbf{K} \mathbf{P} = \frac{EI_z}{70\,l^3} \begin{bmatrix} 0 & 0 & 0 & 0 & 0 & 0 \\ 0 & 0 & 0 & 0 & 0 & 0 \\ 0 & 0 & 6\,l^4 & -30\,l^2 & 8\,l^3 & l^4 \\ 0 & 0 & -30\,l^2 & 1200 & -600\,l & 30\,l^2 \\ 0 & 0 & 8\,l^3 & -600\,l & 384\,l^2 & -22\,l^3 \\ 0 & 0 & l^4 & 30\,l^2 & -22\,l^3 & 6\,l^4 \end{bmatrix} \tag{11}$$

The acceleration of absolute deflection and slope at points P and C are related to the vector $\ddot{\tilde{q}}$ as follows:

$$\ddot{\tilde{q}}_{(1)} = \ddot{y}_P = \mathbf{y}_a.\vec{a}_P \quad \text{and} \quad \ddot{\tilde{q}}_{(2)} = \ddot{\phi}_P = \mathbf{z}_a.\vec{\dot{\omega}}_P$$

and

$$\begin{bmatrix} \ddot{y}_C \\ \ddot{\phi}_C \end{bmatrix} = \begin{bmatrix} \mathbf{y}_a.\vec{a}_C \\ \mathbf{z}_a.\vec{\dot{\omega}}_C \end{bmatrix} = \begin{bmatrix} \underbrace{\begin{matrix} 1 & l \\ 0 & 1 \end{matrix}}_{\tau} & \underbrace{\begin{matrix} 0 & 1 & 0 & 0 \\ 0 & 0 & 1 & 0 \end{matrix}}_{\Phi} \end{bmatrix} \ddot{\tilde{q}}$$

$$= \tau \begin{bmatrix} \mathbf{y}_a.\vec{a}_P \\ \mathbf{z}_a.\vec{\dot{\omega}}_P \end{bmatrix} + \boldsymbol{\Phi}\, \ddot{\tilde{q}}_{(3:6)}\,. \tag{12}$$

where, τ in equation (12) is the coordinate transformation matrix that projects linear and angular accelerations at point P viz. $\ddot{\tilde{q}}_{1:2}$, on point C and Φ is the matrix that maps the double derivatives of relative kinematic parameters i.e. $\ddot{\tilde{q}}_{3:6}$, to the linear and angular accelerations at point C. As the damping matrix $\widetilde{\mathbf{D}}_{(3:6,3:6)}$ is not known at this stage, the undamped equation of motion for the parameters $\tilde{q}$ is considered:

$$\begin{bmatrix} \widetilde{\mathbf{M}}_{(1:2,1:2)} & \widetilde{\mathbf{M}}^T_{(3:6,1:2)} \\ \widetilde{\mathbf{M}}_{(3:6,1:2)} & \widetilde{\mathbf{M}}_{(3:6,3:6)} \end{bmatrix} \begin{bmatrix} \begin{bmatrix} \mathbf{y}_a.\overrightarrow{\ddot{d}}_P \\ \mathbf{z}_a.\overrightarrow{\dot{\omega}}_P \end{bmatrix} \\ \ddot{\tilde{q}}_{(3:6)} \end{bmatrix} + \begin{bmatrix} 0_{2\times 2} & 0_{2\times 4} \\ 0_{4\times 2} & \widetilde{\mathbf{K}}_{(3:6,3:6)} \end{bmatrix} \begin{bmatrix} \begin{bmatrix} * \\ * \end{bmatrix} \\ \tilde{q}_{(3:6)} \end{bmatrix} =$$

$$\begin{bmatrix} -\begin{bmatrix} \mathbf{y}_a.\overrightarrow{F}_{\mathscr{A}/\mathscr{P}} \\ \mathbf{z}_a.\overrightarrow{T}_{\mathscr{A}/\mathscr{P},P} \end{bmatrix} + \tau^T \begin{bmatrix} \mathbf{y}_a.\overrightarrow{F}_{\mathscr{C}/\mathscr{A}} \\ \mathbf{z}_a.\overrightarrow{T}_{\mathscr{C}/\mathscr{A},C} \end{bmatrix} \\ \Phi^T \begin{bmatrix} \mathbf{y}_a.\overrightarrow{F}_{\mathscr{C}/\mathscr{A}} \\ \mathbf{z}_a.\overrightarrow{T}_{\mathscr{C}/\mathscr{A},C} \end{bmatrix} \end{bmatrix} \tag{13}$$

Equations (12) and (13) together can be rearranged in the form of the following state-space representation of the 2-port model of the beam restricted to bending in the $(P, \mathbf{x}_a, \mathbf{y}_a)$ plane:

$$\begin{bmatrix} \dot{\tilde{q}}_{(3:6)} \\ \ddot{\tilde{q}}_{(3:6)} \\ \hline \begin{bmatrix} \mathbf{y}_a.\overrightarrow{\ddot{d}}_C \\ \mathbf{z}_a.\overrightarrow{\dot{\omega}}_C \end{bmatrix} \\ \begin{bmatrix} \mathbf{y}_a.\overrightarrow{F}_{\mathscr{A}/\mathscr{P}} \\ \mathbf{z}_a.\overrightarrow{T}_{\mathscr{A}/\mathscr{P},P} \end{bmatrix} \end{bmatrix} = \left[\begin{array}{c|c} A_{Ty,Rz} & B_{Ty,Rz} \\ C_{Ty,Rz} & D_{Ty,Rz} \end{array}\right] \begin{bmatrix} \tilde{q}_{(3:6)} \\ \dot{\tilde{q}}_{(3:6)} \\ \hline \begin{bmatrix} \mathbf{y}_a.\overrightarrow{F}_{\mathscr{C}/\mathscr{A}} \\ \mathbf{z}_a.\overrightarrow{T}_{\mathscr{C}/\mathscr{A},C} \end{bmatrix} \\ \begin{bmatrix} \mathbf{y}_a.\overrightarrow{\ddot{d}}_P \\ \mathbf{z}_a.\overrightarrow{\dot{\omega}}_P \end{bmatrix} \end{bmatrix} \tag{14}$$

where, having taken into account an arbitrary damping ratio ξ on all the four flexible modes,

$$A_{Ty,Rz} = \begin{bmatrix} 0_{4\times 4} & I_4 \\ -\widetilde{\mathbf{M}}^{-1}_{(3:6,3:6)} \widetilde{\mathbf{K}}_{(3:6,3:6)} & -2\xi\, \mathbf{V}\mathrm{diag}(\omega_i)\mathbf{V}^{-1} \end{bmatrix},$$

$$B_{Ty,Rz} = \begin{bmatrix} 0_{4\times 2} & 0_{4\times 2} \\ \widetilde{\mathbf{M}}^{-1}_{(3:6,3:6)} \Phi^T & -\widetilde{\mathbf{M}}^{-1}_{(3:6,3:6)} \widetilde{\mathbf{M}}_{(3:6,1:2)} \end{bmatrix},$$

$$C_{Ty,Rz} = \begin{bmatrix} \Phi \\ -\widetilde{\mathbf{M}}^T_{(3:6,1:2)} \end{bmatrix} A_{Ty,Rz}(5:8,:),$$

$$D_{Ty,Rz} = \begin{bmatrix} \Phi \widetilde{\mathbf{M}}^{-1}_{(3:6,3:6)} \Phi^T & \left(\tau - \Phi \widetilde{\mathbf{M}}^{-1}_{(3:6,3:6)} \widetilde{\mathbf{M}}_{(3:6,1:2)}\right) \\ \left(\tau - \Phi \widetilde{\mathbf{M}}^{-1}_{(3:6,3:6)} \widetilde{\mathbf{M}}_{(3:6,1:2)}\right)^T & -\widetilde{\mathbf{M}}_{(1:2,1:2)} + \widetilde{\mathbf{M}}^T_{(3:6,1:2)} \widetilde{\mathbf{M}}^{-1}_{(3:6,3:6)} \widetilde{\mathbf{M}}_{(3:6,1:2)} \end{bmatrix},$$

and, $\mathbf{V}$ and $\mathrm{diag}(\omega_i)$ are given by the eigenvalue decomposition of $\widetilde{\mathbf{M}}^{-1}_{(3:6,3:6)} \widetilde{\mathbf{K}}_{(3:6,3:6)}$:

$$\mathbf{V}^{-1} \widetilde{\mathbf{M}}^{-1}_{(3:6,3:6)} \widetilde{\mathbf{K}}_{(3:6,3:6)} \mathbf{V} = \mathrm{diag}(\omega_i^2)\,.$$

This also implies that

$$\mathbf{V}^{-1}\widetilde{\mathbf{M}}^{-1}_{(3:6,3:6)}\widetilde{\mathbf{D}}_{(3:6,3:6)}\mathbf{V} = \mathrm{diag}(2\xi\,\omega_i)\;.$$

2.2.2 Pure Bending in the $(P, \mathbf{x}_a, \mathbf{z}_a)$ Plane

The 2-port model of a beam restricted to pure bending in the $(P, \mathbf{x}_a, \mathbf{z}_a)$ plane, governs the dynamic behavior in translation along $\mathbf{z}_a$ axis and in rotation around $\mathbf{y}_a$ axis and is given by the following state-space representation :

$$\begin{bmatrix} \dot{\tilde{p}}_{(3:6)} \\ \ddot{\tilde{p}}_{(3:6)} \\ \hline \begin{bmatrix} \mathbf{z}_a.\ddot{\vec{d}}_C \\ \mathbf{y}_a.\dot{\vec{\omega}}_C \end{bmatrix} \\ \begin{bmatrix} \mathbf{z}_a.\vec{F}_{\mathscr{A}/\mathscr{P}} \\ \mathbf{y}_a.\vec{T}_{\mathscr{A}/\mathscr{P},P} \end{bmatrix} \end{bmatrix} = \left[\begin{array}{c|c} A_{Tz,Ry} & B_{Tz,Ry} \\ C_{Tz,Ry} & D_{Tz,Ry} \end{array}\right] \begin{bmatrix} \tilde{p}_{(3:6)} \\ \dot{\tilde{p}}_{(3:6)} \\ \hline \begin{bmatrix} \mathbf{z}_a.\vec{F}_{\mathscr{C}/\mathscr{A}} \\ \mathbf{y}_a.\vec{T}_{\mathscr{C}/\mathscr{A},C} \end{bmatrix} \\ \begin{bmatrix} \mathbf{z}_a.\ddot{\vec{d}}_P \\ \mathbf{y}_a.\dot{\vec{\omega}}_P \end{bmatrix} \end{bmatrix} \tag{15}$$

where $\tilde{p}(t)$ is the vector of kinematic parameters, and the matrices $A_{Tz,Ry}$, $B_{Tz,Ry}$, $C_{Tz,Ry}$ and $D_{Tz,Ry}$ are obtained in a procedure similar to the one outlined in Sect. 2.2.1.[1]

2.2.3 Torsion due to Transverse Shear around the $(P, \mathbf{x}_a)$ Axis

The first flexible mode of torsion of the beam $\mathscr{A}$ around the $(P, \mathbf{x}_a)$ axis is modelled by assuming two co-axial inertias, each having a value $\frac{J_x}{2}$ connected together by a torsional spring of stiffness k_x, where $J_x = \rho l I_{px}$ is the mass moment of inertia of the beam around the $\mathbf{x}_a$ axis and and $k_x = \frac{GI_{px}}{l}$ is the torsional stiffness of the beam. The two-port model for the beam torsional dynamic, $R_x(\mathrm{s})$ is derived in analogy with that of two masses connected by a longitudinal spring [2], resulting in the following state-space representation:

$$\begin{bmatrix} \delta\dot{\theta} \\ \delta\ddot{\theta} \\ \hline \mathbf{x}_a.\dot{\vec{\omega}}_C \\ \mathbf{x}_a.\vec{T}_{\mathscr{A}/\mathscr{P},P} \end{bmatrix} = \left[\begin{array}{cc|cc} 0 & 1 & 0 & 0 \\ -\frac{2k_x}{J_x} & -2\xi\sqrt{\frac{2k_x}{J_x}} & \frac{2}{J_x} & -1 \\ \hline -\frac{2k_x}{J_x} & -2\xi\sqrt{\frac{2k_x}{J_x}} & \frac{2}{J_x} & 0 \\ k_x & \xi\sqrt{2J_x k_x} & 0 & -\frac{J_x}{2} \end{array}\right] \begin{bmatrix} \delta\theta \\ \delta\dot{\theta} \\ \hline \mathbf{x}_a.\vec{T}_{\mathscr{C}/\mathscr{A},C} \\ \mathbf{x}_a.\dot{\vec{\omega}}_P \end{bmatrix}$$

$$= \left[\begin{array}{c|c} A_{Rx} & B_{Rx} \\ C_{Rx} & D_{Rx} \end{array}\right] \begin{bmatrix} \delta\theta \\ \delta\dot{\theta} \\ \hline \mathbf{x}_a.\vec{T}_{\mathscr{C}/\mathscr{A},C} \\ \mathbf{x}_a.\dot{\vec{\omega}}_P \end{bmatrix} \tag{16}$$

[1] In equation (11), the term I_z is replaced by I_y and the 4×4 transfer matrix is pre-multiplied and post-multiplied by $\mathrm{diag}([1,\ -1,\ 1,\ -1])$ (since $\mathbf{x}\wedge\mathbf{z} = -\mathbf{y}$), prior to the final form $T_z R_y(\mathrm{s})$ defined by equation (15).

where $\delta\theta = \int(\mathbf{x}_a.\vec{\omega}_C - \mathbf{x}_a.\vec{\omega}_P)dt$, and an arbitrary damping ratio ξ is considered for the flexible mode.

2.2.4 Rigid Body Translation along the $(P, \mathbf{x}_a)$ Axis

When the beam is assumed to be rigid regarding its translation along the longitudinal axis, $(P, \mathbf{x}_a)$, the resulting two-port model $T_x(\mathrm{s})$ is static, as shown below:

$$\begin{bmatrix} \mathbf{x}_a.\vec{a}_C \\ \mathbf{x}_a.\vec{F}_{\mathscr{A}/\mathscr{P}} \end{bmatrix} = \begin{bmatrix} 0 & 1 \\ 1 & -\rho s l \end{bmatrix} \begin{bmatrix} \mathbf{x}_a.\vec{F}_{\mathscr{C}/\mathscr{A}} \\ \mathbf{x}_a.\vec{a}_P \end{bmatrix} = D_{Tx} \begin{bmatrix} \mathbf{x}_a.\vec{F}_{\mathscr{C}/\mathscr{A}} \\ \mathbf{x}_a.\vec{a}_P \end{bmatrix} \quad (17)$$

2.2.5 Two-port Model of Uniform Beam with Integrated Dynamics

Each of the dynamic models, $T_yR_z(\mathrm{s})$, $T_zR_y(\mathrm{s})$, $R_x(\mathrm{s})$ and $T_x(\mathrm{s})$, has two channels - the lower channel being the direct dynamic model at point P i.e. the transfer between the interaction components along the corresponding DOFs and the acceleration of those DOFs at P and the upper channel being the inverse dynamic model at point C i.e. the transfer between the acceleration components of corresponding DOFs and the interaction components at those DOFs at C. All the above dynamics are combined and thereby a dynamic model $[M_{P,C}^{\mathscr{A}}]_{\mathscr{R}_a}(\mathrm{s})$ has been derived and it follows the relation :

$$\begin{bmatrix} [\ddot{\mathrm{x}}_C]_{\mathscr{R}_a} \\ [\mathrm{F}_{\mathscr{A}/\mathscr{P},P}]_{\mathscr{R}_a} \end{bmatrix} = \underbrace{\mathbf{T}^T \begin{bmatrix} T_yR_z(\mathrm{s}) & & & \\ & T_zR_y(\mathrm{s}) & & \\ & & R_x(\mathrm{s}) & \\ & & & T_x(\mathrm{s})) \end{bmatrix} \mathbf{T}}_{[M_{P,C}^{\mathscr{A}}]_{\mathscr{R}_a}(\mathrm{s})} \begin{bmatrix} [\mathrm{F}_{\mathscr{C}/\mathscr{A},C}]_{\mathscr{R}_a} \\ [\ddot{\mathrm{x}}_P]_{\mathscr{R}_a} \end{bmatrix} \quad (18)$$

where $\mathbf{T}$ is the permutation matrix that maps every force and acceleration component in the input and output vectors to the corresponding dynamic model.

3 Application and Validation of Sub-structuring Techniques on the Biomass Study Case

The sub-structuring techniques presented in Sect. 2 have been implemented in SIMULINK in the form of one-port or two-port blocks, so as to include the contribution of flexible appendages, of arbitrary shape or a uniform beam, to the overall dynamics of a structure. The rigid and flexible modes resulting from the modal synthesis of interconnected appendages must be representative of the dynamics of the larger structure itself. The mathematical model underlying the one-port and two-port blocks can be defined by importing the sub-structure-specific mass and modal characteristic parameters like $D_{P_0}^{\mathscr{A}}$, ω_i, ξ_i, L_P and Φ_C from the analysis ouput data of a FEM software like NASTRAN. This is achieved through a SIMULINK-NASTRAN custom interface established for both blocks that will fetch from the user, information including the sub-structure FEM output file names, NASTRAN nodal IDs for connection points, modal truncation, modal damping ratio and the orientation of $\mathscr{R}_a$ with respect to the parent reference frame. Similarly, for the super-element model in equation (18) to be defined, all the terms of vector Θ must be specified at the interface of the corresponding block in SIMULINK. Also blocks that compute uncertain linear state-space models due to uncertainties in the characteristic parameters defining the one-port and two-port models, have been created. The uncertain super-element block, that takes into account uncertainties in the geometrical and mechanical parameters defined by the user, is of prime interest in the co-design step as it permits the global dynamic model to be parametrized in terms of uncertainties in localized sections of the structure.

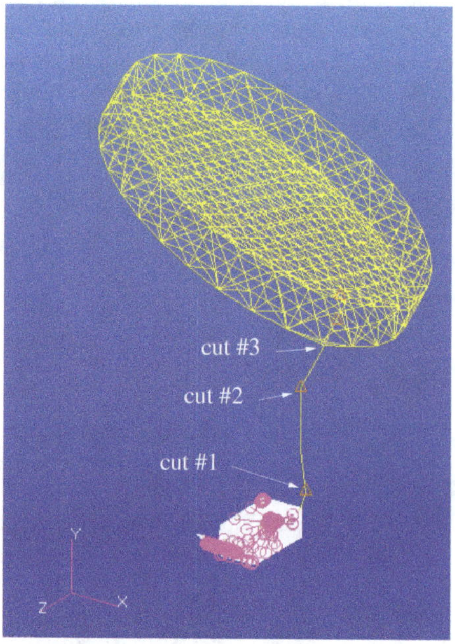

Fig. 5 Biomass satellite - the rigid main body or bus (bottom) is shown in white; the antenna, with arm (attached to bus) and reflector, is shown in yellow. Courtesy by Thales Alenia Space, Italy

The integrated FEM assembly of the Biomass satellite is shown in Fig. 5. The bus $\mathscr{MB}$ is considered rigid, while the flexible antenna $\mathscr{A}_1$ (arm and reflector together) is broken down into a number of combinations of sub-structures, and in each case the applicability of one of the three newly developed sub-structure models has been tested. Three sub-structure combinations to model the antenna are considered:

- single appendage configuration (denoted $\mathscr{A}_1^1$), represented by a 1-port block,
- chain of two appendages (denoted $\mathscr{A}_1^2$), with a cut in the arm portion of antenna, represented by a 2-port block (arm) and a 1-port block (reflector),
- chain of four appendages (denoted $\mathscr{A}_1^3$), with three cuts in the arm (see Fig. 5): the first portion is modelled by a 2-port block, the second and third portions are

uniform beam elements represented by 2 super-element blocks, followed by the portion with the reflector modelled by a 1-port block.

The linear inverse dynamic model of the satellite $[M_B^{\mathscr{M}\mathscr{B}+\mathscr{A}_1}]^{-1}$, between the angular acceleration vector at the bus center of gravity B, $\ddot{x}_B(4:6)$ along the bus reference frame $\mathscr{R}_B$ (output) and a torque vector $F_{ext,B}(4:6)$ along $\mathscr{R}_B$ (input), is considered. The frequency-domain response for this model, computed using the *Satellite Dynamic Toolbox (SDT)* [9], is compared with that computed from the different sub-structure combinations for the antenna.

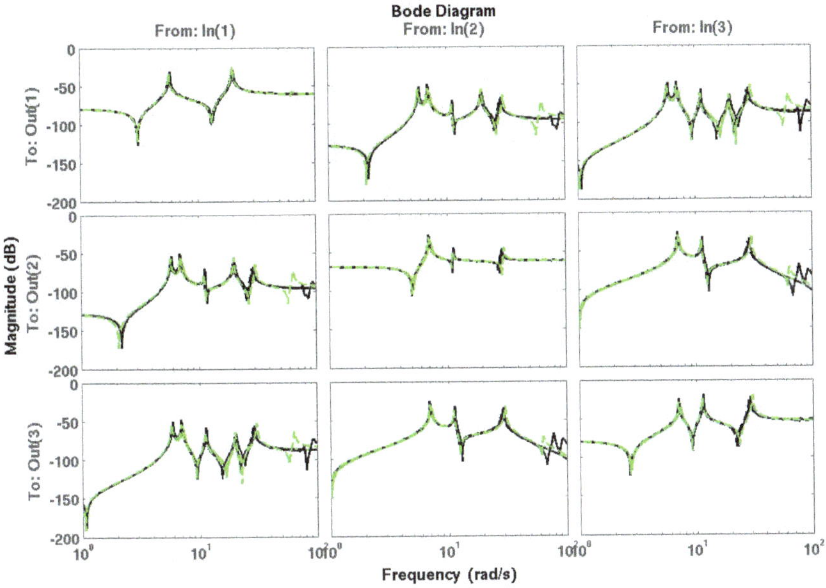

Fig. 6 Frequency-domain responses (magnitude plots) of inverse dynamic models: $[M_B^{\mathscr{M}\mathscr{B}+\mathscr{A}_1}]^{-1}$ (solid green), $[M_B^{\mathscr{M}\mathscr{B}+\mathscr{A}_1^1}]^{-1}$ (dashed black), $[M_B^{\mathscr{M}\mathscr{B}+\mathscr{A}_1^2}]^{-1}$ (dashed green) and $[M_B^{\mathscr{M}\mathscr{B}+\mathscr{A}_1^3}]^{-1}$ (solid black).

Fig. 6 shows a perfect coincidence of the magnitude of Bode plots for the *SDT* model $[M_B^{\mathscr{M}\mathscr{B}+\mathscr{A}_1}]^{-1}$ and the model $[M_B^{\mathscr{M}\mathscr{B}+\mathscr{A}_1^1}]^{-1}$ for the first combination, while there is a very close match with the frequency response when the antenna is modelled as per the second combination ($\mathscr{A}_1^2$) or third combination ($\mathscr{A}_1^3$), till 40 rad/s.

These plots serve to validate the substructuring techniques including single port, double port and super-element models. The mismatch of frequency responses at higher frequencies can be explained by the numerical limit of maximum modal frequency for every substructure posed by the FEM software and the user-defined truncation in SIMULINK. The number of modes considered for the sub-structures should be higher than a minimum value, especially for intermediate appendages, for a more accurate modal synthesis. The torsion modes of the arm alone are at very high frequencies but are driven to very low frequencies when the arm is loaded by the reflector. So these modes appear when the model of the arm and reflector considered as a single appendage (*SDT* or single port model) is truncated, but are ignored when the model of the arm alone (two-port mdoel) is truncated.

In the perspective of co-design, the portion of the arm represented by the two super-element blocks of same section and material properties, but of different lengths l_1 and l_2 in the third sub-structure combination, is assumed to be of a hollow square cross section with thickness t and mean perimeter p (see also Fig. 7), so that the other geometric parameters of the section are functions of the two parameters t and p:

$$s = t\,p \tag{19}$$

$$I_y = \frac{s(\frac{p^2}{8}+2t^2)}{12} = I_z \tag{20}$$

$$I_{px} = 0.125\,s(\frac{p^2}{8}+2t^2) \tag{21}$$

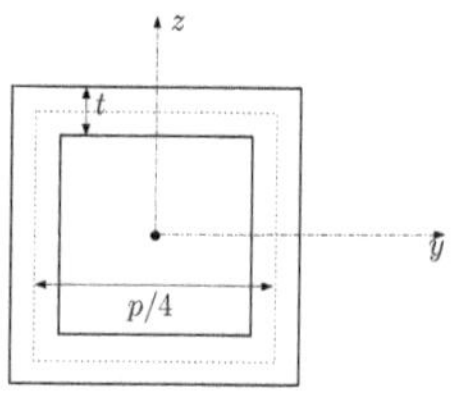

Fig. 7 Parameterization of a square tube section with thickness t and mean perimeter p.

4 Structural-controller Co-design

The primary goal of co-design is to design a reduced order H_∞ attitude controller that will ensure the pointing capability of the Biomass satellite as per the desired performance requirements, and simultaneously optimize the satellite physical parameters and in particular, reduce the structural mass of the satellite components, which will be treated as a secondary criterion. This is a one-step optimization problem that can be solved in MATLAB using the `systune` command [10] where the first criterion is treated as a hard constraint that must be mandatorily met by the solution for decision variables, while the second objective is treated as a soft constraint that will be best satisfied at the end of co-design. However, prior to this step, a H_∞ standard problem has been considered for the design of an attitude controller on the nominal inverse rotational dynamic model of the satellite augmented with sensor and actuator models, without any geometrical uncertainty [6]. This controller is then used to initialize the search while tuning the uncertain geometrical parameters along with the controller parameters, during codesign.

4.1 Satellite Dynamic Model and Avionics

In addition to the antenna $\mathscr{A}_1$, the actual satellite design consists of a flexible solar array $\mathscr{A}_2$ attached to the rigid bus $\mathscr{MB}$. The inverse dynamic model $[M_B^{\mathscr{MB}+\mathscr{A}_1+\mathscr{A}_2}]^{-1}$ of the satellite between a dual force vector input $F_{\mathscr{A}_1/\mathscr{MB},B}$, and a dual acceleration vector ouput $\ddot{x}_B$ at the bus center of gravity B is computed for designing an attitude control law. The inverse dynamic model of the main body and the solar array $[M_B^{\mathscr{MB}+\mathscr{A}_2}]^{-1}(s)$ i.e. the transfer matrix between the dual vectors $F_{ext,B}$ and $\ddot{x}_B$ is obtained using the *Satellite Dynamic Toolbox*. The appendage $\mathscr{A}_1$ is modelled as per the third sub-structure combination ($\mathscr{A}_1^3$) in Sect. 3, with a $\pm 90\%$ variation allowed for the parameters p and t of two adjacent uniform beam portions of the arm, in between the bus and the reflector, that are represented by 2 super-elements $\mathscr{E}_1$ and $\mathscr{E}_2$. The model of a super-element (18) being analytical, it is thus possible to take variations δp and δt on parameters p and t in $\mathscr{E}_1$ and super-elements $\mathscr{E}_2$. Let $\Delta = [\delta p, \delta t]^T$ be the vector of the varying parameters.

The avionics include a star sensor (SST), gyrometers (GYRO) and reaction wheel actuator (RWA) assembly that are used to measure the satellite attitude vector θ, attitude rate vector ω and apply the control torque u on the bus, respectively. They are modelled as low pass filters of first order on the three axes. In Fig. 8, the 6×3 open loop transfer between control input u and measurement output y is denoted by $G(s, \Delta)$ and will be used in the next section for attitude control design. The total inertia of the satellite around the three axes are denoted J_X, J_Y and J_Z.

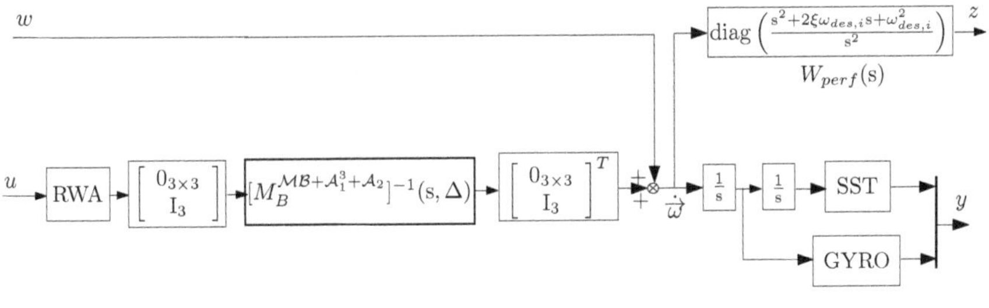

Fig. 8 H_∞ problem $P(s, \Delta)$ for performance (pointing) requirement based on the acceleration sensitivity function.

4.2 Performance Requirements and H_∞ Control Design

Low frequency disturbances (of the order of the orbital frequency of the satellite around the Earth) due to gravity gradient and solar pressure are inevitable in spacecraft dynamics and these disturbances tend to alter the angular orientation of the satellite. An attitude controller designed eventually must ensure that the steady-state error in attitude angles are within a set of limit values $\Delta\theta_i$ (refer Table 1), when constant values of worst-case disturbing torques T_i are applied on each axis $i = X,Y,Z$ of the bus reference frame $\mathscr{R}_b$.

Table 1 Worst-case disturbing torques, pointing requirements, total inertia and required bandwidth for each of the bus reference frame axes

i	X	Y	Z
$T_i^{pert}\,(Nm)$	1.4×10^{-2}	2×10^{-4}	3.9×10^{-3}
$\Delta\theta_i(deg)$	0.03	0.06	0.06
$J_i\,(Kgm^2)$	9287.7	2895.2	8728.4
$\omega_{des,i} = \sqrt{\gamma_{obj}\frac{T_i^{pert}}{J_i\Delta\theta_i}}\,(rd/s)$	0.0657	0.0099	0.0253

In [6], a H_∞ procedure was proposed to design a decentralized, low-order controller $K(\mathrm{s}) = \mathrm{diag}(K_X(\mathrm{s}), K_Y(\mathrm{s}), K_Z(\mathrm{s}))$, so that:

- the pointing requirements for each axis and for the nominal parametric configuration ($\Delta = 0$) are met if the controller $K(\mathrm{s})$ satisfies the H_∞ constraint:

$$\|F_l(P(\mathrm{s},0), K(\mathrm{s})\|_\infty \leq \gamma_{obj}$$

  where $P(\mathrm{s},\Delta)$ is the standard problem based on the acceleration sensitivity function (see Fig. 8). The weighting function $W_{perf}(\mathrm{s})$ is directly computed from pointing requirements and distrurbance magnitudes (see Tab. 1). γ_{obj} is a performance margin (typically $\gamma_{obj} = 1.5$) and $\xi = 0.7$,
- each controller $K_i(\mathrm{s})$ is stable and fit a first order roll-off template $W_{u_i}(\mathrm{s})$ beyond the frequency $\omega_{cut} = 1\,rd/s$ if:

$$\|\frac{1}{W_{u_i}(\mathrm{s})}K(\mathrm{s})\|_\infty \leq \gamma_{obj}, \text{with: } W_{u_i} = \frac{2}{\gamma_{obj}}J_i\omega_{des,i}\omega_{cut}\frac{1+\frac{s}{1000\omega_{cut}}}{s},\ i = X,Y,Z$$

Prior to co-design, a $3\times$ 2-nd order H_∞ controller $K_{nom}(\mathrm{s})$ has been designed on the nominal model ($\Delta = 0$) using such a procedure:

$$K_{nom}(\mathrm{s}) = \arg \min_{K_X(\mathrm{s}),K_Y(\mathrm{s}),K_Z(\mathrm{s})\in\mathscr{S}_2^{2\times1}} \gamma(\Delta, K(\mathrm{s}))$$

where $\mathscr{S}_{nK}^{p\times m}$ is the set of stable m inputs, p outputs, n_K^{th} order minimal linear systems and:

$$\gamma(\Delta, K(\mathrm{s})) = \max\left(\|F_l(P(\mathrm{s},\Delta), K(\mathrm{s}))\|_\infty, \|\frac{1}{W_{u_i}(\mathrm{s})}K_i(\mathrm{s}\|_\infty,\ i = X,Y,Z\right). \quad (22)$$

This controller ensures a stable closed-loop system with good stability margins for the nominal open loop transfer function $-K_{nom}(\mathrm{s})G(\mathrm{s},0)$ (see Table 2).

4.3 Co-design on Uncertain Model

As the analytical two-port model of super-element (eq. (18)) involves only polynomial or fractionnal forms of the parameters t and p, the LFT formalism can be used to split the transfer $[M_B^{\mathscr{M}\mathscr{B}+\mathscr{A}_1^3+\mathscr{A}_2}]^{-1}_{4:6,4:6}(\mathrm{s},\Delta)$ into two units: a nominal model $M_0(\mathrm{s})$ and an uncertainty block Δ_{pt} containing $n_{occ,p}=58$ and $n_{occ,t}=53$ occurences of varying parameters δp and δt along its diagonal. The uncertainty limits of Δ_{pt} is normalized so that the relative variation is always between ± 1 with a nominal value of 0. The nominal model $M_0(\mathrm{s})$ has 114 inputs and outputs, with the first 111 pairs being connected to Δ_{pt} and the remaining three being the torque-acceleration pairs pertaining to the inverse dynamic model (see also Fig. 9). In addition to the three second order controllers $K(\mathrm{s})$, two more decision variables, i.e. the uncertainties δp and δt are considered for the augmented H_∞ standard problem, $P_{cd}(\mathrm{s})$ shown in Fig. 10. The criterion of mass reduction can be implemented by minimizing the norm $\|(1+0.9\,\delta p)(1+0.9\,\delta t)\|_\infty$ which is proportional to the new mass after co-design. However the mass reduction is not limited by any numerical limit and hence it becomes a soft constraint with respect to the criterion to meet performance and roll-off requirements (hard constraints). To summarize, the final control co-design $\{K_{cd}(\mathrm{s}),\delta_{p,cd},\delta_{t,cd}\}$, where $K_{cd}(\mathrm{s})=\mathrm{diag}(K_{X,cd}(\mathrm{s}),K_{Y,cd}(\mathrm{s}),K_{Z,cd}(\mathrm{s}))$, is such that:

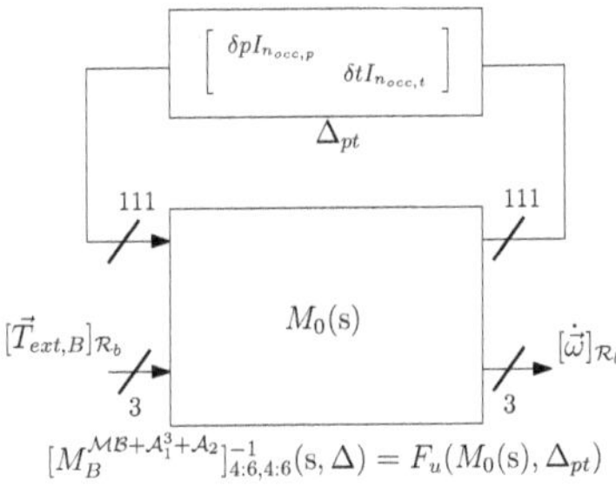

Fig. 9 LFT representation of the inverse dynamic model at bus c.g. with structural parametric uncertainties.

$$\{K_{cd}(\mathrm{s}),\delta p_{cd},\delta t_{cd}\}=\arg\min_{K_X(\mathrm{s}),K_Y(\mathrm{s}),K_Z(\mathrm{s})\in\mathscr{S}_{n_K}^{2\times 1},\delta p,\delta t}(\|(1+0.9\,\delta p)(1+0.9\,\delta t)\|_\infty)$$

$$\text{subject to:}\quad \max(\|F_l(P_{cd}(\mathrm{s}),diag(\Delta_{pt},K_X(\mathrm{s}),K_Y(\mathrm{s}),K_Z(\mathrm{s})))\|_\infty,\ldots$$
$$\|\frac{1}{W_{u_X}(\mathrm{s})}K_X(\mathrm{s})\|_\infty,\|\frac{1}{W_{u_Y}(\mathrm{s})}K_Y(\mathrm{s})\|_\infty,\|\frac{1}{W_{u_Z}(\mathrm{s})}K_Z(\mathrm{s})\|_\infty)\leq\gamma_{obj}. \tag{23}$$

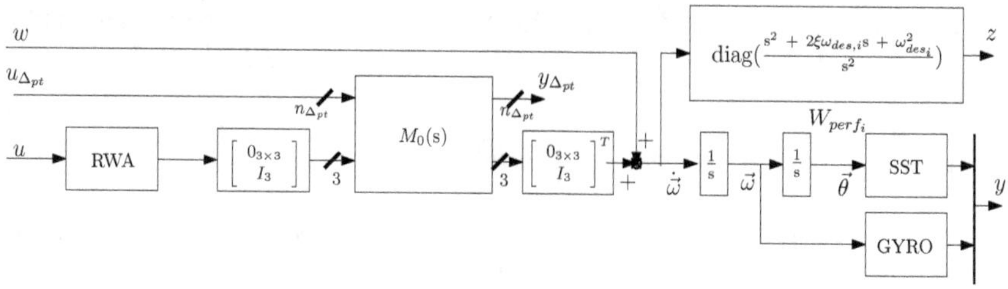

Fig. 10 H_∞ standard problem $P_{cd}(s)$ for codesign showing the control inputs and outputs of the controller as well as the uncertainty matrix Δ_{pt}

Table 2 Comparison of mass, stability and performance indices between the nominal design and the co-design for $\gamma_{obj} = 1.5$

Index/Margin/Metric	Nominal design model	Co-design model
Mass of the two beams of the arm	5.92 kg	1.28 kg
Closed loop LFT norm γ	1.11	1.459
Modulus margin	0.9054	0.6765
Gain margin	20.485 dB	9.8033 dB
Phase margin	$53.83\ ^o$	$39.5423\ ^o$

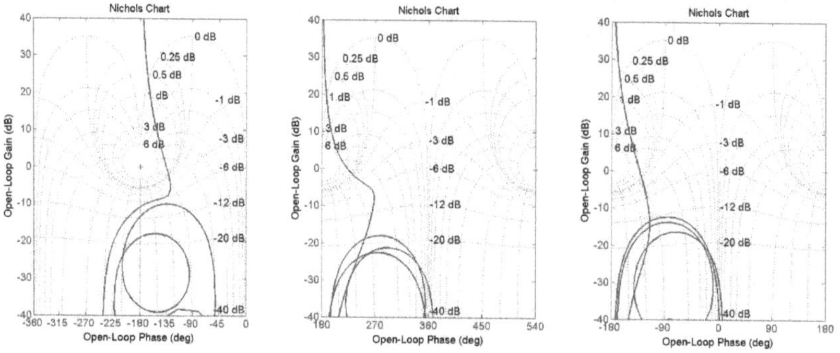

Fig. 11 Nichols plot of $-K_{cd}(s)G(s, [\delta p_{cd}, \delta t_{cd}]^T)$: X(left), Y(center), Z(right) axes

4.3.1 Results from Co-design

With the optimization search initialized from the previously designed H_∞ controller $K_{nom}(s)$, a new 2-nd order controller $K_{cd}(s)$ was found with a 17.23% increase in p, and 81.6% decrease in t. There is a -78.44% reduction of the total mass of the two beams from an original mass of $5.92kg$ at the cost of a small degradation (but still satisfying) of the performance index γ and stability margins (refer Table 2). From the Nichols plot of the open loop transfer $-K_{cd}(s)G(s, [\delta p_{cd}, \delta t_{cd}]^T)$ involving the modified dynamic model after codesign and the controller K_{cd}, the closed loop system is found to be stable (see Fig. 11). Fig.12 shows that the controller $K_{cd}(s)$ follows the template W_u and hence the roll-off requirements are met.

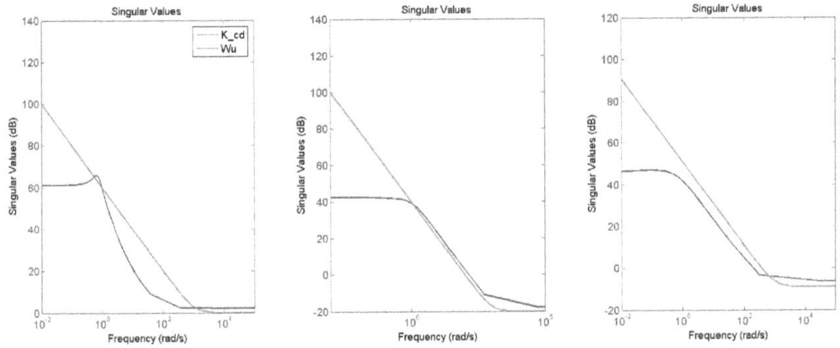

Fig. 12 Singular values vs. frequency for nominal controller K_{cd} (blue) and template W_u (green): X(left), Y(center), Z(right) axes

5 Conclusion

The single and double-port dynamic models used to represent terminal and intermediate sub-structures, and the analytical double-port model used to represent the dynamics of an intermediate uniform beam in an open mechanical chain, have been implemented in MATLAB environment through an interface with the sub-structure finite element model in NASTRAN. The application on Biomass satellite validates these new modelling tools. A particular sub-structure arrangement is used to parameterize the whole satellite model according to some physical design parameters, here: the perimeter and the thickness of the cross section of the boom linking the reflector to the main body. That is certainly the main contribution of this paper. These physical parameters are altered during the co-design so as to save mass with an acceptable performance degradation for the closed-loop system. Thus, mechanical design parameters and attitude control laws are simultaneously optimized to meet pointing requirements.

Future aspects of the study could include the extension of the sub-structuring approaches to model closed mechanical chain and the analysis of the robustness of the co-design model to uncertainties on other structural parameters of the appendages.

References

1. Alazard, D., Cumer, Ch., Tantawi, K.: Linear dynamic modeling of spacecraft with various flexible appendages and on-board angular momentums. In: 7th International ESA Conference on Guidance, Navigation & Control Systems, Tralee, Ireland (2008)
2. Alazard, D., Alvaro Perez, J., Loquen, T., Cumer, C.: Two-input two-output port model for mechanical systems, Submitted to SciTech, Gaylord Palms and Convention Center, Kissimmee, FL (2015)
3. Guy, N., Alazard, D.: Dynamic modeling and analysis of spacecraft with variable tilt of flexible appendages. ASME - Journal of Dynamic Systems Measurement and Control 136(2) (January 2014)

4. Toglia, C., Pavia, P., Campolo, G., Alazard, D., Loquen, T., de Plinval, H., Cumer, C., Casasco, M., Massotti, L.: Optimal Co-Design for Earth Observation Satellites with Flexible Appendages. In: AIAA Guidance Navigation and Control Conference, Boston (MS) (August 2013)
5. Loquen, T., de Plinval, H.: Attitude control of satellites with flexible appendages: a structured H_∞ control design. In: AIAA Guidance Navigation and Control Conference, Minneapolis, MN (August 2012)
6. Alazard, D., Loquen, T., de Plinval, H., Cumer, C.: Avionics/Control co-design for large flexible space structures. In: AIAA Guidance, Navigation and Control Conference, Boston (MS) (August 2013)
7. Girard, A., Roy, N.: Structure Dynamics in Industry, pp. 249-291. Wiley (2008)
8. Chrétien, J.P., Alazard, D.: Modélisation des chaînes polyarticulées flexible en vue de la commande: comparaison de trois techniques de mise en équations, Rapport final n. 2/7735.a, 2/7735.b, 2/7735.c - DERA (1991)
9. Alazard, D., Cumer, C.: Satellite Dynamic Toolbox: Principe, User Guide nd Tutorials (2014), `http://personnel.isae.fr/daniel-alazard/matlab-packages/satellite-dynamics-toolbox.html`
10. Apkarian, P.: Tuning controllers against multiple design requirements. In: American Control Conference (ACC), Washington D.C., pp. 3888–3893 (2013)

Probabilistic Collision Avoidance for Long-term Space Encounters via Risk Selection

Romain Serra, Denis Arzelier, Mioara Joldes, and Aude Rondepierre

Abstract. This paper deals with collision avoidance between two space objects involved in a long-term encounter, assuming Keplerian linearized dynamics. The primary object is an active spacecraft - able to perform propulsive maneuvers - originally set on a reference orbit. The secondary object - typically an orbital debris - is passive and represents a threat to the primary. The collision avoidance problem addressed in this paper aims at computing a fuel-optimal, finite sequence of impulsive maneuvers performed by the active spacecraft such that instantaneous collision probability remains below a given threshold over the encounter and that the primary object goes back to its reference trajectory at the end of the mission. Two successive relaxations are used to turn the original hard chance-constrained problem into a deterministic version that can be solved using mixed-integer linear programming solvers. An additional contribution is to propose a new algorithm to compute probabilities for 3-D Gaussian random variables to lie in Euclidean balls, enabling us to numerically validate the computed maneuvers by efficiently evaluating the resulting instantaneous collision probabilities.

Romain Serra
CNRS, LAAS, 7 avenue du Colonel Roche, F-31400 Toulouse,
Université de Toulouse, INSA, LAAS, 7 avenue du Colonel Roche, F-31400 Toulouse,
e-mail: serra@laas.fr

Denis Arzelier · Mioara Joldes
CNRS, LAAS, 7 avenue du Colonel Roche, F-31400 Toulouse,
Université de Toulouse, LAAS, 7 avenue du Colonel Roche, F-31400 Toulouse,
e-mail {arzelier, joldes}@laas.fr

Aude Rondepierre
Université de Toulouse, Institut de Mathématiques de Toulouse, INSA,
135 avenue de Rangueil, F-31400 Toulouse,
e-mail: aude.rondepierre@insa-toulouse.fr

J. Bordeneuve-Guibé et al. (eds.), *Advances in Aerospace Guidance, Navigation and Control*, DOI: 10.1007/978-3-319-17518-8_39

Introduction

During the last decades, the number of space debris around the Earth has been continuously and rapidly increasing. It now represents the majority of orbital objects and is a real threat for operational spacecraft. To prevent potential collisions on an active satellite, one solution consists in performing one or several evasive maneuvers when the predicted risk is too high according to mission requirements. These requirements also impose to end within given bounds relatively to the reference orbit by the time the threat is gone. This may call for one or several recovery maneuvers. In addition, the design of the overall thrust strategy is driven by the need to minimize fuel consumption in order to preserve the expected lifetime of the satellite.

This study deals with the collision avoidance problem between two orbiting objects involved in a long-term encounter, assuming a Keplerian linearized relative motion. One of the object, called the primary object and denoted by p, is an operational satellite initially set on a reference orbit. It is active in the sense that it is able to perform propulsive maneuvers to change its own trajectory. The second object, denoted by s, is passive: it is typically an uncontrolled space debris. The long-term encounter framework means that the time spent by the two space objects in the encounter region is large enough: it can extend up to a few orbital periods and corresponds to a relative velocity well below the threshold of the km.s^{-1}.

Due to the lack of precision in measurements, the position and the velocity of each object are known with a certain amount of uncertainty, so they can be modeled as random variables. The collision risk is then quantified in terms of probability. The collision avoidance problem is formulated as a joint chance constrained optimization problem: it consists in computing a fuel optimal, finite sequence of impulsive maneuvers performed by the active spacecraft such that the probability of collision between the two objects does not exceed a user-defined threshold and that the primary object goes back to its reference orbit in due time. This paper aims at finding a practically solvable formulation for this optimization problem.

Since the original joint chance-constrained problem is very difficult to solve, simplifying assumptions are usually made to reduce the size of the optimization space or even to avoid the probabilistic formulation. In [15] the execution times and the thrust directions are heuristically fixed *a priori* so that the collision avoidance maneuvers are only optimized in magnitude. In [11], the avoidance strategy is reduced to one single maneuver. The maneuver direction is decoupled from its magnitude which allows to reduce the decision variables from three to one. In the context of satellites in formation flight, [17] proposes a simple strategy consisting in computing one velocity correction to achieve a fixed miss distance between the two objects. Note also the work in [10], where the collision avoidance problem is tackled as a robust optimization problem and in which safe guidance algorithms are designed for the PRISMA mission in the form of linear programs. Most of the time, the proposed strategies do not take any recovery maneuver to the reference orbit into account.

In this paper, the main idea is to relax the initial joint chance constrained optimization problem into a deterministic disjunctive linear program that can be solved for instance with a mixed-integer linear programming solver. To this end, two mechanisms are used: (i) the collision set, usually modeled as a spherical geometrical object [5], is outer approximated here by a convex polyhedron; (ii) the so-called risk selection method, introduced first in [3] for probabilistic path planning. The degree of conservatism of these relaxations will be estimated in the chosen test case presented in the last section of this paper. The first contribution of this article is to propose an effective solution to the collision avoidance problem that takes recovery maneuvers to the reference orbit into account. The second contribution is to propose a new algorithm to efficiently compute the probability for any 3-D Gaussian random variable to lie in a Euclidean ball. This is an extension of the 2-D method described in [16]. This efficient evaluation of the instantaneous probability of collision is used here to analyze *a posteriori* the computed maneuvers and to validate on a finer grid the mitigation of the collision risk.

Notations: Capital letters are dedicated to random variables and small characters to instances of these variables. Let Y be a random vector: y is an occurrence of Y, μ_Y the mean vector of Y and Σ_Y its variance-covariance matrix. Finally, the d-dimensional normal (or Gaussian) distribution of a d-dimensional random vector Y is denoted by: $Y \sim \mathcal{N}_d(\mu_Y, \Sigma_Y)$. $\mathcal{P}(\{\cdot\})$ is the probability of the event $\{\cdot\}$.

1 Probabilistic Model for Collision Avoidance

Two spherical space objects involved in a long-term encounter over a time horizon $[t_0, t_f]$ are considered. This section addresses the mathematical formulation of the problem of a long-term encounter between these two orbiting objects. Let us first introduce the Local Vertical Local Horizontal (LVLH) orbital frame $(O_{\bar{p}}, \mathbf{u_x}, \mathbf{u_y}, \mathbf{u_z})$ attached to the nominal orbit of the primary object (see Figure 1):

- Origin $O_{\bar{p}}$: nominal position of the primary center of mass;
- Z axis (**R-bar**): radial direction (Nadir-Target), oriented towards the center of the Earth;
- Y axis (**H-bar**): perpendicular to the nominal primary orbital plane, pointing opposite the angular momentum;
- X axis (**V-bar**): chosen such that $\mathbf{u_x} = \mathbf{u_y} \times \mathbf{u_z}$.

This frame is used for the linearized equations of motion in what follows.

1.1 Model of the Linearized Dynamics

The long-term encounter assumption means that the time spent by the two objects in the encounter region is significant. One can then reasonably assume that during the encounter the relative distances of the two objects to the reference position are small compared to their distance to the Earth. Under this assumption, the respective equations of motion for each object may be linearized around the reference orbit.

Assuming unperturbed Keplerian motions, the linearized equations are described by the standard Tschauner-Hempel equations [18] and its associated Yamanaka-Ankersen's state transition matrix [20].

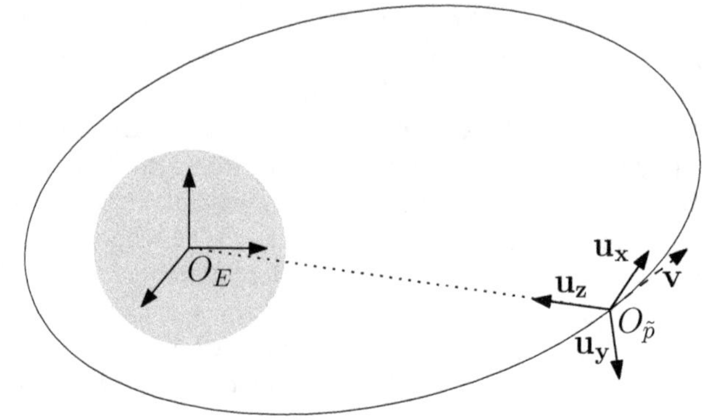

Fig. 1 Local Vertical Local Horizontal (LVLH) frame

Let $\mathbf{x}_p = [\mathbf{r}_p\ \mathbf{v}_p]^T$ and $\mathbf{x}_s = [\mathbf{r}_s\ \mathbf{v}_s]^T$ respectively denote the position and velocity vectors of the primary and the secondary objects. Unlike the secondary object, the primary object can use its thrusters to modify its trajectory. These propulsive maneuvers are assumed to be impulsive. The impulsive approximation for the thrust means that instantaneous velocity increments are applied to the primary object instead of finite-thrust powered phases of finite duration. Let $\hat{T} = \{\hat{t}_1, \hat{t}_2 \ldots, \hat{t}_{N_1}\}$ be a sorted set of dates in $[t_0, t_f]$ when a maneuver can be performed. For each $i \in \{1, \ldots, N_1\}$, the impulsive assumption could be expressed as:

$$\mathbf{x}_p(\hat{t}_i^+) = \mathbf{x}_p(\hat{t}_i^-) + B\Delta V_i, \tag{1}$$

where $B = \begin{bmatrix} 0_3 \\ I_3 \end{bmatrix}$ and ΔV_i is the velocity increment applied at $\hat{t}_i$.

Under the previous assumptions and considering that the primary object p is actuated while s is not, propagation of the state vectors $\mathbf{x}_p$ and $\mathbf{x}_s$ follow equations (2) and (3):

$$\mathbf{x}_p(t) = \Phi(t, t_0)\mathbf{x}_p^0 + \sum_{\substack{i \in \{1,\ldots,N_1\} \\ \hat{t}_i \le t}} \Phi(t, \hat{t}_i) B \Delta V_i \tag{2}$$

$$\mathbf{x}_s(t) = \Phi(t, t_0)\mathbf{x}_s^0, \tag{3}$$

where $\mathbf{x}_p^0 = \mathbf{x}_p(t_0)$, $\mathbf{x}_s^0 = \mathbf{x}_s(t_0)$ and $\Phi(\cdot,\cdot)$ is the Yamanaka-Ankersen transition matrix. The set $\hat{T}$ of dates is assumed fixed *a priori*. Its choice can follow from considerations on heuristics or previous runs of the algorithm. By default, it can simply be a uniform discretization of $[t_0, t_f]$.

It is also assumed that the primary object p is actuated by identical engines rigidly mounted to the body axes of the spacecraft; consequently the fuel consumption will be defined in the sequel as:

$$J = \sum_{i=1}^{N_1} \|\Delta V_i\|_1. \tag{4}$$

Note that a second time grid $T = \{t_0 < t_1 < t_2 < \ldots < t_{N_2} = t_f\}$, uniform and thinner than $\hat{T}$ (i.e. $N_2 > N_1$) is introduced to evaluate the instantaneous probability of collision in the problem of collision avoidance.

1.2 Uncertainty on Positions and Velocities

Due to the current limitations of orbit acquisition systems, information about the objects trajectories and velocities is known with a certain amount of uncertainty. Therefore, positions and velocities of the two objects at initial time t_0 are not known exactly and can be seen as jointly distributed random variables, respectively denoted by $\mathbf{X}_p^0$ and $\mathbf{X}_s^0$, following independent Gaussian distribution laws:

$$\mathbf{X}_p^0 \sim \mathcal{N}_6(\mu_{\mathbf{X}_p^0}, \Sigma_{\mathbf{X}_p^0}), \quad \mathbf{X}_s^0 \sim \mathcal{N}_6(\mu_{\mathbf{X}_s^0}, \Sigma_{\mathbf{X}_s^0}).$$

Under these assumptions, the relative position and velocity vector at time t_0 is also a Gaussian random vector, denoted by $\mathbf{X}^0 = \mathbf{X}_p^0 - \mathbf{X}_s^0$:

$$\mathbf{X}^0 \sim \mathcal{N}_6(\mu_{\mathbf{X}_p^0} - \mu_{\mathbf{X}_s^0}, \Sigma_{\mathbf{X}_p^0} + \Sigma_{\mathbf{X}_s^0}). \tag{5}$$

Let us now focus on the propagation of uncertainties over the time horizon $[t_0, t_f]$. Let $\mathbf{X}_p^k$ and $\mathbf{X}_s^k$ be the random state vectors at time t_k. Due to the linear setting, the Gaussian nature of uncertainty is preserved over the time horizon $[t_0, t_f]$. In other words, at any time t_k, $k \in \{1, \ldots, N_2\}$, one has:

$$\mathbf{X}_p^k \sim \mathcal{N}_6(\mu_{\mathbf{X}_p^k}, \Sigma_{\mathbf{X}_p^k}), \quad \mathbf{X}_s^k \sim \mathcal{N}_6(\mu_{\mathbf{X}_s^k}, \Sigma_{\mathbf{X}_s^k}).$$

Using the propagation equations (2) and (3), the mean and the covariance matrices can be easily expressed as:

$$\begin{aligned}
\mu_{\mathbf{X}_p^k} &= \Phi(t_k, t_0)\mu_{\mathbf{X}_p^0} + \sum_{\substack{i \in \{1,\ldots,N_1\} \\ \hat{t}_i \le t_k}} \Phi(t_k, \hat{t}_i) B \Delta V_i, & \mu_{\mathbf{X}_s^k} &= \Phi(t_k, t_0)\mu_{\mathbf{X}_s^0}, \\
\Sigma_{\mathbf{X}_p^k} &= \Phi(t_k, t_0)\Sigma_{\mathbf{X}_p^0}\Phi(t_k, t_0)^T, & \Sigma_{\mathbf{X}_s^k} &= \Phi(t_k, t_0)\Sigma_{\mathbf{X}_s^0}\Phi(t_k, t_0)^T.
\end{aligned}$$

The relative position and velocity random vector $\mathbf{X}^k = \mathbf{X}_p^k - \mathbf{X}_s^k$ at time t_k is then defined by:

$$\mathbf{X}^k \sim \mathcal{N}_6(\mu_{\mathbf{X}^k}, \Sigma_{\mathbf{X}^k}),$$

where:

$$\begin{aligned}
\mu_{\mathbf{X}^k} &= \Phi(t_k, t_0)(\mu_{\mathbf{X}_p^0} - \mu_{\mathbf{X}_s^0}) + \sum_{\substack{i \in \{1,\ldots,N_1\} \\ \hat{t}_i \le t_k}} \Phi(t_k, \hat{t}_i) B \Delta V_i, \\
\Sigma_{\mathbf{X}^k} &= \Phi(t_k, t_0)(\Sigma_{\mathbf{X}_p^0} + \Sigma_{\mathbf{X}_s^0})\Phi(t_k, t_0)^T.
\end{aligned}$$

Note that the respective probability distributions for position and velocity at any time - as marginals of Gaussian laws - are Gaussian too. Their respective covariance matrices can be simply obtained by taking the adequate 3×3 submatrices of the 6×6 matrices describing the complete states.

1.3 Formulation of the Collision Avoidance Problem

The collision avoidance problem considered in this paper, is to compute a fuel-optimal, finite sequence of impulsive maneuvers performed by the active spacecraft p such that the instantaneous probability of collision between the two objects does not exceed some user-defined tolerance threshold δ. In addition, the primary object has to be back on its reference orbit at final time t_f.

This very last constraint is a key feature of the proposed approach. To ensure that the primary object goes back to its reference orbit at final time t_f, the following boundary condition is imposed: $\mu_{\mathbf{X}_p^{N_2}} = 0$, i.e.:

$$\Phi(t_f,t_0)\mu_{\mathbf{X}_p^0} + \sum_{i=1}^{N_1} \Phi(t_f,\hat{t}_i)B\Delta V_i = 0, \tag{6}$$

which is a linear constraint in the decision variables ΔV_i, $i = 1,\dots,N_1$.

In the literature, the orbiting objects are often assumed to have a spherical shape [5]. This assumption enables to ignore any constraint on the attitude and to give a rough model of objects whose geometry is not very precisely known. Under this assumption, the collision set in the relative position space, referred to as the combined object, is defined as:

$$\{\mathbf{r} \in \mathbb{R}^3 : \|\mathbf{r}\|_2 \leq R\}, \tag{7}$$

where R is the sum of the radii of the two objects. The collision avoidance problem can then be formulated as:

$$\min_{\Delta V_i} \sum_{i=1}^{N_1} \|\Delta V_i\|_1 \ \text{ s.t. } \left| \begin{array}{l} \mathscr{P}(\{\|\mathbf{R}^k\|_2 \leq R\}) \leq \delta,\ k = 1,\dots,N_2 - 1, \\ \Phi(t_f,t_0)\mu_{\mathbf{X}_p^0} + \sum_{i=1}^{N_1} \Phi(t_f,\hat{t}_i)B\Delta V_i = 0. \end{array} \right. \tag{8}$$

where: $\mathbf{R}^k = \begin{pmatrix} I_3 & 0 \end{pmatrix} \mathbf{X}^k$ denotes the relative position random vector.

Remark 1. Rigorously speaking, instead of the instantaneous collision probability, one should consider the probability of collision over the whole time horizon as a constraint. However, despite the simplifying assumptions used here, this quantity is very difficult to manipulate: relative dynamics are still fairly complex and the set of initial conditions leading to a collision occurring before t_f, is difficult to derive.

For any $k \in \{1,\ldots,N_2-1\}$, the instantaneous collision probability at time t_k is the integral of the Gaussian probability density function of $\mathbf{R}^k$ over the ball of radius R centered at the origin. This integral can be computed in a fast and efficient way as shown in Section 2, but is hard to handle as a constraint in problem (8) due to its lack of particular properties with respect to the decision variables.

This section is dedicated to the reformulation of (8) into a more tractable optimization problem. This is done in two steps. Firstly, a conservative polyhedral approximation of the spherical combined object is proposed in Paragraph 1.3.1: the collision set is described by a finite number of equations that are linear in the control vector, and resulting in linear chance constraints that are jointly defined. Then, the so-called risk selection method is used in Paragraph 1.3.2 to replace them by individual probabilistic constraints. The induced individual chance constrained problem is then equivalently reformulated as a deterministic optimization problem.

1.3.1 Polyhedral Approximation of the Collision Set

In this section, the collision set $\{\mathbf{r} \in \mathbb{R}^3 : \|\mathbf{r}\|_2 \leq R\}$ is conservatively replaced by a (convex) polyhedron containing it. Recall that a convex polyhedron in $\mathbb{R}^3$ is defined by a finite number of affine inequalities, or equivalently as the intersection of half-spaces.

The simplest example of a convex polyhedron containing the spherical combined object, is a cube whose faces are tangent to it (see Figure 2).

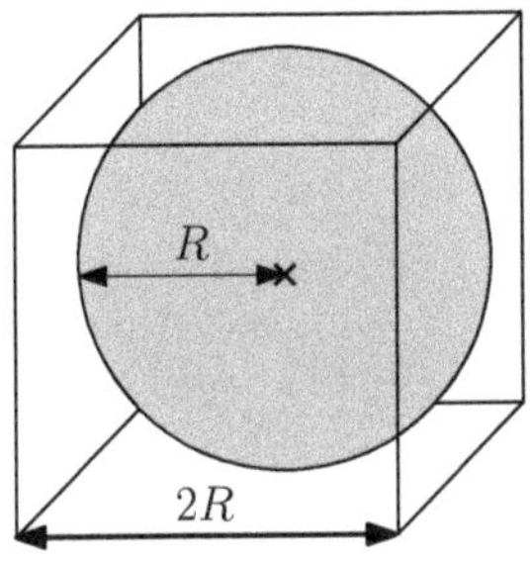

Fig. 2 Combined spherical object included in a cube

More generally, assume that the collision set is included in a given polyhedron:

$$\{\mathbf{r} \in \mathbb{R}^3 : \|\mathbf{r}\|_2 \leq R\} \subset \left\{ \mathbf{r} \in \mathbb{R}^3 : \bigcap_{j=1}^{m} (\mathbf{a}_j^T \mathbf{r} \leq b_j) \right\}, \tag{9}$$

where $\mathbf{a}_j \in \mathbb{R}^3$, $b_j \in \mathbb{R}$, $j = 1,\ldots,m$. Then:

$$\mathscr{P}\left(\left\{\|\mathbf{R}^k\|_2 \leq R\right\}\right) \leq \mathscr{P}\left(\bigcap_{j=1}^{m} \left\{a_j^T \mathbf{R}^k \leq b_j\right\}\right).$$

Using the polyhedral approximation (9), the collision avoidance problem is then replaced by:

$$\min_{\Delta V_i} \sum_{i=1}^{N_1} \|\Delta V_i\|_1 \text{ s.t. } \left| \begin{array}{l} \mathscr{P}\left(\bigcap_{j=1}^{m} \left\{ a_j^T \mathbf{R}^k \leq b_j \right\} \right) \leq \delta, \; k = 1, \ldots, N_2 - 1, \\ \Phi(t_f, t_0)\mu_{\mathbf{X}_p^0} + \sum_{i=1}^{N_1} \Phi(t_f, \hat{t}_i) B \Delta V_i = 0. \end{array} \right. \tag{10}$$

The problem (10) is a joint chance constrained optimization problem which is a relaxation of (8). This approximation is conservative in the sense that any solution of problem (10) is also a solution of (8).

Note that the degree of conservatism is closely related to the tightness of the geometrical approximation between the collision set (7) and its polyhedral approximation (9). There is a clear trade-off between increasing the complexity of the chosen polyhedron by increasing the number of inequalities defining the polyhedron and the computational complexity of the considered optimization problem.

1.3.2 Towards a Deterministic Formulation

In this section, the so-called risk selection method is used to replace the joint chance constraints in (10) by individual chance constraints. This technique was first used in [3] for probabilistic path planning although the name only appeared later (see e.g. [4]). Let us now focus on the probability for the relative position vector to be in the polyhedral collision set at a given time step t_k. One has:

$$\mathscr{P}\left(\bigcap_{j=1}^{m} \left\{ \mathbf{a}_j^T \mathbf{R}^k \leq b_j \right\} \right) \leq \mathscr{P}\left(\left\{ \mathbf{a}_j^T \mathbf{R}^k \leq b_j \right\} \right) \qquad \forall j = 1, \ldots, m, \tag{11}$$

which means that if there exists $j \in \{1, \ldots, m\}$ such that $\mathscr{P}\left(\{ \mathbf{a}_j^T \mathbf{R}^k \leq b_j \} \right) \leq \delta$ then $\mathscr{P}\left(\bigcap_{j=1}^{m} \left\{ \mathbf{a}_j^T \mathbf{R}^k \leq b_j \right\} \right) \leq \delta$. In other words:

$$\bigvee_{j=1}^{m} \left(\mathscr{P}\left(\{ \mathbf{a}_j^T \mathbf{R}^k \leq b_j \} \right) \leq \delta \right) \Longrightarrow \mathscr{P}\left(\bigcap_{j=1}^{m} \left\{ \mathbf{a}_j^T \mathbf{R}^k \leq b_j \right\} \right) \leq \delta, \tag{12}$$

where $\bigvee$ is the disjunction symbol. The original joint chance constraint can thus be conservatively replaced by the following disjunctive set of individual constraints:

$$\bigvee_{j=1}^{m} \left(\mathscr{P}\left(\{ \mathbf{a}_j^T \mathbf{R}^k \leq b_j \} \right) \leq \delta \right). \tag{13}$$

This approach is called risk selection because instead of enforcing the chance for the relative position vector to lie outside the collision set, one enforces it to be in one of the half-spaces outside the convex polyhedron.

After these two successive relaxations (the polyhedral approximation and the risk selection method), the collision avoidance problem (10) is cast as the following individual chance constrained problem:

$$\min_{\Delta V_i} \sum_{i=1}^{N_1} \|\Delta V_i\|_1 \text{ s.t. } \left| \begin{array}{l} \bigvee_{j=1}^{m} \left(\mathscr{P}\left(\{\mathbf{a}_j^T \mathbf{R}^k \le b_j\}\right) \le \delta \right), \; k = 1, \ldots, N_2 - 1, \\ \Phi(t_f, t_0)\mu_{\mathbf{X}_p^0} + \sum_{i=1}^{N_1} \Phi(t_f, \hat{t}_i) B \Delta V_i = 0. \end{array} \right. \tag{14}$$

The new constraint (13) is an individual chance constraint, but remains probabilistic. It can be equivalently reformulated into a deterministic form following [3].

Let Y be a one-dimensional random variable following a Gaussian distribution law: $Y \sim \mathscr{N}_1(\mu_Y, \sigma_Y^2)$. Any chance constraint on Y with a fixed variance and a variable mean, can be translated into a deterministic constraint on the mean μ_Y (see [3, Section III.C]):

$$\mathscr{P}(\{Y < 0\}) \le \delta \Longleftrightarrow \mu_Y \ge \sqrt{2}\sigma_Y \text{erf}^{-1}(1 - 2\delta), \tag{15}$$

where erf is the error function defined by: $\text{erf}(z) = \frac{2}{\sqrt{\pi}} \int_0^z e^{-t^2} dt$. By applying (15) to $Y = \mathbf{a}_j^T \mathbf{R}^k - b_j \sim \mathscr{N}_1(\mathbf{a}_j^T \mu_{\mathbf{R}^k} - b_j, \mathbf{a}_j^T \Sigma_{\mathbf{R}^k} \mathbf{a}_j)$, each individual probabilistic constraint can be equivalently replaced by an affine inequality on the mean:

$$\mathscr{P}\left(\{\mathbf{a}_j^T \mathbf{R}^k \le b_j\}\right) \le \delta \Longleftrightarrow \mathbf{a}_j^T \mu_{\mathbf{R}^k} \ge b_j + c_{jk},$$

where: $c_{jk} = \sqrt{2\mathbf{a}_j^T \Sigma_{\mathbf{R}^k} \mathbf{a}_j} \text{erf}^{-1}(1 - 2\delta)$. In conclusion, the collision avoidance problem (14) can be equivalently reformulated as:

$$\min_{\Delta V_i} \sum_{i=1}^{N_1} \|\Delta V_i\|_1 \text{ s.t. } \left| \begin{array}{l} \bigvee_{j=1}^{m} \left(\mathbf{a}_j^T \mu_{\mathbf{R}^k} \ge b_j + c_{jk} \right), \; \forall k = 1, \ldots, N_2 - 1, \\ \Phi(t_f, t_0)\mu_{\mathbf{X}_p^0} + \sum_{i=1}^{N_1} \Phi(t_f, \hat{t}_i) B \Delta V_i = 0. \end{array} \right. \tag{16}$$

1.4 Towards a Mixed-Integer Linear Program

In the previous section, a tractable deterministic formulation (16) of the collision avoidance problem involving disjunctive constraints, has been derived. Let us now show that the latter is actually a mixed integer linear program which can be solved using dedicated algorithms.

First, in order to get rid of piecewise linear criterion, usual slack variables $\Delta V_i^-, \Delta V_i^+$ belonging to the positive orthant of $\mathbb{R}^3$ are introduced [2]: for all $i \in \{1,\ldots,N_1\}$, the following substitutions are done:

$$\Delta V_i = \Delta V_i^+ - \Delta V_i^-. \tag{17}$$

Note that the original piecewise linear criterion $\sum_{i=1}^{N_1} \|\Delta V_i\|_1$ may be equivalently replaced by the linear criterion $\sum_{i=1}^{N_1} \left(\|\Delta V_i^+\|_1 + \|\Delta V_i^-\|_1\right) = \sum_{i=1}^{N_1}\sum_{j=1}^{3} (\Delta V_i^{+j} + \Delta V_i^{-j})$ since at the optimum $\Delta V_i^{+j} = 0$ or $\Delta V_i^{-j} = 0$ [2].

The disjunctive constraints are then tackled using the classic big **M** method [12]. For the clarity of the presentation, this method is recalled on a one-dimensional example. Let us consider the following two linear disjunctive conditions in the scalar variable y:

$$y \leq y_{min} \text{ or } y \geq y_{max}.$$

If the scalar M is chosen to be sufficiently large, i.e. larger than any other quantity of the problem, the previous conditions are equivalent to the following ones:

$$y \leq y_{min} + qM \quad \text{and} \quad -y \leq -y_{max} + Mq' \quad \text{and} \quad q + q' \leq 1,$$

where q and q' are binary variables: $q, q' \in \{0,1\}$. Any linear program involving such disjunctive constraints, may be recast into a Mixed-Integer Linear Program (MILP). In our case, it leads to an optimization problem with $2 \times 3 \times N_1 = 6N_1$ continuous variables and $m \times (N_2 - 1)$ binary variables:

$$\min_{\Delta V_i^\pm, q_k} \sum_{i=1}^{N_1} \left(\|\Delta V_i^+\|_1 + \|\Delta V_i^-\|_1\right) \text{ s.t. } \left| \begin{array}{l} \bigwedge_{j=1}^{m} \left(\mathbf{a}_j^T \mu_{\mathbf{R}^k} + M q_k^j \geq b_j + c_{jk}\right),\ \forall k = 1,\ldots,N_2 - 1, \\ \Phi(t_f,t_0)\mu_{\mathbf{X}_p^0} + \sum_{i=1}^{N_1} \Phi(t_f,\hat{t}_i) B(\Delta V_i^+ - \Delta V_i^-) = 0, \\ \|\mathbf{q}_k\|_1 \leq m-1,\ k = 1,\ldots,N_2 - 1, \\ \Delta V_i^\pm \geq 0,\ i = 1,\ldots,N_1, \\ \mathbf{q}_k \in \{0,1\}^m,\ k = 1,\ldots,N_2 - 1. \end{array} \right. \tag{18}$$

Note that in the proposed approach, there is no guarantee for the instantaneous collision probability to be below a given threshold between the points of the time grid T. To alleviate this limitation, a precise *a posteriori* analysis of the computed maneuvers is performed by computing the original instantaneous collision probability in (8) over a thinner grid than the one used in (18).

2 Computing the Instantaneous Probability of Collision

In the modeling of the collision avoidance problem, the instantaneous probability of collision $\mathscr{P}(\{\|\mathbf{R}\|_2 \leq R\})$ has to be smaller than a given threshold δ at any instant t_k of a given time discretization of $[t_0, t_f]$. To validate this approach, one needs now to compute the instantaneous probability of collision at each time $t \in [t_0, t_f]$ to numerically verify that it remains below the threshold δ over the whole time interval $[t_0, t_f]$:

$$\forall t \in [t_0, t_f],\ \mathscr{P}(\{\|\mathbf{R}(t)\|_2 \leq R\}) \leq \delta.$$

In that purpose, the conservative polyhedral model is not needed any more and the spherical combined object can be considered again. Under this assumption, the instantaneous probability of collision can be expressed as the integral of a 3-D Gaussian random variable - representing the current relative position of the two objects - on a Euclidean ball. Usually, that kind of integral is evaluated using time-consuming Monte Carlo methods. Here, an analytic formula based on a convergent power series is introduced. It is a 3-D extension to the 2-D method presented in [16]. It is derived by use of Laplace techniques originally developed in [9], and properties of D-finite functions [21, 13]. Computing truncated series should allow for a fast evaluation of the instantaneous probability of collision.

Assuming that $\mathbf{R} \sim \mathscr{N}_3(\mu, \Sigma)$, the instantaneous probability of collision can be expressed as:

$$\mathscr{P}_c = \frac{1}{(2\pi)^{3/2}|\Sigma|^{1/2}} \int_{\mathscr{B}(0,R)} \exp\left(-\frac{1}{2}(r-\mu)^T \Sigma^{-1}(r-\mu)\right) \mathrm{d}r, \tag{19}$$

where $\mu \in \mathbb{R}^3$ and Σ is a 3×3 positive definite matrix. Without loss of generality, the covariance matrix Σ is assumed to be diagonal (this can be always achieved by a classical change of variables). Thus, Σ can be written as:

$$\Sigma = \mathrm{diag}(\sigma_1^2, \sigma_2^2, \sigma_3^2), \tag{20}$$

where $(\sigma_1, \sigma_2, \sigma_3) \in \mathbb{R}_*^{+3}$, and: $\sigma_1 \geq \sigma_2 \geq \sigma_3$. Under these assumptions, the probability $\mathscr{P}_c$ can be rewritten as:

$$\mathscr{P}_c = \frac{1}{(2\pi)^{3/2}\sigma_1\sigma_2\sigma_3} \int_{\mathscr{B}(0,R)} \exp\left(-\frac{1}{2}\sum_{i=1}^{3}\frac{(r_i-\mu_i)^2}{\sigma_i^2}\right) \mathrm{d}r. \tag{21}$$

Both level sets of integrand and domain of integration can be visualized in Figure 3.

Proposition 1.

$$\mathscr{P}_c = \exp\left(-\frac{R^2}{\sigma_3^2}\right) \sum_{k=0}^{+\infty} u_k R^{2k}, \tag{22}$$

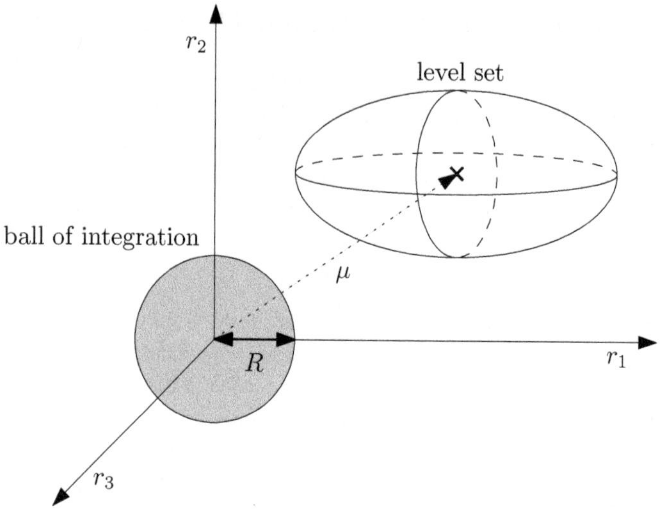

Fig. 3 Geometry of the integral problem

where the sequence $(u_k)_{k\geq 0}$ is positive and explicitly defined by a linear recurrence with polynomial coefficients.

Sketch of the Proof

Let us introduce an intermediate function called g and defined for all $z \in \mathbb{R}^+$ as follows:

$$g(z) = \frac{1}{(2\pi)^{3/2}\sigma_1\sigma_2\sigma_3} \int_{\mathscr{B}(0,\sqrt{z})} \exp\left(-\frac{1}{2}\sum_{i=1}^{3}\frac{(r_i-\mu_i)^2}{\sigma_i^2}\right) \mathrm{d}r,$$

so that the instantaneous probability of collision is given by: $\mathscr{P}_c = g(R^2)$. Using the set-indicator function $\mathbf{1}$, the function g can also be rewritten as:

$$g(z) = \frac{1}{(2\pi)^{3/2}\sigma_1\sigma_2\sigma_3} \int_{\mathbb{R}^3} \mathbf{1}_{\mathscr{B}(0,\sqrt{z})}(r) \exp\left(-\frac{1}{2}\sum_{i=1}^{3}\frac{(r_i-\mu_i)^2}{\sigma_i^2}\right) \mathrm{d}r. \tag{23}$$

The main idea of the proof is to compute the Laplace transform of the function g in closed form, to expand it into a power series in the variable λ^{-1} and to apply the inverse Laplace transform term by term. Finally, a preconditioning with an exponential term is performed in order to obtain a series with positive coefficients.

Let us first apply the Laplace transform to the function g: for all λ whose real part $\Re(\lambda)$ is non-negative, one has:

$$\mathscr{L}_g(\lambda) = \int_0^{+\infty} \exp(-\lambda z) g(z) \mathrm{d}z. \tag{24}$$

The Laplace transform $\mathscr{L}_g$ can be computed in closed form. Starting from:

$$\mathscr{L}_g(\lambda) = \frac{1}{(2\pi)^{3/2}\sigma_1\sigma_2\sigma_3} \int_0^{+\infty} \int_{\mathbb{R}^3} \mathbf{1}_{\mathscr{B}(0,\sqrt{z})}(r) \exp\left(-\lambda z - \frac{1}{2}\sum_{i=1}^{3} \frac{(r_i-\mu_i)^2}{\sigma_i^2}\right) \mathrm{d}r\mathrm{d}z, \tag{25}$$

the Fubini theorem enables us to interchange the two integration symbols:

$$\mathscr{L}_g(\lambda) = \frac{1}{(2\pi)^{3/2}\sigma_1\sigma_2\sigma_3} \int_{\mathbb{R}^3} \int_0^{+\infty} \mathbf{1}_{\mathscr{B}(0,\sqrt{z})}(r) \exp\left(-\lambda z - \frac{1}{2}\sum_{i=1}^{3} \frac{(r_i-\mu_i)^2}{\sigma_i^2}\right) \mathrm{d}z\mathrm{d}r.$$

Now, the integral in the variable z can be computed analytically, so that:

$$\mathscr{L}_g(\lambda) = \frac{1}{(2\pi)^{3/2}\lambda\sigma_1\sigma_2\sigma_3} \int_{\mathbb{R}^3} \exp\left(\sum_{i=1}^{3}\left(-\left(\lambda + \frac{1}{2\sigma_i^2}\right) r_i^2 + \frac{\mu_i r_i}{\sigma_i^2} - \frac{\mu_i^2}{2\sigma_i^2}\right)\right) \mathrm{d}r.$$

The next step consists in doing a change of variables aimed at completing the square in the exponential. More precisely, state:

$$r_i' = r_i - \frac{\mu_i}{2\sigma_i^2}\left(\lambda + \frac{1}{2\sigma_i^2}\right)^{-1} \quad \forall i = 1,2,3. \tag{26}$$

By computation, one finally gets:

$$\mathscr{L}_g(\lambda) = \lambda^{-5/2} \exp\left(-\sum_{i=1}^{3} \mu_i^2 (2\sigma_i^2 + \lambda^{-1})^{-1}\right) \prod_{i=1}^{3} (2\sigma_i^2 + \lambda^{-1})^{-1/2}.$$

Now $\lambda^{5/2}\mathscr{L}_g(\lambda)$ can be seen as a function of the variable λ^{-1}, which can be expanded in a power series $\sum_{k=0}^{+\infty} w_k(\lambda^{-1})^k$. This is valid for $\Re(\lambda) > \frac{1}{2\sigma_3^2}$. The sequence $(w_k)_{k\geq 0}$ follows a linear recurrence with polynomial terms in the variable k. This recurrence can be obtained from the differential equation satisfied by $\lambda^{-1} \to \lambda^{5/2}\mathscr{L}_g(\lambda)$. Indeed this function is D-finite i.e. is solution of a linear differential equation with polynomial coefficients. The Maple package Gfun [14](version 3.65) was used to obtain the recurrence formula. One can now write:

$$\mathscr{L}_g(\lambda) = \lambda^{-5/2} \sum_{k=0}^{+\infty} w_k \lambda^{-k} = \sum_{k=0}^{+\infty} w_k \lambda^{-(k+5/2)}. \tag{27}$$

Let us now apply the inverse Laplace transform to the previous expression. On the left hand side, one gets the function g back. On the right hand side, the inverse Laplace transform is done term by term via the technique described by Widder [19]. It leads to the introduction of the gamma function:

$$g(z) = \sum_{k=0}^{+\infty} \frac{w_k}{\Gamma(k+5/2)} z^{k+3/2} = z^{3/2} \sum_{k=0}^{+\infty} \frac{w_k}{\Gamma(k+5/2)} z^k. \tag{28}$$

In general, this first series is ill-fitted for good numerical results because the terms $\frac{w_k}{\Gamma(k+5/2)} z^k$ have alternating signs and are much higher in absolute value than the actual value of $g(z)$. This problem of evaluating entire functions in finite precision arithmetic was previously addressed in [7, 6] and the same idea is used here. Function g is multiplied by another function h such that both h and hg have explicit power series with positive coefficients - which is good from a computational point of view. Since D-finite functions are closed by multiplication, choosing a D-finite h results in a D-finite product hg, meaning that their power series coefficients satisfy an explicit linear recurrence. Let us do the following preconditioning:

$$\exp\left(\frac{z}{\sigma_3^2}\right) g(z) = \sum_{k=0}^{+\infty} u_k z^k, \tag{29}$$

so that

$$g(z) = \exp\left(-\frac{z}{\sigma_3^2}\right) \sum_{k=0}^{+\infty} u_k z^k. \tag{30}$$

The linear recursive formula for $(u_k)_{k \geq 0}$ can be obtained in closed form using the recurrence defining $(w_k)_{k \geq 0}$. Is was done once again with Gfun. Getting back to the original probability, one immediately gets formula (22).

3 Numerical Results

In this section, the proposed approach is applied on a test case presented by S. Alfano in [1, case 9] where he compares several methods to compute collision probabilities (but offers no avoidance strategy). This is not a real case scenario but it is supposed to be representative of two objects in highly eccentric Earth orbits (HEO).

The reference orbit of the primary object has a mean motion $n = 1.4591 \times 10^{-4}$ s^{-1} and an eccentricity $e = 0.741$. The origin of time $t = 0$ is chosen as the nominal time of closest approach in the unmitigated case. This time corresponds to the instant when the mean relative distance reaches its minimal value if no maneuver is performed. This is typically around this time that the instantaneous collision probability is the highest. The history of instantaneous collision probability, when no maneuver is performed, was computed using the method previously introduced.

Figure 4 shows a close-up around its peak that looks similar to a plot provided by Alfano in [1], although a different model was used to propagate uncertainty in his paper. Clearly, the risk is very high as collision probability reaches near $t = 0$ a maximum of 0.2813 (which is also very close to the value 0.2812 found in [1]).

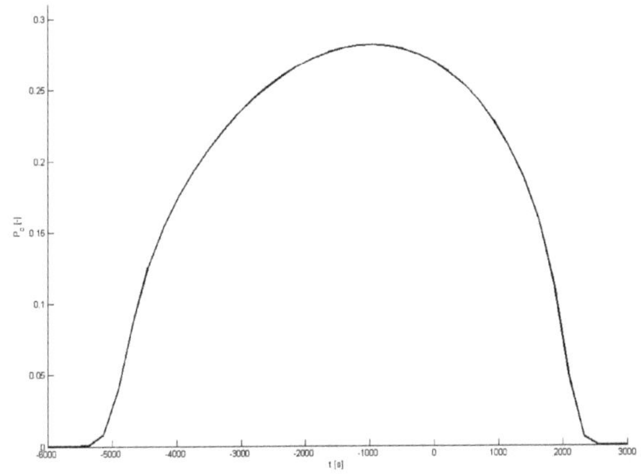

Fig. 4 Instantaneous collision probability as a function of time

For this reason, $t = 0$ is chosen at the midpoint of the interval $[t_0, t_f]$, with $t_0 = -35,000$ s and $t_f = +35,000$ s. The nominal angular position at initial time t_0 is $v_0 = -3.071$ rad. The mean primary position and velocity are assumed to be equal to the nominal one at t_0. Initial conditions for the relative state in the LVLH frame are given in Table 1 (mean vector) and Equation 31 (covariance matrix in meters and seconds). The combined radius of the objects is $R = 6$ m and the risk threshold is $\delta = 10^{-4}$. Both control grid ($N_1 = 50$) and constrained grid ($N_2 = 200$) are uniformly sampled.

Table 1 Mean relative state at time t_0

$\mu_{\mathbf{R}^0}(1)$ (m)	- 46.830
$\mu_{\mathbf{R}^0}(2)$ (m)	0
$\mu_{\mathbf{R}^0}(3)$ (m)	2.986
$\mu_{\mathbf{V}^0}(1)$ (mm.s^{-1})	-0.643
$\mu_{\mathbf{V}^0}(2)$ (mm.s^{-1})	0
$\mu_{\mathbf{V}^0}(3)$ (mm.s^{-1})	- 1.922

$$\begin{pmatrix} 42.9181 & 0 & 2.5144 & 2.607\times 10^{-3} & 0 & 8.0097\times 10^{-3} \\ & 0.0649 & 0 & 0 & -0.0024\times 10^{-3} & 0 \\ & & 0.1549 & 0.1574\times 10^{-3} & 0 & 0.4794\times 10^{-3} \\ & & & 0.1613\times 10^{-6} & 0 & 0.4928\times 10^{-6} \\ & \text{sym} & & & 0.0001\times 10^{-6} & 0 \\ & & & & & 1.5083\times 10^{-6} \end{pmatrix} \quad (31)$$

To limit the number of equations used to define the polyhedral approximation of the collision set and thus the computational time, the spherical combined object is over-approximated by the following cube:

$$\{(r_1,r_2,r_3)\in\mathbb{R}^3 : -r_i \le R,\ r_i \le R \quad \forall i = 1,2,3\}.$$

The overall implementation was done with Matlab© R2014a on an Intel® Xeon® at 3.60GHz. The mixed-integer linear program has been solved using Gurobi 5.6.0 [8] with a big-M of 10^4 m.

The solution is detailed in Table 2 and was obtained after 1.6 s. It consists in only 4 non-negligible impulses. It is noticeable that - except for the first one - every maneuver has only one non-zero component and that there is no out-of-plane maneuvers. The corresponding fuel-cost is $J_1^* = 4.4$ mm/s.

Table 2 Optimal impulses

Number	Instant (s)	LVLH components (mm/s)		
1	-35,000	0.0167	0	-3.0360
2	22,143	0.5167	0	0
3	23,571	0.2777	0	0
4	30,714	-0.5493	0	0

Next Figures 5, 6, 7 and 8 illustrate the effects of the thrusting strategy. The primary mean position and velocity and the nonzero optimal impulses can be seen in Figures 5 and 6: as expected, the primary mean vector goes back to its nominal state at the end of the time horizon.

The history of instantaneous collision probability was also computed in the controlled case, on a tighter grid of 300 points (recall that it was 200 for the optimization algorithm) in 558.9 s. It appears that the maximum risk is very close to the authorized δ-threshold since the highest probability is 1.004×10^{-4}. Note that if the chance constraints are not satisfied on a thinner grid than the one used for the optimization, the algorithm could be run with a bigger value for N_2. Figure 7 depicts one hundred occurrences - randomly sampled - of the uncontrolled relative motion while Figure 8 shows the same trajectories when performing the maneuvers. One can see that the impulses allow to safely avoid the secondary object (in black) for all the trials whereas 33 of them lead to a collision when there is no control.

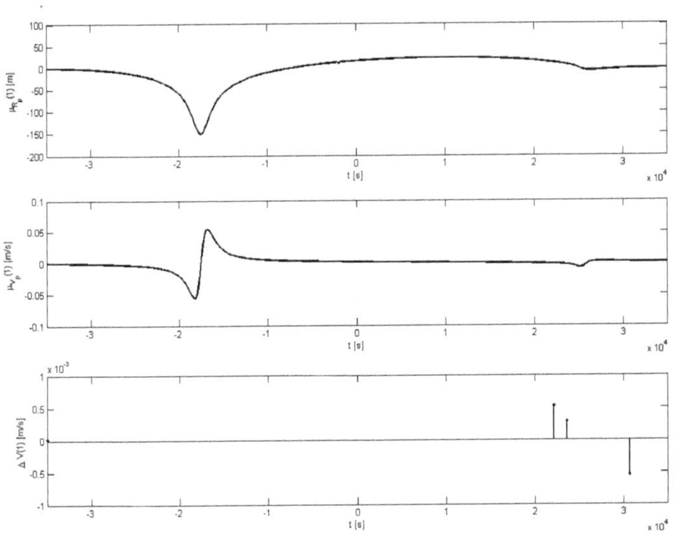

Fig. 5 Primary mean position, mean velocity and optimal impulses along the first LVLH axis versus time

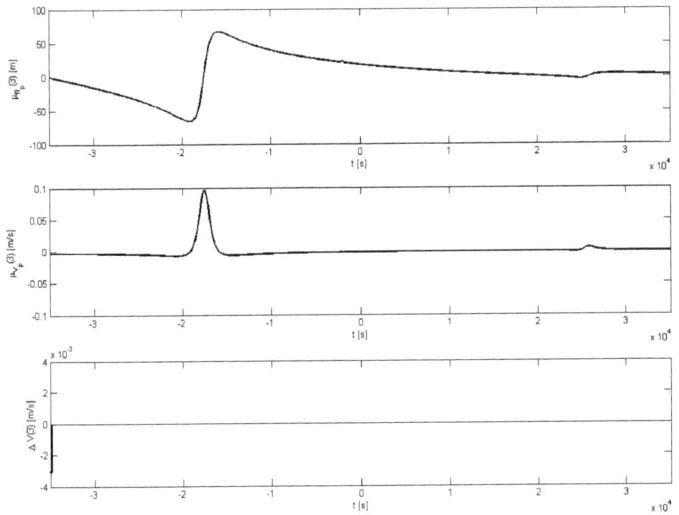

Fig. 6 Primary mean position, mean velocity and optimal impulses along the third LVLH axis versus time

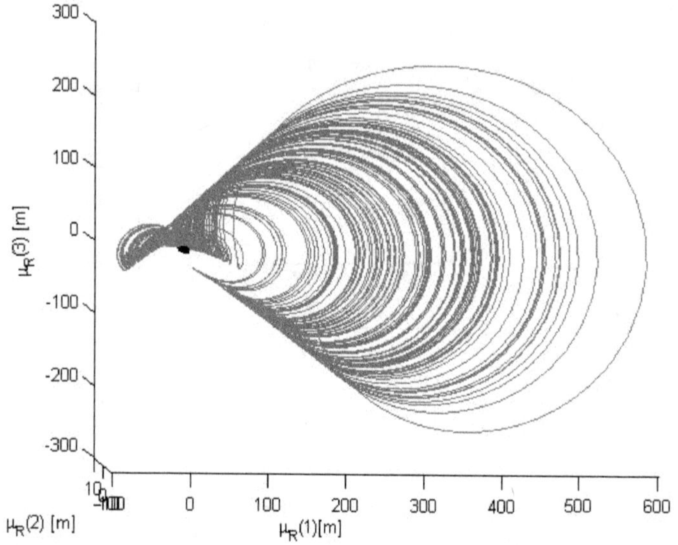

Fig. 7 Examples of uncontrolled relative trajectories of the primary over the encounter. The secondary object is in black.

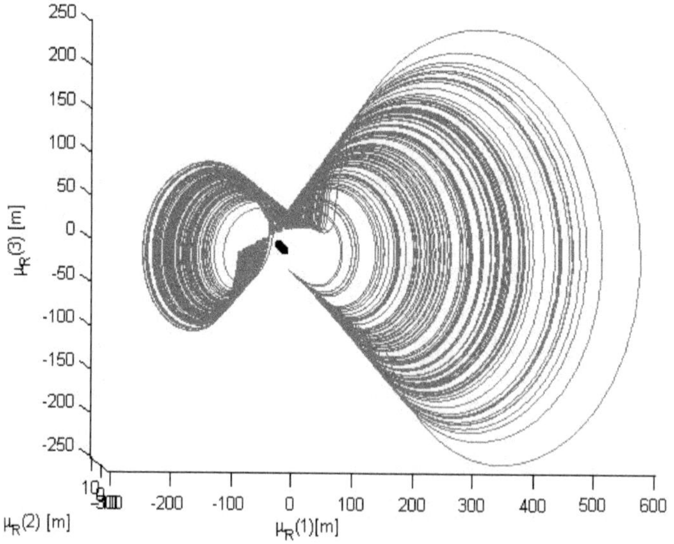

Fig. 8 Examples of controlled relative trajectories of the primary over the encounter. The secondary object is in black.

4 Conclusion

This paper proposes a probabilistic formulation of the collision avoidance problem between an active spacecraft and an orbital debris as a joint chance constrained optimization problem. The framework of this study is restricted to long-term encounters, enabling the linearization of the equations of the relative dynamics. Since the original problem is not tractable, two relaxations are proposed: first the collision set is conservatively approximated by a convex polyhedron containing it, and then the joint chance constrained optimization problem is replaced by an individual chance constrained problem. The latter is equivalently reformulated as a deterministic disjunctive linear program which can be solved using a classical mixed-integer solver. To validate the proposed approach, a new method to efficiently compute 3-D Gaussian integrals on Euclidean balls is presented. It is then used to evaluate the instantaneous probability of collision for a spherical collision set at any time of the mission. The efficiency of the proposed approach is illustrated on a test case by computing the instantaneous probabilities of collision over the whole time horizon. It is numerically checked that the risk remains below the required tolerance threshold. The main limitation of the approach is that the probability of collision over the entire encounter is not directly dealt with in the proposed optimization process. So, there is no theoretical guarantee that the probability of collision remains below the required threshold all along the mission.

As far as the computation of the maneuvers is concerned, improvements can be obtained depending on mission requirements. For instance, an upper bound on the magnitude of the velocity increments can easily be added as linear constraints. The control law could also be restricted to in-plane maneuvers. Another perspective is to extend the collision avoidance problem to multiple active spacecraft (formation flying) and/or multiple debris (orbit safety). This could only be done for a reasonable number of additional objects as complexity would rise accordingly. Concerning the computation of the probability of collision, some improvements will be considered in short term. One perspective will be to find upper bounds on the truncation error and lower bounds on the probability when uncertainty varies. Such results should be of a great interest to retrieve valuable information from a mission point of view on collision probabilities for any type of space encounter.

Acknowledgments. The Authors would like to thank Alexandre Falcoz from Airbus Defence and Space for the grant that partly supports this activity.

References

1. Alfano, S.: A numerical implementation of spherical objet collision probability. Journal of Astronautical Sciences 53(1) (January-March 2005)
2. Bertsimas, D., Tsitsiklis, J.N.: Introduction to linear optimization. AIAA. Athena Scientific, Belmont, Massachusetts (1997)
3. Blackmore, L., Li, H., Williams, B.: A probabilistic approach to optimal robust path planning with obstacles. In: American Control Conference, Minneapolis, MA (June 2006)

4. Blackmore, L., Ono, M., Williams, B.: Chance-constrained optimal path planning with obstacles. IEEE Transactions on Robotics 27(6), 1080–1094 (2011)
5. Chan, F.K.: Spacecraft Collision Probability. AIAA. The Aerospace Press (2008)
6. Chevillard, S., Mezzarobba, M.: Multiple-Precision Evaluation of the Airy Ai Function with Reduced Cancellation. In: Nannarelli, A., Seidel, P.-M., Tang, P.T.P. (eds.) 21st IEEE SYMPOSIUM on Computer Arithmetic, Los Alamitos, CA, April 2013, pp. 175–182. IEEE Computer Society (2013)
7. Gawronski, W., Müller, J., Reinhard, M.: Reduced cancellation in the evaluation of entire functions and applications to the error function. SIAM Journal on Numerical Analysis 45(6), 2564–2576 (2007)
8. Gurobi Optimization, Inc., Gurobi Optimizer Reference Manual (2014), `http://www.gurobi.com`
9. Lasserre, J.B., Zeron, E.S.: Solving a class of multivariate integration problems via laplace techniques. Applicationes Mathematicae (2001)
10. Mueller, J.B., Larsson, R.: Collision avoidance maneuver planning with robust optimization. In: 7th International ESA Conference on Guidance, Navigation and Control Systems (2008)
11. Patera, R.P., Peterson, G.E.: Space vehicle maneuver method to lower collision risk to an acceptable level. Journal of Guidance, Control and Dynamics 26(2) (March-April 2003)
12. Richards, A.R., Schouwenaars, T., How, J.P., Feron, E.: Spacecraft trajectory planning with avoidance constraints using mixed-integer linear programming. Journal of Guidance, Control and Dynamics 25(4) (August 2002)
13. Salvy, B.: D-finiteness: Algorithms and applications. In: Kauers, M. (ed.) ISSAC 2005: Proceedings of the 18th International Symposium on Symbolic and Algebraic Computation, Beijing, China, July 24-27, pp. 2–3. ACM Press (2005) Abstract for an invited talk
14. Salvy, B., Zimmermann, P.: Gfun: a Maple package for the manipulation of generating and holonomic functions in one variable. ACM Transactions on Mathematical Software 20(2), 163–177 (1994)
15. Sánchez-Ortiz, N., Belló-Mora, M., Klinkrad, H.: Collision avoidance manoeuvres during spacecraft mission lifetime: Risk reduction and required δv. Advances in Space Research 38(9), 2107–2116 (2006)
16. Serra, R., Arzelier, D., Lasserre, J.-B., Joldes, M., Rondepierre, A.: A new method to compute the probability of collision for short-term space encounters. In: Proceedings of AAS/AIAA Astrodynamics Specialist Conference, San Diego, California, USA (August 2014)
17. Slater, G.L., Byram, S.M., Williams, T.W.: Collision avoidance for satellites in formation flight. Journal of Guidance, Control, and Dynamics 29(5), 1140–1146 (2006)
18. Tschauner, J., Hempel, P.: Optimale beschleunigungs-programme fur des rendezvous manover. Astronautica Acta 5-6, 296–307 (1964)
19. Widder, D.V.: An introduction to transform theory. Academic Press New York (1971)
20. Yamanaka, K., Ankersen, F.: New state transition matrix for relative motion on an arbitrary elliptical orbit. Journal of Guidance, Control and Dynamics 25(1) (January 2002)
21. Zeilberger, D.: A holonomic systems approach to special functions identities. Journal of Computational and Applied Mathematics 32(3), 321–368 (1990)

Motion Planning and Control of a Space Robot to Capture a Tumbling Debris

Vincent Dubanchet, David Saussié, Daniel Alazard,
Caroline Bérard, and Catherine Le Peuvédic

Abstract. Space robotics has emerged as one of the key technology for on-orbit servicing or debris removal issues. In the latter, the target is a specific point of a tumbling debris, that the « chaser » satellite must accurately track to ensure a smooth capture by its robotic arm. Based on recent works by Aghili, an optimal capture trajectory is presented to match position and speed, but also acceleration of the target. Two controllers are simultaneously synthesized for the satellite and the arm, using the fixed-structure H_∞ synthesis. Their tracking performance is validated for the tumbling target capture scenario. The main goal is to efficiently track the optimal trajectory while using simple PD-like controllers to reduce computational burden. The fixed-structure H_∞ framework proves to be a suitable tool to design a reduced-order robust controller compatible with current space processors capabilities.

1 Introduction

After 60 years of intensive space use, Earth orbits have now reached a shifting point where human intervention is necessary. Known as the « Kessler syndrome », a 200 years forecast states that space access would almost disappear if nothing is done [1, 2], considering the actual debris proliferation in Fig. 1. As shown by the NASA scientist J.C. Liou in [3], at least five massive objects, such as dead satellites or

Vincent Dubanchet · David Saussié
École Polytechnique de Montréal, 2900, boul. Édouard-Montpetit, Montréal, Canada
e-mail: {vincent.dubanchet,david.saussie}@polymtl.ca

Daniel Alazard · Caroline Bérard
Institut Supérieur de l'Aéronautique et de l'Espace, 10,
Avenue Édouard Belin, Toulouse, France
e-mail: {daniel.alazard,caroline.berard}@isae.fr

Catherine Le Peuvédic
Thales Alenia Space, 100, boul. du Midi, Cannes, France
e-mail: catherine.le-peuvedic@thalesaleniaspace.com

J. Bordeneuve-Guibé et al. (eds.), *Advances in Aerospace Guidance, Navigation and Control*, DOI: 10.1007/978-3-319-17518-8_40

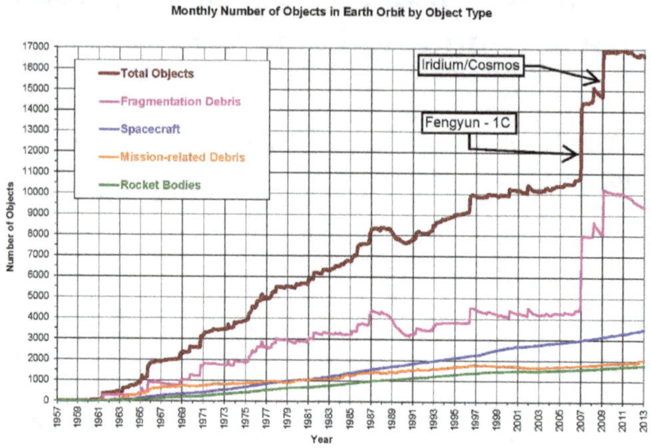

Fig. 1 Debris population evolution during the last 60 years [4]

rocket upper stages, need to be de-orbit each year to reverse the current trend and stabilize the debris population.

Robotics appears as one of the more advanced solution, considering its use in space systems for the last 30 years. It is about to replace astronauts to perform routine tasks and for safety reasons, as impacts in Extra-Vehicular Activities (EVA) become more likely to occur. The first experiment took place with the Canadian robotic arm, called *Canadarm*, attached to the Space Shuttle in the 80's [5]. Many others followed on the *International Space Station* (ISS), like the *DEXTRE* arm, the *European Robotic Arm* or the *JEMRMS*. Apart from operating in micro-gravity, these examples are close to classic terrestrial arms whose base is fixed on the ground. Since the ISS is so massive, it experiences almost no reaction from the arm motion. Challenging missions demonstrated such coupling like the Japanese *Engineering Test Satellite-VII* (ETS-VII) [6], or the American *Orbital Express* project [7]. In the last years, the German space agency DLR[1] has been working on the *DEutsche Orbitale Servicing* (DEOS) mission, and takes advantage of its robotic facilities to test and validate robotic sensors and hardware configurations [8].

Based on a very active and extended literature on the topic, the paper summarizes the main issues on the control of a space robot for debris capture. The focus of literature review is put on path planning methods, optimal control and fixed-structure H_∞ framework in Section 2. Section 3 quickly presents the dynamic model at hand, before solving the optimal path planning problem in Section 4. The controllers are designed and validated in Section 5 with a space robot tracking a tumbling target trajectory. Eventually, Section 6 gathers the paper main results and proposes future research avenues.

[1] *Deutsches zentrum für Luft- und Raumfahrt* (DLR).

2 Literature Review

Modeling, motion planning and control of space robotic manipulators mix spacecraft and robotics theories. As a simplifying hypothesis, it will be considered that the satellite equipped with the robotic arm is floating freely in space such that orbital mechanics can be neglected at this stage[2]. Thereafter, such a system is called a « space robot ».

2.1 *Kinematics and Modeling*

Robot kinematics are of utmost importance to compute dynamic models, and thus, to plan trajectories. The Jacobian matrix $\mathbf{J_m}$ is characterized by the linear dependency between the velocities of the effector and of the joints, as: $\dot{\mathbf{x}}_\mathbf{e} = \mathbf{J_m}(\mathbf{q_m})\,\dot{\mathbf{q}}_\mathbf{m}$, where $\mathbf{x_e}$ is the effector position and attitude, and $\mathbf{q_m}$ the arm configuration (i.e. the set of joint angles). Singularities are then defined as the configurations where the Jacobian matrix becomes rank deficient [10]. For fixed-based arms, singularities only depend on the configuration $\mathbf{q_m}$, and are called *kinematic singularities* [11]. In the case of a free-floating system, the global angular momentum conservation results in a new Jacobian definition. Firstly introduced by Umetani and Yoshida [12], the *Generalized Jacobian Matrix* is expressed as:

$$\mathbf{J}^* = \mathbf{J_m} - \mathbf{J_b}\,\mathbf{I_b}^{-1}\,\mathbf{I_m} \tag{1}$$

where $\mathbf{J_b}$ gives the satellite influence on the effector motion, and $\mathbf{I_b}$ and $\mathbf{I_m}$ are the mass matrices of the base and the manipulator. Since the new Jacobian is now function of the dynamic parameters of both base and arm, additional singularities appear, which are called *dynamic singularities* [13]. Moreover, due to the nonholonomic nature of the angular momentum conservation, they are *path-dependant*. By carefully planning the free-floating robot trajectory to reach a specific configuration, they can be avoided [14, 15].

2.2 *Path Planning*

The kinematics analysis and the modeling phase give a good idea of the robot's workspace. As shown by Vafa in [16], the workspace is drastically reduced for a free-floating system. It even further shrinks when the robot operates in a workspace free of dynamic singularities, called the *Path Independant Workspace* by Papadopoulos [17]. An inertial trajectory planning is proposed in [14] to avoid the path-dependant singularities, by carefully choosing the initial arm's configuration to reach a final inertial position of the effector. This approach is consistent with the arm's ability to re-orient the satellite by cyclic manoeuvers, as it takes advantage of the nonholonomic nature of the system [16]. The base actuation capability is also considered in [18] by the mean of the *Limit Curve*, to choose the effector speed

[2] More insights about the full orbital mechanics of the space robot can be found in [9].

that does not saturate the satellite actuators. In [19], a similar approach is used to maintain the antenna pointing accuracy during the arm's manoeuvers.

An alternative planning method is to consider the base as free-floating (i.e. without active control), and to plan a *reactionless* effector's trajectory. Dubowsky investigated a singular value decomposition of the mass matrix to choose a direction of minimum disturbance [20]. Nenchev also developped the effective *Reactionless Null Space* (RNS) based method; it exploits the arm's redundancy to compute a trajectory without angular momentum transmission from the arm to the satellite [21].

The capture of a tumbling debris by a space robot is only possible if the target trajectory is well-known. Different models can be considered for its dynamics: a linear trajectory is used in [22] and is valid for very small objects, whereas a more generic trajectory has to be computed for a handle on a rotating body [23]. Circular paths are often considered with the hypothesis of a flat-spin motion [24, 25].

Considering this target trajectory, many criteria can be optimized to compute the effector capture trajectory: fuel consumption, actuators saturation, joints position and speed limits, obstacle avoidance, etc. The capture can be realized by matching only the target position, or both position and speed, or even the acceleration to ensure a smoother capture. The first approach leads to an impact at the capture and must be carefully studied to minimize it [26]. To avoid pushing away the debris from the robot, impedance control is also presented in [27]. No impact is expected when position and speed are matched at capture [28, 23]. In this work, trajectories are parametrized as B-splines and the method includes an obstacle avoidance constraint in the trajectory planning optimization problem, eventually solved numerically using the single shooting method. A closed-form solution is also presented in [29] using the Pontryagin's minimum principle. The goal is to minimize a cost function and to match the target position and speed at the instant of capture. In the case of noisy measurements and parametric uncertainties on the debris inertia matrix, a complete scheme is proposed in [30] to estimate and predict its trajectory using a Kalman filter, and thus ensuring a more robust capture.

In the present study, a closed-form solution of the effector trajectory was computed based on [29], but adding the acceleration as a new final constraint. This last requirement avoids any discontinuity in the commanded torques at the joints, and yields a smoother effector trajectory that can track the grapple fixture before closing the capture mechanism.

2.3 Control Issues

Two main strategies exist for the control: the *free-flying* and the *free-floating* mode [31, 32]. The first one couples the *Attitude and Orbital Control System* (AOCS) of the spacecraft with the robotic arm controller, while the other lets the base move freely in reaction of the arm motion. The *free-flying* mode is preferred when the base attitude must be maintained during arm motion and with external disturbances. This is a strong requirement if the robot is tele-operated from the ground, and antennas need to be accurately pointed. Two separate controllers can be designed, as proposed by Oda in [33], but the AOCS can be quickly saturated if the arm moves too fast. As seen above, the trajectory can be adapted to stay within the actuators capabilities. A feedforward controller is also proposed in [34] to improve the overall performance by roughly estimating the disturbance induced by the arm. Eventually, a complete coordinated control is presented in [35] based on an adapted Transposed Jacobian method, which can achieve both position and orientation control of the end-effector and the satellite.

The controller synthesis is performed using the fixed-structure H_∞ synthesis [36]. The classic H_∞ synthesis yields an optimal full-order controller for a given plant, which complies with internal stability and frequency domain performances. As emphasized by Apkarian in [37], as soon as general constraints are specified, like controller order, decentralized control or PID-like structure, optimization techniques are necessary. A nonsmooth optimization method has been introduced by Apkarian and Noll in this paper to overcome these standard H_∞ limitations. A full solution has been added in the `Matlab` *Robust Control Toolbox*, and is called the fixed-structure H_∞ synthesis framework [36, 38]. It offers a convenient tool to specify the controller order and/or structure (e.g., decentralized, PID, observer-based…), to impose frequency domain constraints for performance purpose (e.g., response time, bandwith, disturbance rejection…), and to perform multi-model synthesis. Thus, this approach avoids the reduction of an optimal full order controller obtained through classic H_∞ synthesis solvers.

Moreover, by using the same synthesis framework, key system parameters can be introduced as optimization variables in the design process to perform a co-design study [39].

In this paper, the fixed-structure H_∞ synthesis is used to design a centralized and a decentralized axis-by-axis controller, by solving a multi-model optimization problem. The synthesis is based on the acceleration sensitivity function to perform an efficient disturbance rejection. Still, it will be shown that this method yields good results on the trajectory tracking scenario.

3 Space Robot Models

The dynamic model of a space robot is now given in more detail. A nonlinear model is used to simulate and validate the control laws, whereas a linearized model is used for controller synthesis.

The dynamic model of a general space robot is given in Eq. 2, based on [21]. A modified Newton-Euler algorithm is used to compute the matrices $\mathbf{M}$ and $\mathbf{h}$ for both simulation and control purpose [40]. It combines the dynamic equations of the satellite with the manipulator ones, obtained by a classic Newton-Euler algorithm. The simulation scheme using the $\mathbf{M}$ and $\mathbf{h}$ matrices is detailed in Fig. 2. The base consumption is also evaluated by integrating the base command (i.e., the forces commanded to the thrusters) to compute the consumed fuel mass.

$$\mathbf{M}(\mathbf{q}_m)\begin{bmatrix}\dot{\mathcal{V}}_\mathbf{b}\\ \ddot{\mathbf{q}}_\mathbf{m}\end{bmatrix}+\mathbf{h}(\omega_b,\mathbf{q}_m,\dot{\mathbf{q}}_m)=\begin{bmatrix}\mathbf{F}_\mathbf{b}\\ \tau\end{bmatrix} \quad (2)$$

with

- **M** Global mass matrix
- **h** Nonlinear effects (centrifugal and Coriolis)
- ω_b Base angular rate ($v_b = \left[\mathbf{v_b} \; \omega_b \right]^T$)
- $\mathbf{F_b}$ Efforts applied at the base
- τ Joints torques

Since the H_∞ synthesis applies to linear models, Eq. 2 is linearized around different arm's configurations along the capture trajectory. The linearization is done for a given configuration $\mathbf{q_0}$, and for zero joint velocities. This hypothesis holds well for low speed manoeuvers, but should be revisited for a more agile space robot concept.

Noting $\delta\mathbf{x}$ the variation around the nominal cases, it yields Eq. 3. The nonlinear terms in h vanish because they are quadratic in the velocities [10].

$$\mathbf{M}(\mathbf{q_0}) \begin{bmatrix} \delta\dot{\mathbf{v}}_\mathbf{b} \\ \delta\ddot{\mathbf{q}}_m \end{bmatrix} = \begin{bmatrix} \mathbf{F_b} \\ \tau \end{bmatrix} \tag{3}$$

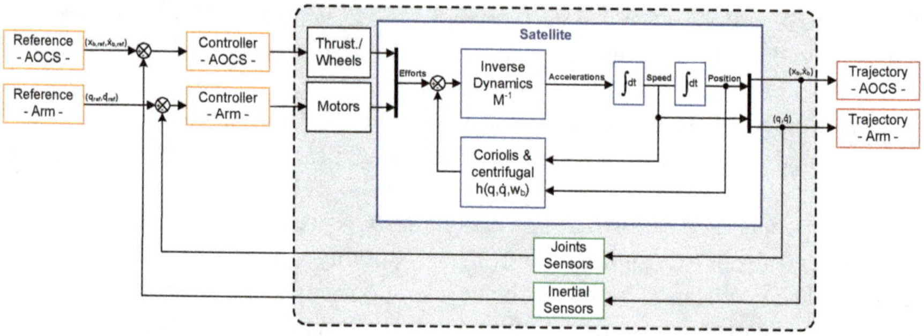

Fig. 2 Nonlinear simulation scheme

4 Optimal Path Planning

The trajectory generation is adapted from [29], by adding the acceleration as a new variable state, to ensure acceleration continuity at capture.

4.1 Target Dynamics

The debris is supposed to rotate freely in space. From the linear momentum conservation, its center of gravity (CoG) has a constant velocity $\dot{\mathbf{r}}_\mathbf{t} = \mathbf{v_0}$, where $\mathbf{r_t}$ is the CoG location. Using the quaternion representation, its dynamics is given by:

$$\frac{d\mathbf{r_t}}{dt} = \dot{\mathbf{r}}_\mathbf{t} \tag{4}$$

$$m_t \frac{d\dot{\mathbf{r}}_\mathbf{t}}{dt} = 0 \tag{5}$$

$$\frac{dQ_t}{dt} = \frac{1}{2}\,\omega_t \otimes Q_t \tag{6}$$

$$\mathbf{I_t}\frac{d\omega_t}{dt} = -\omega_t \times \mathbf{I_t}\omega_t \tag{7}$$

where m_t is the debris mass, $\mathbf{I_t}$ its inertia matrix, Q_t its quaternion of attitude and ω_t its angular rate. $\otimes$ is the quaternion product, as defined in [29]. The time derivative of the angular rate will be denoted by: $\dot{\omega}_t = -\mathbf{I_t}^{-1}(\omega_t \times \mathbf{I_t}\omega_t) \triangleq \phi(\omega_t)$. These equations can be summarized by: $\dot{\mathbf{x}} = \hat{f}(\mathbf{x})$, with $\mathbf{x} = \mathrm{col}\left(\mathbf{r_t}\ \dot{\mathbf{r}}_\mathbf{t}\ Q_t\ \omega_t\right)$.

The grapple fixture location C is now denoted $\mathbf{r_c}$ in the inertial frame, and ρ in the debris local frame. The computation of the position, speed, acceleration and jerk of C are needed to solve the optimal path planning problem afterwards. It yields:

$$\mathbf{r_c} = \mathbf{r_t} + \mathbf{R}(Q_t)\rho \tag{8}$$

$$\dot{\mathbf{r}}_\mathbf{c} = \mathbf{v_0} + \mathbf{R}(Q_t)(\omega_t \times \rho) \tag{9}$$

$$\ddot{\mathbf{r}}_\mathbf{c} = \mathbf{R}(Q_t)(\phi(\omega_t)\times\rho + \omega_t \times (\omega_t \times \rho)) \tag{10}$$

$$\dddot{\mathbf{r}}_\mathbf{c} = \mathbf{R}(Q_t)\left(\dot{\phi}(\omega_t)\times\rho + 2\phi(\omega_t)\times(\omega_t\times\rho) + \omega_t\times(\phi(\omega_t)\times\rho) - (\omega_t^T\omega_t)\,\omega_t\times\rho\right) \tag{11}$$

where $\mathbf{R}(Q_t)$ is the rotation matrix from the debris local frame to the inertial one, also defined in [29]. The derivative of the function ϕ is expressed as:

$$\dot{\phi}(\omega_t) = -\mathbf{I_t}^{-1}\left(\phi(\omega_t)\times\mathbf{I_t}\omega_t + \omega_t\times\mathbf{I_t}\phi(\omega_t)\right) + \omega_t\times\phi(\omega_t) \tag{12}$$

4.2 *Optimal Capture Trajectory*

The end-effector trajectory $\mathbf{r_e}$ is supposed to be generated by a third order derivative:

$$\dddot{\mathbf{r}}_\mathbf{e} = \mathbf{u} \tag{13}$$

By stacking the states under $\chi = \mathrm{col}\left(\mathbf{x}\ \mathbf{r_e}\ \dot{\mathbf{r}}_\mathbf{e}\ \ddot{\mathbf{r}}_\mathbf{e}\right)$, the whole dynamics can be written as:

$$\dot{\chi} = \begin{bmatrix} \hat{f}(\mathbf{x}) \\ \dot{\mathbf{r}}_\mathbf{e} \\ \ddot{\mathbf{r}}_\mathbf{e} \\ \mathbf{u} \end{bmatrix} \triangleq f(\chi,\mathbf{u}) \tag{14}$$

The trajectory is optimized for the cost function $\mathbf{J}$, which includes the time of travel t_f, the weighted acceleration $\ddot{\mathbf{r}}_\mathbf{e}$ and the command $\mathbf{u}$:

$$\mathbf{J} = \int_0^{t_f}\left(1 + w_1\|\ddot{\mathbf{r}}_\mathbf{e}\|^2 + w_2\|\mathbf{u}\|^2\right)dt \triangleq \int_0^{t_f}\mathscr{L}(\chi,\mathbf{u})dt \tag{15}$$

To apply the Pontryagin's minimum principle [41, 42], the system Hamiltonian is defined as follows, where λ are the Lagrange multipliers :

$$\mathbf{H}(\chi,\mathbf{u},\lambda) = 1 + w_1\|\ddot{\mathbf{r}}_\mathbf{e}\|^2 + w_2\|\mathbf{u}\|^2 + \lambda_s^T\hat{f}(\mathbf{x}) + \lambda_{e_1}^T\dot{\mathbf{r}}_\mathbf{e} + \lambda_{e_2}^T\ddot{\mathbf{r}}_\mathbf{e} + \lambda_{e_3}^T\mathbf{u} \tag{16}$$

A final constraint is also added to ensure the continuity of position, speed and acceleration at the capture:

$$\psi(\chi(t_f)) = \begin{bmatrix} \mathbf{r_c}(\mathbf{x}(t_f)) - \mathbf{r_e}(\mathbf{t_f}) \\ \dot{\mathbf{r}}_\mathbf{c}(\mathbf{x}(t_f)) - \dot{\mathbf{r}}_\mathbf{e}(\mathbf{t_f}) \\ \ddot{\mathbf{r}}_\mathbf{c}(\mathbf{x}(t_f)) - \ddot{\mathbf{r}}_\mathbf{e}(\mathbf{t_f}) \end{bmatrix} = 0 \tag{17}$$

By letting the time of travel t_f free, the application of the minimum principle yields the following system of equations describing the optimal solution:

$$\dot{\chi} = f(\chi,\mathbf{u}) \tag{18}$$

$$\dot{\lambda} = -\left(\frac{\partial f}{\partial \chi}\right)^T \lambda - \left(\frac{\partial \mathcal{L}}{\partial \chi}\right)^T \tag{19}$$

$$\frac{\partial \mathbf{H}}{\partial \mathbf{u}} = \left(\frac{\partial f}{\partial \mathbf{u}}\right)^T \lambda + \left(\frac{\partial \mathcal{L}}{\partial \mathbf{u}}\right)^T = 0 \tag{20}$$

$$\chi(0) = \chi_0 \tag{21}$$

$$\lambda(t_f) = \left(\nu^T \frac{\partial \psi}{\partial \chi}\right)^T_{t=t_f} \tag{22}$$

$$\mathbf{H}(t_f) = 0 \tag{23}$$

$$\psi\left(\chi(t_f)\right) = 0 \tag{24}$$

Developing Eq. 19, the differential equations for the Lagrange multipliers λ are:

$$\dot{\lambda}_s = -\left(\frac{\partial \hat{f}}{\partial \mathbf{x}}\right)^T \lambda_s \tag{25}$$

$$\dot{\lambda}_{e_1} = 0 \tag{26}$$

$$\dot{\lambda}_{e_2} = -\lambda_{e_1} \tag{27}$$

$$\dot{\lambda}_{e_3} = -\lambda_{e_2} - 2w_1\ddot{\mathbf{r}}_\mathbf{e} \tag{28}$$

In addition, Eq. 20 results in $\mathbf{u} = \dddot{\mathbf{r}}_\mathbf{e} = -\lambda_{e_3}/(2w_2)$. Combining it with Eqs. 26, 27 and 28, it results in the relation defining the optimal trajectory structure, similar to the one in [29], with $\sigma = \sqrt{w_1/w_2}$:

$$\frac{d^3}{dt^3}\left(\ddot{\mathbf{r}}_\mathbf{e} - \sigma^2\mathbf{r}_\mathbf{e}\right) = \frac{\lambda_{e_1}}{2w_2} \tag{29}$$

Noting that λ_{e_1} is constant, the optimal trajectory takes the following general formulation, where the $\mathbf{k_i} \in \mathbb{R}^3$ are the trajectory parameters :

$$\mathbf{r_e}(t) = \mathbf{k_0} + \mathbf{k_1} t + \mathbf{k_2} \frac{t^2}{2} + \mathbf{k_3} \frac{t^3}{6} + \mathbf{k_4} e^{\sigma t} + \mathbf{k_5} e^{-\sigma t} \tag{30}$$

With the acceleration constraints, the six boundary conditions in Eqs. 21 and 24 allow to compute the parameters $\mathbf{k_i}$:

$$\begin{bmatrix} \mathbb{1} & 0 & 0 & 0 & \mathbb{1} & \mathbb{1} \\ 0 & \mathbb{1} & 0 & 0 & \sigma\mathbb{1} & -\sigma\mathbb{1} \\ 0 & 0 & \mathbb{1} & 0 & \sigma^2\mathbb{1} & \sigma^2\mathbb{1} \\ \mathbb{1} & t_f\mathbb{1} & \frac{t_f^2}{2}\mathbb{1} & \frac{t_f^3}{6}\mathbb{1} & e^{\sigma t_f}\mathbb{1} & e^{-\sigma t_f}\mathbb{1} \\ 0 & \mathbb{1} & t_f\mathbb{1} & \frac{t_f^2}{2}\mathbb{1} & \sigma e^{\sigma t_f}\mathbb{1} & -\sigma e^{-\sigma t_f}\mathbb{1} \\ 0 & 0 & \mathbb{1} & t_f\mathbb{1} & \sigma^2 e^{\sigma t_f}\mathbb{1} & \sigma^2 e^{-\sigma t_f}\mathbb{1} \end{bmatrix} \begin{bmatrix} \mathbf{k_0} \\ \mathbf{k_1} \\ \mathbf{k_2} \\ \mathbf{k_3} \\ \mathbf{k_4} \\ \mathbf{k_5} \end{bmatrix} = \begin{bmatrix} \mathbf{r_e}(0) \\ \dot{\mathbf{r}}_\mathbf{e}(0) \\ \ddot{\mathbf{r}}_\mathbf{e}(0) \\ \mathbf{r_c}(t_f) \\ \dot{\mathbf{r}}_\mathbf{c}(t_f) \\ \ddot{\mathbf{r}}_\mathbf{c}(t_f) \end{bmatrix} \tag{31}$$

Eventually, to fully determine the optimal trajectory, the final time t_f is found with Eqs. 22 and 23. The second one specifies that the Hamiltonian vanishes at final time:

$$\mathbf{H}(t_f) = 0 = \left[1 + w_1 \|\ddot{\mathbf{r}}_\mathbf{e}\|^2 + w_2 \|\dddot{\mathbf{r}}_\mathbf{e}\|^2 + \lambda_s^T \hat{f}(\mathbf{x}) + \lambda_{e_1}^T \dot{\mathbf{r}}_\mathbf{e} + \lambda_{e_2}^T \ddot{\mathbf{r}}_\mathbf{e} + \lambda_{e_3}^T \mathbf{u}\right]_{t=t_f}$$

The unknown variables $\lambda(t_f)$ are defined by the so-called *transversality condition*, with Eq. 22. By eliminating the Lagrange multipliers ν, one obtains :

$$\lambda_s(t_f) = -\left(\frac{\partial \mathbf{r_e}}{\partial \mathbf{x}}\right)_{t_f} \lambda_{e_1}(t_f) - \left(\frac{\partial \dot{\mathbf{r}}_\mathbf{e}}{\partial \mathbf{x}}\right)_{t_f} \lambda_{e_2}(t_f) - \left(\frac{\partial \ddot{\mathbf{r}}_\mathbf{e}}{\partial \mathbf{x}}\right)_{t_f} \lambda_{e_3}(t_f)$$

and using the fact that:

$$\dot{\mathbf{r}}_\mathbf{e} = \left(\frac{\partial \mathbf{r_e}}{\partial \mathbf{x}}\right) \hat{f}(\mathbf{x}) \text{ , and } \ddot{\mathbf{r}}_\mathbf{e} = \left(\frac{\partial \dot{\mathbf{r}}_\mathbf{e}}{\partial \mathbf{x}}\right) \hat{f}(\mathbf{x}) \text{ , and } \dddot{\mathbf{r}}_\mathbf{e} = \left(\frac{\partial \ddot{\mathbf{r}}_\mathbf{e}}{\partial \mathbf{x}}\right) \hat{f}(\mathbf{x})$$

it eventually yields the implicit equation defining the final time of capture t_f, where the $\mathbf{r_e}$ terms are given by Eq. 30 in which the parameters $\mathbf{k_i}$ are also functions of t_f, and the $\mathbf{r_c}$ third derivative is given by Eq. 11.

$$\mathbf{H}(t_f) = 0 = 1 + w_1 \|\ddot{\mathbf{r}}_\mathbf{e}(t_f)\|^2 - w_2 \|\dddot{\mathbf{r}}_\mathbf{e}(t_f)\|^2 + 2w_2 \dddot{\mathbf{r}}_\mathbf{e}(t_f)^T \dddot{\mathbf{r}}_\mathbf{c}(t_f) \tag{32}$$

An example of optimal capture trajectory is given in Figs. 3 and 4.

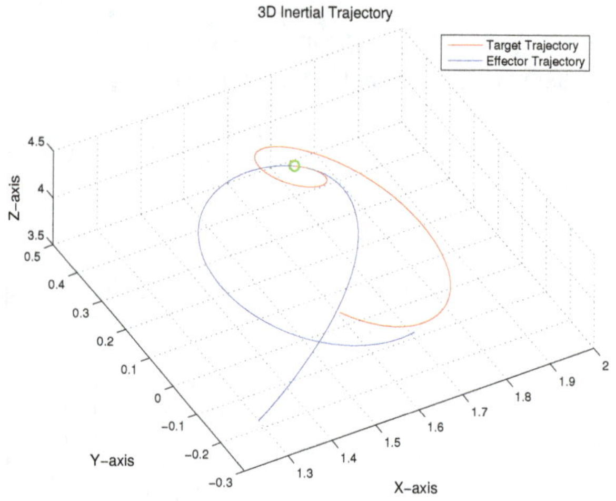

Fig. 3 Example of optimal capture trajectory

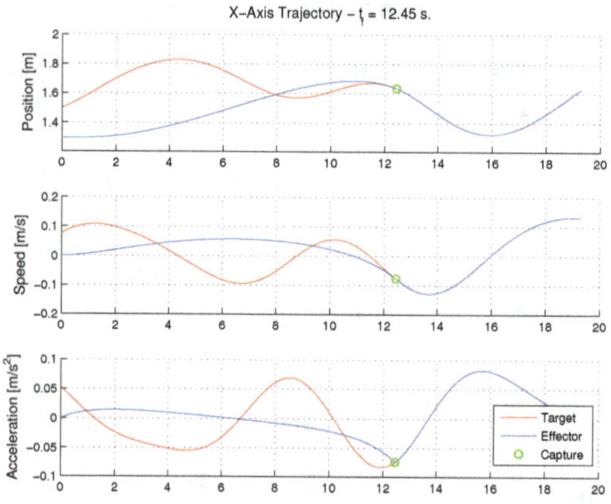

Fig. 4 Example of optimal capture trajectory (X-axis components)

5 Control Design

A control law is now investigated to follow the given effector trajectory. The controller gains are synthesized based on the fixed-structure H_∞ framework. A full space robot equipped with a 6 DoF arm is used to validate this approach.

5.1 *Control Architecture*

The choice between the free-flying or the free-floating mode depends on the processor capabilities and on the need for a centralized or a decentralized control law between the arm and the spacecraft. It is worth recalling that actual space processors are significantly less powerful that terrestrial ones, due to their robustness to the hard environment. A low complexity control architecture will then be preferred in this study to anticipate the limited processing capability of on-board processors. The coordinated control is chosen to extend the cartesian workspace of the robot and to improve its accuracy. Following the approach used in the Japanese mission ETS-VII [33], two separate controllers will be designed based on the spacecraft and the arm's specifications.

The AOCS design is kept very simple with a *Proportional-Derivative* (PD) controller. The global control architecture is illustrated in Fig. 5.

For the arm's controller, two structures are used: one is centralized and gathers all measurements before specifying the joints command, and the other is decentralized on each joint by giving the command torque based on local measurements. It will be shown how a decentralized controller, though much simpler, gives convincing results for trajectory tracking.

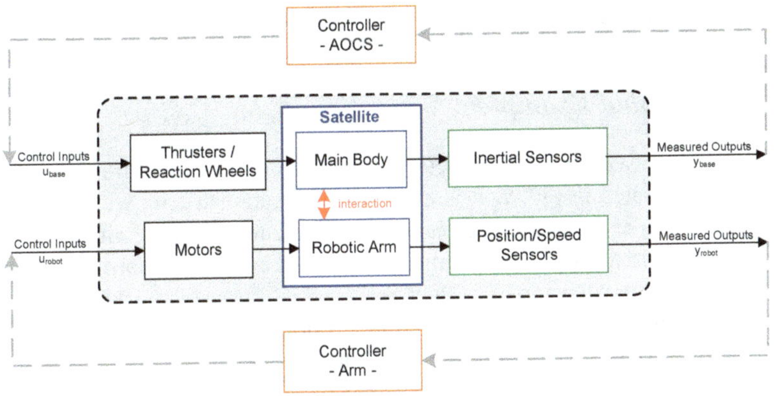

Fig. 5 Decentralized control architecture between AOCS and arm's controller

5.2 *Controller Synthesis*

5.2.1 H_∞ Synthesis Scheme

The synthesis is performed on the acceleration sensitivity function to improve the disturbance rejection [43]. By weighting the acceleration with the desired frequency response given in (33), the resulting H_∞ scheme is illustrated in Fig. 6.

$$W(s) = \frac{s^2 + 2\,\xi_{des}\,\omega_{des}\,s + \omega_{des}^2}{s^2} \tag{33}$$

The weighting function enables to reject efficiently any high-frequency disturbance. They are used for both the base and the arm specifications, with, respectively, $W_b(s)$ defined by $(\omega_b, \xi_b) = (0.01\,rad/s, 0.707)$, and $W_m(s)$ defined by

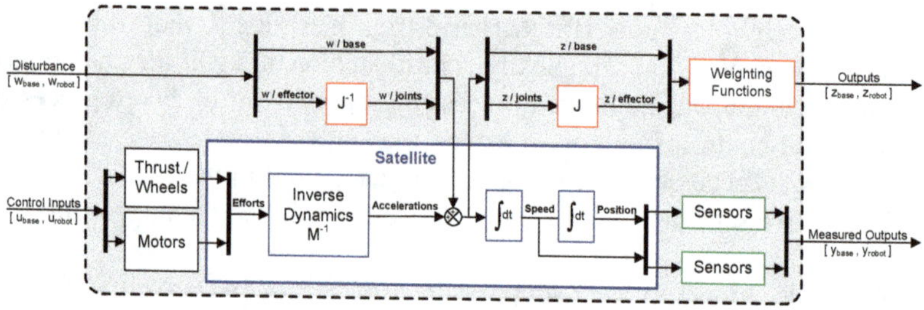

Fig. 6 H_∞ synthesis scheme

$(\omega_m, \xi_m) = (20\,\omega_t, 1)$. The angular rate of the target is fixed to $\omega_t = 13°/s$ in $W_m(s)$. The base's parameters are chosen to reproduce a quasi free-floating system, with a very slow AOCS in order to minimize the base consumption. On the other hand, the manipulator's parameters are taken 20 times greater than the debris tumbling rate ω_t, to ensure a small **tracking error**.

5.3 *Space Robot Example*

The robotic arm used for the complete example has 6 degrees of freedom and its structure is illustrated in Fig. 7. It is inspired from a human arm, with a shoulder, a forearm, an arm and a wrist. The base is a $200\,kg$ cubic satellite with side length of $2\,m$ a side, while the arm weights almost $70\,kg$ and each of its main segments are $2.6\,m$ long for a full extension of $5.7\,m$.

Two strategies are used to control this space robot. The first one is based on a full decoupling controller by means of the mass matrix used in the feedback loop. The final results show that this controller is very efficient for trajectory tracking. The second approach is based on a decentralized structure, with a PD-like controller on each joint. As explained in [44], trajectory tracking can be efficiently done with this type of controller, when the derivative action is sufficiently high. Nevertheless, the tracking error remains around 10 cm since a gain limit has been fixed at 1000 to prevent high command torques. The requirement is fixed to $200\,N.m$ for the joints torque, and to $20\,N.m$ for the base actuators. Each controller is checked on the optimal trajectory obtained for a cylinder debris rotating at $13°/s$, of $0.5\,m$ long and $0.25\,m$ of radius, and weighting $200\,kg$.

5.3.1 Centralized Controller

For the centralized architecture, the arm controller includes the mass matrix $\mathbf{M_0}$ for one configuration $\mathbf{q_0}$ of the trajectory, allowing a full decoupling and ensuring good performance for this very configuration. It takes the following form for a robotic arm of N joints:

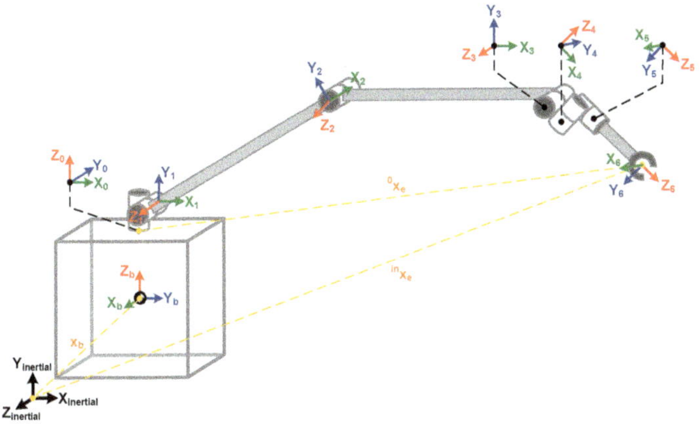

Fig. 7 Space robot example with a 6 DoF arm

$$
K(s) = \mathbf{M_0} \left[\begin{array}{ccccc|ccccc}
K_{p,1} & 0 & \dots & 0 & 0 & K_{d,1} & 0 & \dots & 0 & 0 \\
0 & \ddots & & \vdots & 0 & 0 & \ddots & & \vdots & 0 \\
\vdots & & K_{p,i} & & \vdots & \vdots & & K_{d,i} & & \vdots \\
0 & \dots & & \ddots & 0 & 0 & \dots & & \ddots & 0 \\
0 & 0 & \dots & 0 & K_{p,N} & 0 & 0 & \dots & 0 & K_{d,N}
\end{array} \right] \quad (34)
$$

The synthesis is done using the `systune` function [38]. The weighting functions are still specified as « hard constraints », while the gain limitation and the consumption minimization are set by « soft constraints » [38].

To improve the robust performance of this controller, five different models obtained along the trajectory have been considered at the same time for the synthesis. For all of them, a common controller is synthesized by optimization in `systune`. This multi-model synthesis gives the gain setting in Tab. 1.

The frequency responses of the weighted transfer functions are illustrated in Fig. 8(a). It can be seen that the overall frequency constraints represented by two black curves are well satisfied, for both the base and the manipulator. The trajectory tracking is performed under a 1.5 cm error in Fig. 8(b) and 8(e). It can be noticed that the fuel consumption given in Fig. 8(f) is about 45 mg for the capture trajectory.

Table 1 Gains setting for a **centralized** arm controller

Joint i	1	2	3	4	5	6
$K_{p,i}$	0.4	194.9	11.6	0.5	0.4	942.2
$K_{d,i}$	1.4	35.1	163.6	613.7	602.9	395.2

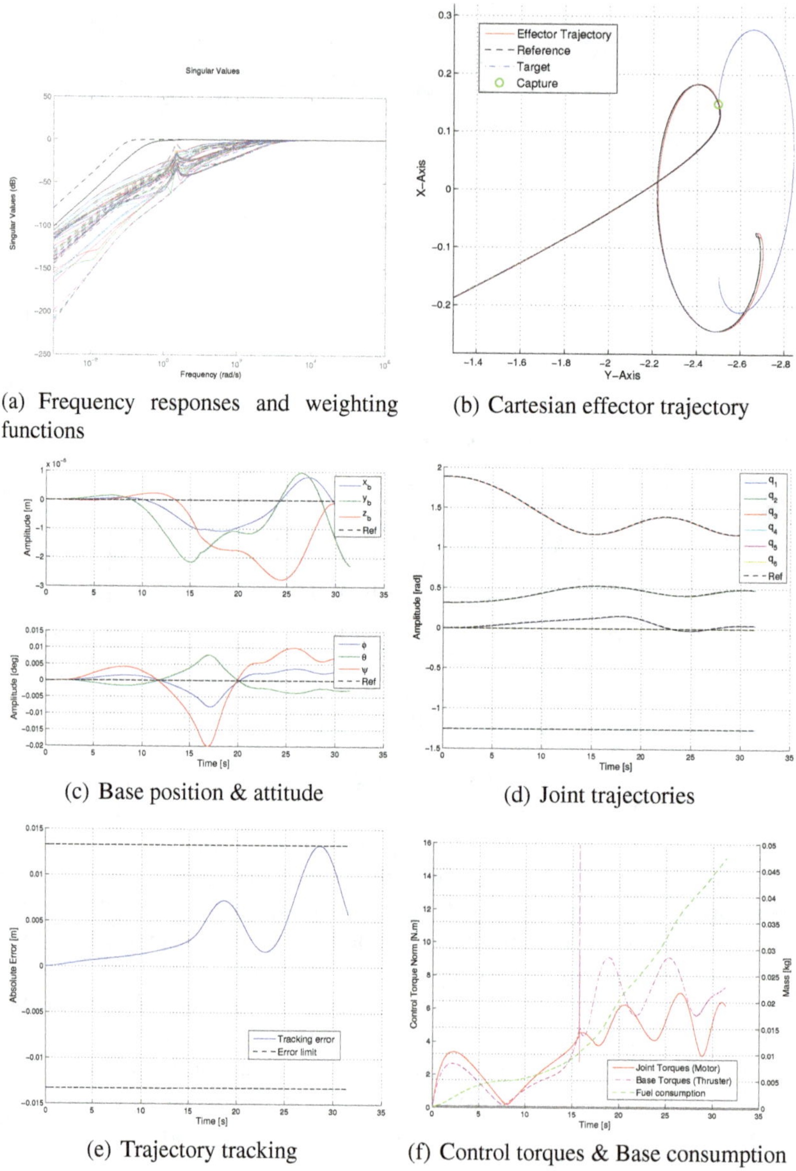

(a) Frequency responses and weighting functions

(b) Cartesian effector trajectory

(c) Base position & attitude

(d) Joint trajectories

(e) Trajectory tracking

(f) Control torques & Base consumption

Fig. 8 Centralized controller - Simulations results for the complete space robot

5.3.2 Decentralized Controller

For the decentralized architecture, the PD controllers for each joint are simultaneously synthesized considering the whole arm dynamics. With this architecture, each of them only uses its local measurement of position and velocity. The coupling between the joints and the base, or between the joints themselves, are actually considered directly in the synthesis since the linearized model is coupled. The synthesized controller is given by:

$$K(s) = \left[\begin{array}{ccccc|ccccc} K_{p,1} & 0 & \dots & 0 & 0 & K_{d,1} & 0 & \dots & 0 & 0 \\ 0 & \ddots & & \vdots & 0 & 0 & \ddots & & \vdots & 0 \\ \vdots & & K_{p,i} & & \vdots & \vdots & & K_{d,i} & & \vdots \\ 0 & \dots & & \ddots & 0 & 0 & \dots & & \ddots & 0 \\ 0 & 0 & \dots & 0 & K_{p,N} & 0 & 0 & \dots & 0 & K_{d,N} \end{array}\right] \quad (35)$$

The synthesis is done again with five different models along the trajectory, to obtain the gains setting given in Tab. 2.

The frequency constraints are mostly fulfilled even if some peaks are degrading the H_∞-norm of some transfer functions, as shown in Fig. 9(a). Even though the controller is much more simpler, the trajectory tracking error only increases to $8\,cm$ of error in Fig. 9(b) and 9(e). Eventually, the fuel consumption rises to almost 50 mg as shown in Fig. 9(f).

Table 2 Gains setting for a **decentralized** arm controller

Joint i	1	2	3	4	5	6
$K_{p,i}$	62.0	161.3	147.2	44.1	1000	3.0
$K_{d,i}$	446.0	1000	1000	1000	945.3	1000

5.3.3 Results

The use of a simple PD controller on each joint can be considered as a good compromise between easiness of implementation and tracking performance. It is clear that a centralized architecture enables a better trajectory tracking, under 1 cm of error, but at the expense of a high computational burden since the mass matrix needs to be evaluated once at the beginning of the trajectory. The tracking error increases to 8 cm for the decentralized structure, but the closed-loop system stays stable and very close to the specified frequency behavior. This last structure remains still attractive with respect to its simplicity. It can be used to approach and to synchronize the effector motion with the target, and then to switch to the centralized architecture to reduce by an order of magnitude the tracking error.

At the consumption level, both structures need nearly the same amount of fuel during the trajectory given in Figs. 8(b) or 9(b). The major difference lies in the base attitude and in the effector positionning. The centralized controller enables to reach an excellent accuracy by maintaining the base position and attitude error almost zero, i.e., less than $0.1\,mm$ in linear displacement, and around 0.01 degree of angular motion, as shown in Fig. 8(c). Conversely, the decentralized structure results in larger displacements, i.e., 10 cm in linear motion and 0.5 degrees in angular one,

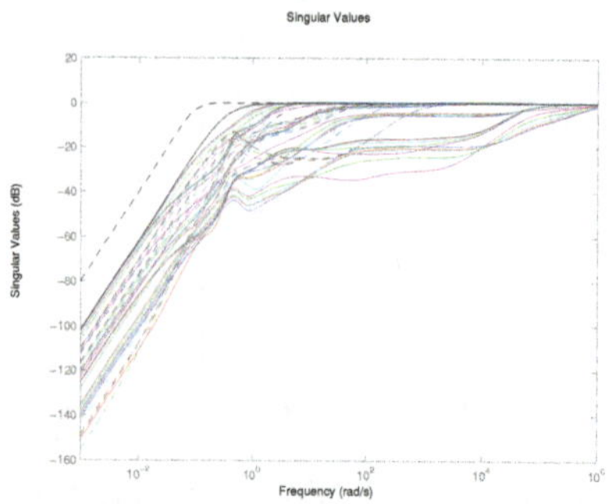

(a) Frequency responses and weighting functions

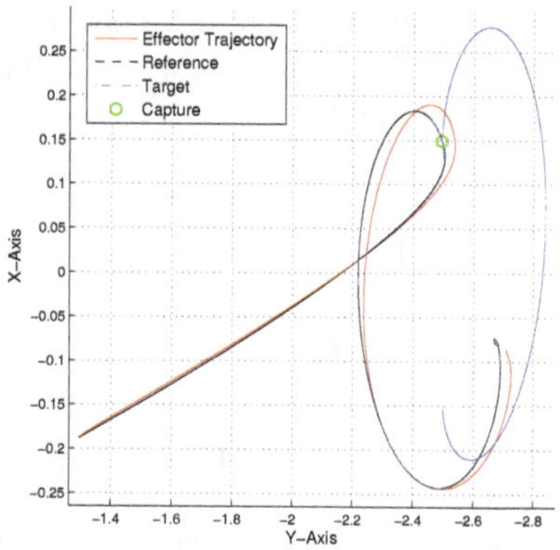

(b) Cartesian effector trajectory

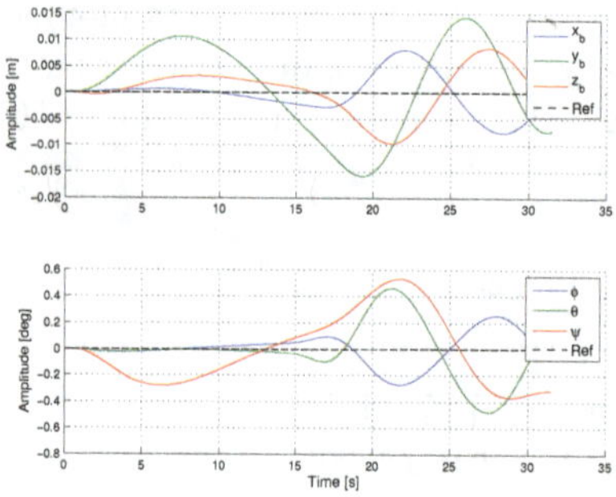

(c) Base position & attitude

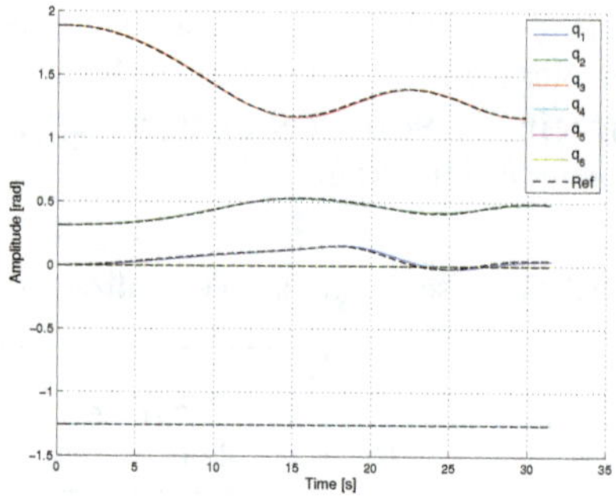

(d) Joint trajectories

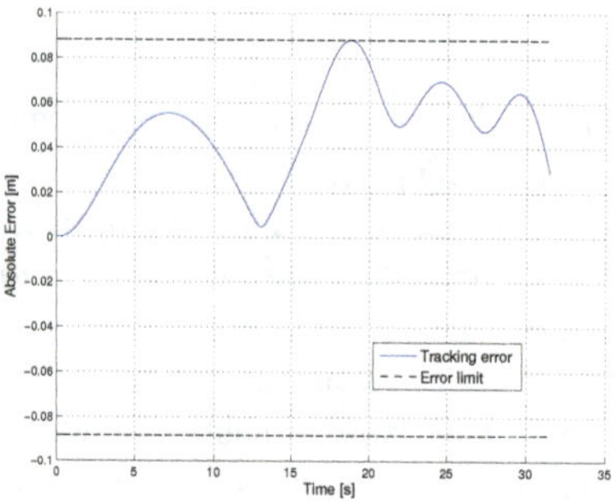

(e) Trajectory tracking

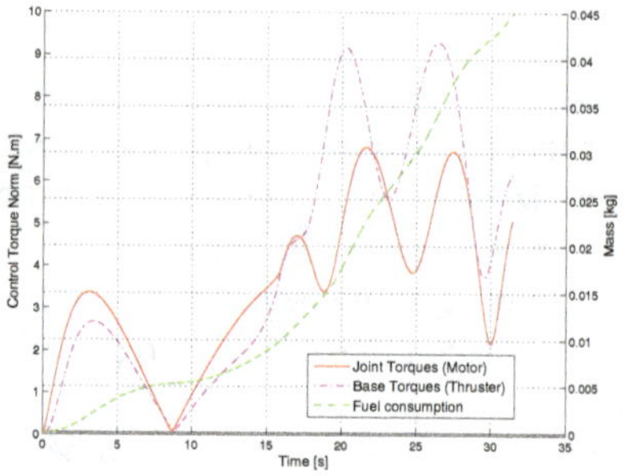

(f) Control torques & Base consumption

Fig. 9 Decentralized controller - Simulations results for the complete space robot

in Fig. 9(c). This difference generates important errors at the effector because of the lever arm of the extended arm, which is around 5 meters in the simulation case.

6 Conclusion

A space robot controller has been presented through this paper. Its main goal was to follow an optimal end-effector trajectory to capture a rotating target. To do so, a first part recalled the main results of space robotics kinematics, before using them to compute the optimal capture trajectory. Then the fixed-structure H_∞ synthesis has been used to synthesize the spacecraft AOCS and the arm controller at the same time. Two main strategies were tested : either a centralized arm controller to improve decoupling on one configuration, or a decentralized one, to keep the structure as simple as possible. These synthesis were performed simultaneously on five distinct models along the trajectory. It has eventually been shown how the decentralized controller can provide an attractive compromise for the trajectory tracking since it does not consume significantly more fuel and is easier to implement.

References

[1] Kessler, D.J., Cour-Palais, B.G.: Collision Frequency of Artificial Satellites: The Creation of a Debris Belt. Journal of Geophysical Research 83(A6), 2637–2646 (1978)

[2] Liou, J.-C.: An active debris removal parametric study for LEO environment remediation. Advances in Space Research 47, 1865–1876 (2011)

[3] Liou, J.-C., Johnson, N.L., Hill, N.M.: Controlling the growth of future LEO debris populations with active debris removal. Acta Astronautica 66(5-6), 648–653 (2010)

[4] NASA Orbital Debris Program Office. Monthly Number of Objects in Earth Orbit by Object Type. Orbital Debris Quaterly News 17(1), 8 (2013)

[5] Aikenhead, B.A., Daniell, R.G., Davis, F.M.: Canadarm and the space shuttle. Journal of Vacuum Science & Technology A 1(2), 126–132 (1983)

[6] Inaba, N., Oda, M.: Autonomous satellite capture by a space robot: world first on-orbit experiment on a japanese robot satellite ETS-VII. In: Proceedings of the 2000 IEEE International Conference on Robotics & Automation, vol. 2, pp. 1169–1174 (2000)

[7] Friend, R.B.: Orbital Express program summary and mission overview. In: Howard, R.T., Motaghedi, P. (eds.) Sensors and Systems for Space Applications II, vol. 6958, pp. 1–11 (2008)

[8] Boge, T., Wimmer, T., Ma, O., Tzschichholz, T.: EPOS - Using robotics for RvD simulation of on-orbit servicing missions. In: AIAA Modeling and Simulation Technologies Conference, Toronto, Canada, vol. 1, pp. 1–15 (2010)

[9] Wang, F., Sun, F., Liu, H.: Space robot modeling and control considering the effect of orbital mechanics. In: 1st International Symposium on Systems and Control in Aerospace and Astronautics, vol. 1, pp. 193–198 (2006)

[10] Spong, M.W., Hutchinson, S., Vidyasagar, M.: Robot Modeling and Control. John Wiley & Sons, New York (2006)

[11] Umetani, Y., Yoshida, K.: Workspace and Manipulability Analysis of Space Manipulator. Transactions of the Society of Instrument and Control Engineers E-1(1), 8 (2001)

[12] Umetani, Y., Yoshida, K.: Resolved motion rate control of space robotic manipulators with generalized jacobian matrix. IEEE Transactions on Robotics and Automation 5(3), 303–314 (1989)

[13] Papadopoulos, E., Dubowsky, S.: On the dynamic singularities in the control of free-floating manipulators. In: ASME Winter Annual Meeting, vol. 15, pp. 45–52 (1989)

[14] Nanos, K., Papadopoulos, E.: On Cartesian Motions with Singularities Avoidance for Free-floating Space Robots. In: IEEE International Conference on Robotics and Automation, vol. 1, pp. 5398–5403 (2012)
[15] Nakamura, Y., Mukherjee, R., Barbara, S.: Nonholonomic path planning of space robots. In: IEEE International Conference on Robotics and Automation, vol. 2, pp. 1050–1055 (1989)
[16] Vafa, Z., Dubowsky, S.: On the dynamics of manipulators in space using the virtual manipulator approach. In: IEEE International Conference on Robotics and Automation, vol. 4, pp. 579–585 (1987)
[17] Papadopoulos, E., Dubowsky, S.: On the nature of control algorithms for free-floating space manipulators. IEEE Transactions on Robotics and Automation 7(6), 750–758 (1991)
[18] Dubowsky, S., Vance, E.E., Torres, M.A.: The Control of Space Manipulators Subject to Spacecraft Attitude Control Saturation Limits. In: NASA Conference on Space Telerobotics, vol. IV, pp. 409–418 (1989)
[19] Oda, M.: Motion control of the satellite mounted robot arm which assures satellite attitude stability. Acta Astronautica 41(11), 739–750 (1997)
[20] Dubowsky, S., Torres, M.A.: Path planning for space manipulators to minimize spacecraft attitude disturbances. In: IEEE International Conference on Robotics and Automation, vol. 3, pp. 2522–2528 (1991)
[21] Nenchev, D.N.: Reaction Null Space of a multibody system with applications in robotics. Mechanical Sciences 4(1), 97–112 (2013)
[22] Yoshida, K., Umetani, Y.: Control of space free-flying robot. In: 29th Conference on Decision and Control, vol. 1, pp. 97–102 (1990)
[23] Lampariello, R.: On Grasping a Tumbling Debris Object with a Free-Flying Robot. In: IFAC Symposium on Automatic Control in Aerospace, pp. 1–6 (2013)
[24] Piersigilli, P., Sharf, I., Misra, A.K.: Reactionless capture of a satellite by a two degree-of-freedom manipulator. Acta Astronautica 66(1-2), 183–192 (2010)
[25] Nguyen-Huynh, T.-C., Sharf, I.: Adaptive reactionless motion for space manipulator when capturing an unknown tumbling target. In: IEEE International Conference on Robotics and Automation, vol. 1, pp. 4202–4207 (2011)
[26] Nenchev, D.N., Yoshida, K.: Impact analysis and post-impact motion control issues of a free-floating space robot contacting a tumbling object. In: Proceedings of the 1998 IEEE International Conference on Robotics & Automation, vol. 1, pp. 913–919 (1998)
[27] Yoshida, K., Nakanishi, H.: Impedance matching in capturing a satellite by a space robot. In: IEEE/RSJ International Conference on Intelligent Robots and Systems, vol. 3, pp. 3059–3064 (2003)
[28] Lampariello, R.: Motion Planning for the On-orbit Grasping of a Non-cooperative Target Satellite with Collision Avoidance. In: International Symposium on Artificial Intelligence, Robotics and Automation in Space, vol. 1, pp. 636–643 (2010)
[29] Aghili, F.: Coordination Control of a Free-Flying Manipulator and its Base Attitude to Capture and Detumble a Noncooperative Satellite. In: IEEE International Conference on Intelligent Robots and Systems, pp. 2365–2372. IEEE (2009)
[30] Aghili, F.: A prediction and motion-planning scheme for visually guided robotic capturing of free-floating tumbling objects with uncertain dynamics. IEEE Transactions on Robotics 28(3), 634–649 (2012)
[31] Dubowsky, S., Papadopoulos, E.: The kinematics, dynamics, and control of free-flying and free-floating space robotic systems. IEEE Transactions on Robotics and Automation 9(5), 531–543 (1993), doi:10.1109/70.258046

[32] Ali, S., Moosavian, A., Papadopoulos, E.: Free-flying robots in space: An overview of dynamics modeling, planning and control. Robotica 25(05), 537–547 (2007)
[33] Oda, M.: On the dynamics and control of ETS-7 satellite and its robot arm. In: Proceedings of the IEEE International Conference on Intelligent Robots and Systems (IROS), vol. 3, pp. 1586–1593 (1994)
[34] Oda, M.: Coordinated control of spacecraft attitude and its manipulator. In: IEEE International Conference on Robotics and Automation, vol. 1, pp. 732–738 (1996)
[35] Papadopoulos, E., Dubowsky, S.: Coordinated manipulator/spacecraft motion control for space robotic systems. In: IEEE International Conference on Robotics and Automation, vol. 2, pp. 1696–1701 (1991)
[36] Gahinet, P., Apkarian, P.: Structured H_∞ synthesis in MATLAB. In: IFAC World Congress, vol. 18, pp. 1435–1440 (2011)
[37] Apkarian, P., Noll, D.: Nonsmooth H_∞ synthesis. IEEE Transactions on Automatic Control 51(1), 71–86 (2006)
[38] Apkarian, P.: Tuning Controllers Against Multiple Design Requirements. In: American Control Conference, vol. 1, pp. 3888–3893 (2013)
[39] Alazard, D., Loquen, T., de Plinval, H., Cumer, C.: Avionics/Control co-design for large flexible space structures. In: AIAA Guidance, Navigation, and Control (GNC) Conference, pp. 1–15. AIAA (2013)
[40] Carignan, C.R., Akin, D.L.: The reaction stabilization of on-orbit robots. IEEE Control Systems 20, 19–33 (2000)
[41] Athans, M., Falb, P.L.: Optimal Control. McGraw-Hill (1966)
[42] Bryson, A.E., Ho, Y.C.: Applied Optimal Control. Hemisphere, New York (1975)
[43] Alazard, D.: Reverse engineering in control design. ISTE / John Wiley & Sons, London / Hoboken (2013)
[44] Kawamura, S., Miyazaki, F., Arimoto, S.: Is a local linear PD feedback control law effective for trajectory tracking of robot motion? In: IEEE International Conference on Robotics and Automation, vol. 3, pp. 1335–1340 (1988)

Cross-Entropy Based Probabilistic Analysis of VEGA Launcher Performance

Anusha Mujumdar, Prathyush Purushothama Menon, Christophe Roux, and Samir Bennani

Abstract. In this paper, a cross entropy based simulation technique is presented for characterising the probability profile of performance of the closed loop Vega launcher model. The paper describes a theoretical development of the technique, develops an algorithm for its use in probability profile generation, and presents the results of the implementation on the Vega model. The results demonstrate the efficacy of the tool, and its benefits in industrial control system analysis.

1 Introduction

One of the greatest challenges in control system design and analysis is the presence of uncertainty. "All models are wrong, but some are useful" [1]. The difference between any physical process and its model is represented in the form of uncertainty. Spacecraft controller validation techniques often rely entirely on models (sometimes coupled with test data). Such knowledge uncertainty is also known as epistemic uncertainty, which is reducible, in theory. Epistemic uncertainty is systematic uncertainty, and describes parameters that have precise values but which are unknown at the time of design. These may or may not

Anusha Mujumdar · Prathyush Purushothama Menon
University of Exeter, United Kingdom EX44QF
e-mail: {apm211,P.M.Prathyush}@exeter.ac.uk

Christophe Roux
ELV, 00187 Rome, Italy
e-mail: christophe.roux@elv.it

Samir Bennani
ESA-ESTEC, 2201 AZ Noordwijk, Netherlands
e-mail: Samir.Bennani@esa.int

J. Bordeneuve-Guibé et al. (eds.), *Advances in Aerospace Guidance, Navigation and Control*, DOI: 10.1007/978-3-319-17518-8_41

be known with precision on launch day. Some examples are the drag coefficient, coupling terms, and engine efficiency. The other class of uncertainty is aleatoric uncertainty, which is statistical, or luck uncertainty, and describes parameters that have unavoidable stochastic variability, such as wind gust. It is important to account for both types of uncertainties in the design and analysis process. It is common to lump both types of uncertainties together, however, this may not be the best strategy and may give rise to overly pessimistic, or conservative results. A good approach is to attempt to reduce all epistemic uncertainties to aleatoric uncertainties [2].

Quantifying the effects of uncertainty is vital to determining spacecraft safety and performance during flight. $\mu-$analysis[3] is an elegant analytical technique for quantifying these effects[4], however it requires the model to strictly be linear. This is often not the case, and so the space industry relies on methods that do not impose assumptions on the model being analysed - such as Monte Carlo techniques [5], and increasingly, on optimisation techniques, for determining the worst case stability and performance criteria[6]-[9].

An important issue of consequence is how likely the occurrence of such a worst case is. Studying this issue requires probabilistic methods, and some research effort has focused on this [10],[11],[12]. Certain recent approaches in this direction are probabilistic gain [13] and randomised analysis techniques [14]. Another relevant recent work is [15], in which the performance of F-16 controller is analysed probabilistically, using the concept of the Wasserstein distance between distributions.

In this paper, the cross entropy (CE) simulation technique [16],[17] is applied to analyse the performance of the European Space Agency's VEGA launcher model [18]. The objective is to generate a "probability profile" of performance. This work builds on previous work published in [19]. CE is a technique that utilises the Kullback-Liebler divergence between two distributions to adaptively arrive at an optimal distribution. In the previous work, the uncertain parameter distributions were simply "shifted" towards the region of higher importance, whereas in the current work, the distributions are shrunk, such that the resulting parameter distributions represent highly accurate regions with low variances. An importance sampling run is hence carried out to determine the probability associated with performance.

The outline of the paper is as follows. First, a mathematical formulation of the problem is presented. The methodology implemented and the cross entropy algorithm used are then discussed. An adaptive initialisation scheme is introduced for efficient computation of the probability profile. The Vega launcher model is described, following which some results are presented.

2 Problem Formulation

For a closed loop dynamical system, let

$$J(\delta) : \mathbb{D} \to \mathbb{R}$$

be a performance criterion, where $\mathbb{D} \subset \mathbb{R}^k$ is an uncertainty domain. A desired threshold for each criterion may be written as $J(\delta) \leq \gamma_{des}$.

In an optimisation-based approach to worst case analysis, the problem is formulated as a maximisation problem subject to the dynamics of the model:

$$\begin{aligned} &\max_{\delta \in \mathbb{R}^k} && J(\delta) \\ &\text{subject to} && \dot{X} = F(X,U,\delta),\ \delta \in \mathbb{D} \end{aligned} \tag{1}$$

Suppose the worst case performance obtained is γ^*.

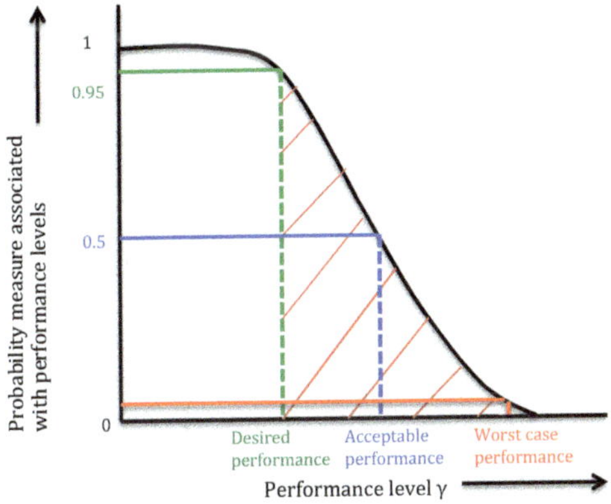

Fig. 1 Probability degradation profile

The interest is to determine the probability of the performance function $J(\delta)$ exceeding a performance level γ, i.e., $P[J(\delta) \geq \gamma]$, as the performance degrades between the desired performance level (γ_{des}) and worst case performance level (γ^*). The characterisation is called the *probability profile*, and is illustrated in Figure 1. As shown, the probability $P[J(\delta) \geq \gamma]$ is determined at the desired performance level γ_{des}, the worst case performance γ^* and performance levels between these two. The worst case performance is interpreted as a rare event [20], and is expected to have a very low associated probability value (10^{-5} or lower).

Several benefits of studying the probability profile exist. Firstly, it indicates the degree of conservatism of the worst case value. If the worst case probability is indeed small, the design need not be changed. It also reveals how feasible the desired performance level is, and whether a more realistic criterion needs to be set. If the performance criteria bounds are to be relaxed, the probability profile gives a good framework for such a relaxation in an informed way. This can lead to a probabilistic method of specifying design criteria.

3 Methodology

Consider the uncertainties δ to be distributed according to the family of probability distribution functions (pdf) $f(\delta; v)$, where v is the parameter in the family. In other words, δ is a random vector such that $\delta \sim f(\delta; v)$. v could be, for example, the mean and standard deviation in the normal distribution. In this work the distribution $f(\delta; v)$ is a normal distribution, and v represents its mean. Finding the probability

$$l = P\left[J(\delta) \geq \gamma\right] \tag{2}$$

requires solving the following non-deterministic polynomial-time hard integration problem:

$$E\left[J(\delta) \geq \gamma\right] = \int_{\mathbb{D}} I_{\{J(\delta)\geq\gamma\}} f(\delta; v)\, d\delta \tag{3}$$

to find the expectation E that $[J(\delta) \geq \gamma]$, where $I_{\{J(\delta)\geq\gamma\}}$ is the indicator function

$$I_{\{J(\delta)\geq\gamma\}} = \begin{cases} 1 \text{ if } J(\delta) \geq \gamma \\ 0 \text{ if } J(\delta) < \gamma \end{cases}. \tag{4}$$

A naive stochastic estimator is Crude Monte Carlo (CMC) simulation [16], in which N samples $\delta_1, \delta_2, ..., \delta_N$ are drawn from the distribution $f(\delta; v)$. Then

$$\hat{l}_{CMC} = \frac{1}{N}\sum_{i=1}^{N} I_{\{J(\delta_i)\geq\gamma\}} \tag{5}$$

gives an unbiased probability estimation. This becomes intractable when $\{J(\delta) \geq \gamma\}$ is a rare event i.e., has a small probability of occurrence (say of the order 10^{-5}) [20]. For rare events we require an estimator that can identify the important parameters. Importance sampling techniques are a class of sampling techniques which work in this fashion [2]. For rare events, CE is particularly suited.

4 The Cross Entropy Method for Rare Event Simulation

CE is an adaptive importance sampling algorithm, developed in [16] to determine the probabilities of rare events. CE adaptively computes the best reference parameter(s), and biases the initial uncertain parameter distribution so that the rare event becomes more likely to occur in the new distribution [17]. Subsequently, an importance sampling run computes the probability of occurrence of the rare event by scaling back from the new distribution to the original distribution, with the help of a likelihood ratio.

Importance sampling seeks to find an *importance sampling density* or the *instrumental density* [2]. Let $g(\delta)$ defined on $\mathbb{D}$ be this unknown distribution, to be found. Suppose that N random samples $\delta_1, ..., \delta_N$ are drawn from $g(\delta)$. Let the initial distribution be $f(\delta; u)$, where u is the initial reference parameter. Then l may be found by modifying equation (3) as follows:

$$E\{J(\delta) \geq \gamma\} = \int_{\mathbb{D}} I_{\{J(\delta) \geq \gamma\}} \frac{f(\delta; u)}{g(\delta)} g(\delta) \, d\delta \tag{6}$$

The stochastic estimator of this is written, similar to (5), as follows:

$$\hat{l} = \frac{1}{N} \sum_{i=1}^{N} I_{\{J(\delta_i) \geq \gamma\}} \frac{f(\delta_i; u)}{g(\delta_i)}. \tag{7}$$

The optimal estimate is then found by having g^* satisfying:

$$g^*(\delta) := \frac{I_{\{J(\delta) \geq \gamma\}} f(\delta; u)}{l} \tag{8}$$

which depends on l and so is unknown.

4.1 Estimation of Rare Event Probabilities

Different IS techniques choose $g(.)$ appropriately. CE is particularly good at estimating small probabilities, i.e., events whose probabilities of occurrence are smaller than 10^{-5}. CE chooses g from within the family of densities $f(\cdot, v)$. The reference parameter v is chosen such that the distance between the densities $g^*(.)$ and $f(\cdot; v)$ is minimum. In the CE method, the Kullback-Leibler divergence, also known as the cross-entropy is the distance between the distributions. For $g^*(\delta)$ and $f(\delta; v)$, the Kullback-Leibler distance is:

$$\begin{aligned} \mathcal{D}(g^*(\delta), f(\delta; v)) = & \int g^*(\delta) \ln(g^*(\delta)) \, d\delta \\ & - \int g^*(\delta) \ln(f(\delta; v)) \, d\delta \end{aligned} \tag{9}$$

Minimising the Kullback-Leibler divergence for $g^*(.)$ and $f(\delta; v)$ is equivalent to minimising the second term $-\int g^*(\delta) \ln(f(\delta; v)) \, d\delta$, which is in turn equivalent to solving the maximisation problem:

$$\max_{v} \int g^{*}(\delta) \ln (f(\delta; v))\, d\delta. \tag{10}$$

Substituting for $g^{*}(.)$ from (8), the optimisation program becomes

$$\max_{v} \int \frac{I_{\{J(\delta) \geq \gamma\}} f(\delta; u)}{l} \ln (f(\delta; v))\, d\delta. \tag{11}$$

If we define the likelihood ratio W as

$$W(\delta; u, v) = \frac{f(\delta; u)}{f(\delta; v)} \tag{12}$$

the solution to the optimisation program may be found by solving for the optimal updating factor v^{*} in the following equation:

$$\frac{1}{N} \sum_{i=1}^{N} I_{\{J(\delta_i) \geq \gamma\}} W(\delta_i; u, v) \nabla \ln (f(\delta_i; v^{*})) = 0. \tag{13}$$

For a gaussian distribution $v = [\mu\ \sigma]$ and $f(\delta; v) = f(\delta; [\mu\ \sigma])$

$$\int I_{\{J(\delta) \geq \gamma\}} W(\delta; [\mu_0\ \sigma_0], [\mu\ \sigma]) f(\delta; [\mu\ \sigma]) \begin{bmatrix} \frac{\partial}{\partial \mu} \ln (f(\delta; [\mu\ \sigma])) \\ \frac{\partial}{\partial \sigma} \ln (f(\delta; [\mu\ \sigma])) \end{bmatrix} d\delta = 0. \tag{14}$$

Substituting $f(\delta; [\mu\ \sigma]) = \frac{1}{\sigma\sqrt{2\pi}} e^{-\frac{(\delta-\mu)^2}{2\sigma^2}}$ in (14) and separating the components we have:

$$\int I_{\{J(\delta) \geq \gamma\}} W(\delta; [\mu_0\ \sigma_0], [\mu\ \sigma]) f(\delta; [\mu\ \sigma]) \frac{\partial}{\partial \mu} \left(\ln \left(\frac{1}{\sigma\sqrt{(2\pi)}} \right) - \frac{(\delta-\mu)^2}{2\sigma^2} \right) d\delta = 0. \tag{15}$$

$$\int I_{\{J(\delta) \geq \gamma\}} W(\delta; [\mu_0\ \sigma_0], [\mu\ \sigma]) f(\delta; [\mu\ \sigma]) \frac{\partial}{\partial \sigma} \left(\ln \left(\frac{1}{\sigma\sqrt{(2\pi)}} \right) - \frac{(\delta-\mu)^2}{2\sigma^2} \right) d\delta = 0. \tag{16}$$

Proceeding to find the updating rule for the mean μ from 15:

$$\int I_{\{J(\delta) \geq \gamma\}} W(\delta; [\mu_0\ \sigma_0], [\mu\ \sigma]) f(\delta; [\mu\ \sigma]) \left(0 - \frac{\partial}{\partial \mu} \frac{(\delta-\mu)^2}{2\sigma^2} \right) d\delta = 0. \tag{17}$$

$$\int I_{\{J(\delta) \geq \gamma\}} W(\delta; [\mu_0\ \sigma_0], [\mu\ \sigma]) f(\delta; [\mu\ \sigma]) \left(\frac{2\delta - 2\mu}{2\sigma^2} \right) d\delta = 0. \tag{18}$$

The left side of the above equation may be written as the expectation $E_v\{I_{\{J(\delta)\geq\gamma\}}W(\delta;[\mu_0\ \sigma_0],[\mu\ \sigma])(\delta-\mu)\}$, where the subscript v indicates that it is taken with respect to the density $f(\delta;v)=f(\delta;[\mu\ \sigma])$. This is expressed stochastically as follows:

$$\frac{1}{N}\sum_{i=1}^{N} I_{\{J(\delta_i)\geq\gamma\}}W(\delta_i;[\mu_0\ \sigma_0],[\mu\ \sigma])(\delta_i-\mu)=0 \tag{19}$$

From the above equation, the updating rule for the mean may be found as follows:

$$\mu^*=\frac{\sum\limits_{i=1}^{N} I_{\{J(\delta_i)\geq\gamma\}}W(\delta_i;[\mu_0\ \sigma_0],[\mu\ \sigma])\,\delta_i}{\sum\limits_{i=1}^{N} I_{\{J(\delta_i)\geq\gamma\}}W(\delta_i;[\mu_0\ \sigma_0],[\mu\ \sigma])} \tag{20}$$

Proceeding to derive the updating rule for the standard deviation from (16):

$$\int I_{\{J(\delta)\geq\gamma\}}W(\delta;[\mu_0\ \sigma_0],[\mu\ \sigma])\,f(\delta;[\mu\ \sigma])\frac{\partial}{\partial\sigma}\left(\ln\left(\frac{1}{\sqrt{2\pi}}\right)+\ln\left(\frac{1}{\sigma}\right)-\frac{(\delta-\mu)^2}{2\sigma^2}\right)d\delta=0. \tag{21}$$

i.e.,

$$\int I_{\{J(\delta)\geq\gamma\}}W(\delta;[\mu_0\ \sigma_0],[\mu\ \sigma])\,f(\delta;[\mu\ \sigma])\left(\left(\frac{-1}{\sigma}\right)+\frac{(\delta-\mu)^2}{\sigma^3}\right)d\delta=0. \tag{22}$$

Separating the additive terms:

$$\int I_{\{J(\delta)\geq\gamma\}}W(\delta;[\mu_0\ \sigma_0],[\mu\ \sigma])\,f(\delta;[\mu\ \sigma])\left(\frac{-1}{\sigma}\right)d\delta = \int I_{\{J(\delta)\geq\gamma\}}W(\delta;[\mu_0\ \sigma_0],[\mu\ \sigma])\,f(\delta;[\mu\ \sigma])\left(\frac{(\delta-\mu)^2}{\sigma^3}\right)d\delta. \tag{23}$$

This gives:

$$\sigma^2=\frac{\int I_{\{J(\delta)\geq\gamma\}}W(\delta;[\mu_0\ \sigma_0],[\mu\ \sigma])\,f(\delta;[\mu\ \sigma])(\delta-\mu)^2}{\int I_{\{J(\delta)\geq\gamma\}}W(\delta;[\mu_0\ \sigma_0],[\mu\ \sigma])\,f(\delta;[\mu\ \sigma])} \tag{24}$$

Drawing N samples of δ from $f(\delta;[\mu\ \sigma])$, we get the stochastic estimator of the above as follows:

$$\sigma^2=\frac{\frac{1}{N}\sum\limits_{i=1}^{N} I_{\{J(\delta_i)\geq\gamma\}}W(\delta;[\mu_0\ \sigma_0],[\mu\ \sigma])(\delta_i-\mu)^2}{\frac{1}{N}\sum\limits_{i=1}^{N} I_{\{J(\delta_i)\geq\gamma\}}W(\delta;[\mu_0\ \sigma_0],[\mu\ \sigma])} \tag{25}$$

This gives the updating rule for the standard deviation as:

$$\sigma = \sqrt{\frac{\sum_{i=1}^{N} I_{\{J(\delta_i)\geq\gamma\}} W\left(\delta; \left[\mu_0\ \sigma_0\right], \left[\mu\ \sigma\right]\right)\left(\delta_i - \mu\right)^2}{\sum_{i=1}^{N} I_{\{J(\delta_i)\geq\gamma\}} W\left(\delta; \left[\mu_0\ \sigma_0\right], \left[\mu\ \sigma\right]\right)}} \tag{26}$$

where $i = 1,...,N$, $j = 1,...,k$. Here, $\delta_{i,j}$ represents the jth element of the ith sample δ_i (recall that each sample of the uncertain parameter belongs to $\mathbb{R}^k$).

4.2 The CE Algorithm

As discussed, the objective is to adaptively compute the best reference parameter, v^*, such that the KL distance is minimum. This is carried out in a multi-stage algorithm by finding the sequence of reference parameters $\{v_t, t \geq 0\}$ and a corresponding sequence of performance levels $\{\gamma_t, t \geq 1\}$ and iterating until $\gamma_t \geq \gamma$. Details of the generalised algorithm may be found in [16]. The principle of the CE algorithm is illustrated in Figure 2. The algorithm biases the initial distribution towards the region in the parameter space that is leads to the higher performances. Then, this distribution is scaled back using the likelihood ratio to determine the probability of occurrence of the performance value in the original performance.

The CE algorithm is shown in Algorithm 1. This is carried out for each performance level of interest γ. Recall that N is the number of evaluations per CE iteration t, ρ is the rarity factor, *maxits* represents the maximum number of CE iterations allowed. N_1 is the number of evaluations for the final importance sampling run. Generally speaking the choice of N and N_1 rely on the model under study, and on the computational budget. In any case both of these are far lower than for naive Monte Carlo. A drawback of importance sampling techniques is that the likelihood ratio becomes unstable for large number of uncertainties.

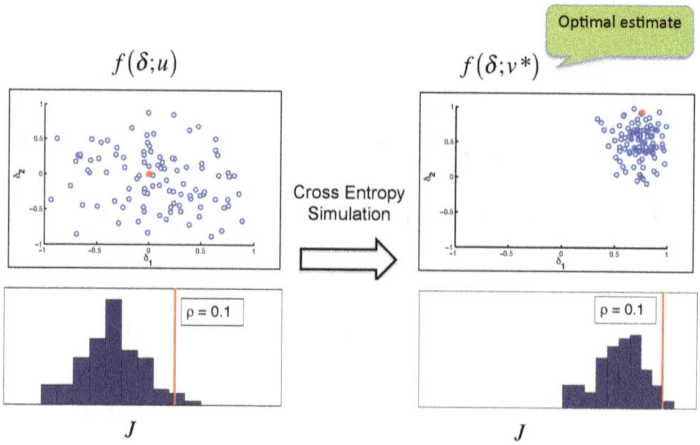

Fig. 2 Principle of the cross entropy method

5 Probability Profile Characterisation

Characterising the evolution of probability as the performance degrades involves finding the probability associated with each performance level, and then plotting these probabilities w.r.t performance. Please refer to [19] for preliminary results on a numerical example and on the Darwin scientific model.

6 Vega Launcher Model

The analysis is performed on a non-linear, six degrees of freedom industrial simulation model of the VEGA launch vehicle[18], a recent European multi-payload launch vehicle developed by ELV under ESA responsibility. The so called ELV VEGA CONTROL simulator is a Matlab C-coded simulator used in industry to prepare and validate the VEGA flight management and flight control system in the atmospheric and exo- atmospheric flight phases. In this study the first flight phase with height between 35m and 60km is considered. The simulation environment has been used as a benchmark model for the ESA research activity "Robust Flight Control System Design Verification and Validation Framework" (ESA AO/1-6322/09/NL/JK) to permit comparison between industrial analysis outcomes based on Monte Carlo simulations with optimization based analysis techniques. Previously, mainly deterministic worst case analysis techniques were assessed upon their merit by measuring the validation gap between traditional Monte Carlo outcomes and worst case analysis results. Subsequently this activity aims at narrowing down the previously determined gap by using probabilistic worst case analysis techniques while associating tight probabilistic risk levels to the requirements under investigation.

The model includes the full set of non-linear equations of motion. The simulator includes a non-linear model of the eletro-mechanical actuator (EMA) dynamics with associated backlash and delays. The simulator also implements the QUASAR Inertial Sensor Unit with its noise and bias characteristics. The propulsion model reflects the dynamics of the P80 solid propulsion system with validated thrust curves that include thrust oscillation effects to assess proper execution of the separation dynamics. The flight mechanic model assembles a fully validated non-linear aerodynamic model for normal force, drag and tree axis moments curves that depends on Mach and angle of attack. Kinematic coupling in all axes are implemented to reflect pitch yaw roll couplings.

A high fidelity structural flexible mode model describing the launcher deformation is included to assess proper filtering and stability properties. The atmosphere model includes also a set of measured sizing wind-gust input models representative for the launch site Kourou.The launcher dynamics are driven by the FPSA ADA/C-flight code reflecting the flight management system for the time line sequence command and execution of associated guidance navigation and control system for thrust vector control (TVC) and roll and attitude control (RACS) and other support functions such as acceleration threshold detection ATD and pyro-valve command for stage separation.

For the purpose of independent validation and verification studies the model has been distributed to various research teams as a compiled executable simulation model. Configuration files have been provided to set boundary conditions and distribution profiles for the uncertainty parameters relevant for the analysis. Access to more than hundred parameters is provided ranging from aerodynamic scatterings, winds, thrust, thrust offset and misalignments, individual mass, center of gravity and inertia variations for all stages and mechanical parts, noise, non-linearities, flex modes uncertainties in frequencies and damping etc. All parameters have normalised deterministic and stochastic dimensionless components.

The overall validation and verification criteria considered in this research activity represent the TVC and RACS technical requirements for the atmospheric phase of flight. Here the major eight requirements reflected as optimization criteria have been extracted from the compliance matrix. These include bounds on loads, lateral deviations as well as limits on separation conditions under the severe wind conditions and in the presence of perturbing roll torques when uncertainties and dispersions are applied.

- $\mathbf{C_1}$ Load requirement boundaries must be limited via the product of dynamic pressure Q and angle of attack α over the entire P80 flight Mach range (criterion C1)
- $\mathbf{C_2}$ to $\mathbf{C_5}$ Drift of position and speed in y- and z-direction within the altitude range of 35m to 60km.
- $\mathbf{C_6}$ Limits on transversal angular acceleration.
- $\mathbf{C_7}$ Transverse angular velocity at separation must be limited to a given value.
- $\mathbf{C_8}$ Total angle of attack at separation must remain bounded and less than a desired value.

In the optimization setting all criteria boundary limits have been normalized to their such that a violation is indicated when the objective function value becomes less than unity. The problem addressed in this paper attempts to assess the requirement on $max(Q\alpha)$ over a reduced parameter combination subset that dominates the requirement degradation.

7 Results

The cross entropy algorithm is applied to the Vega launcher model in order to characterise its probability profile of performance. The performance J considered is the maximum of the dynamic pressure, i.e., $max(Q\alpha)$. Only four of the uncertainties are varied by the algorithm. The uncertainties considered

Data: μ_0,σ_0, *maxits*, ρ, σ^*,α,N and N_1;
Result: The rare-event estimate $\hat{l}$;
Set $t=1$;
Set $\mu(t)=\mu_0$ and $\sigma(t)=\sigma_0$;
while $t \leq maxits$ **do**

1. Generate a sample $\delta_1,...,\delta_N$ from the density $f(.;\mu(t),\sigma(t))$;
2. Compute γ_t, the sample $(1-\rho)$ quantile as follows:
   2a. Calculate the values of the cost $J(\delta_i)$ for each sample δ_i.;
   2b. Arrange them in ascending order. Then find $\gamma_t = J_{[(1-\rho)]}$.;
3. Use the same sample $\delta_1,...,\delta_N$ to find the updating parameter as follows:

$$\mu_j(t+1) = \frac{\sum_{i=1}^{N} I_{\{J(\delta_{i,j})\geq\gamma\}} W\left(\delta_{i,j};[\mu_0,\sigma_0],[\mu(t),\sigma(t)]\right)\delta_{i,j}}{\sum_{i=1}^{N} I_{\{J(\delta_{i,j})\geq\gamma\}} W\left(\delta_{i,j};[\mu_0,\sigma_0],[\mu(t),\sigma(t)]\right)}$$

and

$$\sigma_j(t+1)=\sqrt{\frac{\sum_{i=1}^{N} I_{\{J(\delta_{i,j})\geq\gamma\}} W\left(\delta_{i,j};[\mu_0,\sigma_0],[\mu(t),\sigma(t)]\right)(\delta_{i,j}-\mu_j(t+1))^2}{\sum_{i=1}^{N} I_{\{J(\delta_{i,j})\geq\gamma\}} W\left(\delta_{i,j};[\mu_0,\sigma_0],[\mu(t),\sigma(t)]\right)}}$$

if $min(\sigma(t+1)) \leq \sigma^*$ **then**
$\hat{\mu}_T = \hat{\mu}_(t+1)$;
$\hat{\sigma}_T = \hat{\sigma}_(t+1)$;
Proceed to step 4
end
if $\gamma_t \geq \gamma$ **then**
$\gamma_t = \gamma$;
$\hat{\mu}_T = \hat{\mu}_t$;
$\hat{\sigma}_T = \hat{\sigma}_t$;
Optimal reference parameter found, proceed to step 4;
else
$t=t+1$;
end

end
4. Estimate the rare-event probability l using the likelihood ratio estimator:

$$\hat{l} = \frac{1}{N_1}\sum_{i=1}^{N_1} I_{\{J(\delta)\geq\gamma\}} W\left(\delta_{i,j};[\mu_0,\sigma_0],[\mu_T,\sigma_T]\right);$$

Algorithm 1. The CE Algorithm for Rare Event Simulation

are: "IRSmountingY", "dTc", "SRM_roll" and "air_density_scat". All other uncertainties are fixed at 0.1 (an arbitrary small positive value). These four uncertainties were of interest with respect to the system perspective. CE can handle a reasonable number of uncertainties (so far, we have worked with a maximum of 28 uncertainties in various systems). However, it is a known issue that all importance sampling methods become unstable at high dimensions due to likelihood ratio degeneracy[21].

The initial distributions of these 4 uncertainties are Gaussian distributions with mean 0 and standard deviation 0.4, truncated at [-1,1]. The standard deviation chosen is such that the selected region corresponds to approximately 98.5% of the entire Gaussian (2.5σ). The rationale for such a choice of μ_0 and σ_0 is to obtain a distribution with nominal value at 0 and such that a large number of values lie in the interval [-1,1]. The probability estimation algorithm is applied to find $P[J(\delta) \geq \gamma]$ for every 10,000 Padeg of $max(Q\alpha)$ from $1.8 * 10^5$ Padeg onwards, until a probability of 0 is encountered.

The runs were carried out on Matlab R2012a running on a Windows i7 machine with 8 GB of memory. The algorithmic parameters and constants are listed in table 1.

Table 1 CE algorithm parameters and constants utilised for the run

Name	Symbol	Value
Rarity factor	ρ	0.1
Number of simulations in the adaptive run	N	50
Number of simulations in the final run	N_1	100
Maximum number of iterations of the adaptive run	$maxits$	100
Threshold standard deviation value	$sigma^*$	10^{-5}
Smoothing parameter	α	0.9

The resulting probability profile is shown in Figure 3 (the constituent data is available in Table 2). At the start of the curve, i.e., at performances of 1.8, 1.9 and $2 * 10^5$ Padeg, the associated probability is 1. This indicates that the maximum of $Q\alpha$ will always be at least $2 * 10^5 Padeg$. From here, the probability gradually decreases until the zero probability occurs at $3.1 * 10^5 Padeg$. The tail end of this probability profile is especially interesting, as it provides information about the worst case and performances close to the worst case. Recall from the discussions in the problem formulation section, that the worst case is intuitively perceived to be a rare event. A small probability of $1.13 * 10^{-6}$ is observed to occur at $3 * 10^5$ Padeg. We can infer, therefore, that the worst case value of $max(Q\alpha)$ lies between $3 * 10^5$ and $3.1 * 10^5 Padeg$. In order to further narrow this region, the algorithm was applied to the performance levels $3.06 * 10^5$ and $3.07 * 10^5 Padeg$. These cases gave probabilities of $1.044 * 10^{-7}$ and 0 respectively. The true worst case may therefore can be said to lie in $[3.06 * 10^5, 3.07 * 10^5] Padeg$. Further, the probability of its occurrence is a rare event, i.e., one with a probability lower than 10^{-5}. This agrees with our expectation that the worst case be a rare event.

Apart from the worst case probability, the probability profile offers information on the intermediate performance levels between 2 and $3 * 10^5 Padeg$.

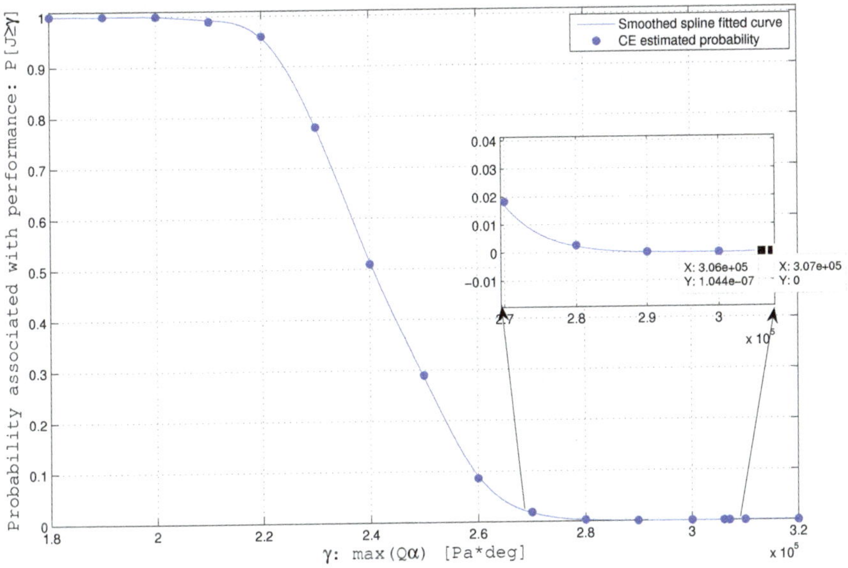

Fig. 3 Probability profile of performance $max(Q\alpha)$. The inset box shows a magnified view of the tail values. The smallest probability obtained is $1.044 * 10^{-7}$ at $3.06 * 10^5$ Padeg. The blue line represents a spline fitted curve, indicating that it is possible to find associated probabilities for several performance values without performing runs each time.

As the performance degrades, the probability gradually reduces. The spline fitted curve in Figure 3 is useful in determining probability values at any performance level of interest. This serves to identify how realistic a performance requirement is. In the current scenario, for example, imposing that the maximum dynamic pressure must be lower than $2.3 * 10^{-5} Padeg$ is not a good choice, since there is a 78% probability that the requirement will be violated. A good requirement is one that has a small probability of violation.

Conversely, the performance associated with some fixed probability may be determined from the probability profile. This approach suggests a new paradigm for specifying performance criteria. For example, a designer may choose to accept a 2% risk of performance violation - this corresponds to a performance value of about $2.68 * 10^5 Padeg$ in the current case, which may then be considered as a performance requirement.

The following plots show some results for the case $\gamma = 3.06 * 10^5$ Padeg, in order to understand the internal working of the CE algorithm. Figure 4 shows the final pdfs of the four parameters, optimised by the cross entropy algorithm. Observe that the resulting variance has been reduced significantly as compared to the initial value.

Table 2 Estimated probabilities associated with various performance levels, and the number of iterations it took the CE algorithm to arrive at the importance sampling distribution

Performance level γ	Estimated probability $P[J \geq \gamma]$	Number of iterations t	Function evaluations $t * N + N_1$
180000	1	1	150
190000	1	1	150
200000	1	1	150
210000	0.99	1	150
220000	0.96	1	150
230000	0.78	1	150
240000	0.51	1	150
250000	0.29	1	150
260000	0.086	2	200
270000	0.018	2	200
280000	0.00266	3	250
290000	0.00014	4	300
300000	$1.13 * 10^{-6}$	6	350
306000	$1.044 * 10^{-7}$	7	400
307000	0	24	1300
310000	0	24	1300
320000	0	24	1300

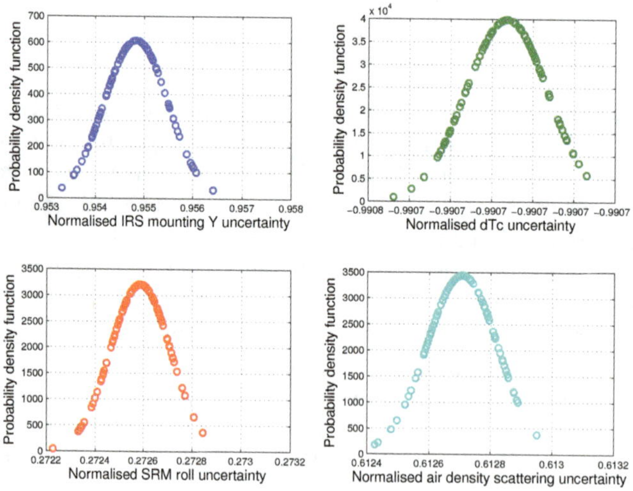

Fig. 4 Optimised probability density functions

Figure 7 shows the evolution of the mean parameter with iterations. Note that while these converge quickly, there is no monotonicity observed in their behaviour. This is a desirable property, and implies that the algorithm is insensitive to the initial conditions and does not get stuck in local minima. Figure 7 shows the evolution of the standard deviation with iterations. Within 7 iterations, the standard deviation achieved is less than 0.1 in all cases. The termination criterion $\gamma_t \geq \gamma$ occurred in 7 iterations, and thus there was no need for the algorithm to fine tune μ and σ any further.

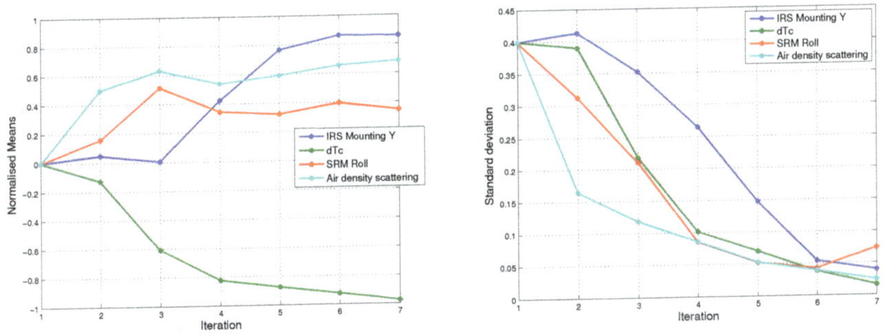

(a) IRSMounting Y (b) dTc

Fig. 5 Evolution of the mean and standard deviation at $\gamma = 3.06 * 10^5 Pa * deg$

The mean values of the uncertainties at the worst case is shown in Figure 6.

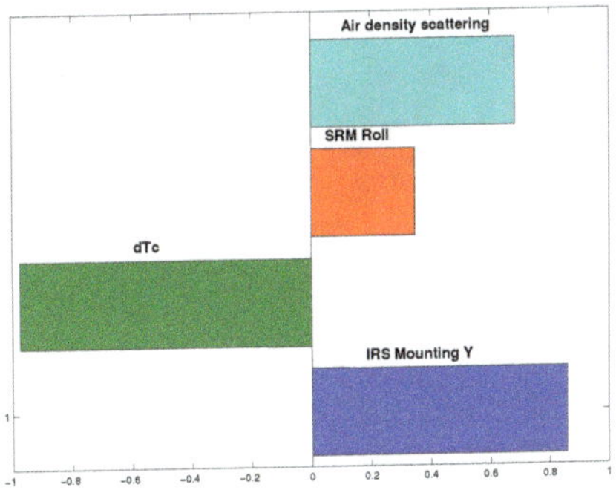

Fig. 6 Mean values of uncertain parameters at worst case

The scatter plots in Figure 7 show the performance distribution as well as the distribution of each of the uncertain parameters.

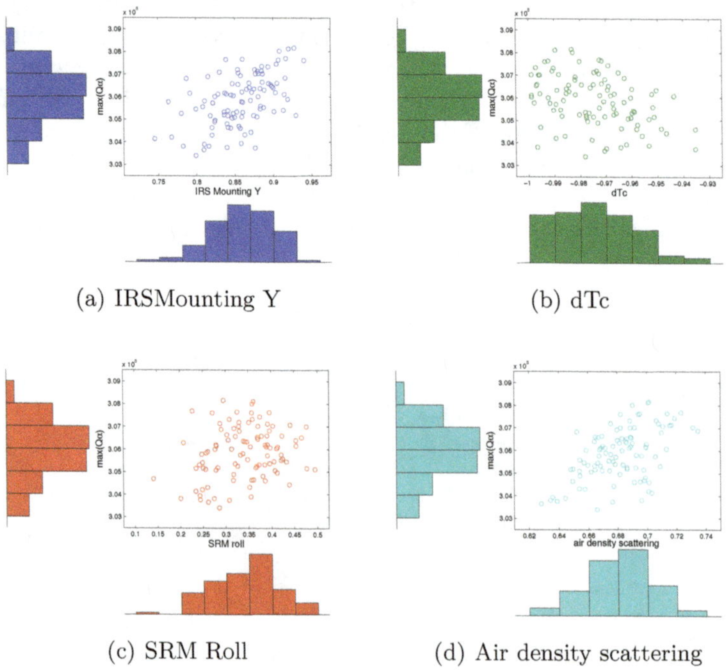

(a) IRSMounting Y (b) dTc

(c) SRM Roll (d) Air density scattering

Fig. 7 Final run with optimised parameter distributions. The costs obtained at different values of the uncertain parameters, along with the parameter and performance distributions are shown here.

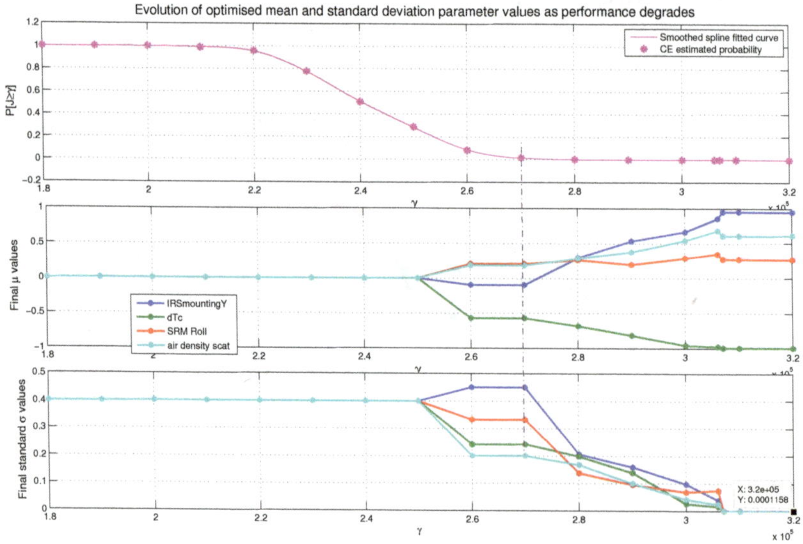

Fig. 8 Variation of μ_T and σ_T over the 17 runs carried out in characterising the probability profile.

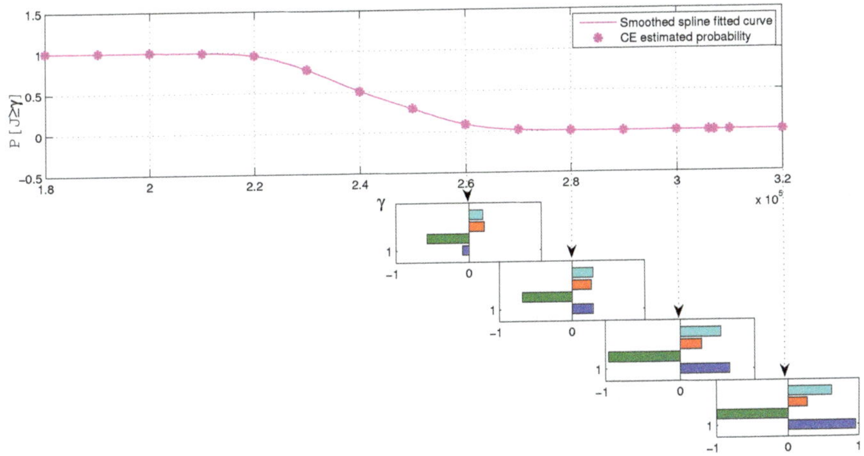

Fig. 9 Probability profile with the μ_T values obtained for each uncertainty. This indicates the directionality of the uncertainties.

It is also interesting to study how $Q\alpha$ varies with the mach number, and also with γ. This is shown in Figure 10. Figure 11 shows how $Q\alpha$ varies with the Mach number over the 17 computations.

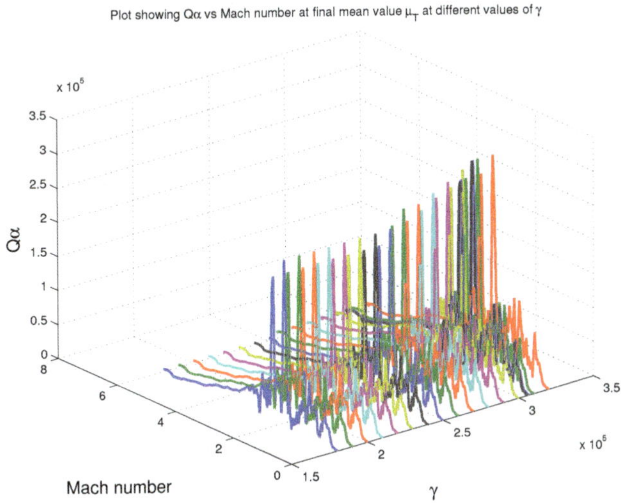

Fig. 10 Plot of $Q\alpha$ vs the mach number and γ. This plot shows the gradual increase in the $Q\alpha$ value obtained as γ increases.

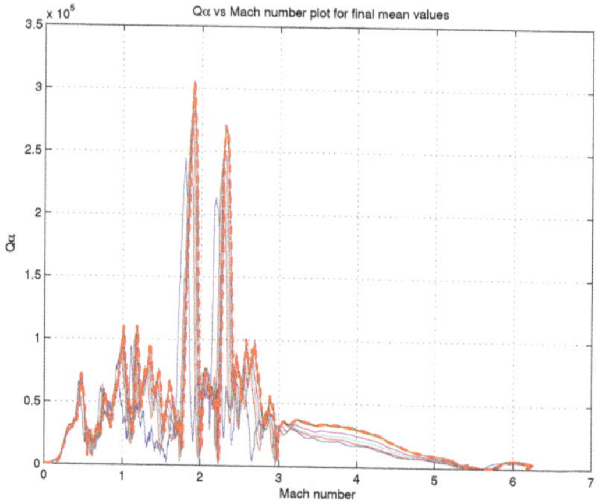

Fig. 11 Plot of $Q\alpha$ vs the Mach number over the 17 computations. The thick red line shows the $Q\alpha$ vs Mach number trajectory at which the peak value of $Q\alpha$ is obtained.

8 Conclusions

The cross entropy simulation method was applied successfully to the Vega launcher model. This demonstrates the efficacy of the technique to characterise the probability profile of industrial control systems. The probability profile is a useful tool for controller analysis, to determine the conservatism of the worst case, and the feasibility of designed performance thresholds.

References

1. Box, G.E.P., Draper, N.R.: Empirical Model Building and Response Surfaces. John Wiley & Sons, New York (1987)
2. Hanson, J.M., Beard, B.B.: Applying Monte Carlo Simulation to Launch Vehicle Design and Requirements Analysis, NASA Marshall Space Flight Center, HuntsvilleTechnical Publication ID 20100038453 (2010)
3. Zhou, K., Doyle, J.C., Glover, K.: Robust and Optimal Control. Prentice Hall, New Jersey (1996)
4. Biannic, J.-M., Roos, C., Knauf, A.: Design and robustness analysis of fighter aircraft flight control laws. European Journal of Control 12(1), 71–85 (2006)
5. Ray, L.R., Stengel, R.F.A.: Monte Carlo Approach to the Analysis of Control System Robustness. Automatica 29, 229–236 (1993)
6. Fielding, C., Varga, A., Bennani, S., Selier, M.: Advanced Techniques for Clearance of Flight Control Laws. Springer (2002)
7. Charbonnel, C.: H_∞ Controller Design and Analysis: Powerful Tools for Flexible Satellite Attitude Control. In: Proceedings of AIAA Guidance, Navigation and Control Conference and Exhibit, Toronto, Canada, August 2-5 (2010)

8. Menon, P.P., Bates, D.G., Postlethwaite, I.: Nonlinear Robustness Analysis of Flight Control Laws for Highly Augmented Aircraft. Control Engineering Practice 15(6), 655–662 (2007)
9. Menon, P.P., Kim, J., Bates, D.G., Postlethwaite, I.: Clearance of Nonlinear Flight Control Laws Using Hybrid Evolutionary Optimization. IEEE Transactions on Evolutionary Computation 10(6), 689–699 (2006)
10. Tempo, R., Giuseppe, C., Dabbene, F.: Randomized Algorithms for Analysis and Control of Uncertain Systems: With Applications. Springer (2013)
11. Khargonekar, P., Tikku, A.: Randomized algorithms for robust control analysis and synthesis have polynomial complexity. In: Proceedings of the IEEE Conference on Decision and Control, vol. 3. IEEE (1996)
12. Calafiore, G.C., Dabbene, F., Tempo, R.: Research on probabilistic methods for control system design. Automatica 47, 1279–1293 (2011)
13. Balas, G.J., Seiler, P.J., Packard, A.: Analysis of an UAV Flight Control System Using Probabilistic μ. In: Proceedings of the AIAA Guidance, Navigation and Control Conference, Minneapolis, Minnesota, August 13-16, pp. 13–16 (2012)
14. Tempo, R., Ishii, H.: Monte Carlo and Las Vegas Randomized Algorithms for Systems and Control. European Journal of Control 13, 189–203 (2007)
15. Halder, A., Lee, K., Bhattacharya, R.: A Probabilistic Method for Nonlinear Robustness Analysis of F-16 Controllers, arXiv preprint arXiv:1402.0147 (2014)
16. De Boer, P.T., Kroese, D.P., Mannor, S., Rubinstein, R.Y.: A tutorial on the cross-entropy method. Annals of operations research 134(1), 19–67 (2005)
17. Rubinstein, R.Y., Kroese, D.P.: Simulation and the Monte Carlo Method, Wiley Interscience (2008)
18. Diaz, E.G.: Design of a robust controller for the VEGA TVC using the μ-synthesis technique, Master's Thesis, ESA (2011)
19. Mujumdar, A., Menon, P.P., Edwards, C., Bennani, S.: Characterising the probability profile of performance degradation of a Darwin satellite controller. In: 2014 IEEE European on Control Conference (ECC), pp. 2514–2519 (June 2014)
20. Bucklew, J.A.: Introduction to Rare Event Simulation. Springer, New York (2004)
21. Rubinstein, R.Y., Glynn, P.W.: How to deal with the curse of dimensionality of likelihood ratios in Monte Carlo simulation. Stochastic Models 25(4), 547–568 (2009)

Author Index

Acosta, Diana 263
Alazard, Daniel 475, 639, 659, 699
Ammann, Nikolaus 605
Andert, Franz 605
Ankersen, Finn 659
Apkarian, Pierre 23
Arzelier, Denis 679

Baier, Thaddäus 3
Bennani, Samir 719
Bérard, Caroline 699
Biannic, Jean-Marc 491
Boada-Bauxell, Josep 401
Bolting, Jan 491
Bordeneuve, Joël 475
Bouadi, Hakim 455
Boubakir, Ahsene 151
Burlion, Laurent 23, 401, 531

Casasco, Massimo 571
Chaix, Jérémie 343
Chambon, Emmanuel 23
Chriette, Abdelhamid 401
Chu, Qiping P. 81, 187, 363
Cieslak, Jérome 67, 323
Cumer, Christelle 639

Dávila, J. 323
Defaÿ, François 491
Delahaye, Daniel 521
de Loza, Alejandra Ferreira 67, 323
Denieul, Yann 475
de Plinval, Henry 531
de Visser, C.C. 187

Döll, Carsten 283
Drouin, Antoine 455
Dubanchet, Vincent 699

Ehlers, Jana 433

Fezans, Nicolas 433
Fichter, Walter 135, 151, 571
Fridman, Leonid 67

Gibert, Victo 401
Grzymisch, Jonathan 571

Hayoun, Shmuel Y. 227
Heller, Matthias 3
Hellmundt, Fabian 99
Henry, David 67, 323
Holzapfel, Florian 99

Joldes, Mioara 679
Joos, Hans-Dieter 53
Jouhaud, Frank 119

Kampen, Erik-Jan van 81, 363
Kampen, E. van 187
Kaneshige, John 263
Kim, Mingu 39
Kim, Youdan 39
Klöckner, Andreas 509
Kownacki, Cezary 415

Lamp, Maxim 549
Lee, Jin-Ik 247
Lee, Seokwon 39
Lee, Yongwoo 39

Le Peuvédic, Catherine 699
Lizy-Destrez, Stéphanie 343
Lombaerts, Thomas 263
Loquen, Thomas 639
Losa, Damiana 571
Lovera, Marco 205
Lu, Peng 81, 363
Luckner, Robert 549

Maass, Bolko 605
Maier, Rudolf 99
Mangion, David Zammit 381
Massotti, Luca 659
Menon, Prathyush Purushothama 719
Mora-Camino, Felix 455
Mujumdar, Anusha 719
Murali, Hari Hara Sudhan 659

Ołdziej, Daniel 415
Osterhuber, Robert 99
Oumer, Nassir W. 625

Panin, Giorgio 625
Panomruttanarug, Benjamas 455
Panza, Simone 205
Perez, Jose Alvaro 639
Pittet, Christelle 591, 639
Plestan, Franck 401
Prieur, Pascal 591
Puechmorel, Stephane 521

Rondepierre, Aude 679
Roux, Christophe 719

Sakai, Shin-ichiro 303
Sanfedino, Francesco 343
Saussié, David 699
Scardino, Marco 343
Schuet, Stefan 263
Serra, Romain 679
Shima, Tal 167, 227
Shin, Hyo-Sang 247
Soken, Halil Ersin 303
Souanef, Toufik 135, 151
Stolle, Martin 283

Tal, Ezra 167
Taquin, Gilles 475
Theuma, Kevin 381
Toglia, Chiara 659
Tol, H.J. 187
Toussaint, Clément 475
Tsourdos, Antonios 247

Van Eykeren, Laurens 363

Wahid, Mastura Ab 455
Watanabe, Yoko 283
Wildschek, Andreas 99

Zolghadri, Ali 67, 323